INHALED PARTICLES VIII

Proceedings of an International Symposium
on Inhaled Particles organised by the
British Occupational Hygiene Society

26–30 August 1996

Editors

N. CHERRY
T. OGDEN

Supplement to the
Annals of Occupational Hygiene

PERGAMON

U.K.	Elsevier Science Ltd, The Boulevard, Langford Lane, Kidlington, Oxford OX5 1GB, U.K.
U.S.A.	Elsevier Science Inc., 655 Avenue of the Americas, New York, NY 10010, U.S.A.
JAPAN	Elsevier Science Japan, Tsunashima Building Annex, 3-20-12 Yushima, Bunkyo-ku, Tokyo 113, Japan

First edition 1997

Published as a Supplement to the journal *The Annals of Occupational Hygiene*

ISBN 0 08 0427405
ISSN 0003–4878

Printed in Great Britain by BPC-AUP Aberdeen Ltd.

INHALED PARTICLES
VIII

Illustration of actual workplace conditions in oilfield sandblasting in Texas in the early 1990s.

CONTENTS

Contents

Contents   vii

SECTION 4: COAL, SILICA AND OTHER DUSTS

SECTION 5: DEPOSITION, CLEARANCE AND MODELLING

SECTION 6: OTHER DUSTS AND TOPICS

SECTION 7: INDOOR AIR

PREFACE

INHALED PARTICLES VIII was held in Cambridge on 26–30 August 1996. The meeting, the latest in a series of international symposia organised by the British Occupational Hygiene Society since 1960, carried the title "Occupational and Environmental Implications for Human Health", reflecting the Scientific Committee's wish to consider health as the central theme of the conference. To this end, the call for abstracts specified eight types of agent known to affect human health, and invited papers that would enable delegates to integrate research findings from basic mechanisms to the epidemiological impact on human health. Although this attempt at integration was more successful in some sessions than in others, delegates responded warmly to this approach which may form the basis of future meetings in this series.

This volume collected together extended abstracts, reviewed and edited by ourselves, of all but eight of the 135 papers and posters presented in Cambridge. The papers are shorter than those published in previous proceedings and have not been subjected to full peer review. However need for timely publication of important contributions from international centres precluded the lengthy process of review and revisions that has delayed previous publications. We trust that you will agree that the decision to proceed with a rapid publication was the correct one.

In all events of this sort there are many whose efforts risk going unrecorded. We would like to note here particularly the contributions of other members of the Scientific Committee:

Dr M. R. Bailey	NRPB, Didcot
Dr P. Baxter	University of Cambridge
Dr R. C. Brown	Independent consultant
Dr L. C. Kenny	Health and Safety Laboratories, Sheffield
Prof. J. C. McDonald	National Heart and Lung Institute, London
Dr J. Pritchard	Glaxo Group Research Limited, Ware

The BOHS Secretariat was represented by the committee by Mrs P. Blythe who, together with Mrs T. Bull, provided sterling support throughout the planning and organisation of the meeting.

The structure of the meeting included both formal presentations and posters, with discussion sessions integrating the two. This format made unusually heavy demands on the Chairs of the sessions and we would like to thank all those who agreed to take this on, namely: M. D. Attfield, M. R. Bailey, P. Baxter, J. Bignon, R. Brown, G. J. Burdett,

J. W. Cherrie, A. Churg, C. G. Collier, K. Donaldson, G. W. Gibbs, J. Heyder, J. Leigh, R. L. Maynard, J. C. McDonald, H. Muhle, D. Norback, R. F. Phalen, J. Pritchard, O. G. Raabe, C. Rice, A. Rogers, C. E. Rossiter, A. Searl and J. H. Vincent.

Finally, two of our speakers warrant special mention. Dr Paul Schulte from NIOSH, kindly agreed to give the Colt Lecture on the "Molecular Epidemiology of Occupational and Environmental Lung Disease". We are grateful both to him and to the Colt Foundation which sponsored the event. The other guest speaker, Mr Henry Waltons was particularly welcome as an organiser of the first, 1960, meeting and Editor of many of the Proceedings. His history of the meetings is included as the opening paper in this volume.

Nicola Cherry   Trevor Ogden
Centre for Occupational Health,   *Independent Consultant*
University of Manchester

Pergamon

Ann. occup. Hyg., Vol. 41, Supplement 1, pp. xvii–xix, 1997
British Occupational Hygiene Society
Published by Elsevier Science Ltd
Printed in Great Britain
0003–4878/97 $17.00 + 0.00
Inhaled Particles VIII

INHALED PARTICLES VIII: HISTORY OF THE SYMPOSIA

In the late 1950s the lung dust diseases, silicosis and coal-workers' pneumoconiosis dominated the industrial hygiene scene; asbestos was but a cloud on the horizon. New research results and new ideas for better hygienic measurements abounded, for example the concept of aerodynamic selective sampling, but people in the corridors of power were reluctant to change established practices. So the BOHS announced "a Symposium on the physical, chemical and physiological factors governing the entry of harmful substances into the body via the respiratory system" to be held in 1960. (Note: no mention of clearance, biological response or epidemiology.) The Society believed "that this Symposium would meet a real need in providing an opportunity for the comprehensive discussion of recent research in this field and that the published proceedings would be of value not only to research workers but also to those concerned with the practical problems of establishing and maintaining safe environments". The bringing together into one volume of information normally scattered over diverse literature was an important objective of the enterprise. It was an ambitious undertaking for a small society barely 7 years old, with meagre financial resources. The announcement received a good response. Several VIPs gave their active support. Most of the experts worldwide that we hoped to attract joined in when they learned that their peers were doing likewise. Governmental and other institutions, and industry, gave generous financial help. The Symposium, held in Oxford, in March 1960, was attended by 251 delegates from 20 countries (50% more than are present here). Thirty-eight papers were presented. Consecutive interpretation was provided for the discussion, which was recorded. The registration fee was £4.00 and the event was an undoubted success. American visitors found the experience of unheated student accommodation in historic Balliol College "something to talk about when we get home". Subsequent symposia were held during summer vacations! (Was anyone other than present speaker there? No! Who was not born then? No response. . . A mature company!) The proceedings, "Inhaled Particles and Vapours", edited by Norman Davies, were published by Pergamon Press in 1961. The Society's journal, *The Annals of Occupational Hygiene*, had been launched in the previous year. Pergamon's proprietor, Robert Maxwell, took on both of these when no other publisher was interested. John Gilson, President of BOHS in 1961 reported on the proceedings: "There is no doubt that this work which is unique in the history of the subject, will be found in laboratories and libraries all over the world. It is carrying the name of the Society with it and giving us something to live up to." The 1960 meeting had been thought of as a one-off event. But its success—and continuing interest in inhaled particles—prompted BOHS to organize Inhaled Particles and Vapours II, in Cambridge, in 1965. This time there was no difficulty in attracting papers or people. Forty-seven papers (a 25% increase) were presented to a capacity audience of 286 delegates from 21 countries. Some late applicants had to be turned away. Simultaneous interpretation in French and German was provided. (I think only Charles Rossiter of the present company, other than myself, was there.) The inclusion of 'vapours' in the first two symposia was to meet the wishes of those BOHS Council Members whose interests lay there rather than with particles: the 'Wets' (chemists) vs. the 'Drys' (physicists). Few 'vapour' papers were offered, and these did not tie in well with the particles theme. By the late 1960s some results from important epidemiological studies of silicosis and CWP were emerging, and parties concerned suggested that the BOHS might arrange an international conference. This developed into Inhaled Particles III (no vapours) held in London in September 1970. There was again an enormous expansion: 339 delegates came from 25 countries, including Russia for the first time. Ninety-one papers were accepted and the programme was extended to 2 weeks. The European Community kindly provided interpreters for simultaneous translation, free of charge. (Two contributors to the 1970 meeting, Otto Raabe and Graham Gibbs, are here presenting papers again.) The two-week experiment was not wholly successful for either delegates or organizers, and so future symposia were restricted to one 5-day week. The two-fold increase in the number of papers also created publication problems. The status of the Symposia for attracting prime material was linked with that of the published Proceedings. Papers were peer-reviewed and edited to the standard of regular scientific journals, and came to be accepted as such by referencing journals and libraries. Transcripts of the discussion—and any addition-al written questions—were referred back to authors for considered response and included in the Proceedings. The increased size of Inhaled Particles and rising production costs caused Pergamon Press to ask for camera-ready copy, which we rejected because it would restrict editorial review. So we looked for another publisher, and Inhaled Particles III was published as two volumes and 1090 pages, in letter-press, by Unwin Brothers, in 1971. Vapours had been squeezed out of the 1970 agenda. So the 'wets' arranged a separate meeting on Inhaled Gases and Vapours in 1971. This had limited response and was not repeated. But we were asked by BOHS Council to re-instate vapours in the next, 1975, symposium. Edinburgh 1975 was the next venue, and the peak occasion, with 434 delegates from 29 countries. One hundred and sixty-five papers were offered but only 66 (40%) could be accepted for presentation in the 5 days available. Sixty-one 'particle' papers with discussions were published, by Pergamon Press once again, as Inhaled Particles IV, in two Parts. Five 'vapour' papers were printed separately in *The Annals of Occupational Hygiene*. The two topics were not again combined. Edinburgh 1975 was followed by Inhaled Particles V, held at Cardiff in 1980. Again offers of papers far exceeded the capacity of a 5-day meeting; 69 were accepted for presentation. The attendance, 317, was somewhat less than at Edinburgh. The local civic authorities, City and Region, laid on generous receptions for our international audience. We survived a protocol crisis when the respective top persons of these local authorities discovered they had not received reciprocal invitations from their opposite numbers! The Proceedings of the 1975 meeting, Inhaled Particles V, were again published by Pergamon Press in the customary letter-press style. But they issued it as an extra numbered Volume of *The Annals*, instead of a supplement, and billed all subscribers accordingly. Not surprisingly, this brought protests. By the time of the VIth Symposium (Cambridge 1985), veterans of the original organizing teams, and the editor, had been succeeded by new people with new ideas. Poster presentations were introduced so that fewer would-be contributors need be disappointed, and a Workshop on Lung Dosimetry was added in parallel sessions. This more than doubled the total presentations, to 144, and registrations rose to near the 1975 peak. It also doubled the editorial and printing load (equalling more than 2 years of *The Annals*) and Pergamon Press again asked for camera-ready (or computer-ready) copy. The European Community, which had generously provided multi-language interpretation at earlier symposia, agreed to re-type the edited texts for computerized printing. This did not work smoothly. There was a 3-year delay before publication—though the final result was fine—and preparations for Symposium V II were set back by a year, to 1991.

This hiatus did not diminish support for the VIIth Symposium, held in Edinburgh, 1991. Two hundred and fifty papers were offered and 140 were accepted, divided equally between oral and poster presentation. One hundred and twenty-six reached publication. The editors, still determined to maintain the standards of previous proceedings, were over-burdened. Publication again took 3-years. This elicited an official apology from the BOHS and an assurance of timely publication for any future venture of the same kind. The series was clearly in jeopardy. I was not privy to the BOHS Council's discussions but am told that some members fought for another go, led by Nicola Cherry. So here you are again! The programme and attendance show continued support, and that Inhaled Particles is still a live subject. There are oral presentations, parallel sessions, and posters. All are to be published rapidly as shortened papers, with a little editorial tidying-up, and as a supplement to *The Annals*. Professionalism has come to Head Office! I have many memories of the earlier symposia. I particularly remember the first meeting 36 years ago in Oxford when, never having been President of anything before (or since), I had to address the dinner guests from the daunting height (I was going to say 'pinnacle of Academe' then remembered that we are now in Cambridge) of the Master's Chair at the High Table in the Great Hall of Balliol. I recall quoting some words from Shakespeare's *Hamlet* to the assembled company—and I can apply them again to you now, "What a piece of work is a man! How noble in reason! How infinite in faculty! . . . this quintessence of dust." You are the quintessence of dust, but I trust that in future years you can look back on the proceedings of this Symposium as noble in reason!

Finis.

W. Henry Walton

THE COLT LECTURE

Ann. occup. Hyg., Vol. 41, Supplement 1, pp. 1–6, 1997
British Occupational Hygiene Society
Published by Elsevier Science Ltd
Printed in Great Britain
0003–4878/97 $17.00 + 0.00
Inhaled Particles VIII

PII: S0003–4878(96)00125–1

MOLECULAR EPIDEMIOLOGY OF OCCUPATIONAL AND ENVIRONMENTAL LUNG DISEASE

P. A. Schulte

National Institute for Occupational Safety and Health, 4676 Columbia Parkway,
Cincinnati, OH 45226, U.S.A.

BIOMARKERS AND MOLECULAR EPIDEMIOLOGY

Advances in molecular biology and analytical chemistry are providing potentially useful tools for the epidemiologic study of occupational and environmental lung disease (NRC, 1989; Paoletti, 1995; Schulte, 1996). These tools are biological indicators, or biomarkers, of events that represent exposure to a xenobiotic, effects of exposure and susceptibility to effects of exposure. Taken together, these biomarkers can be used to describe a continuum of events following a xenobiotic exposure and proceeding through resultant diseases. Using a continuum concept expands the classic epidemiologic model that involves statistically inferring an association between an exposure and disease. This classic approach has yielded many useful findings for public and occupational health with regard to pulmonary diseases such as coal workers' pneumoconiosis, beryllium disease, asthma from wood dust and lung cancer from asbestos, cigarette smoking, bis-chloromethyl-ether and uranium. The epidemiologic assessment involving these lung diseases generally relied on statistical inference rather than mechanistic insight to associate exposure with disease; but that is not to say that investigators conducting such studies lacked biological hypotheses for such associations.

What differs from these previous investigations is the potential for identifying various events in a continuum between an exposure and disease (NRC, 1989). These events are represented by biomarkers. Biomarkers are generally biological measurements representing or correlating with an event, serving as the event, or predicting the event. Biomarkers can range from those on the macroscale, such as dysfunction, disease and death down to those on the microscale (such as the cell, gene or molecule). The middle of this range can include biomarkers such as various anatomic, physiologic and functional changes.

Biomarkers can serve as dependent or independent variables in classical epidemiological study designs such as cohort, case–control, cross-sectional and intervention studies. As a dependent variable, a biomarker of effect, [such as C8 oxidation of deoxyguanosine in coal workers (Schins, 1995) or 4-hydroxy-2-noncnal protein adducts in individuals exposed to ozone (Li *et al.*, 1996)] could, if validated, serve as surrogates for lung disease, or earlier stage events that could be associated with later disease, in an epidemiologic study. For independent variables, biomarkers of exposure can be used. For example, in addition to using a job title or a few exposure measurements as an independent variable for exposure, it may be possible to use a biologic measure of cumulative dose for better defining biologically important exposure. Biomarkers may also be used to evaluate the factors that mediate between an exposure and disease. In epidemiology, these are called effect modifiers. For example, in people with the null GSTM1 allele, the risk of lung cancer given exposure to polycyclic aromatic hydrocarbons, was approximately two-fold greater than for those with the other prominent alleles given the same exposure level (Tang *et al.*, 1995).

These two uses of biomarkers, to define a continuum and to serve as variables and effect modifiers in epidemiologic research, have been termed as "molecular epidemiology" (Schulte and Perera, 1993). "Molecular epidemiology of lung disease" is a heuristic phrase that represents the natural confluence of powerful developments in basic biomedical sciences and the field-tested methods of epidemiology to study lung disease. Although molecular epidemiology can be viewed as an evolutionary step in epidemiology, a supplemental set of tools, or even a separate discipline, it does not represent a shift in the basic paradigm of epidemiology (Schulte, 1993). However, the term can be used to stand for the application of the whole range of biomarkers used in epidemologic investigations (Rothman *et al.*, 1995). These include molecular, genetic, cellular, histologic and physiologic markers. The literature discussing molecular epidemiology contains a difference of opinion about this term. The definition used in this paper emphasises epidemiology and treats the molecular descriptor as a way to define the variables to use in epidemiologic investigations. Others focus on the use of molecular tools to gain mechanistic insight but not necessarily to enhance understanding of epidemiologic associations. These molecular tools are most useful in epidemiology and risk assessment when the mechanistic information is used in conjunction with the established principles of epidemiology. Ultimately, the term "molecular epidemiology" should be seen as a signpost pointing to a more biological basis for epidemiology.

MOLECULAR EPIDEMIOLOGY AND LUNG DISEASE

Except for immunological markers, biomarkers (particularly cellular and molecular markers) have been infrequently used in the study of occupational and environmental lung disease. A TOXLINE (U.S. National Library of Medicine) title-field search was conducted of all the literature between January 1986 and July 1996 that could be identified with a search strategy that included various occupational and environmental lung diseases synonyms plus the terms "human and epidemiologic" plus "cellular", "molecular", "genetic", "DNA", or "biomarkers". Approximately 5221 citations (both human and epidemiologic) were identified, 70 of these were actually published investigations involving biomarkers in studies of occupational or environmental lung disease. When these 70 studies were assessed according to the established biomarker categories, the following percentage distribution was identified: internal dose, 9.4%; biologically effective dose, 20.2%; early biological effects, 25.5%; alternations of structure and function, 10.8%; clinical disease, 9.7%; and susceptibility, 20.2%. This distribution roughly indicates how often the markers were used and would probably change slightly with different search strategies and classification logic. Nonetheless, the exercise provides a glimpse of how biological markers have been used to study occupational and environmental lung disease. In general, these studies provide no information about unknown causes of lung disease. However, they serve other important functions: they validate markers as indicators of exposure, effect, or susceptibility and they elucidate the causal pathway and biological plausibility of previously identified associations.

Two types of studies involved internal dose biomarkers: those measuring a xenobiotic in lung tissue indirectly by imaging and those using other biologic specimens such as blood to measure metabolites (e.g. of PAHs) or antibodies (e.g. of tuberculosis). Biomarkers of biologically effective dose were generally assessed in surrogate tissues such as peripheral blood lymphocytes. Most of these studies were exposure-selective, cross-sectional studies that assessed the relation of various DNA adducts with exposure. Occasionally, protein adducts were assessed. Case–control studies were also frequently used to study markers of biologically effective dose. At least one study compared cancer patients on the basis of smoking history and looked for marker–exposure associations based on tissue measurements. Many of the studies involving DNA adducts also assessed susceptibility factors such as various genotypes for p450 enzymes.

The category "early biologic effects" is the largest. However, the inclusion of a marker in this category is subjective and dependent on the state of the science. If a marker is a biologic effect that is known to be directly on the causal pathway to lung disease and is not reversible, it may be rightfully placed in the next category, "alteration of structure and function". At this time, for most markers, the extent to which a marker is a contributing causal factor in lung disease is not known so markers, such as oncogenes, tumour suppressor genes or proteins, are included as markers of early biologic effects. The most frequent marker found in this category is the p53 tumour suppressor gene. The studies involving altered structure and function are similarly subjective and are based on the extent to which a marker predicts disease. Many of the studies in this category, as well as in the clinical disease category, involved immunologic markers such as tumour necrosis factor alpha or IgE in asthma studies. In no cases were molecular changes used as dependent variables to represent disease in epidemiologic studies.

Studies with biomarkers of susceptibility represent the second largest category. Generally, these involve phase 1 (e.g. CYP1A1) and phase II (GSTM1) metabolic polymorphisms. The existing knowledge about the causes of occupational and environmental lung disease has resulted from clinical, pathological and epidemiological investigations. These investigations did not involve molecular markers, but in a broader sense, they involved a range of biological markers at the histologic, physiologic and individual levels. The exceptions are immunopathology studies and assessments of cytotoxicity and cellular pathology. Thus, for diseases like coal workers' pneumoconiosis, silicosis, asbestosis and lung cancer from inhaled carcinogens, the major etiologic questions have been answered previously by classic epidemiologic methods. The contribution of molecular biomarkers in understanding these diseases probably involves issues of medical screening and surveillance (i.e. what early effects can be identified for therapeutic intervention); identification of those workers likely to progress to more serious disease (e.g.

progressive massive fibrosis in coal workers' pneumoconiosis) and contribution to risk assessments (e.g. what are the risks at low levels of exposure?).

FUTURE EFFORTS

In the future, molecular biomarkers might be useful in the area of new or undefined hazards. For example, many countries are wrestling with the question of potential health effects from manmade mineral fibres, such as refractive ceramic fibres. Are these fibres fibrogenic or carcinogenic? In some cases, these fibres have been produced for relatively short periods and with low airborne concentrations. The question of concern is what the health risk is at these concentrations. Biomarkers of effect such as those found for more widely studied fibres like asbestos, could be used to determine whether there are any exposure–response relationships in workers exposed to a new fibre of concern. A research study could be designed to compare workers exposed to asbestos with those exposed to refractive ceramic fibres for assays of identified biomarkers. Using appropriate controls and covariate adjustments, it would be possible to determine if the exposed workers were similar to each other or greater than the controls. This could be a signal to restrict exposures. If such a relationship was identified and risk could be inferred then precautions could be taken.

Similarly, with mixtures of substances that cause oxidative stress, understanding molecular-level events could provide useful surrogates for exposures or early effects. For example, it has been shown that 4-hydroxy-2-nonenal protein adducts may be associated with apoptosis and necrosis and may serve as a secondary toxic messenger for acute ozone injury from exposures to low concentrations. This marker could be used as a dependent variable in a population study.

Molecular markers may also be useful in distinguishing between etiological pathways. Consider for example, the mutation spectrum of the p53 tumour suppressor gene. Different carcinogens seem to cause different characteristic mutations. For example, lung cancers in cigarette smokers are different from lung cancers in nonsmokers. Cigarette smokers have more G:C to T:A transversions (Harris, 1996). The p53 mutation spectrum can be influenced by both the type and location of the promutagenic lesion and understanding these characteristics can provide etiological information. Understanding this kind of information in both humans and animals potentially increases our ability to assess cancer risks accurately. Cancer risk assessment has been forced to provide conservative risk estimates because in addition to gaps in classical exposure and disease data, limited knowledge exists about complex pathobiological processes during carcinogenesis, differences in metabolism of carcinogens, different DNA repair capacities, variable genomic stability among animal species, and variation among individuals with inherited susceptibilities (Sutter, 1995). Molecular biomarkers measured in animals and humans according to the established parallelogram approach used in genetic toxicology can provide useful data for low-dose assessments of risk (Sutter, 1995).

At this time, the lack of significant contributions of molecular markers to the epidemiologic study of occupational and environmental lung disease should not be viewed with concern about their utility. Most of the markers thus far identified have been in the developmental/validation stage. The challenge now is to use categories of markers and assays that have been found with known carcinogens and toxicants to study and control substances that are suspected pulmonary hazards to fill in gaps in the data. Validation efforts are needed in the laboratory and the field before a marker can be used to assess a suspicious pulmonary toxicant. Nonetheless, as biomarkers are found to be useful, they need to be used proactively to identify potential hazards before widespread exposures occur. In the area of occupational and environmental lung disorders, biomarkers could be applied to topics such as: fibrosis and cancer from man-made mineral fibres, carcinogen risk assessment for diesel emissions, early effects from air pollution and exposure to oxidative stressors. Finally, more attention should be given to using molecular biomarkers to assess prevention and intervention efforts.

Successful applications of molecular markers will not be rapid if a passive approach is taken by the occupational and environmental communities. The mapping of the human DNA (and various other genomes) is yielding a vast amount of raw information that (unless interpreted) will not provide tangible tools for research and control. Occupational and environmental researchers and risk assessors need to increase their knowledge and access to the raw data. Computer-driven "bioinformatics" coupled with the ability to rapidly obtain partial sequences of expressed genes portend revolutionary changes in scientific research and the practice of occupational medicine and in the collation of data for useful regulatory and prevention purposes (Rodbell, 1996). Assuming the irreversibility of gene patterns indicates hazard to the cell and possibly to the human body, dose–response relationships, under appropriate cell/tissue culture and exposure conditions can be accurately determined. The important difference from other toxicological approaches is that the method involves the total expressed gene pattern. Epidemiologists and industrial hygienists need to envision investigations that involve the laboratory screening of chemicals followed by field research (Rodbell, 1996).

More research is also needed to assess the relationship of biomarkers in blood to pathologic changes in lungs. Linkage of these two specimens with each other and with lung imaging technology should enhance the utility of biomarkers.

Finally, investigators need to integrate molecular epidemiologic approaches with other approaches for etiologic, intervention, and clinical research on lung diseases. Such integration can be accomplished in part by encouraging banking of biological specimens when possible.

The literature review showed that only a small number of biomarkers were used in epidemiological studies of lung disease. Generally, clusters of studies used the same markers (e.g. p53). This small number is good in one sense, since widespread use of unvalidated markers is unwarranted. Nonetheless, a large and diverse array of biomarkers have been identified for various lung diseases and their causal pathways. More of these markers need to be field tested and used in epidemiologic studies.

The use of molecular biological markers to study and control occupational and environmental lung disease can raise various ethical, legal and social issues. Researchers need to anticipate these issues and the impacts on participants in studies and make efforts to protect them.

REFERENCES

Harris, C. (1996) p53 tumour suppressor gene: at the crossroads of molecular carcinogenesis, molecular epidemiology and cancer risk assessment. *Environ. Health Perspectives* **104**, 435–39.

Li, Li, Hamilton, R. F., Jr, Kirichenko, A. and Holian, A. (1996) 4-hydroxy-2-nonenal as a mediator for oxidative injury. *Toxicol. appl. Pharmacol.* **139**, 135–143.

National Research Council (1989) *Biological Markers in Pulmonary Toxicology*. National Academy Press, Washington, D.C.

Paoletti, P. (1995) Application of biomarkers in population studies for respiratory non-malignant diseases. *Toxicol.* **101**, 99–105.

Rodbell, M. (1996) Bioinformatics: An emerging means of assessing environmental health. *Environ. Health Perspectives* **104**, 136.

Rothman, N., Stewart, W. F. and Schulte, P. A. (1995) Incorporating biomarkers into cancer epidemiology: a matrix of biomarker and study design categories. *Cancer Epid. Biomarkers Prev.* **4**, 301–311.

Schins, R. P. F., Schilderman, A. E. L. and Borm, P. S. A. (1995) Oxidative DNA damage in peripheral blood lymphocytes in coal workers. *Int. Arch. occup. Environ. Health* **67**, 153–157.

Schulte, P. A. (1993) A conceptual and historical framework for molecular epidemiology. In *Molecular Epidemiology: Principles and Practices* (Edited by P. A. Schulte and F. P. Perera), pp. 3–44. Academic Press, San Diego.

Schulte, P. A. (1996) Use of biomarkers to investigate occupational and environmental lung disorders. *Chest* **109**, 95–123.

Sutter, T. R. (1995) Molecular and cellular approaches to extrapolation for risk assessment. *Environ. Health Perspective* **103**, 386–89.

Tang, D. L., Chiamprasert, S., Santella, R. M. and Perera, F. P. (1995) Molecular epidemiology of lung cancer: carcinogen—DNA adducts, GSTM1 and risk. [Abstract] Annual Meeting of the American Association for Research, Toronto.

SECTION 1

ENVIRONMENTAL EXPOSURES

Pergamon

Ann. occup. Hyg., Vol. 41, Supplement 1, pp. 7–13, 1997
© 1997 British Occupational Hygiene Society
Published by Elsevier Science Ltd. All rights reserved
Printed in Great Britain
0003–4878/97 $17.00 + 0.00
Inhaled Particles VIII

PII: S0003–4878(96)00126–3

PRO-INFLAMMATORY EFFECT OF PARTICULATE AIR POLLUTION (PM10) *IN VIVO* AND *IN VITRO*

W. MacNee,* X. Y. Li,* P. S. Gilmour† and K. Donaldson†

*Unit of Respiratory Medicine, Department of Medicine, University of Edinburgh, Royal Infirmary, Lauriston Place, Edinburgh EH3 9YW; and the †Department of Biological Sciences, Napier University, Edinburgh, U.K.

INTRODUCTION

Numerous epidemiological studies have shown that particulate air pollution, especially small particles with an aerodiameter of ≤ 10 μm (PM10) are associated with increased morbidity and mortality from airways diseases, such as asthma and chronic obstructive pulmonary disease (COPD) (Schwartz, 1994; Pope *et al.*, 1995). However, the mechanism for the adverse effects of PM10 at such low airborne mass concentrations and the association with cardiovascular deaths is not understood. These effects have been shown in diverse geographical loations (Schwartz, 1994), which suggests that the exact composition of the particulate air pollution is not critical. We hypothesised (Fig. 1) that PM10 particles have oxidant properties, related to their size, resulting in airway inflammation, increased airspace permeability and interstitialisation, producing an enhanced inflammatory response (Seaton *et al.*, 1995). We believe these effects are due to the oxidant properties of PM10 and result in exacerbations of airways diseases and a change in the rheological and coagulation properties of the blood, through both local and systemic effects. This latter effect may be critical in precipitating cardiovascular events and hence deaths in an already compromised and susceptible population. The purpose of this study was to begin to test this hypothesis by measuring pro-inflammatory potential and the oxidant activity of PM10. We have also tested the contention that the size rather than the composition of the particles is important, by comparing the effects of PM10 with those of other fine and ultrafine particles.

METHODS

Particles

PM10 particles were collected on filters from the Edinburgh monitoring station of the Enhanced Urban Network. Since the extraction procedure produced a suspension of PM10 contaminated with small numbers of filter fibres (0–10 per light microscopic field at × 80 magnification), filter fibre suspension (FFS, > 300 fibres per light microscopic field at × 80 magnification), was prepared by sonicating an unused filter in PBS. We estimated that between 50–125 μg of PM10 in 0.2 ml PBS was instilled into rat lungs and was compared in some experiments with instillation

Local, vascular and systemic effects of pulmonary deposition of PM10 and ultrafine particles

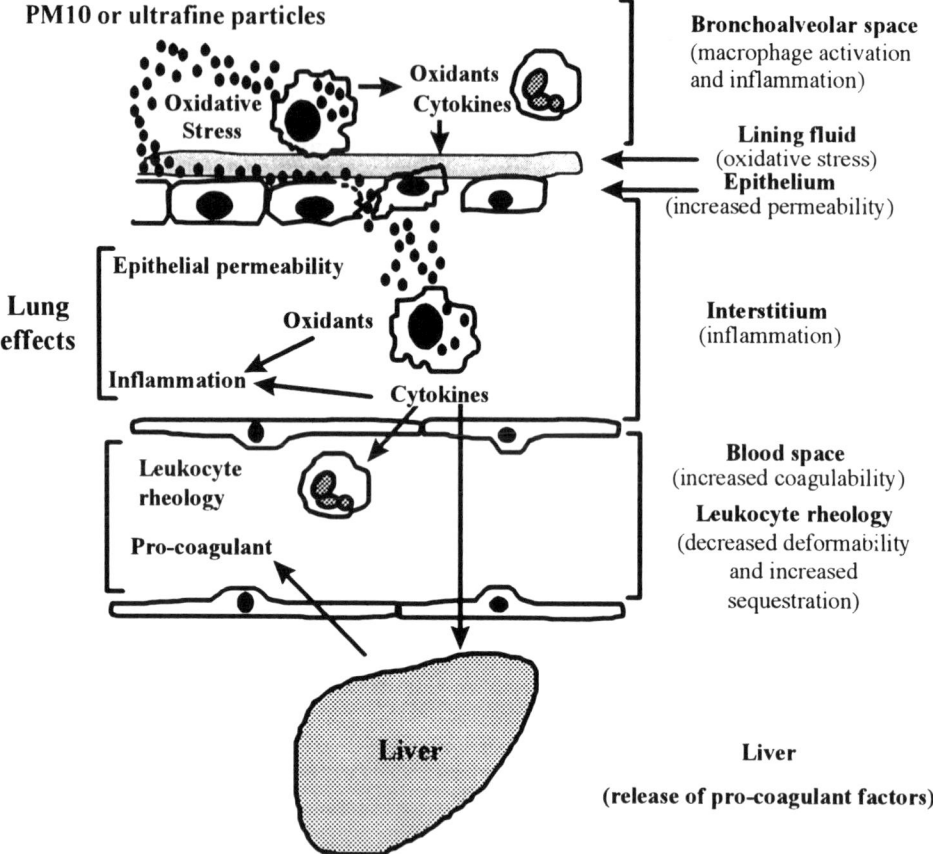

Fig. 1. Diagrammatic representation of a hypothesis for the mechanism of the harmful effects of particulate air pollution (PM10).

of the same mass (125 µg) of fine (CB, 200–500 nm diameter) and ultrafine carbon black (ufCB, 20 nm diameter). In limited experiments rats inhaled 1 mg.m^{-3} of ultrafine titanium dioxide (ufTiO$_2$, 96 nm diameter) for 7 h.

Particle instillation/bronchoalveolar lavage (BAL)

Syngeneic Wistar-derived rats of the HAN strain were instilled intratracheally with 0.2 ml of PBS-particle suspension. Control animals were not instilled, or instilled with 0.2 ml of PBS or FFS alone. Six hours after instillation of the particle suspensions, rats were sacrificed, the lungs removed, BAL was performed and lung homogenate prepared. Supernatant from BAL leukocytes (1×10^6 per ml cultured for 24 h) was collected for the measurement of nitrite (Ding *et al.*, 1988), tumour necrosis factor (TNG, Li *et al.*, 1995) and lactated dehydrogenase (LDH, Bergmeyer *et al.*, 1965).

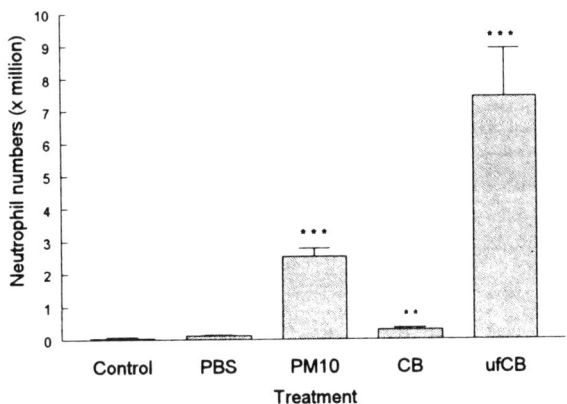

Fig. 2. The number of neutrophils in bronchoalveolar lavage (BAL) from rats 6 h after no instillation (control) or intratracheal instillation with PBS, PM10, fine (CB) or ultrafine (ufCB) carbon black (n = 3–5, **p < 0.01, ***p < 0.001 compared with PBS).

Epithelial permeability in vivo and in vitro

Rat lung epithelial permeability was assessed as the total protein concentration in BAL fluid (Li *et al.*, 1995). A549 human type II alveolar epithelial cells were cultured as a monolayer on tissue inserts and the permeability of the monolayer was determined as the passage of BSA across the cell monolayer (Li *et al.*, 1994).

GSH and GSSG

GSH/GSSG concentrations were measured by the GSSG-reductase-DTNB recycling procedure (Li *et al.*, 1994). In the inhalation experiments with ufTiO$_2$, systemic oxidant stress was measured as Trolox Equivalent Antioxidant Capacity (TEAC) of plasma (Rahman *et al.*, 1996).

Statistical analysis

The results were expressed as mean ± SEM. Differences between mean values were assessed by analysis of variance.

RESULTS

Intratracheal instillation of PM10 caused neutrophil influx into the airspaces 6 h after instillation (Fig. 2). CB installation produced a small, but significant neutrophil influx. However the greatest inflammatory cell influx occurred following instillation of ufCB (Fig. 2).

BAL leukocytes obtained 6 h after PM10 instillation produced greater amounts of TNF and NO in culture (Fig. 3). Although BAL leukocytes from PM10-instilled animals had a greater potential to produce TNF and NO in culture, TNF and NO in BALF were not significantly different 6 h after PM10 instillation, compared with control animals (data not shown).

PM10 increased airspace epithelial permeability 6 h after instillation (Fig. 4). As with the influx of inflammatory leucocytes the greatest increase in airspace epithelial permeability occurred following instillation of ufCB.

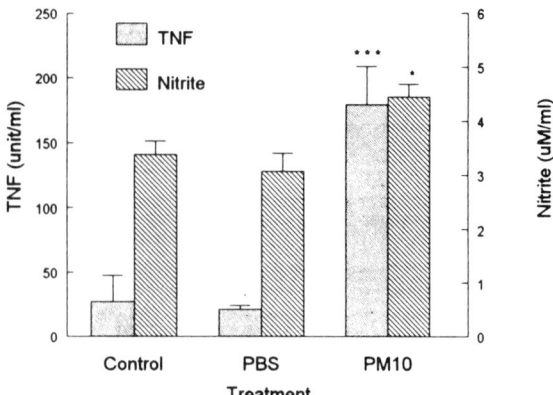

Fig. 3. The effect of intratracheal instillation of PM10 on TNF and NO production by BAL leukocytes in culture ($n = 3$, *$P < 0.05$, ***$P < 0.001$ compared with PBS).

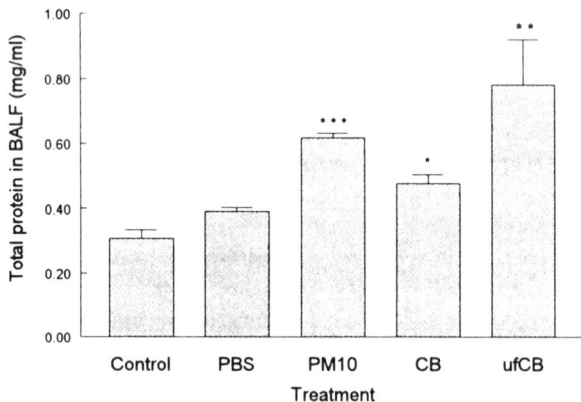

Fig. 4. The effect of intratracheal instillation of PM10, fine (CB) and ultrafine (ufCB) carbon black on rat lung epithelial permeability *in vivo*, measured as total protein values in BALF 6 h after instillation ($n = 3.6$, *$P < 0.05$, **$P < 0.01$ and ***$P < 0.001$ compared with PBS).

The addition of PM10 to A549 type II alveolar epithelial monolayers *in vitro* increased their permeability to BSA (control 0.10 ± 0.02 mg.ml^{-1}, PM10 0.20 ± 0.02 mg.ml^{-1}, $P < 0.01$), which was not due to cell death since monolayers of A549 cells incubated with PM10 for 6 h did not release increased amounts of LDH (control 23.5 ± 2.5, PM10 21.0 ± 7.1 units 10^{-6} cells, $P > 0.05$).

Intratracheal instillation of PM10 decreased GSH without any significant change in GSSG in BAL fluid 6 h after instillation (Fig. 5).

Instillation of filter fibre suspension alone (FFS) did not significantly alter BAL neutrophil counts (control 0.68 ± 0.28, FFS $0.58 \pm 0.15 \times 10^6$ neutrophils, $P > 0.05$), nor increase epithelial permeability, or decrease GSH levels in BAL fluid (data not shown). Preliminary experiments showed that following inhalation of ufTiO$_2$ rat lung epithelial permeability increased (Fig. 6), associated with a decrease in lung GSH and a fall in the antioxidant capacity of plasma (Fig. 7).

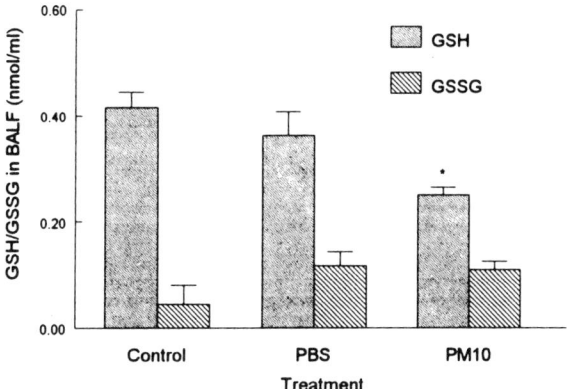

Fig. 5. The effect of intratracheal instillation of PM10 on GSH and GSSG concentrations in BALF 6 h after instillation in rat lungs ($n = 3$, $^*P < 0.05$ compared with PBS).

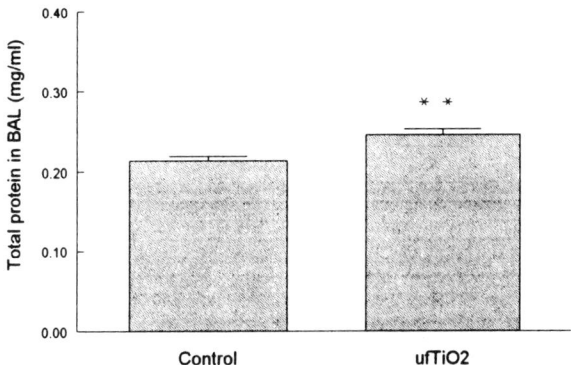

Fig. 6. The effect of inhalation of ultrafine TiO$_2$ (ufTiO$_2$ 1 mg m^{-3} for 7 h) on rat lung epithelial permeability ($n = 3$, $^{**}P < 0.01$).

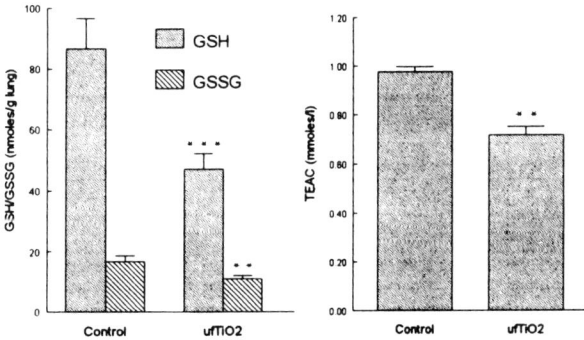

Fig. 7. The effect of inhalation of ultrafine titanium dioxide (ufTiO$_2$ 1 mg m^{-3} for 7 h) on rat lung GSH (left) and plasma trolox equivalent antioxidant capacity (TEAC, right); $n = 3$, $^{**}P < 0.01$ and $^{***}P < 0.001$.

DISCUSSION

This study supports our hypothesis (Seaton *et al.*, 1995) that PM10 causes airway inflammation and increased epithelial permeability. The PM10 suspensions were contaminated with a small number of fibres derived from the preparation procedure. However instillation of FFS alone, containing 30 times as many fibres as the PM10 suspension, did not produce pro-inflammatory effects.

Evidence in support of our contention that it is the oxidant properties of PM10 which is responsible for its biological activity *in vivo* is shown by changes in the important lung antioxidant glutathione. Our hypothesis that the epithelial injury was due to the free radical activity of PM10 is also supported by other studies in our laboratory which demonstrate that PM10 particles produce plasmid DNA scission, a sensitive assay for detecting the ability of particles to cause free radical injury (Gilmour *et al.*, in press).

We explored the contention that the size, rather than the composition of PM10 is important in its pathogenicity. The exact composition of our PM10 sample is not available. However, analysis of PM10 obtained from other sources indicates that carbonaceous material makes up 50% of the mass (Clarke *et al.*, 1984). Hence we compared the effects of instillation of both fine (CB) and ultra fine carbon black (ufCB) particles, with those of a similar mass of PM10. These studies show that ufCB, rather than CB, produce similar qualitative, although greater quantitative pro-inflammatory effects to those of PM10 in the rat lung. The greater inflammatory effect of ufCB was anticipated from our hypothesis, since ufCB is composed entirely of ultra fine particles, which is not the case for PM10, which by definition has only 50% of its particles with an aerodiameter of < 10 μm.

The association between an oxidant-induced decrease in lung GSH and an increase in epithelial permeability has been suggested from our work cigarette smoke (Li *et al.*, 1994, 1996). Other candidate inflammatory mediators may be involved in increasing epithelial permeability, such as TNF and nitric oxide. Although data from the present study does not show increased NO or TNF in BAL fluid, following instillation of PM10, *in vitro* BAL leukocytes from PM10-treated rats produced significantly more NO and TNF in culture than control BAL cells. The absence of detectable TNF and NO levels in BALF from PM10-treated rats compared with control animals is likely to result from the presence of inhibitors in the BAL fluid.

The results of this study are preliminary and are limited, due to the lack of the availability of large quantities of PM10, to an animal model of instillation, rather than the prefered inhalation model, which would be more relevant to environmental exposures. However, preliminary studies of the effects of acute inhalation of ultrafine titanium dioxide (ufTiO$_2$ 96 nm diameter), which is not inflammogenic as larger diameter particles (583 nm diameter, Driscoll *et al.*, 1990), indicate a similar influx of neutrophils into the airspaces and increased epithelial permeability, as we showed with instillation of PM10. In these inhalation experiments there was evidence of both local, lung oxidant stress, (decrease in lung GSH) and also systemic oxidant stress (decrease in the plasma TEAC). Interestingly a decrease in plasma antioxidant capacity also occurs after cigarette smoking in healthy volunteers and in exacerbations of both asthma and chronic obstructive pulmonary disease (Rahman *et al.*, 1996). We consider that the additional oxidant stress, produced by PM10 could act in susceptible populations, whose antioxidant defences are already compromised, to precipitate exacerbations of their airways disease. Furthermore, systemic oxidant stress may change neutrophil rheology, hence increasing neutrophil sequestration in the lungs, as has been shown during cigarette smoking (MacNee *et al.*, 1989). Increased neutrophil sequestration in the microcirculation of other organs such as the heart may contribute to the development of cardiovascular events (Fig. 1). This hypothesis is being tested in on going experiments.

In conclusion, this study provides evidence that PM10 has free radical activity and causes an inflammatory response and lung epithelial injury. These data provide support for our hypothesis (Seaton *et al.*, 1995) of the mechanism of the harmful effects of PM10 in exacerbating airways diseases.

Acknowledgements—Supported by the Medical Research Council, the British Lung Foundation and 3M Pharmaceuticals, U.K.

REFERENCES

Bergmeyer, H. U., Bernt, E. and Hess, B. (1965) *Lactate Dehydrogenenase. Method of Enzymatic Analysis* (Edited by H. U. Bergmeyer). Verlag Chemie, New York.

Clarke, A. G., Willison, M. J. and Zeki, E. M. (1984) A comparison of urban and rural aerosol composition using dichotomous samples. *A. J. Mos. Environ.* **18**, 1707–1775.

Ding, A. H., Nathan, C. F. and Stuehr, D. J. (1988) Release of reactive nitrogen intermediates and reactive oxygen intermediates from mouse peritoneal macrophages. *J. Immunol.* **141**, 2407–2412.

Driscoll, K. E., Lindenschmidt, R. C., Maurer, J. K., Higgins, J. M. and Ridder, G. (1990) Pulmonary response to silica or titanium dioxide: Inflammatory cells, alveolar macrophage-derived cytokines and histopathology. *Am. J. Respir. Cell Mol. Biol.* **2**, 381–390.

Gilmour, P. S., Brown, D. M., Lindsay, T. G., Beswick, P. H., MacNee, W. and Donaldson, K. Adverse health effects of PM10: involvement of iron in the generation of hydroxyl radical. *Occ. Environ. Med.* (in press).

Li, X. Y., Donaldson, K., Brown, D. and MacNee, W. (1995) The role of tumour necrosis factor in increased airspace epithelial permeability in acute lung inflammation. *Am. J. Respir. Cell Mol. Biol.* **13**, 185–195.

Li, X. Y., Donaldson, K., Rahman, I. and MacNee, W. (1994) An investigation of the role of glutathione in increased epithelial permeability induced by cigarette smoke *in vivo* and *in vitro. Am. J. Respir. Crit. Care Med.* **149**, 1518–25.

Li, X. Y., Rahman, I., Donaldson, K. and MacNee, W. (1996) Mechanisms of cigarette smoke induced increased airspace permeability. *Thorax* **51**, 465–471.

MacNee, W., Wiggs, B., Belzberg, A. S. and Hogg, J. C. (1989) The effect of cigarette smoking on neutrophil kinetics in human lungs. *N. Engl. J. Med.* **321**, 924–928.

Pope, C. A., Bates, D. V. and Raizenne, M. E. (1995) Health effects of particulate air pollution: time for reassessment? *Environ. Health Perspect.* **103**, 472–480.

Rahman, I., Morrison, D., Donaldson, K. and MacNee, W. (1996) Systemic oxidative stress in asthma, COP and smoking. *Am. J. Respir. Crit. Care Med.* **154**, 1055–1060.

Schwartz, J. (1994) Air pollution and daily mortality: a review and meta-analysis. *Environ. Res.* **64**, 36–52.

Seaton, A., MacNee, W., Donaldson, K. and Godden, D. (1995) Particulate air pollution and acute health effects. *Lancet* **345**, 176–178.

Pergamon

Ann. occup. Hyg., Vol. 41, Supplement 1, pp. 14–18, 1997
© 1997 Published by Elsevier Science Ltd on behalf of BOHS
Printed in Great Britain. All rights reserved
0003–4878/97 $17.00 + 0.00
Inhaled Particles VIII

PII: S0003–4878(96)00127–5

LACK OF CORRELATION BETWEEN PM$_{10}$ MEASUREMENTS AND UPPER RESPIRATORY TRACT DOSE

D. Fisher*† and M. McCawley‡

*West Virginia University, 1095 Willowdale Rd, Rm 111, Morgantown, WV 26505, U.S.A.;
†Current address: National Institutes of Health, Bethesda, MD, U.S.A.; ‡National Institute for
Occupational Safety and Health, 1095 Willowdale Rd, Rm 111, Morgantown, WV 26505, U.S.A.

INTRODUCTION

A size-selective air intake that has a 50% cut point at 10 μm (PM$_{10}$) was developed to monitor those ambient air particulate concentrations that might cause lung disease. The underlying assumption in establishing the current U.S. ambient air monitoring system was that PM$_{10}$ samples would provide equal protection regardless of the size distribution of the particulate matter. The size fraction, PM$_{10}$, is based on the aerodynamic diameter of particles that are capable of penetrating to, but not necessarily depositing in, the thoracic region of the respiratory system. Epidemiologic studies suggest that exposure to particulate air pollution may cause increased mortality and morbidity from respiratory and cardiovascular disease (Pope *et al*, 1996). Epidemiological studies seek a dose–response relationship to understand the etiology of disease. These analyses seek to find the relationship between particulate air pollution and mortality. However, questions about the usefulness of the PM$_{10}$ estimates have emerged. Previous results suggested that there was a questionable relationship between particulate dose and the PM$_{10}$ measurement.

For occupational environments, McCawley (1993) showed that there is a bias in the relationship between exposure (penetration) and dose (deposition) due to the variations in size distribution in occupational environments. The amount of bias associated with the use of the penetration criteria is dependent upon the particle size distribution. Since deposition is a measurement of dose and as such is the quantity of particulate crossing the physical boundaries of the person and entering the tissue, deposition should relate more directly to response than should measurements such as PM$_{10}$ which only measure penetration.

A high percentage of particulate mass in the atmosphere can occur in particle sizes below 1 μm in diameter (Fig. 1). These submicrometer dusts can be metal or organic fumes, many of which are suspected carcinogens. Not only does the submicrometer fraction contain carcinogens but, the likelihood of finding a correlation between penetration and deposition criteria is severely decreased when there is a substantial submicrometer contribution to the overall mass. The PM$_{10}$ technique when used to estimate submicrometer particulate levels is, for all practical purposes, collecting 100% of the particles.

Diseases such as asthma, chronic bronchitis or bronchial carcinomas are examples of responses initiated by an effective dose, presumably at the critical site in the

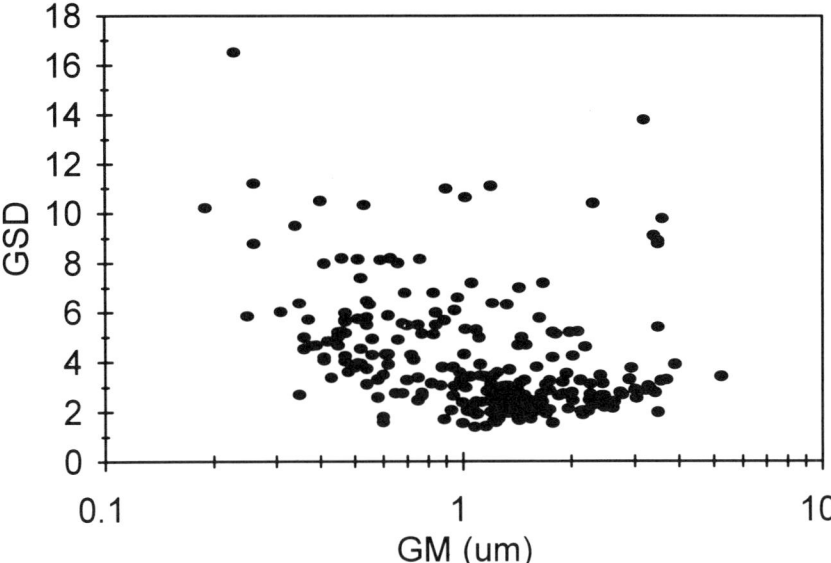

Fig. 1. A plot of all 250 size distributions' geometric means (GM) and geometric standard deviations (GSD) used for the meta analysis, showing 126 GMs less than 1 μm.

upper respiratory tract, that is, the tracheobronchial region. However, PM$_{10}$, which is similar to the thoracic particulate mass fraction, is that particulate that represents exposure for the whole lung (Boubel *et al.*, 1994). Depending on the particle size, the tracheobronchial fraction may be only a small and variable component of that measurement.

The question raised here is whether the PM$_{10}$ criterion is a good indicator of the dose of particulate responsible for diseases associated with the tracheobronchial region regardless of the size distribution from which the sample is drawn. The hypothesis to be tested is that the slope of the regression line for measurements of tracheobronchial dose vs PM$_{10}$ penetration is not significantly different than zero at a confidence level of 95%. The question being addressed is—"does the same lack of correlation between penetration and deposition, resulting from varying particle size distributions in occupational environments, exist for ambient air?" This paper will assess this relationship by analysing data on size distributions of ambient air particles from published studies.

METHODS

From articles on ambient air studies, a variety of particle size distributions was found. A meta analysis of eight studies (Kadowaki, 1976; Lee and Patterson, 1969; Lee and Goranson, 1970; Lee *et al.*, 1972; Lee and Smith, 1972; Lundgren, 1970; Mainwaring and Harsha, 1976; Patterson and Wagman, 1977) with 250 particle size distributions was performed. For each of the size distributions, the geometric mean (GM) and geometric standard deviation (GSD) was used as given or as estimated from figures to calculate the penetration and deposition. The PM$_{10}$ criterion was used to calculate the percentage of the distribution that contributes to penetration.

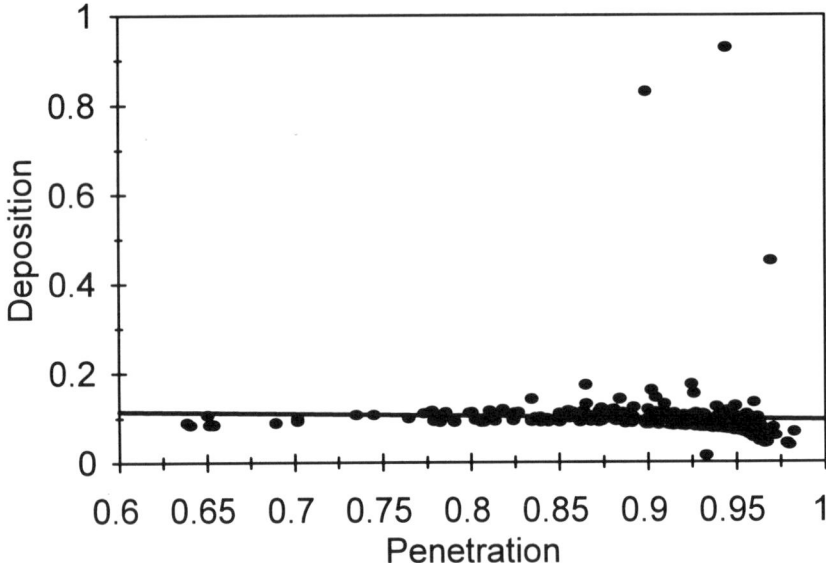

Fig. 2. Two-hundred and fifty data points derived from all the size distributions, comparing estimates of the fraction of the total distribution equivalent to the PM_{10} (penetration) and the fraction of the same distribution that would deposit in the tracheobronchial region of the lung. The slope of the regression line was found to not be significantly different than 0 ($\alpha = 0.05$) indicating that there would be no functional relationship between the two measures.

Deposition was estimated using Stahlhofen *et al.* (1989), for a breathing rate of seven breaths per minute and a minute volume of 3500 ml. Results of studies in which the size distributions of the mass fraction of particular chemical constituents were given are also included. As noted above, the toxic properties of the particulate may relate more to the chemical constituents, the organics or metals content for example. Thus the size distribution of those mass fractions may be relevant, if mass can be assumed to be relevant at all.

RESULTS AND DISCUSSION

Figure 2 shows the regression line for all the calculated values of deposition and penetration in those studies. It should be noted that the slope (-0.05) was not significantly different than 0 ($\alpha = 0.05$), indicating a lack of any measurable correlation. PM_{10}, therefore, does not by itself well represent dose to the upper respiratory tract as was previously seen in occupational environments. However subtraction of the $PM_{2.5}$ fraction (which is a parallel criterion to the PM_{10}, but with the 50% collection point at 2.5 μm) did not improve the correlation for ambient air even though a similar technique had worked to improve correlations for occupational environments.

If there is not a good relationship between the penetration and deposition, then additional variability will be introduced into the dose–response analysis. This lessens the chance of observing a statistically significant exposure–response relationship and calls into question the general applicability of the PM_{10} sampling network. It is vital that a relationship between measurement and dose be closely

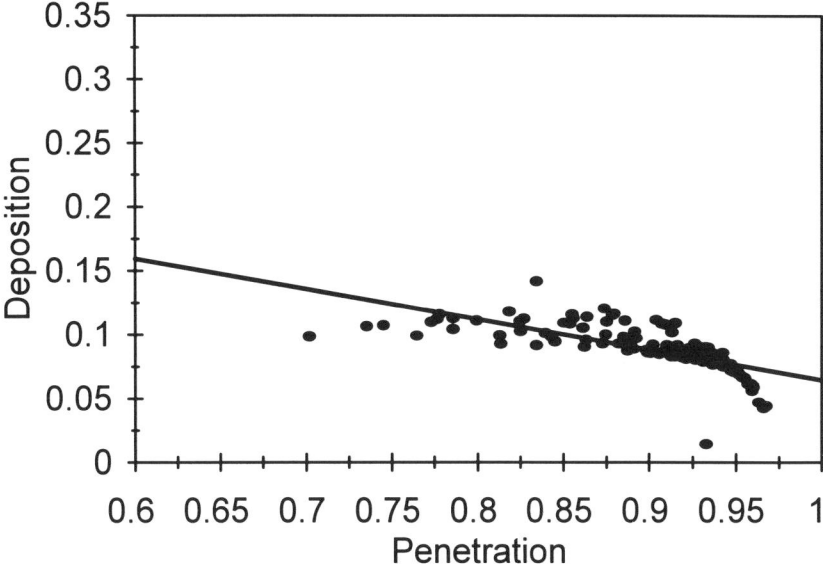

Fig. 3. One hundred and twenty-four data points for only those size distributions with GM > 1.0 µm and GSD < 4.5. As in Fig. 2, there was no positive functional relationship found between PM$_{10}$ (penetration) and tracheobronchial deposition.

correlated in order to set standards that are applicable regardless of the particle size distribution. At the same time, it is acknowledged that PM$_{10}$ is not meant to represent the dose to the lung of submicrometer aerosols.

All the size distributions with GMs greater than 1 µm and having a GSD less than 4.5 were examined. This was done to eliminate those distributions that might have substantial amounts of submicrometer particle mass, since there would be no association between deposition and penetration in that region for PM$_{10}$. When the PM$_{10}$ is compared with the tracheobronchial deposition a negative slope is seen (Fig. 3). This implies improbably that as the exposure increases the dose decreases. However, using the subtraction method (PM$_{10}$ minus PM$_{2.5}$), increased the correlation. Figure 4 shows the results of the regression calculation for this data, which had a correlation coefficient of 0.81.

CONCLUSION

Interpretation of size distribution information noted here indicates that PM$_{10}$ is not a good measure of dose to the upper respiratory tract. While there could be temporal stability to the size distributions the spatial differences make it difficult to support any finding that the PM$_{10}$ offers equal protection at all locations for a mass dose to the upper respiratory tract. This offers an explanation for some of the lack of observed correlation between exposure measurement and health effects seen in epidemiology studies.

It has been demonstrated in an occupational and now in an ambient air study that there can exist a lack of a consistent relationship between penetration measurement and dose due to the variations in size distribution. The PM$_{10}$ criterion and the

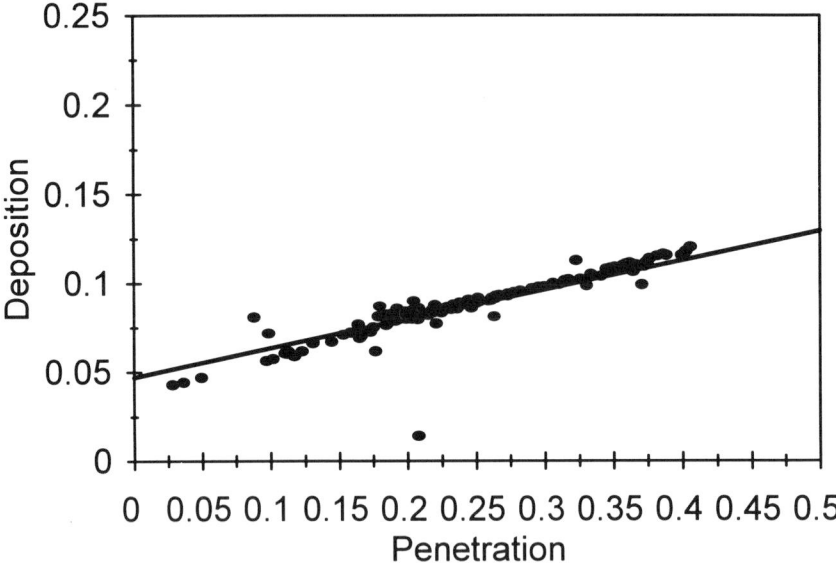

Fig. 4. The same 124 size distributions used in Fig. 3, but the measure of penetration is now PM_{10}–$PM_{2.5}$ compared with tracheobronchial deposition. The statistically significant positive relationship between the two measures is lessened if the submicrometer size distribution are included in the analysis.

thoracic criterion because of its similarity, are poor indicators of dose to the tracheobronchial region.

REFERENCES

Boubel, R. W., Fox, D. L., Turner, D. B. and Stern, A. C. (1994) *Fundamentals of Air Pollution*, p. 374. Academic Press, New York.

Kadowaki, S. (1976) Size distribution of atmospheric total aerosols, sulfate, ammonium and nitrate particulate in the Nagoya area. *Atmos. Environ.* **10**, 39–43.

Lee, R. E. Jr and Goransons, S. (1970) National air surveillance cascade impactor network, III. Variation in size of airborne particulate matter over a three-year period. *Environ. Sci. Tech.* **10**, 1022–1027.

Lee, R. E. Jr, Goranson, S., Enrione, R. and Morgan, G. (1972) National air surveillance cascade impactor network II. Size distribution measurements of trace metal components. *Environ. Sci. Tech.* **6**, 1025–1030.

Lee, R. E. Jr and Patterson, R. K. (1969) Size determination of atmospheric phosphate, nitrate, chloride, and ammonium particulate in several urban areas. *Atmos. Environ.* **3**, 249–255.

Lee, R. E. and Smith, C. F. (1972) Size distribution of suspended particulate from lignite combustion. *Environ. Sci. Tech.* **6**, 929–930.

Lundgren, D. A. (1970) Atmospheric aerosol composition as a function of particle size and of time. *J. Air Pollut. Control. Assoc.* **19**, 795–801.

McCawley, M. (1993) Caveats in the use of particle size-selective sampling criteria. *Proc. 2nd Int. Conf. on Occupational Health and Safety in the Minerals Industry*, Perth, Australia.

Mainwaring, S. and Harsha, S. (1976) Size distribution of aerosols in Melbourne city air. *Atmos. Environ.* **10**, 57–60.

Patterson, R. and Wagman, J. (1977) Mass and composition of an urban aerosol as a function of particle size for several visibility level. *J. Aerosol Sci.* **8**, 269–279.

Pope, C. A., Dockey, D. W. and Schwartz, J. (1996) Review of epidemiologic evidence of health effects of particulate air pollution. *Inhalation Toxicol.* **7**, 1–18.

Stahlhofen, W., Rudolf, G. and James, A. C. (1989) Intercomparison of experimental regional aerosol deposition data. *J. Aerosol Med.* **2**, 285–308.

 Pergamon

Ann. occup. Hyg., Vol. 41, Supplement 1, pp. 19–23, 1997
© 1997 Published by Elsevier Science Ltd on behalf of BOHS
Printed in Great Britain. All rights reserved
0003–4878/97 $17.00 + 0.00
Inhaled Particles VIII

PII: S0003–4878(96)00128–7

COMPARISON OF THE NUMBER OF ULTRA-FINE PARTICLES AND THE MASS OF FINE PARTICLES WITH RESPIRATORY SYMPTOMS IN ASTHMATICS

A. Peters,* H. E. Wichmann,* T. Tuch,†‡ J. Heinrich* and J. Heyder†

*GSF-Institute of Epidemiology, Postfach 1129, Oberschleissheim, D-85758, Neuherberg, Germany;
†GSF-Institute for Inhalation Biology, Neuherberg, Germany; and
‡Ludwig-Maximilians University, Munich, Germany

INTRODUCTION

Recently it has been recognised that ambient particles might play an important role in pollution-induced respiratory responses (Dockery and Pope, 1994; Pope *et al.*, 1995a,b; Brunekreef *et al.*, 1995; Bascom *et al.*, 1996). However, it is unclear which properties of the particles are responsible for the observed health effects. Fine and even ultra-fine particles might be more "toxic" than coarse particles and the "physical" toxicity of these particles might even exceed their "chemical" toxicity. Consequently more insight into health related aspects of particulate air pollution will be obtained by correlating respiratory responses with mass and number concentration of ambient particles. Therefore, the role of fine and ultra-fine particles in eliciting respiratory health effects was studied in adults with a history of asthma in Erfurt, Eastern Germany (Peters *et al.*, in press). The present paper presents more detailed analyses on the temporal relationship between particles and respiratory symptoms.

METHODS

The aerosol size spectrometer consisted of two sensors covering different size ranges. In the range 0.01–0.3 μm ambient particles were classified with an electrical mobility analyses (TSI, model 3071) according to their volume-equivalent diameter and counted with a condensation particle counter (TSI, model 3760). In the size range 0.1–2.5 μm particles were classified by an optical particle counter (PMS, model LAS-X): particles smaller than the wavelength of the applied laser light according to their volume-equivalent diameter; larger particles according to their cross-section illuminated by the laser beam. However, while the electrical classification is independent of the chemical properties, the optical classification is not. Therefore, the response function of the optical counter requires adjustment to the refraction index of the particles to be classified. For this adjustment, monodisperse particles selected from the ambient aerosol by electrical classification were frequently used in this study (Tuch *et al.*, submitted). The number distribution can be converted into a particle volume distribution, which was used to calculate the particle mass distribution assuming an average density of the aerosol particles. When daily total particle volume concentrations and occasional $PM_{2.5}$ measure-

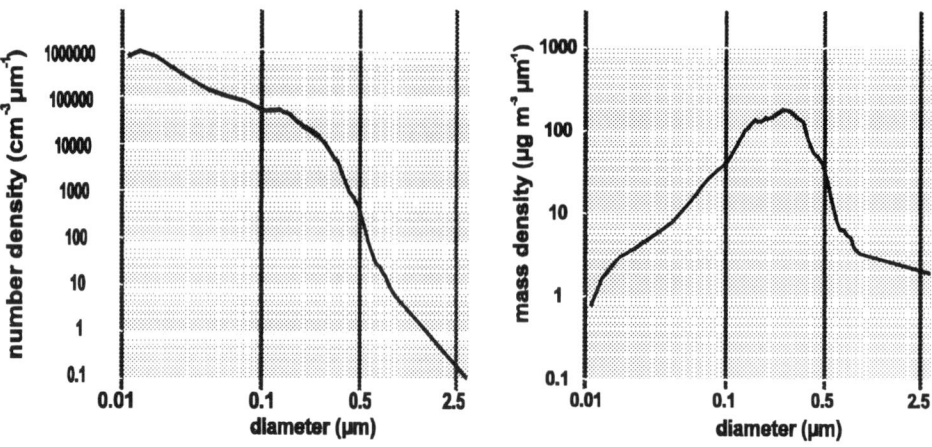

Fig. 1. Mean number and mass distribution of ambient particles in Erfurt, 1991/92 (Tuch *et al.*, submitted).

ments were compared an apparent particle density of 1500 kg m^{-3} could be determined for Erfurt. The derived particle mass distributions were used to calculate integral mass concentrations of particles between selected cut-off diameters. The following abbreviations will be used consequently: NC$_{0.01-0.1}$ for the number concentration of the ultra-fine particles, NC$_{0.01-2.5}$ for the total number concentration of fine particles, MC$_{0.1-0.5}$ for the mass concentration of fine particles with a diameter between 0.1 and 0.5 μm and MC$_{0.01-2.5}$ for the mass concentration of the fine particles.

During the winter 91/92 27 non-smoking adults with a history of asthma participated in a panel study (Peters *et al.*, 1996, in press). They were 44–80 years old in 1992 and 63% of them were women. The panelists recorded their respiratory symptoms daily in a diary. The analysis was restricted to the winter period (from September through March). Data on fine and ultra-fine particles was only available after 1 October, 1991. Data were obtained on 145 days. Regression analyses of population averaged time-series were used to control for possible confounding by time-varying influences on respiratory symptoms as has been described previously (Peters *et al.*, 1996, in press). The prevalence of feeling ill during the day and cough was analysed using a logistic regression model adjusting for a linear trend, temperature, weekend and an autocorrelated error structure of order one. A 5-day mean was used to estimate the cumulative impact of air pollution. It was calculated by averaging the exposure on the current day and 4 days prior. In case of missing data the remaining measurements were taken. All regression coefficients were expressed as effects associated with a change of the exposure for one inter-quartile range.

RESULTS

Daily mean number and mass distributions were obtained with the mobile aerosol spectrometer. Figure 1 presents the mean distribution of the number and

Table 1. Odds ratios (OR) for the symptom feeling ill during the day in association with the fine and ultra-fine particles

	IQR [# cm^{-3}]	$NC_{0.01-2.5}$ OR	95% Cl	IQR	$NC_{0.01-0.1}$ OR	95% Cl
same-day	15 120	1.29	(1.05, 1.58)	12 000	1.21	(0.98, 1.50)
lag 1	15 120	1.36	(1.11, 1.66)	12 000	1.30	(1.07, 1.58)
lag 2	15 120	1.27	(1.04, 1.54)	12 000	1.26	(1.04, 1.52)
lag 3	15 120	1.11	(0.90, 1.37)	12 000	1.06	(0.87, 1.29)
lag 4	15 120	1.22	(1.00, 1.48)	12 000	1.20	(1.00, 1.45)
lag 5	15 120	1.23	(1.02, 1.49)	12 000	1.19	(0.99, 1.43)
5 day mean	10 508	1.39	(1.15, 1.68)	9 200	1.44	(1.15, 1.81)

	[µg m^{-3}]	$MC_{0.01-2.5}$ OR	95% Cl	[µg m^{-3}]	$MC_{0.1-0.5}$ OR	95% Cl
same-day	57	1.24	(1.09, 1.41)	47.5	1.23	(1.09, 1.40)
lag 1	57	1.27	(1.09, 1.49)	47.5	1.23	(1.07, 1.41)
lag 2	57	1.15	(0.97, 1.33)	47.5	1.12	(0.97, 1.30)
lag 3	57	1.13	(0.96, 1.33)	47.5	1.11	(0.97, 1.27)
lag 4	57	1.14	(0.98, 1.33)	47.5	1.12	(0.98, 1.28)
lag 5	57	1.18	(1.02, 1.38)	47.5	1.16	(1.01, 1.32)
5 day mean	50	1.21	(1.06, 1.38)	33.7	1.19	(1.05, 1.35)

the mass of the particles during the winter 91/92. Seventy-three per cent of the particles were smaller than 0.1 µm in diameter. Eighty-two per cent of the mass concentration were associated with particles in the size range 0.1–0.5 µm (Tuch *et al.*, submitted; Peters *et al.*, in press). Since the number concentration of the particles smaller than 0.1 µm was not highly correlated with the mass concentration of particles with a diameter between 0.1 and 0.5 µm ($r = 0.51$) (Peters *et al.*, 1996), these two fractions and the total number and the total mass of the fine particles were chosen to evaluate the impact of the fine and ultra-fine particles on respiratory symptoms.

Analyses were adjusted for a linear trend, mean daily temperature and weekend, but only a negative linear trend showed to be statistically significant. A stronger association with the prevalence of feeling ill during the day was observed for the number concentration of particles ($NC_{0.01-2.5}$) than for their mass concentration ($MC_{0.01-2.5}$) (Table 1). The exposure on previous days appeared to contribute to the prevalence of feeling ill concurrently (Peters *et al.*, in press). This observation was supported both by estimating the impact of previous days alone or by using a 5 day mean (Table 1). While no strong association between the particle number concentrations on the same day and the prevalence of cough was observed, some indication was found that prolonged exposure to elevated number concentrations of ultra-fine particle might be associated with increasing reporting of cough (Table 2). The difference between the odds ratios for the 5 day means of $NC_{0.01-0.1}$ and $MC_{0.1-0.5}$ was statistically significant ($p < 0.05$). Both the mass of the fine particles $MC_{0.01-2.5}$ and the mass of the particles between 0.1 and 0.5 µm $MC_{0.1-0.5}$ showed only associations with cough when the same-day concentrations were considered.

Table 2. Odds ratios for the symptom cough in association with the fine and ultra-fine particles

	IQR [# cm^{-3}]	NC$_{0.01-2.5}$ OR	95% CI	IQR	NC$_{0.01-0.1}$ OR	95% CI
same-day	15 120	1.16	(0.98, 1.37)	12 000	1.12	(0.95, 1.33)
lag 1	15 120	1.06	(0.89, 1.26)	12 000	1.04	(0.88, 1.23)
lag 2	15 120	1.10	(0.94, 1.29)	12 000	1.12	(0.97, 1.31)
lag 3	15 120	1.07	(0.91, 1.27)	12 000	1.08	(0.92, 1.27)
lag 4	15 120	1.06	(0.90, 1.25)	12 000	1.09	(0.94, 1.28)
lag 5	15 120	0.96	(0.82, 1.13)	12 000	0.97	(0.83, 1.14)
5 day mean	10 580	1.17	(1.01, 1.37)	9 200	1.26	(1.06, 1.50)

	[μg m^{-3}]	MC$_{0.01-2.5}$ OR	95% CI	[μg m^{-3}]	MC$_{0.1-0.5}$ OR	95% CI
same-day	57	1.19	(1.07, 1.32)	47.5	1.18	(1.06, 1.31)
lag 1	57	1.07	(0.92, 1.25)	47.5	1.06	(0.93, 1.21)
lag 2	57	0.93	(0.80, 1.09)	47.5	0.94	(0.82, 1.08)
lag 3	57	0.97	(0.84, 1.12)	47.5	0.97	(0.86, 1.10)
lag 4	57	0.88	(0.76, 1.03)	47.5	0.90	(0.75, 1.03)
lag 5	57	0.95	(0.82, 1.09)	47.5	0.95	(0.84, 1.08)
5 day mean	50	1.02	(0.91, 1.15)	39.7	1.02	(0.91, 1.14)

DISCUSSION

The measured total particle number concentration was determined by ultra-fine particles in Erfurt during the winter 91/92. The mass of the fine particles was dominated by particles between 0.1 and 0.5 μm in diameter. The relatively low correlation between the mass and the number concentrations of the particles enabled us to analyse their contribution to health effects in adults with a history of asthma (Peters *et al.*, in press).

Increases in the prevalence of feeling ill during the day and of cough were observed both for the mass and the number concentrations of the particles. The general health of the panellists meliorated in associations with elevated levels of particles on the same day as well as on previous days. The strongest association with the prevalence of feeling ill during the day was found for the 5-day mean of the number concentrations of ultra-fine particles. Similarly, the prevalence of cough increased in association with the 5-day mean of the number concentrations of ultra-fine particles. Therefore, the increases in symptoms and the decrease in lung function (Peters *et al.*, in press) might be caused by an inflammation in the alveoli as a reaction to ultra-fine particles, as has been hypothesised recently by Seaton *et al.* (Seaton *et al.*, 1995).

REFERENCES

Bascom, R., Bromberg, P. A., Costa, D. A., Devlin, R., Dockery, D. W., Frampton, M. W., Lambert, W., Samet, J. M., Speizer, F. E. and Utell, M. (1996) Health effects of outdoor air pollution. *Am. J. Resp. Crit. Care Med.* **153**, 3–50.
Brunekreef, B., Dockery, D. W. and Krzyzanowski, M. (1995) Epidemiologic studies on short-term effects of low levels of major ambient air pollution components. *Environ. Health Perspect.* **103**, 3–13.
Dockery, D. W. and Pope, C. A. (1994) Acute respiratory effects of particulate air pollution. *Annu. Rev. Public Health* **15**, 107–132.

Peters, A., Goldstein, I. F., Beyer, U., Franke, K., Heinrich, J., Dockery, D. W., Spengler, J. D. and Wichmann, H. E. (1996) Acute health effects of exposure to high levels of air pollutants in eastern Europe. *Am. J. Epidemiol.* **144**, 570–581.

Peters, A., Wichmann, H. E., Tuch, T., Heinrich, J. and Heyder, J. Respiratory effects are associated with the number of ultra-fine particles *Am. J. Resp. Crit. Care Med.* (in press).

Pope, C. A., Bates, D. V. and Raizenne, M. E. (1995a) Health effects of particulate air pollution: time for reassessment? *Environ. Health Perspect.* **103**, 472–480.

Pope, C. A., Dockery, D. W. and Schwartz, J. (1995b) Review of epidemiological evidence of health effects of particulate air pollution. *Inhal. Toxico.* **7**, 1–18.

Seaton, A., MacNee, W., Donaldson, K. and Godden, D. (1995) Particulate air pollution and acute health effects. *Lancet* **345**, 176–178.

Tuch, T., Brand, P., Wichmann, H. E. and Heyder, J. Application of a mobile aerosol spectrometer for characterizing ambient aerosol particles in Eastern Germany (submitted).

Pergamon

Ann. occup. Hyg., Vol. 41, Supplement 1, pp. 24–31, 1997
© 1997 British Occupational Hygiene Society
Published by Elsevier Science Ltd. All rights reserved
Printed in Great Britain
0003–4878/97 $17.00 + 0.00
Inhaled Particles VIII

PII: S0003–4878(96)00129–9

GENOTOXIC EFFECTS INDUCED BY AIRBORNE PARTICULATES ON TRACHEOBRONCHIAL EPITHELIAL CELLS *IN VITRO*

C. Hornberg, L. Maciuleviciute and N. H. Seemayer

Medical Institute of Environmental Hygiene, Gurlittstr. 53, D-40225 Düsseldorf, Germany

INTRODUCTION

Airborne particulate matter adsorbs chemicals, forming a locus for chemical interactions and a vehicle for transmission. Chemical analyses have shown that airborne particulate matter from heavily industrialised areas contains more than 500 chemical substances, among them putative and known carcinogens, particularly polycyclic aromatic hydrocarbons, acridines, traces of heavy elements etc. (Goldstein, 1983). It has been recognised for many years that high levels of particulate air pollution can lead to increased mortality and morbidity. Dockery and Schwartz (1995) reported that daily mortality was associated with particulate air pollution measured as total suspended particulates (TSP). Similar analysis have shown positive correlations between particulate air pollution and daily mortality in six communities in the United States (Dockery *et al.*, 1993; Schwartz, 1994). Other epidemiologic studies have highlighted an association between inhalable particulates with an aerodynamic diameter of less than 10 μm and especially fine-particulates below 2.5 μm incriminated to contribute to excess morbidity and mortality in certain U.S. cities (Pope *et al.*, 1995).

The respiratory tract is the major site of exposure to airborne particulates (Heyder, 1993) and genotoxic activity of airborne particulates leading to mutation and cancer has been well documented (Epstein *et al.*, 1977; Chrisp and Fisher, 1980). An important target of aerosols, especially of inhalable particulate matter with a particle size smaller than 10 μm are the tracheobronchial epithelial cells. The most common cancer in man, the bronchogenic carcinoma, originates from these cells (Tomatis, 1990). Several *in vitro* models have been developed for the study of respiratory toxicology, carcinogenesis and mutagenesis utilising rodent and human tracheal epithelial cells in culture (Nettesheim and Barrett, 1984; Harris 1987). In previous studies it has been reported that organic extracts of airborne particulates are cytotoxic, mutagenic and carcinogenic in a number of short-term bioassays using rodent and human tissue culture cells (Seemayer *et al.*, 1988; Hornberg *et al.*, 1993; Hornberg and Seemayer, 1995a, b; Motykiewicz *et al.*, 1996). In this study we present data of genotoxic effects of a sample of airborne particulates collected in Duisburg, a city in the highly industrialised Rhine–Ruhr district, Germany. As target cells we utilised tracheal epithelial cells of the golden Syrian hamster, the rat and human bronchial epithelial cells of the line BEAS-2B (Reddel *et al.*, 1988).

Cytotoxicity was determined by reduction of plating efficiency. As a sensitive criterion of genotoxicity testing we determined the induction of sister chromatid exchanges (SCE) in the three test systems.

MATERIAL AND METHODS

Airborne particulate sampling and extraction

Sample no. 69 of airborne particulates was collected in spring 1991 in the city of Duisburg, Germany, a highly industrialised area and the centre of iron, steel and coke production. For collection we used a high volume sampler HVS 150 (Ströhlein Instruments) equipped with glass fibre filters. The filters were extracted with dichlormethane as described previously (Tomingas et al., 1977; Seemayer et al., 1990). Global extracts (GEX) were quantitatively transferred to dimethyl sulphoxide (DMSO) for biological testing in cell culture experiments. The final concentrations of DMSO in medium did not exceed 0.5%. Air volumes of collection of airborne particulates were measured and extractable substances are presented as cubic metres (cbm) ml^{-1} medium. The benzo(a)pyrene content of the extract was measured as described previously in detail (Tomingas et al., 1977). Sample of GEX no. 69 containing extractable substances of airborne particulates from 1557 cbm of air dissolved per ml DMSO, revealed a benzo(a)pyrene concentration of 1640 ng ml^{-1} DMSO.

Isolation and cultivation of tracheal epithelial cells

The isolation of tracheal epithelial cells from the golden Syrian hamster and the rat has been described in detail previously (Seemayer et al., 1994; Hornberg and Seemayer, 1995b). Briefly, excised tracheas were longitudinally opened along the cartilaginous part, spread and the inner epithelial layer covered with Eagle's minimum essential medium containing 0.25% protease (type 14, Sigma). Following digestion for 24 h at 37 or 4°C, respectively, the dissociated epithelial cells were flushed out with medium from the inner layer of tracheas and suspended in a complex mixture of media. After centrifugation the cell pellet was resuspended in "complete medium". Complex tissue culture medium "complete medium" contains 3 media in equal parts containing Ham's F-12 medium, CMRL 1066 medium and conditioned medium of confluent NIH 3T3 cells supplemented with insulin, transferrin, hydrocortisone, epidermal growth factor, bovine serum albumin without or with 1.5% bovine serum as previously described in detail (Hornberg et al., 1993). Penicillin and streptomycin were added to all media. Cultivation of tracheal epithelial cells was performed on a fibronectin matrix in plastic flasks at 37°C and at pH 7.4 in a humidified atmosphere of 5% CO_2 in air.

The human bronchoepithelial cell line BEAS-2B obtained by courtesy of Curt C. Harris, National Cancer Institute Bethesda from Dr Lang (Hamburg), was cultivated in serum-free "complete medium". After reaching nearly confluence cells were detached with a trypsin-EDTA-solution. Thereafter, to abrogate the trypsin effect, trypsin inhibitor solution (soyabean trypsin inhibitor, Sigma) was added, followed by centrifugation at 1000 rpm for 5 min. Cell pellet was resuspended in "complete medium" for further cultivation.

Detection of "sister chromatid exchanges (SCE)" in tracheal epithelial cell cultures from rodents and in human bronchoepithelial cells of line BEAS-2B

Rodent tracheal epithelial cells at passage 1–4 were detached with a trypsin-EDTA-solution, resuspended in fresh "complete medium" and seeded into plastic dishes (Quadriperm, Hereaus) on fibronectin precoated slides at a cell concentration of 1.25×10^5 cells. Following a cultivation period of 24 h extract no. 69 was added to the cultures in declining 5–7 concentrations in presence of bromodeoxyuridine (15 µg m^{-1}). Cultivation was continued for 48 h, the last 3 h in presence of demecolcine (Colcemid, 10 µg ml^{-1}, Sigma) or nocodazole (20 ng ml^{-1}, Janssen, Belgium). After hypotonic treatment and fixation of cells, chromosomes were stained by the fluorescent plus Giemsa (FPG) technique as earlier reported (Hornberg *et al.*, 1993). Human bronchoepithelial cells of line BEAS-2B were seeded on glass slides into Quadriperm dishes at a cell concentration of 70 000 cells. After a cultivation period of 72 h at 37°C cell cultures were refed with "complete medium" and bromodeoxyuridine (15 µg ml^{-1}) as well as various concentration of extract no. 69 were added. Three hours before the end of a 72 h cultivation period demecolcine (Colcemid 0.2 µg ml^{-1}) was added. The further steps as well as the staining of chromosomes by the fluorescent plus Giemsa (FPG) technique was identical with the treatment of hamster or rat tracheal epithelial cells.

Reduction of "plating efficiency"

Logarithmically growing cells of the human cell line A 549 (pneumocyte type II) were seeded at a concentration of 100 cells per tissue culture flasks of 25 cm^2 using Dulbecco's modification of Minimum Essential Medium supplemented with 10% fetal calf serum and antibiotics. 24 h after seeding to cell cultures various concentrations of extract no. 69 were added and incubated for 120 h. Thereafter cell cultures were rinsed with phosphate buffered saline (PBS), supplied with fresh medium and further cultivated for 6 days. Following fixation with Bouin's solution and staining with crystal violet "plating efficiency" was determined by scoring colonies of more than 50 cells (Seemayer *et al.*, 1990).

Statistical analysis

Experimental data were computerised and mean values, limits of confidence and standard deviations determined. Bartlett's test for equal variances, one way analysis of variance and Student's *t*-test were performed.

RESULTS

The rodent tracheal epithelial cells and the human bronchial epithelial cell line BEAS-2B used in this study show a morphology of typical epithelial cells containing cytokeratin demonstrable by indirect immunoflurescence (Seemayer *et al.*, 1994; Hornberg *et al.*, 1996). The BEAS-2B cell line was established by Reddel *et al.* (1988) by transformation with an Adenovirus 12-SV40 hybrid virus. These cells express the SV 40 specific T-antigen and reveal predominantly a near diploid chromosome number.

The cytotoxic effect of extract no. 69 of airborne particulates was evaluated by reduction of "plating efficiency" of human cell line A-549. While no reduction of

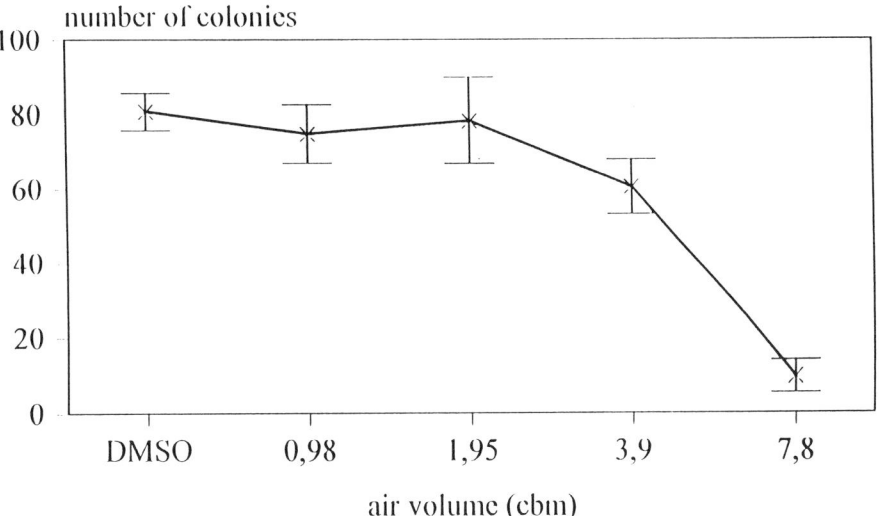

Fig. 1. Reduction of "plating efficiency" of human cells of line A-549 in presence of various concentrations of extract (GEX) no. 69 of airborne particulates. Air volumes of collection are presented in cubic metre (cbm). Values are means and limits of confidence are presented.

"plating efficiency" was observed in presence of extractable substances from 0.98 cbm of air, a reduction to nearly 60% of control values occurred by extractable substances from 3.90 cmb of air (Fig. 1). Higher concentrations of extract no. 69 disclosed strong dose-dependent reduction of "plating efficiency".

We compared the genotoxic activity of extract no. 69 of airborne particulates utilising cultures of hamster and rat tracheal epithelial cells and human bronchial epithelial cell line BEAS-2B. The dose-dependent induction of SCE by GEX no. 69 in tracheal epithelial cell cultures of the rat and of the Syrian golden hamster is depicted in Figs 2 and 3. On the abscissa, increasing concentrations of GEX no. 69 are outlined corresponding to air volumes of particulate collection in the range of 0.25 up to 15.57 cbm air. On the ordinate, SCE/chromosome are shown. Rodent tracheal epithelial cells reveal a strong dose-related increase of SCE in the presence of GEX no. 69. Up to 2.5-fold values of SCE can be seen in the presence of the extract from 15 cbm air; substances from 0.25 cbm air are also effective. The concentration of benzo(a)pyrene in the experiment with GEX no. 69 was in the range of 0.25–16 ng ml^{-1} medium. The lowest effective dose leading to a significant increase of SCE in the presence of GEX no. 69 was observed with rat tracheal epithelial cells at a concentration of 0.49 cbm ml^{-1} medium ($P < 0.001$) and with hamster tracheal epithelial cells at 0.98 cbm ml^{-1} medium ($P < 0.001$).

The human bronchial epithelial cell line BEAS-2B revealed a strong dose-related induction of "sister chromatid exchanges" in presence of increasing concentrations of extract no. 69 (Fig. 4). A more than two-fold increase of SCE was observed in presence of extractable substances from 2.01 cbm of air. It is remarkable that substances from as little as 0.25 cbm of air caused a highly significant ($P < 0.001$) increase of "sister chromatid exchanges". The steepest dose-related increase of

C. Hornberg *et al.*

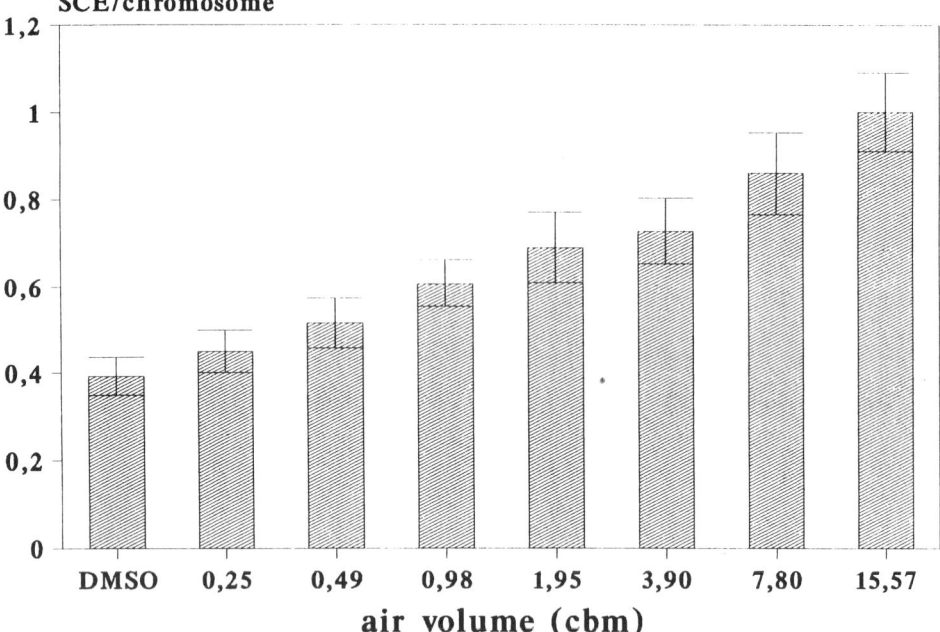

Fig. 2. Dose-related induction of SCE in tracheal epithelial cells of the golden Syrian hamster *in vitro* in the presence of various concentrations of extract (GEX) no. 69 of airborne particulates in cubic metre (cbm). Values are means and limits of confidence are presented.

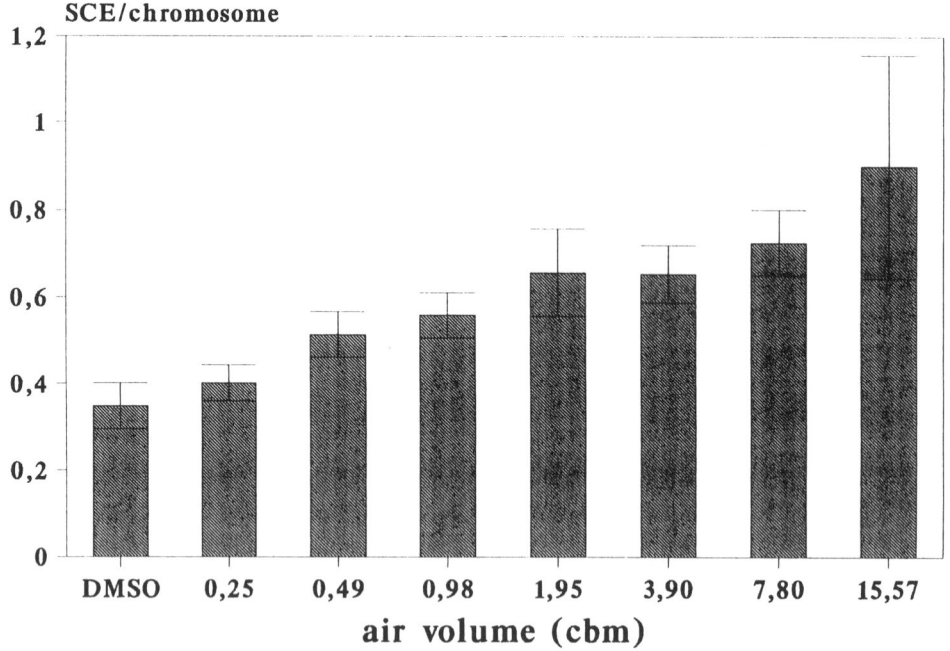

Fig. 3. Dose-related induction of SCE in tracheal epithelial cells of the rat *in vitro* in presence of various concentrations of extract (GEX) no. 69 of airborne particulates in cubic metre (cbm). Values are means and limits of confidence are presented.

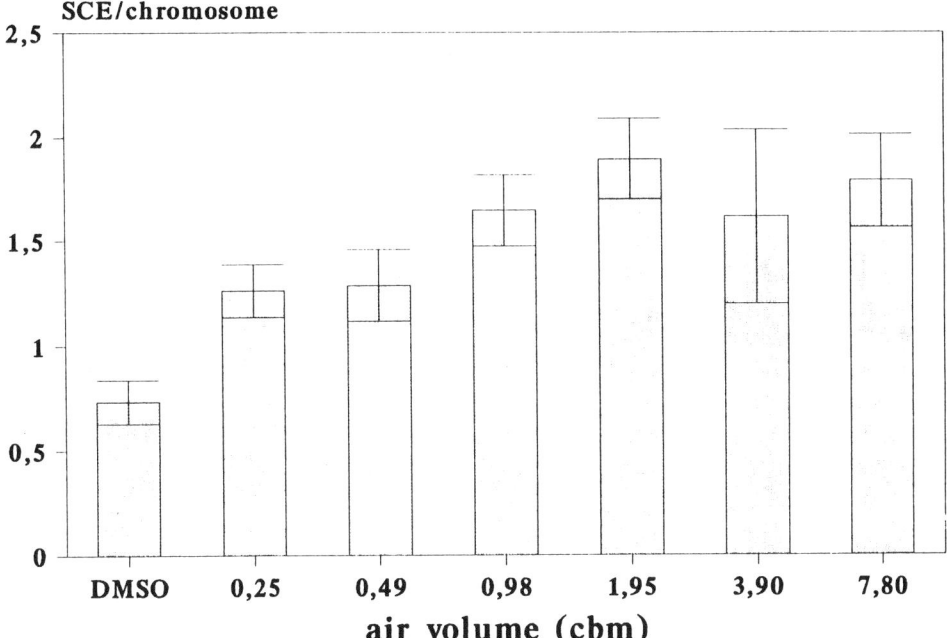

Fig. 4. Dose-related induction of SCE in human bronchial epithelial cells of line BEAS-2B *in vitro* in presence of various concentrations of extract (GEX) no. 69 of airborne particulates in cubic metre (cbm). Values are means and limits of confidence are presented.

"sister chromatid exchanges" disclosed the human bronchial epithelial cell line BEAS-2B, followed by rodent tracheal epithelial cells.

DISCUSSION

Airborne particulates pose, by inhalation, a health risk to humans by their cytotoxic, mutagenic and carcinogenic activity. Recent results demonstrate an association between airborne particulates and increasing mortality rate in some U.S. cities (Dockery *et al.*, 1993; Pope *et al.*, 1995). Genotoxic damage in humans from air pollutants have been reported (Perera *et al.*, 1992; Motykiewitcz *et al.*, 1996) in a biomonitoring survey in Poland utilising six endpoints of molecular biology and cytogenetics. Results provided a link between exposure and genetic alterations relevant to cancer induction.

Earlier reports and results presented show that rodent and human tissue culture cells can be used successfully to evaluate cytotoxic, mutagenic and carcinogenic potency of airborne particulates and exhibit special advantages for testing biological effects of air pollutants (Seemayer *et al.*, 1984). The culture of human airway epithelial cells especially plays an important role in advancing the understanding of the metabolic and molecular mechanisms underlying normal function and disease pathology of airway epithelial cells (Gruenert *et al.*, 1995).

Despite a reduction of airborne particulates in the Rhine–Ruhr district since 1969 from 100–200 μg cbm^{-1} of air to values of 30–50 μg cbm^{-1} of air today, cytotoxic, mutagenic and carcinogenic activity of airborne particulates calculated per volume of air remained nearly unchanged (Hornberg *et al.*, 1993; Hornberg and Seemayer, 1995, a, b; Seemayer *et al.*, 1984, 1990, 1994). Our results demonstrate a cytotoxic effect of sample no. 69 of airborne particulates corresponding to extractable substances from 4.1 cbm of air upwards. In contrast, genotoxic activity was already demonstrable with extractable substances from 0.25–0.50 cbm of air by induction of "sister chromatid exchanges" in rodent tracheal epithelial cells and human bronchial epithelial cells.

A dose related increase of sister chromatid exchanges in exposed rodent and human tracheobronchial epithelial cells reveals a serious risk for humans by genotoxic activity of airborne particulates. For health risk evaluation we have to take into consideration a daily respiratory ventilation in man of 12–14 cbm of air or the inhalation of 4000–5000 cbm of air per year. Furthermore, we must take into account the lowest effective genotoxic dosage demonstrable by a significant increase of sister chromatid exchanges. Repeated inhalation of genotoxic substances from airborne particulates could exceed body's ability to eliminate and detoxify (Repace, 1982) and therefore pose a health risk of intoxication and cancer development.

REFERENCES

Chrisp, C. E. and Fisher, G. L. (1980) Mutagenicity of airborne particles. *Mutat. Res.* **76**, 143–164.

Dockery, D. W., Pope III, C. A., Xu, X., Spengler, J. D., Ware, J. H., Fay, M. E., Ferris, B. J. and Speizer, F. E. (1993) An association between air pollution and mortality in six U.S. cities. *N. Engl. J. Med.* **329**, 1753–1759.

Dockery, D. W. and Schwartz, J. (1995) Particulate air pollution and mortality: more than the Philadelphia story. *Epidemiology* **6**, 629–632.

Epstein, S. S., Fujii, K. and Asahina, S. (1977) Carcinogenicity of a composite organic extract of urban particulate atmospheric pollutants following subcutaneous injection in infant mice. *Environ. Res.* **19**, 163–176.

Goldstein, P. D. (1983) Toxic substances in the atmospharic environment. *J. Air. Poll. Control Assoc.* **33**, 454–467.

Gruenert, D. C., Finkbeiner, W. W. and Widdicombe, J. H. (1995) Culture and transformation of human airway epithelial cells. *Am. J. Physiol.* **268**, L347–L360.

Harris, C. C. (1987) Human tissue and cells in carcinogenesis research. *Cancer Res.* **47**, 1–10.

Heyder, J. (1993) Regional deposition of inhaled particles in the human respiratory tract. In *Advances in Controlled Clinical Inhalation Studies* (Edited by U. Mohr *et al.*), pp. 103–107. Springer, Berlin.

Hornberg, C., Seemayer, N. H., Hadnagy, W. and Ivanfy, K. (1993) Tracheal epithelial cells of the golden Syrian hamster *in vitro* as a tool for detection of genotoxic activity of airborne particulates. *J. Aerosol Sci.* **24**, 91–92.

Hornberg, C. and Seemayer, N. H. (1995a) Tracheal epithelial cells *in vitro* as a model to study genotoxicity of airborne particulates. *Toxic. In Vitro* **9**, 397–402.

Hornberg, C. and Seemayer, N. H. (1995b) Induction of sister chromatid exchanges in rodent tracheal epithelial cells as a sensitive bioassay for detection of genotoxic activity of airborne particulates. *Exp. Toxic Pathol.* **47**, 241–243.

Hornberg, C., Maciuleviciute, L. and Seemayer, N. H. (1996) Sister chromatid exchanges in rodent tracheal epithelium exposed in vitro to environmental pollutants. *Toxicol. Letters* **88**, 55–64.

Motykiewicz, G., Perera, F. P., Santella, R. M., Hemminki, K., Seemayer, N. H. and Chorazy, M. (1996) Assessment of cancer hazard from environmental pollution in Silesia. *Toxicol. Letters* **88**, 169–173.

Nettesheim, P. and Barrett, J. C. (1984) Tracheal epithelial cell transformation: a model system for studies on neoplastic progression. *CRC Critical Reviews in Toxicology 12* (Edited by G. Klein and S. Weinhouse), pp. 215–239. CRC Press, Boca Raton, FL.

Perera, F. P., Hemminki, K., Grzyzbowska, E., Motykiewicz, G., Michalska, J. Santella, R. M., Young, T. L., Dickey, C., Brandt-Rauf, P., DeVivo, I., Blaner, W., Tsai, W.-Y. and Chorazy, M.

(1992) Molecular and genetic damage in humans from environmental pollution in Poland. *Nature* **360**, 117–129.

Pope III, C. A., Bates, D. V. and Raizenne, M. E. (1995) Health effects of particulate air pollution: time for reassessment? *Environ. Health Perspect.* **103**, 472–480.

Reddel, R. R., Ke Y., Gerwin, B. I., McMenamin, M. G., Lechner, J. F., Su, R. T., Brash, D. E., Park, J.-B., Rhim, J. U.S. and Harris, C. C. (1988) Transformation of human bronchial epithelial cells by infection with SV 40 or adenovirus-12 SV40 hybrid virus, or transfection via strontium phosphate coprecipitation with a plasmid containing SV 40 early region genes. *Cancer Res.* **48**, 1904–1909.

Repace, J. L. (1982) Indoor air pollution. *Environ. Int.* **8**, 21–36.

Schwartz, J. (1994) Total suspended particulate matter and daily mortality in Cincinnati, Ohio. *Environ. Health Perspect.* **102**, 186–189.

Seemayer, N. H., Manojlovic, N. and Schürer, C. C. (1984) Cell cultures as a tool for detection of cytotoxic, mutagenic and carcinogenic activity of airborne particulate matter. *J. Aerosol Sci.* **15**, 426–430.

Seemayer, N. H., Hadnagy, W., Behrend, H. and Tomingas, R. (1988) Indicators of potential health risks by airborne particulates: cytotoxic, mutagenic and carcinogenic effects on mammalian cells *in vitro*. *Environmental Hygiene* (Edited by N. H. Seemayer and W. Hadnagy), pp. 54–59. Springer, Berlin.

Seemayer, N. H., Hadnagy, W. and Tomingas, R. (1990) Evaluation of health risks by airborne particulates from *in vitro* cyto- and genotoxicity testing on human and rodent tissue culture cells: a longitudinal study from 1975 until now. *J. Aerosol. Sci.* **21** (Suppl. 1), 501–504.

Seemayer, N. H., Hornberg, C. and Hadnagy, W. (1994) Comparative genotoxicity testing of airborne particulates using rodent tracheal epithelial cells and human lymphocytes *in vitro*. *Toxicol. Letters* **72**, 95–103.

Tomatis, L. (1990) Air pollution and cancer: an old and new problem. Air pollution and human cancer (Edited by L. Tomatis), pp. 1–7. Springer, Berlin.

Tomingas, R., Voltmer, G. and Bednarik, H. (1977) Direct fluorimetric analysis of aromatic polycyclic hydrocarbons on thin-layer chromatograms. *Sci. Total Environ.* **7**, 216–267.

Pergamon

Ann. occup. Hyg., Vol. 41, Supplement 1, pp. 32–38, 1997
© 1997 British Occupational Hygiene Society
Published by Elsevier Science Ltd. All rights reserved
Printed in Great Britain
0003–4878/97 $17.00 + 0.00
Inhaled Particles VIII

PII: S0003–4878(96)00130–5

SURFACE FREE RADICAL ACTIVITY OF PM$_{10}$ AND ULTRAFINE TITANIUM DIOXIDE: A UNIFYING FACTOR IN THEIR TOXICITY?

P. Gilmour*, D. M. Brown*, P. H. Beswick*, E. Benton*, W. MacNee†
and K. Donaldson,

*Biomedicine Group, Department of Biological Science, Napier University, 10 Colinton Rd, Edinburgh
EH10 5DT, U.K.; and †Unit of Respiratory Medicine, University of Edinburgh, Edinburgh, U.K.

INTRODUCTION

Epidemiological evidence has been accumulating showing a strong relationship between particulate environmental air pollution (PM$_{10}$) and end-points of respiratory ill health such as attacks of asthma, COPD, diminished lung function and cardio-vascular deaths (Pope *et al.*, 1995). To date there has been no plausible biological hypothesis to explain this relationship at the very low airborne mass concentrations of particulate air pollution that are found (< 50 μg ml^{-1}). We recently hypothesised (Seaton *et al.*, 1995) that an ultrafine (< 100 nm diameter) component of PM$_{10}$ is responsible for its adverse effects. This is based on the initial studies of Oberdorster and colleagues (Feirin *et al.*, 1992) who demonstrated that titanium dioxide in the ultrafine form (20 nm diameter) was highly inflammogenic to the lungs of rats compared to fine (200 nm diameter) TiO$_2$ particles at the same airborne mass concentration.

We now hypothesise that the adverse effects of PM$_{10}$ on the lung result from free radical activity at the surface of an ultrafine fraction. We further hypothesise that the interstitialisation that was seen with UFTiO$_2$ (Ferin *et al.*, 1992) could similarly occur with the ultrafine component of PM$_{10}$. If the ultrafine material has free radical activity then the increased surface area that is presented to the epithelial surface by a relatively small mass of ultrafine particles could compromise epithelial integrity leading to interstitialisation.

We demonstrate here that ultrafine TiO$_2$ and PM$_{10}$ both have hydroxyl radical activity and that UFTiO$_2$ is capable of causing hydroxyl radical-mediated membrane damage to erythrocytes; fine TiO$_2$ has much less of these properties. Additionally PM$_{10}$ hydroxyl radical activity is either in the ultrafine fraction or is released in soluble form.

MATERIALS AND METHODS

Particles

PM$_{10}$ was collected from the Edinburgh sampling site of the enhanced urban network and removed from the filters by sonication; a blank filter, similarly

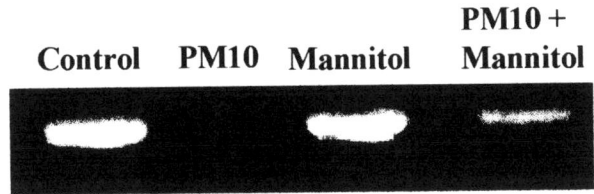

Fig. 1. Image of an agarose gel showing depletion of the supercoiled DNA band by PM$_{10}$ and protection by mannitol.

treated, was used to control for any free radical activity released from the filters themselves. Fine TiO$_2$ was obtained from Tioxide Ltd with average particle diameter of 583 nm and UFTiO$_2$ from Degussa (P25) with average particle diameter of 96 nm.

Plasmid assay for hydroxyl radical activity

The ability to break strands of supercoiled plasmid DNA was the sensitive assay used here for detecting particle surface-associated free radicals (Gilmour *et al.*, 1995). Briefly φ X174 RF plasmid DNA was incubated with the differing quantities of particles. All treatments were carried out in a volume of 20 μl, achieved by the addition of filtered ultra-pure distilled water. Each treatment or control sample was incubated at 37°C in a water bath for 8 h. The three plasmid forms (super-coiled, relaxed coil and linearised plasmid) were separated by electrophoresis for 16 h at 30 V on a 0.8% agarose gel. The proportion of the plasmid forms, which provide a measure of the free radical damage to the plasmid, were quantified by scanning laser densitometry and free radical damage to DNA was expressed as depletion of the supercoiled DNA band. To confirm the role of hydroxyl radicals in the DNA damage, mannitol was used as a scavenger.

Damage to the plasma membrane of sheep erythrocytes

As a model membrane target, we utilised the sheep erythrocyte membrane, and ascertained the ability of TiO$_2$ samples to cause direct membrane damage. Washed sheep erythrocytes were uncubated for 30 min with differing concentrations of particles and membrane damage was assessed as the absorbance at 540 nm (haemoglobin) in supernatants of centrifuged samples. Again, mannitol was used to confirm the role of hydroxyl radicals and DSF-B was used to investigate the role of iron.

RESULTS

PM$_{10}$

PM$_{10}$ had free radical activity that caused scission depletion of supercoiled plasmid DNA as shown in Fig. 1. This could be protected against with mannitol, clearly implicating hydroxyl radical in the injury.

We quantified the amount of DNA in the supercoiled ban using scanner laser densitometry and image density analysis software (Fig. 2) revealing that a blank filter yielded very little free radical activity whilst the PM$_{10}$ loaded filter gave up to

P. Gilmour *et al.*

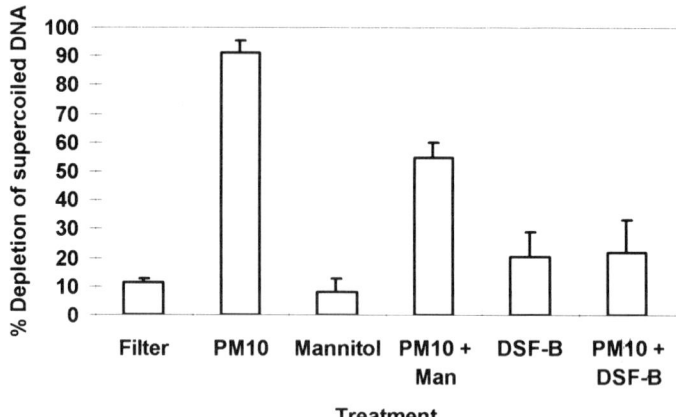

Fig. 2. Free radical activity of PM_{10} material and amelioration by mannitol (Man) and by desferoxamine-B (DSF-B); results are mean $\pm$ sem of at least three separate experiments.

90% depletion of the supercoiled band. Confirmation of the role of hydroxyl radical and iron were shown by the blockage of the free radical injury by mannitol and desferoxamine-B, respectively, in Fig. 2.

Evidence that the free radical activity emanates from an ultrafine fraction comes from studies where the PM_{10} was centrifuged to clarity and was found to have the same free radical activity as the standard PM_{10} material (Fig. 3). Another interpretation of this result is of course that the PM_{10} releases a soluble component that is responsible for the hydroxyl radical; this involves iron as evidenced by the inhibition with DSF-B shown in Fig. 3.

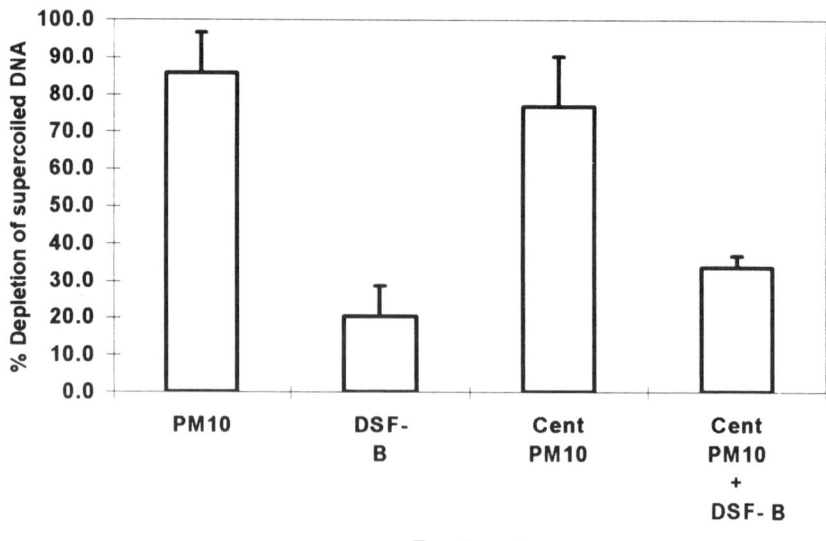

Fig. 3. Free radical activity, as shown in the plasmid assay, of PM_{10} material and PM_{10} material centrifuged to clarity (Cent PM_{10}) and the blocking effect of Desferoxamine-B (DSF-B).

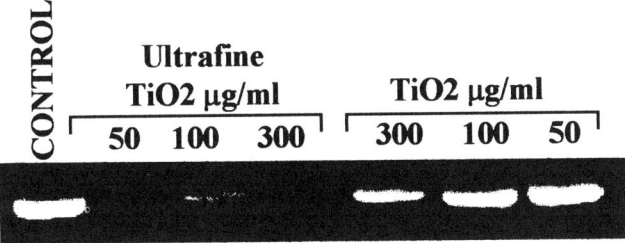

Fig. 4. Image of a supercoiled DNA band in an agarose gel to show depletion of the supercoiled DNA band by UFTiO$_2$ and relative lack of free radical activity of TiO$_2$.

Ultrafine TiO$_2$

When UFTiO$_2$ and TiO$_2$ were studied in the plasmid assay for hydroxyl radical activity the UFTiO$_2$ was found to have substantially more activity than TiO$_2$ (Figs. 4 and 5). As an alternative assay for the free radical–injuring activities of UFTiO$_2$ and TiO$_2$ we measured the abilities of the two samples to cause haemolysis of sheep erythrocytes. As shown in Fig. 6, only UFTiO$_2$, not TiO$_2$, was able to cause haemolysis in a concentration dependent manner. Figure 7 shows that the haemolytic effect of UFTiO$_2$ was blockable with mannitol at higher concentrations, confirming the role of hydroxyl radical in the haemolysis caused by UFTiO$_2$.

Iron involvement in the haemolytic effect of UFTiO$_2$ could not be demonstrated using DSF-B, but we cannot yet discount interference from haemoglobin as a confounding factor.

DISCUSSION

We demonstrate here that PM$_{10}$ material and UFTiO$_2$ both have hydroxyl radical activity that is detectable in a supercoiled plasmid DNA assay and that UFTiO$_2$ also has the ability to cause hydroxyl radical-mediated haemolysis of sheep erythrocytes. Unfortunately, because of the small amounts of PM$_{10}$ currently

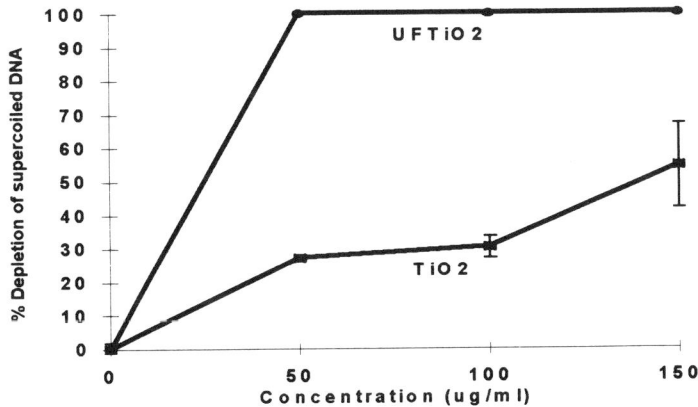

Fig. 5. Depletion of supercoiled DNA by UFTiO$_2$ and TiO$_2$ as quantified by scanning laser densitometry of the supercoiled bands.

P. Gilmour *et al.*

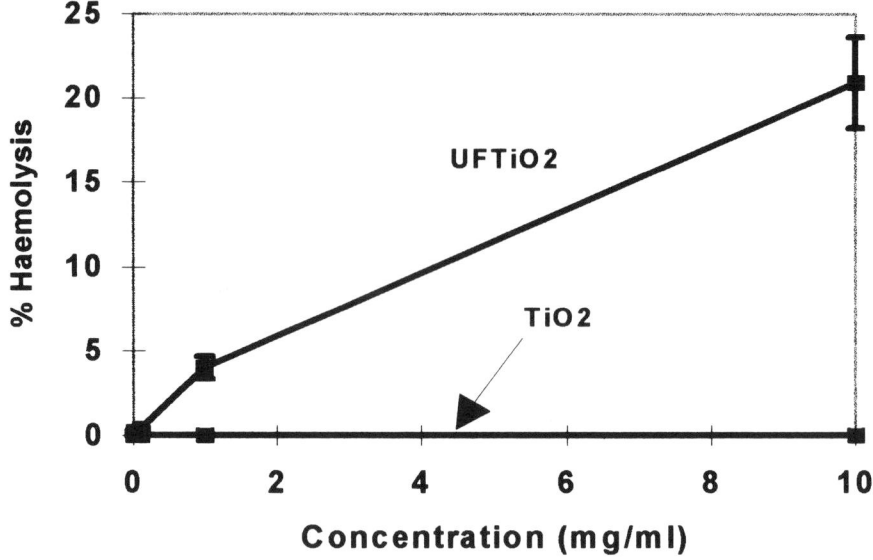

Fig. 6. Haemolytic activity of UFTiO$_2$ and TiO$_2$: results are mean ± sem of three experiments.

available we have been unable to test the haemolytic activity of PM$_{10}$. TiO$_2$ has barely detectable free radical activity in the plasmid assay and is unable to cause haemolysis at equivalent mass to UFTiO$_2$. Quartz has been reported to cause haemolysis by an oxidative mechanism, blockable with catalase (Razzaboni and Bolsaitis, 1990), but catalase had no effect on the haemolysis caused by UFTiO$_2$ here (data not shown).

The particle size of the PM$_{10}$ material used here is unknown but the size sampling characteristics of the tapered element oscillating microbalance that is used in the Edinburgh sampling site of the enhanced urban network follows the PM$_{10}$ convention and collects 10 μm aerodynamic diameter particles with 50% efficiency.

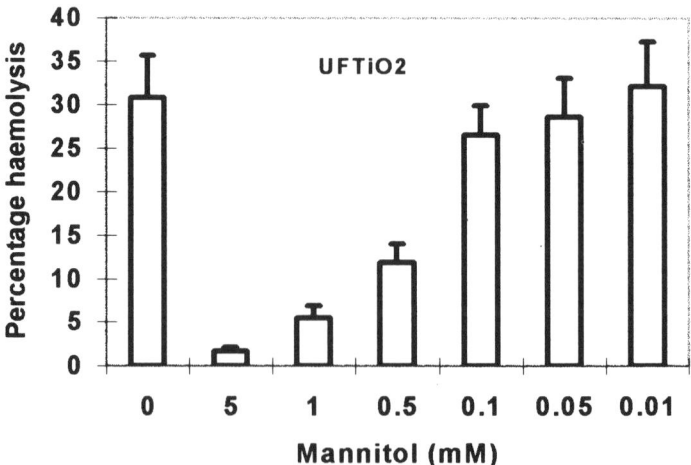

Fig. 7. Amelioration of UFTiO$_2$-mediated haemolysis with mannitol; mean ± sem of four experiments.

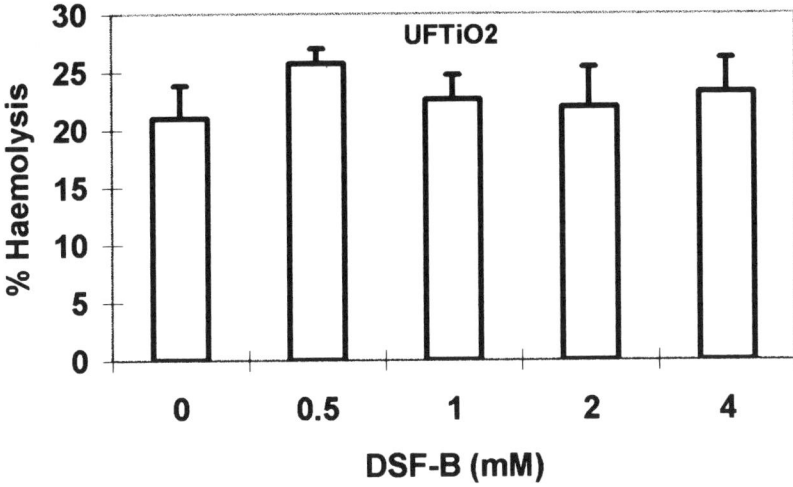

Fig. 8. Effect of iron chelation with DSF-B on UFTiO$_2$-mediated haemolysis.

It also, however, collects a smaller proportion, by number, of larger particles and collects particles down to the ultrafine range, very likely in large numbers (Quality of Urban Air Review Group, 1996). We have no information on the harvesting efficiency of our method for retrieving PM$_{10}$ from the filters for different particle sizes. However, we believe that there is a large proportion, in particle number and surface area terms, of ultrafine particles in the PM$_{10}$ used in these experiments. We also hypothesise that the hydroxyl radical activity resides in the ultrafine fraction of the PM$_{10}$. This was supported by the fact that in a sample of clear supernatant from which PM$_{10}$ particles had been removed by centrifugation, there was the same amount of free radical activity as in the whole suspension. However, an alternative explanation is that the "hydroxyl generating system" which involves iron, as shown by DSF-B studies, diffuses from the PM$_{10}$ particles into solution.

The marked difference in hydroxyl radical activity of UFTiO$_2$ and TiO$_2$ could underlie the difference in pathogenicity found with these two materials by Ferin *et al.* (1992) and the differential expression of anti-oxidant genes in the lungs of rats inhaling UFTiO$_2$ or TiO$_2$ (Janssen *et al.*, 1994).

The hydroxyl radical-mediated haemolysis caused by UFTiO$_2$ could not be inhibited with DSF-B, ruling out Fenton chemistry as the source of the hydroxyl radical for this particle but this could be a result of free iron from haemoglobin binding to desferrioxamine and confounding the assay. In preliminary studies however, using the plasmid assay, there was no protective effect of chelation with desferoxamine on the free radical damage caused by UFTiO$_2$.

The involvement of iron in the hydroxyl radical-generating activity of PM$_{10}$ but not in the case of UFTiO$_2$, if this is confirmed, is a clear difference between the two particulate materials. UFTiO$_2$ had an average particle diameter of 96 nm and TiO$_2$ had an average particle diameter of 583 nm. Particle surface area calculations indicate that the maximum difference in surface area is 7 times more surface area of UFTiO$_2$ compared to an equal mass of TiO$_2$. Consideration of the data given above show that the differences in the hydroxyl radical activity between TiO_2 and $UFTiO_2$ cannot be explained by differences in surface area because in excess of a sevenfold increase in TiO_2 still did not cause the same degree of plasmid-damaging or haemolytic activity as $UFTiO_2$.

In conclusion, the data shown here reveal that both PM_{10} and ultrafine TiO_2 have hydroxyl radical-generating activity. Preliminary studies with $UFTiO_2$ suggests that the source of the free radical activity is different for these two particles, involving iron in the case of PM_{10} but not in the case of $UFTiO_2$. However, the latter finding could be artefact due to the presence of haemoglobin in the haemolysis assay.

The data do not establish that the hydroxyl radicals emanate from an ultrafine component in the PM_{10} but generally support this contention. An alternative explanation is that the hydroxyl radical generating system, which involves iron, diffuses into solution from the PM_{10} particles. The latter would be likely to occur in the lung and the modifying effect of components of the lung lining fluid on this process are under investigation.

Acknowledgements—We acknowledge the financial assistance of the Colt Foundation and the British Occupational Hygiene Research Foundation.

REFERENCES

Ferin, J., Oberdorster, G. and Penney, D. P. (1992) Pulmonary retention of ultrafine and fine particles in rats. *Am. J. respir. Cell Mol. Biol.* **6**, 535–542.
Gilmour, P., Beswick, H. P. and Donaldson, K. (1995) Surface free radical activity of a range of respirable industrial fibres assessed using ϕx 174 RF1 plasmid DNA. *Carcinogenesis* **16**, 2973–2979.
Janssen, Y. M., Marsh, J. P., Driscoll, K. E., Borm, P. J., Oberdorster, G. and Mossman, B. T. (1994) Increased expression of manganese-containing superoxide dismutase in rat lungs after inhalation of inflammatory and fibrogenic minerals. *Free Radical Biol. Med.* **16**, 315–322.
Pope, C. A., Bates, D. V. and Raizenne, M. E. (1995) Health effects of particulate air pollution: time for reassessment?. *Environ. Health Perspect.* **103**, 472–480.
Quality of Urban Air Review Group (1996) Airborne particulate matter in the United Kingdom, 3rd Report, 176 pp.
Razzaboni, B. L. and Bolsaitis, P. (1990) Evidence of an oxidative mechanism for the hemolytic activity of silica particles. *Environ. Health Perspect.* **87**, 337–341.
Seaton, A., MacNee, W., Donaldson, K. and Godden, D. (1995) Particulate air pollution and acute health effects. *Lancet* **345**, 176–178.

Ann. occup. Hyg., Vol. 41, Supplement 1, pp. 39–42, 1997
© 1997 British Occupational Hygiene Society
Published by Elsevier Science Ltd. All rights reserved
Printed in Great Britain
0003–4878/97 $17.00 + 0.00
Inhaled Particles VIII

PII: S0003–4878(96)00164–0

HEALTH EFFECTS OF SULPHUR-RELATED ENVIRONMENTAL AIR POLLUTION—THE ROLE OF ACIDIC AEROSOLS

J. Heyder, I. Beck-Speier, B. Busch, G. A. Ferron, E. Karg, W. G. Kreyling, A.-G. Lenz, K. L. Maier, H. Schulz, S. Takenaka and A. Ziesenis

GSF-National Centre for Environment and Health, Institute for Inhalation Biology, D-85758 Oberschleissheim, Germany

INTRODUCTION

The contribution of gaseous and particulate pollutants to the ambient air pollution mixture varies with time and geographic location. Often sulphur dioxide and particles are considered together because they are the primary products of fossil fuel combustion processes and are usually present together as components of a complex mixture. These air pollution scenarios can be simulated by considering mixtures of gaseous sulphur IV (sulphur dioxide), particle-associated sulphur IV (neutral sulphite particles) and particle-associated hydrogen ions.

In a first attempt to study respiratory responses induced by sulphur-related air pollution, dogs were exposed to gaseous and particulate sulphur IV over a period of 290 days (Heyder *et al.*, 1992). No indication of chronic bronchitis was observed, but there were functional and morphometrical indications of early stages of lung emphysema, hyperplastic changes in the nasal cavity and disturbances of ciliated cell development in the trachea (Tables 1 and 2).

A more complex air pollution mixture was used in the study presented here. Over a period of 13 months particle-associated hydrogen ions were delivered to canine lungs in addition to gaseous and particulate sulphur IV. Since the size distributions of particles carrying H^+ and sulphur IV were almost identical and since particles carrying H^+ grew in the intrapulmonary moist air almost as much as particles carrying sulphur IV, the intrapulmonary sites of H^+ and sulphur IV deposition was very similar. Therefore, in all likelihood interactions between respiratory responses induced by sulphur IV and H^+ occurred and synergy was anticipated.

METHODOLOGY

The long-term exposure study with 16 young adult beagle dogs housed in whole body chambers at minimum restraint (23°C, 0.41 relative humidity, 8 mm s^{-1} air flow) was designed as a longitudinal study, characterized by repeated observations of individual pulmonary response patterns to inhaled combustion-related air pollution at near-ambient levels.

To establish baseline data for biological markers characterizing pulmonary

Table 1. Pollution-induced functional response patterns of the canine respiratory system

Respiratory and nonrespiratory lung function	Biological marker	Response to	
		sulphur IV	sulphur IV and hydrogen ions
Lung mechanics			
lung volumes		—	—
lung elasticity	static lung compliance	→*	—
	dynamic lung compliance	—	—
Alveolar–capillary barrier			
gas transfer	diffusing capacity	→*	—
permeability	albumin concentration	←*	—
Macrophage-mediated particles clearance	detection of intrapulmonary activity after inhalation of radiolabeled insoluble particles	←*	→*
Macrophage-associated defence capacity			
oxidative defence	release of superoxide	→*	—
phagocytic capacity	phagocytosis index	→*	→*
	fraction of phagocytosing cells	→*	←*
lysosomal activity	β-N-acetylglucosaminidase concentration	←*	→*
intracellular particle dissolution	extracellular activity of radiolabel from phagocytised radiolabeled particles of moderate solubility	←*	←
Type II cell function	alkaline phosphatase concentration	—	←
Oxidant status of extracellular proteins	methionine sulphoxide concentration	→*	—
	carbonyl concentration	→*	—
Cell injury	lactate dehydrogenase concentration	—	—

— no change detected, * statistical significant changes detected.

Table 2. Pollution-induced structural response patterns of the canine respiratory system

Target site	Response to	
	sulphur IV	sulphur IV and hydrogen ions
Nasal cavity	hyperplastic changes	—
Larynx and trachea	loss of cilia	—
Bronchi and nonrespiratory bronchioles	—	increased volume density of glands
Acini	enlargement of airspaces	type II cell proliferation

response patterns and pollution-induced changes of these data, eight dogs were repeatedly examined while the two chambers housing the dogs were ventilated over a period fo 16 months with unpolluted air (control period of the study) and then over a period of 13 months with polluted air (exposure period of the study). Thus each individual served as its own control, eight dogs housed in two chambers ventilated with unpolluted air over a period of 29 months served as controls.

The exposure system was designed as a dynamic system with continuous horizontal, unidirectional, low-turbulent displacement airflow and continuous particle injection into the airflow (Karg *et al.*, 1992). The animals were exposed for 16.5 h d^{-1} to a neutral sulphite aerosol at a mean particle mass concentration of 1.52 mg m^{-3} (1.02 µm mass median aerodynamic diameter, 2.2 geometric standard deviation, 0.33 mg m^{-3} particle-associated sulphur IV, 0.15 mg m^{-3} gaseous sulphur IV) and for 6 h d^{-1} to an acidic sulphate aerosol at a mean particle mass concentration of 5.66 mg m^{-3} (1.08 µm mass median aerodynamic diameter, 2.0 geometric standard deviation, 15 µmole m^{-3} particle-associated hydrogen ions). The particles were produced by nebulization of aqueous solutions of sodium bisulphite (Na_2SO_3) and sodium hydrogen sulphate ($NaHSO_4$) into the chamber-supplying conditioned airflow. Within the exposure chambers the sulphur IV-carrying sulphite particles assumed solid state while the hydrogen ion-carrying sulphate particles could be considered saturated droplets. Within the moist air of the canine lungs sulphite and sulphate particles increased about thrice in diameter.

As a result of microbial action on animal excreta gaseous ammonia was released into the chamber atmospheres which could neutralize the acidic particles. The analysis of eluates of intrachamber filter samples indicated that 50% neutralization occurred with ammonia so that the dogs were exposed to a mean H$^+$-concentration of 15 µmole m^{-3}. The extent of intrapulmonary neutralization with endogenous ammonia is less known. Measurements of ammonia concentration in the air exhaled by the animals suggested that intrapulmonary neutralization was lower than that occurring in the exposure chambers.

To assess pulmonary responses, respiratory lung function of the dogs was tested before the onset and at the end of the exposure period of the study, and serial segmental lung lavages were repeatedly performed during control and exposure period to obtain epithelial lining fluid from the lungs for analysis of cellular contents, cell function and biochemical indicators of lung injury. At the end of the study the lungs of all animals were morphologically and morphometrically examined.

RESULTS AND DISCUSSION

In addition to responses observed as a result of sulphur IV exposure of canine lungs, functional and structural responses observed in this study are listed in Tables 1 and 2. They indicate that interactions between responses induced by sulphur IV and hydrogen ions occurred. But antagonism rather than synergism was observed. The responses produced by sulphur IV were either less pronounced, not detectable or even reversed when particle-associated hydrogen ions were also delivered to the canine lungs. On the other hand, the proximal alveolar region responded to the exposure to sulphur IV and hydrogen ions in contrast to the solely exposure to sulphur IV. In the epithelial lyning fluid the concentration of alkaline phosphatase was elevated and, at the same time, type II epithelial cells proliferated.

CONCLUSIONS

(1) The pulmonary response pattern to an inhaled pollutant can be altered considerably by the additional inhalation of another pollutant.

(2) In this study, no synergy of pulmonary responses could be associated with inhaled neutral sulphite and acidic sulphate particles.

(3) Acidic aerosols may be less hazardous than suspected.

(4) A final interpretation of the response pattern induced by sulphur IV and hydrogen ions can only be given when the pulmonary response pattern induced by hydrogen ions is known. The corresponding study is in progress.

REFERENCES

Heyder, J., Beck-Speier, I., Ferron, G. A., Heilmann, P., Karg, E., Kreyling, W. G., Lenz, A.-G., Maier, K., Schulz, H., Takenaka, S. and Tuch, T. (1992) Early responses of the canine respiratory tract following long-term exposure to a sulfur (IV) aerosol at low concentration. *Inhal. Toxicol.* **4**, 159–174.

Karg, E., Tuch, T., Ferron, G. A., Haider, B., Kreyling, W. G., Peter, J., Ruprecht, L. and Heyder, J. (1992) Design, operation and performance of whole body chambers for long-term aerosol exposure of large experimental animals. *J. Aerosol Sci.* **23**, 279–290.

 Pergamon

Ann. occup. Hyg., Vol. 41, Supplement 1, pp. 43–48, 1997
© 1997 British Occupational Hygiene Society
Published by Elsevier Science Ltd. All rights reserved
Printed in Great Britain
0003–4878/97 $17.00 + 0.00
Inhaled Particles VIII

PII: S0003–4878(96)00131–7

A NOVEL TEST STAND FOR THE GENERATION OF DIESEL PARTICULATE MATTER

D. Dahmann

Institut für Gefahrstoff-Forschung der Bergbau-BG, Waldring 97, Bochum 44789, Germany

INTRODUCTION

Todays industry uses a wide variety of diesel engines. Especially in mineral salt and potash mining, they have become the most important production factor since their introduction in the early 60s. IARC classified diesel particulate matter (DPM) as probably carcinogenic (IARC, 1989). NIOSH recommended that DPM be regarded as possibly carcinogenic (NIOSH, 1988). In Germany, discussions about possible health hazards due to DPM exposure resulted in 1990 in its official classification as cancerogenic (MAK Werte, 1990). The respective limit value (Technische Richtkonzentration (TRK)) was published one year later (MAK Werte, 1991).

Several problems caused by DPM have not been solved yet. Internationally for instance, relatively few quantitative exposure data are available. Moreover, the inherent characteristics of the cancerogenic effect are obviously still under discussion among medical experts although an agreement about the general cancerogenic properties of DPM seems to exist. In addition, the measuring technique to be referred to for DPM determination (sampling and analytical steps) is internationally still inconsistent. However, to measure the particulate proportion of diesel exhaust, i.e. especially "elementary carbon" (EC) is undisputed. To separate this substance from coarse dust, sampling of the respirable dust fraction according to EN 481 (EN 481, 1993) is necessary.

Even though the German TRK value is still based on the so-called "total carbon" (TC) concept (carbon dioxide in the respirable dust fraction due to combustion determined by coulometric measurement and recalculation as carbon), a procedure to determine elementary carbon was developed in Germany and put into practice. The introduction of specific limit values on the basis of elementary carbon is scheduled to take place in the autumn of 1996. (See procedure description by Dahmann *et al.* (1996) and in ZH 1/120.44 (1995).) This method also requires respirable dust sampling on quartz fibre filters. Using inert gas, the filters are subjected to desorption/decomposition under high temperatures. The compounds to be desorbed (so-called "organic carbon" (OC)) are removed from elementary carbon and also determined after catalytic oxidation to CO_2. As usual, the remaining elementary carbon is determined by oxygen combustion and subsequent coulometric titration. Since the introduction of this method, a great number of measuring values has been obtained. The paper by Dahmann *et al.* (1996) includes a summary of these data.

The problems still to be solved are:

(1) A "primary standard" for diesel particulate matter or elementary carbon aiming at analytical quality management in laboratories has not been provided yet.
(2) Regarding its parameters, the exact method to determine elementary carbon in diesel soot is internationally inconsistent. Particularly from experts working in environmental protection, further steps to refine the analytical method are required, "thermo-optical method") (Birch and Carey, 1996).
(3) The exact composition of diesel soot in samples of ambient air, mainly with regard to specific cancerogenic components like PAHs and nitro PAHs, has not been sufficiently investigated yet. Existing information is almost exclusively based on "freshly generated" diesel soot from engine test stand trials.
(4) It is at present almost impossible to validate different sampling techniques by exact direct comparisons. The reason is that no suitable sources for the provision of diesel particulate matter in ambient air concentrations are available.
(5) The very nature of the analytical procedure has the effect that the determination of elementary carbon in coal mines is not feasible at the moment. Investigations on the influence of hard coal dust on diesel particulate matter in workplaces, especially its analytical determination, should have priority.

In general, existing diesel test stands are designed for the quantitative determination of emissions of diesel particulate matter from particular engines. For instance, particle emissions after dilution of the exhaust gas and cooling to about 50°C are only gravimetrically determined. Sampling is performed near the engine and the exhaust gas is not allowed to age sufficiently. These problems induced us to design, to build and to operate a test stand for the generation of diesel particulate matter in ambient air. In the following, I am going to present interim results.

DESCRIPTION OF THE TEST CHANNEL

Figure 1 shows the design principle of the test channel. A ventilator having a maximum performance of 250 m^3 min^{-1} sucks the air and conducts it through a system of ducts. This system is Y-shaped and has the described dimensions. Dosage of diesel particulate matter is performed via a branch of the system (see Fig. 1). The diesel engine has a performance of 45 kW at 4200 rpm. It is equipped with a water-cooled eddy current brake permitting the adjustment of varying engine loads. Almost any kind of aerosol or gas can be added via the second branch of the duct. This includes for example hard coal dust and/or water vapour. The mixture is obtained as described after a distance of 11 m. Subsequently, the aerosols are directed into an expansion chamber of about 20 m^3 and from there into the atmosphere via a purification device. The test stand permits the sampling of aerosols and gases at various sites. Isokinetic probe measurements can be performed in each branch as well as in the test channel proper (MP3). In addition, sampling of ambient air in the man-sized expansion chamber by common air

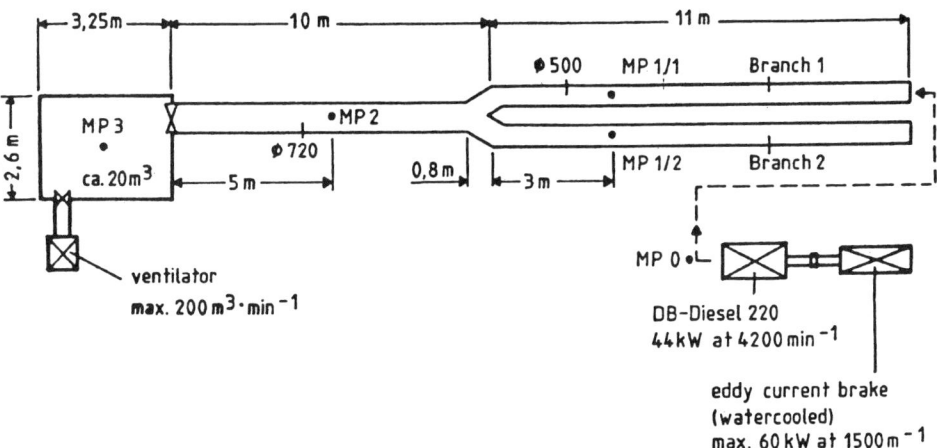

Fig. 1. Diesel particulate matter: IGF test stand.

samplers is possible. The ventilator is progressively adjustable. During a typical engine test at about 1500 rpm, approximately 0.2 m^3 exhaust gas per minute are fed into the Y-duct. With a closed second branch and a ventilator suction performance of about 35 m^3 min^{-1}, a dilution factor of approximately 1:17.5 is obtained. Temperatures are as follows: the fresh exhaust gas of the diesel engine has a temperature of about 120°C while entering the channel. Within the expansion chamber, a negligible temperature increase of about 1°C during sampling experiments over several hours occurs. The application of above-mentioned test parameters results in an exhaust gas velocity in the first part of the duct of about 3 m s^{-1} and in the second part of 1.5 m s^{-1}. Under these conditions, exhaust gases have a time of flight of about 11 s from entry into the test channel to discharge into the expansion chamber. In the mixing section behind the convergence of both Y-branches, the time of flight of aerosols until their discharge into the expansion chamber amounts to approximately 7 s.

PERFORMED EXPERIMENTS AND RESULTS

Some of our experiments and their outcome demonstrating the principle of the test device and its possibilities shall be outlined. They also serve to optimise the test channel.

Reproducibility of dust concentrations in the channel

To check the homogeneity of dust concentrations in the test channel under otherwise identical conditions (suction performance, operational data of the engine), a series of repetitive measurements was conducted. Particularly isokinetic measurements at measuring point 2 (see Fig. 1) with branch 2 being closed were performed.

The results of a series of 12 repetitive measurements at 1500 rpm and without use of the brake may serve as example. The typical fuel with a comparatively low sulphur content (approx. 0.05%) for potash mining was used. The extracted air volume of about 0.7 m^3 at a mean dust concentration of 2.6 mg m^{-3} contained about 1.10 mg m^{-3} of elementary carbon. The respective standard deviations were found to be 8.99 or 7.1 resp. These data are relevant if the emission of an engine and the influence of its operation on ambient air aerosols are to be investigated. In this way, filter samples of a roughly pre-determined concentration range can be achieved. Especially, the influence of different fuel types or engine loads can be studied without difficulty.

In this connection, we wanted to find out in a secondary experiment whether, and if possible, how the heating of coated filters can reduce the organic carbon content to a reproducible level. A test series showed that the OC content could be reduced to a constant value of about 3% (referred to EC) after 2 h at $250°$ (in the drying stove). Further temperature increases were not effective.

Homogeneity of filter coating

To provide the analytical laboratory with large filters (150 mm diameter) for quality management, quartz fibre filters were coated with DPM using the instrument VC25 (Ströhlein, Kaarst) which samples the inhalable dust fraction. Sampling was performed in the expansion chamber (MP 3). The instruments were operated for 20–30 min (exhaust volume 6–14 m^3) under the conditions described under A. In this way, EC concentrations of about 1 mg m^{-3} were prepared. The filters were punched and filter areas were individually investigated. Filter area diameters of 1 cm usually yield relative repetitive standard deviations between 3 and 6% ($N = 15$). The filters thus obtained are applicable to quality management in the laboratory. Similar filters were meanwhile successfully used in the framework of a European inter-laboratory test for EC/OC determination (Guillemin et al., 1996).

Comparison of different sampling methods

For method comparisons, several samplers were used in parallel in the expansion chamber (MP 3). The prerequisite is a homogeneous DPM distribution in the chamber. To ascertain homogeneity, several grid measurements were performed. Six different samplers were positioned in parallel on three levels of 60, 100 and 180 cm, respectively. We used personal dust samplers with Casella cyclone pre-separators having a sampling volume flow of 2 l min^{-1}. The stationary instruments were positioned in the centre of six rectangles with identical shapes and areas for about 3 h. Operational conditions of the engine largely corresponded to those described above.

Figure 2 shows the result of a typical grid measurement (concentrations in mg m^{-3}). By neglecting the outcome at point 6, the average results at the remaining five measuring points show the following relative standard deviations:

OC: 4.21% for a mean concentration of 1.148 mg m^{-3}
EC: 3.07% for a mean concentration of 2.29 mg m^{-3}.

Similar results were also achieved on the remaining two measuring levels. Standard deviations of 5.32% (OC) or 4.76% (EC) at 60 cm and 5.38% (OC) or 4.43% (EC) at 120 cm were observed. The relative standard deviation of these three results increases between 10 and 30% if measuring point 6 is included. This surprisingly high homogeneity of samples in large parts of the measuring chamber at this early

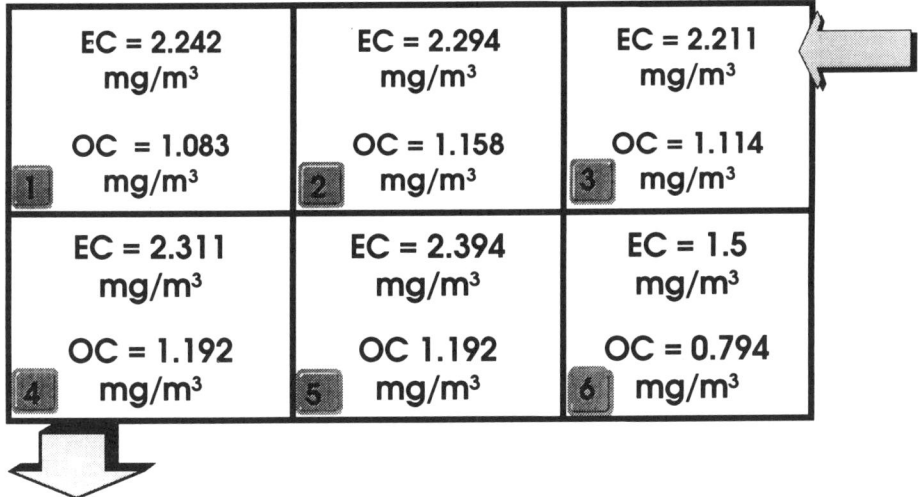

Fig. 2. Comparison of different sampler types. In a typical experiment, six different personal dust samplers (21 cyclone) were used simultaneously on three different levels above ground and positioned at equal distances within the chamber (MP 3). Example of Results at 180 cm.

stage justifies our expectation that comparative sampling in the chamber will be successful.

SUMMARY AND DISCUSSION

A novel test stand for the generation of diesel particulate matter was developed and successfully tested. The device offers the following advantages:

- Availability of large coated filters which, after the punching of filter areas, can be used for quality management in laboratories or comparative measurements.
- It could be proven that diesel emissions of a relatively constant composition over a long period of time can be provided.
- First trials showed that the test stand will also enable the comparison of different sampling techniques. This outcome has to be confirmed by further tests.

At present, filters in the test stand are investigated for the chemical composition of diesel soot. Main emphasis will be put on investigating the characteristics with regard to the dosage of non-diesel aerosols or gases via the second branch of the test stand.

REFERENCES

Birch, M. E. and Carey, R. A. (1996) Elemental carbon-based method for occupational monitoring of particulate diesel exhaust. *Analyst*, **121**.
Dahmann, D., Fricke, H.-H. and Bauer, H.-D. (1996) Diesel engine emissions in workplace-atmospheres in Germany. *Occ. Hyg.* **3**, 255–262.

EN 481 (1993) EN 481 Workplace atmospheres: size fraction definitions for measurement of airborne particles. CEN European Committee for Standardization, Brussels, Belgium.

Guillemin, M. *et al.* (1996) International round robin tests on the measurement of carbon in diesel exhaust particulates. *Int. Arch. occ. Environ. Health.*

IARC (1989) *IARC Monographs on the Evaluation of Carcinogenic Risks to Humans*, Vol. 46, 458 pp. Diesel and gasoline exhausts and some nitroarenes. International Agency for the Research on Cancer, World Health Organisation, Lyon, France.

ISO 8178-1 (1993) ISO 8178 Reciprocating internal combustion engines—exhaust emission measurement—part 1 testbed measurement of gaseous and particulate emissions. International Organisation for Standardisation.

MAK-Werte (1990) Maximale Arbeitsplatzkonzentrationen und Biologische Arbeitsplatztoleranzwerte (TRGS 900). *Bundesarbeitsblatt*, p. 35. Verlag W. Kohlhammer, Stuttgart, Germany.

MAK-Werte (1991) Maximale Arbeitsplatzkonzentrationen und Biologische Arbeitsplatztoleranzwerte (TRGS 900), *Bundersarbeitsblatt*, p. 36. Verlag W. Kohlhammer, Stuttgart, Germany.

NIOSH (1988) Current Intelligence Bulletin no. 50: carcinogenic effects of exposure to diesel exhaust, DHHS (NIOSH) Publication no. 88–116. National Institute for Occupational Safety and Health, Cincinnati, OH.

ZH 1/120.44 (1995) Verfahren zur Bestimmung von Kohlenstoff im Feinstaub—anwendbar für partikelförmige Dieselmotor. *Emissionen in Arbeitsbereichen*. Carl Heymanns, Verlag KG, Köln.

Pergamon

Ann. occup. Hyg., Vol. 41, Supplement 1, pp. 49–53, 1997
© 1997 British Occupational Hygiene Society
Published by Elsevier Science Ltd. All rights reserved
Printed in Great Britain
0003–4878/97 $17.00 + 0.00
Inhaled Particles VIII

PII: S0003–4878(96)00132–9

ESTIMATING RESIDENTIAL POLYCYCLIC AROMATIC HYDROCARBON (PAH) RELATED LUNG CANCER RISKS USING OCCUPATIONAL DATA

G. W. Gibbs

Safety Health Environment International Consultants Corp., 14–51221 Rae Rd 265, Spruce Grove, Alberta, Canada T7Y 1E7

INTRODUCTION

Increased lung cancer risks have been reported in gasworkers (Doll *et al.*, 1965; Pike, 1983; Berger and Manz, 1992), coke-oven workers (Redmond, 1983; EPA, 1984; Pott, 1985; WHO, 1986; Pott and Heinrich, 1990), roofers (Hammond *et al.*, 1976), aluminium smelter workers (Gibbs and Horowitz, 1976; Armstrong *et al.*, 1994) and persons exposed to Chinese smoky coal (Tuomisto and Jantunen, 1987; He *et al.*, 1989). Several of these studies, using benzo[a]pyrene (B[a]P) as a surrogate, have been used to derive public PAH related lung cancer risk estimates. In their case–cohort study of aluminium smelter workers (Armstrong *et al.*, 1994) lung cancer mortality increased with cumulative B[a]P exposure. The objective of the present study was to estimate the PAH-related lung cancer risk for persons living in the vicinity of the aluminium smelter complex (Armstrong *et al.*, 1994) and to examine factors which might affect the validity of such estimates. Some preliminary findings are summarised below.

METHOD

In order to examine risks based on the aluminium study with those from other coal tar pitch volatile (CTPV) derived PAH studies, published lung cancer risk estimates were compiled. Using the same studies, new lifetime estimates of (B[a]P) related lung cancer risks were calculated assuming residents inhaled 23 m^3 of air per day and were continuously exposed for 50 years. A linear non-threshold model was assumed. Factors likely to affect the validity of estimates were listed and data gathered to evaluate their impact.

RESULTS AND DISCUSSION

Lung cancer risk estimates without any consideration of validity are shown in Table 1. In spite of study design differences, often poor quality exposure information and many assumptions, they fall in a relatively narrow range. This may be due, in part, to the similarity in the crude estimated exposures in the coke and gasworks analyses—top side, approximately 30 $\mu g\ m^{-3}$. In spite of uncertainties, the risk estimates from several independent studies suggest that the "carcinogenic

G. W. Gibbs

Table 1. Lifetime lung cancer risk estimates per ng m^{-3} B[a]P per 100 000 persons exposed

PAH source	New/ recalculated Public***	Published Worker*	Published Public**	Ref@
Gasworks				
United Kingdom	9.5	nc	nc	Doll *et al.* (1965)
			43.0	Pike (1983)
			4.0	WHO (1986) Pike recalc.
Germany	4.2	nc	nc	Berger and Manz (1992)
Coke ovens				
	5.8	1.4c	nc	Redmond (1983)
	7.8	nc	nc	Costantino *et al.* (1995)
	—	1.0	5.0	Pott (1985)
(top) 30 000 ng m^{-3}	7.2			Pott and Heinrich (1990)
(side) 3000 ng m^{-3}	6.6			Pott and Heinrich (1990)
	—	nc	8.7	WHO (1986)
	—	—	5.0c	EPA (1984)
Aluminum smelter	4.4	1.0	nc	Armstrong *et al.* (1994)
Roofing	0.3	nc	nc	Hammond *et al.* (1976)
Smoky coal				
10 ng m^{-3}	—	—	0.1	He *et al.* (1989) nl
1000 ng m^{-3}	—	—	2.5	He *et al.* (1989) nl
	2.6	—	—	He *et al.* (1989) l
	—	—	6.5	Tuomisto and Jantunen (1987)

* All males except for smoky coal estimates. All working populations except smoky coal which involved indoor domestic exposure. ** Public risk estimates were based on a variety of different assumptions. *** Assumes 50 years of exposure, breathing rate of 23 m^{-3} per day for the public and 10 m^{-3} per day for workers. Assumes 9% lifetime lung cancer risk for males when using case–control or case–cohort studies. @ Either paper in which risk estimate was published or data source for estimate. nl—non-linear estimates; l—assuming linearity; nc—not calculated; c—calculated value converted to B[a]P equivalent using 0.71% factor.

potency" of coal tar derived PAHs from the studied sources are similar. Factors with potential to influence the validity of extrapolated risk estimates can be divided into: (a) those affecting the original occupational risk estimates; (b) those (e.g. smoking) distinguishing residents from workers; and (c) those relating to qualitative differences in exposure between the workplace and community.

(a) A critical review of the aluminium smelter study is not the aim of this report. However some factors were identified as important. These included the B[a]P exposure estimates (e.g. past benzene soluble matter (BSM) concentrations were converted to B[a]P based on recent B[a]P/BSM ratios); the use of hybrid arithmetic and geometric mean concentrations to assess exposure; the effect of respirator usage on actual exposures; smoking data reliability, use of Provincial rather than local rates and the estimated absolute risks used in the risk determination. The use of mid–late 1980s Canadian and Quebec rates to predict absolute risks in hypothetical populations exposed at various B[a]P levels (Armstrong *et al.*, 1994) ignores local mortality factors. It also assumes that these rates are the most appropriate for deriving future risk estimates. When male lung cancer mortality rates in two municipalities providing the smelter workforce (1951–1973; 1975–1989) were examined, they were about 17% (observed (O) = 355; expected (E) = 302.5; $P < 0.05$)) and 35% ($O = 405$; $E = 299.0$; $P < 0.05$) higher than provincial rates. The extent to which this is due to work in the smelter is not known. However, female lung cancer rates were 32% ($O = 89$; $E = 67.3$; $P < 0.05$) and 19% ($O = 80$; $E = 67.1$) higher, respectively, and women were not "PAH exposed" in the smelter. These results suggest that worker lung cancer risks are probably overestimated, perhaps by as much as 35%. As male smoking rates are declining, the future risks for workers will be overestimated even more. Converting hybrid geometric/arithmetic to arithmetic means would increase exposure estimates lowering actual risk further. It is evident that this, like any other study, should be critically evaluated prior to any attempt at risk extrapolation.

(b) When community male smoking rates were compared with those of smelter workers, the latter appeared to smoke more, but smoking effects were probably small. The percentage of the 696 men in the Armstrong subcohort who survived to 1983 were 40 years and over and had ever smoked was 84.5%, compared to 70.8% (CV 10.7%) in the community. As the cohort and community sample mean ages were 66 and 57, the difference could be due to age. There were 46.6% (CV 17.3%) of female residents who had ever smoked. Thus male risk estimates of lung cancer based on the Armstrong study may greatly overestimate female risks, due to smoking differences, if the interaction between smoking and B[a]P exposure is multiplicative. Whether additive or multiplicative is not known and present data are inadequate to accurately evaluate the risk for non-smokers. Changing smoking habits in the community will also be important.

(c) The factors influencing environmentally induced disease are pollutant composition and properties, level, duration and pattern of exposure, period since first exposure, individual factors (e.g. smoking) and susceptibility factors (e.g. pre-existing diseases, genetics, age). As increasing B[a]P exposures relate to increasing lung cancer risk in the smelter, it has been assumed that the B[a]P index reflects the concentrations of the main "potentially carcinogenic agents" in the workplace air. Extrapolation to the community will be invalid if (a) the profile or relative concentrations of the smelter "potentially carcinogenic agents" differ from those in community air or if (b) other substances increase or reduce lung cancer risks. In the smelter, high B[a]P exposures also involve exposures to fluorides in various forms, alumina dust, electric and magnetic fields, heat and other agents. Residents are not so exposed. Particle overload effects may have occurred in the smelter, virtually impossible for residents. The role of other agents in carcinogenesis is not known and data comparing smelter and community air are as yet inadequate to confirm completely the qualitative and quantitative similarity of smelter and environmental PAH exposures.

Another problem is the shape of the exposure–response curve. The above estimates assumed a linear model through zero. However, the shape is uncertain and uncertainty increases as the concentration decreases well below measured risk levels. Armstrong *et al.* reported a worker lifetime lung cancer risk of 20.2% at a concentration of 20 000 ng m^{-3} B[a]P. The lowest concentration category for which risks were measured in workers was 100 ng m^{-3}. One of the highest levels of B[a]P in the area surrounding the smelter in 1992, based on results from a sampling station where the profile of PAHs was similar to that of the smelter emission was

7.5 ng m^{-3} (annual arithmetic mean) and 1.8 ng m^{-3} (annual geometric mean) (Violette *et al.*, 1993).

The major determinant of cumulative exposure at low levels of exposure such as encountered by the public will be duration of exposure, while for workers, level of exposure is often more important. A lifetime exposure of 400 ng m^{-3}—years B[a]P for a member of the public at 10 ng m^{-3} involves 40 years of 24 h exposures. A smelter worker could theoretically acquire such an exposure in less than 1 month of regular work at 20 000 ng m^{-3} B[a]P. Extrapolation assumes that long term low level exposure carries the same risk as short term higher exposures even at very low exposures. Other assumptions used in community risk estimations include breathing rates for the general public and workers, duration of exposure (50 years was used in the above calculations because the peak of lung cancer deaths occurs about 10 years later). Opinions differ concerning the lag period. Real durations of exposure time (residency times) are likely to be much less. The general public, today, is not exposed for 24 h per day to the external pollutants as many work in buildings or homes with air cleaners and also leave the region for holidays or business. In the aluminium smelter study, exposure in the last 10 years prior to death did not give rise to a significant excess of lung cancer, while exposure in the last 20 years and especially in the last 40 years did. This suggests a latency of at least 10 years. Applying a lag period would reduce, for workers, the exposure contributing to risk, increasing their risk per unit B[a]P exposure. Thus latency choices will affect estimated risk. While the extent to which lung cancer risks might be influenced is not known, the general population includes people whose general health status is considerably worse than that of workers and competing risks may be quite different and must be considered in risk extrapolation.

CONCLUSIONS

Although subject to uncertainty, lung cancer risks per unit B[a]P exposure derived from various studies are remarkably similar, suggesting that the potency of coal-tar derived PAHs from the various industries are also similar. While tempting to extrapolate such risks to the general population, assumptions and uncertainties must first be quantified. Estimates of community lung cancer risks based on the experience of aluminum smelter workers are likely to have little validity unless adequate consideration is given to the factors shown or suspected of influencing extrapolated risk estimates.

Acknowledgements—Statistics Canada and Dr Ben Armstrong kindly provided data to facilitate some of the evaluations made in this study. The study was supported by Alcan Smelters and Chemicals.

REFERENCES

Armstrong, B., Tremblay, C., Baris, D. and Theriault, G. (1994) Lung cancer mortality and polynuclear aromatic hydrocarbons: A case-cohort study of aluminum production workers in Arvida, Quebec, Canada. *Amer. J. Epidem.* **139**, 250–262.
Berger, J. and Manz, A. (1992) Cancer of the stomach and the colon–rectum among workers in a coke gas plant. *Amer. J. ind. Med.* **22**, 825–834.

Costantino, J. P., Redmond, C. K. and Bearden, A. (1995) Occupationally related cancer risk among coke oven workers: 30 years of follow-up. *JOEM* **37**, 597–604.

Doll, R., Fisher, R. E. W., Gammon, E. J., Gunn, W., Hughes, G. O., Tyrer, F. H. and Wilson, W. (1965) Mortality of gasworkers with special reference to cancers of the lung and bladder, chronic bronchitis, and pneumoconiosis. *Brit. J. ind. Med.* **22**, 1–12.

EPA (1984) Carcinogen Assessment of Coke Oven Emissions, U.S. EPA, 1984, Washington, DC. Final report no. EPA-600/6-82-003F.

Gibbs, G. W. and Horowitz, I. (1979) Lung cancer mortality in aluminum reduction plant workers. *J. occup. Med.* **21**, 347–353.

Hammond, E. C., Selikoff, I. J., Lawther, P. L. and Seidman, H. (1976) Inhalation of benzpyrene and cancer in man animals. *N.Y. Acad. Sci.* **271**, 116–124.

He, X., Chen, W. and Chen, H. (1989) An epidemiological study on dose–response relationship between B[a]P concentrations in indoor air and lung mortality in Xuanwiei, China. In *Present and Future of Indoor Air Quality: Proceedings of the Brussels Conference* (Edited by C. J. Bieva *et al.*) pp. 227–233. Elsevier Science, Amsterdam.

Pike, M. C. (1983) Human cancer risk assessment. In *Polycyclic Aromatic Hydrocarbons Evaluation of Sources and Effects*. National Academy Press, Washington, DC. C1–C28.

Pott, F. (1985) Pyrolytic emissions, profiles of polycyclic aromatic-hydrocarbons and lung-cancer risk—data and evaluation. *Staub Reinhaltung Der Luft* **45**(7–8), 369–379.

Pott, F. and Heinrich, U. (1990). Relative significance of different hydrocarbons for the carcinogenic potency of emissions from various incomplete combustion processes. In *Complex Mixtures and Cancer Risk* (Edited by H. Vainio *et al.*), pp. 288–297. IARC, Lyon.

Redmond, C. K. (1983) Cancer mortality among coke oven workers. *Environ. Health Perspect.* **52**, 67–73.

Tuomisto, J. and Jantunen, M. (1987) A simple way of comparing carcinogenic effects of chemical and radioactive emissions NPHI A/2 1987, Kuopio, Finland. Kuopion yliopiston painatustkeskus, p. 18. National Public Health Institute.

Violette, R., Gariepy, R., Proulx, A. L. *et al.* (1993) PAH emissions reduction program at a horizontal stud Soderberg plant at Jonquiere, Quebec, Canada and the reduction of B[a]P in ambient air (1985–1992). Paper 93-FA-163.03, 86th Annual meeting and exhibition, June 13–18, 1993, pp 1–24. Air and Waste Management Association.

WHO (1986) *Polynuclear Aromatic Hydrocarbons (PAH) Air Quality Guidelines*, pp. 105–117. World Health Organization, Geneva.

 Pergamon

Ann. occup. Hyg., Vol. 41, Supplement 1, pp. 54–59, 1997
© 1997 British Occupational Hygiene Society
Published by Elsevier Science Ltd. All rights reserved
Printed in Great Britain
0003–4878/97 $17.00 + 0.00
Inhaled Particles VIII

PII: S0003–4878(96)00133–0

CHARACTERISATION OF PARTICULATE MATTER PM10 AND PM2.5 IN NORTHRHINE WESTPHALIA, SAXONIA AND LITHUANIA—FIRST RESULTS

E. Kainka,* G. Kramer† and J. Dudzevicius‡

*Medical Institute of Environmental Hygiene at the University, Auf'm Hennekamp 50,
D-40225 Düsseldorf, Germany, †Institute of Hygiene at the University of Leipzig, Germany,
and ‡Medical Academy of Kaunas, Lithuania

INTRODUCTION

The measurement of equal concentrations of total suspended particulate matter (TSP) in different areas may not cause equal effects on human health because of differences in the quantity of PM10 and PM2.5 as well as variations with regard to the components. Epidemiological studies in U.S. cities emphasised the importance of PM10 measurement and especially the measurement of PM2.5 instead of TSP. The correlation between daily mortality and PM10 concentration or especially PM2.5 concentration in urban areas was very high. A lower correlation was attributed to the coarse fraction of PM10. Therefore, more specific studies should concentrate on the finer fractions of airborne particulate matter. That is the reason why the international and especially the European discussion tends more and more to the measurement of PM10 and/or PM2.5. The United States has done measurements of PM2.5 since the beginning of 1996.

The PM10 and PM2.5 fractions are different compositions of several classes of pollutants. They differ also in their sources, amount, size, in physical and chemical behaviour and even in ability to penetrate from outdoors to indoors. The last point makes PM2.5 measurements more important in interpreting community exposure.

This paper will analyse and characterise the PM10 measurements and particle components of the coarse and fine fraction in several urban areas of Northrhine Westphalia, Saxonia and Lithuania as well as of special sources of pollutants like road traffic, industry and heating. Electron microscopical studies, chemical and crytallographical analysis of single particles and collections of particles characterise the components of several fractions and study areas. The identification of components may give hints on toxic effects.

METHODS

Study areas

One of three study areas in Northrhine Westphalia is in Düsseldorf, with a measuring point at a testing container in Düsseldorf Mörsenbroich that is situated at a multiple cross roads. Another study area is Duisburg. Mostly industrial air pollutants are measured at this testing container in Duisburg Walsum. The third study area is the small town of Borken situated north of the Ruhr area. Nearly clean air conditions are measured at a testing container in Borken. The study area in Saxonia is Leipzig, with measurements at a testing container in the South of Leipzig, where specially air pollution due to heating can be measured. The three study areas of Lithuania are situated in Kaunas at testing containers or stations. At station 1 mainly the traffic air pollution is measured, at station 2 the heating air pollution and at station 3 industrial air pollution.

Sampling method

The sampling instrument used is a low volume PM10 dichotomous sampler by Graseby Andersen. The volume air flow is 1 cubic metre per hour. This impactor divides the finer suspended particles in the two fractions PM10 coarse and PM10 fine (= PM2.5). These samplers are exposed near to testing containers in the study areas.

During several 24 h measurements, particulates of the coarse and fine fraction are collected on glass fibre filters to determine the concentrations and also for later extraction and chemical analysis. During 2 h measurements, particulates of the two fractions are collected on poly-carbonate filters with a poresize of 2 μm that will be used for scanning and transmission electron microscopical analyses. These discontinuous measurements and sampling methods are planned to cover each season of a year, to get better knowledge of special seasonal influences.

Electron microscopy

Particle analyses are done on scanning and transmission electron microscopy using EDX and diffraction methods.

RESULTS

Concentration of PM10 coarse and fine (PM2.5)

In general, variations of the concentrations of PM10 coarse and fine can be observed during a week. Examples of maximum and minimum daily amount of PM10 and its coarse and fine fraction in the study areas of Northrhine Westphalia at wintertime are shown in Figs 1 and 2. In Leipzig (south) four continuous 24 h measurements are done in winter on a Thursday up to a Sunday. The daily amount of PM10 and its coarse and fine fraction can be seen in Fig. 3. The measured concentrations of the Lithuanian stations are not yet available.

Particulate components

Coarse particulates collected at the Düsseldorf–Mörsenbroich station (traffic air pollution) are dominated by large soot agglomerates in addition of tyre abrasion and geogenic clay minerals. Fine particulates are mainly single soot particles or small soot agglomerates.

Coarse particulates collected at the Duisburg–Walsum station (industrial air pollution) are slag (mostly from blast furnaces), xenomorphic silica (changed by melting), several salt particles, gypsum and metallic particles. In the fine particulates soot and fly ash dominate.

Coarse particulates collected at the Borken station (nearly clean air) are mostly

E. Kainka *et al.*

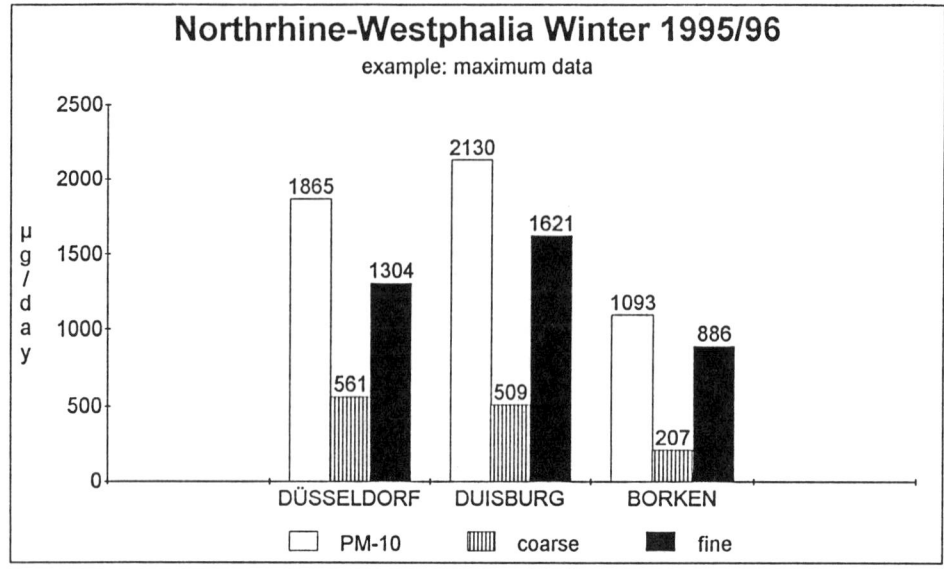

Fig. 1. Example of the maximum daily amount of total PM10 and of the coarse and fine fraction of Düsseldorf–Mörsenbroich (Tu, 23.01.96), Duisburg–Walsum (Th, 08.02.96) and of Borken (Tu, 13.02.96).

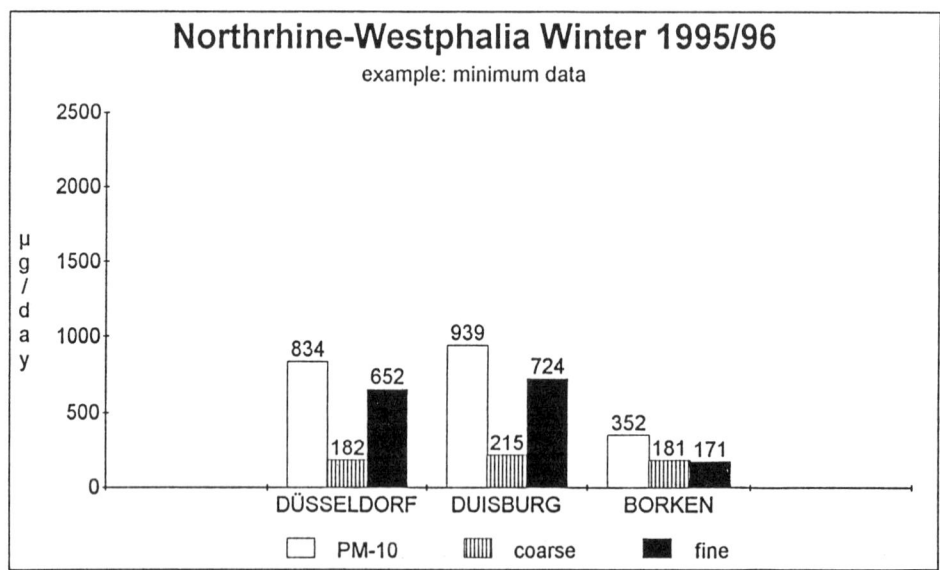

Fig. 2. Example of the minimum daily amount of total PM10 and of the coarse and fine fraction of Düsseldorf–Mörsenbroich (Sa, 20.01.96), Duisburg–Walsum (Sa, 10.02.96) and of Borken (Sa, 02.03.96).

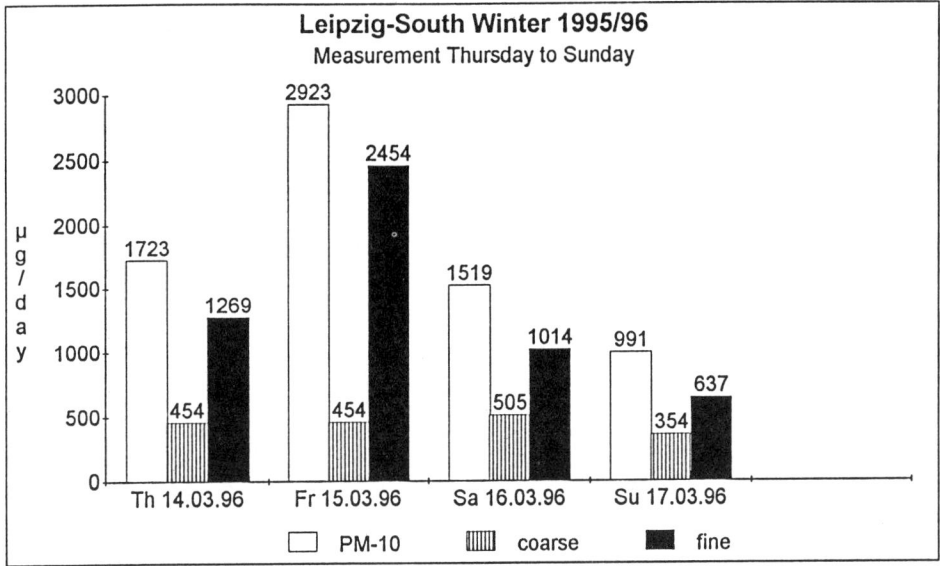

Fig. 3. Daily amount of total PM10 and of the coarse and fine fraction of four continuous measurements in Leipzig–South.

of biogenic and geogenic origin. Clay minerals and quartz indicate signs of weathering. The fine particulates are silicates and in addition some soot particles.

Coarse particulates collected at the Leipzig–South station (heating air pollution) are typical particles of combustion (in addition to particles of traffic air pollution) like soot agglomerates, sometimes with ball-like morphology, charcoal, big and small balls of fly ash, xenomorphic silicates and sulfates like gypsum. The fine particulates are dominated by soot agglomerates and small balls of fly ash (Fig. 4).

Coarse particulates collected at the Kaunas station 1 (traffic air pollution) are soot agglomerates, fly ash and clay minerals. Fine particulates are mostly soot particles. Coarse particulates collected at the Kaunas station 2 (heating air pollution) are soot agglomerates, fly ash, charcoal, xenomorphic silicates and sulphates. Fine particulates are soot particles and agglomerates added by small balls of fly ash (Fig. 4).

Coarse particulates collected at the Kaunas station 3 (industrial air pollution) are silicate particles, metallic particles and several salts. The fine particulates consist of soot particles, small agglomerates, fly ash and silicates.

DISCUSSION

The discontinuous measurements of PM10 which have started in Northrhine Westphalia, Saxonia and Lithuania are done by the above-mentioned low volume dichotomous sampler with a sampling rate comparable to the human breathing volume. The sampling of the two fractions of PM10 collects the bimodal aerosols in ambient air, as shown by Wilson *et al.* (1995).

Unlike the measurements of total suspended particles, the separate sampling of

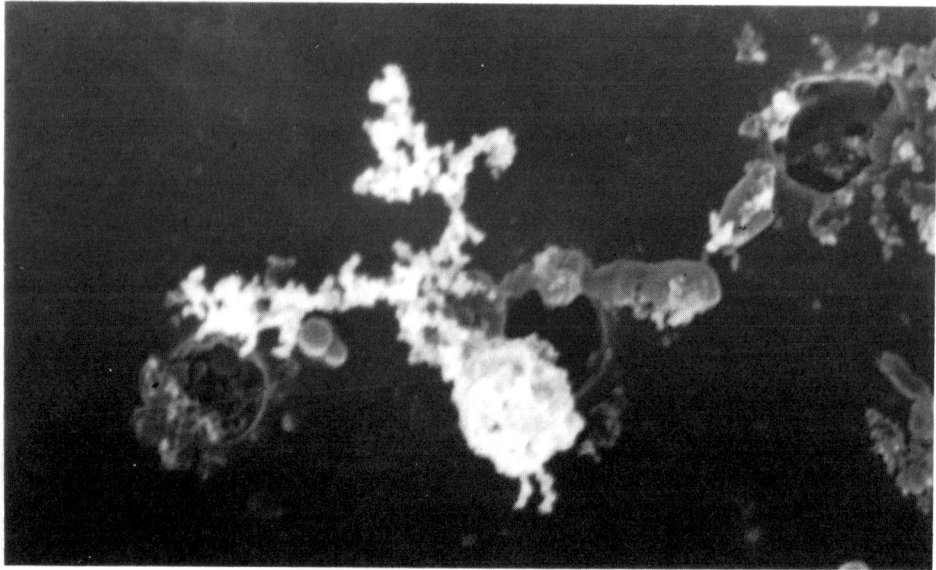

Fig. 4. Scanning electron micrograph of the PM2.5 fraction: soot agglomerates and fly ash, collected in Leipzig–South at wintertime 1996, magnification: 8000 ×.

the coarse and fine fraction of PM10 gives much more information, especially in the components and the amount of the PM2.5 fraction. The toxicity of this fraction is reported by epidemiologic studies in U.S. cities. Dockery *et al.* (1989, 1993) and Wilson *et al.* (1995) documented high correlations between daily mortality and the daily concentration of PM2.5 in several U.S. cities. Also Monn (1994) reported PM10 concentrations in eight regions of Switzerland.

The first results of our measurements show for all stations and kinds of air pollution that there is a day by day variation of the PM10 concentration and the coarse and fine fraction during a week (Figs 1–3). The main variation is in the PM2.5 fraction. In a clean area region, the measured amount of PM2.5 can be smaller than the amount of the coarse fraction of PM10. It can be mentioned that the Lithuanian PM10 concentration in Kaunas is lower than in Germany. The immense reduction of industry in recent time is obvious.

With the aim of doing further chemical analyses and to apply cytotoxical and genotoxical tests on extracts of the fractions, we started to analyse the particles and to characterise particles collections of the single fractions and stations according to their source. In all samples of the fine fractions soot particles and fly ash are more or less present. Single soot particles of 10 nm size or, even small soot agglomerates, may reach the alveoli. This is also true for the small balls of fly ash, because of their low density. The high content of heavy metals in this amorphous silica material may be the toxic component. Geogenous particles, for example silicas, derive from mechanical abrasion and are found mostly in coarse fractions. The weathered surface of these particles makes them less toxic than a changed and recrystallised surface of a silicate by combustion.

The particulate air pollution caused by traffic is highly concentrated in the fine fraction. This air pollution cannot just be documented at a typical station with traffic exhaust. The influence of traffic air pollution exists also in the particulate matter of areas with industrial or heating air pollution like Leipzig–South or even in so called clean air areas like Borken. The variations during a week are obvious. On the Lithuanian air pollution it can be mentioned that the air pollution by traffic exhaust is low in contrast to Germany. The traffic density and the number of diesel cars is lower. There are many electric buses in Kaunas and other buses that use normal petrol. The measurements of PM10 and PM2.5 in connection with the single particle analyses continue to develop a detailed and source revealed description of the particulate composition of ambient air in urban and rural areas in Western and Eastern Europe.

REFERENCES

Dockery, D. W., Speizer, E., Stram, D. O., Ware, J. H., Spengler, J. D., Benjamin, G. and Ferris, J. R. (1989) Effects of inhalable particles on respiratory health of children. *Am. Rev. Respir. Dis.* **139**, 587–594.

Dockery, D. W., Pope, C. A., Xu, X., Spengler, J. D., Ware, J. H., Fay, M. E., Ferries, B. G. and Speizer, F. E. (1993) An association between air pollution and mortality in six U.S. cities. *N. Engl. J. Med.* **329** (24), 1753–1759.

Monn, C. (1994) PM10 concentrations and aerosol particle size distribution in outdoor air in eight regions of Switzerland. *J. Aerosol Sci. suppl.* **1**, 159–160.

Wilson, W. E. and Suh, H. H. (1995) Differentiating fine and coarse particles: definitions and exposure relationships relevant to epidemiological studies. *Trends in Aerosol Research IV* (Seminar in Duisburg on the 27 January 1995), pp. 57–71.

 Pergamon

Ann. occup. Hyg., Vol. 41, Supplement 1, pp. 60–64, 1997
© 1997 British Occupational Hygiene Society
Published by Elsevier Science Ltd. All rights reserved
Printed in Great Britain
0003–4878/97 $17.00 + 0.00
Inhaled Particles VIII

PII: S0003–4878(96)00072–5

MODELLING HYGROSCOPIC PARTICLE GROWTH IN HUMAN LUNG AIRWAYS

G. M. Schum and R. F. Phalen*

California Environmental Protection Agency, Sacramento, CA 95812–0806; and
*Community and Environmental Medicine, University of California, Irvine, CA 92697–1825, U.S.A.

INTRODUCTION

An inhaled particle deposition program has been modified to take hygroscopic aerosol growth into account. The new model uses empirical or theoretical growth curves to calculate aerosol particle deposition in each airway generation. A substantial body of research on hygroscopic growth of particles has evolved which uses thermodynamic modelling to relate postulated temperature and humidity conditions in the lung to particle growth rate as a function of time (Ferron and Busch, 1996). Our approach differs from these models by simply using empirical or theoretical aerosol growth curves. Using this approach, we show that predicted inhaled particle growth and regional deposition patterns are similar to those predicted by other models.

METHODS

Using the analysis of Ferron and Busch (1996) for a NaCl particle growing from an initial size of 1 μm MMAD in airways that are at 99.5% relative humidity (RH), the particle grows to about 4 μm in approximately 10 s. The growth curves cited in Ferron and Busch for the 1 μm initial size have a strong theoretical foundation. These curves generally all have some initial rapid growth phase as dry particles encounter the saturated humidity conditions in the thoracic airways. This rapid growth phase is nearly always less than one second in duration. The particles grow in a semi-log linear fashion until they become nearly saturated, usually on a time scale of 1–5 s, followed by a saturation plateau which may or may not be reached during the time of a single breathing cycle (generally less than 6–8 s for normal resting ventilation).

Figure 1 shows the NaCl theoretical growth curves for 1 μm particles exposed to 99.5% RH and 96% RH environments. We used three functions to fit the particle growth curves:

$$(1) \text{ Logistic:} \quad y \text{ (μm)} = a_0 + a_1/(1 + \exp(-(\text{time}-a_2)/a_3))$$
$$(2) \text{ Sigmoidal:} \quad y \text{ (μm)} = a_0 + a_1/(1 + (\text{time}/a_2)\,\hat{}\,a_3)$$
$$(3) \text{ Exponential:} \quad y \text{ (μm)} = a_0 + a_1{}^* \exp(-\text{time}/a_2).$$

For the 99.5% RH curve, the logistic function fitted to 12 data points provided a nearly exact reconstruction. The sigmoidal function was also used to predict growth

NaCl Particle Growth Curves
Dashed lines are curve fits to data

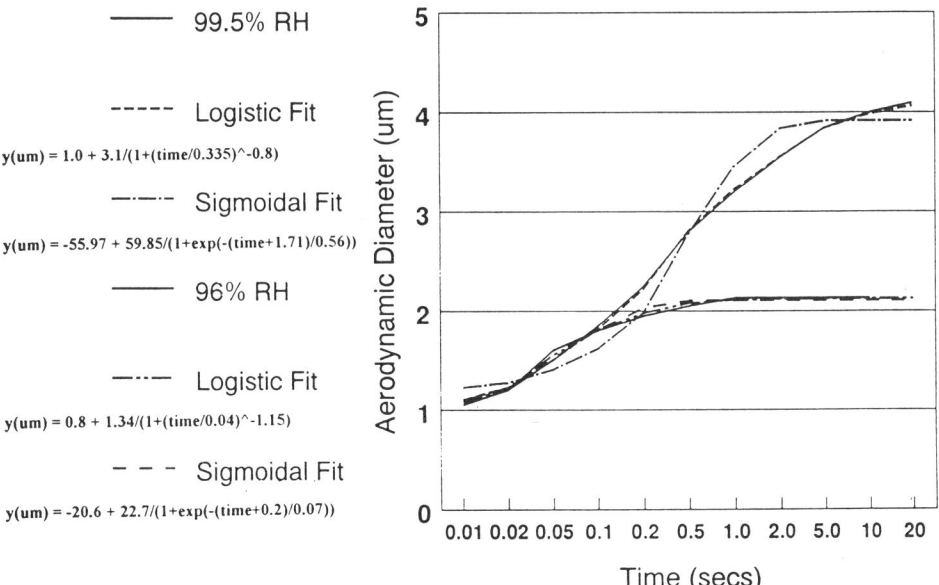

—— 99.5% RH

- - - - Logistic Fit

y(um) = 1.0 + 3.1/(1+(time/0.335)^-0.8)

—··— Sigmoidal Fit

y(um) = -55.97 + 59.85/(1+exp(-(time+1.71)/0.56))

—— 96% RH

—··— Logistic Fit

y(um) = 0.8 + 1.34/(1+(time/0.04)^-1.15)

– – – Sigmoidal Fit

y(um) = -20.6 + 22.7/(1+exp(-(time+0.2)/0.07))

Fig. 1. Theoretical growth of a 1 μm particle at 99.5% RH and 96% RH (data from Ferron) compared with fitted logistic and sigmoidal functions.

and deposition, but the results (not shown) were nearly the same as those predicted by the logistic growth model. The fitted sigmoidal and exponential models were very similar. This suggests that the specific function selected is not critical, as long as it matches the overall shape of the curve.

We used the age-scaleable computational particle deposition model of the National Council of Radiological Protection (Yeh *et al.*, 1996). This model uses the Yeh and Schum (1980) mathematical equations to predict deposition in the tracheobronchial and pulmonary regions. For particle growth, we assumed a 99.5% RH in all thoracic airways. While there is uncertainty in this assumption, we feel it is a good initial estimate.

To illustrate the applicability of our approach, we computed particle growth and regional deposition for both an adult and a young child for resting and moderate exercise respiratory patterns. It is important to accurately estimate breathing frequency in our model, since particle growth depends on residence time, which is a direct function of frequency. It is also important to note that children, because of their higher specific ventilation rates, will have much higher breathing frequencies than an adult for a given level of activity, reducing the time particles have to grow.

For simplicity, we present the results for mouth breathing, assuming that transit time and particle growth in the oropharynx are negligible. For nasal breathing, additional research is needed.

Hygroscopic Particle Growth
Logistic Model, Resting Ventilation

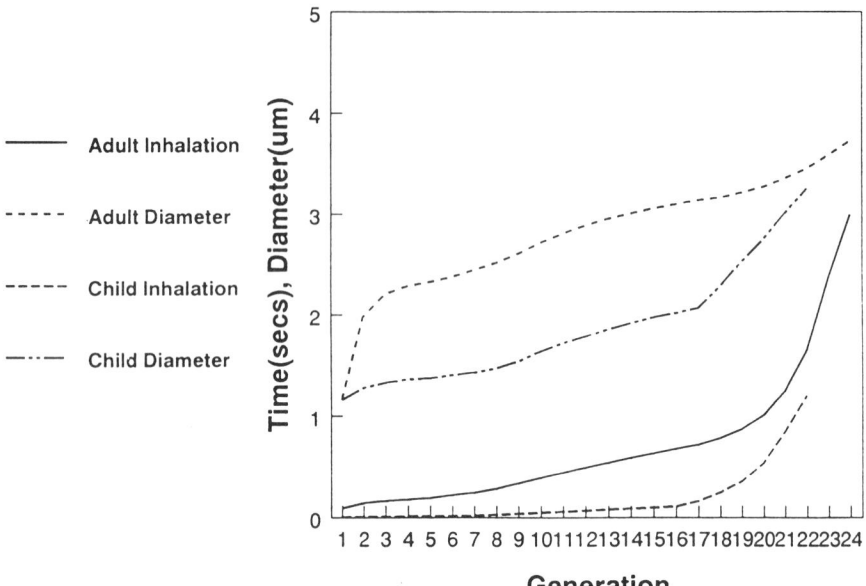

Fig. 2. Growth of a 1 μm NaCl particle in the thoracic airways of an adult and child lung model using a logistic growth function fit to data from Ferron for 99.5% RH. Adult: 10 bpm, 750 cc TV; Child: 25 bpm, 300 cc TV.

RESULTS

Figure 2 shows the relationship between cumulative residence time on inspiration and logistic particle growth for the adult and child during resting ventilation. Note that particle growth rates in the conducting airways of the adult and child are markedly different. However, once the particles reach the pulmonary region (generation 17), the growth curves will rapidly converge so that deposition, primarily by sedimentation, will dominate the results.

The predicted regional deposition values are shown in Table 1 for the adult and child in resting and exercise activities. Also shown is predicted deposition for a non-hygroscopic unit density 1 μm particle. For our representative case of a 1 μm MMAD particle with a logistic growth curve similar to the theoretical growth curve for 99.5% RH NaCl cited in Ferron and Busch (1996), the following conclusions may be reached:

(1) Depending on the respiratory frequency, strongly hygroscopic particles will generally double to triple in size in the conducting airways of an adult.
(2) Growth rates in the conducting airways are greater in an adult than in a young child, which results from the higher respiratory rates of children for similar activity levels.
(3) Growth to nearly equilibrium sizes during the total time spent in an acinar unit (inhalation time + exhalation time) is achieved in both the adult and the child.

Table 1. Calculated deposition for 1 μm NaCl particle

	Adult resting		Adult exercise		Child resting		Child exercise	
Lung model variables								
Height (cm)	180		180		100		100	
TLC (cc)	5564		5564		1755		1755	
FRC (cc)	2887		2887		702		702	
FRC + 1/2 TV (cc)	3262		3637		852		952	
Breathing pattern variables								
BPM (frequency)	10		30		25		50	
TV (cc)	750		1500		300		500	
Q (lmin⁻¹)	7.5		45		7.5		25	
Pause	—		—		7.5		—	
	Time (s)	Aerodynamic diameter	Time (s)	Aerodynamic diameter	Time (s)	Aerodynamic diameter	Time (s)	Aerodynamic diameter
Growth timing variables								
Inspiration through trachea	0.10	1.16	0.017	1.16	0.006	1.16	0.002	1.16
through primary bronchi	0.15	1.00	0.025	1.42	0.010	1.24	0.003	1.21
through lobar bronchi	0.17	2.21	0.029	1.51	0.012	1.33	0.003	1.23
through terminal bronchiole	0.68	3.10	0.12	2.07	0.12	2.02	0.34	1.56
end inspiration alveoli	3.00	3.81	1.99	3.14	1.20	3.44	0.60	3.07
Expiration through acinus	5.32	3.96	1.89	3.64	2.30	3.71	1.17	3.43
through trachea	6.00	3.98	2.00	3.66	2.40	3.73	1.20	3.44

Logistic growth model: $y \, (\mu m) = 1.0 + 3.1/(1 + (time/0.335)\char`\^ -0.8)$

	Adult resting		Adult exercise		Child resting		Child exercise	
Deposition fraction								
TOTAL		0.54		0.39		0.38		0.29
NOPL		0.001		0.016		0.007		0.037
TB		0.11		0.055		0.066		0.085
PUL		0.43		0.31		0.31		0.166
No growth model								
Deposition fraction								
TOTAL		0.18		0.13		0.12		0.15
NOPL		0.001		0.016		0.007		0.037
TB		0.033		0.024		0.034		0.06
PUL		0.15		0.09		0.08		0.05

(4) Little additional growth occurs in the conducting airways on exhalation in either the adult or child.

(5) Hygroscopic particle growth generally leads to about a threefold increase in total deposition in both the adult and child, with the largest relative increase due to deposition in the pulmonary compartment. This deposition, for these size particles (2–4 μm after growth), is due to the larger sedimentation efficiencies for the distal airways.

(6) Using mathematical functions to simulate theoretical or experimental particle growth curves shows the same general trends as those predicted by other, more elegant models. Our model, however, provides lower total deposition than some other published values (Ferron and Busch, 1996).

(7) For the case of adult resting ventilation, our results are close to those measured by Anselm *et al.* (1986), which showed that inhaled dry 0.7 μm NaCl particles were exhaled as 3.85 μm particles. Our model calculations for 1 μm particles indicate they would be exhaled as 3.98 μm MMAD particles, suggesting our growth model is reasonably accurate.

Acknowledgements—Supported by NHLBI grant R01 HL39682 to R. F. Phalen. The assistance of M. J. Oldham and R. Mannix is duly appreciated.

REFERENCES

Anselm, A., Gebhart, J., Heyder, J. and Ferron, G. (1986) Human inhalation studies of growth of hygroscopic particles in the respiratory tract: Effects of variations in respiratory pattern. In *Aerosols: Formation and Reactivity. 2nd Int. Aerosol Conf.*, Berlin, pp. 252–255. Pergamon Journals Ltd, Oxford.
Ferron, G. A. and Busch, B. (1996) Deposition of hygroscopic aerosol particles in the lungs. In *Aerosol Inhalation: Recent Research Frontiers* (Edited by J. C. M. Marijnissen and L. Gradon) pp. 143–152. Kluwer Academic Publishers, Netherlands.
Yeh, H. C. and Schum, G. M. (1980) Models of human lung airways and their application to inhaled particle deposition. *Bull. Math. Biol.* **42**, 461–480.
Yeh, H. C., Cuddihy, R. G., Phalen, R. F. and Chang, I-Y. (1996) Comparisons of calculated respiratory tract deposition of particles based on the NCRP/ITRI and the new ICRP66 Model. *Aerosol Sci. Technol.* **25**, 134–140.

SECTION 2

RADIOACTIVE PARTICLES

Pergamon

Ann. occup. Hyg., Vol. 41, Supplement 1, pp. 65–69, 1997
© 1997 British Occupational Hygiene Society
Published by Elsevier Science Ltd. All rights reserved
Printed in Great Britain
0003–4878/97 $17.00 + 0.00
Inhaled Particles VIII

PII: S0003–4878(96)00073–7

LONG-TERM BEHAVIOUR OF URANIUM IN THE LUNGS OF BABOONS AFTER INHALATION OF URANIUM DIOXIDE AND OCTOXIDE

J. L. Poncy, P. Massiot, C. Frot, S. Matton, G. Rateau, G. Grillon, D. Hoffschir and P. Fritsch

Laboratoire de Radiotoxicologie, CEA/DSV/DRR/SRCA, Laboratoire de Radiotoxicologie, BP 12, 91680 Bruyères-le-Châtel, France

INTRODUCTION

Uranium oxides (UO_2, U_3O_8) represent one of the most important intermediates of the manufactured fuel used in nuclear plants. During fuel processing, the predominant route of intake of uranium oxide powder for potentially exposed workers is inhalation. For this reason, the long-term behaviour of inhaled uranium particles is a fundamental parameter in lung retention and incidence of long-term lung pathology.

The biokinetic behaviour of uranium in humans has been estimated by comparing rodent data on the translocation of uranium deposited in the lungs to blood, with mechanical clearance measured in humans exposed to particles labelled with ^{85}Sr or ^{88}Y (Stradling *et al.*, 1988, 1989). However, a 1 year experimental study after inhalation of two ceramic forms of uranium oxides in baboons has shown that this animal species was the most appropriate to assess distribution of uranium in the lungs and other organs including the different parts of the skeleton (Métivier *et al.*, 1992).

For the short- and middle-term assessment of occupational exposure to uranium oxides, external chest monitoring and urinary assays remain the main procedure. Nevertheless, beyond 1 year post acute exposure, it was difficult to appreciate the long-term uranium retention in the deep lung by chest monitoring for low initial lung deposit and no experimental data are available. Other methods could be used such as analysis of particles extracted by broncho-alveolar lavage (BAL).

The aim of this study was to compare the long-term clearance of two industrial uranium oxides after inhalation in baboons by chest monitoring for 850 days and to characterize particles in alveolar macrophages extracted by segmental pulmonary lavage using energy dispersive X-ray spectroscopy (EDS) analysis.

MATERIAL AND METHODS

Two enriched uranium oxides (3.1% ^{235}U) as UO_2 and U_3O_8, obtained from nuclear fuel manufactures were used. A respirable fraction of the powders was prepared just before aerosol exposure by sedimentation in plasmagel (Bellon,

France) to obtain a polydisperse particle suspension with a count geometric mean diameter of 0.37 μm (δg 3.4) for UO_2, and 0.52 μm (δg 4) for U_3O_8. Two baboons (*Papio papio*), weighing between 10.8 and 13.2 kg were exposed to each oxide using our specific inhalation facility as previously described (André *et al.*, 1989). Before inhalation, animal health and lung integrity was assessed by X-ray examinations of the lungs and biochemical analysis of blood and urine. Prior to inhalation, injection of diazepam (Valium, Roche, 0.5 mg kg^{-1} body wt) was used to tranquilize animals and then the baboons were anaesthetized by an intramuscular injection of ketamine (Imalgene 500, Iffa Mérieux, 10 mg kg^{-1} body wt). Dried food (UAR, Villemoisson, France) supplemented with fresh fruits and water were freely available throughout the experiment. All experiments on animals were performed by scientists certified by the French Ministry of Agriculture for carrying out these procedures.

Gamma spectrometric measurements of the 17, 62, 92 and 185 keV emissions were used for the assessment of uranium in the chest of each animal. Measurements were performed immediately after inhalation to determine the initial lung deposit and until 850 days after exposure. The detection system used two sodium iodide crystals (type 121, SM 51 Scintiflex, Q and S, Ltd) protected by a thick beryllium window. The limit of detection corresponded to 170 μg of uranium.

About 5 years after exposure, baboon alveolar macrophages were collected by limited subsegmental lung lavage that was performed by repeated instillation of 5 ml 0.9% NaCl until a total volume of 20 ml had been collected. Cytospin cell samples were prepared on slides coated with collodium and then EDS analysis of particles was performed in the entire alveolar macrophages as previously described (Massiot *et al.*, 1996).

RESULTS

The estimated initial uranium lung deposit of each animal was 9.4 and 11.2 mg for UO_2 and, 4.3 and 3.4 mg for U_3O_8. Figure 1 shows the evolution of lung retention of uranium, expressed as a percentage of the initial lung deposits as a function of time following exposure. Two main lung clearance periods can be distinguished, the early one was observed the first week after exposure and involved for each animal a similar fraction of the initial lung deposit, about 20% and the late one from about 100–850 days that fit to a single exponential as a function of time. For a same powder, similar clearance parameters were measured, and the half-time of the early clearance was about the same for UO_2 and U_3O_8. By contrast, significant difference was observed between the half-time of late clearance: about 300 days and more than 850 days for UO_2 and U_3O_8, respectively.

Five years after inhalation, EDS analysis demonstrated that particles of uranium were still in alveolar macrophages collected by BAL. These particles corresponded to less than 10% of the total phagocytosed particles observed that were mostly SiO_2 and Al silicate. The spectrum of a particle of uranium after either UO_2 or U_3O_8 exposure with the specific U X-rays is showed in Fig. 2. Other specific X-rays were encountered such as P Kα that was observed for all the particles of uranium analysed and as Si Kα and/or Fe Kα that was observed for most of them. The Cu Kα and Kβ X-rays were due to the copper grids used.

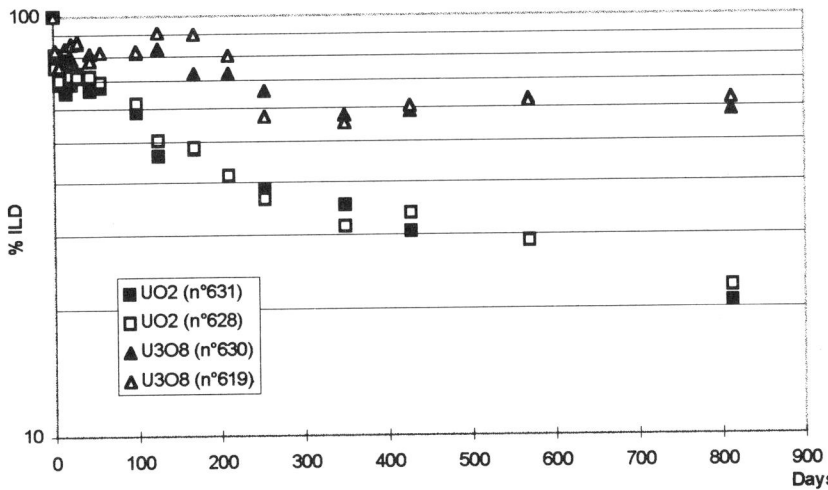

Fig. 1. Evolution of uranium lung retention after inhalation of UO_2 and U_3O_8 in baboons.

DISCUSSION

Recommendations of the International Commission on Radiological Protection (ICRP 54, 1988) assign UO_2 and U_3O_8 to transportability class Y, with an half-life time in the lungs of about 500 days. These recommendations were based on human (West *et al.*, 1979; Price, 1989; Stradling *et al.*, 1989) and animal (Stradling *et al.*, 1988; Métivier *et al.*, 1992; Ansoborlo *et al.*, 1994) data.

Taking into account the animal weight, the initial lung deposits of our baboons were approximately 20 times those calculated in human lungs after acute exposure to the ALI of class Y compounds of natural uranium as recommended by ICRP 61 (1991), for UO_2 and eight times for U_3O_8. This calculation assumes the lung deposit parameters used by the new respiratory tract model (ICRP 66, 1994) in an adult man after inhalation of particles with an AMAD of 1 μm.

For each uranium oxide, the lung clearance cannot be described using similar functions as previously described (Métivier *et al.*, 1992). Thus, in order to compare the results, we have defined a late lung clearance from 100 to 850 days after exposure. The clearances of UO_2 were very similar, about 300–350 days, whereas, those for U_3O_8 were very different, about 200 days (Métivier *et al.*, 1992) and more than 850 days. The different process lines of the two powders might explain such differences.

In fact, little information is available on the distribution of uranium in the different compartments of the respiratory tract after aerosol exposure. The two major sites of accumulation of uranium after long-term chronic exposure to natural UO_2 particles were lung interstitium in which particles could be encountered as clusters, bones and tracheobronchial lymph nodes (TLN) as observed by Leach *et al.* (1970, 1973) in different animal species. Unlike lungs, during the 4 years of chronic exposure in beagles and rhesus monkeys, the amount of uranium content in TLN increased continuously. In the case of baboon, the evolution of the U distribution could be similar to that reported for lungs and TLN after PuO_2 exposure (Métivier *et al.*, 1989). Further studies are in progress to characterize, the

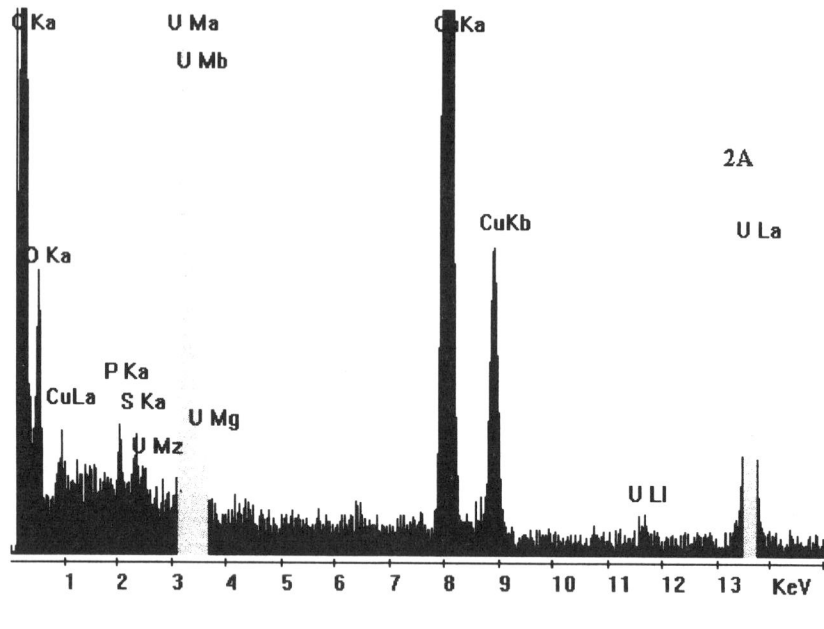

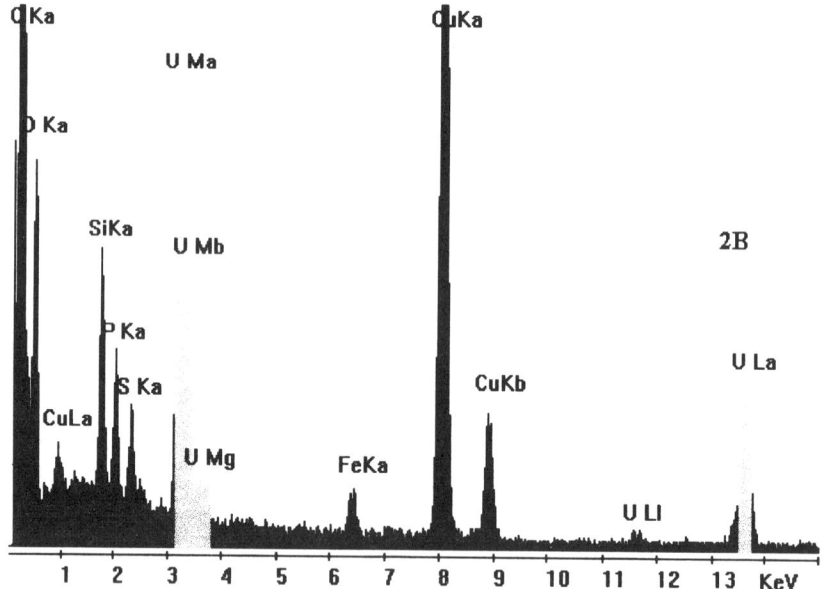

Fig. 2. X-ray spectrum of phagocytosed particle in baboon alveolar macrophages after inhalation of uranium oxides as UO_2 (2A) and U_3O_8 (2B).

long-term relative distribution of uranium in alveolar macrophages, lung interstitium and TLN.

Using electron energy loss spectrometry analysis, presence of uranium particles was observed in alveolar macrophages of rat, a few days after intratracheal instillation of uranium oxides (Henge-Napoli *et al.*, 1994). Most needles present in macrophages were associated with the oxide particles and characterised as poorly soluble uranyl phosphate. This rapid chemical evolution of the uranium oxides in macrophages might explain the long-term retention of uranium in the deep lung. Even at 5 years post-inhalation, we have observed some particles of uranium in alveolar macrophages extracted by limited subsegmental lung lavage. The method used (Massiot *et al.*, 1996), allows us a quantitative estimate and could be potentially applied to human. We observed the specific P Kα X-ray associated with the particles of uranium and, for some of them, other X-rays such as that of Si or Fe. Further studies are in progress to characterize the evolution of the chemical composition of uranium oxide particles after digestion of the biological matrix which will provide more information on the dissolution process and on the role of elements present in the phagolysosomal compartment (Matton *et al.*, submitted).

REFERENCES

André, S., Charruau, J., Rateau, G., Vavasseur, C. and Métivier, H. (1989) Design of a new inhalation device for rodents and primates. *J. Aerosol Sci.* **20**, 647–656.

Ansoborlo, E., Henge-Napoli, M. H., Donnadieu-Claraz, M., Roy, M. and Pihet, P. (1994) Industrial exposure to uranium aerosols at laser enrichment processing facilities. *Radiat. Prot. Dosim.* **53**, 163–167.

Henge-Napoli, M. H., Ansoborlo, E., Donnadieu-Claraz, M., Berry, J. P., Gibert, R. and Pradal, B. (1994) Solubility and transferability of several industrial forms of uranium oxides. *Radiat. Prot. Dosim.* **53**, 157–161.

International Commission on Radiological Protection (1988) Individual monitoring for intakes of radionuclides by workers: design and interpretation. *ICRP Publication 54*, Pergamon Press, Oxford.

International Commission on Radiological Protection (1991) Annual limits on intake of radionuclides by workers based on the 1990 recommendations. *ICRP Publication 61*, Pergamon Press, Oxford.

International Commission on Radiological Protection (1994) Human respiratory tract model for radiological protection. *ICRP Publication 66*, Pergamon Press, Oxford.

Leach, L. J., Maynard, E. A., Hodge, H. C., Scott, J. K., Yuile, C. L., Sylvester, G. E. and Wilson, H. B. (1970) A five-year inhalation study with natural uranium dioxide (UO₂) dust -I.Retention and biological effect in the monkey, dog and rat. *Health Phys.* **18**, 599–612.

Leach, L. J., Yuile, C. L., Hodge, H. C., Scott, J. K., Sylvester, G. E. and Wilson, H. B. (1973) A five-year inhalation study with natural uranium dioxide (UO₂) dust—II. Post-exposure retention and biological effects in the monkey, dog and rat. *Health Phys.* **25**, 239–258.

Massiot, P., Fritsch, P., Le Naour, H., Rateau, G., Lizon, C. and L'Hullier, I. (1996) Electron probe X-rays microanalysis of inhaled (UO2, PuO2) mixed oxides in alveolar macrophages observed *in toto. Int. Congr. on Radiation Protection 9*, Vienna, Hoftburg, Austria. Vol. 2, pp. 508–510.

Matton, S., Massiot, P., Le Naour H., Lizon C., Rateau G. and Fritsch P. EDS analysis of inhalable cerium oxide particles after *in vitro* dissolution assay and after intratracheal injection. *Ann. occ. Hyg.* (submitted).

Métivier, H., Masse, R., Rateau, G., Nolibé, D. and Lafuma, J. (1989) New data on the toxicity and translocation of inhaled ²³⁹PuO₂ in baboons. *Radiat. Prot. Dosim.* **26**, 167–172.

Métivier, H., Poncy, J. L., Rateau, G., Stradling, G. N., Moody, J. C., Gray, S. A. (1992) Uranium behaviour in the baboon after the deposition of a ceramic form of uranium dioxide and uranium octoxide in the lungs: implications for human exposure. *Radioprotection* **27**, 263–281.

Price, A. (1989) Review of methods for assessment of intake of uranium by workers at BNFL Springfields. *Radiat. Prot. Dosim.* **26**, 35–42.

Stradling, G. N., Stather, J. W., Gray, S. A., Moody, J. C., Hodgson, A., Sedgwick, D. and Cooke, N. (1988) The metabolism of ceramic and non-ceramic forms of uranium dioxide after deposition in the rat lung. *Human Toxicol.* **7**, 133–139.

Stradling, G. N., Stather, J. W., Price, A. and Cooke, N. (1989) Limits of intake and interpretation of monitoring data for workers exposed to industrial uranium bearing dusts. *Radiat. Prot. Dosim.* **26**, 83–87.

West, C. M., Scott, L. M. and Schultz, N. B. (1979) Sixteen years of uranium personnel monitoring experience-in retrospect. *Health Phys.* **36**, 665–670.

Pergamon

Ann. occup. Hyg., Vol. 41, Supplement 1, pp. 70–76, 1997
British Occupational Hygiene Society
© 1997 NRPB. Published by Elsevier Science Ltd
Printed in Great Britain
0003–4878/97 $17.00 + 0.00
Inhaled Particles VIII

PII: S0003–4878(96)00074–9

BIOKINETICS OF A URANIUM OCTOXIDE-BEARING AEROSOL FORMED IN A NEW LASER ENRICHMENT PROCESS: IMPLICATIONS FOR HUMAN EXPOSURE

A. Hodgson, S. A. Hodgson, G. N. Stradling, A. Birchall, T. Fell, M. H. Hengé-Napoli* and E. Ansoborlo*

National Radiological Protection Board, Chilton, Didcot, U.K.; and *Institut de Protection et de Surete Nucleaire, Pierrelatte, France

INTRODUCTION

This study forms part of a programme of work to investigate the biokinetics of uranium in aerosols formed during laser isotopic separation at a new uranium enrichment processing facility (Ansoborlo *et al.*, 1994). These aerosols can consist of UO_2, U_{metal} and U_3O_8 in different proportions and contain an ultrafine component. A previous study (Hengé-Napoli *et al.*, 1994) identified an aerosol consisting of 85% UO_2/U_{metal} and 15% U_3O_8 with up to 20% of the activity associated with an ultrafine component. The ultrafine fraction was shown to consist of small clusters of 0.1 μm particles. This material had similar absorption characteristics to the default values recommended in the *International Commission on Radiological Protection (ICRP) Publication 66* (ICRP, 1994) for compounds with moderate absorption rates (Type M). Workers can be exposed to two different types of uranium aerosols which in normal conditions are a mixture of $UO_2 + U_{metal}$. However, due to the high pyrophoricity of uranium metal, a fire could ignite which would alter the initial composition. Although considerable information exists on the biokinetics of U_3O_8 (Stradling *et al.*, 1989) there is a paucity of data available on aerosols formed during uranium fires.

The aim of this study was to use material-specific physiochemical data and biokinetic data, obtained from animal experiments, to provide a basis for deriving dose limits and assessing monitoring procedures. The study employed the most recent recommendations, models and methodologies of the ICRP *ICRP Publications 60* (ICRP, 1991), *66* (ICRP, 1994), *68* (ICRP, 1994) and *69* (ICRP, 1995). All computer codes used were developed specifically to implement the latest ICRP models.

MATERIALS AND METHODS

The material was obtained from the AVLIS pilot facility during cleaning operation, specifically on structures where ignition can occur. Individual personal air samples were taken using a Gilair pump in compliance with French AFNOR standard NF X 43256. The sampling rate was $1 \, l.m^{-1}$ using a 0.8 μm pore size filter.

Activity collected on the filter was determined by α-counting and the particle size distribution was obtained using an 8-stage Anderson Mark II cascade impactor operating at a flow rate of 28 l.m^{-1}. The density of the aerosol was measured with a pycnometer and the specific surface area determined by the BET (Brunauer, Emmet and Teller) analytical method using absorption of N_2 gas. Further physico–chemical analysis was carried out using X-ray microanalysis and i.r. techniques (Ansoborlo et al., 1994).

The respirable fraction, particles less than 5μm activity median aerodynamic diameter (AMAD), was obtained by sedimenting the dust in a column of ethanol. Ethanol was used as the sedimentation medium in preference to water to minimise leaching of uranium. Before administration of the particles to the rats, the ethanolic suspension was evaporated to incipient dryness and the particles resuspended in 0.9% (w/v) saline containing 0.01% (v/v) of the dispersing agent Renex 78 [Atlas Chemical Industries (U.K.) Ltd, London]. The animals used were female rats of the HMT strain (NRPB, Chilton) about 16 weeks old and weighing 200 g. Food and water were always available and all animal procedures were conducted in compliance of the Animals (Scientific Procedures) Act, 1986. Animals were first anaesthetised using 3% halothane in oxygen, using a Fluotek MkIII vaporiser (Cyprene Ltd, Keighley, West Yorkshire). The saline suspension of the particles (0.1 ml; 80 μg uranium) was administered to 35 rats by intratracheal instillation using a procedure described previously (Ellender, 1987). Some groups of animals were maintained in metabolism cages for the separate collection of urine. The animals were killed, in groups of five, between 1 and 168 days after exposure to determine the lung retention and tissue distribution of the uranium. Samples collected included the lung, liver, kidneys and carcass (excluding the gastrointestinal tract). All samples were ashed at 500°C in a muffle surface to reduce bulk and the uranium content was determined by delayed neutron counting. Greater than 90% of the uranium present in the carcass is presumed to be in the skeleton.

Interpretation of the data

The human respiratory tract model in *ICRP Publication 66* provides comprehensive detail on the deposition and clearance of inhaled material. In the model, it is assumed that: the clearance rates due to particle transport and absorption from lungs to blood are independent, competing processes; that particle transport is species-dependent but independent of the material while absorption is independent of species but material dependent; and that absorption is the same from each region of the respiratory tract except the anterior region of the nose from which there is no absorption. By adopting this approach, material specific absorption parameters, obtained from animal studies, information can be used in human biokinetic and dosimetric models to derive dose coefficients and to provide on tissue retention and excretion rates (Birchall et al., 1995).

Material specific absorption kinetics for uranium were evaluated from the animal data by the program GIGAFIT, developed at the NRPB. GIGAFIT is a parameter fitting program that fits a function, or model, with up to 30 variable parameters to 12 independent data sets simultaneously. GIGAFIT was configured to implement a biokinetic model of the HMT rat lung (Birchall et al., 1995). Particle deposition in the human lungs was then calculated according to the ICRP human respiratory tract model with the personal computer program LUDEP (Jarvis and Birchall, 1993). A modified version of LUDEP, which incorporates the most recent ICRP systemic model for uranium, was developed to compute the dose coefficients for the different isotopic compositions of uranium (^{234}U, ^{235}U and ^{238}U) for each particle size distribution. The final dose coefficient reflects the proportion that each isotope and each particle size fraction contributes towards the total activity of the aerosol. The corresponding material specific annual limit on intake (ALI) was derived from the expression

$$\text{ALI} = \frac{0.02}{e(50)},$$

where 0.02 Sv is the annual effective dose limit, currently recommended by ICRP and e(50) is the effective dose coefficient (committed effective dose per unit intake) SvBq^{-1} committed over 50 years. The same version of LUDEP was used to predict the retention of uranium in the lung and its subsequent excretion via urine following acute and chronic exposure. The modified version of LUDEP was quality assured by comparing output against the computer code PLEIADES (Program for LinEar Internal Age-dependent DosES) for default particle sizes of uranium. PLEIADES, also developed by Board staff, was used to calculate dose coefficients for *ICRP Publications 68, 69* and *71*.

RESULTS

Uranium concentrations in air over 22 individual sampling measurements ranged from 0.2 to 18.5 Bq.m^{-3}. The particle size measured at the workplace was 8 μm AMAD ($\sigma_g = 2.4$), with 8% of the activity less than 1 μm. The density of the dust was 6.8 g.cm^{-3} and the specific surface area was 1 m^2.g^{-1}. X-ray microanalysis and i.r. techniques showed the compound comprising 70% U$_3$O$_8$ and 30% UO$_2$ + U$_{metal}$.

The tissue distribution and excretion data obtained from the animal study are shown in Table 1. The results show that the uranium is moderately transportable: 9% of the initial lung deposit is absorbed to blood by 1 day increasing to almost 20% by 168 days. The lung retention and uranium absorption to blood data were used with GIGAFIT to derive material specific absorption parameters, suitable for estimating dose coefficients in humans, and these are shown in Table 2.

DISCUSSION

After an intake by inhalation, the models most likely to be used to predict the radiological consequences are those currently recommended by ICRP. The *ICRP Publication 66* human respiratory tract model recommends three default types for absorption (fast [F], moderate [M] and slow [S]) and an aerodynamic particle size of 5 μm ($\sigma_g \simeq 2.5$) for occupational exposure. However, the ICRP advocates the use of material-specific data whenever possible. The biokinetic behaviour of uranium after absorption to blood is assumed to be independent of the chemical form on intake.

Table 1. Tissue distribution and retention of uranium

| Days | Lungs | Kidneys | Carcass | % Initial lung deposit[a] | | To blood[c] | Faeces + GIT[d] |
				Total body[b]	Urine		
1	71.7 ± 4.1	0.79 ± 0.08	3.32 ± 0.36	75.8 ± 4.2	4.76 ± 0.55	8.87 ± 0.66	19.4 ± 4.2
3	61.7 ± 3.8	0.92 ± 0.06	3.86 ± 0.45	66.5 ± 3.8	6.98 ± 0.66	11.8 ± 0.8	26.5 ± 3.8
7	48.4 ± 3.2	0.72 ± 0.09	4.36 ± 0.43	53.5 ± 3.2	9.25 ± 0.75	14.3 ± 0.9	37.3 ± 3.3
14	37.6 ± 2.6	0.46 ± 0.03	4.84 ± 0.52	42.9 ± 2.7	10.2 ± 0.9	15.5 ± 1.0	46.9 ± 2.9
28	29.1 ± 2.0	0.32 ± 0.02	4.43 ± 0.42	33.9 ± 2.1	11.5 ± 1.0	16.3 ± 1.1	54.6 ± 2.4
84	18.6 ± 1.3	< 0.28[e]	3.37 ± 0.30	22.0 ± 1.3	13.7 ± 0.9	17.1 ± 0.9	64.3 ± 1.6
168	12.1 ± 1.4	< 0.28[e]	3.04 ± 0.32	15.2 ± 1.4	16.2 ± 1.2	19.2 ± 1.2	68.6 ± 1.9

[a] Mean ± SE. five animals per group, initial lung deposit = 80 μg natural uranium.
[b] Sum of uranium in tissues, excluding liver (undetectable).
[c] Kidneys + carcass + urine; absorption from gastrointestinal tract (GIT) was ignored.
[d] Initial lung deposit − (total body + urine).
[e] Minimum detectable activity; all livers were in this category.

Table 2. Material specific absorption parameters (determined by GIGAFIT)

Compound	Fraction dissolved rapidly (F_r)	Rapid dissolution rate (S_r)	Slow dissolution rate (S_s)
Dust	0.089	100	0.0026
M[a]	0.1	100	0.005
S[a]	0.001	100	0.0001

[a] Absorption type defined in *ICRP Publication 66*.

Table 3. Material specific dose coefficient (Sv.Bq^{-1})

Isotope	0.1 µm (σg = 1.6) AMAD	8 µm (σg = 2.4) AMAD
^{234}U	1.895×10^{-5}	2.557×10^{-6}
^{235}U	1.731×10^{-5}	2.220×10^{-6}
^{238}U	1.635×10^{-5}	2.028×10^{-6}
3.5% Enriched[a,b]		3.74×10^{-6}
ALI (Bq)		5348
Mass (mg)		74

[a] Sum of the contribution of each isotope. Activity distribution: ^{234}U (0.80), ^{235}U (0.04), ^{238}U (0.16).
[b] Particle size contribution 0.1 µm AMAD (0.08), 8 µm AMAD (0.92).

The dose coefficient was calculated assuming that 92% of the activity was associated with the large particles and 8% with the ultrafine component. The uranium isotopic composition was 80% ^{234}U, 4% ^{235}U and 16% ^{238}U by activity. The calculated values are shown in Table 3 with the derived dose coefficient (3.7 × 10^{-6} Sv Bq^{-1}) and the corresponding ALI (5350 Bq or 74 mg of uranium) with the isotopic composition specified. These values were derived assuming a specific activity of 72 000 Bq.g^{-1} and a committed effective dose of 20 mSv.

Implications for monitoring

The predicted lung retention and daily urinary excretion rate of uranium following acute and chronic intakes are shown in Fig. 1. The daily urinary excretion rate at time *t* is defined as the amount of uranium excreted between $(t - 1)$ and *t* day. Intakes are expressed as a fraction of unit intake for acute exposure and as a fraction of the daily intake for chronic exposure. For assessing chronic intakes, the method adopted is that described by *ICRP Publication 54* (1988). Similar calculations, performed using identical particle size parameters for default absorption Type M and S uranium compounds, are also shown in the figure for comparison.

After an acute inhalation exposure about 10% of the intake (55% of the 0.1 µm AMAD particles and 5% of the 8 µm AMAD) would initially deposit in the lung. The remainder would be either deposited higher in the respiratory tract or be exhaled. Assuming ^{235}U would be used for chest monitoring, with a minimum detectable activity (MDA) of 3 Bq (ICRP 54, 1988), it can be inferred that acute exposure at the reporting level (0.1 ALI) would not be detected. Although chest monitoring can detect larger intakes, it is inappropriate as acute intakes should be

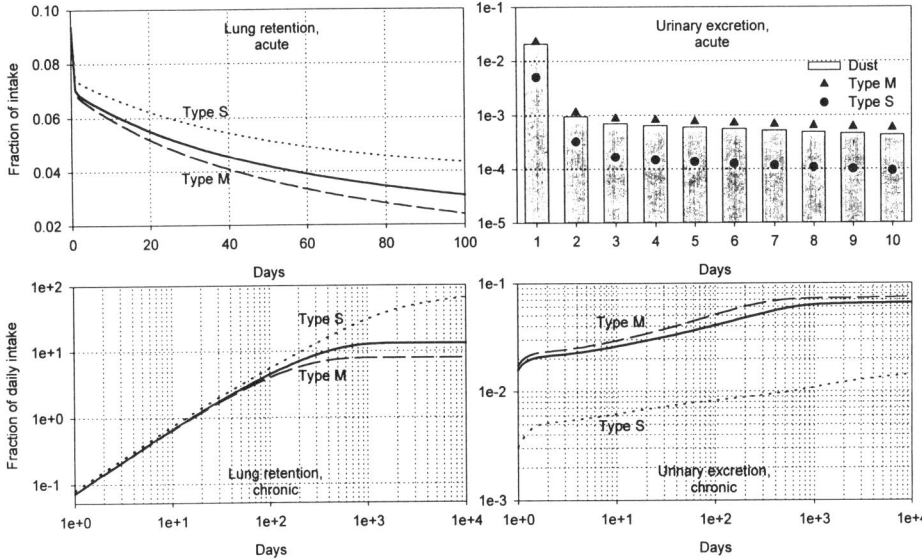

Fig. 1. Predicted lung retention and urinary excretion rates for dust and Type M and S uranium compounds after acute and chronic intake.

restricted to 2.5 mg of uranium (or 180 Bq of this dust), the permissible daily intake limit for chemical toxicity (OJEC, 1980). Chronic daily intakes at the ALI level would never be detected at the MDA quoted. In conclusion, chest monitoring is not applicable except to indicate intakes well in excess of toxic limits.

Approximately 2% of an acute intake is excreted via the urine within the first 24 h period. The daily excretion rate reduces more than 20-fold over the second 24 h period and 40-fold by 1 week after exposure. Therefore, for accurate assessment of an acute intake, it is essential that sampling commences as soon as practicable and that the exposure history is known. Techniques capable of detecting fractions of the daily intake, initially 0.015 rising to 0.065 per unit intake at near equilibrium, are required to assess chronic exposure. To monitor at the reporting level (0.1 ALI/365) demands sensitivity better than 0.02 Bq or 0.2 μg.d^{-1} at early times. This value increases to 0.1 Bq or 0.9 μg at times later than 5000 days.

REFERENCES

Ansoborlo, E., Hengé-Napoli, M. H., Donnadiu-Claraz, M., Roy, M. and Pihet, P. (1994) Industrial exposure to uranium aerosols at laser enrichment processing facilities. *Radiat. Prot. Dosim.* **53**, 163–167.

Birchall, A., Bailey, M. R. and Jarvis, N. S. (1995) Application of the new ICRP respiratory tract model to inhaled plutonium nitrate using experimental biokinetic data. In *Proc. Int. Conf. On Radiation Dose Management in the Nuclear Industry*, Windermere, U.K. Oct 9–11, pp. 216–223.

Ellender, M. (1987) The clearance of uranium after deposition of the nitrate and bicarbonate in different regions of the rat lung. *Human Toxicol.* **6**, 479–482.

Hengé-Napoli, M. H., Ansoborlo, E., Gray, S. A., Hodgson, A. and Stradling, G. N. (1995) The biokinetics of a uranium aerosol formed in a new laser enrichment process: implications for human exposure. In *Proc. Int. Conf. On Radiation Dose Management in the Nuclear Industry*, Windermere, U.K., Oct 9–11, pp. 224–229.

International Commission on Radiological Protection (1988) Individual monitoring for intakes of radionuclides by workers: design and interpretation. ICRP Publication 54. *Ann. ICRP*, Vol. 19, Nos 1–3. Pergamon Press, Oxford.

International Commission on Radiological Protection (1991) 1990 recommendations of the International Commission on Radiological Protection. ICRP Publication 60. *Ann. ICRP*, Vol. 21, Nos 1–3. Elsevier Science Ltd, Oxford.

International Commission on Radiological Protection (1994) Human respiratory tract model for radiological protection. ICRP Publication 66. *Ann. ICRP*, Vol. 24, Nos 1–3. Elsevier Science Ltd, Oxford.

International Commission on Radiological Protection (1994) Dose coefficients for intakes of radionuclides by workers (replacement of ICRP Publication 61). ICRP Publication 68. *Ann. ICRP*, Vol. 24, No. 4. Elsevier Science Ltd, Oxford.

International Commission on Radiological Protection (1995) Age-dependent doses to members of the public from intakes of radionuclides: part 3 ingestion dose coefficients. ICRP Publication 69. *Ann. ICRP*, Vol. 25, No. 1. Elsevier Science Ltd, Oxford.

Jarvis, N. S. and Birchall, A. (1993) LUDEP 1.0. Personal Computer program for calculating internal doses using the new ICRP respiratory tract model. NRPB-SR264.

OJEC (1980) Official Journal of European Communities. L246, Vol. 23, 17 September 1980. Office for Official Journal of the European Communities, Luxembourg.

Stradling, G. N., Stather, J. W. and Gray, S. A. (1989) The metabolic behaviour of U_3O_8 bearing residues after their deposition in the rat lung: implication for human exposure. *Exp. Pathol.* **37**, 76–82.

Ann. occup. Hyg., Vol. 41, Supplement 1, pp. 77–81, 1997
© 1997 British Occupational Hygiene Society
Published by Elsevier Science Ltd. All rights reserved
Printed in Great Britain
0003–4878/97 $17.00 + 0.00
Inhaled Particles VIII

Pergamon

PII: S0003–4878(96)00075–0

HETEROGENEITY OF A (U, Pu) MIXED OXIDE POWDER AT THE LEVEL OF INHALED PARTICLES: AN EDS ELECTRON MICROSCOPY STUDY IN THE RAT

P. Massiot, H. Le Naour, S. Matton, C. Lizon, G. Rateau, I. L'Hullier,
G. Grillon and P. Fritsch

Laboratoire de Radiotoxicologie, CEA/DSV/DRR/SRCA, BP 12, 91680 Bruyères le Châtel, France

INTRODUCTION

Early studies using energy dispersive X-ray spectrometry (EDS) by transmission electron microscopy have shown that mixed (U, Pu) oxide containing 25% of Pu was a homogeneous solid solution, $(U, Pu)O_{1.96}$ (Eidison and Meiwhinney, 1983). We have recently published preliminary EDS results on the chemical composition of a $(U, Pu)O_2$ powder containing 5.3% Pu in rat alveolar macrophages after inhalation exposure (Massiot *et al.*, 1996). This inhaled powder appeared heterogeneous, most of the particles being "pure" UO_2 whereas, a few of them were "pure" PuO_2.

The aim of this study was to provide semi-quantitative results in order to estimate the range of heterogeneity of this inhaled powder. This was performed on entire alveolar macrophages obtained by pulmonary lavage, a method which provides more information about distribution of particles in the alveolar macrophage population (Massiot *et al.*, 1996) than the usual method, after digestion of the biological matrix (Pairon *et al.*, 1994).

MATERIAL AND METHODS

Three month old male Sprague–Dawley rats were exposed to a $(U, Pu)O_2$ powder containing 5.3% of Pu. The Pu used for the mixed oxide fabrication contained 1.1% of [238]Pu, 8.0% of [241]Pu and, at the time of the experiment, 11% of the total α activity was due to [241]Am. The aerosol was obtained by nebulisation of the powder, suspended in ethanol, using our inhalation facility as previously described (André *et al.*, 1989). On day 3 after exposure, four animals were killed under pentobarbital anaesthesia (40 mg kg⁻¹). Pulmonary lavage was performed by repeated instillation of 5 ml 0.9% NaCl until a total volume of 20 ml had been collected. Cytospins of the cell suspension were prepared on slides and cells were fixed for 30 min in 70% ethanol.

Collodium coated slides were used for electron microscopy study. In this case, the collodium film was put on copper grids and coated with carbon. Observations were performed with a CM 200 Philips microscope using the scanning transmission mode at 200 kV. EDS analyses were performed with an EDAX X-ray detector (super ultra thin window). The particle diameter distribution was measured using standard image analysis software developed in our laboratory. The shape of particles was estimated at different tilts with respect to the beam direction under high-angle annular-dark field (HAADF) observation mode (Treacy, 1981).

The EDS was performed either by scanning a rectangular field including the whole particle or by a static analysis using a 14 nm nanoprobe. Quantitative analysis was performed with an EDAX software which was based on Cliff and Lorimer equations (Cliff and Lorimer, 1972). For a few particles, these spectrum analyses were completed by a specific Pu and U Lα X-ray mapping to provide element imaging. Semi-quantitative results were obtained by image analysis.

Uncoated slides were used for autoradiography observed by light microscopy using standard methods (nuclear emulsion, Ilford K5).

The total lung burden was determined by scintillation counting from the results obtained for the lavaged lungs and an aliquot of the lavage fluid. The samples were mineralized and the amounts of total alpha emitters were measured using a scintillation medium adapted for acid solution (Packard, Instagel+).

RESULTS

In our experimental conditions the retention of the total α emitters in lungs was 4.3 kBq (SD = 0.7) and about 30% of the lung retention was extracted by pulmonary lavage. Particles were easily identified at low magnification within the entire cells using the HAADF mode. About 40% of alveolar macrophages contained particles and the mean number of particles per loaded macrophage was about three. The median actual diameter of the particles was 0.2 μm (σg = 2) but no particle with a diameter of more than 0.6 μm was encountered over several hundred observed ones. Figure 1 shows the same particle at tilts from +40° to −60°. This particle corresponded to an aggregate of small particles with a diameter of about 50 nm. In fact, most of the observed particles were aggregates.

Figure 2 shows the spectrum of a particle emitting both Pu and U specific X-rays obtained by scanning over the entire particle. Because the overlapping of the U and Pu M X-rays, only the Lα-rays could be used to determine accurately the relative amount of U and Pu. In this special case, expressed as metal content, the particle contained about 11% Pu.

Figure 3 shows, for the same particle as Fig. 2, spectra obtained on two different areas with a static 14 nm nanoprobe and the U and Pu Lα X-ray mapping of the particle. The static analysis (Fig. 3) showed that some areas within the particle were composed of "pure" UO_2 and "pure" PuO_2 without any significant mixing. This heterogeneous composition was confirmed by U and Pu X-ray mapping (Fig. 3).

Of the 34 particles analysed with a scanning rectangular field analysis, 10 particles contained both Pu and U whereas 24 contained only Pu. However, after element mapping only two of them might correspond to actual mixed oxide that always contained more than 25% Pu. Figure 4 shows autoradiography of alveolar macrophages. For this 1 day exposure time, most alveolar macrophages did not contain any α emitters but, when α tracks were observed, their number per particle varied within a range from 1 to more than 500.

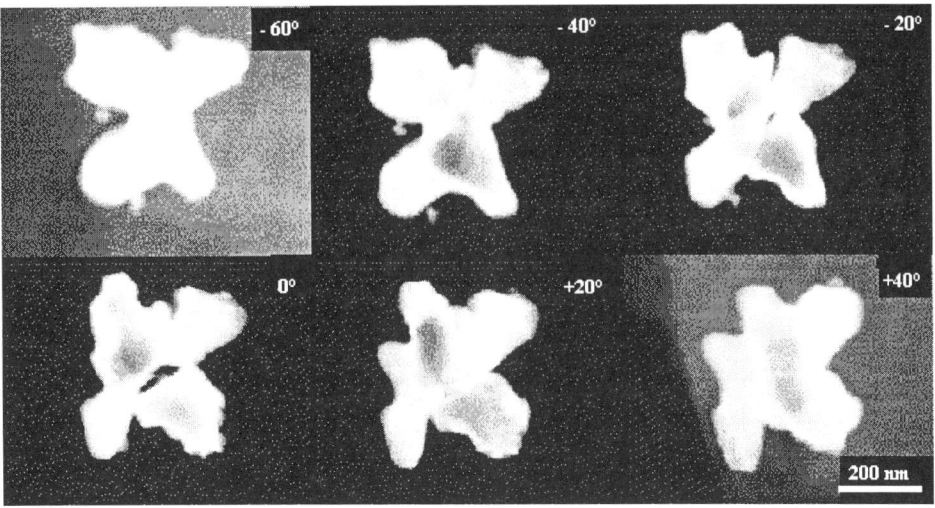

Fig. 1. Mixed oxide particle observed at different tilt angles.

DISCUSSION

The U and Pu Lα X-rays have a similar high energy, 13.612 and 14.276 keV, respectively, that presented a negligible absorption within the sample, in the range of particle size observed. For this reason, quantitative estimate of the U/Pu ratio could be directly determined from the ratio of the net intensity of these rays. Thus, the method used (EDS of particles included within a biological matrix—the entire alveolar macrophage) appears quite appropriate for quantitative study of the relative amounts of Pu and U in $(U, Pu)O_2$ particles after lung deposition following inhalation. Moreover, it provides data on the mean number of particles per cell and on the shape and size of the particles. This method cannot be used to study particle interaction with some cellular elements such as phosphorus for example. For that purpose, EDS has to be performed after digestion of the cellular matrix and

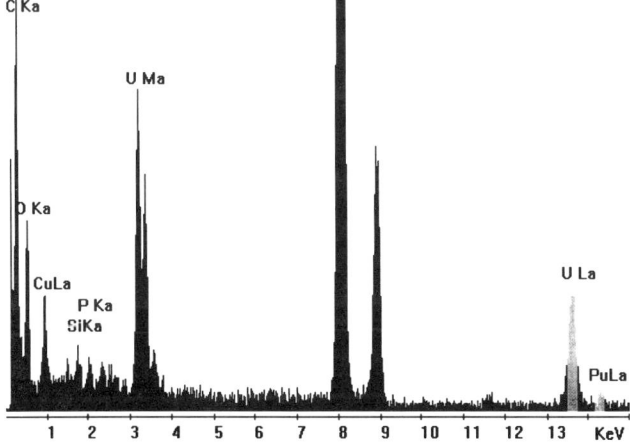

Fig. 2. X-ray spectrum obtained after scanning over the entire particle.

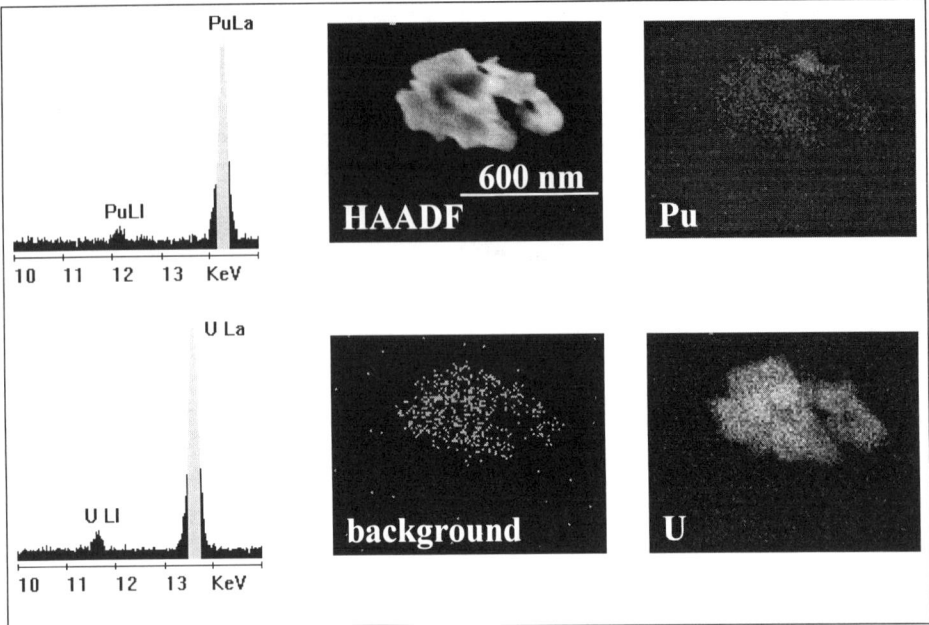

Fig. 3. Spectra from spot analysis showing "pure" UO_2 and "pure" PuO_2 areas and Lα X-ray U and
Pu mapping after subtraction of background obtained at a similar energy. Same particle as Fig. 2.

deposition of the particles on a filter coated with carbon (Pairon *et al.*, 1994;
Matton *et al.*, 1996).

Using EDS, the detection limit for Pu was estimated at 0.5–1% expressed as
Pu/(U + Pu) ratio for counting times from 500 to 2000 s. From the spectrum
analysis performed after scanning over the entire particle, we have demonstrated
that the mixed oxide studied corresponds to a heterogeneous powder ranging from

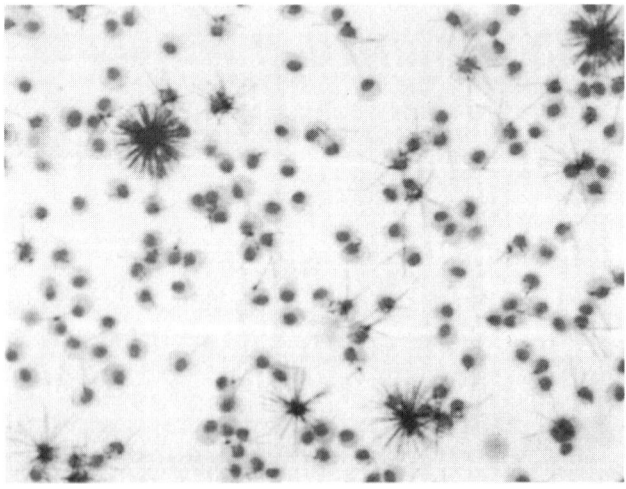

Fig. 4. Autoradiography of cells extracted by pulmonary lavage after a 1-day exposure time.

"pure" UO_2 to "pure" PuO_2. However, the homogeneous composition within particles or aggregates can be only characterized by element mapping. With this method, very few actual (U, Pu)O_2 particles have been encountered (2/34). Because of the overlapping of individual particles within aggregates, the actual presence of (U, Pu)O_2 solid solution is not actually demonstrated and further mapping experiments are needed.

Results obtained by autoradiography confirm those obtained with EDS. We have calculated the number of α tracks emitted per day, on 2 π, from homogeneous mixed oxide and pure PuO_2 according to their chemical and isotopic compositions. The values were 1.2 and 21 for the median actual diameter measured (0.2 µm) and 30 and 570 for the maximal observed diameter (0.6 µm) for homogeneous mixed oxide and pure PuO_2, respectively. Thus, the range of the number of tracks observed by autoradiography, after a one day exposure, could only be obtained if pure PuO_2 particles were present. Most of the few tracks associated with particles could be due to small PuO_2 particles in aggregates but not to actual (U, Pu)O_2.

We conclude that, after inhalation of the (U, Pu)O_2 studied, most of the dose delivered to the lungs comes only from a few particles that induce a heterogeneous irradiation of the lungs. Further studies, that combine autoradiography and EDS, are in progress to characterize the physico-chemical properties of (U, Pu)O_2 according to their mean Pu content and their fabrication process.

REFERENCES

Andre, S., Charuau, J., Rateau, G., Vavasseur, C. and Métivier, H. (1989) Design of a new inhalation device for rodents and primates. *J. Aerosol Sci.* **20**, 647–656.

Cliff, G. and Lorimer, G. W. (1972) Quantitative analysis of thin metal foils wing EMMA-4, the ratio technique. In *Proc. 5th Europ. Congr. on E.M.* Manchester, pp. 140–141. Pub. The Institute of Physics, Bristol and London.

Eidison, A. and Meiwhinney, J. (1983) *In vitro* dissolution of respirable aerosols of industrial uranium and plutonium mixed-oxide nuclear fuels. *Health Phys.* **45**, 1023–1027.

Massiot, P., Fritsch, P., Le Naour, H., Rateau, G., Lizon, C. and L'Hullier, I. (1996) Electron probe X-ray microanalysis of inhaled (UO_2, PuO_2) mixed oxides in alveolar macrophages observed *in toto*. *Int. Cong. on Radiation Protection 9*, April 14–19, 1996, Vienna, Hofburg, Austria, Vol. 2, pp. 508–510.

Matton, S., Massiot, P., Le Naour, H., Lizon, C., Rateau, G. and Fritsch, P. (1997) EDS electron microscopy of inhalable cerium oxide particles after *in vitro* dissolution assay and after intratracheal injection (submitted).

Pairon, J. C., Billon-Galland, M. A., Iwatsubo, Y., Bernstein, M., Gaudichet, A., Bignon, J. and Brochard, P. (1994). Biopersistence of non fibrous mineral particles in the respiratory tracts of subjects following occupational exposure. *Environ. Health Persp.* **102** (suppl 5), 269–275.

Treacy, M. M. J. (1981) Imaging with Rutherford scattered electrons in the scanning electron microscope. *Scanning Electron Microsc.* **1**, 185–197.

 Pergamon

Ann. occup. Hyg., Vol. 41, Supplement 1, pp. 82–85, 1997
© 1997 British Occupational Hygiene Society
Published by Elsevier Science Ltd. All rights reserved
Printed in Great Britain
0003–4878/97 $17.00 + 0.00
Inhaled Particles VIII

PII: S0003–4878(96)00076–2

INHALATION OF (U, Pu)O$_2$ IN THE RAT: PRELIMINARY RESULTS ON THE TRANSURANIUM ELEMENTS BEHAVIOUR AND ON THE EFFICACY OF A DTPA TREATMENT

B. Ramounet, P. Fritsch, C. Lizon, J. L. Poncy, G. Grillon, M. Verry and G. Rateau

Laboratoire de Radiotoxicologie, CEA/DSV/DRR/SRCA, BP 12, 91680 Bruyères le Châtel, France

INTRODUCTION

A few studies have concerned the behaviour of actinides after inhalation of (U, Pu)O$_2$ powders containing from 1 to 30% Pu which were regarded as homogeneous solid solutions (Stanley *et al.*, 1982; Talbot and Baker, 1989; Morgan and Black, 1989; Lataillade *et al.*, 1995). The results suggested that transfer of Pu to the extra-pulmonary organs increased as the U/Pu or the ^{238}Pu/total Pu ratio increased.

We have recently shown that inhaled (U, Pu)O$_2$ from a mixed oxide powder containing 5.3% Pu was heterogeneous at the nanometric level in terms of U/Pu ratio. Most of the inhaled particles from an aerosol of this powder corresponded to "pure" UO$_2$ and "pure" PuO$_2$ (Massiot *et al.*, 1996a,b).

The aim of this work was to provide experimental results on the biological behaviour of transuranium elements in lungs and their retention in extra-pulmonary organs after inhalation of this "heterogeneous" (U, Pu)O$_2$ powder. Moreover, an estimate of the efficacy of a DTPA treatment on the Pu retention values has been performed after a repeated daily treatment for one week and, the chemical, and the isotopic compositions have been determined in the different organs studied.

METHODS

Two groups of 20 male Sprague–Dawley rats, referred to as group I and group II, were exposed at 3 months of age to the (U, Pu)O$_2$ aerosol containing 5.3%. The Pu used for the mixed oxide fabrication contained 1.1% of ^{238}Pu, 8.0% of ^{241}Pu and, at the time of the experiment, 11% of the total α activity was due to ^{241}Am. The aerosol was obtained by nebulization of the powder using our specific device as previously described (André *et al.*, 1989). Two hours later, six animals from group II received an intramuscular injection of DTPA (30 μmol kg^{-1}) and this treatment was repeated daily for 6 days.

Rats were killed under pentobarbital anaesthesia (40 mg kg^{-1}) at different times after the aerosol exposures. The amounts of Pu and Am in lungs, liver, femurs and kidneys were measured by α- and γ-ray spectrometry. For some animals, *in vivo* counting was performed by X- and/or γ-ray spectrometry using an argon methane proportional counter or a NaI detector. Liquid scintillation measurements were performed on mineralized samples with or without specific solvent extraction of Pu (Keough and Powers, 1970) and for total α, with a scintillation medium adapted for

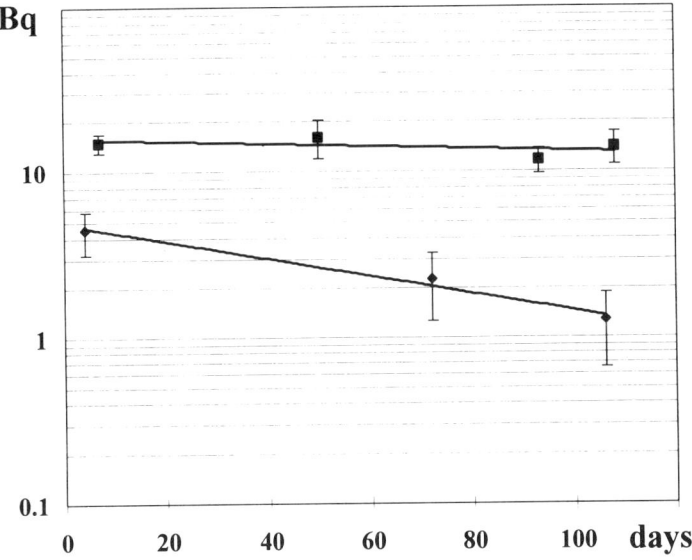

Fig. 1. Evolution of lung retention of total α emitters as a function of time following aerosol exposure for the two groups of rats studied. Mean value for at least four animals per point, bars: SD.

acid solution (Packard, instagel+). ^{241}Pu amounts were estimated by liquid scintillation assuming a counting yield similar to that of tritium. Alpha spectrometry was performed after actinide precipitation with lanthane fluoride.

RESULTS

The initial lung burdens measured on day 4 after exposure for group I and on day 7 after exposure for group II. These burdens, expressed as total α emissions were 4.3 kBq ($SD = 0.7$, $n = 4$) and 15.0 kBq ($SD = 6.0$, $n = 6$) for groups I and II, respectively.

The lung clearances of the two animal groups for the first 110 days after exposure are shown in Fig. 1. A single exponential function with a half life of about 60 days was observed for the lowest initial lung burden (group I), whereas the clearance process appeared almost entirely inhibited for the highest initial lung burden (group II). In this latter animal group, four rats of the five animals remaining alive after 110 days died before day 160 post-exposure.

On day 7 post-exposure, the DTPA treatment performed on group II did not modify significantly the pulmonary retention of Pu as compared to the untreated controls. However, this DTPA treatment induced a significant decrease of the amount of Pu retained in liver and kidneys that corresponded to at least a 50% decrease as compared to controls (Table 1).

Table 2 shows, in group II, that the chemical and isotopic compositions of the actinide retentions in lungs, liver, kidneys and femurs, measured 7 days after exposure, were very similar to those provided for this (U, Pu)O^2 powder (Table 3). On day 50 post-exposure, preliminary results suggested that the cumulated retention of actinides in the extra-pulmonary organs could correspond to more than 1.3% of the initial lung burden.

Table 1. Pu retention values in controls and DTPA treated animals. Mean values for six rats measured on day 7 post-inhalation ± SD

	DTPA treated rats	Untreated controls
Lungs	13 831 Bq ± 5558	14 013 Bq ± 1741
Liver	4 Bq ± 2	16 Bq ± 8
Kidneys	0.25 Bq ± 0.2	0.59 Bq ± 0.65

Table 2. Isotopic composition of the actinide retention in different organs. Mean values for six rats measured on day 7 post-inhalation ± SD

	$^{239+240}$Pu/ total α	^{238}Pu + ^{241}Am/ total α
Liver	29.5% ± 0.8	70.5% ± 0.9
Kidneys	30.7% ± 2.1	69.4% ± 2.2
Lungs	29.4% ± 0.3	70.6% ± 0.3

Table 3. Comparative isotopic composition of the initial lung burden and the crude powder. Mean values for six rats measured on day 7 post-inhalation

	Crude powder	Initial lung deposit
^{241}Am/total α	0.11	0.08
β ^{241}Pu/total α	26	28
$^{239+240}$Pu/total α	0.28	0.30

DISCUSSION

Although the inhaled (U, Pu)O$_2$ particles were heterogeneous, the isotopic and chemical compositions of the initial lung deposit were very similar to those provided for the mixed oxide powder. Further studies are in progress to characterize these compositions in the different organs as a function of time following inhalation.

Two different clearance processes have been described during the first weeks following inhalation of "insoluble" particles in the rat. The first one, observed during the first 7–10 days post-exposure, corresponds to the mucociliary clearance of the upper airways. The second one, observed for 200–300 days, corresponds to alveolar migration to the upper respiratory tract via the bronchial tree lumen. After inhalation of ^{239}PuO$_2$, a gradual inhibition of the first clearance process has been reported in the range of initial lung burdens from 0.7 to 9 kBq (Sanders *et al.*, 1993). Alteration of the second process was induced at higher initial lung burden with a nearly total inhibition for initial lung burden at 3.7 kBq (Sanders, 1972). In our experimental conditions, we could only study this second process. The clearance was nearly entirely inhibited for an initial lung burden at 15 kBq, whereas the half life observed at 4.3 kBq was quite similar to the value we have reported in the same strain of rat after exposure to a (U, Pu)O$_2$ aerosol containing 20% of Pu or industrial PuO$_2$ (Lataillade *et al.*, 1995). However, this half life was about twice that measured after inhalation of Fe$_2$O$_3$ (Fritsch and Masse, 1992) or after inhalation of ^{239}PuO$_2$ for low initial lung burden (Lundgren *et al.*, 1995). Such a difference might be explained by the high radiotoxicity due to the heterogeneity of the (U, Pu)O$_2$ particles used (Massiot *et al.*, 1996a,b). Further studies are in progress to determine if the alveolar macrophage clearance parameters could be altered, even at a low initial lung burden, as a function of the specific alpha activity of the inhaled particles.

To our knowledge, only one study has concerned the effect of a DTPA treatment on Pu extra-pulmonary retention after oxide inhalation (Fukuda *et al.*, 1996). In this study, the optimal treatment corresponded to intraperitoneal administration (150 μmol kg^{-1}) performed one day after exposure that was repeated daily for 5 days and then addition of DTPA to the drinking water. In this case, no significant diminution of Pu extra-pulmonary retention could be observed. In our study, a significant decorporation was observed that could mainly be due to the early treatment performed 2 h after exposure. Thus, our results demonstrate the efficacy of an early DTPA treatment to decrease Pu retention in extra-pulmonary organs after inhalation of (U, Pu)O$_2$.

Our preliminary results on the transfer of transuranium elements to the extra-pulmonary organs were in the range of those reported in the rat after inhalation of (U, Pu)O$_2$ (Stanley *et al.*, 1982; Lataillade *et al.*, 1995).

In conclusion, these preliminary results provide new information on the behaviour of transuranium elements after exposure to (U, Pu)O$_2$ aerosols. New questions, especially about lung clearance alterations as a function of the radiotoxicity of the inhaled particles and their heterogeneity, and about the efficacy of a DTPA treatment following inhalation of actinide oxides, arise. This justifies the need for further studies to provide new experimental data.

REFERENCES

Fritsch, P. and Masse, R. (1992) Overview of pulmonary alveolar macrophage renewal in normal rats and during different pathological processes. *Environ. Health Perspect.* **97**, 59–67.

Fukuda, S., Iida, H., Yamada, Y., Koizumi, A., Sato, H., Isigure, N., Nakano, T. and Enomoto, H. (1996) Chelating agent, DTPA, can not remove effectively inhaled-plutonium oxide in rats. *Int. Congr. on Radiation Protection 9*, April 14–19, 1996, Vienna, Hofburg, Vol. 2, pp. 460–462.

Keough, R. F. and Powers, G. J. (1970) Determination of plutonium in biological materials by extraction and liquid scintillation counting. *Analyt. Chem.* **42**, 419–421.

Lataillade, G., Verry, M., Rateau, G., métivier, H. and Masse, R. (1995) Plutonium solubility in rat and monkey lungs after inhalation of industrial plutonium oxide and mixed uranium and plutonium oxide. *Int. J. Radiat. Biol.* **67**, 373–380.

Lundgren, D. L., Haley, P. J., Hahn, F. F., Diel, J. H., Griffith, W. C. and Scott, B. R. (1995) Pulmonary carcinigenicity of repeated inhalation exposure of rats to aerosols of ^{239}PuO$_2$ *Radiat. Res.* **142**, 39–53.

Massiot, P., Fritsch, P., Le Naour, H., Rateau, G., Lizon, C. and L'Hullier, I. (1996a) Electron probe X-ray microanalysis of inhaled (UO$_2$, PuO$_2$) mixed oxides in alveolar macrophages observed *in toto*. *International Congress on Radiation Protection 9*, April 4–19, 1996, Vienna, Hofburg, Austria, Vol. 2, pp. 508–510.

Massiot, P., Le Naour, H., Matton, S., Lizon, C., Rateau, G., Grillon, G. And Fritsch, P. (1196b) Heterogeneity of (U, Pu)O$_2$: an EDS electron microscopy study of inhaled particles in the rat (submitted).

Morgan, A. and Black, A. (1989) Lung retention and translocation of Pu in mice following inhalation of ^{238}PuO$_2$ and ^{239}PuO$_2$ fired at 550–1250°C and (U, Pu)O$_2$ fired at 1400 and 1600°C. *Radiat. Prot. Dosim.* **26**, 297–301.

Sanders, C. L. (1972) Deposition patterns and the toxicity of transuranium elements in lung. *Health Phys.* **22**, 607–615.

Sanders, C. L., Lauhala, K. E., McDonald, K. E. and Sanders, G. A. (1993) Lifespan studies in rats exposed to ^{239}PuO$_2$ aerosol. *Health Phys.* **64**, 509–521.

Stanley, J. A., Edison, A. F. and Mewhinney, J. A. (1982) Distribution, retention and dosimetry of plutonium and americium in the rat, dog and monkey after inhalation of an industrial-mixed uranium and plutonium oxide aerosol. *Health Phys.* **43**, 521–530.

Talbot, R. J. and Baker, S. T. (1989) Solubility of reactor fuels in the mouse lung with respect to their U/Pu and ^{238}Pu/^{239}Pu ratios. *Radiat. Prot. Dosim.* **26**, 207–210.

 Pergamon

Ann. occup. Hyg., Vol. 41, Supplement 1, pp. 86–91, 1997
British Occupational Hygiene Society
© 1997 AEA Technology plc. Published by Elsevier Science Ltd
Printed in Great Britain
0003–4878/97 $17.00 + 0.00
Inhaled Particles VIII

PII: S0003–4878(96)00077–4

EARLY CELLULAR RESPONSES IN RATS EXPOSED TO RADON AND RADON PROGENY

C. G. Collier,* M. Bisson,‡ S. T. Baker,* T. Eldred,* P. Fritsch,† J-P. Morlier†
and G. Monchaux†

*AEA Technology Plc, 551 Harwell, Didcot OX12 0PX, U.K.; †CEA-DSV/DPTE, bp 6, 92265,
Fontenay-aux-Roses, Cedex, France; and ‡INERIS, BP2, F-60550, Verneuil en Halatte, France

INTRODUCTION

Domestic exposure to radon/radon progeny is estimated to account for 5–10% of lung tumours, and may contribute to the 95% related to smoking Doll (1992). These estimates are based on extrapolation from mining exposures to the domestic situation. However, as considerable differences exist between the two situations of terms of radon levels, dose rates, exposure duration, populations exposed and the presence of other pollutants, direct extrapolation from mining to domestic exposures may not be appropriate. Animal studies permit the effects of these different conditions on lung tumour risks to be assessed.

Within the current CEC programme on the risk to the population from inhalation exposure of radon decay products (progeny), two groups (AEA, U.K. and CEA, France) are conducting life-span studies in rats to determine the effect of dose and dose rate on lung tumour induction by radon/radon progeny. Whilst most of the animals exposed are followed for life-span, a few from each group are taken from early effects studies, to assess the induction of nuclear aberrations in alveolar macrophages and proliferation in lung cells (epithelial and macrophage). These early effects may be useful as *in vivo* dosimeters for different lung regions and as predictors of long-term effects.

The early effects peak at approximately 14 days post exposure regardless of exposure duration (2–15 days), (Taya *et al.*, 1994; Bisson *et al.*, 1994), so both groups studied effects at this time. Radon/radon progeny exposure [440 working level months (WLM) (1.54 J h^{-1} m^{-3})] doubled epithelial cell proliferation regardless of animal age, and proliferation levels were up to 40× higher in young animals (3–12 weeks old), Collier *et al.* (1996). For this reason all studies used adult animals. Taya *et al.* (1994) showed that over the cumulative dose range 120–990 WLM at a fixed dose rate of 1200–1300 working levels (WL, 0.025–0.027 J m^{-3}), there was a linear dose–response for both alveolar and bronchial epithelial cell proliferation. For micronucleus induction in alveolar macrophages, Johnson and Newton (1994) showed a linear dose–response over the range 70–1020 WLM following exposure over a 3–5 h period (dose rates 4000–35 000 WL). However, at doses rates of 1000 WL, Collier *et al.* (1996) showed that the dose–response was not linear over the range 175–3200 WLM, with the response either peaking and then declining or peaking and then remaining steady, depending on the effect being

Table 1. Exposure conditions, total dose, dose rate and duration in early effects studies

Study number	Cumulative exposure WLM (range of different exposure groups)	Exposure concentration (PAEC), WL	Particle concentration particles/cc	'Unattached' fraction, %	Exposure duration (days)
AEA	0	0	10^4	—	18 d
Study 1	200, 400, 800, 1600,	1000	10^4	2%	2–18d
AEA	0	0	10^4	—	18 d
Study 2	1000	250, 500, 1000,	10^4	2%	3.75–27.5d
CEA	0	0	6×10^3	—	0
Study 3	300	2143	6×10^3	8%	1d
	1000	7143	6×10^3	8%	1d

studied. The dose at which the peak response occurred depended on the effect being studied. Comparison of the early effects following radon progeny exposure with those of low LET radiations *in vivo* showed that at dose rates of 1000 WL, 1 WLM was equivalent to 8–9 mGy ^{60}Co in terms of micronucleus induction in alveolar macrophages (Bisson *et al.* 1994). For high LET radiations *in vitro*, exposure of cultured alveolar macrophages to alpha particles from ^{238}Pu gave an equivalence to radon exposure *in vivo* of 1 WLM:10 mGy for micronucleus induction, Johnson and Newton (1994).

These early effects provide a sensitive and, at least over a range of lower doses, reliable indicator of the dose received within any one study. However, direct comparison between studies is complicated by the use of different dose rates and exposure conditions. This paper reviews the results from early effects studies at AEA Technology over a range of total doses (100–3200 WLM) and reports new data on the effect of dose rate (range 250–2000 WL). New results from the CEA on the effect of cumulative dose on alveolar macrophage numbers, proliferation and nuclear aberration incidences in alveolar macrophages following radon/radon progeny exposure are reported. Results from the two groups are compared and preliminary conclusions drawn on the effect of dose and dose rate on the early response of lung cells to radon/radon progeny exposure.

MATERIALS AND METHODS

All exposures at both facilities were whole body and used male Sprague–Dawley rats (AEA–Charles River, U.K., 3–12 months old; CEA–OFA Iffa Credo, France, 2 months old). At AEA Technology, animals were exposed in a purpose built facility (Strong and Walsh, 1990). The animals were exposed continuously for periods of up to 7 days in two chambers with closed air circuits for recirculation of the radon exposure atmosphere. Carnuba wax aerosol (100–250 nm diameter) was generated as a carrier for attachment of radon progeny. Accurate measurement of the potential alpha energy concentration (PAEC), radon progeny concentration and "unattached" fraction were performed continuously during exposure. At CEA, the exposures were conducted for 8.5 h d^{-1}, in 10 m^3 inhalation chambers at the Razes facility (COGEMA) as described previously (Charneaud *et al.*, 1982). Measurements of equilibrium factor, "unattached" fraction, and particle concen-

C. G. Collier *et al.*

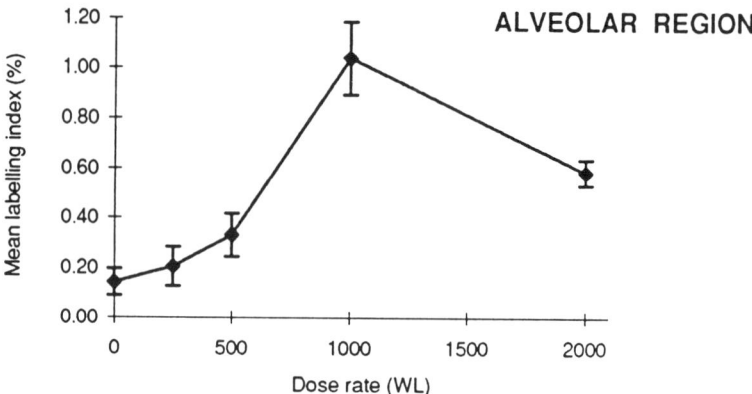

Fig. 1. Proliferation in the alveolar epithelium following radon progeny exposure at constant doses and varying dose rates. Mean ± SE.

tration were made. Exposure conditions and the durations for each study (2 at AEA, 1 at CEA), are given in Table 1.

In all studies, 14 days after exposure, groups of exposed animals (AEA $n = 2-6$ depending on exposure level, CEA $n = 4$) received intraperitoneal injections of 5-bromo-2'-deoxyuridine (4 ml, 5 mg ml^{-1}) (BrdU, Sigma), 4 h before death at AEA and 18 and 12 h before death at CEA. Animals were killed by intraperitoneal injection of sodium pentabarbitone (Sagatal, RMB Animal Health Ltd, Dagenham, 400 mg kg^{-1}). Free cells were recovered from the lungs by lavage with sterile saline (0.15 M). The number of cells present in the recovered lavage fluid was determined using a Coulter counter (Coulter, Luton, U.K.). Cytospin slides (Shandon) were prepared from the lavage fluid for differential cell counts and measurement of the incidence of nuclear aberrations (binucleated cells and cells containing micronuclei or fragmented nuclei). At CEA, the Cytospin preparations were immunocytochemically stained for BrdU and the incidence of stained macrophages was measured to give alveolar macrophage proliferation incidence. At AEA, paraffin wax sections were prepared from the left lobe of each lung and pairs of serial sections were stained alternately with haematoxylin and eosin and with an immunocytochemical stain for BrdU. The incidence of proliferating epithelial cells was determined as the ratio of BrdU stained cells to all cells (from cell counts on the H and E stained sections).

RESULTS

Cell proliferation

Inhalation exposure of rats to radon/radon progeny induced cell proliferation, (as determined by BrdU incorporation) in the epithelial cells and in alveolar macrophages. The magnitude of the response was determined by dose and dose rate.

In study 1 at AEA using a constant dose rate, proliferation in the alveolar region of the lung (Fig. 1), increased with dose up to a cumulative exposure of 1000 WLM. Beyond this dose, proliferation levels decreased markedly to levels below those for control animals. Analysis of variance showed significant differences between the responses for different groups. In the bronchial epithelial cells proliferation was much lower than in the alveolar region (peak levels of 0.13% compared with more than 1% in the alveolar region) and hence the statistical power of the results is much weaker. However, cumulative doses of 200 and 400 WLM did result in significantly higher proliferation in the bronchial region over control levels. Proliferation within alveolar macrophages (study 3 at CEA) was significantly elevated above control levels, at cumulative exposures of 300 WLM. At the higher dose of 1000 WLM, proliferation was similar to that observed in control animals.

At constant doses (study 2), the lowest dose rate (250 WL) showed only slightly elevated alveolar epithelial proliferation levels over controls. At progressively increasing dose rates this effect became more marked up to a peak at a dose rate of 1000 WL. There was a marked decline in response at the highest dose rate. Analysis of variance showed significant differences in response between groups. In the bronchial region, very low proliferation incidences were seen at dose rates below 1000 WL. Above this level a marked increase in proliferation with increasing dose rate was observed. Regression analysis showed a significant dose rate response.

Lavagable cell numbers

At constant dose rates, study 1 at AEA showed no significant differences in the total number of lavagable cells recovered from the lungs. The number of macrophages recovered in both study 1 and study 3 showed a significant decrease in the number of macrophages with dose. Differential cell counts conducted for other cell types in study 1, all showed significant positive correlation with dose indicating an inflammatory response at higher doses (1600 WLM or higher). In study 2, no significant differences were found for total cell number, macrophage number, or differential cell counts either between exposed animals and controls or between exposure groups. Indicating that exposure at this level had not resulted in inflammation and dose rate did not affect this response.

Nuclear aberrations

The incidence of micronucleated cells showed a significant increase with dose in study 3 (Fig. 2) and in study 1 up to a maximum dose of 1600 WLM, beyond which the incidence decreased. Binucleated cells also showed a significant increase with dose in study 1, up to a maximum dose of 1000 WLM, beyond which the response levelled off. Multiple nucleated cells (or cells containing multiple nuclear fragments) showed a significant correlation with dose in both study 1 (Fig. 3) and 3 over the entire range of doses studied (up to 3200 WLM).

In study 2, there was no significant effect of dose rate on the incidence of micronucleated alveolar macrophages. The incidence of both binucleated cells and multinucleated cells showed a significant positive correlation with dose rate.

DISCUSSION

The effect of total cumulative dose at a fixed dose rate (1000 WL giving 200–3200 WLM,) and on dose rate giving a fixed cumulative dose (250–2000 WL giving 1000 WLM) on early effects in the rat lung following inhalation of radon and radon progeny were studied at AEA. In addition the CEA group studied the effect of total cumulative exposure (300 and 1000 WLM) achieved with differing dose rates.

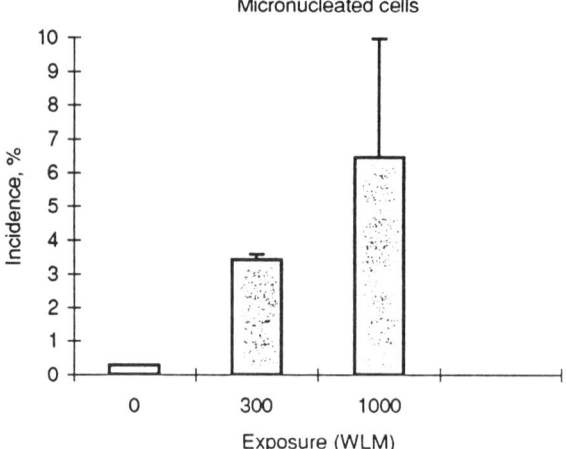

Fig. 2. The incidence of micronucleated cells following radon progeny exposure. Mean ± SE.

The effects observed are summarised in Table 2. Most effects studied showed a positive dose–response relationship, with a peak response beyond which the response either declined or reached a plateau, probably due to cell killing.

Only proliferation in the alveolar and bronchial epithelium and the induction of binucleated and micronucleated cells showed a positive response to dose rate. For alveolar epithelial cell proliferation, the response peaked at 1000 WL. At fixed dose rates, the peak responses for alveolar and bronchial epithelium proliferation occurred before the peaks in the nuclear aberrations responses. Hence proliferation appears to be more sensitive to cell killing and it may be assumed that the decline in alveolar proliferation response at high dose rates is also due to higher cell killing.

Analysis of cohorts of miners who had received doses of up to 2000 WLM concluded that the excess lifetime lung cancer mortality was 350 deaths/10^6/WLM, BEIR IV (1988). The doses used in studies 1–3 were in a comparable range and the increasing effects seen with increasing dose are consistent with the miner data. The

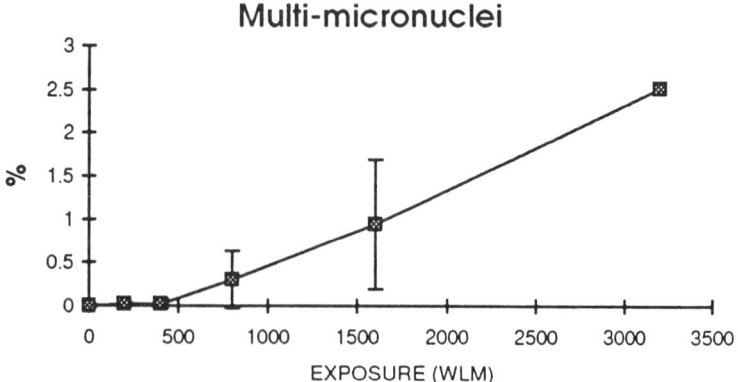

Fig. 3. Incidence of cells containing multiple or fragmented nuclei following exposure to radon progeny at constant dose rate, Mean ± SE.

Table 2. Summary of effects found (+ve indicates significant positive correlation)

Effect studied	Dose (100–3200 WLM)	Dose rate (250–2000 WL)
Cell proliferation		
Alveolar epithelial cell	+ve (< 1000 WLM)	+ve (< 1000 WL)
Bronchial epithelial cell	elevated response at 200 and 400 WLM	+ve
Alveolar macrophages	elevated response at 300 WLM	Not studied
Macrophage cell number	−ve	No effect
Inflammatory cell number	+ve (> 1600 WLM)	No effect
Nuclear aberrations		
Micronuclei	+ve (< 1600 WLM)	No effect
Binucleated cells	+ve (< 1000 WLM)	+ve
Multiple/fragmented	+ve	+ve

miner data showed an inverse response with dose rate, BEIR IV (1988). For the estimation of risk from domestic exposure, when data from miners exposed at the lower dose rates (< 20 WLM y^{-1}) were used, the estimated risk from radon exposure was found to be 50–100% higher, Darby and Doll (1990). The positive dose rate response observed for some early effects in this study is not inconsistent with the inverse dose rate response observed in the miners, as the dose rates in this study (250–2000 WL) were very high compared to those in the mines (< 10 WL).

The usefulness of the early effects to predict long-term tumour incidences will not be known until completion of the life span studies at AEA and CEA. Should one or a combination of the effects prove reliable indicators of long-term effects, future studies may be conducted using early effects only as end points, reducing the duration and costs of studies to determine the effect of exposure conditions on lung tumour induction by radon progeny. Further refinement of the techniques may also allow doses to particular regions of the lung to be determined. When comparing responses (both early effects and tumour induction) between studies, it is important to consider the dose rate as well as the total cumulative exposures.

REFERENCES

BEIR IV (1988) Health risks of radon and other internally deposited alpha emitters. National Research Council. National Academy Press, Washington D.C.

Bisson, M., Collier, C. G., Poncy, J-L., Taya, A., Morlier, J. P., Strong, J. C., Baker, S. T., Monchaux, G. and Fritsch, P. (1994) Biological dosimetry in different compartments of the respiratory tract after inhalation of radon and its daughters. *Rad. Prot. Dos.* **56** (1–4) 89–92.

Chameaud, J. R., Perraud, J. R., Chrétien, J., Masse, R. and Lafuma, J. (1982) Lung carcinogenesis during *in vivo* cigarette smoking and radon daughters exposure in rats. *Recent Results in Cancer Research* **82**, 11–20.

Collier, C. G., Baker, S. T., Eldred, T. M. and Strong, J. C. (1996) Early effects of radon progeny exposure in animals. Presented at *9th International Radiation Protection Association Congress*, Vienna, April 1996.

Darby, S. C. and Doll, R. (1990) Radiation and exposure rate. *Nature* **344**, 824.

Doll, R. (1992) Risks from radon. *Rad. Prot. Dos.* **42** (3) 149–153.

Johnson and Newton (1994) Estimation of radon progeny dose to the peripheral lung and the effect of radon progeny exposure on alveolar macrophages. *Rad. Res.* **139**, 163–169.

Strong, J. C. and Walsh, M. (1990) A facility for studying the carcinogenic and synergistic effects of radon daughters and other agents in rodents. In *Indoor Radon and Lung Cancer, Reality or Myth Proceedings of 24th Hanford Symposium on Health and the Natural Environment.* Battelle Press, Richland.

Taya, A., Morgan, A., Baker, S. T., Humphreys, J. A. H., Bisson, M. and Collier, C. G. (1994) Changes in the rat lung after exposure to radon and its progeny. *Rad. Res.* **139** 170–177.

Pergamon

Ann. occup. Hyg., Vol. 41, Supplement 1, pp. 92–98, 1997
© 1997 British Occupational Hygiene Society
Published by Elsevier Science Ltd. All rights reserved
Printed in Great Britain
0003–4878/97 $17.00 + 0.00
Inhaled Particles VIII

PII: S0003–4878(96)00134–2

INHALATION AND RETENTION OF THORIUM DUSTS BY MINERAL SANDS WORKERS

G. S. Hewson

Department of Minerals and Energy, 100 Plain Street, East Perth, WA 6004, Australia

INTRODUCTION

Heavy mineral sands consisting of ilmenite, rutile, zircon and monazite, have been mined and processed in Western Australia since 1956. The industry is a significant one; in 1995 it mined 37.4 million tonnes of sand and processed 2.6 million tonnes of heavy mineral concentrate to produce 2.1 million tonnes of the individual mineral sands, at a value of $550 million. A specific occupational health concern associated with this industry is radiation exposure arising out of the presence of thorium and, to a lesser extent, uranium with all the heavy minerals. Monazite, a rare earth phosphate, is radiologically the most significant mineral, containing typically between 5 and 7% thorium and 0.1 and 0.3% uranium. Although monazite is a low volume product, comprising only about 0.5% of total mineral sand production, it tends to preferentially concentrate in airborne dust because it is softer than the titanium and zirconium bearing minerals. This is of particular concern during the processing of mineral sands because the minerals are subjected to a variety of vigorous physical treatment processes, such as screening and magnetic, electrostatic and gravity separation. Without the application of appropriate dust control technology, considerable airborne dust (and consequently radioactivity) concentrations may be experienced by workers who operate and maintain the separation plant.

While the presence of radioactivity in this industry has been long appreciated, early protective and regulatory measures were focussed on control of external radiation exposure in circuits where monazite was being concentrated and bagged. Intake of radioactive dust was only recognised as a potentially significant source of exposure in the early 1980s, following national review and acceptance of ICRP Publication 30 (ICRP, 1979). This resulted in derived air concentration (DAC) values for thorium an order of magnitude lower than those previously applied to the industry (Hewson and Terry, 1995).

The issue of intake of radioactive dust prior to the mid 1980s is important because there are many long-term workers in the industry and it is well recognised that each of the five dry separation plants were very dusty. In addition, early work practices, such as use of compressed air to blow down equipment, and sweeping, banging and brushing of floors and plant equipment, coupled with limited use of respiratory protective equipment, exacerbated the potential for dust inhalation by workers operating and maintaining the separation plants. Industry management and employees, through a concern about the likely long-term health implications of exposure to radioactive dust, have both been anxious to ascertain the extent of past exposure and have supported a range of bioassay research endeavours.

This paper examines the inhalation exposure estimates obtained from bioassay studies and compares them with estimates from personal air sampling and a retrospective assessment of intake of radioactivity. One objective of such a comparison is to assess the veracity of recommended biokinetic models. Another is to investigate the feasibility of using an alternative, cost-effective exposure assessment technique, such as exhaled thoron, as a routine monitoring tool.

METHODS

As thorium oxides have a very long half-time of clearance from the lung, bioassay measurements, apart from analyses of faeces, will reflect long-term chronic inhalation. To relate the results from bioassay to air sampling requires knowledge of annual average daily intake over the exposure period.

For each year since 1977 an estimate of the average airborne alpha activity concentration across the industry was made as follows:

$$X = (6).(20).(0.06).D_g.SA_{Th}.M_{HMC}$$

where X is the average alpha activity, Bq m^{-3}

     6   is the number of alpha particles emitted per decay of ^{232}Th in ore dust

     20   is the concentration factor for monazite in airborne dust (Hewson and Terry, 1995)

     0.06   is the average thorium content of monazite

     D_g is the arithmetic mean gravimetric dust concentration, mg m^{-3}

     SA_{Th} is the specific activity of ^{232}Th, 4.1 Bq mg^{-1}

     M_{HMC} is the average monazite content of heavy mineral concentrate.

The mean gravimetric dust concentration for separation plant workers was obtained from the Department's atmospheric contaminant exposure database, known as CONTAM. The number of personal air samples has varied from 165 in 1977 to in excess of 2581 in 1989 and the number of samples per "designated" radiation worker is now typically in the range of 6–10 per year. Details of the dust sampling measurements are summarised in Hewson and Terry (1995).

The monazite content of heavy mineral concentrate feedstock (M_{HMC}) to the dry separation plant was determined by reviewing historical mineral sands production data reported to the Department of Minerals and Energy by the mineral sands companies.

For the years 1975, 1976 and the period prior to 1975, it was assumed that the mean personal dust concentration was 15 mg m^{-3}, based on measurements in the late 1970s. This assumption maybe overly conservative as the two new plants which commenced operations in the mid to late 1970s were larger than the three existing plants and were considered very dusty by the inspectorate. An average M_{HMC} content prior to 1975 of 0.40% was calculated by examining total production figures to that time.

The estimates derived from this analysis were used to assign pre-1986 values for the annual intake of radioactive dust (and hence internal radiation dose) for those workers who had participated in bioassay studies. This required knowledge of the employment history of the individual and average annual working hours; information which was readily available from company records. The intakes were subsequently converted to committed (over 50 years) effective doses using conversion factors derived from the new ICRP 66 lung model (ICRP, 1994). Procedural details are described by Hewson and Terry (1995) and Terry and Hewson (1994).

Results of recent bioassay studies were reviewed, including: (1) the concentration of thorium in blood serum and urine (Hewson and Fardy, 1993); (2) the concentration of thorium in faeces (Terry *et al.*, 1995); (3) the concentration of thoron exhaled in breath; and (4) the amount of thorium decay chain radionuclides deposited in lungs and other organs (Terry and Hewson, 1994, 1995).

The predicted concentration of thorium in urine, blood serum and lungs was calculated using the ICRP 66 lung model (ICRP, 1994) and ICRP 30 biokinetic model for thorium (ICRP, 1979) and assuming workers were chronically exposed to relatively insoluble thorium ore dust with a median particle size (or activity median aerodynamic diameter—AMAD) of 10 μm.

RESULTS

At the five individual separation plant sites the average monazite content was found to vary between less than 0.1% and about 2.0%. The average monazite content across the industry has varied from 0.11% in 1994 to 1.13% in 1983, with the latter figure corresponding to annual monazite production of 15 606 tonnes. The value for 1994 is biased low as the industry substantially curtailed monazite production in 1994 as a result of a drastic fall off in demand. Thus, the above retrospective analysis cannot be applied to the 1994 and 1995 dust data.

CONTAM records show that mean dust concentrations in the dry separation plants have progressively declined from 16.8 mg m^{-3} in 1977 to 1.4 mg m^{-3} in 1991, although a substantial level of non-compliance with the general dust standard of 10 mg m^{-3} was evident up to the mid 1980s. The average alpha activity derived from the retrospective assessment procedure for the period < 1975–1993 is shown in Fig. 1, together with the average monazite content and the actual average alpha activity, routinely determined by the mineral sands industry since 1986. It can be seen that there is reasonably good correlation between the actual and estimated alpha activities since 1986, which provides confidence in the reasonableness of the estimates in the period before 1986. The increased measured alpha activity in 1989 corresponds to a period when large scale engineering control work was being undertaken in some of the plants and it is assumed that the additional maintenance activity was resulting in increased resuspension of settled dust.

Depending on the site (or monazite content of feed material), pre-1986 doses to separation plant workers were estimated to vary between 9 and 90 mSv per annum (cf. current ICRP recommended dose limit of average 20 mSv per annum). For other workers, annual doses of one-tenth or one-third these values were chosen, depending on nature of task and extent of dust exposure (Hewson and Terry, 1995). It is emphasised that the estimates relate to a "typical" plant producing about 2000–3000 tonnes monazite per annum or with about 0.5–0.7% monazite in

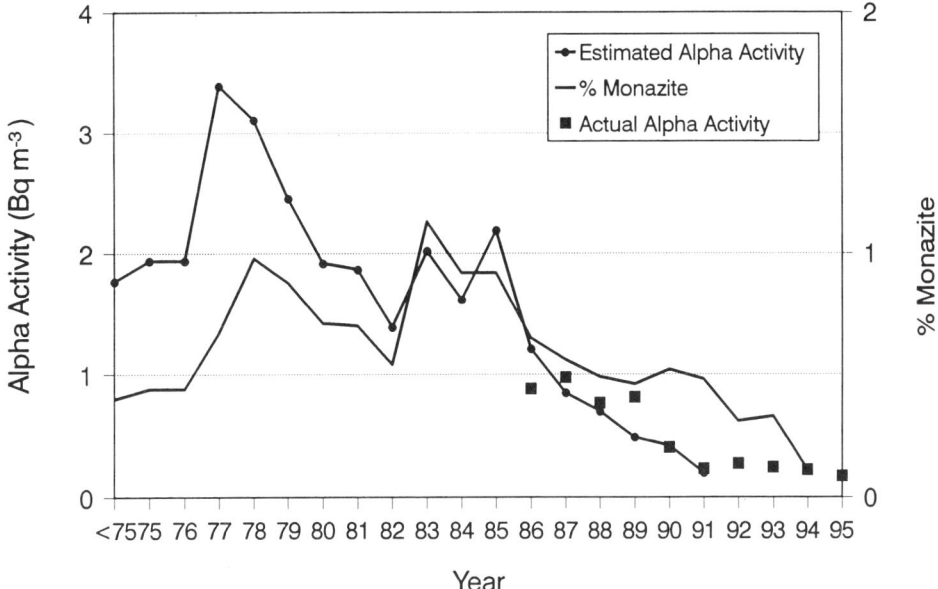

Fig. 1. Mean airborne radioactivity concentrations, expressed as total alpha activity, in Western Australian mineral sands separation plants (1975–1995).

heavy mineral concentrate. Adjustments for plants producing more or less monazite, or with more or less monazite in feed, are described elsewhere (Hewson and Terry, 1995).

A summary of the bioassay results is presented in Fig. 2. In this analysis the results from bioassay have been normalised against the expected average daily intake of ^{232}Th as estimated from company air sampling programs and the retrospective assessment described earlier.

DISCUSSION

The urine and serum results indicate that absorption of thorium is considerably less than predicted by the application of air sampling measurements to the recommended biokinetic and dosimetric models. This implies that the intake is much lower than expected and/or the biokinetic model is inappropriate. As the concentration of thorium in urine and serum reflects the long term accumulation of thorium in the body and it has been established that long term workers were exposed to high dust concentrations, it is suspected that the major discrepancy is due to the model assumption in relation to dissolution of deposited dust particles to blood.

The faeces results were based on measurements on only two workers and, while inter-worker variations were evident, the average result was broadly consistent with that expected on the basis of personal air sampling measurements. Note that in the case of faeces, the thorium concentration reflects recent thorium intake. The daily excretion rate of thorium in faeces reaches a constant value in about 7 days

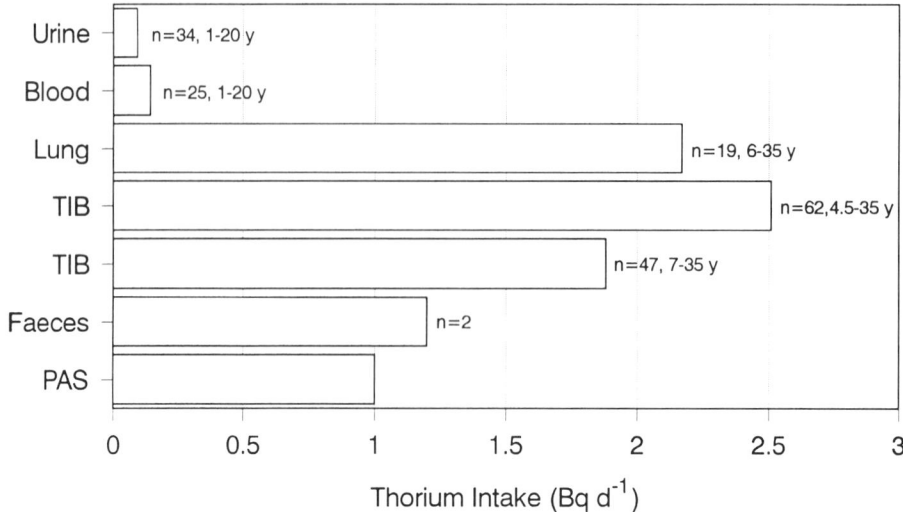

Fig. 2. Estimated intake of ^{232}Th by mineral sands workers using various bioassay techniques, normalised against personal air sampling measurements. The longer thoron in breath (TIB) bar refers to 62 workers with employment between 4.5 and 35 years, and the shorter to a subset of 47 workers with employment longer than 7 years.

from the start of intake (and is in equilibrium with daily inhalation rate) and reaches negligible amounts in about 7 days from the cessation of intake.

The thoron in breath results show that, for those employees above the minimum detection limit (MDL), the thorium lung burdens are significantly higher than predicted on the basis of applying intake data (from personal air sampling data) to the new ICRP 66 lung model (ICRP, 1994). However, the results are biased by the large number of workers (81 out of 145 workers tested) below the MDL of the technique. Of interest is that 75% of short-term (employment duration < 7 years) workers are below the MDL and that the average intake of short-term and long-term workers is 0.15 and 0.54 Bq d^{-1} ^{232}Th, respectively. Thus, it is evident that there has been improvement in the exposure status of mineral sands workers over the last 7–10 years.

The breath measurements revealed that 25 workers (17%), including three short term workers were estimated as receiving an average annual dose of 20 mSv or more over their total employment period. One of the highest lung burdens was measured on a retiree, who had been out of the industry for 10 years and other substantial burdens were measured on long-term workers who had not been working in dusty jobs for about 10 years. These results underline: (1) that considerable intakes of radioactive dust were incurred by some workers in the 1960s and 1970s; and (2) that once thorium ore dust lodges in the lungs it is avidly retained.

The elevated lung burdens recorded against the three short-term workers are of concern, as their predicted intake is significantly above their work category colleagues. Differences in an individual's results arise through personal factors such as work habits, use or non-use of a respirator, mode (i.e. mouth or nose) and rate of breathing and smoking habit. Such factors may result in the actual exposure being either higher or lower than estimated from personal air sampling. The discrepancy may also be due to a combination of incorrect intake data, inappropriate metabolic data or incorrect assumptions about dust characteristics, such as particle size. As the metabolic data reflect the latest scientific knowledge it is more likely that the air sampling data is not providing representative coverage of high exposure tasks. However, the discrepancy could be explained if the AMAD was 5 μm (not 10 μm) or if the worker was a mouth breather.

The range of ratios of predicted thorium lung burden (from air sampling) to estimated burden (from breath measurements) varied from 0.43 to 13.5, confirming that there is substantial inter-worker variability in intake, even for workers with similar employment periods in the same work category. It is interesting to note, however, that the average ratio of measured to predicted lung burdens for long term workers is within a factor of 2, which provides some confidence in the reasonableness of the retrospective assessment. Many health physicists might consider this as good agreement.

The results indicate that thoron in breath measurements provide data that are complementary to air sampling data and are able to identify workers whose work habits require closer investigation. One disadvantage of the breath measurements is that the MDL is relatively high and therefore the technique is primarily restricted to long-term workers or short-term workers with significant intake. The Department of Minerals and Energy has recently developed and commissioned an alternative breath measurement device, utilising an electrostatic collection chamber and this has resulted in a 50% improvement in the MDL. The Department has provided this device to industry and it is currently being trialled on selected workers using a measurement strategy based on tests at the start, middle and end of the year. Preliminary results indicate that the technique is sufficiently sensitive and cost effective and could replace the comprehensive and expensive regime of personal air sampling that is currently used by the industry. The major advantage of the technique over air sampling is that it identifies individual workers at elevated risk and enables suitable investigation and intervention strategies to be implemented. While the air sampling regime is considered comprehensive, in comparison to sampling conducted in other industry sectors, it should be recognised that the number of air samples on an individual worker represents only about 5% of the worker's annual working shifts.

CONCLUSIONS

Reasonably good correlation between estimates of thorium intake derived from air sampling and retrospective assessment and estimates from bioassay measurements was found for long-term workers. Bioassay measurements on recent workers, although higher than expected, confirm the increasing improvement in industrial hygiene conditions and work practices since the early 1980s. Periodic thoron-in-breath measurements could replace the comprehensive regime of personal air sampling which currently exists in the industry, thereby providing a relatively sensitive and cost effective diagnostic monitoring tool.

Continued study of the mineral sands workforce could yield important information about the inhalation and retention of insoluble dust particles. Mineral sands dust has the advantage of a "radioactive signature", which allows direct *in vivo* measurements of the dust in the lungs and other organs and the measurement of thoron exhaled in breath. Periodic measurements of workers who are no longer employed in dusty tasks will enable an assessment of the clearance of dust from the lungs.

REFERENCES

Hewson, G. S. and Fardy, J. J. (1993) Thorium metabolism and bioassay of mineral sands workers. *Health Phys.* **64**, 147–156.

Hewson, G. S. and Terry, K. W. (1995) Retrospective assessment of radioactivity inhaled by mineral sands workers. *Radiat. Prot. Dosim.* **59**, 291–298.

International Commission on Radiological Protection (1979) Annual limits on intake for workers. ICRP Publication 30, Pergamon Press, Oxford.

International Commission on Radiological Protection (1994) Human respiratory tract model for radiological protection. ICRP Publication 66, Pergamon Press, Oxford.

Terry, K. W., Hewson, G. S. and Meunier, G. M. (1995) Thorium excretion in faeces by mineral sands workers. *Health Phys.* **68**, 105–109.

Terry, K. W. and Hewson, G. S. (1995) Thorium lung burdens of mineral sands workers. *Health Phys.* **69**, 233–242.

Terry, K. W. and Hewson, G. S. (1994) Thorium lung burdens of workers in the mineral sands industry. Report of research commissioned by the Titanium Minerals Committee of the Chamber of Mines and Energy. Department of Minerals and Energy, Perth Western Australia.

Pergamon

Ann. occup. Hyg., Vol. 41, Supplement 1, pp. 99–103, 1997
© 1997 British Occupational Hygiene Society
Published by Elsevier Science Ltd. All rights reserved
Printed in Great Britain
0003–4878/97 $17.00 + 0.00
Inhaled Particles VIII

PII: S0003–4878(96)00078–6

EDS ANALYSIS OF INHALABLE CERIUM OXIDE PARTICLES AFTER *IN VITRO* DISSOLUTION ASSAY AND AFTER INTRATRACHEAL INJECTION

S. Matton, P. Massiot, H. Le Naour, C. Lizon, G. Rateau and P. Fritsch

Laboratoire de Radiotoxicologie, CEA/DSV/DRR/SRCA, BP 12, 91680 Bruyères le Châtel, France

INTRODUCTION

Several *in vitro* assays have been developed to measure the solubilisation rates of mineral particles for blood transfer estimate (Kanapilly *et al.*, 1973). Different solutions could be used which simulate different biological compartments such as interstitial fluid, gastric content, and also phagolysosome which corresponds to the cellular compartment involved after phagocytosis of inhaled particles by alveolar macrophages (Helfinstine *et al.*, 1992). The solubilized fraction from the powder has only been considered, but to our knowledge, no data are available on the physico-chemical modifications of the particles related to the composition of the stimulating solution which might control the particle solubilisation kinetics.

After phagocytosis by alveolar macrophages, EDS analysis performed on thin sections has demonstrated the presence of uranyl phosphate needles in phagolysosomes containing uranium oxide particles (Henge-Napoli *et al.*, 1994). This could show interactions between uranium from the oxide and the phosphate ions of this cell compartment.

The aim of this work was to develop methods, based on energy dispersive X-ray spectrometry (EDS), to compare the physico-chemical evolution of cerium oxide inhalable particles after *in vitro* dissolution processes for 24 h in different phagolysosome-simulating media and after intratracheal instillation in the rat. Cerium oxide has been studied because its chemical properties might be similar to those of some actinide oxides for which many results on dissolution and physico-chemical properties are available *in vitro* and *in vivo* after lung deposition (Kanapilly and Goh, 1973; André *et al.*, 1989; Ansoborlo *et al.*, 1990; Henge-Napoli *et al.*, 1994).

MATERIALS AND METHODS

Cerium hydroxide was obtained by adding 5 M NaOH to a 2.5 g l.$^{-1}$ CeCl$_3$ solution, under constant stirring, until no more precipitation occurred. Cerium phosphate was prepared similarly, by adding 0.1 M NaH$_2$PO$_4$ to a 2.5 g l.$^{-1}$ CeCl$_3$ solution prepared in H$_3$PO$_4$ 3 M. The hydroxide was then rinsed with distilled water and heated at 600°C for 48 h, the phosphate was air dried at 60°C. The inhalable fraction from the cerium oxide powder was obtained by sedimentation in a 2% collodium solution in isoamyl acetate for 5 min. Several rinses with acetone were carried out and the powder was air dried before preparing each aqueous suspension.

The solutions used for the *in vitro* experiments corresponded to a Tris–HCl buffer, 0.02 M at pH 5.5, containing 0, 3, 60, 300 and 1000 mM of phosphate ions. The oxide powders were stirred in these solutions for 24 h and were then collected on carbon-coated filters (Nucleopore, HTTP, 0.4 μm), rinsed with distilled water, coated again with carbon and put on 300 mesh copper grids.

Intratracheal injections of a suspension of cerium oxide in the buffer without phosphate were carried out on 3 month old Sprague–Dawley male rats. Animals were killed under pentobarbital anaesthesia (40 mg kg^{-1}), 7 days after the administration. Alveolar macrophages were extracted by pulmonary lavage after repeated instillations of 5 ml NaCl 0.9% until a total volume of 20 ml had been obtained. The cell suspension was centrifuged (1000 **g**) for 10 min and the cell pellet was treated by sodium hypochlorite (48°) for 2 h to digest the biological matrix (Pairon *et al.*, 1994). The particles were then put on grids as previously described.

EDS analysis was performed with an EDAX silicon detector (super ultra thin window) associated with an electron microscope (CM 200 Philips). The electron beam was obtained at 200 kV. Spectra were recorded at the center of the particles using a spot size of 14 nm. Chemical compositions were estimated with an EDAX software based on Cliff and Lorimer equations (Cliff and Lorimer, 1972).

RESULTS

Electron microscopy has shown that most of the oxide powders corresponded to aggregates. Such aggregates of 50–100 nm diameter particles were also observed after incubation *in vitro*, after intratracheal injection and for the cerium phosphate powder.

Figure 1 (A) shows one X-ray spectrum obtained on the initial cerium oxide particles in the range of 0–7 keV. The specific Ce Lα, Lβ, Lγ, Lℓ, Mα, Mγ and Lλ X-rays were emitted from the particle. The spectrum obtained out of the particle [Fig. 1 (B)] shows that, except for the O Kα X-ray due to the oxide, in this range of energies, the other significant X-rays, Cu Lα, C Kα and Cl Kα, are due to the grid, the carbon layer and the chloroform treatment used to disolve the filter, respectively.

Figure 2 shows a X-ray spectrum from a cerium phosphate particle with the specific P Kα X-ray. Preliminary results clearly demonstrated a decrease of the P Kα/Ce Lα net X-ray intensity ratio as the particle diameter increased. The calculated mass concentration P/Ce was 1.32 ($SD = 0.27$, $n = 11$) for 250–500 nm diameter particles and 0.63 ($SD = 0.32$, $n = 10$) for 750–1500 nm diameter particles.

Figure 3 shows the X-ray spectrum of a particle, after a 24 h incubation in TRIS-HCl containing 3 mM of phosphate ion. A clear P Kα X-ray is also observed and in this case, the P/Ce ratio was about 0.61. Such specific P Kα X-rays were also encountered on most of the particles for each phosphate ion concentration studied. For particles of about the same diameter, the P/Ce ratio seemed to increase as the phosphate ion concentration increased.

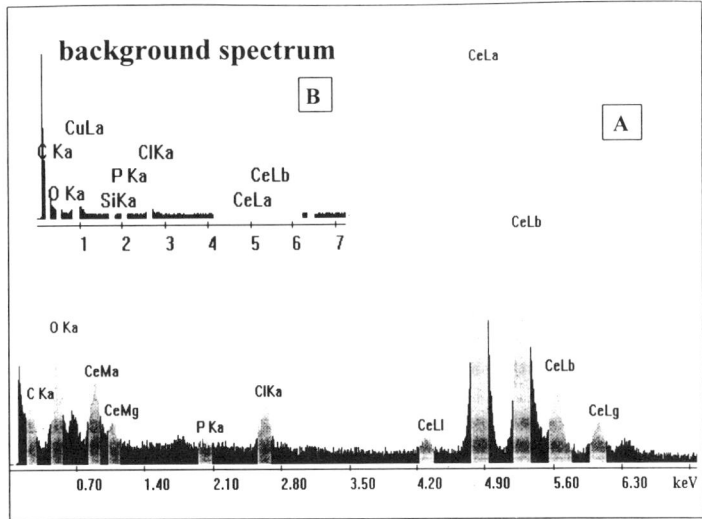

Fig. 1a. X-ray spectrum obtained on the initial cerium oxide particles.
Fig. 1b. Corresponding background.

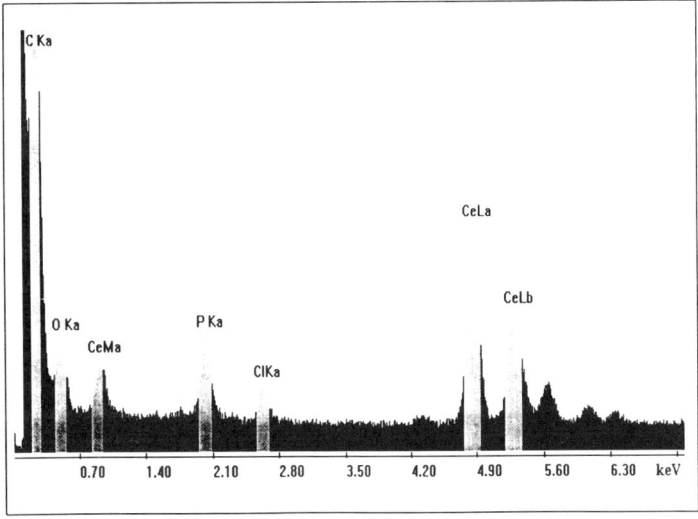

Fig. 2. X-ray spectrum from a cerium phosphate particle.

Figure 4 shows the X-ray spectrum of a particle obtained from lavaged alveolar macrophages after the intratracheal instillation. As after *in vitro* incubation, P Kα X-ray is observed and a P/Ce ratio of this particle is 0.20. However, another X-ray is encountered, the Si Kα ray. Such a X-ray was present in most of the particles analysed within a wide range of intensities.

DISCUSSION

This study shows that EDS could be a powerful tool to characterize the evolution

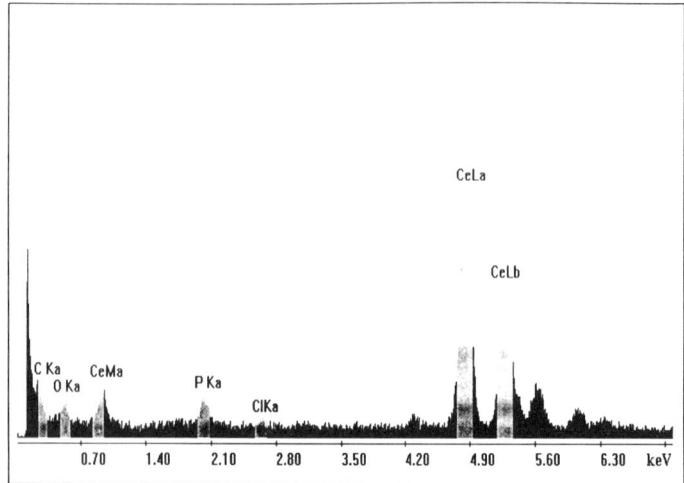

Fig. 3. X-ray spectrum of a particle, after a 24 hour incubation in TRIS-HC1 containing 3 mM of phosphate ion.

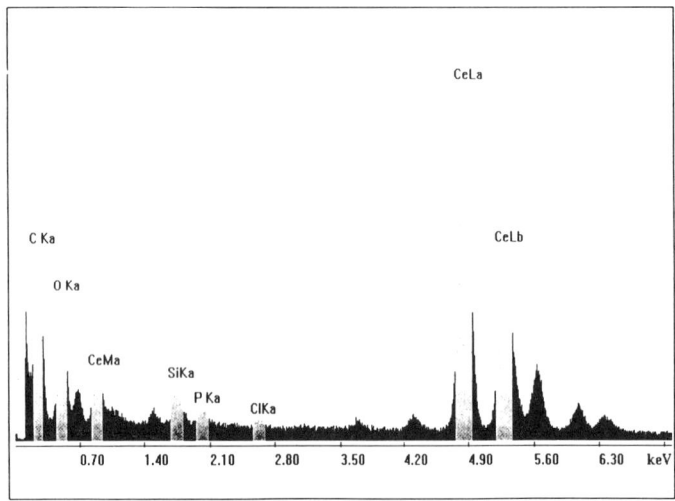

Fig. 4. X-ray spectrum of a particle obtained from lavaged alveolar macrophages after intratracheal instillation.

of the physico-chemical form of poorly soluble particles during their dissolution process occuring either *in vitro*, in media simulating different biological compartments or *in vivo*, after phagocytosis by alveolar macrophages.

Qualitatively, similar results were obtained for the occurrence of P Kα X-ray *in vivo* and *in vitro*. This seems to demonstrate an interaction of cerium from the oxide particles with phosphates from the surrounding compartment. The observed phenomenon could visualize the transformed state of particle that is taken into account in the model proposed by the ICRP (Moss and Kanapilly, 1980; ICRP 66, 1994) for dissolution kinetics of inhaled particles. In our study, the cerium oxide seemed to be trapped within a cerium phosphate matrix. However, the phenomenon observed *in vitro* might appear different from that actually occuring *in vivo*. The presence of other elements *in vivo*, such as silicon, could involve specifically the phagolysosome compartment as a function of the nature of the previously phagocytosed particles. Thus, a potential interaction between phagocytosed particles during, or even after, their dissolution proccess has to be taken into account. Studies are in progress to characterize, as a function of time, the chemical composition of particles and/or phagolysosomes *in vivo* and *in vitro*. Moreover, these studies will include element mapping to determine the homogeneity of the distribution of these elements.

Because the low energy of P Kα X-ray (2.013 keV), it can be significantly absorbed within the sample. Thus, accurate quantitative analysis needs absorption corrections. This was clearly visualized by the decrease of P Kα/Ce Lα X-ray ratio as the particle diameter increased. To provide actual measurement of the P/Ce ratio during the dissolution process of cerium oxide, studies are still in progress with pure cerium phosphate powders to establish an absorption calibration curve as a function of particle diameter. Potentially, our experimental approach on entire particles would provide much more quantitative informations than that performed on thin sections (Henge-Napoli *et al.*, 1994).

In conclusion, this study suggests that the transformed state of particle during their dissolution process occuring *in vivo* or *in vitro* could be characterized by EDS analysis on entire particles. The *in vivo* dissolution process appears as a complex phenomenon that involves the contents of the phagolysosomal compartment. Thus, the previous phagocytosed particles have to be taken into account.

REFERENCES

André, S., Metivier, H., Auget, D., Lantenois, G., Boyer, M. and Masse, R. (1989) Assessment of uranium tetrafluoride dissolution in the lung by *in vivo* and *in vitro* methods. *Radiat. Prot. Dosim.* **26**, 75–81.

Ansoborlo, E., Chalabresse, J., Escallon, S. and Hengé-Napoli, M. H. (1990) *In vitro* solubility of uranium tetrafluoride with oxidizing medium compared with *in vivo* solubility in rats. *Int. J. Radiat. Biol.* **58**, 681–689.

Cliff, G. and Lorimer, G. W. (1972) Quantitative analysis of thin metal foils wing EMMA-4, the ratio technique. In *Proc. 5th Europ. Congr. on E. M.*, Manchester, pp. 140–141. Pub. The Institute of Physics, Bristol and London.

Helfinstine, S. Y., Guilmette, R. A. and Schlapper, G. A. (1992) *In vitro* dissolution of curium oxide using a phagolysosomal simulant solvent system. *Environ. Health. Persp.* **97**, 131–137.

Henge-Napoli, M. H., Ansoborlo, E., Donnadieu-Claraz, M., Berry, J. P., Gilbert, R. and Pradal, B. (1994) Solubility and transferability of several industrial forms of uranium oxides. *Radiat. Prot. Dosim.* **53**, 157–161.

International Commission on Radiological Protection. (1994) *ICRP Publication 66*. Pergamon Press, Oxford.

Kanapilly, G. M. and Goh, C. H. T. (1973) Some factors affecting the *in vitro* rates of dissolution of respirable particles of relatively low solubility. *Health Phys.* **25**, 225–237.

Moss, O. R. and Kanapilly, G. M. (1980). Dissolution of inhaled aerosols. *Generation of Aerosols and Facilities for Exposure Experiments* (Edited by K. Willeke), pp. 105–124. Ann Arbor, Michigan

Pairon, J. C., Billon-Galland, M. A., Iwatsubo, Y., Bernstein, M., Gaudichet, A., Bignon, J. and Brochard, P. (1994) Biopersistence of non fibruous mineral particles in the respiratory tracts of subjects following occupational exposure. *Environ. Health Persp.* **102** (suppl 5), 269–275.

 Pergamon

Ann. occup. Hyg., Vol. 41, Supplement 1, pp. 104–110, 1997
British Occupational Hygiene Society
© 1997 NRPB. Published by Elsevier Science Ltd
Printed in Great Britain
0003–4878/97 $17.00 + 0.00
Inhaled Particles VIII

PII: S0003–4878(96)00124–X

BIOKINETICS OF RECYCLED URANIUM TRIOXIDE AND IMPLICATIONS FOR HUMAN EXPOSURE

J. C. Moody,* A. Birchall,* G. N. Stradling,* A. R. Britcher†
and W. P. Battersby

*National Radiological Protection Board, Chilton, Didcot, Oxon OX11 0RQ; and †British Nuclear
Fuels Ltd, Sellafield, Cumbria CA20 1PG, U.K.

INTRODUCTION

One aim of reprocessing nuclear fuel is to recover uranium for future fabrication. This involves the formation of uranium trioxide (UO_3), to which personnel can be exposed by inhalation during handling procedures. The study described here investigates the biokinetic behaviour of uranium after deposition of UO_3 in rat lungs. This provides an experimental basis for assessing appropriate limits on intake for workers and guidance on bioassay interpretation. These judgements consider the implications of recent recommendations by the International Commission on Radiological Protection (ICRP) on dose limits, biokinetic models and methodologies as well as current legislation and limits based on chemical toxicity.

MATERIALS AND METHODS

A sample of UO_3 dust was collected in the work area, the isotopic composition by activity was determined as 50.6% ^{234}U, 1.2% ^{235}U and 48.2% ^{238}U. For use in animal experiments a particle fraction of small size was obtained using a micronising mill. After sedimentation in ethanol, a particle fraction 1.57 μm AMAD σ_g 1.62 was retrieved, as determined by an Amherst Process Instruments Aerosizer. The animals were 20-week-old female HMT rats of about 220 g to which food and water were made freely available. All procedures were in accord with U.K. legislation— the *Animal (Scientific Procedures) Act*, 1986.

The principal aim was to assess absorption of uranium from lungs to blood. In order to quantify lung clearance at early times ($t < 7$ d), UO_3 was administered by intratracheal instillation as a suspension in saline. The initial lung deposit (91.9 ± 2.3 ng ^{235}U, ~ 24 μg total U) was assessed from five aliquots of the same volume of the suspension instilled (100 μl). Groups of five rats were subsequently killed at intervals from 30 min to 42 days after instillation. The uranium content of selected ashed tissues was determined by delayed neutron counting of ^{235}U and corrected for endogenous uranium by analysing tissues from unexposed rats. Supplementary studies involving intragastric gavage of UO_3 and intravenously injected ^{233}U nitrate were performed to assess absorption of uranium from the gut. These studies lasted 7 days.

Table 1. Tissue distribution of uranium in rats after instillation of UO_3

Time	Lungs	Kidneys	Carcass	Urine	Faeces[d]	To blood
			% of initial lung deposit,[a] mean $\pm$ SE, $n = 5$			
30 min	46.7 $\pm$ 3.1	3.73 $\pm$ 0.20	20.45 $\pm$ 1.04	nd[c]	29.1 $\pm$ 3.3	24.18 $\pm$ 1.06
1 hour	34.7 $\pm$ 6.5	5.01 $\pm$ 0.78	20.00 $\pm$ 1.55	6.63 $\pm$ 1.95	33.7 $\pm$ 7.0	31.64 $\pm$ 2.61
3 hours	24.0 $\pm$ 4.4	3.70 $\pm$ 0.64	13.01 $\pm$ 1.22	19.0 $\pm$ 3.5	40.3 $\pm$ 5.8	35.71 $\pm$ 3.76
6 hours	23.4 $\pm$ 0.56	4.94 $\pm$ 0.92	13.17 $\pm$ 1.08	22.3 $\pm$ 3.8	36.2 $\pm$ 4.1	40.41 $\pm$ 4.06
1 day	15.1 $\pm$ 1.1	5.53 $\pm$ 0.33	9.73 $\pm$ 0.41	30.7 $\pm$ 1.9	38.9 $\pm$ 2.3	45.96 $\pm$ 1.97
3 days	7.25 $\pm$ 0.65	3.45 $\pm$ 0.84	9.09 $\pm$ 0.69	38.0 $\pm$ 2.8	42.2 $\pm$ 3.1	50.54 $\pm$ 3.00
7 days	5.85 $\pm$ 0.66	< 2[b]	8.35 $\pm$ 0.54	44.0 $\pm$ 1.1	41.8 $\pm$ 1.4	52.35 $\pm$ 1.23
14 days	4.86 $\pm$ 0.34	< 2	8.18 $\pm$ 0.76	39.8 $\pm$ 0.8	47.2 $\pm$ 1.2	47.98 $\pm$ 1.10
42 days	2.54 $\pm$ 0.16	< 2	8.48 $\pm$ 0.41	nd	—	—

(a) Initial lung deposit: 0.092 $\pm$ 2.2 μg ^{235}U, 23.9 $\pm$ 0.6 μg U.
(b) Activities at or below the MDA, 1.9 ng ^{235}U.
(c) Not determined.
(d) Determined by difference.
Activities in the liver were at or below the MDA so not reported separately.
Activity in the faeces obtained by difference.

RESULTS

Lung retention and tissue distribution data after instillation of UO_3 are given in Table 1. The amounts of uranium absorbed from the lungs were obtained by summing systemic content and urine. The lung content fell rapidly with 50% being cleared within 30 min and 75% by 3 h after exposure. Only 2.5% was retained at 42 d. Dissolution and absorption into blood followed by excretion in urine was an important route of clearance accounting for 36 and 46% cleared by 3 h and 1 d, respectively. The calculated f_1 for UO_3 was in the range 0.02–0.05 compared with ICRP recommendations, of 0.02 for Type F and M materials (ICRP, 1994b).

The experimental data were analysed using GIGAFIT (graphically interactive general algorithm for fitting—Birchall et al., 1995), a simulation modelling software package, to derive material specific absorption parameters for the ICRP human respiratory tract model (ICRP, 1994a). The parameters which best fitted the intratracheal instillation data are given in Table 2 with values for Types F and M. These parameters may be used directly in code implementing the human respiratory tract model such as LUDEP (Jarvis, 1994).

Table 2. Material specific absorption parameters for use with ICRP 66 respiratory tract model calculated for UO_3 and default values

Expressed as rapid and slow dissolution fractions				Expressed as initial and "transformed" particles			
Parameter	UO_3	Type F	Type M	Parameter	UO_3	Type F	Type M
f_r	0.638	1	0.1	s_p (d^{-1})	63.83	100	10
s_r (d^{-1})	100	100	100	s_{pt} (d^{-1})	36.17	0	90
s_s (d^{-1})	0.035	—	0.005	s_t (d^{-1})	0.035	—	0.005
f_b	0.488	0	0	f_b	0.488	0	0
s_b (d^{-1})	3.259	—	—	s_b (d^{-1})	3.259	—	—

The relationship between these alternative representaions is defined in ICRP Publication 66 as:
(i) $s_p = s_s + f_r (s_r - s_s)$; (ii) $s_{pt} = (1 - f_r) (s_r - s_s)$; (iii) $s_t = s_s$.

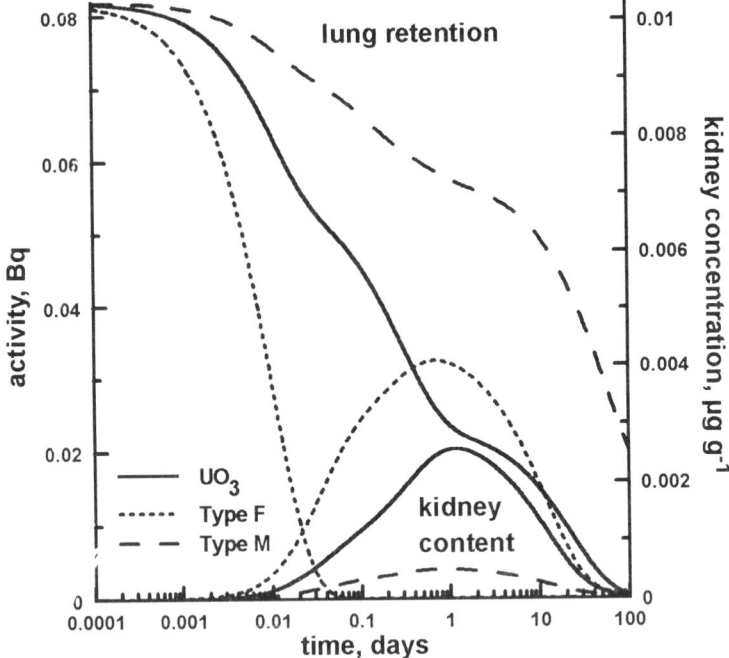

Fig. 1. Predicted lung and kidney retention of uranium in humans after acute intake of 1 Bq, 5 μm AMAD UO_3.

DISCUSSION

The main purpose of the animal experiments was to quantify absorption rates of uranium encountered in the workplace for extrapolation to man. Emphasis here is placed on lung clearance, urinary excretion and kidney concentration. Using standardized criteria described in ICRP Publication 71 (ICRP, 1995) together with the data obtained in the study, this UO_3 can broadly be described as Type F. In this same publication, ICRP assign UO_3 to Type M although a caveat identifies that pure forms may be more appropriately described as Type F. In calculations, a particle size of 5 μm AMAD σ_g 2.5 was assumed which closely resembled airborne material in the work place.

After an acute intake by humans, the lung content is predicted to fall rapidly over the first day with only 25% of that deposited in the lower respiratory tract retained by 1 day (Fig. 1). There is very little retention at times approaching 100 days after exposure. Interestingly, the kidney content (Fig. 1) rises sharply to a peak of 25% of the lower respiratory tract deposit at 1 day after exposure, returning to a very small amount by 100 days. Instantaneous urinary excretion (Fig. 2) falls progressively from 0.3 Bq d^{-1} at 0.1 d to 0.02 and 1.0 10^{-4} Bq d^{-1} at 1 and 100 days after exposure, respectively.

Dose coefficients were calculated using the computer code PLEIADES (program for linear internal age-dependent doses) which was also used to calculate dose coefficients in ICRP Publications 68 (ICRP, 1994b) and 72 (ICRP, 1996). The dose coefficient for 5 μm AMAD particles is 1.0 10^{-6} Sv Bq^{-1}, which is equivalent to an

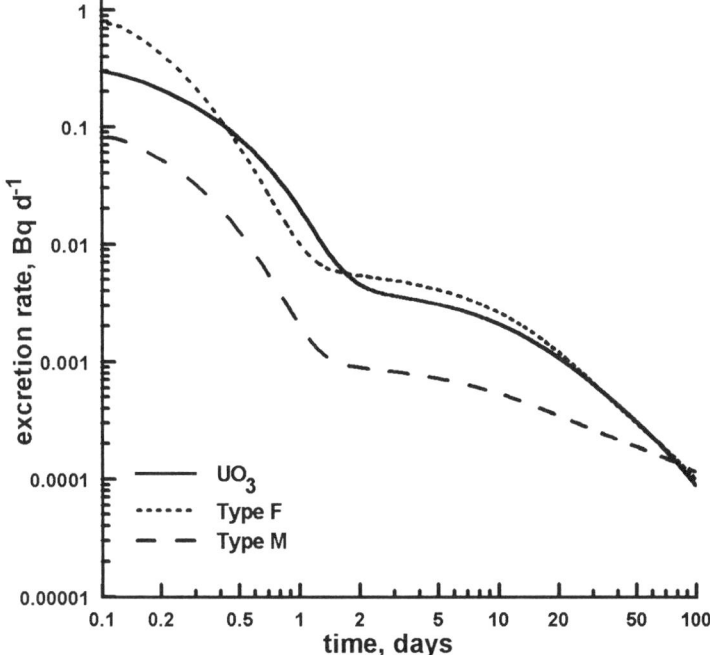

Fig. 2. Predicted instantaneous urinary excretion of uranium in humans after acute intake of 1 Bq, 5 μm AMAD UO_3.

annual limit of intake (ALI) of 20 kBq (20 mSv annual dose limit) or 780 mg for this UO_3 preparation. It has long been recognised, however, that uranium is chemically toxic to the kidneys and this has given rise to toxicologically based limits on intake. ICRP recognised in Publications 30 (ICRP, 1979) and 54 (ICRP, 1988) that for soluble forms of depleted, natural and low-enrichment uranium, the limit on intake should be based on chemical toxicity and that radiologically based annual limits on intake were entirely inappropriate for such materials.

Current EC legislation stipulates a toxicity based daily limit of 2.5 mg (0.26 mg m^{-3}, 6.7 Bq m^{-3}) originating from ICRP Publication 6 (ICRP, 1964). This value is based on an equilibrium concentration in the kidney of 3 μg g^{-1} considered to be a threshold for kidney damage, but calculated using a simplistic metabolic model. However, in the U.K., the Health and Safety Executive's (HSEs) Guidance Notes on Occupational Exposure Limits, 1996, set a long-term time weighted average (TWA) exposure limit of 0.2 mg m^{-3} (8 h TWA—5.13 Bq m^{-3}) for soluble and insoluble uranium compounds. Alternatively, with the experimentally derived dose coefficient, a value of 8.7 Bq m^{-3} (0.33 mg m^{-3}—5 μm AMAD) can be calculated. It can be seen, as ICRP identified, that a limit based on chemical toxicity will override even with a 20 mSv annual dose limit. Moreover, in their comprehensive reviews, Diamond (1989) and Leggett (1989) highlight the uncertainties associated with the 3 μg g^{-1} level and suggest that there is good reason for the limit to be an order of magnitude lower.

It is interesting to note that the values adopted by HSE are based on those from the American Conference of Government Industrial Hygienists (ACGIH). However, other safety organisations such as the U.S. Department of Labor, Occupational Safety and Health Administration (OSHA) and the U.S. National Institute for Occupational Safety and Health (NIOSH) distinguish between soluble (0.05 mg m^{-3}) and insoluble (0.2 mg m^{-3}) compounds.

After an acute intake, predicted uranium excretion in urine a few days after exposure is very low, 0.0026 Bq d^{-1} per Bq intake at 7 days after exposure (Fig. 2). Although dependent on the analytical technique, this is clearly asking a great deal of any routine monitoring programme.

To provide guidance on the assessment of a chronic intake, ICRP Publication 54 (ICRP, 1988) takes a hypothetical case where the daily intake during exposure is equal to ALI/365. This is cited for comparison against monitoring results. Alternatively, the derived air concentration (DAC) assumes an exposure regime representing 50 × 40 h working weeks, at a breathing rate of 1.2 m^3 h^{-1} (2400 m^3 y^{-1}). As an exposure regime consistent with a DAC will mean an intake more than ALI/365 during exposure days, it follows that this will be accompanied by elevated urine levels at certain times compared with the hypothetical case. This may lead to false identification of overexposure. To address this inconsistency, a more realistic, intermittent intake regime compatible with the principle of the DAC was modelled. It was assumed that chronic exposure occurred 8 h a day at 1.2 m^3 h^{-1} with no weekend exposure. From this, the instantaneous kidney concentrations, and daily urinary excretion were calculated. For completeness, a constant chronic exposure regimen was also modelled. Both scenarios assumed identical intakes of 7 Bq per week.

Not surprisingly, periodicity in the assumed intermittent chronic regimen is mirrored in the calculated kidney concentration (Fig. 3). Although the kidney concentration during the first two weeks of exposure is higher than the constant chronic case, the two cases overlie, and are within 10%. Similar calculations undertaken for daily urinary excretion (Fig. 4—upper plots) demonstrate a dramatic fall in urine excretion during weekend breaks. After only a short period of exposure during the susbequent week, the daily urinary excretion rate returns to the higher level. This problem of underestimating intakes from urine analysis would be compounded by typical "post-break-monitoring" procedures where samples are typically taken before work on Monday morning, and may only represent 2 h excretion. This too was modelled (Fig. 4—lower plots). It can be seen that when assuming a constant chronic exposure, intakes can be underestimated by 300–400%. It is well known that sampling immediately after a shift is likely to result in false positives from contamination, but consideration could be given to showering before collection on Fridays or any other procedure appropriate to the shift pattern. After an identified potential intake sampling will obviously commence immediately as is current practice.

CONCLUSIONS

For the protection of workers encountering such UO_3 there are many points of reference. For radiological protection, this UO_3 can be described as consistent with an absorption Type F compound, which is in line with studies on exposed workers

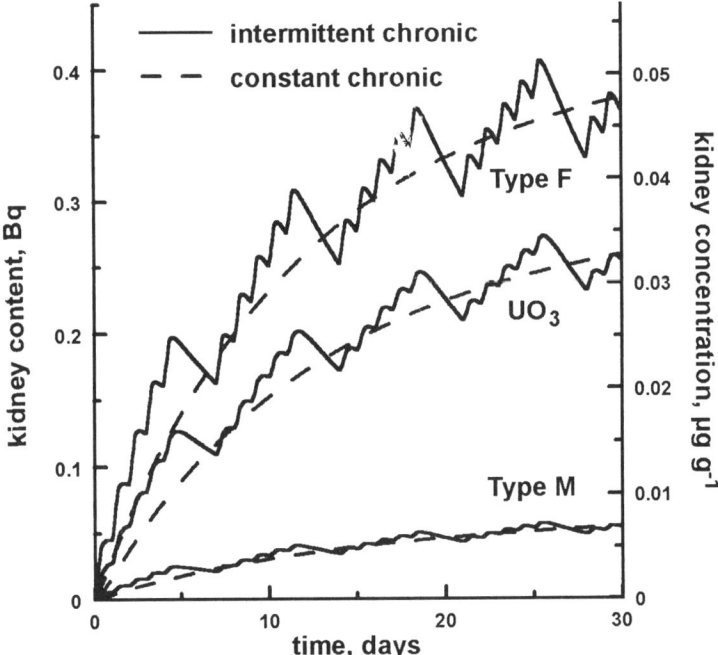

Fig. 3. Predicted uranium content of human kidneys after chronic exposure at 7 Bq per week, 5 μm AMAD UO₃.

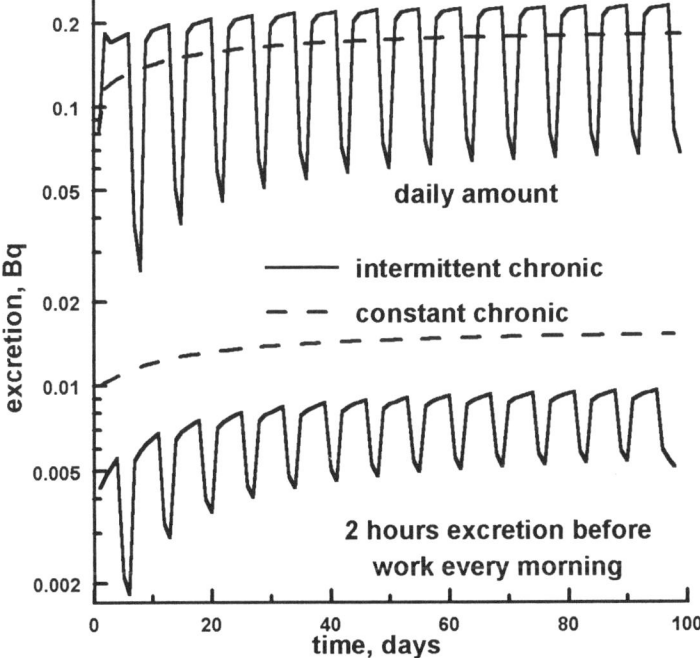

Fig. 4. Predicted urinary excretion of uranium in humans after chronic exposure to 7 Bq per week, 5 μm AMAD UO₃.

as well as other animal experiments undertaken at NRPB. In addition, this study has provided specific absorption parameters for dosimetric and biokinetic calculations. However, to protect workers in such a facility, due regard should be given to toxicologically based exposure limits as this will be more restrictive. With occupational protection now being more prospective, exposure limits set by OSHA and NIOSH might be considered more widely which would limit air concentration to 0.05 mg m^{-3}. Interestingly, neither recent ICRP publications nor forthcoming EC legislation consider this aspect of protection. Central to toxicity based limits is the concentration of uranium in kidneys. A maximum concentration of 3 µg g^{-1} in the kidneys has been the limiting concentration in kidneys for some time, and has been regarded as offering adequate protection. Moreover, as the derivation of this value relied on ICRP models now superseded, with the advent of ICRP's new models and modern thinking on toxicology, this value needs to be reassessed.

In demonstrating compliance with safety standards by using personal monitoring data, post break monitoring may not be satisfactory for a routine programme and may not detect acute intakes that were not identified during the previous week. Indeed, consideration should be given to sample collection before the break commences. Probably high volume and personal air samplers may be of more use for demonstrating compliance and measurements augmented with detailed particle size data. Alternatively, consideration should be given to respiratory protection to control exposure rather than intake.

REFERENCES

Birchall, A., Bailey, M. R. and Jarvis, N. S. (1995) Application of the new ICRP respiratory tract model to inhaled plutonium nitrate using experimental biokinetic data. In *Proc. Int. Conf. On Radiation Dose Management in the Nuclear Industry*, Windermere, U.K., Oct 9–11, pp. 216–223.

Diamond, G. L. (1989) Biological consequences of exposure to soluble forms of natural uranium. *Radiat. Prot. Dosim.* **26**, 23–33.

Jarvis, N. S. and Birchall, A. (1994) LUDEP 1.0, A personal computer program to implement the new ICRP respiratory tract model. In proc. workshop—intakes of radionuclides: detection, assessment and limitation of occupational exposure, Bath, 13–17 September 1993. *Radiat. Prot. Dosim.* **53** (1–4), 191–193.

International Commission on Radiological Protection. (1964) Recommendations of the International Commission on Radiological Protection. ICRP Publication 6. Pergamon Press Ltd, Oxford.

International Commission on Radiological Protection. (1979) Limits for intakes of radionuclides by workers. ICRP Publication 30. *Ann. ICRP* **2** (3–4), Elsevier Science Ltd, Oxford.

International Commission on Radiological Protection. (1988) Individual monitoring for intakes of radionuclides by workers: design and interpretation. ICRP Publication 54. *Ann. ICRP* **19** (1–3). Elsevier Science Ltd, Oxford.

International Commission on Radiological Protection. (1994a) Human respiratory tract model for radiological protection. ICRP Publication 66. *Ann. ICRP* **24** (1–3). Elsevier Science Ltd, Oxford.

International Commission on Radiological Protection. (1994b) Dose coefficients for intakes of radionuclides by workers; replacement of ICRP Publication 61. ICRP Publication 68. *Ann. ICRP* **24** (4). Elsevier Science Ltd, Oxford.

International Commission on Radiological Protection. (1995) Age-dependent doses to members of the public from intakes of radionuclides: part 4 inhalation dose coefficients. ICRP Publication 71. *Ann. ICRP* **25** (3–4). Elsevier Science Ltd, Oxford.

International Commission on Radiological Protection. (1996) Age-dependent doses to members of the public from intake of radionuclides: part 5 compilation of ingestion and inhalation dose coefficients. ICRP Publication 72. Elsevier Science Ltd, Oxford (in press).

Leggett, R. W. (1989) The behavior and chemical toxicity of U in the kidney: a reassessment. *Health Phys.* **57**, 365–383.

Pergamon

Ann. occup. Hyg., Vol. 41, Supplement 1, pp. 111–115, 1997
© 1997 British Occupational Hygiene Society
Published by Elsevier Science Ltd. All rights reserved
Printed in Great Britain
0003–4878/97 $17.00 + 0.00
Inhaled Particles VIII

PII: S0003–4878(96)00090–7

ALVEOLAR MACROPHAGE LETHALITY INDUCED BY α IRRADIATION *IN VITRO*

C. Lizon, I. Bailly, F. Guezingar, F. Jouanny, J. L. Poncy and P. Fritsch

Laboratoire de Radiotoxicologie, CEA/DSV/DRR/SRCA, BP 12, 91680 Bruyères le Châtel, France

INTRODUCTION

After γ irradation, rodent alveolar macrophages (AM) have been shown to be very radioresistant (Kubota *et al.*, 1994). However, only a few studies have concerned functional and genetic alterations induced in AM after α irradiation.

After inhalation of ^{239}PuO$_2$, a decrease of AM motility (Nolibé *et al.*, 1974), an alteration of the AM phagocytosis ability and an increase of the number of cells with micronuclei and multinucleated cells (Morgan and Talbot, 1992) have been reported. After addition of ^{241}AmO$_2$ particles to the culture medium, dose–effect relationships have been determined *in vitro* on alterations of phagocytic ability and viability of AM (Taya and Mewhinney, 1992). In all these studies, the actual dose delivered to the cells was unknown, but heterogeneously distributed, depending on the size distribution of particles and on the number of particles per cell.

Homogeneous irradiation can be performed *in vitro* using electrodeposited sources of α emitters. To our knowledge, only one study has been performed on AM that measured micronuclei induction (Johnson and Newton, 1994).

The aim of this study was to determine rat AM survival after a homogeneous *in vitro* α irradiation to relate AM death to the number of α tracks crossing the entire cell or the cell nucleus and to the delivered dose.

METHODS

Three month-old male Sprague–Dawley rats were used. Animals were killed under pentobarbital anaesthesia (40 mg kg^{-1}). Lungs were removed and AM were extracted by pulmonary lavage after repeated instillations of 5 ml NaCl 0.9% until a total volume of 20 ml had been collected. The cell suspension was centrifuged (200 **g**) for 10 min and the pellet suspended in culture medium. The culture medium was RPMI containing 10% of either fetal calf serum (FCS) or a substitute (BMS, Polylabo). AM were plated at a cell density of 4.10^4 per cm^{-2} on 6 μm thick mylar foil maintained by a special ring on the culture vial. Three ^{239}Pu electrodeposited sources were used with flux at $0.7.10^{-4}$, $1.2.10^{-4}$ or $1.4.10^{-4}$ α μm^{-2} s^{-1}.

To determine adherent cell number per unit of area, cultures were observed either unstained, with Hoffman contrast and video-image recording, or after paraformaldehyde fixation followed by standard staining with Mayer's hemalun or Giemsa. Viability was measured 1 h after the end of irradiation. It was expressed as percent of unirradiated cell density using several control cultures.

The mean AM total area and the mean nucleus AM area were measured on fixed and stained cells using a standard image analysis software developed in the laboratory. Confocal microscopy observations (Biorad, MRC 1024) were performed after paraformaldehyde fixation and propidium iodide staining.

RESULTS

After 3 days in culture, AM density of unirradiated cells appeared about three times larger for BMS than for FCS supplemented medium. Moreover, this density remained nearly constant for 5 days in BMS medium whereas it gradually decreased in FCS medium. Thus, irradiation experiments have been performed using the BMS medium.

Large variations have been observed for measurements of the control AM densities after plating on mylar, depending on the culture vial, even for cells obtained from the same animal. For a given survival value, the mean of the standard deviations measured in controls (at least three culture vials) was as high as 33% of mean.

Figure 1 shows AM survival as a function of α fluence. The survival curve could fit a single exponential function:

$$Y = 121.6.\ e^{(-7.38\ .10^{-2}.X)} \qquad\qquad R^2 = 0.742.$$

Thus, a 50% decrease of AM density was observed for a fluence at 12 α μm^{-2}.

Figure 2 shows the distribution of total and nucleus AM areas. From these normal distributions, the mean AM area was 127.41 μm^2, SD = 39.83 and the mean nuclear area was 36.17 μm^2, SD = 7.8. The AM shape evaluation has been performed using confocal microscopy. Figure 3 shows a vertical section of three macrophages observed on day one in culture. At this time, a spherical shape could be considered.

Preliminary results have been obtained by video-microscopy on living AM using Hoffman contrast. In controls, for the same vial, the number of adherent AM remained nearly constant for 5 days in culture. Thus, in these experimental conditions, the same vial can be used for measuring control survival just before irradiation.

DISCUSSION

In our experimental conditions BMS serum substitute improved the AM survival compared to FCS. This result confirms our previous observations using another serum substitute (Harper *et al.*, 1996). Thus, the medium supplemented with BMS is more adapted for AM survival studies than the usual FCS one.

After homogeneous α irradiation, the dose–effect relationship of AM lethality fit to a single exponential function of α fluence. Because no significant decrease of AM survival was observed for fluences below 3 α μm^{-2} and because the calculated survival was higher than 100% in controls, a threshold or a shoulder could be expected.

Assuming a spherical shape of the cell, as suggested by confocal microscopy observations, actual fluence to two targets, the cell nucleus and the whole cell, can

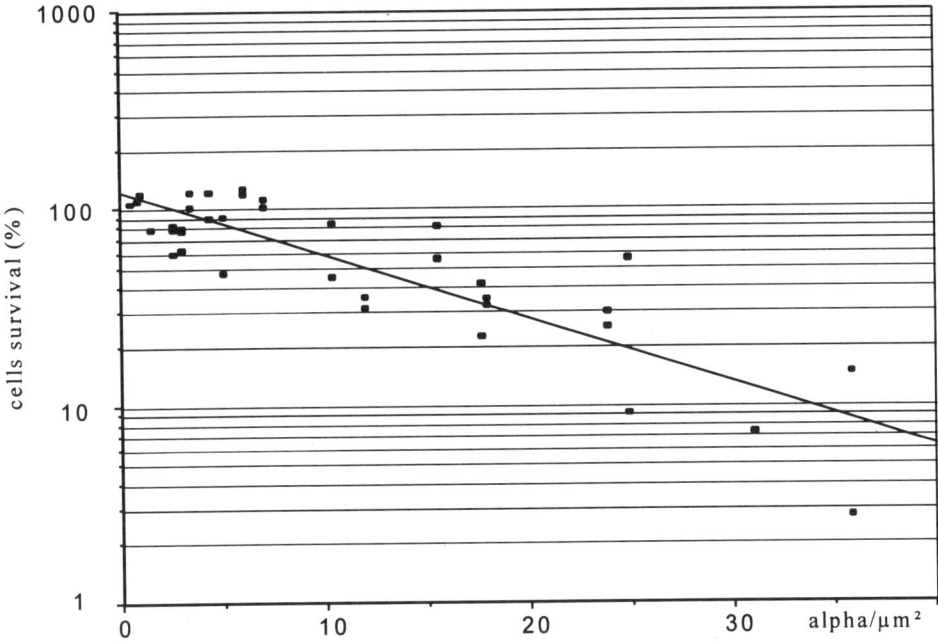

Fig. 1. Evolution of AM survival as a function of α fluence (individual values).

be calculated according to Crawford-Brown and Shyr (1987). The number of α tracks needed to kill 50% of the AM population are 262 and 928 α for the nucleus and the whole cell, respectively. This corresponds to a mean delivered dose at 155 Gy for the cell nucleus and at 152 Gy for the whole cell.

To overcome problems related to variability in the AM density values according to culture vial, we have followed the individual behaviour of living AM. This could be obtained by video microscopy using Hoffman contrast. The preliminary results obtained on controls showed a smaller variability than those observed after

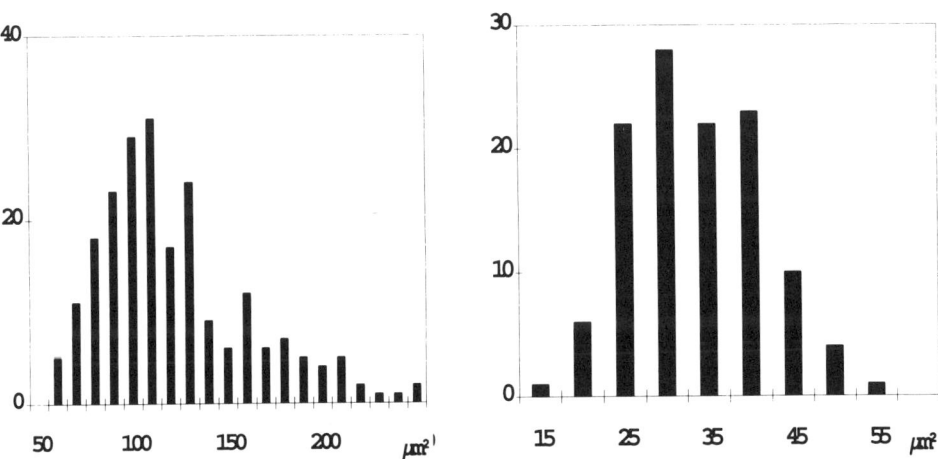

Fig. 2. Distribution of total and nucleus AM areas, one day after plating on mylar foil.

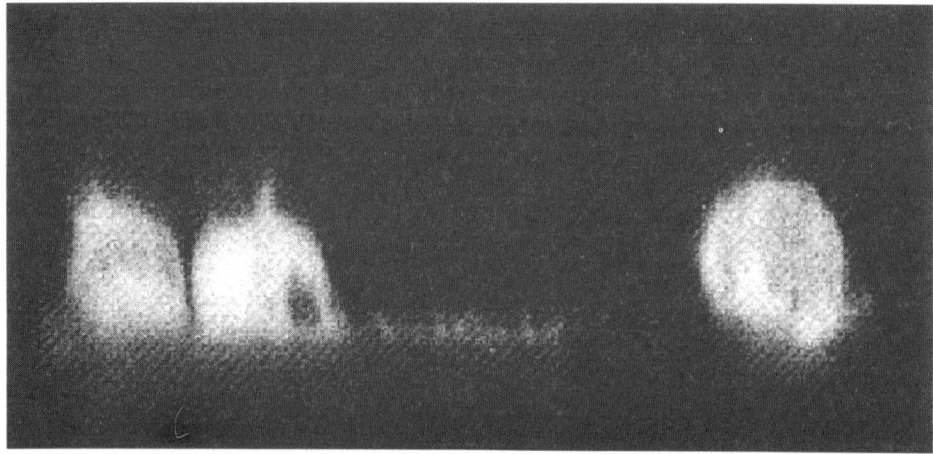

Fig. 3. Vertical section of three AM 1 day after plating on mylar, obtained by confocal microscopy.

fixation. Thus, this method will be used in further studies to characterise the expected threshold or shoulder in the dose effect relationship.

By contrast to the previous studies, AM death induced after phagocytosis of poorly soluble α emitters can be estimated from our results according to the specific α activity and the particle diameter. Table 1 shows, for different inhaled actinide oxides with AMAD of 1 and 5 μm as considered for aerosol outside or inside the working place (ICRP, 1994), the time needed to kill 50% of AM. The calculation was performed assuming that each AM contained one particle and that no spontaneous cell death occurs. *In vivo*, the half-life of control rat AM has been

Time needed to kill 50% AM	UO_2	$^{237}NpO_2$	$^{239}PuO_2$	$(U, Pu)O_2$ with ^{238}Pu and ^{241}Am *	$^{238}PuO_2$
1 μm AMAD	14400 years	7 years	28 days	5.5 days	2.5 hours
5 μm AMAD	115 years	20 days	5.5 hours	1 hour	1.2 minutes

Fig. 4. Time needed to kill 50% of AM after phagocytosis of actinides oxides (1 or 5 μm AMAD, 1 particle per cell), assuming no AM renewal. *Heterogeneous inhalable powder with 1.1% ^{238}Pu and 11% of total α due to ^{241}AM, characterised by Massiot *et al.*, 1997 and used for lung clearance measurement by Ramounet *et al.*, 1997.

estimated to 7–10 days (Fritsch *et al.*, 1992).Thus, modifications of AM population that could alter lung clearance parameters might be expected for particles inducing 50% AM death within less than 1 week. From Table 1, a significant death could be induced by inhalable particles with a larger α specific activity than that $^{239}PuO_2$ such as Pu oxide containing ^{238}Pu and/or ^{241}Am and $^{238}PuO_2$. In the case of $^{239}PuO_2$, only large particles could induce AM death. Several experimental results agree with such hypothesis (Sanders, 1972; Morgan *et al.*, 1988; Ramounet *et al.*, 1997). Further studies are in progress to measure other physiological end points after α irradiation such as motility or cytokines release and to characterise the actual targets for the radioinduced alterations (cytoplasm, nucleus).

REFERENCES

Crawford-Brown, D. J. and Shyr, L. J., (1987) The relationship between hit probability and dose for alpha emissions under selected geometrics. *Radiat. Protec. Dosim.* **20**, 155–168.

Fritsch, P. and Masse, R. (1992) Overview of pulmonary alveolar macrophage renewal in normal rats and during different pathological processes. *Environ. Health Perspect.* **97**, 59–67.

Harper, R., Stirling, C., Patrick, G., Hoffschir, D., Poncy, J. L. and Kreyling, W. G. (1996) The survival and function of nondividing alveolar macrophages under standard culture conditions. *Inhal. Toxicol.* **8**, 405–422.

International Commission on Radiological Protection (1994) *ICRP Publication 66.* Pergamon Press, Oxford.

Johnson, N. F. and Newton, G. J. (1994) Estimation of the dose of radon progeny to the peripheral lung and the effect of exposure to radon progeny on the alveolar macrophage. *Radiat. Res.* **139**, 163–169.

Kubota, Y., Takahashi, S. and Sato, H. (1994) Effect of γ-irradiation on the function and viability of alveolar macrophages in mouse and rat. *Int. J. Radiat. Biol.* **65**, 335–344.

Massiot, P., Le Naour, H., Matton, S., Lizon, C., Rateau, G., Grillon, G. and Fritsch, P. (1997) Heterogeneity of (U, Pu)O_2: an EDS electron microscopy study of inhaled particles in the rat (submitted).

Morgan, A., Black, A., Knight, D. and Moores, S. R. (1988) The effect of firing temperature on the lung retention and translocation of Pu following the inhalation of $^{238}PuO_2$ and $^{239}PuO_2$ by CBA/H mice. *Health Phys.* **54**, 301–310.

Morgan, A. and Talbot, R. (1992) Effects of inhaled alpha-emitting actinides on mouse alveolar macrophages. *Environ. Health Persp.* **97**, 177–184.

Nolibé, D., Métivier, H., Masse, R. and Lafuma, J. (1974) Motilité des macrophages et épuration alvéolaire chez différentes espèces animales (mobility of macrophages and alveolar phagocytosis in various animal species). *Rev. Fran. Mal. Respir.* **2**, 128–132.

Ramounet, B., Fritsch, P., Lizon, C., Poncy, J. L., Grillon, G., Verry, M. and Rateau, G. (1997) Inhalation of (U, Pu)O_2 in the rat: preliminary results on the transuranium element behaviour and on the efficacy of a DTPA treatment (submitted).

Sanders, C. L. (1972) Deposition patterns and the toxicity of transuranium elements in lung. *Health Phys.* **22**, 607–615.

Taya, A. and Mewhinney, J. A. (1992) Cytotoxicity, uptake and dissolution of $^{241}AmO_2$ particles in dog alveolar macrophages *in vitro*. *Int. J. Radiat. Biol.* **62**, 81–88.

 Pergamon

Ann. occup. Hyg., Vol. 41, Supplement 1, pp. 116–122, 1997
British Occupational Hygiene Society
© 1997 NRPB. Published by Elsevier Science Ltd
Printed in Great Britain
0003–4878/97 $17.00 + 0.00
Inhaled Particles VIII

PII: S0003-4878(96)00064-6

DESIGN OF A VOLUNTEER STUDY TO DETERMINE THE CLEARANCE OF INHALED PARTICLES FROM THE HUMAN NASAL PASSAGE

J. R. H. Smith, M-D. Dorrian, M. I. Youngman, M. R. Bailey, G. Etherington and A. L. Shutt

National Radiological Protection Board, Chilton, Didcot, Oxfordshire OX11 0RQ, U.K.

INTRODUCTION

The National Radiological Protection Board, (NRPB), U.K., is conducting a comprehensive volunteer study on the clearance of inhaled particles from the human nasal passage. Inhalation is the major route of intake of radioactive material for workers, with much inhaled material depositing in the nose, typically > 40% for particles of aerodynamic diameter (d_{ac}) > 2 µm. However, there is little information on the clearance from the human nasal passage of particles deposited during breathing for the 12 h following exposure (Fry and Black, 1973; Lippmann, 1970) and none on the clearance of the significant fraction that remains in the nasal passage at longer times. The new respiratory tract model adopted by the International Commission on Radiological Protection (ICRP, 1994) represents retention of most of the particles deposited in the extra-thoracic (ET) airways by two compartments: (a) particles deposited in the skin lined anterior nasal passage are cleared extraneously (by nose blowing) with a half-time of about one day; and (b) particles deposited in the ciliated posterior nasal passage, mouth and larynx clear to the gastro-intestinal (GI) tract with a half-time of 10 min. It is assumed that in all conditions material deposited in the nose is equally divided between the anterior and posterior nasal passages. However, it is recognised (ICRP, 1994) that there is considerable uncertainty associated with these values, especially with regard to the partition of deposition between the anterior and posterior nasal passages, which would be expected to depend on particle size and breathing pattern and with the clearance pattern from the anterior nasal passage. It is the aim of this study to address these uncertainties. It will investigate the effects of particle size and breathing pattern on the pattern of aerosol deposition in the respiratory tract and particle clearance from the nasal passage over ranges typical of occupational exposure. The degree of inter-subject variation and the efficency of nose blow sampling will also be studied. Preliminary results from three pilot experiments are presented.

ETHICAL APPROVAL AND INFORMED CONSENT

Ethical approval for the study was ȯbtained from the Central Oxford Research

Ethics Committee who required evidence that subjects would give informed consent. Subjects for the study must have no chronic ailments or serious abnormalities of the respiratory tract. An independent medical examiner established that volunteers gave informed consent.

The participation of each volunteer will be controlled so that the committed effective dose equivalent received in any twelve month period of the study will not exceed 0.01 mSv. This is well within the annual limit of 0.5 mSv that the World Health Organisation recommends for Catetory I projects, WHO (1977).

METHOD

The study is composed of a series of experiments in which volunteers inhale, through the nose, an insoluble, radio-labelled, monodisperse aerosol in the size range 0.6 μm > 30 μm d_{ae}. The particles are generated using either a spinning top aerosol generator (STAG Mk II, Research Engineers, London, U.K.: 0.6 to ~ 6 μm) or a vibrating orifice aerosol generator (VOAG, model 3450, TSI Inc., St. Paul, MN, U.S.A.: ≳ 6 μm) (Youngman, in press). During administration the volunteer breathes according to one of three regimes: naturally through the nose whilst sitting at rest; naturally through the nose whilst exercising at a workload of 80 watts, defined as "light exercise" (ICRP, 1994); or to a controlled breathing pattern. The first two regimes are representative of occupational breathing conditions, the latter will be used for comparison with computational model predictions. Volunteers pedal an ergometer to control their work rate during the "light exercise" experiments. Pilot experiments using ^{99m}Tc (half-life: 6.02 h) labelled particles will be performed for every particle size and breathing regime combination before main experiments, using ^{111}In (half-life: 2.83 days) labelled aerosols, are conducted on at least three volunteers, so enabling inter-subject variation to be determined.

The administration of the radio-labelled aerosol to the volunteer is controlled to give an initial nasal deposit (IND) of about 500 Bq ^{99m}Tc for pilot experiments and 1 kBq ^{111}In for main experiments. The required intake is determined from the activity of the aerosol particles and the fraction of inhaled particles expected to deposit in the ET airways; with the deposition fraction calculated using LUDEP 2.0 (Jarvis et al., 1996), a PC program which implements the ICRP (1994) human respiratory tract model, using the volunteer's tidal volume and respiratory frequency for the required breathing regime.

Administration methods vary with particle size. Large particles (> 10 μm d_{ae}) will be directly inhaled from the resuspension chamber. Particles less than 10 μm d_{ae} are inhaled from the chamber through computer controlled valves (Fig. 1); which automatically switch as the volunteer inhales and exhales. A 100 μl aliquot of the radio-labelled particle suspension is dispersed into the chamber using a compressed air nebuliser. Following resuspension the volunteer inhales the aerosol from the chamber at the start of the next inhalation. The aerosol concentration is controlled so that the administration will take approximately four breaths. The laser photometer (Fig. 1) measures the concentration of the inhaled aerosol. The pneumotachograph measures the volunteer's flow rate. Thus the number of particles inhaled is determined and the administration automatically stopped once

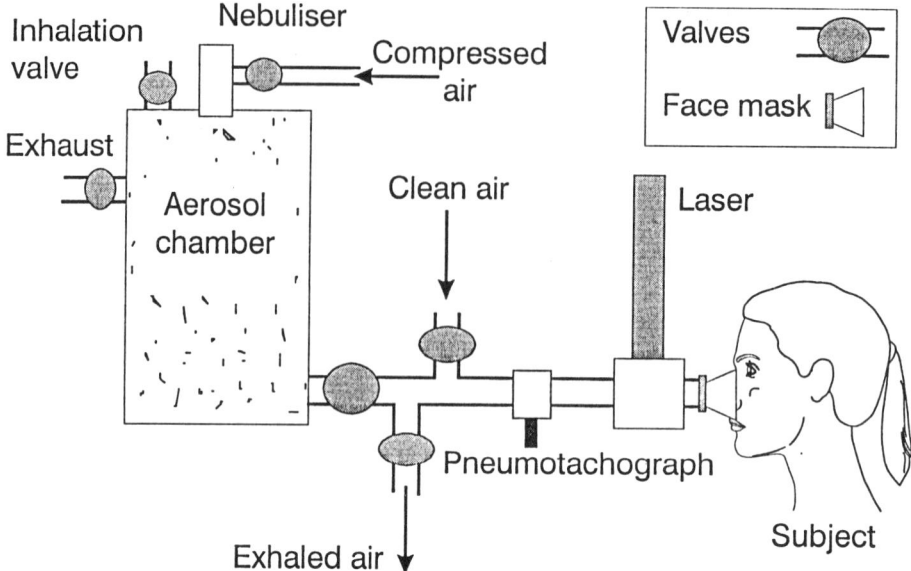

Fig. 1. Administration apparatus for aerosols < μm d_{ae} (draft diagram).

the required activity has been inhaled. The administration apparatus is then removed from the laboratory so that the activity remaining in it will not interfere with the *in vivo* measurements on the volunteer.

The clearance of the deposited particles is determined by *in vivo* measurements of γ-rays emitted by the radio-label, with additional information being gained from nose blow samples. On occasion, urine samples are measured to check the insolubility of the radio-label. All biological samples are measured by γ-ray spectrometry. The use of ^{99m}Tc and ^{111}In as the radio-labels means that clearance can be followed for up to 1 day after administration in pilot experiments and for about 5 days for main experiments, greatly expanding the time span of available clearance data. The start of *in vivo* measurements is synchronised with the start of aerosol administration. The volunteer inhales the aerosol seated within an array of five 110 mm diameter sodium iodide [NaI(T1)] detectors, collimated with lead side shields. Two detectors measure activity in the head, two are positioned against the upper back to measure lung activity, and a fifth is positioned in front of the stomach (Fig. 2). A computer controlled data acquisition system performs a sequence of 1 and 5 min measurements for the first 45 min after administration, so measuring any initial rapid particle clearance from the nasal airways.

At the end of the administration activity deposited on the face is removed by face wipes. The face wipes and nose mask are measured for activity. For most experiments volunteers may blow their noses at will, but are provided with a kit bagging and labelling such samples. The activities of these samples are measured as soon as reasonably practicable. The volunteer is also asked to keep a diary of events that might affect nasal clearance (for example eating, drinking, bathing).

A low background *in vivo* measurement detector array, housed in a steel room (Sumerling *et al.*, 1985), is used to make high accuracy measurements of the activity

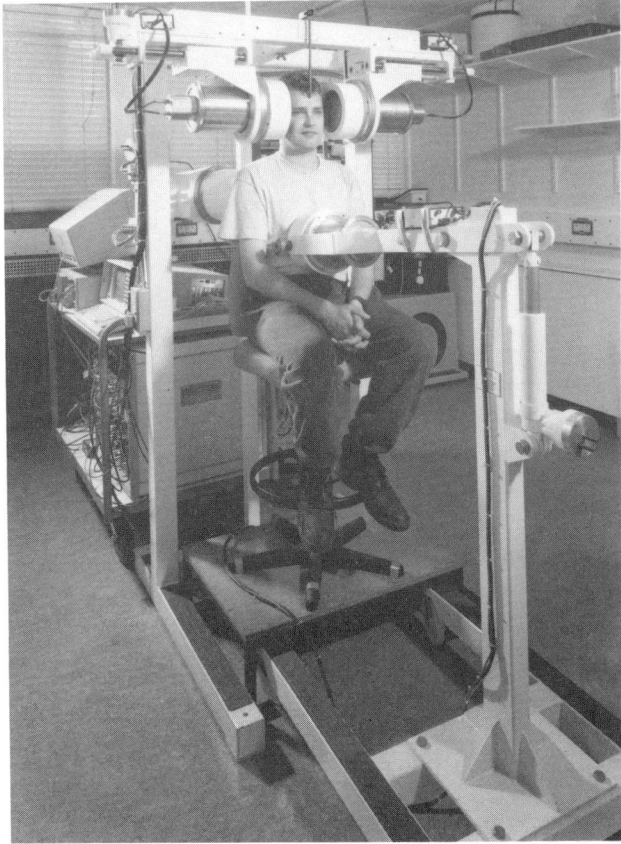

Fig. 2. The rapid particle clearance NaI(T1) detector *in vivo* measurement array.

in the head (nasal passage), chest (lungs) and abdomen (GI tract) of the volunteer from approximately 1 h after intake (Fig. 3). Each measurement is made using four 150 mm diameter NaI(T1) detectors collimated with lead side shields. Measurements are repeated at intervals for several hours after administration and then at about 24 h after intake and, for the main experiments, over several following days until the activity cannot be measured with sufficient accuracy. Measurements made with both *in vivo* measurement systems are corrected for contributions from activity elsewhere in the body.

RESULTS

To date, three administrations have been made of ^{99m}Tc-labelled 3 μm d_{ae} particles to volunteers sitting at rest. Particle clearance from the nasal passage was measured for up to 23 h after intake, doubling the period for which data were previously available (Fig. 4). All three volunteers provided complete nose blow samples for this period. Two volunteers provided urine samples which confirmed

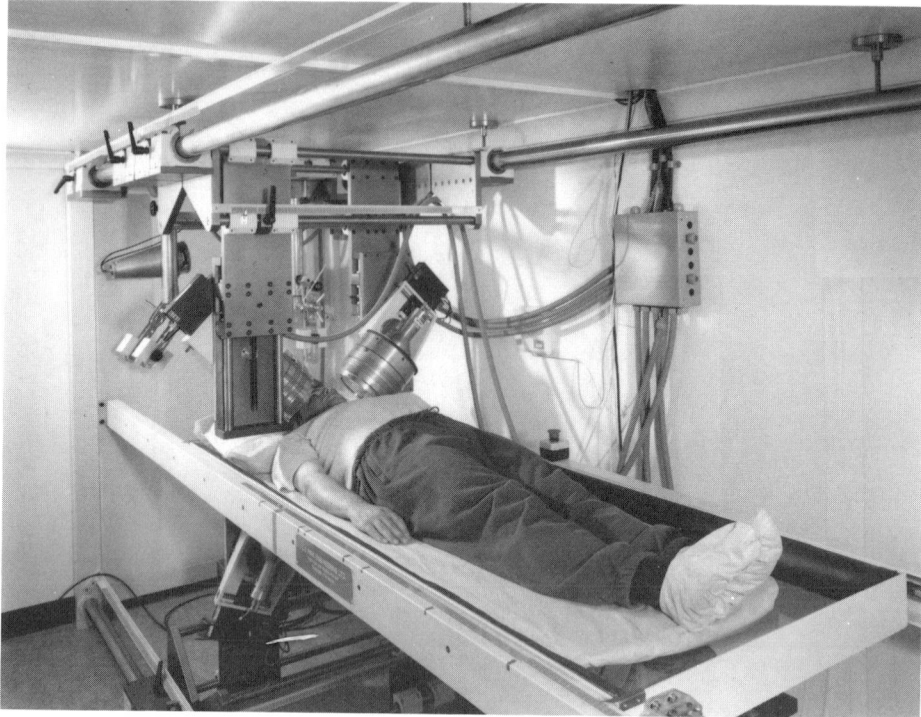

Fig. 3. Head measurements being performed using the low background NaI(T1) detector *in vivo* measurement array.

that ^{99m}Tc can be regarded as an insoluble radio-label. Full analysis of data obtained is in progress and the results presented below are provisional.

Respiratory tract deposition

The ratios of particles deposited in the nasal passage to particles deposited in the lungs seem to be markedly lower than the values predicted using the ICRP (1994) model in all three experiments. Preliminary analysis indicates that nasal deposition fractions are similar to the model estimates, suggesting that lung deposition is higher for nasal inhalation than estimated by the model under these conditions.

Particle clearance from the nasal passage

The initial activity deposited in the nasal passage was estimated by summing the activities measured in the head and abdomen (nasal passage and GI tract) directly after the administration rig had been removed from the laboratory. Activity in the GI tract at that time was assumed to have cleared rapidly from the pharynx. Nasal clearance shows a strong degree of inter-subject variation. However, for all volunteers approximately 90% IND had cleared at 20 h after intake. Over the period of measurement approximately 5, 16 and 20% IND was cleared by nose blows by volunteers A, B and C, respectively.

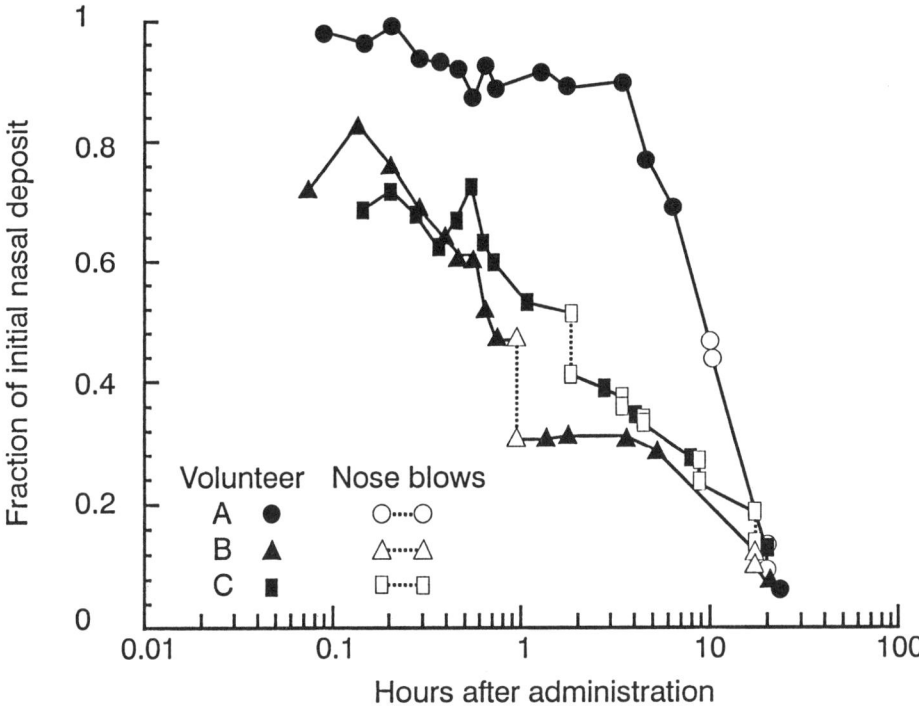

Fig. 4. Particle clearance from the nasal passage measured as a fraction of the initial nasal deposit (draft graph).

Table 1. Comparison of measured ratio of nasal passage to lung deposition for inhaled aerosols with estimates made using ICRP Publication 66 human respiratory tract model, (ICRP, 1994)

	ICRP (1994) default subject	Volunteers A	B	C
Sex, age	Male	Male, 43	Male, 37	Male, 47
Smoking history	Non-smoker	Ex-smoker	Non-smoker	Non-smoker
Tidal volume, cm³	750	490	600	560
Respiratory frequency (breaths per minute)	12	11.2	13.4	15.2
Particle size, μm	3.0	3.1	3.1	3.1
Geometric standard deviation, σ_g	1.2	1.1	1.03	1.1
Measured ratio of nose/lung deposition*	N/A	1.0 ± 0.2	0.9 ± 0.1	0.7 ± 0.2
ICRP (1994) estimated ratio of nose/lung deposition	2.2	1.5	2.2	2.6

*The uncertainties quoted for the measured nasal passage to lung deposition ratios are the 95% confidence intervals from counting statistics only. There is additional uncertainty due to geometric factors.

DISCUSSION

The experimental methodology for the study has been established. The results of the three pilot experiments show consistency with the results of the two earlier studies (Fry and Black, 1973; Lippmann, 1970). Preliminary analysis indicates that lung deposition was higher than estimated by the ICRP (1994) model under the conditions of the experiment. However, it should be borne in mind that this is a provisional analysis which has not considered all factors, such as the measurement uncertainties due to geometrical factors, which may affect these findings. Particle clearance from the nasal passage has strong inter-subject variation, but 90% clearance occurred within 24 h in all three experiments. Nose blow samples do not account for all long-term nasal clearance. These results indicate that the study will yield important information for the biokinetic modelling of particle clearance from the nasal passages and for the assessment of doses from intakes of inhaled radioactivity.

Acknowledgements—This work is part-funded by the Commission of the European Communities under Contract no. F14P CT950026 "Inhalation of Radionuclides". Mr M. V. Holding is thanked for his assistance to the study in developing and constructing equipment for the study.

REFERENCES

Fry, F. A. and Black, A. (1973) Regional deposition and clearance of particles in the human nose. *Aerosol Sci.* **4**, 113–124.

Jarvis, N. S., Birchall, A., James, A. C., Bailey, M. R. and Dorrian, M-D. (1996) LUDEP 2.0: Personal computer program for calculating internal doses using the ICRP Publication 66 Respiratory Tract Model. NRPB-SR287, National Radiological Protection Board, Chilton, U.K.

ICRP (1994) Human respiratory tract model for radiological protection. ICRP Publication 66. *Annals of the ICRP*, Vol. 24 (1–3). Elsevier Science Ltd, Oxford.

Lippmann, M. (1970) Deposition and clearance of inhaled particles in the human nose. *Ann. Otol. Rhinol. Laryngol.* **79**, 519–528.

Sumerling, T. S., McClure, D. R. and Massey, D. K. (1985) Measurement of total body radioactivity: the procedures used at the board. NRPB-R188, National Radiological Protection Board, HMSO, London.

Youngman, M. J. Production of polystyrene particles for human nasal deposition and clearance studies. 10th Annual Conf. Aerosol Society: aerosols, their generation, behaviour and applications. *J. Aerosol Sci.* (in press).

World Health Organisation (1977) Use of ionizing radiation and radionuclides in human beings for medical research, training and nonmedical purposes. A report of a WHO expert committee. Technical report series 611. World Health Organisation, Geneva.

SECTION 3

FIBRES

 Pergamon

Ann. occup. Hyg., Vol. 41, Supplement 1, pp. 123–128, 1997
© 1997 British Occupational Hygiene Society
Published by Elsevier Science Ltd. All rights reserved
Printed in Great Britain
0003–4878/97 $17.00 + 0.00
Inhaled Particles VIII

PII: S0003–4878(96)00107–X

TRENDS IN OCCUPATIONAL GROUPS AND INDUSTRIES ASSOCIATED WITH AUSTRALIAN MESOTHELIOMA CASES 1979–1995

A. J. Rogers†, P. Yeung‡, A. Johnson§, J. Leigh‡ and P. Davidson‡

†Alan Rogers OH&S, PO Box 2128, Clovelly, NSW 2031, Australia; ‡Worksafe Australia, PO Box 58, Sydney, NSW 2001, Australia; and §NSW Workers' Compensation (Dust Diseases) Board, 82 Elizabeth St., Sydney, New South Wales, Australia

INTRODUCTION

In the mid 1970s concern was raised by a group of occupational health professionals, compensation authorities and government regulators about the rising numbers of mesothelioma cases occurring in the Australian community. Asbestos exposure was known at this time to be a major contributor to the cases and certain industries such as crocidolite mining at Wittenoom, local asbestos cement production, war-time gas mask construction and the manufacture of crocidolite covered welding rods had contributed to local cohorts. Most states by this stage had specific asbestos regulations designed to prevent health effects principally asbestosis in the mining and manufacturing workforce with compensation authorities conducting regular inspection and health audits.

Despite the introduction of these contemporary control strategies mesothelioma cases have continued to rise in line with the world wide trend (Peto *et al.*, 1995). Although this increase can be attributed to the long latency effects of the historical heavy exposure to asbestos especially amphiboles, there is still considerable concern in the community in regard to the use of bound chrysotile products and to the legacy of asbestos products such as in public and private buildings. This study provides an analysis of the trends that have developed in Australian mesotheliomas as a means of predicting future outcomes.

MATERIALS AND METHODS

Australian cases of mesothelioma have been collected by some state compensation authorities for more than 30 years and by a national registration scheme since 1979.

The Australian Mesothelioma Surveillance Program (1979–1985)

This research project centred in the Commonwealth Department of Health was set up by a diverse group of occupational health professionals to investigate amongst other things the incidence and aetiology of mesothelioma in Australia (Ferguson *et al.*, 1987). Detailed occupational and environmental histories were obtained by direct interview with patients or next of kin and, failing this, work associates. All histories were independently examined by two occupational hygienists who had considerable experience in investigating exposures in a range of asbestos manufacturing and user industries.

The Australian Mesothelioma Register (1986–1995)

The comprehensive collection and reporting of cases continued post 1985 (Leigh *et al.*, 1991, 1996). Owing to a number of factors the data collected on each case became less substantive with reliance placed on local rather than central diagnosis, case histories were reliant on information provided from clinical records with the consequence of considerable variability of details, quality and interpretation of information. Central register data was reviewed and coded by two experienced occupational hygienists one of which was involved with the coding of the 1979–1985 cases.

The NSW Worker's Compensation (Dust Diseases) Board (1985–1995)

As part of the requirements of the "no fault" compensation scheme for dust diseases operating in the State of New South Wales (NSW), data is collected on asbestos related diseases associated with employment in that State (DDB, 1995). Detailed occupational histories are obtained by direct interview which is followed up by industrial inspectors who conduct field inspection of premises and employer records and prepare a report to the independent medical panel.

RESULTS

In the period 1979–1995, 3538 cases were collected nationwide with an increase of 6–7% p.a. although this varies considerably across different industries and occupations (Tables 1 and 2). Five industries are typical of the differing trends found during the period of study and show the considerable influence of exposure to amphiboles.

The building industry produced the largest number of cases and also the largest increase ($n = 559$, 448%) across a wide range of occupations. Some cases recalled being present when "Limpet" (mainly amosite and crocidolite in Australia) was sprayed by insulation workers. Trades such as electricians and plumbers indicated that they had disturbed a variety of asbestos products whilst carrying out repairs, renovation or installation of new services. Many carpenters and labourers indicated that they had worked with asbestos cement products which were commonly used in domestic and commercial construction in the 1950s, 1960s and 1970s.

Shore based shipbuilding, repair and demolition accounted for the second largest industry group. Case numbers appeared to have remained steady with a variable but slight increase of around one case per year. The majority of these were involved in boiler and engine repairs which involved the disturbance of asbestos lagging. Other trades such as shipwrights, welders, plumbers and electricians indicated working with asbestos however and only a small proportion of cases listed their occupation as ships laggers.

The local asbestos cement production industry ranks third in overall cases with an increase of approximately 5% p.a. although there is some evidence from Fig. 1 that this rate may have plateaued in the last 5 years. Almost all cases have arisen

Table 1. Industry distribution of mesothelioma cases in three different periods

Industry	Mesothelioma Program 1979–85	Mesothelioma Register 1986–90	1991–95
Primary asbestos industries			
Wittenoom crocidolite mining/milling	49	66	49
Asbestos cement production	53	55	68
Insulation manufacture/installation	27	16	26
Stevedoring	22	27	34
Primary asbestos user industries			
Ship construction/repair/demolition	88	87	132
Building	65	203	291
Railway loco fabrication/maintenance	45	29	63
Coal fired power station construction and maintenance	29	32	47
Secondary user/product contact			
Engineering fabrication/repair	15	50	88
Metal smelting	10	12	27
Chemical plants	10	18	7
Motor vehicle repair	7	16	24
Other industries	301	341	271
Insufficient information/unknown	137	218	383
Total	858	1170	1510

Table 2. Occupational distribution of mesothelioma cases in three different periods

Occupation	Mesothelioma Program 1979–85	Mesothelioma Register 1986–90	1991–95
Building trades			
Carpenters	63	107	158
Electricians	25	30	59
Plumbers	13	29	39
Builders labourers	0	34	35
Painters	6	25	24
Plasterers	3	4	11
Bricklayer (insulation/furnace)	3	5	1
Bricklayer (non-insulation)	0	5	7
Metal and engineering trades			
Boiler makers & assistants	50	76	69
Welders	15	7	15
Car mechanics	7	16	24
Shipping			
Ableseamen	2	6	17
Stokers	7	3	7
Marine engineers	1	10	13
Wharf labourers	21	19	33

from employees who commenced in the era when dust levels were relatively high and crocidolite was used (Boyle and Rogers, 1993). This industry had been operational from 1916 to 1987 and used at its peak over 90% of all asbestos fibre mainly in the form of roofing and cladding. By 1954 this accounted for 25% of new housing and a world ranking of four in per capita consumption. Imported

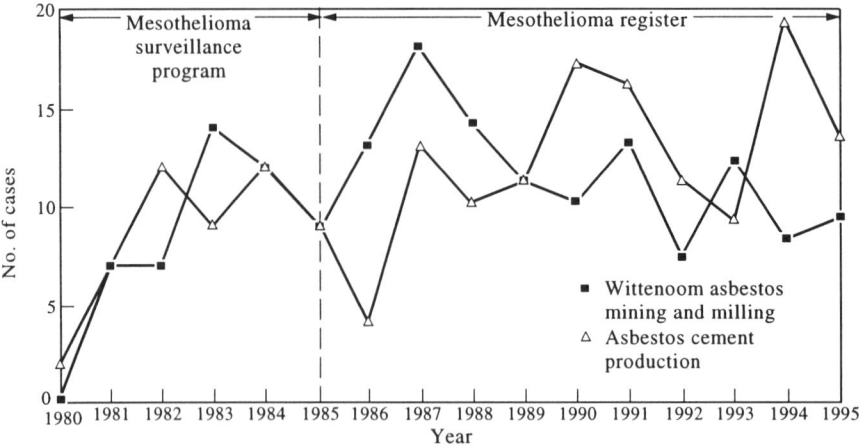

Fig. 1.

crocidolite, amosite and chrysotile and local sources of crocidolite and chrysotile were used in various combinations throughout the history of the industry.

Crocidolite mining, milling and associated activities at Wittenoom which ran in a number of phases from 1937 to 1966 ranked fourth and there is clear evidence that the numbers have been gradually decreasing over the last 7 years. Most of the cases were as a direct exposure at the various mines and mills whilst others were involved in transportation or the use of tailings for local road and concrete construction. Environmental cases from non mine employees such as women, children and town visitors are not included in this study having been reported elsewhere (Rogers and Nevill, 1995). Although the infamy of the Wittenoom mine gives the impression that Australia was a major producer of asbestos, more than 95% was imported from Canada and South Africa (Boyle and Rogers, 1993). The stevedoring industry which was responsible for unloading both imported and locally produced crocidolite and chrysotile shows a steady increase of approximately 3–4% in cases.

A group of primary asbestos user industries such as railway loco construction and repair, various engineering and metal fabrication groups and coal fired electric power stations have combined contributed 11% of cases and the number of cases from each industry has increased more than 50% since 1980. All of these industries involved the large scale use of both bulk crocidolite and bulk amosite along with chrysotile products such as rope, cloth and board. A steady rate of approximately 60 cases per year arose from a wide variety of both large and small industries. These involved jobs where the potential contact with amphibole or chrysotile asbestos ranged from negligible to highly likely.

DISCUSSION

Although a steady increase in mesothelioma over all industries and occupations cases is observed, there appear to be at least two separate groups. Cases arising from primary asbestos industries appear to have plateaued and for Wittenoom they are on the decline. This is in line with predictions made using proportional latency

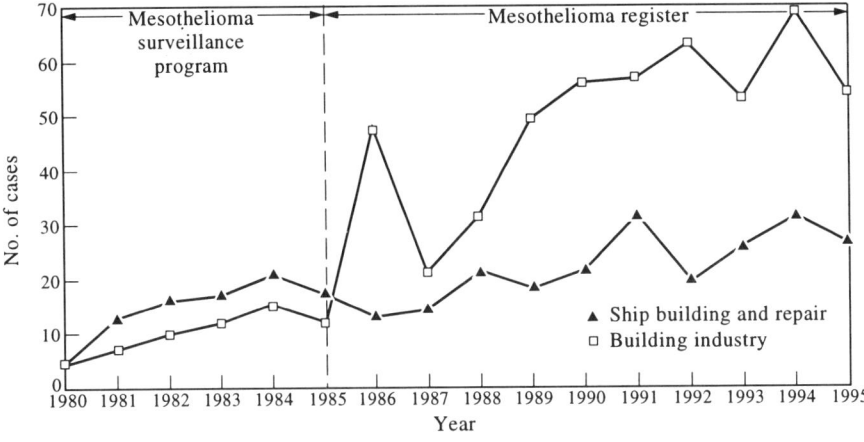

Fig. 2.

calculations made for these industries which ceased contact with amphiboles, particularly crocidolite, some 20–30 years ago (Boyle and Rogers, 1993; Rogers and Nevill, 1995). Data from the census of U.K. asbestos workers employed pre- and post the 1969 Asbestos Regulations, also indicates a decline in mesothelioma numbers for those manufacturing industries such as asbestos cement and asbestos compounding that traditionally used amphibole asbestos (Hutchings *et al.*, 1995).

The above group is in sharp contrast to those industries who were users of the primary asbestos products and in trades where contact was made with *in situ* asbestos products. These industries were not included in early asbestos regulations which were usually specificly designed to prevent disease in asbestos mining and product manufacturing. Of particular concern are the trades associated with the building industry which were not covered by specific legislation in the largest State of NSW until 1983. For post 1985 cases collected in the national Register, there was usually insufficient information in the occupational history to determine if asbestos products were used and if so to what extent exposure occurred. To overcome this difficulty reference was made to the NSW DDB cases where considerable details on asbestos contact had been gathered. Building industry cases shown in Figs 2 and 3 indicate that the ratio of national to NSW cases has risen from approximately 2:1 to 4 or 5:1 since the change in programme to register information. This is higher than the ratio of 3:1 across all industries (Leigh *et al.*, 1996). The differences urge caution in assigning of all cases from the building trade being associated with asbestos exposure in that industry although it had the potential for continued intermittent contact during renovation and repair, introduced the use of power tools which created higher potential for dust exposure and had considerable growth in the workforce particularly during the post war industrial expansion and later immigration programs.

In examining the industry trends against the overall increasing numbers of mesothelioma it is possible to observe how the implementation of good occupational hygiene practice such as the technical prohibition on the use of amphiboles, the control of dust emissions, adequate legislation and enforcement and removal of

A. J. Rogers *et al.*

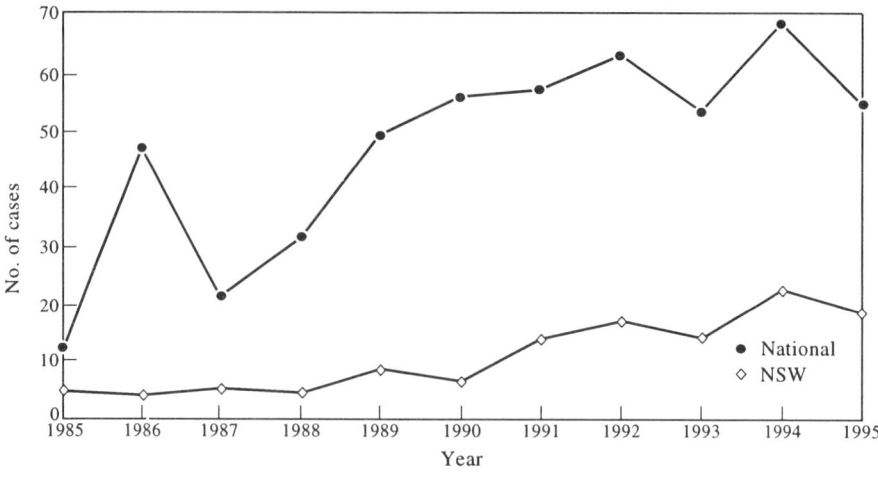

Fig. 3.

asbestos from buildings and ships has lead to a reduced mesothelioma risk in the workforce albeit due to latency effects it has taken some 30–40 years for the results to show. Controls introduced later in the asbestos user industries will not show up in reduced mesothelioma numbers till around 2010.

In addition to data presented in this study, the reporting of cases not associated with significant occupational exposure has increased and this has resulted in conflicting reports as to the significance of intensity of exposure, trace environmental, low level occupational and chrysotile exposure or non-asbestos background mesotheliomas as contributing factors to mesothelioma in Australia.

REFERENCES

Boyle, M. and Rogers, A. (1993) Mesothelioma cases—the experience of the Australian asbestos cement manufacturing industry. *Proceedings of 1993 Annual Conference of the Australian Institute of Occupational Hygienists*, pp 131–135. AIOH PO Box 1205, Tullamarine Vic 3043, Australia.

1995 Report of the NSW Workers' Compensation (Dust Diseases Board) (1995) 82 Elizabeth St., Sydney 2000, Australia.

Ferguson, D. A., Berry, G., Jelihovsky, T., Andreas, S., Rogers, A., Chung Fung, S., Grimwood, A. and Thomson, R. (1987) The Australian mesothelioma surveillance program 1979–1985. *Med. J. Aust.* **147**, 166–172.

Hutchings, S., Jones, J. and Hodgson, J. (1995) *Occupational Health—Decennial Supplement*, Chapter 9, pp. 127–152. Asbestos related diseases, UK Health and Safety Executive, HM Stationery Office, London.

Leigh, J., Corvalan, C., Grimwood, A., Berry, G., Ferguson, D. A. and Thompson, R. (1991) The incidence of malignant mesothelioma in Australia, 1982–1988. *Am. J. ind. Med.* **20**, 643–655.

Leigh, J., Hull, B. and Davidson, P. (1996) The incidence of mesothelioma in Australia 1992–1994 and Australian mesothelioma register report, 1996. National Institute of Occupational Health and Safety, PO Box 58, Sydney, NSW 2001, Australia.

Peto, J., Hodgson, J., Matthews, F. and Jones, J. (1995) Continuing increase in mesothelioma mortality in Britain. *Lancet,* **345**, 535–539.

Rogers, A. and Nevill, M. (1995) Occupational and environmental mesotheliomas due to crocidolite mining activities in Wittenoom, Western Australia. *Scand. J. Work Environ. Health* **21**, 259–264.

 Pergamon

Ann. occup. Hyg., Vol. 41, Supplement 1, pp. 129–133, 1997
© 1997 British Occupational Hygiene Society
Published by Elsevier Science Ltd. All rights reserved
Printed in Great Britain
0003–4878/97 $17.00 + 0.00
Inhaled Particles VIII

PII: S0003–4878(96)00163–9

MESOTHELIOMA MORTALITY IN BRITAIN: PATTERNS BY BIRTH COHORT AND OCCUPATION

J. T. Hodgson,* J. Peto,† J. R. Jones* and F. E. Matthews†

*Epidemiology and Medical Statistics Unit, Health and Safety Executive, Room 241, Magdalen House, Stanley Precinct, Bootle, Merseyside L20 3QZ, U.K.; and †Section of Epidemiology, Institute of Cancer Research, 15 Cotswold Road, Belmont, Surrey SM2 5NG, U.K.

OBJECTIVES

To assess the impact of asbestos exposure on successive cohorts of the British population and on different occupational groups.

METHODS

Data were taken from the HSE's mesothelioma register, which has records of all death certificates in Great Britain since 1968 on which mesothelioma was recorded. Death rates by age and birth cohort were modelled by Poisson regression. Occupational effects were assessed by calculating proportional mortality ratios, based on the occupations recorded on death certificates. The main cohort and occupational analyses reported here are based on data up to 1991. The data for 1992–1994 have recently been released and the final set of figures compares the observed data for these three years with expected values based on projections of the age-cohort model fitted to the date up to 1991.

COHORT ANALYSIS

Male mesothelioma death rates show a clear pattern by age and cohort (see Fig. 1 for clarity, only alternate 5 year cohorts are shown and the youngest observation point for each cohort, which has zero deaths for most cohorts, has also been omitted).

Rates increased steeply with age and with a similar age-specific pattern for each cohort. At all ages for which a direct comparison can be made, rates increase from cohort to cohort up to the 1940s. The multiplicative model:

$$r_{ac} = A_a C_c, \text{ where } r_{ac} \text{ is the rate for age group a in cohort c,}$$
$$\{A_a\} \text{ are age factors common across cohorts and}$$
$$\{C_c\} \text{ are cohort factors common across ages}$$

provides a very close statistical fit to the observed rates. The details of the fitting process have been described elsewhere (Peto *et al.*, 1995). The estimated lifetime risk of mesothelioma implied by this model for each cohort rises over 20-fold from the cohort born around 1900 to that born around 1945, for which the lifetime risk is

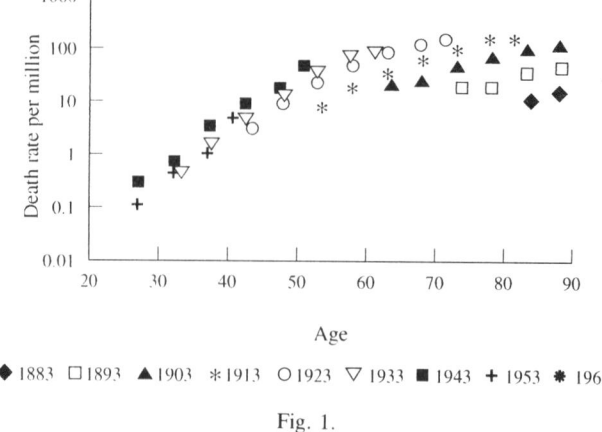

◆ 1883 □ 1893 ▲ 1903 * 1913 ○ 1923 ▽ 1933 ■ 1943 + 1953 * 1963

Fig. 1.

about 1.3%. Rates then fell and the 1950 and 1955 cohorts had risks 30 and 50% lower, respectively, than those for the 1945 cohort. The pattern for females was similar, though the age and cohort slopes were less steep.

OCCUPATIONAL ANALYSIS

The occupations with the highest PMRs are shown in Table 1. All have clearly raised potential for asbestos exposure, and building workers, especially plumbers and gas fitters, carpenters and electricians are the largest high risk group. The groups with risks clearly raised above the average only account for about half the recorded deaths.

In order to display PMRs across the whole range of occupations without the picture being confused by the imprecisely determined PMRs of small job groups, Fig. 2 shows the PMRs for 27 major job groups. These job groups were selected initially by choosing each job group which accounted for more than 1% of total (all cause) male deaths. There were 28 such job groups. Two of these groups (sales representatives and sales managers nec) had very similar PMRs and have been combined into a single category. Two job groups which fell below the 1% of total deaths boundary but which had natural links with job groups above that boundary were combined with their larger partner and treated as a single group: face trained coal miners with other coal miners and teachers in higher education with teachers nec. The final result is a set of 27 job groups which together account for 63% of all deaths.

The PMRs for these 27 groups show a wide spread from 0.25 for farmers up to 4.45 for plumbers and gas fitters, an 18-fold difference. As one moves up the spectrum of risk three broad groupings can be distinguished. Between the lowest two or three jobs and the highest dozen or so jobs the PMRs are very closely bunched between 0.5 and 0.9. Coal miners and farmers have PMRs which fall clearly below this range and motor mechanics (although they have a wide confidence limit) are closer to these two lowest risk groups than to the next highest group. The 11 job groups with PMRs above 0.9 (they are in fact all above 1) show a

Table 1. Proportional mortality ratios (PMRs) of men aged 16–74 from mesothelioma in England and Wales 1979–80, 1982–90

Job group	PMR (all men = 1)	Number	Percent	Cumulative percent
Metal Plate Workers	7	110	2.5	2.5
Vehicle Body Builders	6.19	35	0.8	3.2
Plumbers and Gas Fitters	4.43	201	4.5	7.7
Carpenters	3.66	258	5.8	13.5
Electricians	2.91	161	3.6	17.1
Upholsterers	2.83	19	0.4	17.5
Construction Workers	2.56	187	4.2	21.7
Boiler Operators	2.54	39	0.9	22.5
Electrical Plant Operators	2.54	18	0.4	22.9
Chemical Engineers and Scientists	2.48	18	0.4	23.3
Sheet Metal Workers	2.33	48	1.1	24.4
Scaffolders	2.26	11	0.2	24.7
Production Fitters	2.16	304	6.8	31.4
Professional Engineers nec	2.11	105	2.3	33.8
Plasterers	2.03	27	0.6	34.4
Welders	2.03	70	1.6	35.9
Managers in Construction	1.97	40	0.9	36.8
Dockers and Goods Porters	1.95	69	1.5	38.4
Electrical Engineers (so described)	1.87	39	0.9	39.2
Technicians nec	1.72	24	0.5	39.8
Builders and Handymen	1.64	98	2.2	42
Laboratory Technicians	1.64	27	0.6	42.6
Draughtsmen	1.61	28	0.6	43.2
Machine Tool Operators	1.33	179	4	47.2
Painters and Decorators nec	1.31	100	2.2	49.4

steady and quite steep gradient of increasing risk with no obvious break points. All these jobs have a clear potential for above-average asbestos exposure.

There are no indications in the PMR values of the central group of jobs for making any sub-grouping; however, with two exceptions (production and maintenance managers and lorry drivers) the manual occupations within this group lie at

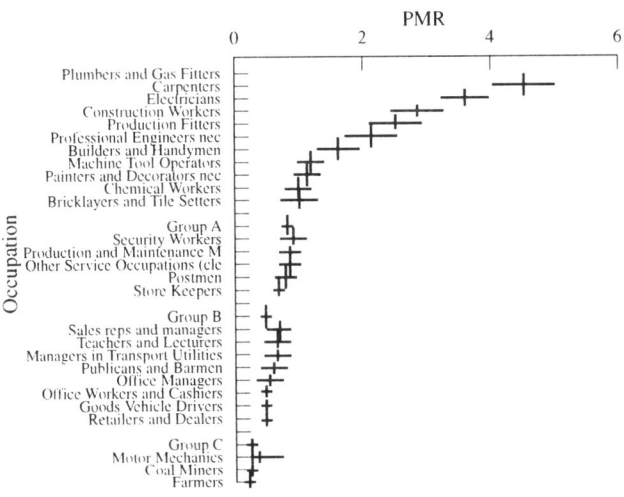

Fig. 2.

Table 2. Observed male mesothelioma deaths up to 1994 by age group and birth cohort, and (in brackets) difference from numbers predicted on the basis of projections based on data up to 1991

Cohort: men born in period	Age group						Total observed– projected difference for cohort
	30–34	35–39	40–44	45–49	50–54	55–59	
July 1938–June 1943	5	18	63	178 (+ 16)	282 (+ 24)	21 (− 3)	37
July 1943–June 1948	7	33	90 (+ 10)	135 (− 13)	20 (+ 5)	—	2
July 1948–June 1953	4	23 (+ 2.4)	39 (− 3.6)	1 (− 4.9)	—	—	− 6.1
July 1953–June 1958	4 (+ 0.9)	7 (− 1.8)	2 (+ 0.7)	—	—	—	− 0.2

the higher risk end and the non-manual occupations at the lower risk end. If the high and low ends of this middle group are aggregated in this way the average PMRs for the two sub-groups are significantly different: 0.8 (95% CI, 0.72, 0.89) for the higher (mainly manual) group; and 0.57 (95% CI, 0.52, 0.62) for the lower (mainly non-manual) group. If the lowest three job groups are combined, their PMR is significantly lower again, 0.28 (95% CI, 0.23, 0.35). These groups are shown on Fig. 2 as Groups A, B and C (labelled in order of decreasing risk).

Of course, the boundaries between these groups are to some extent arbitrary. However, this grouping does capture the main features of the data in terms of the distribution of PMRs and (apart from group C) shows a good degree of coherence in the kinds of occupation within the groups. It is worth noting that none of the 167 smaller job groups in the full occupational classification (data not shown here) has a PMR significantly lower than that of group C, so this group can reasonably be taken to define the lowest risk group for mesothelioma (in the period of observation). In relation to this group, group B has about twice the mesothelioma risk and group A about three times the risk.

The relative risks between these groups have been fairly stable over time and overall death rates for the three higher risk groups have increased about 6-fold between 1968–73 and 1987–91. The low risk group's rate has increased 3-fold over the same period.

COHORT PATTERNS IN THE LATEST DATA

The results described above are all based on data up to 1991. After a delay while the national mortality statistics system was being computerised, the mesothelioma data for the three years 1992–1994 have now been released. It is of interest to examine whether the new data are consistent with the data up to 1991 and in particular whether the age-specific rates for the more recently born cohorts (for whom asbestos exposure has almost certainly fallen, as controls have been more strictly applied, and usage has fallen) continue to rise in line with the pattern seen in earlier cohorts. The comparisons are shown in Table 2.

These data show that the observed numbers in the most recent 3 years have been slightly higher than projected for the earliest of the four cohorts shown, close to the projected values for the 1943–48 cohort and below the projected values for the two later cohorts.

CONCLUSIONS

Asbestos exposure sufficient to cause mesothelioma has been widespread in the male population. Building workers have been at high risk and should continue to be alert to the dangers of working on asbestos materials.

REFERENCE

Peto, J., Hodgson, J. T., Matthews, F. E. and Jones, J. R. (1995) Continuing increase in mesothelioma mortality in Britain. *Lancet* **345**, 535–39.

 Pergamon

Ann. occup. Hyg., Vol. 41, Supplement 1, pp. 134–136, 1997
© 1997 British Occupational Hygiene Society
Published by Elsevier Science Ltd. All rights reserved
Printed in Great Britain
0003–4878/97 $17.00 + 0.00
Inhaled Particles VIII

PII: S0003–4878(96)00108–1

CROCIDOLITE, RADIOGRAPHIC ASBESTOSIS AND SUBSEQUENT LUNG CANCER

N. H. de Klerk,* A. W. Musk,† J. J. Glancy,‡ S. C. Pang,§ H. G. Lund,*
N. Olsen† and M. S. T. Hobbs*

*Department of Public Health, University of Western Australia, Western Australia 6970;
Departments of †Respiratory Medicine, ‡Radiology, Sir Charles Gairdner Hospital, Nedlands,
Australia; and §Health Department of Western Australia.

INTRODUCTION

Historically, excess lung cancer rates were first noted in subjects with asbestosis. This raised the question of whether asbestosis was a necessary step in the causal path from asbestos exposure to lung cancer. Answering this question is fundamental to an understanding of the way in which asbestos acts to produce lung cancer with substantial significance for the shape of the asbestos–lung cancer exposure–response relationship at low doses and therefore the setting of exposure limits. On the medico-legal side, awards of compensation for lung cancer in persons exposed to asbestos have been decided on whether or not they have asbestosis. Previous evidence regarding this question has been discussed at length elsewhere (Henderson *et al.*, 1996).

The aims of this study were to estimate the exposure–response relationships between crocidolite and lung cancer in subjects both with and without radiographic evidence of parenchymal fibrosis.

SUBJECTS

All known cases of lung cancer occurring up to December 1994, in the cohort of 6910 former Wittenoom crocidolite workers and from 1 to 61 controls per case were matched on year of birth. Controls were alive, cancer free and had ceased exposure at Wittenoom at the time of the case diagnosis. Case diagnosis was based on cancer registry and death certificate diagnoses throughout Australia. All this information was obtained from the ongoing cohort study of former Wittenoom crocidolite workers (Armstrong *et al.*, 1988).

METHODS

X-rays

Radiographs were selected as close as possible to the date of diagnosis of the case (up to 5 years before and 3 months after) and scored according to the ILO Classification of Radiographs for the Pneumoconioses (ILO, 1980). For 390 subjects they were scored by three independent readers and the median reading

Table 1. Smoking categories—number (%); asbestos exposure—geometric mean (range); and X-ray
severity—number (%) in major ILO categories of profusion of small opacities

	Cases ($n = 55$)	Controls ($n = 841$)
Smoking category		
Current smokers	38 (69)	346 (41)
Ex-smokers	15 (27)	337 (40)
Never smoked	2 (4)	158 (19)
Asbestos exposure		
Days worked	385 (5–4712)	162 (1–4628)
Cumulative fibres ml^{-1} years^{-1}	17.1 (0.1–736)	8.1 (0.1–431)
Average fibres ml^{-1}	16.2 (1–110)	17.4 (1–130)
Major ILO category		
0	29 (53)	653 (78)
1	15 (27)	161 (19)
2	6 (11)	22 (3)
3	5 (9)	5 (1)

used in the analyses. The remainder were scored by two of these three readers and the reading from the "best" reader, as judged by agreement with the median reading for the 390 X-rays that had been read by all three, was used in analyses.

Exposure

Duration and intensity of exposure at Wittenoom were obtained from employment records and a survey of airborne respirable fibres > 5 microns in length (Armstrong *et al.*, 1998). These estimates have recently been validated by reference to quantitative lung fibre analysis (de Klerk *et al.*, 1996). Smoking histories were obtained from a 1979 questionnaire or from hospital case notes (de Klerk *et al.*, 1991).

Statistical analysis

Conditional logistic regression analyses modelled the relative risk of lung cancer incidence: covariates considered were duration and intensity of exposure, smoking category, weeks from X-ray and level of profusion of small opacities (or any profusion $\geq 1/0$, or any profusion $\geq 1/1$).

RESULTS

X-rays were found from periods less than 5 years before the onset of disease for 55 cases and 841 controls. There were 88 cases from the cohort who had no X-rays located and a further 53 for whom the only X-rays available were taken more than 5 years before diagnosis (the majority of these were X-rays taken at the start of employment). The prevalence of profusion of small opacities $\geq 1/0$ on the ILO scale was greater among cases than among selected control films (Table 1).

Smoking was the strongest predictor of lung cancer and current smokers had the highest risk. After adjustment for smoking, both the presence of asbestosis on the X-ray and the level of cumulative exposure to crocidolite significantly increased the risk of lung cancer (Table 2). The interaction between fibrosis and exposure was

Table 2. Lung cancer incidence, exposure to asbestos and smoking

	RR	95% CI
Smoking habit		
Non-smoker	1.0	—
Ex-smoker (> 1 y)	3.6	1.7, 7.4
Current	9.8	2.2, 42.4
Asbestos exposure		
Log (fibres ml^{-1} years^{-1})	1.25	1.06, 1.48
Asbestos on X-ray		
(1/0 or worse)	2.54	1.39, 4.64

not significant ($P = 0.36$), but there was some evidence that, although the risk was higher, the exposure–response slope was less steep in the presence of fibrosis.

CONCLUSIONS

This study has demonstrated significant exposure–response relationships between the amount of exposure to asbestos and incidence of lung cancer in all subjects as well as in subjects with no radiographic evidence of asbestosis. The level of radiographic fibrosis conferred additional risk beyond that associated with level of exposure. At least in this cohort, radiographic asbestosis does not appear to be a pre-requisite of asbestos-associated lung cancer.

Acknowledgements—This study was funded by the Australian National Health and Medical Research Council. The authors are very grateful to Jan Eccles, Jan Sleith, the Staff of the Perth Chest Clinic and Naomi Hammond for invaluable assistance with retrieval of X-rays, entering of the X-ray data onto computer and general clerical assistance.

REFERENCES

Armstrong, B. K., Klerk, N. H. de, Musk, A. W. and Hobbs, M. S. T. (1988) Mortality in miners and millers of crocidolite in Western Australia. *Br. J. Ind. Med.* **45**, 5–13.

Henderson, D. W., de Klerk, N. H., Hammar, S. P., Hillerdal, G., Huuskonen, M. S., Leigh, J., Pott, F., Roggli, V., Shilkin, K. B., Tossavainen, A. (1997) Asbestos and lung cancer: an old controversy revisited. In *Tumours and Tumour-like Disorders of the Lungs* (Edited by B. Corrin). Churchill Livingstone, London.

ILO (1980) Guidelines for the use of ILO international classification of radiographs of pneumoconioses. International Labour Office Occupational Safety and Health Series, No. 22, Geneva.

Klerk, N. H. de, Armstrong, B. K., Musk, A. W. and Hobbs, M. S. T. (1991) Smoking, exposure to crocidolite, and the incidence of lung cancer and asbestosis. *Br. J. ind. Med.* **48**, 412–417.

Klerk, N. H. de, Musk, A. W., Williams, V. M., Filion, P. R., Whitaker, D. and Shilkin, K. B. (1996) Comparison of measures of exposure to asbestos in former crocidolite workers from Wittenoom Gorge, W. Australia. *Am. J. ind. Med.* **30**, 579–587

 Pergamon

Ann. occup. Hyg., Vol. 41, Supplement 1, pp. 137–141, 1997
© 1997 Published by Elsevier Science Ltd on behalf of BOHS
Printed in Great Britain. All rights reserved
0003–4878/97 $17.00 + 0.00
Inhaled Particles VIII

PII: S0003–4878(96)00109–3

AN EXPOSURE–RESPONSE ANALYSIS OF RESPIRATORY DISEASE RISK ASSOCIATED WITH OCCUPATIONAL EXPOSURE TO CHRYSOTILE ASBESTOS

L. Stayner,* R. Smith,* A. J. Bailer,*† S. Gilbert,* K. Steenland,* J. Dement,‡
D. Brown§ and R. Lemen*¶

*NIOSH, 4676 Columbia Parkway, Robert Tafts Labs, Cincinnati, OH 45226, U.S.A.;
†Miami University, Oxford, ‡Duke University, Durham, §NIEHS, RTP and ¶3495 Highgate Hills
Drive, Deluth, GA 30136, U.S.A.

INTRODUCTION

There has been considerable debate in the scientific literature concerning the significance of the risks associated with exposure to chrysotile asbestos (Mossman *et al.*, 1990; Stayner *et al.*, 1996). This paper presents the findings from exposure–response and risk analyses of lung cancer and asbestosis mortality based on a cohort mortality study of U.S. textile workers exposed to chrysotile asbestos.

MATERIAL AND METHODS

Data were used from a recent update of a cohort mortality study of workers exposed to chrysotile asbestos in a South Carolina textile factory (Dement *et al.*, 1994). The analysis was restricted to include workers employed in the textile production operations for at least 1 month between 1 January, 1940 and 31 December, 1975. Follow-up of this cohort for vital status was until 31 December, 1990. Chrysotile exposure levels by areas of the plant, specific jobs and calendar years have been previously estimated and were used with work history information to estimate individual cumulative exposures for this analysis.

Exposure–response analyses were conducted for cancers of the trachea, bronchus and lung ("lung cancer") and for asbestosis and pneumoconiosis ("asbestosis"). The underlying cause of death was used for lung cancer (ICD9 = 162) and a multiple cause of death approach (Steenland *et al.*, 1992) was used for "asbestosis" (ICD9 = 501 and 505). Based on these definitions, there was a total of 126 lung cancer and 45 cases of asbestosis available for analysis.

Poisson regression methods were used to analyze the exposure–response relationship between chrysotile asbestos exposure and respiratory disease mortality. For lung cancer, the person–years and observed deaths were restricted to include only those with at least 15 years of time since the date of first exposure.

A wide variety of parametric models were evaluated including additive, log–linear, log–quadratic, additive relative rate and power function models (Stayner *et al.*, 1995). The fit of these models was contrasted by comparing their deviances and by graphically comparing them with a categorical model, and a restricted cubic

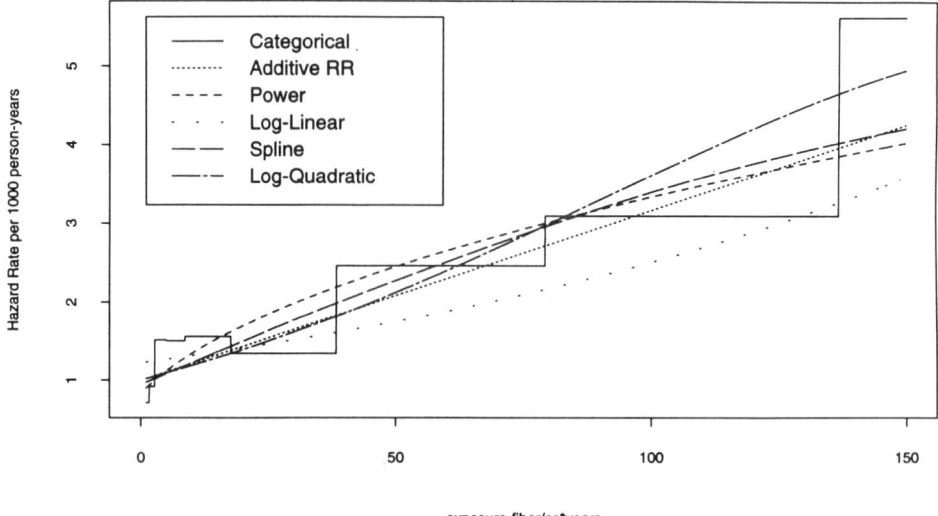

Fig. 1. Lung cancer mortality rates as a function of cumulative asbestos exposure predicted by alternative models for white males, age 50 in 1940–1969.

spline model (Herndon and Harrell, 1995). Finally, a "threshold" model (Ulm, 1990) was evaluated to test whether the fit of the model was improved by including a threshold parameter.

Estimates of excess lifetime risk of dying from lung cancer and asbestosis were developed for varying levels of chrysotile asbestos exposure based upon an actuarial method, which accounts for the influence of competing risks (BEIR IV, 1988).

RESULTS

The results from fitting alternative Poisson regression models to the lung cancer rates are illustrated in Fig. 1. Exposure was a highly significant predictor ($P < 0.001$) of lung cancer mortality in all of the models evaluated. The additive relative rate (ARR) model gave the best fit to the data and provided similar estimates of the rate as the spline model and the categorical model.

A significant interaction was found between cumulative exposure and time since first exposure ($P = 0.04$), and an additive relative rate model with separate slopes for cumulative exposures with 15 to < 30, 30 to < 40 and $\geq$ 40 years of latency was chosen as the final model for estimating lifetime risks.

The results from fitting alternative Poisson regression models for asbestosis are illustrated graphically in Fig. 2. The exposure–response relationship was found to be highly statistically significant ($P < 0.001$) in all of the models evaluated. The power model was found to provide the best fit to the data of all of the parametric models, and produced similar estimates of the hazard rate as the categorical mode and somewhat lower estimates than the spline model. The fit of the models for lung cancer and asbestosis were not improved by the inclusion of a threshold parameter, and thus there was no evidence for a "threshold" type response for these outcomes.

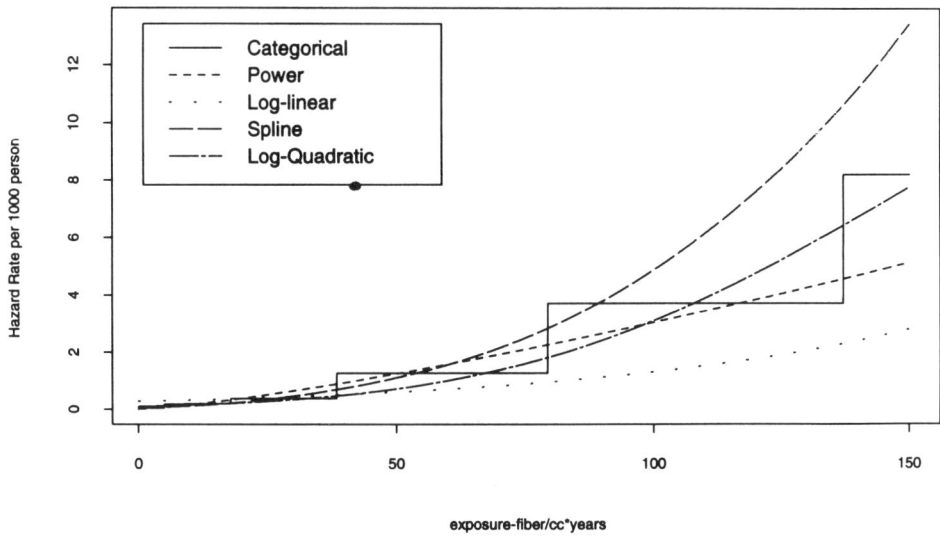

Fig. 2. Asbestosis mortality rates as a function of cumulative asbestos exposure predicted by alternative models for white males, age 50 in 1940–1969.

Predicted lifetime excess risks of lung cancer and asbestosis assuming 45 years of exposure to varying chrysotile asbestos exposure levels are presented in Fig. 3 for lung cancer and in Fig. 4 for asbestosis. The risks vary by gender and race because of differences in the background rates used in the models. The predicted risks for asbestosis are less than those for lung cancer at low exposure levels (e.g. < 0.5). At

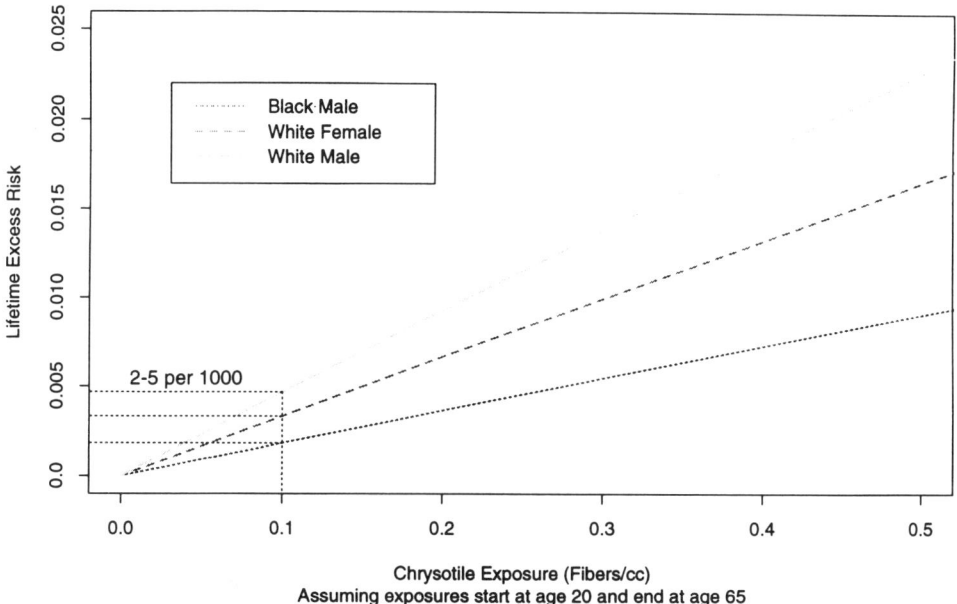

Fig. 3. Lifetime excess risk for lung cancer assuming 45 years of exposure to varying concentrations of chrysotile asbestos.

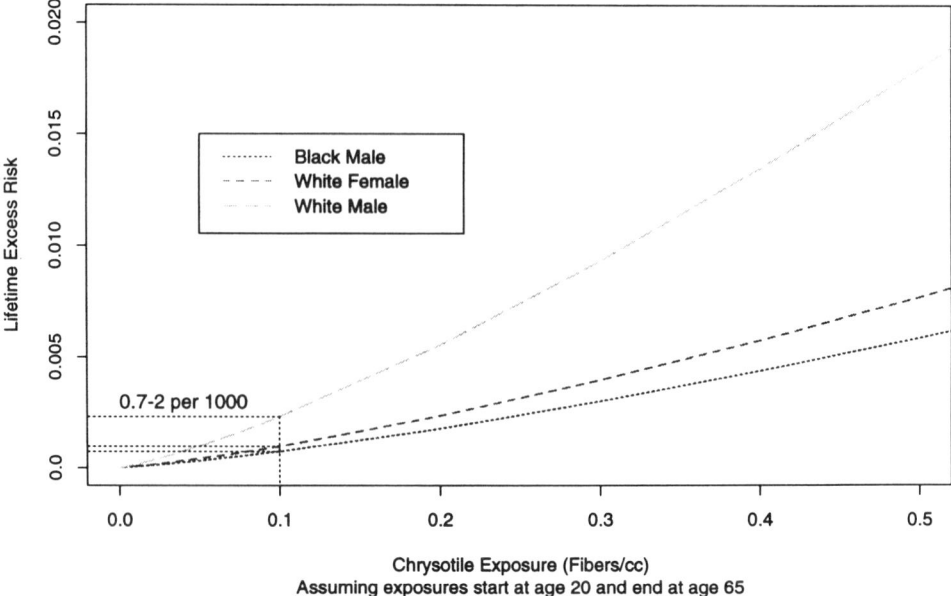

Fig. 4. Lifetime excess risk for asbestosis assuming 45 years of exposure to varying concentrations of chrysotile asbestos.

higher exposure levels this pattern is reversed with the predicted risks for asbestosis being higher than those for lung cancer.

DISCUSSION

The results from these analyses clearly demonstrate a strong exposure–response relationship between chrysotile exposure and mortality from asbestosis and lung cancer, which is not surprising given the results from previous studies. The exposure–response relationship for lung cancer appeared to be linear on a multiplicative scale, whereas the exposure–response relationship for asbestosis appeared to be sub-linear.

There was no statistical evidence for a threshold in either the lung cancer or asbestosis model. Thus the results from this analysis fail to provide any support for arguments that have been made for a threshold for the effects of chrysotile asbestos on lung cancer and asbestosis risks (Browne, 1986). Based on this analysis, the predictions of excess lifetime risk for white males exposed for 45 years at the recently revised OSHA standard of 0.1 fibers cc^{-1} was predicted to be approximately 5 per 1000 for lung cancer and 2 per 1000 for asbestosis. The lung cancer risks estimated in this analysis were substantially higher than what was observed in previous analyses of Quebec chrysotile miners and millers (McDonald *et al.*, 1980, 1994). The reasons for these widely varying results are not known.

REFERENCES

Biological Effects of Ionizing Radiation (BEIR) IV (1988). Health risks of radon and other internally deposited alpha-emitters. Committee on the Biological Effects of Ionizing Radiation, Board of Radiation Effects Research, Commission on Life Sciences, National Research Council. National Academy Press, Washington, D.C.

Browne, K. (1986) A threshold for asbestos related lung cancer. *Br. J. ind. Med.* **43**, 556–558.

Dement, J. M., Brown, D. P. and Okun, A. (1995) Follow-up study of chrysotile asbestos textile workers: cohort mortality and case-control analyses. *Am. J. ind. Med.* **26**, 431–447.

Herndon, J. E., and Harrell, F. E. (1995) The restricted cubic spline as baseline hazard in the proportional hazards model with step function time-dependent covariables. *Stat. Med.* **14**(19), 2119–29.

McDonald, J. C., Liddell, F. D. K., Dufresne, A. and McDonald, A. D. (1993) The 1891–1920 birth cohort of Quebec chrysotile miners and millers: mortality 1976–88. *Br. J. ind. Med.* **50**, 1072–1081.

McDonald, J. C., Liddell, F. D. K., Gibbs, G. W., Eyssen, G. E. and McDonald, A. D. (1980) Dust exposure and mortality in chrysotile mining, 1910–75. *Br. J. ind. Med.* **37**, 11–24.

Mossman, B. T., Bigman, J., Corn, M., Seaton, A. and Gee, J. B. L. (1990) Asbestos: scientific developments and implications for public policy. *Science* **24**, 294–301.

Stayner, L. T., Dankovic, D. A. and Lemen, R. A. (1996) Occupational exposure to chrysotile asbestos and cancer risk: a review of the amphibole hypothesis. *AJPH* **86**(2), 179–186.

Stayner, L., Smith, R., Bailer, J., Luebeck, E. G. and Moolgavkar, S. H. (1995) Modeling epidemiologic studies of occupational cohorts for the quantitative assessment of carcinogenic hazards. *Am. J. ind. Med.* **27**, 155–70.

Steenland, K., Nowlin, S., Ryan, B. and Adams, S. (1992) Use of multiple-causes mortality data in epidemilogic analyses. *Am. J. Epidemiol.* **136**, 855–862.

Ulm, K. W. (1990). Threshold models in occupational epidemiology. *Math. comput. Modeling* **14**, 649–52.

 Pergamon

Ann. occup. Hyg., Vol. 41, Supplement 1, pp. 142–147, 1997
© 1997 British Occupational Hygiene Society
Published by Elsevier Science Ltd. All rights reserved
Printed in Great Britain
0003–4878/97 $17.00 + 0.00
Inhaled Particles VIII

PII: S0003–4878(96)00110–X

DEPOSITION OF FIBROUS AEROSOL IN A MODEL OF A HUMAN LUNG BIFURCATION UNDER CYCLIC FLOW CONDITIONS

T. Myojo

National Institute of Industrial Health, 21-1, Nagao 6-chome, Tama-ku, Kawasaki, 214, Japan

INTRODUCTION

Various lung diseases associated with exposure to fibrous materials have been pointed out by many authors. Pott (1978) proposed the well-known hypothesis that carcinogenic potency depends upon the length, diameter and aspect ratio of inhaled fibres. However, only a small number of experiments have been conducted on the deposition process of fibrous aerosols in airways (Timbrell, 1972; Petra, 1979, Kahn, 1982; Sussman *et al.*, 1991) because of difficulties in the generation and measurement of fibrous aerosols compared to compact aerosols.

Myojo (1987, 1990) observed deposited fractions of fibrous aerosols and presented profiles of deposition velocity on a simplified model of bifurcating tubes. The dimensions of the model bifurcation were based on the third and fourth generation of Weibel's lung model A (Weibel, 1963) and a steady inspiratory flow of glass fibre aerosol was employed. Deposited fraction and deposition velocity for a fibre length range were obtained by the direct observation of fibres deposited on the model bifurcation.

The current paper presents additional data obtained using a unique breathing simulator (Myojo, 1989) to generate cyclic flow in the model bifurcating tubes. The relationships between fibre diameter, fibre length and deposited fractions were determined for cyclic and steady flow conditions.

EXPERIMENTAL METHODS

Figure 1 shows the general arrangement of the fibrous aerosol generator, the model bifurcation and the flow control units. It is essentially the same as that used in the previous work (Myojo, 1987, 1990) except for cyclic flow generation. The fibrous aerosols generated from the fluidized bed were led to an aerosol neutralizer 11 of [85]Kr and then inlet 12. At inhalation, the aerosols passed through the model bifurcation 1 which were made of brass and corresponded approximately to the bifurcation between the third and fourth generations of Weibel's lung model A. The fibres which did not deposit on the bifurcation were collected on nucleopore filters 3 (pore size 0.4 μm). At exhalation, clean air was supplied from slit/cam valve 7 to mixer 2 then to bifurcating tube 1 and nucleopore filters 3.

The operating principle of the breathing simulator is shown in Fig. 2. Air

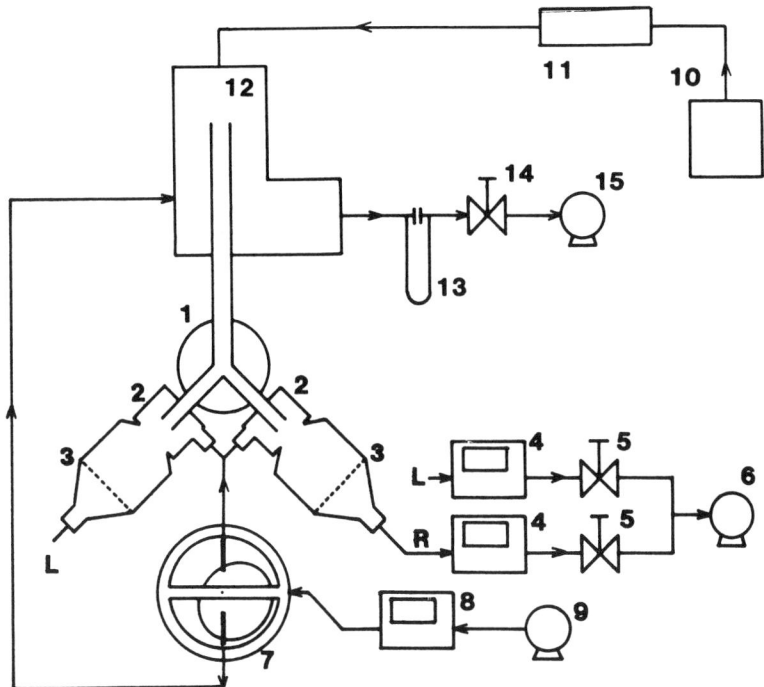

Fig. 1. Experimental apparatus. (1) Bifurcating tubes, (2) mixer, (3) filter holder, (4) massflowmeter, (5) valve, (6) vacuum pump, (7) slit/cam valve, (8) massflow controller, (9) compressor, (10) fibrous aerosol generator, (11) neutralizer, (12) aerosol inlet, (13) orifice flowmeter, (14) valve, (15) vacuum pump.

pumped from the compressor passes through the valve and into the slit/cam valve. The slit/cam valve opens and closes continuously to generate a sinusoidal flow pattern. Therefore airflow at the mixer connected to the slit/cam valve outlet has same pattern; Q_3. The vacuum pump draws air at a constant flow rate; $Q_1 + Q_2$. Cyclic flow is generated in the bifurcating tube, with air expired from the bifurcating tube ($Q_3 > Q_1 + Q_2$) and with air inspired through the tube toward the nucleopore filters ($Q_3 < Q_1 + Q_2$). The main advantage of this system is that whole fibers passing through small bifurcating tubes are captured at the nucleopore filter, making it very similar to steady flow conditions.

At steady flow conditions, the flow rate in each daughter tube was 15.6 and 31.2 cm^3 s^{-1} corresponding to a breath of 500 and 1000 cm^3, respectively, inspired at a constant rate for 2 s. A maximum flow rate of 24.5 cm^3 s^{-1} in the daughter tubes was selected to obtain the same average flow rate of 15.6 cm^3 s^{-1}. The instantaneous flow rate curves were monitored with a pneumotachograph connected to a pressure transducer and a chart recorder. Air volume of one breath inspired and expired from bifurcating tubes were measured by a glass cylinder with soap bubble.

The wall surface of the daughter tube from carina to 4 mm in an axial direction and from carina to 90° in a circumferential direction was directly observed under the scanning electron microscope (Hitachi HS-2B). The fibres were divided into three length groups, i.e. 10–20 μm, 20–40 μm and 40–80 μm. More than 350 fibres collected on the nucleopore filter were also measured and the number concentra-

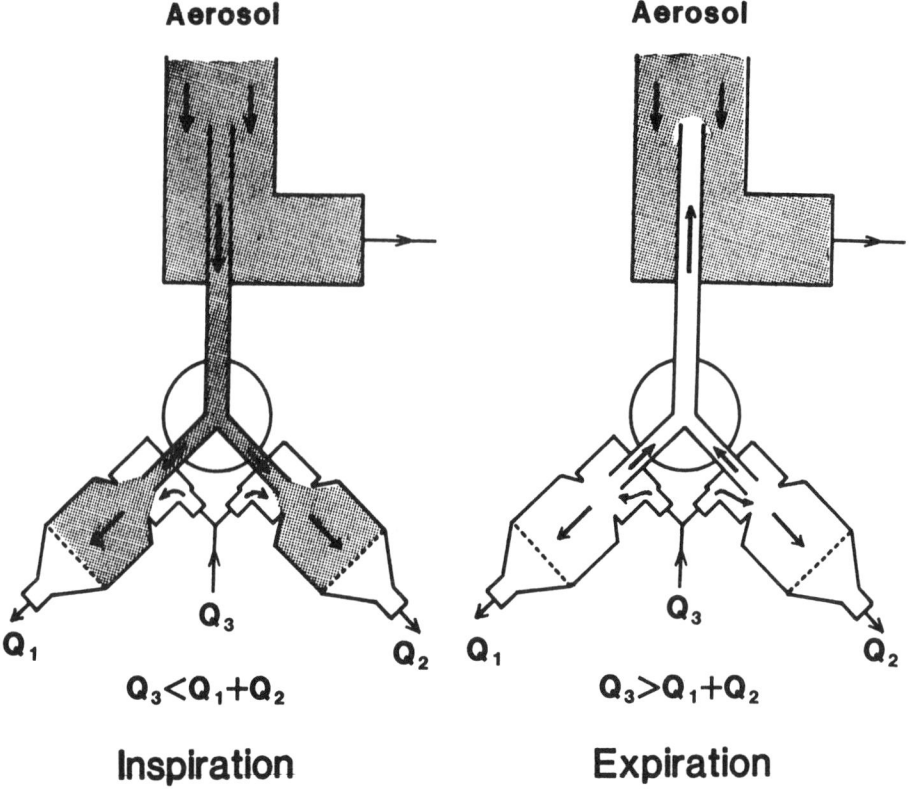

Fig. 2. Generation of cyclic flow in model bifurcation. Q_1, Q_2: flow rate through the filter (constant flow). Q_3: flow rate from slit/cam valve (pusatile flow).

tions of the aerosols were determined. From these data, the deposited fraction $Fd(L1,L2)$ of the bifurcation was calculated for the range of fibre with length from $L1$ to $L2$. Both length and diameter of each fibre were also measured to determine the values of means, variances and correlation from the micrographs of the fibre.

Bivariate lognormal distribution of fibrous aerosol has been used to characterize the length and the diameter of asbestos and man-made mineral fibre (Timbrell, 1982; Schneider and Holst, 1983; Cheng, 1986). The deposition fraction for the length range from $L1$ to $L2$ and for the diameter range from D to $D + \Delta d$, $Fd(D,L1,L2)$ is given by

$$Fd(D,L1,L2) = Fd(L1,L2) \cdot K \qquad (1)$$

where $Fd(L1,L2)$ is the experimentally obtained deposition fraction by the method above-mentioned and K is defined as

$$K = \frac{F1(D + \Delta D,L1,L2) - F1(D,L1,L2)}{F2(D + \Delta D,L1,L2) - F2(D,L1,L2)} . \qquad (2)$$

Here, $F1$ and $F2$ are the cumulative undersize distribution of D for the length range between $L1$ and $L2$. $F1$ and $F2$ are the distributions F for deposited fibres and inlet

Table 1. Deposited fractions under cyclic and steady flow conditions

Run no.	Fiber length Flow pattern	10–20 μm			20–40 μm			40–80 μm		
		Fd	K	Fd(1–2 μm)	Fd	K	Fd(1–2 μm)	Fd	K	Fd(1–2 μm)
1	steady	4.9×10^{-4}	1.24	6.1×10^{-4}	1.6×10^{-3}	0.96	1.5×10^{-3}	4.0×10^{-3}	0.81	3.3×10^{-3}
4	cyclic (15 breath min^{-1})	4.6×10^{-4}	1.31	6.0×10^{-4}	1.9×10^{-3}	1.08	2.1×10^{-3}	5.9×10^{-3}	0.94	5.5×10^{-3}
5	cyclic (30 breath min^{-1})	6.6×10^{-4}	1.26	8.3×10^{-4}	2.6×10^{-3}	0.86	2.2×10^{-3}	6.7×10^{-3}	0.69	4.6×10^{-3}
2	steady	1.2×10^{-3}	1.36	1.6×10^{-3}	5.7×10^{-3}	1.00	5.7×10^{-3}	1.1×10^{-2}	0.80	8.8×10^{-3}

Fd = total deposited fraction.
K = factor defined by equation (2) ranging from 1 to 2 μm.
Fd(1–2 μm) = deposited fraction averaging from 1 to 2 μm (total × factor).

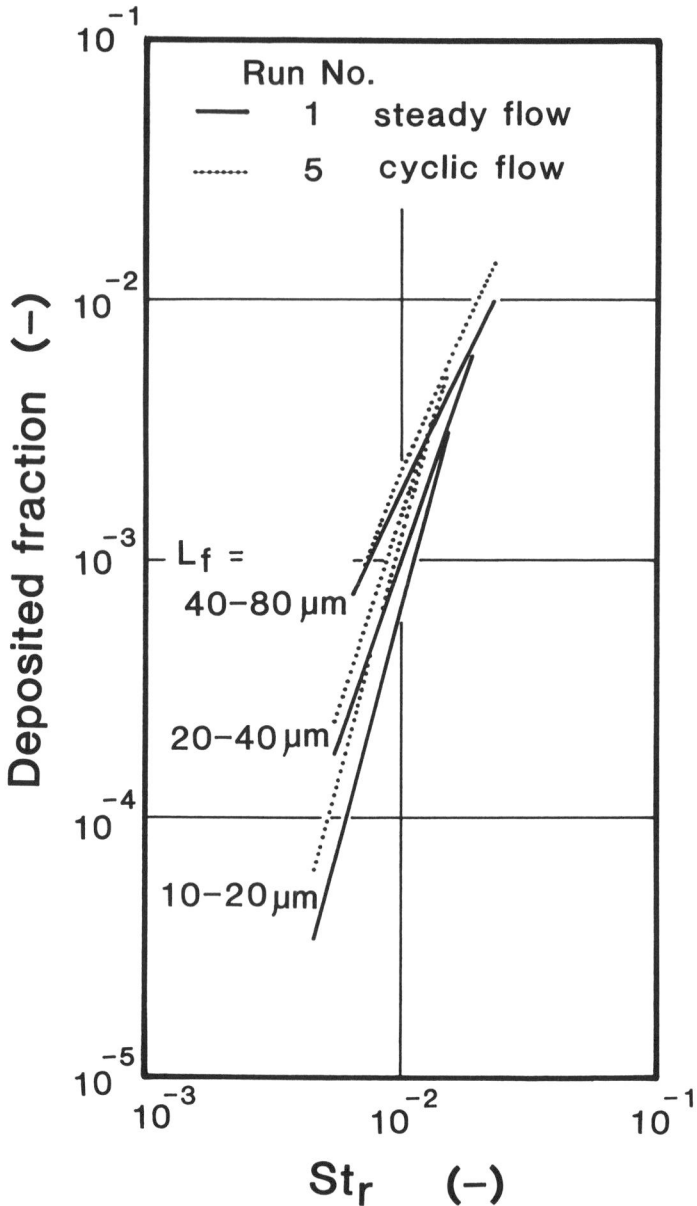

Fig. 3. Relationship between Stokes No. St_r for randomly-oriented fibres and deposited fraction at steady and cyclic flow conditions.

fibres, respectively. We applied the bivariate lognormal distribution to obtain the function F. The details of these calculations were published in previous paper (Myojo, 1990).

RESULTS AND DISCUSSION

Cyclic flow patterns measured by pneumotachographs agreed well with sinusoidal curves. Ideal volume per breath of this system is 62.5 cm^3 with a 4 s cycle and 31.3 cm^3 with a 2 s cycle. However, the measured volume of inspiration is 64 cm^3 with 4 s and 31 cm^3 with 2 s. Also, volume of expiration is 59 cm^3 with 4 s and 29 cm^3 with s 2. Incomplete zero adjustment and fluctuation of flow rate caused small volume differences between inspiration and expiration.

The deposited fraction Fd for each fibre length, the factor K defined by equation (2) and the deposited fraction averaged over 1.0–2.0 μm in fibre diameter are shown in Table 1. The average flow rates for run No.1, No.4 and No.5 are identical, but the values of Fd and Fd(1–2 μm) increase as the frequency of cyclic flow and the fibre length increase. Consequently, the deposited fractions at 30 breaths min^{-1} were 1.4–1.7 times greater than those from the steady flow with the same average flow rate. Figure 3 shows the relationship between Stokes number and the deposited fraction. In the figure, Stokes number St for randomly-oriented fibres by Harris and Fraser (1976) is chosen as the abscissa. The lines are drawn from 1.0 to 2.0 μm in diameter as the valid diameter range.

Figure 3 shows that the deposited fraction increases steeply with St. The deposited fractions at cyclic flow are higher than those at steady flow. In particular, deposited fractions at large Stokes number are influenced by cyclic flow. The deposition of aerosol particles in the lung during pulsatile flow was higher than the deposition during steady flow conditions (Bell, 1978; Gurman *et al.*, 1984; Kim and Garcia, 1991). Our system, which can generate complete cyclic flows, can also be applied for the measurement of ordinary aerosol deposition.

REFERENCES

Bell, K. A. (1978) *Recent Developments in Aerosol Science*, pp. 97–134. Wiley, New York.
Cheng, Y. S. (1986) *Aerosol Sci. Tech.* **5**, 359–368.
Gurman, J. L., Lippmann, M. and Schlesinger, R. B. (1984) *Aerosol Sci. Tech.* **3**, 245–252.
Harris, R. L. and Fraser, D. A. (1976) *Am. Ind. Hyg. Assoc. J.* **37**, 73–89.
Kahn, R. A. (1982) Ph.D. thesis, University of Pittsburgh.
Kim, C. S. and Garcia, L. (1991) *Aerosol Sci. Tech.* **14**, 302–315.
Myojo, T. (1987) *J. Aerosol Sci.* **18**, 337–347.
Myojo, T. (1989) *Am. ind. Hyg. Assoc. J.* **50**, 240–244.
Myojo, T. (1990) *J. Aerosol Sci.* **21**, 651–659.
Petra, A. L. (1979) Ph.D. thesis, North Carolina State University.
Pott, F. (1978) *Staub Reinhaltung der Luft* **38**, 486–490.
Schneider, T. and Holst, E. (1983) *J. Aerosol Sci.* **14**, 139–146.
Sussman, R. G., Cohen, B. S. and Lippmann, M. (1991) *Inhal. Toxicol.* **3**, 145–160.
Timbrell, V. (1982) *Ann. occup. Hyg.* **26**, 347–369.
Timbrell, V. (1972) *Assessment of Airborne Particles* (Edited by T. T. Mercer, P. E. Morrow and W. Stober), pp. 429–445. Springfield, Illinois.
Weibel, E. R. (1963) *Morphometry of the Human Lung*. Springer, Berlin.

 Pergamon

Ann. occup. Hyg., Vol. 41, Supplement 1, pp. 148–153, 1997
© 1997 British Occupational Hygiene Society
Published by Elsevier Science Ltd. All rights reserved
Printed in Great Britain
0003–4878/97 $17.00 + 0.00
Inhaled Particles VIII

PII: S0003–4878(96)00111–1

CLEARANCE OF RESPIRABLE PARA-ARAMID FROM RAT LUNGS: POSSIBLE ROLE OF ENZYMATIC DEGRADATION OF PARA-ARAMID FIBRILS

A. Searl

Institute of Occupational Medicine, 8 Roxburgh Place, Edinburgh EH8 9SU, U.K.

INTRODUCTION

Para-aramid (marketed as Kevlar* and Twaron[†]) is a strong organic polymer fibre of high thermal stability used in protective clothing, industrial fabrics, tyre cords, friction products, high strength cables and high performance composites. The relative toxicity of different fibre types after inhalation is related to their durability within lung tissue (Davis, 1994). The aim of this study was to assess the biopersistence of respirable para-aramid fibrils in rat lungs. This paper addresses the possible role of enzymes in the clearance of para-aramid from lung tissue.

METHODS

Groups of 40 rats were exposed by inhalation to the same target concentration (700 fibres ml^{-1}) for each of three fibre types, para-aramid, chrysotile and code 100/475 glass microfibres for 10 days. The lung fibre burden was determined for groups of five animals at 0, 3 and 7 days and 1, 3, 6, 12 and 18 months after the end of exposure. The recovered fibres were characterised using scanning electron microscopy (SEM). Para-aramid fibrils have a ribbon-like rather than cylindrical morphology and lie flat on filter surfaces after filtration from aqueous suspension. The dimensions measured were the length and ribbon diameter, but not the thickness perpendicular to the filter surface.

The chrysotile and code 100/475 glass fibres were recovered from lung tissue using NaOCl bleach (Sebastien *et al.*, 1989). Two tissue digestion methods were used for the para-aramid fibrils. Samples of lung tissue from all the para-aramid animals were digested in a 11% solution of ethanolic KOH for 4–6 h at 60°C (Kelly *et al.*, 1985). These samples were further treated with diluted Clorox (a U.S. brand of NaOCl bleach) to reduce the amount of lung residue present. In addition, samples of lung tissue at 0, 6 and 12 months after exposure were digested using an enzyme mixture: collagenase, papain, DNA-ase and lipase at 37°C followed by a brief treatment with a detergent (10% sodium dodecyl sulphate) to reduce the amount of lung residue present. The lung tissue digestion methods were validated by exposing respirable samples of the test fibres to the reagents used for digestion and by digesting samples of lung tissue spiked with known quantities of fibre.

---

* Dupont's registered trade mark.
† Akzo Nobel's registered trade mark.

Table 1. Effects on tissue digestion methods on the dimensions, numbers and volumes of respirable para-aramid fibrils as determined by SEM analysis of fibrils reacted directly with the reagents used for digestion, fibrils recovered from lung tissue spiked with known amounts of para-aramid and fibrils recovered from time zero of the biopersistence experiments.

	GML (μm)	GMD (μm)	Recovery fibre numbers	Recovery volume	No. fibril sized	No. sample
Respirable para-aramid	7	0.26	(100%)	(100%)	400	6
Respirable para-aramid in suspension						
Enzyme digestion	8.0	0.26	128%	150%	202	2
Enzymes followed by SDS	8.6	0.26	78.7%	159%	201	2
Respirable para-aramid recovered from spiked lung						
Enzyme + SDS	5.3	0.26	(96%)*	29%	200	1
KOH + Clorox (45 min)	7.5	0.26	(88%)*	95%	205	4
Para-aramid recovered at time zero in inhalation experiment						
Enzyme + SDS	6.7	0.21	7.2×10^7	0.016 mm^3	401	2
KOH + Clorox (45 min)	9.9	0.27	4.4×10^7	0.017 mm^3	602	3
KOH + Clorox (4 h)	10.4	0.30	8.9×10^7	0.043 mm^3	801	4

* Assumes same ratio of fibre number to mass as in suspended respirable para-aramid sample. The enzyme digestions used a mix of collagenase, papain, DNAase and lipase. The KOH digestions involved incubating samples for 45 min of 4 h with 11% KOH in 80% ethanol in water, followed by removal of residual lung tissue by treatment with 10% Clorox for 10 min. SDS—1% solution of sodium dodecyl sulphate. GML—geometric mean length, GMD—geometric mean diameter.

VALIDATION OF TISSUE DIGESTION METHODS FOR PARA-ARAMID

The ethanolic KOH–Clorox method: respirable para-aramid reacts slowly with ethanolic KOH to form a potassium salt. Para-aramid also dissolves in undiluted Clorox, but more slowly than in U.K. retail bleach of equivalent NaOCl content. Experiments with lung tissue indicated that 45 minutes incubation with ethanolic KOH followed by a 10 min treatment with 10% Clorox (= 0.5% NaOCl) produced a satisfactory preparation for SEM sizing and had a negligible effect on fibril dimensions or numbers (Table 1). The lung tissue appeared to have a protective effect on the para-aramid fibrils. With the para-aramid exposed lungs however, it was necessary to extend the digestion time to 4 h for time points up to 6 months and 6 h for the later time points. A comparison of data for 45 min and 4 h incubation with ethanolic KOH suggested that the longer incubation had little effect on fibril dimensions but was associated with a doubling in fibril numbers (Table 1). Given the ribbon-like morphology of fibrils, it was impossible to determine whether this increase in fibril numbers was due to length-wise splitting during the prolonged incubation in the KOH solution.

The enzyme method

The enzyme mix had little effect on the dimensions or numbers of respirable para-aramid fibrils. Experiments with lung tissue however indicated that enzyme digestion led to a marked reduction in the number of long fibrils present (Table 1).

Conclusion. Neither digestion method was ideal, but the ethanolic KOH digestion appeared to have less effect on recovered fibres than the enzyme digestion.

RESULTS FROM BIOPERSISTENCE STUDY

The lung burden analyses for all three fibre types show large reductions in the overall number and volume of retained fibres during the 16 months following exposure (Fig. 1). The apparent half-time of clearance of each of the fibre types depends on the size fraction and time interval examined (Searl, 1996). The data for the para-aramid fibrils recovered using KOH show rapid clearance of the longest fibres (> 15 μm) combined with an initial increase in the numbers of shorter fibres (Fig. 1). The length distribution of para-aramid fibrils (recovered using KOH) after 6 months is similar to that of the respirable fibres recovered from lung tissue using the enzyme digestion method at time zero (Fig. 2). The chrysotile data show a more rapid reduction in the numbers of retained short fibres than of long fibres (Fig. 1).

The para-aramid lung burden data from the enzyme digestions show a marked reduction in fibril numbers through time but no change in fibril dimensions (Fig. 2). The total lung burden, at time zero as determined using the enzyme digestion method, was similar in terms of estimated volume to that determined using the KOH method with a 45 min incubation, but less than half of that determined using the KOH method with a 4 h incubation.

DISCUSSION

In terms of total lung burden, the para-aramid fibrils had a similar biopersistence to the chrysotile and code 100/475 glass fibres over the 16 months. The change in length distribution of the para-aramid fibrils is consistent with the disintegration of the longest respirable fibrils into shorter fragments that are subsequently more readily cleared by macrophages. In contrast, the pattern of clearance of chrysotile is consistent with macrophage clearance of short fibres and minimal transverse breakage of fibres. Overall the long para-aramid fibrils are much less durable than long fibres of chrysotile. Given that fibres that are too long to be removed by macrophages and are also durable may be more hazardous than shorter fibres, respirable para-aramid is likely to be less hazardous than chrysotile.

The similarity of the length distributions of para-aramid fibrils recovered using KOH after 6 months and fibrils recovered using the enzyme mix at time zero suggests that similar enzymes may play an important role in promoting the degradation of para-aramid fibrils in lung tissue. The similarity in length distributions combined with the reduction in breakage rate of long para-aramid fibrils through time also suggests that there might be a limited number of sites along the length of para-aramid fibrils open to the processes of chemical degradation that led to disintegration. Para-aramid is extremely resistant to dissolution in the inorganic salt systems normally used to simulate lung fluids (Minty *et al.*, 1995). It is however readily biodegradable both in lung tissue (Warheit *et al.*, 1992) and when used to construct artificial ligaments in sheep (Dauner *et al.*, 1990). The para-aramid structure contains amide bonds analogous to those hydrolysed during the enzymatic

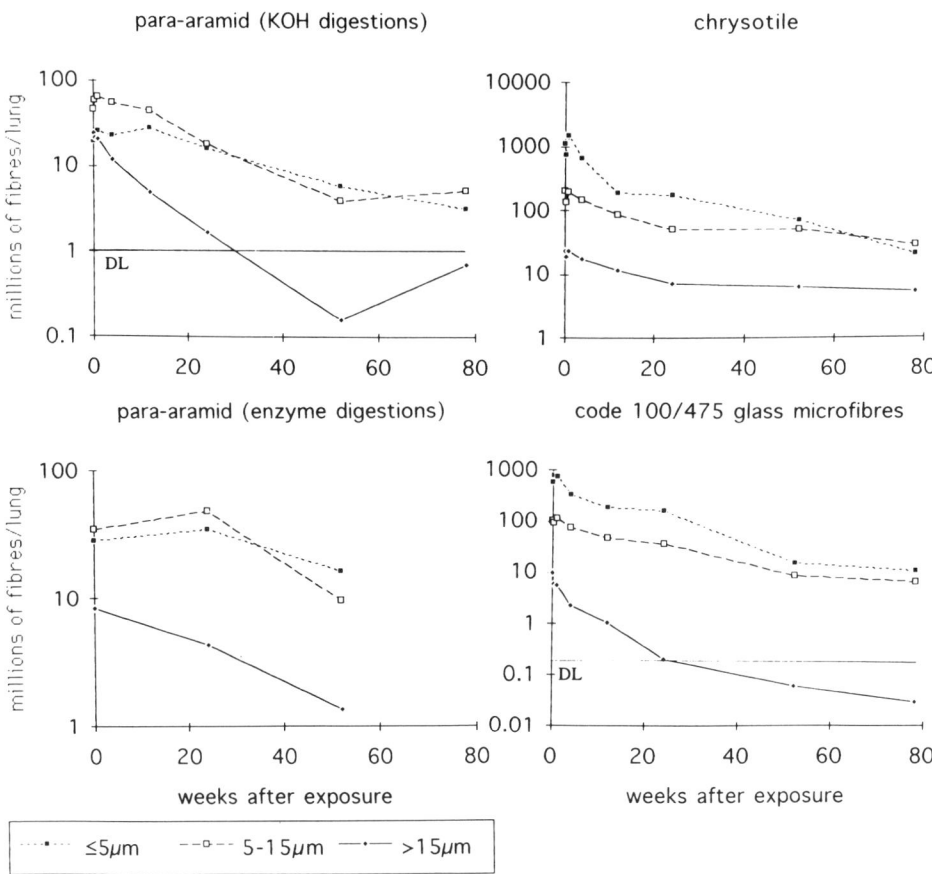

Fig. 1. Change in mean numbers of fibres per lung as determined by SEM for the three test fibres through time following exposure. At least 200 fibres/animal were sized and there were 5 rats per time point. Two sets of data are shown for para-aramid. The main data set was derived using ethanolic KOH to digest samples of lung tissue. Some further samples of lung tissue were digested using an enzyme mix (collagenase, papain, DNAase and lipase). DL—Detection Limit.

breakdown of proteins. Most of these bonds are protected by adjacent benzene rings that would restrict any interaction between the amide bond and bulky molecules such as enzymes. Deformation of the polymer chain at a kink band or other discontinuity within the structure could open the amide bond to enzyme attack. The length of fibrils created in the lung through the disintegration of longer fibrils may be governed by the spacing of intercrystalline discontinuities within the para-aramid structure. Yang (1994), for example, illustrates kink bands in para-aramid at a similar 5–7 μm spacing to the modal length class of fibrils recovered from lung tissue in this study. The spacings of similar discontinuities that are susceptible to hydrolysis may vary in fibrils from different sources.

Acknowledgements—This study was funded by Akzo Nobel and DuPont. The author is grateful to Dr D. Warheit and M. Hartsky of DuPont who provided help and advice with the development of the fibre

A. Searl

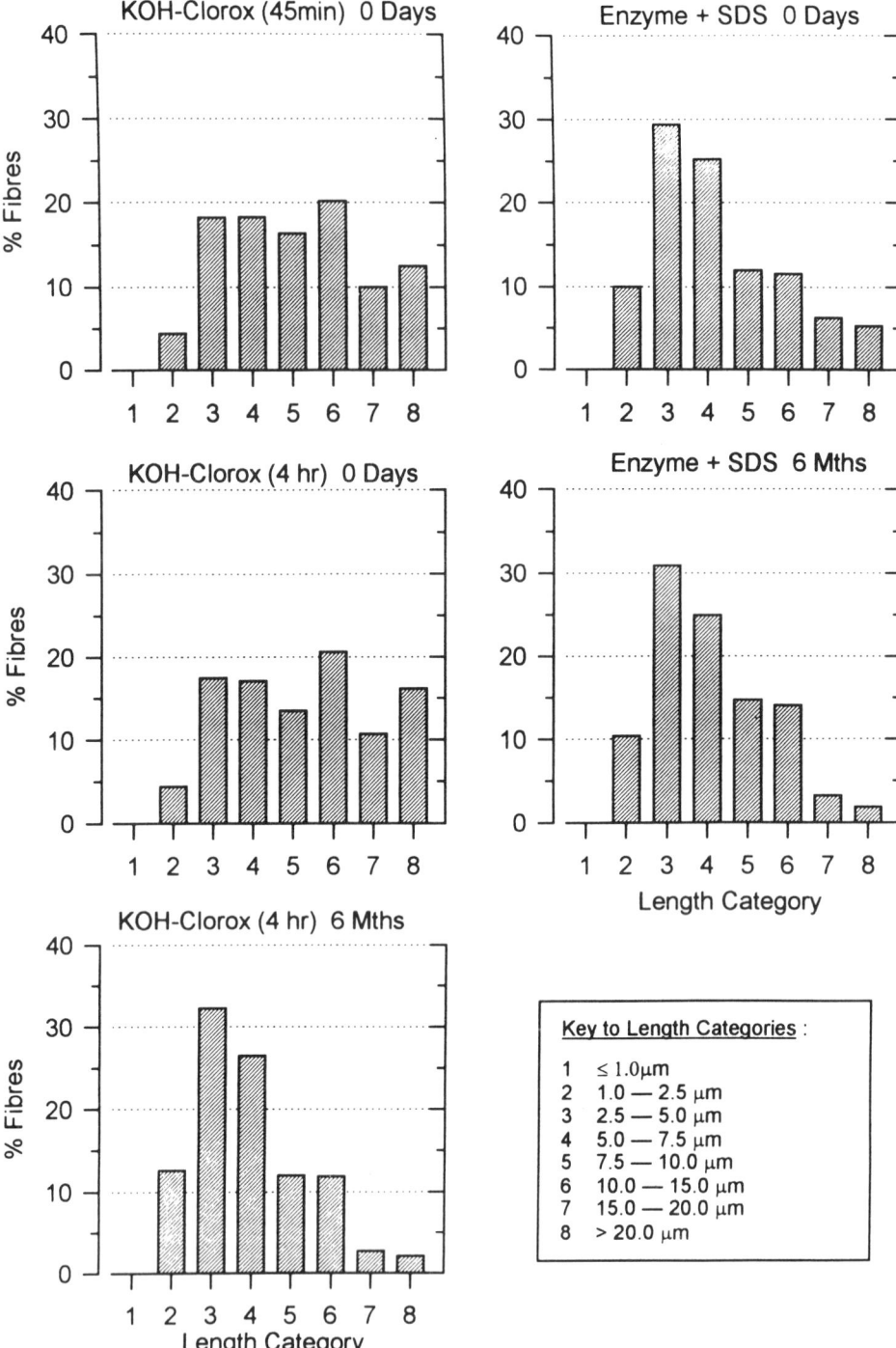

Fig. 2. Comparison of length distributions of para-aramid fibrils recovered at zero and six months after the end of inhalation using the ethanolic KOH digestion method and the enzyme digestion method.

cloud generation and the lung digestion techniques. The inhalation exposures and animal care were performed by R. Cullen, S. Clark and M. Whittington. The lung burden analysis was performed by C. MacGonagle and R. McCue and data analysis was performed by D. Buchannan.

REFERENCES

Davis, J. M. G. (1994) The role of clearance and dissolution in determining the durability or biopersistence of mineral fibers. *Environ. Hlth Perspect.* **102**, 113–117.

Dauner, M., Plank, H., Syre, I. and Dittel, K.-K. (1990) Para-aramid fiber for artificial ligament. *Advances in Biomaterials* **9**, 445–449.

Kelly, D. P., Merriman, E. A., Kennedy, G. L. and Lee, K. P. (1993); Deposition, clearance and shortening of Kevlar para-aramid fibrils in acute, subchronic and chronic inhalation studies in rats. *Fundam. appl. Toxicol.* **21**, 345–354.

Minty, C. A., Meldrum, M., Phillips, A. M. and Ogden, T. L. (1995) P-aramid respirable fibres: criteria document for an occupational exposure limit. Health and Safety Executive, EH65, pp. 30.

Searl, A. (1996) A comparative study of the clearance of respirable para-aramid, chrysotile and glass fibres from rat lungs. *Ann. occ. Hyg.* (in press).

Sebastien, P., McDonald, J. C., McDonald, A. D., Case, B. and Harley, R. (1989) Respiratory cancer in chrysotile textile and mining industries: exposure inferences from lung analysis. *Br. J. ind. Med.* **46**, 180–189.

Warheit, D. B., Kellar, K. A. and Hartsky, M. A. (1992) Pulmonary cellular effects in rats following aerosol exposures to ultrafine Kevlar aramid fibrils: evidence for biodegradability of inhaled fibrils. *Toxicol. appl. Pharmacol.* **116**, 225–239.

Yang, H. H. (1994) *Kevlar Aramid Fiber* Wiley, Chichester.

 Pergamon

Ann. occup. Hyg., Vol. 41, Supplement 1, pp. 154–160, 1997
© 1997 British Occupational Hygiene Society
Published by Elsevier Science Ltd. All rights reserved
Printed in Great Britain
0003–4878/97 $17.00 + 0.00
Inhaled Particles VIII

PII: S0003–4878(96)00112–3

THE SYNTHESIS AND CHARACTERISATION OF FIBRES STRUCTURALLY RELATED TO THE MINERAL ERIONITE

A. P. Rood,* J. A. Hoskins† and L. R. Hibbs†

*Davy Faraday Laboratory, The Royal Institution, 21 Albemarle Street, London W1X 4BS, U.K.;
and †MRC Toxicology Unit, Hodgkin Building, University of Leicester, Lancaster Road,
Leicester LE1 9HN, U.K.

INTRODUCTION

The fibrous zeolite erionite has been identified as the most probable reason for the high levels of human mesotheliomas reported from villages in Cappadocia, Turkey (Barris, 1978). Inhalation studies in rats have confirmed it as a potent carcinogen and this taken together with the epidemiological studies suggests that erionite fibres are more potent, on a fibre for fibre basis, than similarly sized fibres of crocidolite.

We have synthesised chemically homogeneous fibres structurally related to erionite in an effort to understand more of the underlying properties that make this mineral so biologically active. Fibres with controlled morphology, based on aluminium phosphate, have been prepared by growing them with an organic template under hydrothermal conditions in the laboratory. Such fibres have the same size and the same zeolite pore structure, as erionite. We also report a new method of growing mesoporous films of silica on the surface of fine glass fibres by mixing them with the gel precursor in the synthesis and using the glass as a substrate for nucleation and growth.

Aluminium phosphate fibre samples with a range of compositions have been prepared with various amounts of silicon, which replaces phosphorous and also iron and cobalt, to replace some of the aluminium. Fibre size distributions have been determined by SEM, chemical compositions by EDX and crystal structures by XRD. The size distributions of the fibres show a major proportion are within the WHO range for potentially biologically active fibres.

METHOD AND MATERIALS

Samples of fibrous erionite from Oregon, Nevada, California and from Arizona, were obtained from the historic collection of the Royal Institution.

Synthetic analogues of the erionite structure were prepared using the method of Lohse *et al.* (1994). In this preparation, water and aluminium isopropoxide are stirred to form a homogeneous suspension to which is added silica and phosphoric acid, together with a cyclohexylamine template and the whole boiled for 10 min. The resulting gel was treated with aqueous hydrogen fluoride and the mixture heated overnight in an autoclave at 200°C. Fibres containing iron or cobalt were

(a)

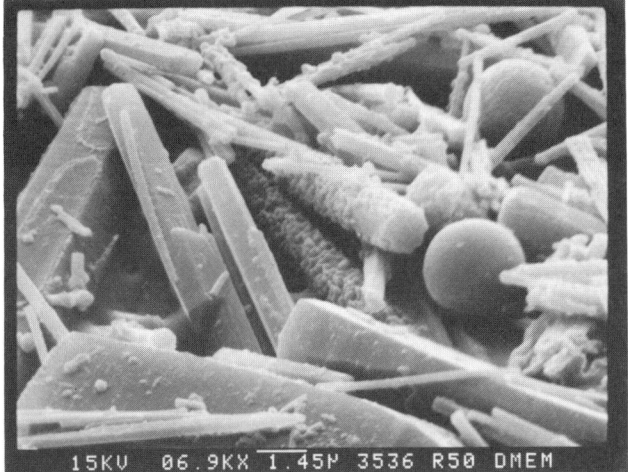

(b)

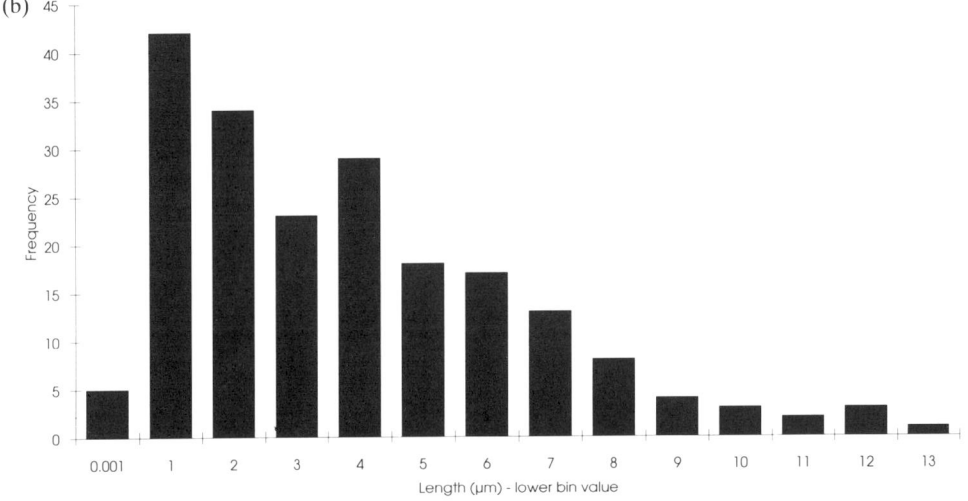

Fig., 1 (a), (b).

prepared by adding these cations at the early stages of the preparation. The resulting crystals were filtered off, washed with distilled water and dried in an oven.

Mesoporous crystalline films of silica were prepared from template surfactant assemblies on the surfaces of fine glass fibres, using the method of Ozin *et al.* (1996).

Template molecules in both the above preparations were removed by calcining in oxygen at up to 500°C. Samples were characterised structurally by XRD, chemically by EDX and fibre size distribution determined by SEM measurements on gold coated dispersions.

RESULTS

In Fig. 1 the SEM image the EDX trace and the fibre length distribution of the

R50 UNCOATED

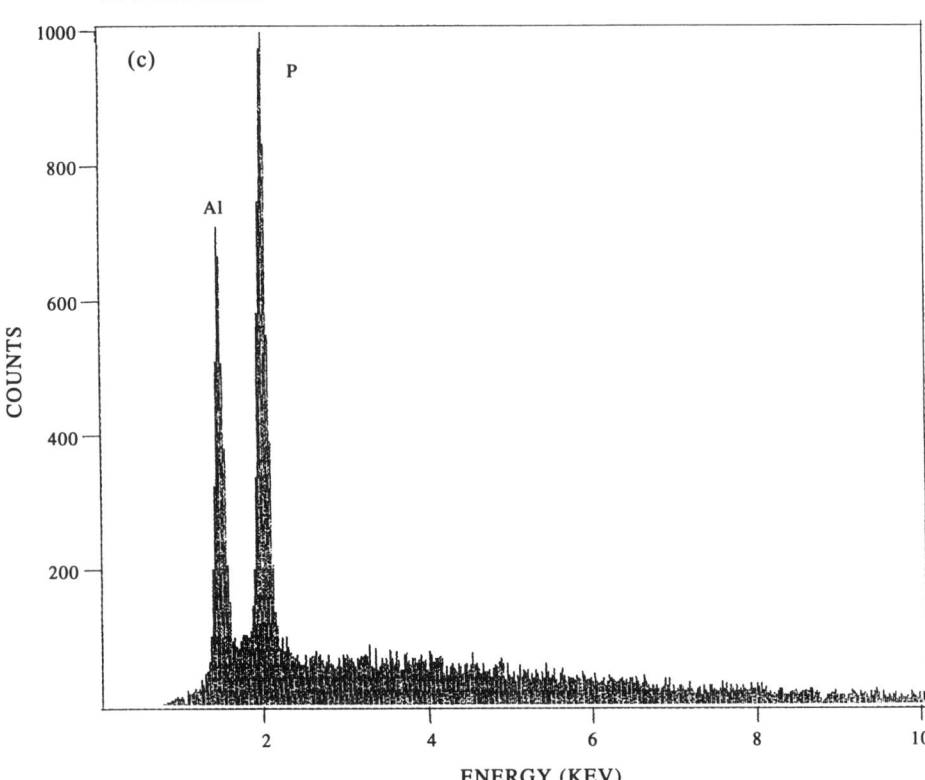

Fig. 1. SEM image, (b) length distribution and (c) EDX trace of synthetic aluminium phosphate fibres.

aluminium phosphate analogues of erionite are shown. The basic morphology of the fibres can be seen together with non-fibrous debris which can be a minor component in some preparations. The XRD trace from such material is shown in Fig. 4, on which is superimposed the lines from the characteristic d-spacings of natural erionite.

In Fig. 2 a sample of natural erionite from Eureka County (Nevada) is similarly shown, together with the corresponding size distribution. A number of natural erionites have been measured in this way, with a statistically significant number of fibres being measured (typically over 200).

The growths of microporous crystalline silica on the surface of fine glass fibres is also shown in Fig. 3. This material has been previosuly characterised as Mobil catalytic material type 41 and is known to have large pores of approximately 3 nm in diameter.

DISCUSSION

The most important determinant of a fibre's potential to produce mesothelioma, when implanted into animals, is the size of that fibre. Work by Wagner *et al.* (1973) and by Stanton *et al.* (1981) has indicated that fibres less than 0.3 μ in diameter and

(a)

(b)

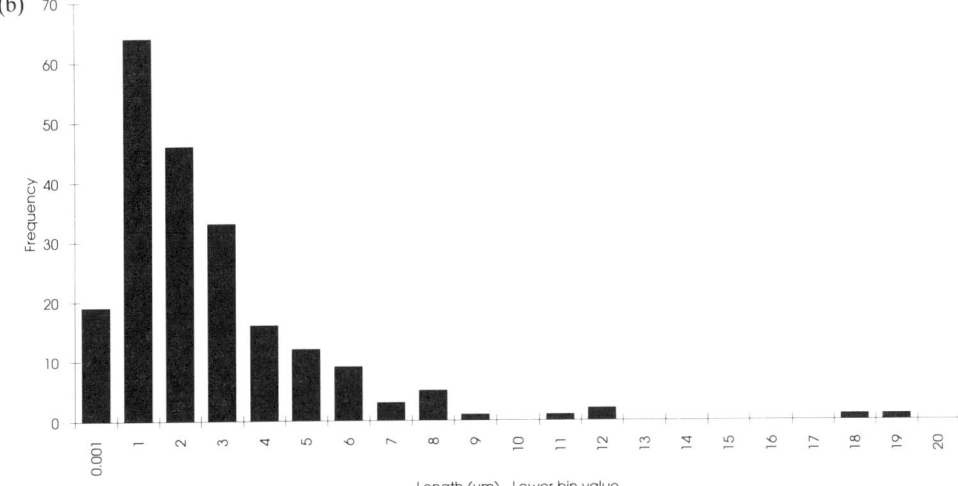

Fig. 2. (a), (b).

greater than 8 μ in length, have the greatest potential for introducing mesothe-liomas after intrapleural or intraperitoneal injection into rats.

Size on its own does not account for all observations, the biopersistence of the fibre is critical. This is particularly so when the route of entry is by inhalation, where chemical and physical clearance is active. Also other factors, as yet not well understood, may involve the surface characteristics of the fibres, Kane (1991).

Erionite illustrates the size effects well, although samples vary from geological location to location, the fibres of this mineral have median diameters of 0.3–0.5 μ, median lengths between 3.5–5 μ and with aspect ratios around 10:1.

Our structural analogues of erionite have the same internal pore structures of a large diameter (0.6 nm) internal channel linked by 0.35 nm diameter ports. Fibres of sample R50 shown in Fig. 2 have a 0.5 micro median diameter and 4.4 micron median length and are in the range of the natural erionites. The internal chemical

EUREKA UNCOATED

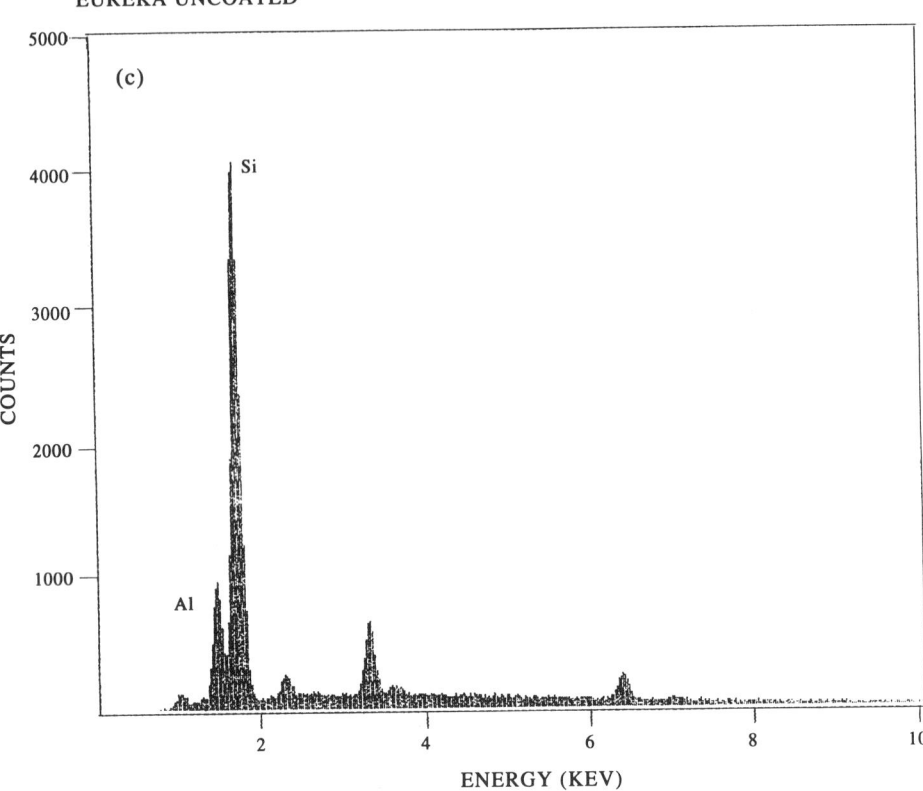

Fig. 2. (a) SEM image, (b) length distribution and (c) EDX trace of natural erionite.

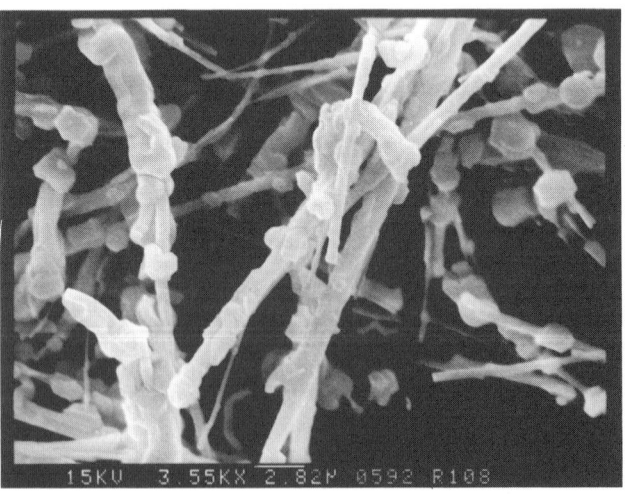

Fig. 3. SEM image of mesoporous silica on glass fibres.

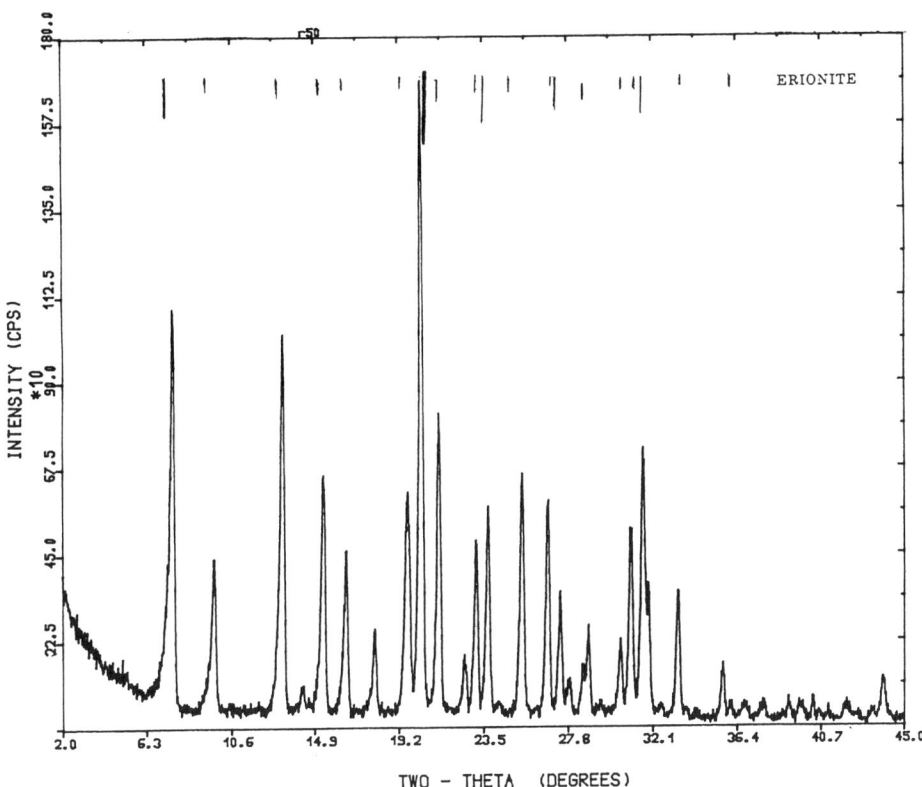

Fig. 4. XRD trace of synthetic aluminium phosphate fibres with erionite shown for comparison.

environment, and the surfaces of these synthetic fibres will be distinct from natural erionite. The most obvious difference is the presence of phosphorous, which is not found in erionite and the low levels of silicon possible in the aluminium phosphate fibres. Up to 12% of silicon is the maximum we have managed to introduce into the synthetic material before losing the fibrous habit. Natural erionite contains up to 27% of silicon and cations such as calcium, sodium and potassium are also present in the structure.

If chemistry as well as structure is important in understanding the erionite activity, variation can be played in this model system by the ability to introduce atoms such as iron or other transition metals into the lattice. The same opportunity exists to tether active systems internally on the MCM-41 surface by novel synthetic techniques being developed by Maschmeyer *et al.* (1995).

REFERENCES

Baris, Y. I., Sahin, A. A., Ozesmi, M., Kerse, I., Ozen, E., Kolacan, B., Altinors, M. and Goktepeli, A. (1978) An outbreak of pleural mesothelioma and chronic fibrosing pleurisy in the village of Karin/Urgup in Anatolia. *Thorax* **33**, 181.

Kane, A. B. (1991) Fibre dimensions and mesothelioma: a reappraisal of the Stanton hypothesis. In *Mechanisms in Fibre Cacinogenisis* (Edited by R. C. Brown, J. A. Hoskins and N. F. Johnson), pp. 131–141. Plenum Press, New York.

Lohse, U., Jancke, K., Loffler, E. and Scaller, T. (1994) The influence of HF on the incorporation of silicon into the erionite-like ALPO-17 molecular sieve. *Cryst. Res. Tech.* **29**, 237–45.

Maschmeyer, T., Rey, F., Sankar, G. and Thomas, J. M. (1995) Hetrogeneous catalysts obtained by grafting metallocene complexes onto mesoporous silica. *Nature* **378**, 159.

Ozin, G. A., Yang, H., Coombs, N. and Sokolov, I. (1996) Free-standing and orientated mesoporous silica films grown at the air–water interface. *Nature* **381**, 589–92.

Stanton, M. F., Layard, M., Tegeris, A., Miller, E., May, M., Morgan, E. and Smith, A. (1981) Relation of particle dimensions to carcinogenicity in amphibole asbestoses and other fibrous minerals. *J. Natl. Canc. Inst.* **67**, 965–75.

Wagner, J. C., Berry, G. and Timbrell, V. (1973) Mesothelioma in rats after inoculation with asbestos and other minerals. *Br. J. Cancer* **28**, 173–85.

Pergamon

Ann. occup. Hyg., Vol. 41, Supplement 1, pp. 161–167, 1997
© 1997 British Occupational Hygiene Society
Published by Elsevier Science Ltd. All rights reserved
Printed in Great Britain
0003–4878/97 $17.00 + 0.00
Inhaled Particles VIII

PII: S0003–4878(96)00113–5

MALIGNANT MESOTHELIOMA IN AUSTRALIA (1945–1995)

J. Leigh, B. Hull and P. Davidson

National Institute of Occupational Health and Safety, GPO Box 58, Sydney 2001, Australia

INTRODUCTION

In Australia, more chrysotile than amphibole asbestos was mined until 1939. With the commencement of mining at Wittenoom, Western Australia in 1937, crocidolite dominated production, until final closure in 1966. New South Wales, the first State to mine asbestos, also produced the largest tonnages of chrysotile (until 1983) as well as smaller quantities of amphibole (until 1949). With the closing of the crocidolite mine at Wittenoom in 1966, Australian asbestos production and exports declined. Imports of chrysotile also started to decline. The main sources of raw asbestos imports were Canada (chrysotile) and South Africa (crocidolite and amosite).

In addition to imports of asbestos fibre, Australia also imported many manufactured asbestos products, including asbestos cement articles, asbestos yarn, cord and fabric, asbestos joint and millboard, asbestos friction materials and gaskets. The main sources of supply were the United Kingdom, U.S.A., Federal Republic of Germany and Japan. In Australia over 60% of all production and 90% of all consumption of asbestos fibre was used by the asbestos cement manufacturing industry. From about 1940 to the late 1960s all three types of asbestos were used in this industry, crocidolite then being phased out. Much of this industry output remains in service today in the form of "fibro" houses and water and sewerage piping. By 1954 Australia was number four in the Western world in gross consumption of asbestos cement products, after U.S.A., U.K. and France, and clearly first on a per capita basis. From World War II to 1954, 70 000 asbestos cement houses were built in the state of New South Wales alone (52% of all houses built). In Australia as a whole, until the 1960s, 25% of all new housing was clad in asbestos cement.

Exposures in the past were very high in some industries and jobs [e.g. 25 million particles per cubic foot (150 fibres ml^{-1}) in asbestos pulverisors and disintegrators in the asbestos cement industry; up to 600 fibres ml^{-1} in baggers at Wittenoom]. Australia still imports about 2000 tonnes a year of chrysotile fibre and about $A13.5m worth of asbestos products a year, over half as friction material but also fabricated yarn, fabric, jointing, gaskets, millboard and asbestos cement products. Handling of asbestos in place and removal operations are subject to a strict National Code of Practice. A series of regulations adopted in the late 1970s and early 1980s by the various states now impose exposure limits of 0.1 fibre ml^{-1} for crocidolite, amosite and mixtures and 0.1–1.0 fibre ml^{-1} for chrysotile (TWA 8 h membrane filter method light microscopy, WHO fibres).

With this background, it was almost certain that Australia would suffer a mesothelioma epidemic of a severe nature. The first reported case from Wittenoom was in 1962. Retrospective search identified 658 cases (535 male, 123 female) occurring in Australia from 1945 to 1979 (Musk *et al.*, 1989).

METHODS

The Australian Mesothelioma Surveillance Program (Ferguson *et al.*, 1987) began on 1 January 1980. Formal voluntary notification of cases was actively sought from a network of respiratory physicians, pathologists, general and thoracic surgeons, medical superintendants, medical records administrators, State and Territory departments of occupational health, cancer registries, compensation authorities or any other source. A full occupational and environmental history was obtained for each case, either from the patient or next-of-kin. The history taking was non-directive but included specific questions on asbestos exposure at the end.

Occupational and environmental exposure was based on the opinions of two experienced hygienists, who were, however, not independent nor blinded as to disease status. The diagnosing pathologist was requested to provide slides and/or tissue specimens. These were circulated among a pathology expert panel for confirmation of diagnosis. Post-mortem examination was actively sought for in every case in order to confirm diagnosis and to obtain lung tissue free of tumour for lung fibre content analysis.

Since 1 January 1986, a less detailed notification system has operated, with a short questionnaire history, which is followed up by mail. Only histologically confirmed cases are accepted but there is no pathology panel diagnosis confirmation. This is now known as the Australian Mesothelioma Register, but is a continuation of the Program.

RESULTS AND DISCUSSION

Table 1 shows notifications up to end 1995. Up to end 1995, a total of 4129 notifications had been received*. Notifications show a continuing upward trend (Figs 1 and 2). Both male and female rates have increased but the male rate is over seven times the female rate. In 1995, the male crude rate was 6.7 per 100 000 per year and the female crude rate 1.1 per 100 000 per year. These are the highest reported rates in the world. The incidence is now similar to Hodgkins lymphoma or liver cancer and the mortality greater than that of cervical cancer. Western Australia has the highest incidence but contributes only 15% of the total cases. Wittenoom contributes about 5% of the Australian cases. Most of the cases come from the two most populous and industrialised states, New South Wales and Victoria (Leigh *et al.*, 1991).

In 93.2% of all cases the mesothelioma was pleural in site, 6.5% peritoneal and only four cases were in other sites. Among men 94.3% were pleural, 5.3% peritoneal; among women 86.3% pleural, 13.7% peritoneal. Of the cases that underwent pathology panel review 96% were confirmed as mesothelioma (73%

---

*Total notifications to 20/12/96 = 4585. 1996 notifications 1/1/96–20/12/96 = 456.

Table 1. Mesothelioma notifications in Australia 1980–1995

	NSW	VIC	QLD	WA	SA	TAS	NT	ACT	TOTAL
1980	16	1	0	0	0	0	0	0	17
1981	54	3	19	11	5	5	0	0	97
1982	89	20	10	10	21	2	0	1	153
1983	58	22	28	47	19	6	0	0	180
1984	79	41	21	26	14	1	1	2	185
1985	72	40	27	30	19	1	0	1	190
1986	46	34	38	32	18	2	1	1	172
1987	54	40	26	28	32	0	0	2	182
1988	57	27	45	23	36	1	0	2	191
1989	124	25	35	44	22	3	0	1	254
1990	111	82	43	26	25	1	0	1	289
1991	105	44	46	66	56	10	0	2	329
1992	117	45	40	37	39	3	1	1	283
1993	99	34	42	47	25	5	0	0	252
1994	152	41	75	32	30	8	0	1	339
1995	124	92	51	33	44	10	1	3	358
ALL	1357	591	546	492	405	58	4	18	3471
%	39.0	17.0	15.7	14.2	11.7	1.7	0.1	0.5	100

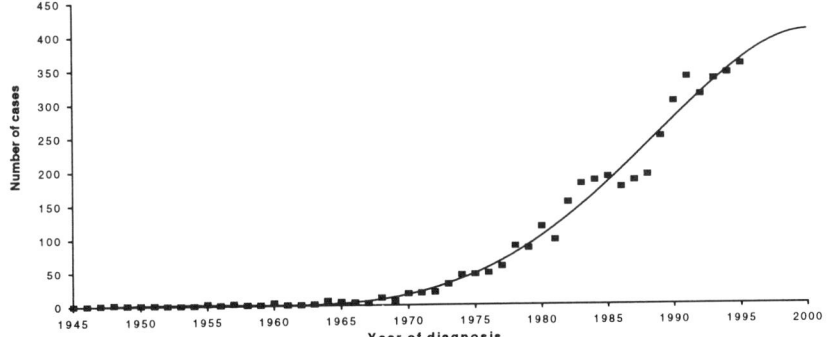

Fig. 1. Incident cases of malignant mesothelioma in Australia 1945–1995.

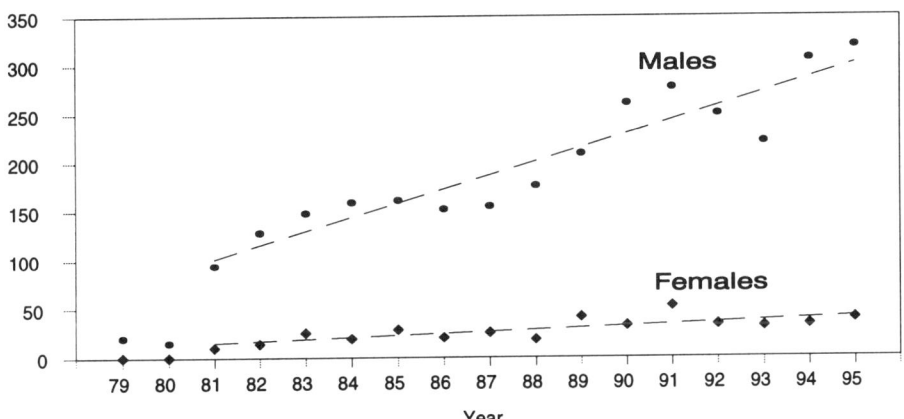

Fig. 2. Mesothelioma cases notified per year.

Table 2. Australian mesothelioma register asbestos exposures as documented in the register from 1 January 1986 to 31 March 1995

Circumstance of exposure	No. exposed (with no other exposure) (single)	No. exposed (with other exposures) (multiple)	Total exposed
Acoustic engineer	1	—	1
Airconditioning	7	8	15
Aircraft	7	—	7
Armed forces/wartime	14	6	20
Asbestos bagging (not Wittenoom)	2	—	2
Asbestos bags—handled which had contained	3	—	3
Asbestos clothing worn	4	1	5
Asbestos covers for cooking	1	—	1
Asbestos dwelling/fence—built/renovated	42	6	48
Asbestos dwelling—lived in	9	6	15
Asbestos products factory—lived near	6	2	8
Asbestos products factory—worked near	1	—	1
Asbestos mine—worked/lived near (not Wittenoom)	7	1	8
Asbestos product handled in the workplace	31	6	37
Asbestos product manufacturer—worked	87	33	120
Asbestos product part of workplace or surrounds	8	6	14
Asbestos tailings—played on as a child	1	—	1
Asbestos/or products worker—lived with/washed clothes	28	3	31
Bakery (ovens)	2	—	2
Boilermaker/cleaner/attendant/installer/welder	59	35	94
Brake linings—made/repaired	37	9	46
Bricklayer	7	1	8
Brickworks	8	3	11
Builder/builder's labourer	105	22	127
Carpenter/joiner	131	28	159
Cement factory worker	4	—	4
Chemical engineer	1	—	1
Concreting	2	—	2
Construction worker	5	1	6
Civil engineer	1	—	1
Demolition	3	2	5
Design engineer	—	1	1
Diesel engineer	—	1	1
Dockyard worker	16	15	31
Electrical engineer	3	1	4
Electrical fitter	10	3	13
Electrical mechanic	2	—	2
Electrician	30	6	36
Electroplater	—	1	1
Engineer	8	—	8
Fireproofing	3	—	3
Firedoors	3	—	3
Firefighter	2	2	4
Fitter/turner	27	11	38
Foundry	2	1	3
Furnace	4	—	4
Glassworks/glaziers	2	—	2
Industrial chemist	4	—	4
Industrial engineer	1	—	1
Instrument technician	1	—	1
Insulation	6	1	7
Jeweller	1	—	1

continued

Table 2. *continued*

Circumstance of exposure	No. exposed (with no other exposure) (single)	No. exposed (with other exposures) (multiple)	Total exposed
Laboratory technician	1	—	1
Labourer	15	3	18
Lagger	21	12	33
Lagging in workplace	11	1	12
Laundry/drycleaners	9	2	11
Linesman	4	—	4
Locksmith	1	—	1
Machine fitter	1	1	2
Machine inspector	2	—	2
Machine operator	1	—	1
Machinist	2	—	2
Maintenance carpenter	1	—	1
Maintenance electrician	2	—	2
Maintenance engineer	2	—	2
Maintenance fitter	10	3	13
Maintenance mechanic	1	—	1
Maintenance worker	7	2	9
Marine engineer	5	6	11
Mechanical engineer	1	—	1
Mechanical fitter	2	2	4
Metal fabrication	1	—	1
Metallurgy	1	—	1
Navy	60	39	99
Painter/decorator	30	5	35
Panelbeater	3	—	3
Papermill	2	1	3
Patternmaker	2	1	3
Pipes—handled/cut/stored/drilled	12	2	14
Plasterer	9	5	14
Plumbing	44	13	57
Power station worker	59	39	98
Pressure pak manufacturer	1	—	1
Printing	7	—	7
Railways	63	31	94
Renovations/maintenance/lagging in workplace	13	2	15
Roofing	2	2	4
Sheetmetal	5	9	14
Ships—building/repairing/on	60	47	107
Shop fitter	1	—	1
Site visits/inspections	4	2	6
Smelting	1	—	1
Steelworks	7	2	9
Storeman	10	—	10
Stoves	2	—	2
Sugar mill	6	4	10
Tannery	1	—	1
Telephone technician	5	2	7
Tiler	3	—	3
Toolmaker	3	1	4
Trades assistant	8	—	8
Transporting asbestos	7	3	10
Transporting asbestos product	6	1	7
Tyre factory	7	4	11

continued

Table 2. *continued*

Circumstance of exposure	No. exposed (with no other exposure) (single)	No. exposed (with other exposures) (multiple)	Total exposed
Waterside worker	54	7	61
Weighing trucks	1	—	1
Welder	13	5	18
Whitewash—Greece/Cyprus	2	1	3
Wittenoom	114	18	132
Wood machinist	1	—	1
Asbestos exposure			
single	1478		
multiple	280		
No apparent asbestos exposure	361		
No response to questionnaire	254		
Total cases from 1/1/86–31/3/95	2373		
Proportion of respondents with asbestos exposure	1758		
	2119	= 83%	

definite, 17% probable and 6% possible). In the cases reported with detailed histories (up to end 1985), the most common occupational exposures were repair and maintenance of asbestos materials (18%), shipbuilding (11%), asbestos cement production (7%), asbestos cement use (7%), railways (6%), Wittenoom crocidolite mining/mining (6%), insulation manufacture/installation (4%), wharf labouring (3%), power stations (3%), boilermaking (2%), para-occupational, hobby and environmental (15%). The definition of exposure was extended to slight exposures such as working with asbestos cement in home construction and maintenance, visiting Wittenoom, or living close to an asbestos factory or mine. (These categories had been categorized as "no known history" in some earlier reports.) Only 18% had no known history of exposure according to the extended definition. Of this "no known history" group, 81% had fibre counts detected in the lungs, 30% with more than 10^6 fibres $g^{-1} > 2\,\mu$ including "long" ($> 10\,\mu$ fibres) suggesting that nearly all cases have been exposed to some level of asbestos. Indeed absence of fibres in the lungs does not negate exposure as fibres may have initiated mesothelioma and then been cleared before death. Mean latency from first exposure to presumptive diagnosis was 37.4 years (Ferguson *et al.*, 1987).

In the cases reported since 1 January 1986, when less detail of history of exposure was sought, 86% of males responding to questionnaire and 46% of females gave a history of asbestos exposure (overall 83%) (non-response 11%). Table 2 shows the range of circumstances (combination of occupations and industries and environmental setting) in which exposure occurred.

Some common exposure histories were: repair and maintenance of asbestos materials (13%), shipbuilding (3%), asbestos cement production (4%), railways (3%), powerstations (3%), boilermaking (3%), Wittenoom (5%), wharf labour (2%), para-occupational, hobby, environmental (4%), carpenter (4%), builder (6%), navy (3%), plumber (2%), brake linings (2%), multiple (12%). The pattern of exposure is shifting away from the older traditional industries towards product,

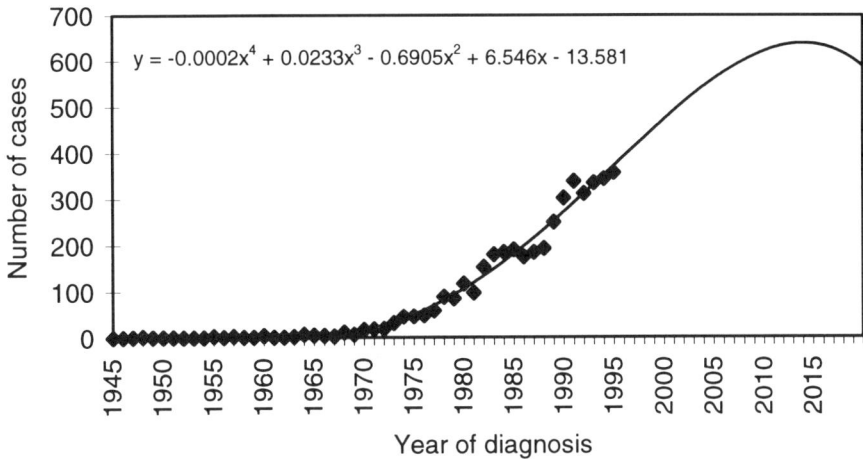

Fig. 3. Incident cases of malignant mesothelioma in Australia 1945 to 1995 and extrapolation to 2020 assuming maximum at 2010.

domestic and environmental exposure. The incidence is still increasing and assuming peak amphibole exposure occurred about 1965 and peak chrysotile about 1975, peak incidence is not expected until about 2010. The expected total number of cases from 1945 to 2020 is estimated to be about 18 000, based on models by Berry (1991) and de Klerk *et al.*, (1989) for Wittenoom, extrapolated for Australia as a whole (assuming Wittenoom contributes 5% of cases), and direct extrapolation from the best fit to the empirical incidence curve, constrained to have a maximum value at 2010, following a 40 year latency from the time of maximum exposure (1970) (Fig. 3).

REFERENCES

Berry, G. (1991) Prediction of mesothelioma, lung cancer and asbestosis in former Wittenoom asbestos workers. *Br. J. ind. Med.* **48**, 793–802.

Berry, G. (1995) Environmental mesothelioma incidence, time since exposure to asbestos and level of exposure. *Environometrics* **6**, 221–228.

Ferguson, D. A., Berry, G., Jelihovsky, T., Andreas, S. B., Rogers, A. J., Fung, S. C., Grimwood, A. and Thompson, R. (1987) The Australian Mesothelioma Surveillance Program 1979–1985. *Med. J. Aust.* **147**, 166–172.

Klerk, N. H. de, Armstrong, B. K., Musk, A. W. and Hobbs, M. S. T. (1989) Predictions of future cases of asbestos related disease among former miners and millers of crocidolite in Western Australia. *Med. J. Aust.* **151**, 616–620.

Leigh, J., Corvalan, C. F., Grimwood, A., Berry, G., Ferguson, D. A. and Thompson, R. (1991) The incidence of malignant mesothelioma in Australia 1982–1988. *Am. J. ind. Med.* **20**, 643–655.

Musk, A. W., Dolin, P. J., Armstrong, B. K., Ford, J. M., de Klerk, N. H. and Hobbs, M. S. T. (1989) The incidence of malignant mesothelioma in Australia 1947–1980. *Med. J. Aust.* **150**, 242–246.

 Pergamon

Ann. occup. Hyg., Vol. 41, Supplement 1, pp. 168–171, 1997
© 1997 British Occupational Hygiene Society
Published by Elsevier Science Ltd. All rights reserved
Printed in Great Britain
0003–4878/97 $17.00 + 0.00
Inhaled Particles VIII

PII: S0003–4878(96)00114–7

RECONSTRUCTION OF INHALATION EXPOSURE AND UPTAKE FOR ASBESTOS FIBRES

J. W. Cherrie

University of Aberdeen, and Institute of Occupational Medicine, 8 Roxburgh Place,
Edinburgh EH8 9SU, U.K.

INTRODUCTION

In epidemiological studies and risk assessments involving inhalation of hazardous substances, it is conventional to estimate cumulative exposure as the product of exposure level (i.e. concentration in the worker's breathing zone) and exposure duration. In some cases the time component is implicit, e.g. where exposure levels are expressed as 8 h or 15 min time-weighted averages. However, a third variable is required if we wish to estimate the quantity of material inhaled into the body: the worker's breathing rate.

The importance of breathing ventilation rate has long been recognised. Oldham and Roach (1952) discussed their particle concentrations in coal mines in relation to cumulative exposure and the product of concentration and ventilation rate, which they termed "dose". Similarly, in laboratory studies of people exposed to solvent vapours the impact of work rate and hence ventilation rate, is well known (Opdam, 1989; Wallen *et al.*, 1985; Fiserova-Bergerova, 1985; Bowes *et al.*, 1995); with increasing breathing rate leading to increased concentrations of the substance or its metabolites in the body. Jones *et al.* (1981) have also published a field study of the differences in ventilation rate among coalminers underground.

In this study the cumulative exposure of a group of workers is compared with an alternative exposure variable: the uptake, which is defined as the product of exposure level (e.g. measured in fibres m^{-3}), exposure duration (e.g. in hours) and breathing rate (e.g. in $m^3\,h^{-1}$). For asbestos, uptake is expressed as the number of fibres inhaled, while cumulative exposure is presented as fibre ml^{-1}.years.

METHODS

The study group comprised 21 maintenance workers who were employed in three buildings where asbestos-containing materials were present. Twenty-six separate tasks were identified where exposure to asbestos may have occurred. A questionnaire was used to obtain details of the workers' duration of employment in each task, along with the frequency and duration of the work. Each worker was also interviewed to obtain descriptions of the tasks they had carried out.

The exposure level to respirable fibres was estimated for each task using a semi-quantitative algorithm which has been validated in separate studies (Cherrie *et al.*, 1996; Cherrie and Schneider, in preparation). Estimates of the subject's ventilation rate during a task were derived by firstly asking each worker to rate the physical effort required for a task into five categories (from "rest"—metabolic rate 115 W to "very high"—metabolic rate 520 W). Examples of the type of work activities in each category were shown to the worker to guide their choice. Ventilation rate was estimated using the method outlined in BSI (1994), assuming body height (1.7 m), body weight (70 kg) and the percentage oxygen exhaled (16.2%). Cumulative exposure for each task was estimated as the product of estimated task exposure level and total task duration. Uptake was estimated as the product of the task cumulative exposure and average breathing ventilation rate. An individual's cumulative exposure and uptake were obtained by summing the contributions from each task they had carried out.

RESULTS

Estimates of exposure level were made for 30 tasks (four of the original tasks being subdivided on the basis of the interview responses) as shown in Fig. 1. The estimated exposure levels ranged from < 0.01 to 1.5 fibres ml^{-1}, while the average daily task durations were between 0.03 and 8 h. The highest exposure levels were assessed for cleaning areas contaminated with asbestos or direct work with asbestos-containing materials. In almost all cases the workers were exposed to amosite asbestos (Cherrie, 1996). Breathing ventilation rate for each task was allocated to one of four categories between 12 and 28 l min^{-1}, based on the average response from the workers who had carried out these activities.

Estimated cumulative exposure ranged between 0.002 and 0.93 fibre ml^{-1}.years, while uptake ranged from 4.5×10^6–1.9×10^9 fibres. Figure 2 shows the estimated uptake and cumulative exposure for each individual in the study. The diagonal line on the diagram shows the expected relationship if the ventilation rate were constant at 24 l. min^{-1}. From this it can be seen that there is some evidence that uptake and cumulative exposure were not linearly related, with uptake being lower than expected at higher exposures.

DISCUSSION

The procedures used to estimate workers' exposure have been validated in a previous study (Cherrie and Schneider, in preparation). In that study we predicted asbestos exposure levels for simulated work activities in a contaminated building and then completed the tasks, measuring the personal exposure of those involved. The correlation between the log-transformed estimated and measured exposure levels was high ($r^2 = 0.86$), with a small positive bias in the estimated values (+ 45%).

In the present study it has been shown to be possible to reconstruct past asbestos exposure and uptake for building maintenance workers, based on information collected during interviews. The application of these procedures may provide a way of estimating the likely risks to health which may occur in specific working populations. For example, based on the present exposure assessments it has been estimated that on average for these maintenance workers there is a lifetime risk of dying from mesothelioma of approximately one or two per thousand (Cherrie, 1996).

J. W. Cherrie

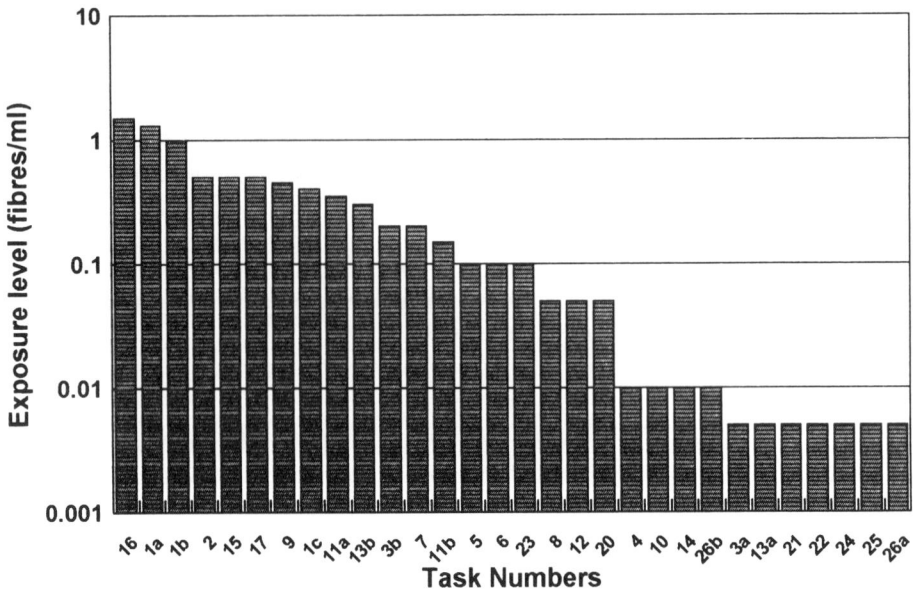

Fig. 1. Estimated exposure levels for each task.

Task	Description
1	Working near dusty work when building being completed (Building A)
2	Sweeping or cleaning sub-basement prior to painting (Building A)
3	Cleaning or changing air filters in sub-basement (Building A)
4	Cable running in ducts (Building A)
5	Cable running above false ceiling (Building A)
6	Maintenance on document conveyor system (Building A)
7	Brushing out ventilation ducts (Building A)
8	Maintenance on ventilation boxes (Building A)
9	Boiler inspection including removal of asbestos rope (Building A)
10	Removal of riser panels (Building C)
11	Cable running above false ceiling (Building B)
12	Removal of light fittings (Building B)
13	Maintenance on ventilation ducts (Building B)
14	Taking insulation samples for asbestos analysis (Building A)
15	Cable running (Building C)
16	Drilling through asbestos panel (Building A)
17	Replacing fire dampers in ventilation system (Building A)
18	Boiler inspection: pipes and asbestos lagging (Building B)
20	Cable running in floor ducts (Building C)
21	Ceiling work (Building C)
22	Washing light fittings and diffusers (Building C)
23	Cable running in floor ducts (Building B)
24	Cable running in boiler house (Building B)
25	Working in attic area above kitchen (Building B)
26	Replacing lamps (Building A)

There is some evidence that estimated uptake may not be linearly related to cumulative exposure, because of the association of lower estimated ventilation rates with dustier tasks. The majority of the activities with higher estimated exposure levels involved bystander exposure during the cleaning, by others, in the course of building construction. Differences in ventilation rate, either between

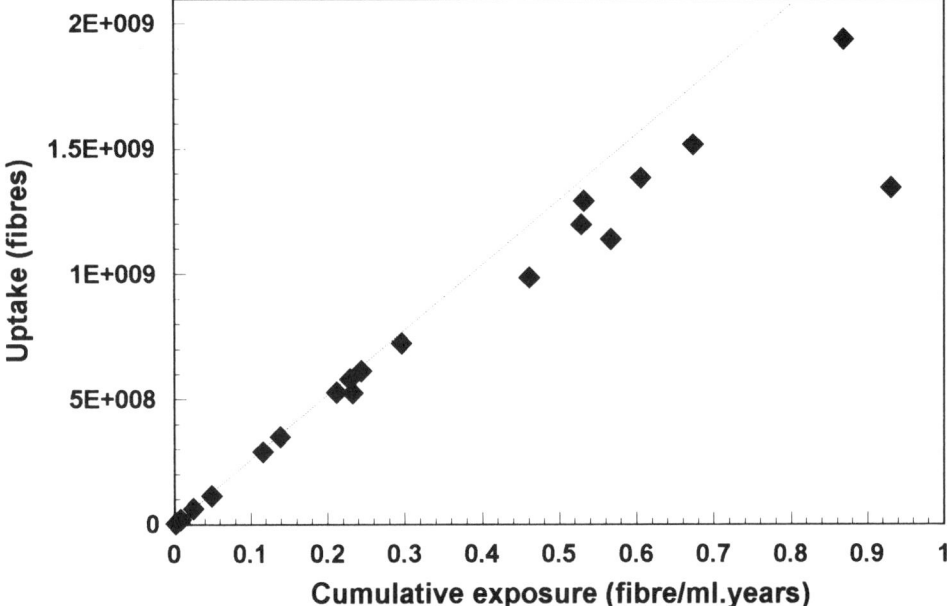

Fig. 2. Comparison of estimated uptake and cumulative exposure for the maintenance workers. Diagonal line shows expected relationship for a constant breathing ventilation rate of 24 l.min^{-1}.

different jobs or different individuals, could be as large as five times, although in this study the differences were on average much smaller. In most epidemiological studies and risk assessments it is assumed that differences in breathing rate between individuals are small and lead to non-differential missclassification of workers' exposures. However, the results from this study show that this may not always be true and this may have implications for the reliability of risk assessments based solely on estimates of cumulative exposure.

REFERENCES

Bowes, S. M., Frances, M., Laube, B. L. and Frank, R. (1995) Acute exposure to acid fog: influence of breathing pattern on effective dose. *Am. ind. Hyg. Assoc. J.* **56**, 143–150.

British Standards Institution (1994) *Ergonomics. Determination of Metabolic Heat Production.* BSEN 28996. BSI, London.

Cherrie, J. W. (1996) Reconstructing the past: estimating exposure to hazardous substances in occupational epidemiology. PhD thesis, University of Aberdeen, U.K.

Cherrie, J. W. and Schneider, T. Validation of a new method for structured subjective assessment of past concentrations (in preparation).

Cherrie, J. W., Schneider, T., Spankie, S. and Quinn, M. (1996) A new method for structured subjective assessments of past concentrations. *Occ. Hyg.* **3**, 75–83.

Fiserova-Bergerova, V. (1985) Toxicokinetics of organic solvents. *Scand. J. Work Environ. Health* **11** (Suppl 1), 7–21.

Jones, C. O., Gauld, S., Hurley, J. F. and Rickmann, A. M. Personal differences in the breathing patterns and volumes and dust intakes of working miners. IOM Report number TM/81/11. Institute of Occupational Medicine, Edinburgh.

Oldham, P. D. and Roach, S. A. (1952) A sampling procedure for measuring industrial dust exposure. *Br. J. ind. Med.* **9**, 112–119.

Opdam, J. J. G. (1989) Respiratory input in inhalation experiments. *Br. J. ind. Med.* **46**, 145–156.

Wallen, M., Holm, S. and Byfalt-Nordqvist, M. (1985) Co-exposure to toluene and p-xylene in man:uptake and elimination. *Br. J. ind. Med.* **42**, 111–116.

 Pergamon

Ann. occup. Hyg., Vol. 41, Supplement 1, pp. 172–177, 1997
© 1997 British Occupational Hygiene Society
Published by Elsevier Science Ltd. All rights reserved
Printed in Great Britain
0003–4878/97 $17.00 + 0.00
Inhaled Particles VIII

PII: S0003–4878(96)00115–9

OCCUPATIONAL EXPOSURE FOR TUNGSTEN OXIDE FIBRES, FIBRE–DOSE, DURING HARD METAL PRODUCTION

W. Sahle, S. Krantz, B. Christensson and I. Laszlo

National Institute for Working Life, Department of Work Organisation and Technology,
S-171 84 Solna, Sweden

INTRODUCTION

Hard metal is manufactured by a process of powder metallurgy from tungsten metal and carbon with cobalt as a binder. In addition, chromium, molybdenum, vanadium, titanium, tantalum and nickel may be added. Therefore, hard-metal workers may be exposed to several potentially toxic materials and mixed dust inhalation is very common. In view of the mixed dust exposures and the variable histological appearances of lung tissues from hard metal workers, the term "mixed dust pneumoconiosis" was suggested by Rüttner *et al.* (1987). This was before the presence of tungsten oxide fibres in the environment was known.

Following the discovery of airborne tungsten oxide fibres in working environment (Sahle, 1992), static and personal samples were collected at two leading hard-metal production plants in Sweden. A total of 174 static samples; 137 for total dust, 37 for respirable dust and 68 personal samples were studied. Personal samples were collected for whole shifts on workers at different working days and also short sampling times were used for monitoring specific work task. The samples were studied by gravimetric method and the tungsten oxide fibres with an aspect ratio $\geq$ 5 and a diameter $\leq$ 3 µm for fibre length $\geq$ 5 and 0.5 µm were studied by transmission electron microscope. Details of sampling procedures and a preliminary personal exposure data were presented in a previous paper (Sahle *et al.*, 1996).

The preliminary report showed that dust/fibre exposure of workers varied from day-to-day. Exposure levels are raw material and task dependant. The real exposure of workers should, therefore, take into considerations the time spent in the different work tasks and the use of respirator. Assuming full protection with respirator, an empirical occupational exposure-dose per year and person was evaluated for hard metal workers involved in production of tungsten metal. The fibre dose-estimation was calculated based on personal sample values and from estimated duration, frequency and task-share among workers for different activities for a duration of a year. This is presented below.

METHODOLOGY OF ANALYSIS

The overall occupational duration of exposure-time for a worker in a year, D_y, was approximated to 1600 h (43 weeks/year, 215 working shifts per year, each of 480 min and 30 min of pause). Workers alternate between the following activities;

material transport, charging of raw material, functional inspection of furnaces, vessels exchange, sieving, grinding and mixing of unreduced material, cleaning of factory floor, furnace maintenance and so forth. Some of the activities were done more than once per shift, while other activities were done once per- shift, day, or week etc. The personal exposure levels were, therefore, task frequency dependant and during cleaning of cyclone and restoration of furnaces workers used respirator.

The accumulated duration of exposure-time for a specific task per person and year, D_i, was normalised as follows:

$$D_i = \frac{d_i}{D_y} \tag{1}$$

where d_i, was an estimated exposure time in hours in a duration of a year for a task and person, and D_y was the total working hours per person and year. The normalisation gives

$$\sum_{i=1}^{n} D_i = \sum_{i=1}^{n} \frac{d_i}{D_y} = 1. \tag{2}$$

The exposure-dose for a task is given as a product of dust/fibre concentration for a task (C_i) and the normalised duration of the task per year and person (D_i). The overall exposure dose for a worker is that estimated as shown in eq (3), by summation of the sub-tasks exposure-dose levels. The exposure time for activities done under respirator was estimated to zero hours of exposure per year.

$$\text{Exposure-dose per person and year} = \sum_{i=1}^{n} (D_i * C_i). \tag{3}$$

RESULTS

Three different initial materials were used for production of tungsten metal. The raw material charged varied in amounts. In one of the factories studied, referred as factory 1 below, the reduction was conducted in two stages and ammonium-paratungstet or tungsten blue oxide were used as raw materials. The second factory-studied used tungsten trioxide and a single reduction system. The personal samples investigated showed variations in airborne dust content and concentrations depending on raw materials used. Figure 1 shows the mean concentrations for personal samples obtained when the following raw materials were charged; ammonium paratungstet (APT), blue tungsten oxide (BTO) and yellow tungsten trioxide (YTO), respectively. For details see Sahle et al. (1996).

At the raw material charging site in factory one, there is a local exhaust. Therefore, the total dust collected on the personal samples amounted to 1–2 mg m^{-3}. The YTO is used as raw material in the second factory. As the raw material charging site does not have local exhaust, the total dust level of personal samples was about 8 mg m^{-3}. Raw material charging took about 10% of the total work activities in both factories studied. The raw materials APT and YTO were non-fibrous, while the blue oxide was a fibrous material. Therefore, as clearly seen in Fig. 1, the airborne fibre concentration values of personal samples during BTO charging showed higher fibre exposure level.

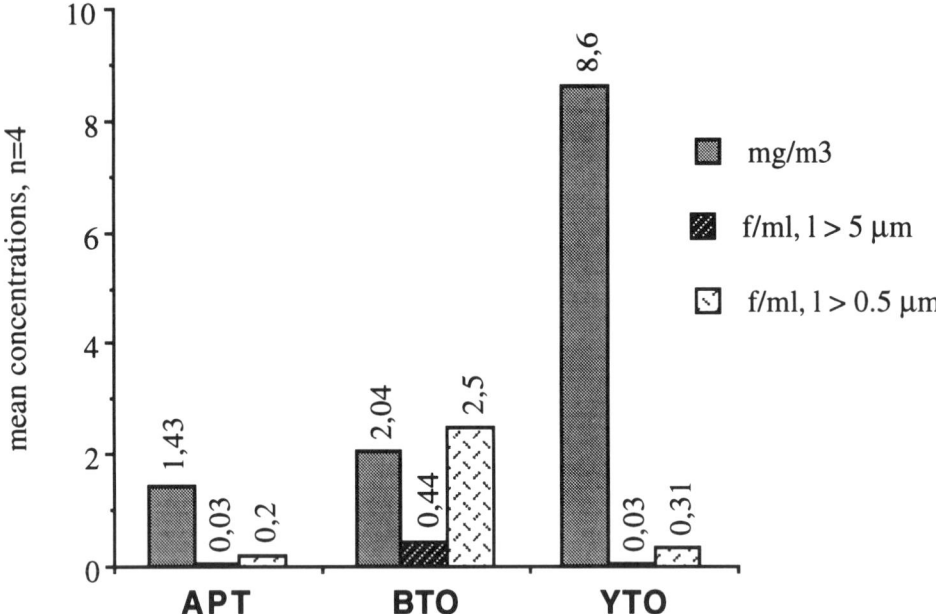

Fig. 1. Mean concentrations for personal samples during raw material charging.

The variability of the personal exposure levels for different tasks and the predominance of short fibres over fibres longer than 5 μm is illustrated in Fig. 2. Excluding samples collected during activities conducted with respirator, higher level of personal exposure for tungsten oxide fibres occurs during handling of unreduced material and during charging of the blue tungsten oxide as raw material. The overall accumulated dust/fibre exposure-dose per year and person is shown in Fig. 3. These results represent the actual dust and fibre exposure-dose of workers during tungsten metal production. The total dust exposure level respective of the raw material used was about 1.2–1.4 mg m^{-3}. Although the total dust level was approximately the same for both factories, the fibre exposure levels for fibres $1 \geq 5$ μm was about 10-fold during the use of APT and BTO compared to the YTO raw material. This implies that the choice of raw material affects the level of fibre exposure for these industries. Since BTO is formed by APT calcination, the overall personal exposure values does not differ between the use of the BTO or the APT as raw materials.

DISCUSSION

Since the pioneer work of Jobs and Ballhusen (1940) on hard-metal disease, several occupational health problems, such as interstitial fibrosis, occupational asthma and extrinsic allergic alveolites, are recognised in the hard-metal production industries. The extensive literature has been reviewed several times (Fairhill *et al.*, 1947; Bech *et al.*, 1962; Cugell 1992). The cause of occupational asthma is well established, but the cause of lung fibrosis is still debated.

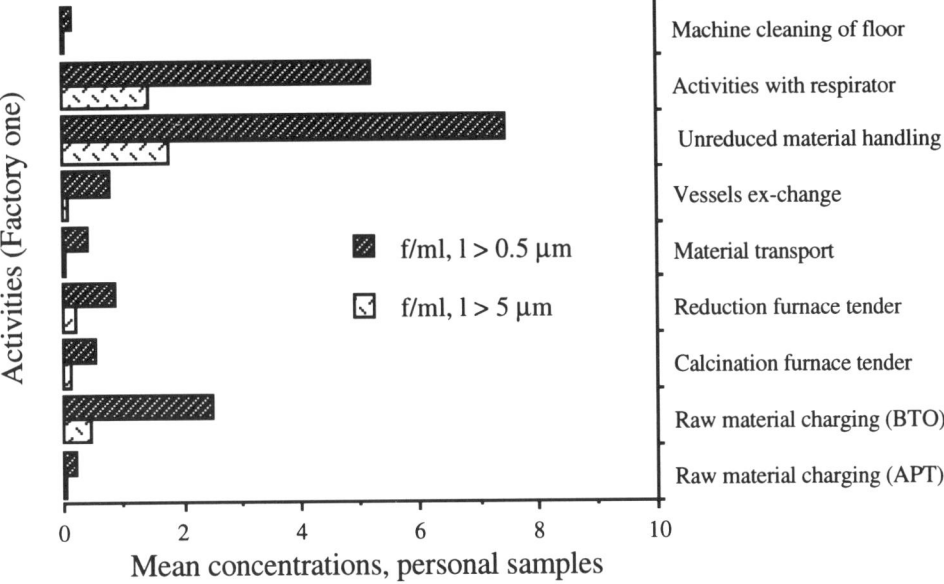

Fig. 2. Illustration of the task dependance of personal fibre exposure levels.

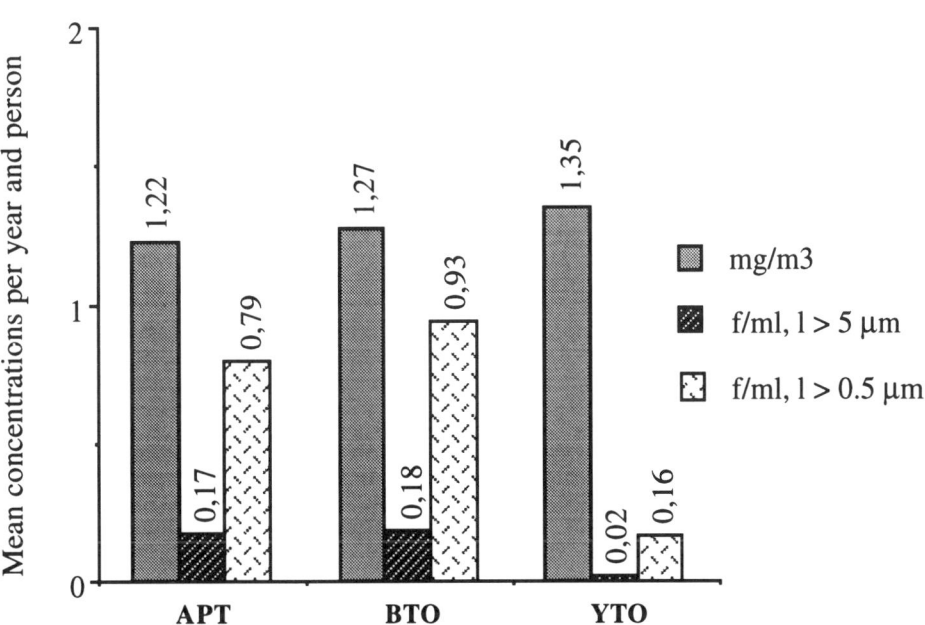

Fig. 3. The actual dust/fibre exposure estimate of workers during tungsten metal production using the raw material as bases.

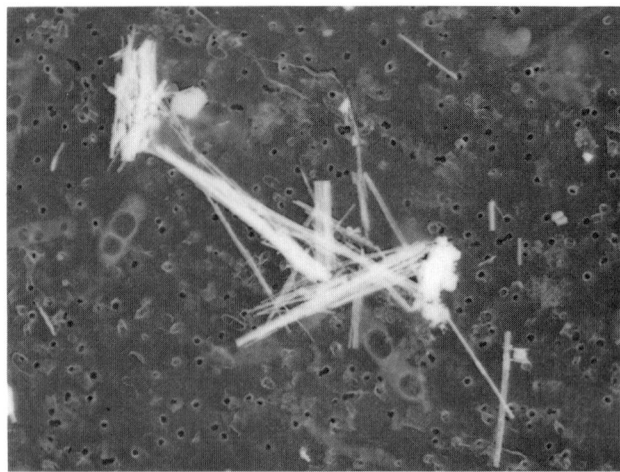

Fig. 4. Transmission electron micrograph of airborne personal tungsten oxide fibres, collected on 0.2 μm nuclepore filters.

The airborne dust that resulted in fatal lung fibrosis among hard-metal workers is a heterogeneous mixture consisting of multiple components. Tungsten has been regarded as an inert element with regard to inducing pulmonary effects, but tungsten oxide fibres may be carcinogenic (Leanderson and Sahle, 1995). Tungsten oxide fibres are found to generate hydroxyl radicals and are cytotoxic to human lung cells, therefore, the fibres are potentially harmful when inhaled. Furthermore, the fibres are freshly formed when released into the ambient air. Therefore, its toxic activity is possibly higher than that reported.

On the one hand, tungsten metal production is always associated with formation of intermediary fibrous oxides. The airborne tungsten oxide fibres, whose morphology is illustrated in Fig. 4, are very thin and most of these can not be seen under phase contrast optical microscope. Most of the fibres are shorter than 5 μm, but with very high aspect ratio (Sahle and Laszlo, 1996).

On the other hand, hard-metal workers with occupationaly caused lung diseases have always showed very large amounts of tungsten in their lungs. Compared with controls, the exposed workers had 70 000 times higher levels of tungsten in their lungs (Rizzato *et al.*, 1986). Although, the durability of such tungsten oxide fibres in a biological environment is as yet unknown, its casual associations in terms of source, ambient concentration and exposure–dose relations are important information to link occupational exposure and clinical outcome.

In conclusion, this report demonstrates that reliable information on dust exposure levels should be derived from personal measurement, but the formal reporting of exposure data ought to take in to considerations the duration of the different work activities. Furthermore, the study clearly manifests that a single reduction stage not only reduces dustiness and the number of furnaces demanding maintenance, but also the tungsten trioxide raw material as it is a non-fibrous material is preferable with respect to reducing workers exposure to the fibrous tungsten oxide.

Acknowledgement—The authors dedicate this article to the late Professor Arne Magnéli, the sole of Magnéli phases, and among other matters for his outstanding contributions in the tungsten oxide system. The study was partially supported by the Swedish council for Work Life Research, contract number 93–0077.

REFERENCES

Bech, A. O., Kipling, M. D. and Heather, J. C. (1962) Hard metal disease. *Br. J. ind. Med.* **19**, 239–252.

Cugell, D. W. (1992) The hard-metal diseases. *Clinics in Chest Med.* **13**, 269–279.

Fairhall, L. T., Castberg, H. T., Carrozzo, N. J. and Brinton, J. P. (1947) Industrial hygiene aspects of the cemented tungsten carbide industry. *Occup. Med.* **4**, 371–379.

Jobs, H. and Ballhausen, C. (1940) Powder metallurgy as a source of dust from the medical and technical standpoint. *Vertravenargt* **5**, 142–148.

Leanderson, P. and Sahle, W. (1995) Formation of hydroxyl radicals and toxicity of tungsten oxide fibres. *Toxic. in Vitro* **9**, 175–183.

Rizzato, G., Lo-Cicer, S., Barberis, M., Torre, M., Pietra, R. and Sabbioni, E. (1986) Trace of metal exposure in hard-metal lung disease. *Chest* **89**, 101–106.

Rüttner, J. R., Spyche, M. A. and Stolkin, I. (1987) Inorganic particulates in pneumoconiotic lungs of hard-metal grinders. *Br. J. Ind. Med* **44**, 657–660.

Sahle, W. (1992) Possible role of tungsten oxide whiskers in hard-metal pneumoconiosis. *Chest (Lett.)* **102**, 1310.

Sahle, W. (1995) Formation and emission of tungsten oxide fibres during hard-metal production. In *New Materials and the Working Environment* (Edited by U. Midtgård), pp. 77–97. National Institute of Occupational Health, Denmark, Copenhagen.

Sahle, W., Krantz, S., Christensson, B. and Laszlo, I. (1996) Preliminary data on hard-metal workers exposure to tungsten oxide fibres. *Sci. Total Environ.* **191**, 153–167.

Sahle, W. and Laszlo, I. (1996) Airborne inorganic fibre level monitoring by TEM: comparison of direct and indirect sample transfer methods. *Ann. occup. Hyg.* **40**, 29–44.

 Pergamon

Ann. occup. Hyg., Vol. 41, Supplement 1, pp. 178–183, 1997
© 1997 British Occupational Hygiene Society
Published by Elsevier Science Ltd. All rights reserved
Printed in Great Britain
0003–4878/97 $17.00 + 0.00
Inhaled Particles VIII

PII: S0003–4878(96)00091–9

REVERSIBILITY OF FIBROTIC LESIONS IN RATS INHALING SIZE-SEPARATED CHRYSOTILE ASBESTOS FIBRES FOR 2 WEEKS

K. E. Pinkerton,* A. A. Elliot,* S. R. Frame† and D. B. Warheit†

*University of California, Davis; and †DuPont Haskell Laboratory, PO Box 50, Elkton Rd, Newark, DE19714–0050, U.S.A.

INTRODUCTION

Inhaled asbestos fibres cause lung disease in both humans and animals. Fibre inhalation results in the development of cellular injury, inflammation and tissue remodelling. Epithelial proliferation and excessive collagen deposition are major features associated with this process. Alterations in lung tissues at alveolar duct bifurcations have been noted following exposures as short as 1 h to aerosolized asbestos fibres (Brody et al., 1981; Warheit et al., 1984; Chang et al., 1988). One consequence of this pattern of fibre deposition is a highly localized and active region of epithelial injury and repair with the proliferation of alveolar type II cells and interstitial cells. These changes progress during the first 48 h following exposure and continue to be evident 1 month after fibre exposure has ended.

The long-term consequences of limited exposure to asbestos fibres are not as well defined. Although a number of studies have examined exposure to asbestos fibres and the sequelae over a long period of recovery for lung tumor incidence, there have been no studies which have quantitatively assessed tissue changes occurring during a long postexposure period or following short-term exposures to chrysotile asbestos fibres. The purpose of this study was to examine the consequences of short-term exposure to aerosolized chrysotile asbestos fibres over a 1 year period of recovery in the lungs of rats. This study specifically addresses the issue of whether the initial changes seen immediately following asbestos exposure continue to persist or to progressively worsen with longer periods of recovery.

METHODS

Groups of male Crl:CDBR rats (7–8 weeks old, Charles River Breeding Laboratories, Kingston, New York) were exposed by inhalation 6 h day^{-1}, 5 days week^{-1} for 2 weeks to mean fibre concentrations of 458 and 782 fcc^{-1} and assessed at several postexposure time periods. After completion of exposures, the lungs of chrysotile-exposed animals and aged-matched sham controls were assessed for asbestos-induced histopathological effects immediately after exposure, as well as 5 days, 1, 3, 6 and 12 months postexposure.

Animals were anesthetised with an intraperitoneal injection of sodium pentobar-

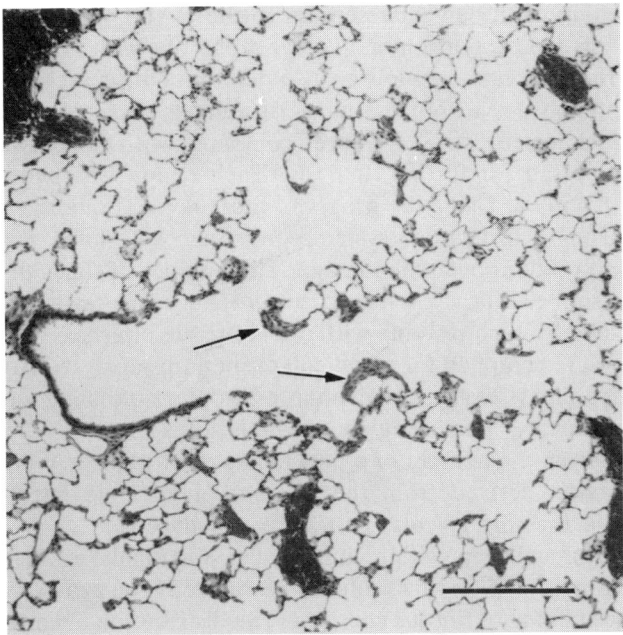

Fig. 1. Bronchiole–alveolar duct junction (BADJ) in longitudinal profile. The terminal bronchiole and a pair of alveolar ducts with an intervening tissue ridge or first alveolar duct bifurcation is present. BADJs and the central acini from the lungs of rats 1 week after a 2 week exposure to asbestos fibres. Thickening of alveolar septal tips and walls is illustrated (arrows) and was most extensive through 4 weeks, but showed a reduction at 13 weeks which continued through 1 year postexposure. Scale bar is 250 μm.

bital and the lungs fixed by intratracheal instillation of glutaraldehyde at a hydrostatic pressure of 20 cm of fixative. Within 48 h, tissue slices from the lungs were embedded in paraffin. Sections 5–6 μm thick were cut with a rotary microtime and stained with hematoxylin and eosin, or sirius red for histologic examination. Tissue sections were examined microscopically and bronchiole-alveolar duct junctions (BADJs) were identified. Selection of BADJs was based on the criterion of a terminal bronchiole opening directly onto the longitudinal profile of an alveolar duct (Fig. 1). In many instances, the first alveolar duct bifurcation formed by the tissue ridge separating two alveolar ducts arising from the same terminal bronchiole was also present.

Tissue sampling and morphometric analysis

Morphometric analysis was carried out only on the high dose (782 fcc^{-1}) animals. Each BADJ selected was examined for the presence of delicate, lacy tissues forming the alveoli of the duct wall along with regions showing marked thickening. Alveolar wall thickening was defined as an increase in the cellularity and volume of tissues forming the alveolar septal tip, the opening forming the entrance from individual alveoli to the alveolar duct lumen (Fig. 1).

To determine the extent of changes occurring within the central acinus following exposure to fibres, the distance from the most distal portion of the alveolar duct showing thickening of the alveolar septum to the BADJ was measured for each isolation. The relative proportion of alveolar septal tips with thickening of 50% or greater compared to control tissue was also determined for each isolation.

Alveolar septal tissues within the lungs of animals exposed to fibres that had increased in thickness by 50% or more, compared with lung tissues of control animals were selected for further analysis. Each region was photographed at a magnification of 1500×. Using the program, Stereology Toolbox, each tissue field was analyzed using a test lattice system consisting of 21 lines, each 2.25 cm in length, placed at random over each image. The number of times the ends of each line fell over tissue, capillary, or alveolar macrophage was recorded. The number of intercepts made by each test line with the air–tissue interface was also counted. Measurements were compiled for all alveolar septal tips analyzed from a minimum of 4 BADJs per animal. Tissue and alveolar macrophage volumes were normalized to the alveolar surface area using the formula, $V_v S_v$, where V_v is the volume fraction derived from the number of points falling on structures of interest and S_v is the surface fraction derived from the number of test line intercepts with the air-to-tissue surface within alveolar tissue areas analyzed.

The presence of collagen within lung tissues was analyzed in tissue sections stained with sirius red in picric acid solution. Sirius red is an azo dye which is highly selective for the staining of fibrous collagen. The distribution of collagen fibrils and bundles throughout parenchymal tissues and thickened alveolar septal tips was determined at each time period examined following exposure to fibres.

RESULTS

Repeated exposure to asbestos fibres over a 10-day period resulted in significant changes occurring in alveolar ducts immediately beyond the terminal bronchiole. These changes under the exposure regimen used in our study were not confined to only first alveolar duct bifurcations, but extended along the length of the alveolar duct. The BADJ could be easily identified along with numerous prominent thickenings of alveolar septal tips and bifurcations along the length of the alveolar duct forming the pulmonary acinus (Fig. 1). The most prominent changes within the alveolar duct noted as increased cellularity and septal wall thickening, was evident during first month after fibre exposure had ended. With greater postexposure time, the prominence of septal tips with increases in the cellularity and thickness of alveolar walls, decreased significantly (Fig. 2).

The volume of alveolar tissues formed by septal tips and walls and normalized to alveolar surface area is shown in Fig. 3 for control and asbestos-exposed animals. Measurements taken from the lungs of rats exposed to asbestos fibres were confined to only those septal tips and alveolar walls found to be increased in cellularity and thickness. Thickening of alveolar septal tissues was almost doubled compared with that seen in control animals immediately following as well as 5 days and 1 month following completion of the exposure. By 3 months postexposure, a significant reduction was noted in the volume of affected alveolar tissues in the lungs of exposed animals. Although alveolar septal tips and walls were on average thicker than those seen in control animals, the differences between control and treated animals were not statistically significant. Alveolar tissue volume, even in

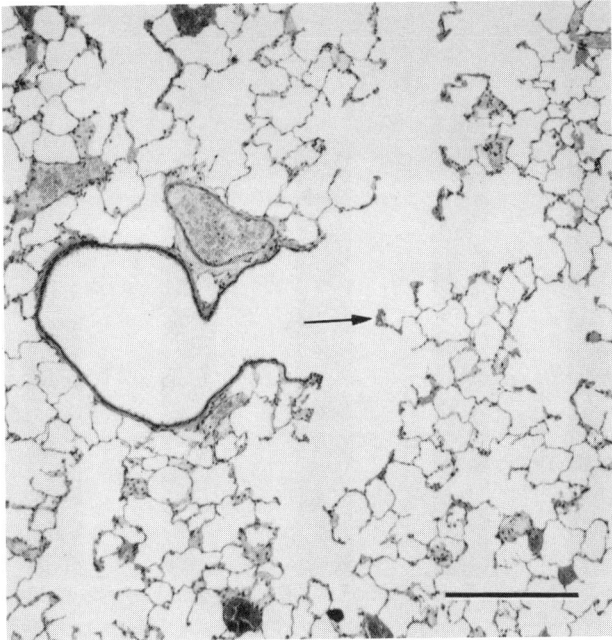

Fig. 2. BADJs and the central acini from the lungs of rats following 52 weeks postexposure to asbestos fibres. The lesion characterised by septal wall thickening in Fig. 1 is essentially reversible (arrows) at 1 year postexposure. Scale bar is 250 μm.

the most affected regions of the lungs of animals exposed to fibres did not change significantly from 3 months to 1 year after the end of exposure. Over this same period of time, tissue volumes in control animals did not change significantly.

The volume of macrophages normalised to alveolar surface area was significantly elevated through the first 3 months following exposure to aerosolised fibres (Fig. 4). A five-fold elevation above control values was noted 1 month after the end of exposure. Following 3 months recovery, macrophage volume was significantly decreased. No significant differences were noted in the volume of macrophages 26 or 52 weeks after fibre exposure compared with control values.

DISCUSSION

This study illustrates that the inhalation of asbestos fibres over a short period of time leads to significant changes within the alveolar tissues of the lung. These changes consist of increased cellularity and thickening of the alveolar walls and septal tips which form the alveolar duct immediately beyond the BADJ. These alterations were most prominent immediately following exposure, but by 1 month postexposure, changes had reached their maximum extent and had begun to resolve. This pattern had been noted in earlier studies with prominent changes persisting 1 month after the end of exposure. From our studies we found that further recovery time from exposure to the fibres resulted in a significant reduction

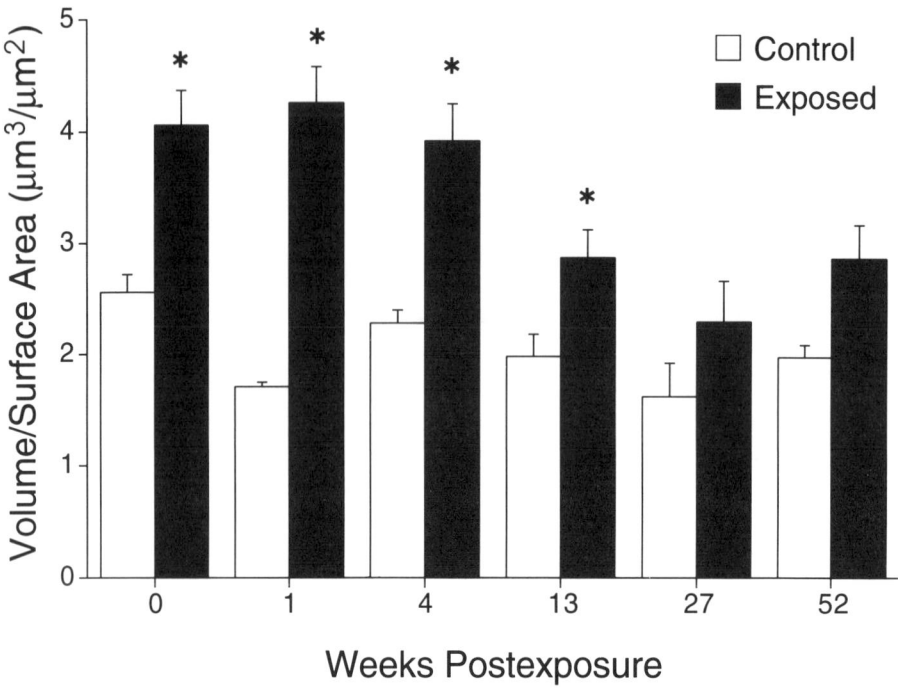

Fig. 3. Alveolar tissue volume (mean ± SEM) following exposure to asbestos fibres. All measurements are normalised to the alveolar surface area for comparison. An asterisk denotes a statistically significant difference ($P < 0.05$) compared with control.

and some resolution of the injury and remodelling of the tissues associated with fibre inhalation. These consisted of a reduction in the cellularity of the alveolar tissues, particularly those areas that were most affected initially by exposure to asbestos fibres. Although there remained a slight thickening of septal tissues, these were not statistically significant from that noted in control animals. As early as 3 months following exposure, tissue changes had shown resolution with no further progression 1 year after the end of fibre exposure. The distance into the alveolar duct in which tissue changes were seen did not change (data not presented). The proportion of the alveolar duct wall involved in tissue changes was significantly reduced as early as one month after the end of exposure. These findings suggest that with the inhalation of asbestos fibres, there is extensive involvement of the lung parenchyma that is not confined to the first alveolar duct bifurcation. These changes are typical of what has been reported in the past. This study demonstrates that with repeated exposures to high levels of respirable fibres over the short-term does not lead to a progression and a worsening of injury over the long-term period of recovery as originally speculated. There is significant resolution of the initial tissue changes that were noted up to 1 month following the end of fibre exposure. This resolution appears to occur as early as 3 months following exposure and does not show progression to a more severe state 1 year after the end of fibre exposure.

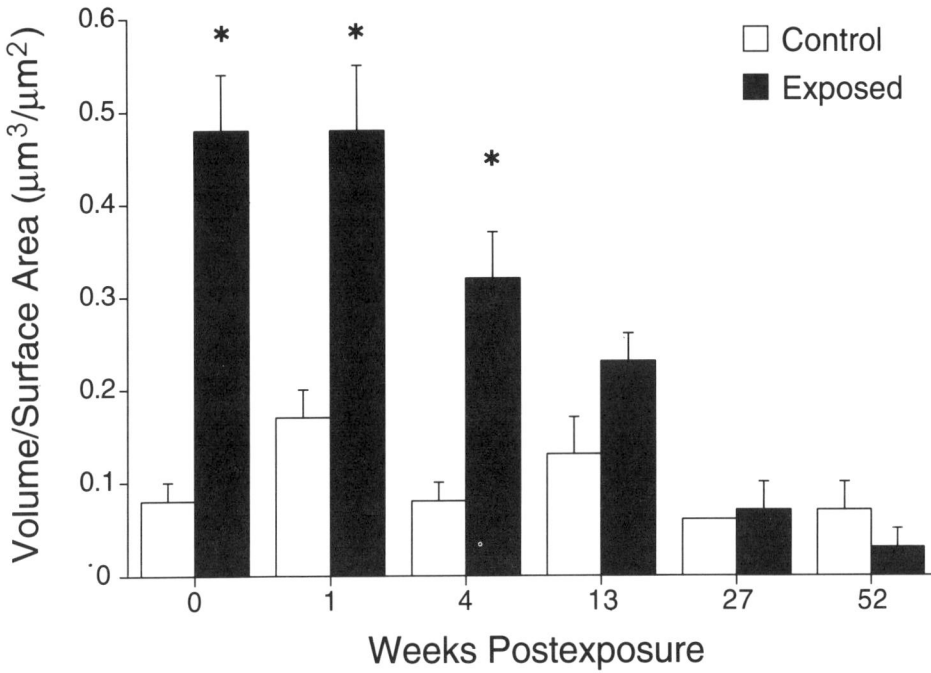

Fig. 4. Alveolar macrophage volume (mean ± SEM) following exposure to asbestos fibres. All measurements are normalised to the alveolar surface area. An asterisk denotes a statistically significant difference ($P < 0.05$) compared with control.

Acknowledgments—This study was supported in part by a grant from the NIEHS Center in Environmental Health—ES 05707 and a base grant to the California Regional Primate Research Center RR00169. The authors acknowledge the technical support of Mark A. Hartsky and Suzanne I. Snajdr.

REFERENCES

Brody, A. R., Hill, L. H., Adkins, Jr, B. and O'Connor, R. W. (1981) Chrysotile asbestos inhalation in rats: deposition pattern and reaction of alveolar epithelium and pulmonary macrophages. *Am. Rev. Respir. Dis.* **123**, 670–679.
Chang, L. Y., Overby, L. H., Brody, A. R. and Crapo, J. D. (1988) Progressive lung cell reactions and extracellular matrix production after a brief exposure to asbestos. *Am. J. Pathol.* **131**, 156–170.
Warheit, D. B., Chang, L. Y., Hill, L. H., Hook, G. E. R., Crapo, J. D. and Brody, A. R. (1984) Pulmonary macrophage accumulation and asbestos induced lesions at sites of fiber deposition. *Am. Rev. Respir. Dis.* **129** 301–310.

Pergamon

Ann. occup. Hyg., Vol. 41, Supplement 1, pp. 184–188, 1997
© 1997 Published by Elsevier Science Ltd on behalf of BOHS
Printed in Great Britain. All rights reserved
0003–4878/97 $17.00 + 0.00
Inhaled Particles VIII

PII: S0003–4878(96)00116–0

INVESTIGATION OF THE DURABILITY OF CELLULOSE FIBRES IN RAT LUNGS

H. Muhle, H. Ernst and B. Bellmann

Fraunhofer Institute of Toxicology and Aerosol Research, Nikolai-Fuchs-Str. 1,
D-30625 Hannover, Germany

INTRODUCTION

Inhalable cellulose fibres can be released in production processes in the paper industry or during spraying of specially prepared chips of recycled newspapers which are used for thermal insulation of walls in houses. In the latter process, fibre concentrations may amount up to 50×10^6 fibres m^{-3} in building sites (Tiesler and Schnittger, 1992). For the definition of a "critical" fibre the same criteria as for inorganic fibres were used (fibre length ≥ 5 μm, fibre diameter < 3 μm, aspect ratio $\geq 3:1$).

Milton *et al.* (1990) instilled cellulose dust intratracheally in hamsters by a single dose of 0.75 mg/100 g body weight. Hamsters were killed after 8 weeks. Lungs of animals showed a significant number of granulomata and thickened interalveolar septae.

In a 28-day inhalation study with cellulose in rats, Davis (1993) reported the appearance of alveolitis and granulomata. The test material originated from thermal insulation products. It is an open question, whether cellulose fibres can accumulate in lungs or whether these fibres will show a degradation under physiological conditions. The goal of this investigation was to analyse quantitatively the biodurability of cellulose fibres in lungs of rats.

MATERIALS AND METHODS

Chemical pure microcrystalline wood cellulose fibres (type Avicel PH 105 Serva, Heidelberg, Germany) and the commercially available product Isofloc originating from recycled newspaper were classified to obtain a respirable fraction of these materials, Isofloc is used as a thermal insulating product and was treated with 12% borax and 8% boric acid. Airborne fibres were sized in a high-volume heavy-gain impaction sampler (Bellmann and Muhle, 1994). The obtained rat-inhalable fractions were named Cellulose-F (reference substance) and Isofloc-F. The size distribution was determined by electron microscopic investigations. For Isofloc-F the median fibre length was 7.6 μm, the median fibre diameter was 0.50 μm. For Cellulose-F the corresponding values were 4.2 and 0.87 μm, respectively (see Table 1).

Test materials were suspended in saline and 2 mg per animal was instilled intratracheally into 10-weeks-old female Wistar rats (30 animals per group). This

Table 1. Percentiles of the frequency distribution of the length and diameter of fibres in the sized materials

Material	Fibre length [μm]				Fibre diameter [μm]			
	10% <	50% <	90% <	99% <	10% <	50% <	90% <	99% <
Isofloc-F	3.8	7.6	22.9	40.6	0.15	0.50	1.49	2.90
Cellulose-F	1.9	4.2	8.4	14.2	0.36	0.87	1.49	2.56

amount of mass leads to dust overload conditions in lungs. Macrophage-mediated particle clearance is reduced at this lung burden (Muhle *et al.*, 1990). A third group of 30 animals was treated by saline only. At 2 days, 1, 3, 6 and 12 months after treatment at each date six animals per group were sacrificed and lungs were isolated. Tissues of four lungs per group were digested by 26% hypochlorite for 4 h at 4°C on a roller. The suspension was centrifuged for 25 min at 26 900 **g**. The pellet was resuspended in water and filtered on a nuclepore filter. The treatment of test materials by hypochlorite under these conditions did not change the size distribution of the fibres. The wood Cellulose-F was analysed by a scanning electron microscope. Fibres originating from recycled newspapers were thinner and, therefore, investigated in a transmission electron microscope. For further details of methods of fibre analysis see Bellmann and Muhle (1994). Two lungs were taken for histopathological examination after H and E stain.

For calculation of clearance kinetics a regression analysis of logarithm of number or mass of fibres per lung of individual animals versus time after instillation was done (Muhle *et al.*, 1994).

RESULTS

Lung weight

Lung weight was significantly increased in the Isofloc-F-treated group up to half a year after treatment. In the Cellulose-F-exposed group only a mild increase of lung weight was observed. After one year no statistically significant differences were seen in both treatment groups.

Fibre retention

One year after treatment of the experimental animals cellulose fibres of both materials were present in lungs. Fibre number of the reference material Cellulose-F was almost unchanged in the lungs (see Fig. 1). The calculated half-time of the fibre clearance was in the range of 1000 days (see Table 2). This means that this fibre type shows a considerable biopersistence. Up to half a year after treatment for Isofloc-F fibres a splitting into thinner fibrils were observed (see Fig. 2). After one year the evaluation by transmission electron microscopy with regard to fibre number and fibre diameter was difficult for Isofloc-F because branching of fibrils and a low contrast in the electron microscope; however, the presence of these fibres could be demonstrated unequivocally.

Because of the splitting of Isofloc-F fibres no half-time can be calculated for the fibre number. On the basis of fibre mass which was calculated from the fibre size distribution and fibre numbers a half-time of 72 days was obtained for the first half year (see Table 2).

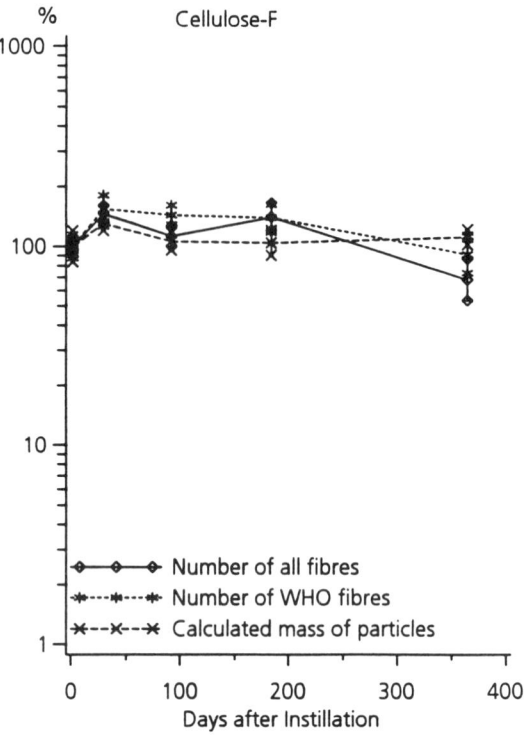

Fig. 1. Retention of microcrystalline wood cellulose fibres and of their calculated particle mass in rat lungs.

Histopathology

Lungs treated with Cellulose-F and Isofloc-F showed almost identical changes. At 2 and 30 days after treatment mild multifocal granulomatous inflammation was observed. Cellulose fibres were phagocytised by alveolar macrophages. After 3 and 6 months fibre associated granulomata, slight interstitial fibrosis, alveolar histiocytosis, alveolar lipoproteinosis and alveolar cell hyperplasias were found. After one year the severity of lesions reported at earlier sacrifice dates increased.

DISCUSSION

The objective of this paper was to investigate the biodurability of the cellulose containing dust Isofloc-F. As this material originates from recycled newspaper dust it contains also other constituents of wood like lignin. Therefore, as a reference

Table 2. Calculated half-times of the clearance with 95% confidence limits

	Half-times in days calculated on the basis of		
Group	Number of fibres	Number of WHO fibres	Mass of fibres
Isofloc-F	—	—	72 (53–113)
Cellulose-F	564 (274–∞)	1046 (351–∞)	> 1000 (676–∞)

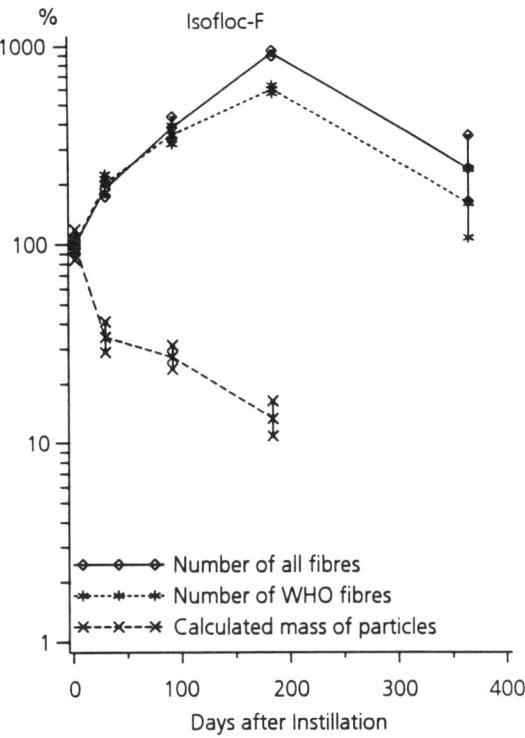

Fig. 2. Retention of fibres from recycled newspaper and of their calculated particle mass in rat lungs.

material chemical pure microcrystalline wood cellulose was used. After sizing these materials rat-inhalable fractions were obtained.

The lung burden which was given to the animals was in the range of "dust overload" of lungs (Muhle *et al.*, 1990). The reason for dose selection was to prevent to a great extent the macrophage-mediated fibre clearance. This type of clearance physically removes fibres from lungs. This could possibly interfere with the kinetics of the degradation of the cellulose fibres. Retention data showed that in rat lungs fibres of microcrystalline wood cellulose are quite persistent. Fibres originating from recycled newspapers splitted in lungs. Obviously no significant enzymatic digestion of cellulose fibres takes place in lungs.

It is concluded that cellulose fibres show a higher biodurability in lungs than ceramic fibres tested by the same protocol (Bellmann and Muhle, submitted). Therefore, cellulose fibres have the potential to accumulate in lungs. Histopathological results and published data show inflammatory reactions and fibrotic lesions in rat lungs. It is recommended that a long-term inhalation study is performed to investigate chronic effects.

Acknowledgements—This project was funded by: Hauptverband der gewerblichen Berufsgenossenschaften, Arbeitsgemeinschaft der Bau-Berufsgenossenschaften and Ökologische Bautechnik Hirschhagen GmbH.

REFERENCES

Bellmann, B. and Muhle, H. Biopersistence of various types of mineral fibres in the rat lung after intratracheal application (submitted).

Bellmann, B. and Muhle, H. (1994) Investigation of the biodurability of wollastonite and xonotlite. *Environ. Health Perspect.* **102** (Suppl. 5) 191–195.

Davis, J. M. G. (1993) The need for standardising testing procedures for all products capable of liberating respirable fibres: the example of materials based on cellulose. *Br. J. ind. Med.* **50**, 187–190.

Milton, D. K., Godleski, J. L., Feldman, H. A. and Greaves, I. A. (1990) Toxicity of intratracheally instilled cotton dust, cellulose, and endotoxin. *Am. Rev. Respir. Dis.* **142**, 184–192.

Muhle, H., Bellmann, B., Creutzenberg, O., Heinrich, U. and Mermelstein, R. (1990) Dust overloading of lungs: investigations of various materials, species differences and irreversibility of effects. *J. Aerosol Med.* **3**, S111–S128.

Muhle, H., Bellmann, B. and Pott, F. (1994) Comparative investigations of the biodurability of mineral fibres in the rat lung. *Environ. Health Perspect.* **102** (Suppl. 5) 163–168.

Tiesler, H. and Schnittger, J. (1992) Untersuchungen zur Belastung durch faserförmige Stäube bei der Verarbeitung von Cellulose-Dämmstoffen. *Zbl. Arbeitsmed.* **42**, 278–285.

 Pergamon

Ann. occup. Hyg., Vol. 41, Supplement 1, pp. 189–193, 1997
© 1997 British Occupational Hygiene Society
Published by Elsevier Science Ltd. All rights reserved
Printed in Great Britain
0003–4878/97 $17.00 + 0.00
Inhaled Particles VIII

PII: S0003–4878(96)00062–2

MESOTHELIOMA AFTER ENVIRONMENTAL CROCIDOLITE EXPOSURE

J. Hansen,* N. H. de Klerk,† A. W. Musk,* J. L. Eccles† and M. S. T. Hobbs†

*Department of Respiratory Medicine, Sir Charles Gairdner Hospital; and †Department of Public Health, University of Western Australia, Nedlands, 6907 Western Australia, Australia

INTRODUCTION

It is well-known that occupational exposure to asbestos, especially crocidolite, causes mesothelioma (Dement *et al.*, 1994; McDonald *et al.*, 1993; Newhouse and Sullivan, 1989; Armstrong *et al.*, 1988) and that the rate of mesothelioma increases with both level of exposure and time since first exposure (de Klerk *et al.*, 1989; Peto, 1985). That non-occupational, or environmental, exposure to asbestos results in an increased risk of mesothelioma is no longer in doubt (Magnani *et al.*, 1993; Joubert *et al.*, 1991; Reid *et al.*, 1990; Wagner *et al.*, 1960). It appears that exposure to crocidolite, particularly by living in the vicinity of a mine or mill, confers the greatest risk (Reid *et al.*, 1990; Siemiatycki, 1982). No previous study of environmental exposure to asbestos and risk of mesothelioma has been able to assess exposure levels which would make the derivation of exposure–response relationships possible (Dodoli *et al.*, 1992; Reid *et al.*, 1990; Siemiatycki, 1982; Hammond *et al.*, 1979).

Nearly 7000 people were employed by the Australian Blue Asbestos (ABA) Company at the Wittenoom crocidolite mine and mill between 1943 and 1966. They experienced an increased risk of mesothelioma which was dose-dependent (Armstrong *et al.*, 1988). A cohort of former residents of Wittenoom, who lived there for at least one month between 1943 and 1993 and were not directly employed in the mining and milling of crocidolite, has been established (Hansen *et al.*, 1993). These residents include: the wives and families of former ABA employees; government employees, such as teachers, hospital staff and police; people who worked for other mining companies which used Wittenoom as a base camp for exploration of the surrounding area; people who were self-employed; and families of these people. Information was obtained from numerous sources, including the local school register, hospital attendances, the WA electoral roll, birth certificates, information provided by Wittenoom crocidolite workers who answered a mailed questionnaire in 1979 and participants in a cancer-prevention programme using vitamin A dietary supplements. This study aims to estimate exposure–response relationships between mesothelioma and environmental exposure to crocidolite in these subjects.

METHODS

All subjects who claimed to have worked with crocidolite at Wittenoom, were excluded from the cohort. Before proceeding with death and disease searches, all females not traced to a current address were sought in the WA marriage register, commencing the year she was last known to be alive, or, for those not known to have married, the year she turned 15. Computerised record linkage procedures were used to search the annual death index for WA from 1969 to 1993, to determine the numbers and causes of deaths to arise in this cohort. Prior deaths were found by manually searching of records from 1943 to 1968. The coded cause of death was routinely extracted from the death tapes for deaths since 1969, or coded by an experienced nosologist for other deaths. Records of all cases of malignant mesothelioma in WA are registered by the WA Mesothelioma Registry. The Australian Mesothelioma Surveillance Programme is notified of other cases occurring elsewhere in Australia. Records of both these registries were inspected to obtain cases arising in this cohort.

The current address of people who were not known to have died and who were not participating in the vitamin A cancer prevention programme, was obtained from the WA Electoral Roll, the Commonwealth Electoral Roll or through relatives. They were mailed a questionnaire which gathered information on the time they spent at Wittenoom, smoking and medical histories and demographic data. Participants of the vitamin A programme provided this information when they enrolled in the programme.

The level of individual exposure was estimated for each subject in the cohort (Hansen *et al.*, 1996). Briefly, subjects not working for ABA directly with crocidolite at Wittenoom were estimated to have an intensity of exposure of 1.0 fml^{-1} from 1943 to 1957 (when a new mill was commissioned and the town was moved from being adjacent to the mill to a site 10 km away) and then 0.5 fml^{-1} between 1958 and 1966, when the mining operations ceased. Since then, exposure of residents in the town continued because of the widespread use of tailings around the town, and individual exposures were assigned by interpolation from periodic surveys from 0.5 fml^{-1} in 1966 to 0.014 fml^{-1} in 1992. Duration of residence was combined with intensity of exposure to give a measure of cumulative exposure for each person, which was adjusted to account for the continuous nature of exposure experienced by these subjects.

Mesothelioma incidence rates were standardised to the World Population. Cox regression (Cox, 1972) was used to examine the separate and combined effect of age, sex, time since first residence at Wittenoom, estimated levels of exposure and age at first exposure on the incidence of mesothelioma. Length of stay at Wittenoom and cumulative exposure were treated as time-dependent covariates, and age at first exposure, sex, calendar period of first exposure, whether a subject lived with an asbestos worker, or washed the clothes of an asbestos worker, were included as fixed covariates. Data were censored at 31 December, 1993 or age 85, whichever occurred first and the "survival time" variable used was time since first residence at Wittenoom. The best transformation for each variable, where appropriate, (categorical, log or untransformed) was assessed by linearity of trends and likelihood ratio criteria and included in the final model if they improved the fit ($P < 0.05$).

RESULTS

One hundred and fifty one subjects claimed to have worked with crocidolite at Wittenoom and were excluded from the cohort. To the end of 1993, 27 cases of malignant pleural mesothelioma had been diagnosed and histologically confirmed by either the WA Mesothelioma Registry or the Australian Mesothelioma Surveillance Programme. A further five cases arose in subjects excluded from this cohort who worked with asbestos at Wittenoom. Another 17 cases of mesothelioma were reported to the WA Mesothelioma Registry who claimed to have had exposure to crocidolite at Wittenoom. They have not been included in this study because their residence was short and was not independently established by methods previously described (Hansen *et al.*, 1993).

Of the 18 female and 9 male cases, 25 arose in people who also experienced "domestic" exposure as they had lived with men who worked with crocidolite at Wittenoom. So far, only 1 case has arisen in a resident who first went to Wittenoom after the mining operations ceased in 1966. Sixteen of the cases (59%) had occurred since 1989. No cases arose within 20 years of first residence at Wittenoom.

When compared with the rest of the cohort, mesothelioma cases stayed longer at Wittenoom (mean stay 65 months vs. 33 months, $P < 0.001$), had a higher average intensity of exposure (mean value 0.3 fml^{-1} vs. 0.5 fml^{-1}, $P < 0.001$), and a higher cumulative exposure to crocidolite (16.3 fy ml^{-1} vs 5.4 fy ml^{-1}, $P < 0.001$). The standardised incidence of mesothelioma to 1993 in this cohort was 207 per million person-years (pmpy) and was similar for males and females (186 and 210 pmpy, respectively). The rate increased significantly with time from first exposure, duration of exposure and cumulative exposure. Incidence rose from 180 pmpy 20–29 years since first exposure, to 791 pmpy 30–39 years since first exposure and 999, 40 or more years since first exposure. Over all times from first exposure, the incidence of mesothelioma rose from 57 pmpy for those who lived at Wittenoom for less than 1 year, to 143 pmpy for residents with duration of residence between 1 and 5 years, to 720 pmpy for those who lived at Wittenoom longer than 5 years. The incidence of mesothelioma was 87 pmpy for subjects whose estimated cumulative exposure was less than 7 fy ml^{-1}, rising to 327 and 1544 pmpy for estimated cumulative exposure of 7–20 fy ml and over 20 fy ml^{-1}, respectively.

In all regression models, age at first exposure, sex, year of first exposure, intensity of exposure, whether a subject lived with an asbestos worker or washed his clothes, had no significant effect on mesothelioma incidence ($P > 0.05$). Relative risks of mesothelioma were significantly increased to 2.5 (exposure duration: 1–5 years) and 6.2 (exposure duration > 5 years), compared to subjects who stayed at Wittenoom for less than 1 year ($P < 0.001$); and to 2.5 (cumulative exposure: 7–20 fy ml^{-1}) and 6.3 (cumulative exposure > 20 fy ml^{-1}), when compared to subjects whose cumulative exposure was less than 7 fy ml^{-1} ($P < 0.001$).

DISCUSSION

This study of former residents of Wittenoom who were environmentally exposed to crocidolite shows that the incidence of mesothelioma increased significantly with increasing time from first residence at Wittenoom and with increasing level of exposure, whether assessed by duration of residence or by cumulative exposure. The incidence of mesothelioma increased from about 180 per million person-years (pmpy) at 20–29 years since first exposure, to 1000 pmpy at 40 or more years from first exposure. No figures are available from studies of other environmentally exposed subjects, but the corresponding figures for the Wittenoom workers' cohort were approximately 900 and 7000 pmpy and typically about 1000 pmpy and 5000 pmpy for other occupational cohorts (de Klerk and Armstrong, 1992).

Regression analysis also showed evidence of an exposure–response relationship with mesothelioma incidence in this cohort. When duration of residence and estimated intensity of exposure were included in the model together, the association between rate of mesothelioma and intensity of exposure was not significant. This result was not unexpected as intensity of exposure was measured with much more error than duration of residence (Hansen *et al.*, 1996), and the amount of error introduced into the estimate of cumulative exposure by the estimate of intensity of exposure could explain why the model using just duration of residence appeared a better fit than the one using cumulative exposure.

This is the first study to show exposure–response relationships between incidence of mesothelioma and environmental exposure to any form of asbestos. The study of a birth cohort of exposed subjects near South African crocidolite mines has been unable to determine the intensity and duration of exposure for members of the cohort (Reid *et al.*, 1990).

Wittenoom residents in this study experienced a standardised incidence rate of mesothelioma of around 210 pmpy, which is substantially higher than the Western Australian rate in 1988 of 50 pmpy for men and 8 pmpy for women (de Klerk and Armstrong, 1992). It is exceeded in Australia only by that found among Aboriginal residents of the Pilbara region of WA, who have a crude rate of 250 pmpy (Musk *et al.*, 1995), and is one of the highest population rates in the world (de Klerk and Armstrong, 1992). Other studies of environmental asbestos exposure have reported high rates among specific populations, including residents near an asbestos-cement factory in Italy who experienced age-standarised incidence rates of 114 pmpy for males and 73 pmpy for females (Magnani *et al.*, 1993).

Most population studies of mesothelioma report differences in the incidence rates for men and women with the rates for males being typically 5–9 times greater than those for females. This is thought to be due to a much greater proportion of male cases having prior occupational exposure to asbestos than females (Leigh *et al.*, 1991). Females are more likely to obtain their exposure environmentally. Similar mesothelioma incidence rates for males and females shown in this study reflect the nature of exposure: both males and females were exposed to crocidolite by living in Wittenoom and men with known occupational exposure to crocidolite at Wittenoom were excluded from the analysis.

In conclusion, this study has shown that at the "low" levels of crocidolite exposure experienced by former residents of Wittenoom, there is a significantly increased risk of mesothelioma, which is dose-dependent.

REFERENCES

Armstrong, B. K., de Klerk, N. H., Musk, A. W. and Hobbs, M. S. T. (1988) Mortality in miners and millers of crocidolite in Western Australia. *Br. J. ind. Med.* **45**, 5–13.

Cox, D. R. (1972) Regression models and life-tables. *J. R. Stat. Soc. B.* **34**, 187–220.

de Klerk, N. H. and Armstrong, B. K. (1992) The epidemiology of asbestos and mesothelioma. In *Malignant Mesothelioma* (Edited by D. W. Henderson, K. B. Shilkin, D. Whitaker and S. L. P. Langlois) pp. 223–250. Hemisphere, New York.

de Klerk, N. H., Armstrong, B. K., Musk, A. W. and Hobbs, M. S. T. (1989) Cancer mortality in relation to crocidolite at Wittenoom Gorge in Western Australia. *Br. J. ind. Med.* **46**, 529–536.

Dement, J. M., Brown, D. P. and Okun, A. (1994) Follow-up study of chrysotile textile workers: cohort mortality and case-control analyses. *Am. J. ind. Med.* **26**, 431–447.

Dodoli, D., Del Nevo, M., Fiumalbi, C., Iaia, T. E., Cristaudo, A., Comba, P., Vita, C. and Battista, G. (1992) Environmental household exposures to asbestos and occurrence of pleural mesothelioma. *Am. J. ind. Med.* **21**, 681–687.

Hammond, E. C., Garfinkel, L., Selikoff, I. J. and Nicholson, W. J. (1979) Mortality experience of residents in the neighbourhood of an asbestos factory. *Ann. NY Acad. Sci.* **330**, 417–422.

Hansen, J., de Klerk, N. H., Musk, A. W. and Hobbs, M. S. T. Individual exposure levels in people environmentally exposed to crocidolite. *Ann. occup. Environ. Hyg.*(in press).

Hansen, J., de Klerk, N. H. Eccles, J. L., Musk, A. W. and Hobbs, M. S. T. (1993) Malignant mesothelioma after environmental exposure to blue asbestos. *Int. J. Cancer* **54**, 578–581.

Joubert, L., Siedman, H. and Selikoff, I. J. (1991) Mortality experience of family contacts of asbestos factory workers. *Ann. NY Acad. Sci.* **643**, 416–418.

Leigh, J., Corvalan, C. F., Grimwood, A., Berry, G., Ferguson, D. A. and Thompson, R. (1991) The incidence of malignant mesothelioma in Australia 1982–1988. *Am. J. ind. Med.* **20**, 643–655.

McDonald, J. C., Liddell, F. D. K., Dufresne, A. and McDonald, A. D. (1993) The 1891–1920 birth cohort of Quebec chrysotile miners and millers: mortality 1976–88. *Br. J. ind. Med.* **50**, 1073–1081.

Magnani, C., Bellis, D., Borgo, G., Botta, M., Ivaldi, C., Mollo, F. and Terracini, B. (1993) Incidence of mesotheliomas among people environmentally exposed to asbestos. *Eur. Respir. Rev.* **3**, 105–107.

Musk, A. W., de Klerk, N. H., Eccles, J. L., Hansen, J. and Hobbs, M. S. T. (1995) Mesothelioma in Pilbara Aborigines. *Aust. J. Publ. Health* **19**, 520–522.

Newhouse, M. L. and Sullivan, K. R. (1989) A mortality study of workers manufacturing friction material: 1941–86. *Br. J. ind. Med.* **46**, 176–179.

Peto, J. (1985) Some problems in dose-response estimation in cancer epidemiology. In *Methods of Estimating Risk of Chemical Injury: Human and Non-Human Biota and Ecosystems* (Edited by V. B. Vouk, G. C. Butler, D. G. Hoel and D. B. Peakell), pp. 361–380.

Reid, G., Keilkowski, D., Steyn, S. D. and Botha, K. (1990) Mortality of an asbestos-exposed birth cohort: a pilot study *South Afr. Med. J.* **78**, 584–586.

Siemiatycki, J. (1982) Health effects on the general population: mortality in the general population in asbestos mining areas. In *Proc. World Symp. on Asbestos*, Montreal, pp. 337–348. Canadian Asbestos Information Centre.

Wagner, J. C., Sleggs, C. A. and Marchand, P. (1960) Diffuse pleural mesothelioma and asbestos exposure in North West Cape Province. *Br. J. ind. Med.* **17**, 260–271.

 Pergamon

Ann. occup. Hyg., Vol. 41, Supplement 1, pp. 194–196, 1997
© 1997 British Occupational Hygiene Society
Published by Elsevier Science Ltd. All rights reserved
Printed in Great Britain
0003–4878/97 $17.00 + 0.00
Inhaled Particles VIII

PII: S0003–4878(96)00061–0

ASBESTOS FIBRES IN LUNG TISSUE AND DEGREE OF INTERSTITIAL FIBROSIS IN SUBJECTS EXPOSED TO CROCIDOLITE

A. W. Musk,* N. H. de Klerk,† V. M. Williams,‡ P. R. Filion,‡ D. Whitaker‡ and K. B. Shilkin‡

*Department of Respiratory Medicine, Sir Charles Gairdner Hospital, Nedlands, 6009, Western Australia; †Department of Public Health, University of Western Australia; and ‡Department of Anatomical Pathology, Western Australian Centre for Pathology and Medical Research (PathCentre), QEII Medical Centre, Nedlands, Western Australia, Australia

INTRODUCTION

Exposure–response relationships between both prevalence and incidence of asbestosis, and duration and intensity of exposure to crocidolite have been established (Cookson *et al.*, 1986a; de Klerk *et al.*, 1989). Radiographic progression of asbestosis after onset is, however, only weakly influenced by past exposure (Cookson *et al.*, 1986b). The extent of tissue fibrosis appears to be more strongly related to fibre surface area than to fibre concentration (Timbrell *et al.*, 1988).

The aims of this study were to examine the relationship between fibre levels in lung tissue and the degree of diffuse interstitial fibrosis in subjects exposed to crocidolite and to determine the effect of exposure to tobacco smoke on the relationship between lung fibre level and degree of interstitial fibrosis.

SUBJECTS

One hundred and forty subjects were included in this study on whom smoking information was available for 78 (Table 1).

Table 1. Study subjects—Exposure and smoking

Asbestos exposure history	Number of subjects
No known exposure	17
Exposure to asbestos at Wittenoom	42
Other occupational exposure to asbestos	40
Unknown exposure history	41
TOTAL	140
Smoking history	
Never	10
Ex	19
Current	49
Unknown	62

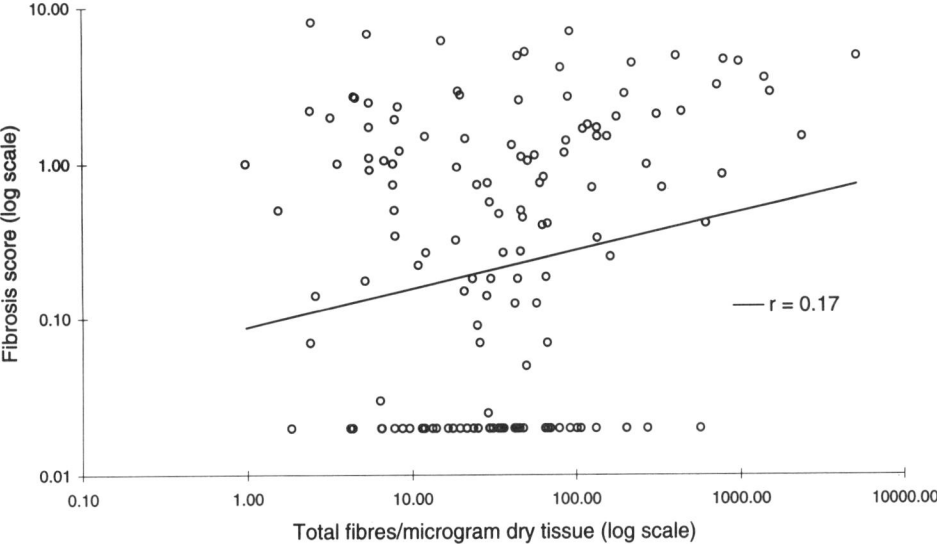

Fig. 1. Fibre concentration and fibrosis. All subjects ($n = 140$).

METHODS

Asbestos exposure and smoking histories were determined where available from clinical notes and records of the Wittenoom follow-up studies (de Klerk *et al.*, 1994). Fibre-counting and sizing on excised or post-mortem lung specimens used TEM with EDAX to identify fibre types. Fibre surface area was estimated from the diameter and length of the fibre size distributions.

Interstitial fibrosis was scored by consensus by three pathologists concurrently (Ashcroft *et al.*, 1988): the mean score from 20 fields was used for analysis.

The relationships between log fibre counts and surface areas, age, sex, smoking and the log (fibrosis score + 0.01) were assessed using linear regression.

RESULTS

There was a weak but statistically significant association between total fibre concentration and fibrosis score ($r = 0.17$, $P = 0.05$), which was approximately linear (Fig. 1). The relationship with fibre surface area was similar ($r = 0.17$, $P = 0.05$). The associations of fibrosis with asbestos fibres only and crocidolite fibres only were weaker and insignificant.

The fibrosis score increased by 0.001 (95% CI 0.0005–0.0014) on average for each additional million fibres per gram of lung tissue. No effect of sex or age on the fibrosis score was observed. The fibrosis score was additionally associated with smoking, although not significantly so, in that ex-smokers had a lower average score than current smokers who had a lower average score than non-smokers.

CONCLUSIONS

While the amount of exposure to crocidolite and all fibres (and hence the lung tissue fibre counts) affect the risk of asbestosis, there is only a weak association between fibre count or surface area and the level of fibrosis. This is consistent with previous observations on the association between degree of exposure to asbestos and radiographic progression of fibrosis in subjects with asbestosis (Cookson *et al.*, 1986b).

Acknowledgements—This study was funded by the Australian National Health and Medical Research Council. The authors are very grateful to Jan Eccles and Naomi Hammond for invaluable clerical assistance.

REFERENCES

Ashcroft, T., Simpson, J. M. and Timbrell, V. (1988) Simple method of estimating severity of pulmonary fibrosis on a numerical scale. *J. Clin. Pathol.* **41**, 467–470.

Cookson, W. O. C. M., de Klerk, N. H., Musk, A. W., Armstrong, B. K., Glancy, J. J. and Hobbs, M. S. T. (1986a) The prevalence of radiographic asbestosis in crodidolite miners and millers at Wittenoom, Western Australia. *Br. J. ind. Med.* **43**, 450–457.

Cookson, W. O. C. M., de Klerk, N. H., Musk, A. W., Glancy, J. J., Armstrong, B. K. and Hobbs, M. S. T. (1986b) The natural history of asbestosis in former crocidolite workers of Wittenoom Gorge. *Am. Rev. Respir. Dis.* **133**, 994–998.

de Klerk, N. H., Armstrong, B. K., Musk, A. W. and Hobbs, M. S. T. (1989) Predictions of future asbestos-related disease cases among former miners and millers of crocidolite in Western Australia. *Med. J. Aust.* **151**, 616–620.

de Klerk, N. H., Musk, A. W., Armstrong, B. K. and Hobbs, M. S. T. (1994) Diseases in miners and millers of crocidolite from Wittenoom, Western Australia: a further follow-up to December 1986. *Ann. occup. Hyg.* **38** (Supp 1), 647–655.

Timbrell, V., Ashcroft, T., Goldstein, B., Heyworth, F., Meurmann, L. O., Rendall, R. E. G., Reynolds, J. A., Shilkin, K. B. and Whitaker, D. (1988) Relationships between retained amphibole fibres and fibrosis in human lung tissue specimens. *Ann. occup. Hyg.* **32** (suppl. 1), 323–340.

 Pergamon

Ann. occup. Hyg., Vol. 41, Supplement 1, pp. 197–202, 1997
© 1997 British Occupational Hygiene Society
Published by Elsevier Science Ltd. All rights reserved
Printed in Great Britain
0003–4878/97 $17.00 + 0.00
Inhaled Particles VIII

PII: S0003–4878(96)00060–9

A MODEL TO PREDICT DEPOSITION OF MAN-MADE VITREOUS FIBRES IN THE HUMAN TRACHEOBRONCHIAL REGION

M. M. Quinn,* M. J. Ellenbecker,* T. J. Smith,† D. H. Wegman* and E. A. Eisen*

*University of Massachusetts Lowell, Work Environment Dept, 1 University Avenue, Lowell,
MA 01854, U.S.A.; and †Harvard University School of Public Health, Boston, Massachusetts, U.S.A.

INTRODUCTION

Biologically-based indices of fibre exposure

In previous work, we proposed an approach to construct indices of fibre exposure (Quinn, 1992; Quinn *et al.*, 1996) for a study of lung cancer based on three basic conditions for biologic activity: (1) the fibres must be able to cause the cellular changes associated with tumour formation; (2) the fibres must be expected to deposit in the target tissues; and (3) the fibres must persist at the target tissue long enough to cause carcinogenic cellular changes. Specific hypotheses regarding these three conditions relative to fibre length and diameter are represented in a series of matrices. The matrices are multiplied to produce a single value that is an index of the fibre exposure hypothesised to be related to lung cancer risk. The work presented here was developed for the matrix which represents the second biologic condition: fibre deposition at the target tissue as a function of the fibre lengths and diameters measured from a particular air sample.

The target tissue for human lung cancer

To develop a hypothesis regarding the target tissue for human lung cancer, a review of the medical and experimental literature was conducted. A summary of the review indicated that:

(1) Greater than 90% of all lung tumours are found in the bronchial wall.

(2) Tumours in the bronchiolar and alveolar walls are rare, less than 2–3% (Parkes, 1982).

(3) About 75% of all lesions originate in the first-, second- or third-order bronchi, the rest are mostly in the segmental bronchi, generations 4–5 (Robbins and Cotran, 1979).

(4) Among asbestos fibre exposed cases of lung cancer, the distribution of tumour types is similar to that among lung cancer cases not known to have had an asbestos exposure (Kannerstein and Churg, 1972). There is little information regarding the location of lung tumours for other types of fibre exposure.

It was concluded that it is reasonable to assume that the target tissue for human lung cancer is that region where the tumours occur and that the first five generations of the tracheobronchial tree appear to include most of the target tissue

Table 1. Air sampling results matrix. The numbers of fibres counted and sized
from an air sample and tallied in a matrix adapted from the WHO/Euro Reference
Scheme for MMMF

Length (μm)	Diameter (μm)						
	0.05–0.1	0.1–0.2	0.2–0.6	0.6–1.0	1–3	3–5	5–7
2–5	1	2	5	4	1	0	0
5–10	1	1	10	4	6	2	0
10–20	3	3	13	14	18	2	1
> 20	2	2	11	25	69	13	4

Most air samples from the MMVF industry would not actually contain fibres
in every length and diameter category; these hypothetical data are used only
for illustration. The zero values occur because particles having these
diameters and lengths do not meet the WHO fibre definition which requires
a 3:1 aspect ratio.

for human lung cancer induced by inhaled fibres. In contrast, the target tissue for
lung cancer is often explicitly or implicitly assumed to be the alveolar region. For
example, "respirable" fibres, defined as penetrating with high efficiency to the
alveolar region, usually are used as the fibre measure in epidemiological studies of
fibres and lung cancer risk. Respirable fibres may have diameters up to 3 μm and
any length greater than 5 μm.

OBJECTIVE

The purpose of this study was to determine, for a given airborne fibre exposure,
the efficiencies with which the fibres of different lengths and diameters deposit in
the tracheobronchial (TB) region, specifically generations 1–5, of the human lung.

METHODS

An air sample was collected in an environment where MMVF were aerosolised.
The air sample was analysed using scanning electron microscopy (SEM) according
to the WHO/Euro Reference Scheme (WHO, 1985). The fibre lengths and
diameters from the air sample were tallied in a matrix adapted from the WHO
Scheme. An example of a typical air sampling results matrix is given in Table 1. A
model of fibre deposition efficiencies in the TB region (Weibel generations 1–5
with the trachea equal to generation 1) was developed using a set of empirically-
derived deposition efficiency equations in an iterative procedure to calculate the
cumulative deposition efficiency for fibres having the lengths and diameters
represented by each cell in the air sampling results matrix (Table 1). The results of
this proposed model then were compared to results using conventional methods.

The proposed method

A model of fibre deposition in the human TB region was developed using an
iterative procedure to calculate the cumulative deposition efficiency in generations
1–5 for fibres having the lengths and diameters represented in each cell of Table 1.
As a starting point, the model used a set of equations (Sussman, 1988; Sussman *et*

al., 1991a,b) that predict the deposition efficiency (D_{eff}), the fraction of fibres having a particular diameter and length (in four length categories: 2–5 μm, 5–10 μm, 10–20 μm, > 20 μm) depositing in each of the first five Weibel generations of the lung (trachea = generation 1). A series of linked spreadsheets was used to build the model. Cumulative deposition efficiencies were predicted for three different respiratory flow rates: 15, 30 and 60 l. min^{-1}. The Sussman (1988) equations (17)–(20) were derived from experiments in which casts of the human lung were exposed to fibrous aerosol. The experiments were performed using cyclic flow through a larynx attached to the casts of human lungs in an artificial thorax. The test aerosol was crocidolite having a density of 2.5 g cm^{-3}; the density of glass is similar. The equations account for the fibre diameter, length and density, the air flow (l. min^{-1}) in a particular generation of the lung, the radius of a particular airway and an interception term based on fibre length and whether the flow in the airway is turbulent or transitional.

The model calculated the cumulative deposition efficiency for fibres having the midpoints of the length and diameter intervals represented by each cell of Table 1. In the first step of the model, the deposition efficiency for airway generation 1, D_{eff}, was calculated using the Sussman equation that corresponds to the fibre length interval for a specific cell. The air flow (l. min^{-1}) and radius of the airway used in the equation were for the first airway generation. The interception term varied depending on the overall respiratory flow rate. Flow was transitional at 15 and 30 l. min^{-1}; it was turbulent at 60 l. min^{-1}. The deposition efficiencies for airway generations 2–5 were then calculated using the same Sussman equation as in step 1, except that the air flow and airway radius used in the equation were for each of the specific airway generations. Finally, the cumulative deposition efficiency for all five generations was obtained for the fibres in the particular fibre length and diameter cell (Table 1) by calculating the product of each generation's D_{eff} and (1 − D_{eff}) from the previous generation [for generation 2: $D_{eff2} \times (1 - D_{eff1})$]. For an actual air sample, the model repeats these steps for each length and diameter cell in Table 1.

The conventional method

The aerodynamic equivalent diameter, D_{ae}, was calculated for each fibre in the air sample. These were compared to models for lung regional penetration based on spherical particles of unit density (British Medical Research Council, BMRC and the American Conference of Governmental Industrial Hygienists, ACGIH). The Stöber equation for calculating the D_{ae}, D_{aeS}, was used: $D_{aeS} = 2.19 \, D_{f}\beta^{0.171}$, where D_{aeS} = Stöber aerodynamic equivalent diameter of a fibre, D_{f} = fibre diameter and β = aspect ratio (length to diameter) of the fibre. Other equations for the fibre D_{ae} produced similar results.

RESULTS

Table 2 summarises the D_{aeS} for the fibres having the particular diameters and lengths in the air sample. Examination of the Stöber equation shows that the calculation of D_{aeS} weights fibre diameter much more than length. For a given fibre diameter, such as in the 0.2–0.6 μm diameter column (Table 2), there is little variation in the D_{aeS} across fibre length categories. For example, fibres of diameter

M. M. Quinn *et al.*

Table 2. The Stöber aerodynamic equivalent diameter (D_{aeS}) for a fibre having the average length and diameter of each cell

Length interval (μm)	Average fibre length (μm)	Average fibre diameter (μm)						
				Diameter intervals (μm)				
		0.05–0.1	0.1–0.2	0.2–0.6	0.6–1.0	1–3	3–5	5–7
		0.075	0.15	0.40	0.80	2.0	4.0	6.0
				D_{aeS} (μm)				
2–5	3.5	0.32	0.56	1.3	2.2	4.8	8.6	12.0
5–10	7.0	0.36	0.63	1.4	2.5	5.4	9.6	13.5
10–20	15.0	0.41	0.72	1.6	2.9	6.2	11.0	15.4
> 20	30	0.46	0.81	1.8	3.3	7.0	12.4	17.3

0.4 μm and lengths of 3.5, 7.0, 15.0 and 30.0 μm have D_{aeS} values of 1.3, 1.4, 1.6 and 1.8 μm, respectively.

The conventional practice would compare these D_{aeS} to the BMRC or ACGIH curves for particle penetration. Such a comparison indicates that greater than 90% of fibres having these D_{aeS} would pass through the tracheobronchial (TB) region and penetrate to the alveolar region. In contrast, the proposed model predicts significantly different TB deposition efficiencies for the fibres longer than 10 μm with these D_{aeS} values. The predicted cumulative deposition efficiencies in generations 1–5 are: 0.03, 0.10, 0.21 and 0.46, respectively.

The predicted values of the deposition efficiencies in generations 1–5 vary considerably for fibres of the same diameter across the length categories. It is seen that, for a given fibre diameter, the cumulative deposition efficiency increases considerably as the fibre length increases (Table 3). This is because the model predicts that fibre interception with the airway wall increases nearly exponentially as fibre length changes from 2–5 μm to greater than 20 μm. Consider, for example, fibres having a diameter of 0.2–0.6 μm (Table 3). Cumulative deposition efficiency increased when either fibre diameter or length increased (Fig. 1), but length, not diameter, had the most significant effect on deposition. This contradicts the conventional method which minimizes the effect of fibre length on regional lung deposition. In addition, fibre deposition in the TB region increased as the respiratory flow rate increased. For example, for fibres with diameter 0.4 μm and

Table 3. Predicted cumulative deposition efficiency in generations 1–5. Respiratory flow rate: 30 1. min^{-1}. Fibres in each cell were assigned the corresponding length and diameter midpoints

Length (μm)	Diameter (μm)						
	0.05–0.1	0.1–0.2	0.2–0.6	0.6–1.0	1–3	3–5	5–7
2–5	0.011	0.019	0.029	0.036	0.046	NA	NA
5–10	0.047	0.071	0.104	0.126	0.156	0.177	NA
10–20	0.106	0.148	0.206	0.244	0.293	0.328	0.348
> 20	0.262	0.352	0.463	0.532	0.612	0.665	0.693

NA = Not applicable. Particles having these lengths and diameters do not meet the WHO fibre definition which requires a 3:1 aspect ratio.

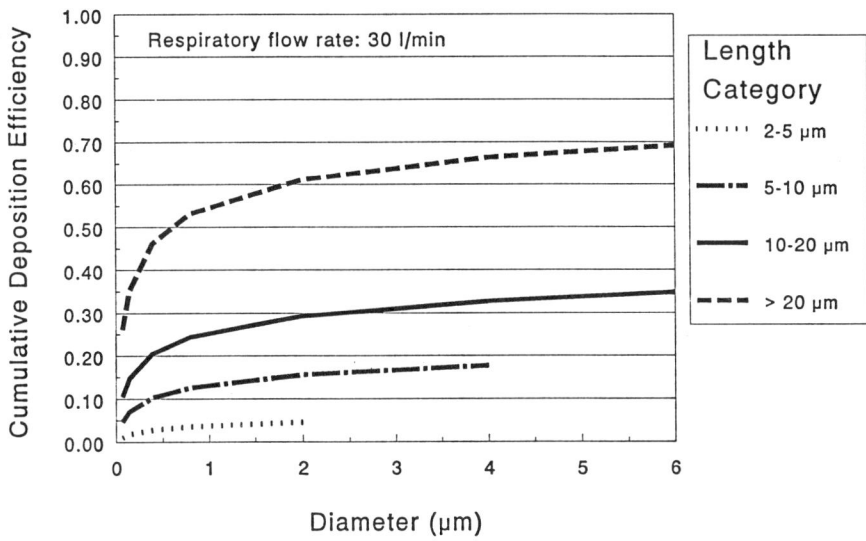

Fig. 1. Predicted cumulative deposition efficiency.

Note. 3:1 aspect ratio limits diameters within length categories.

length 15 μm, the cumulative deposition efficiency in generations 1–5 was 0.16, 0.21 and 0.23 at respiratory flow rates of 15, 30 and 60 l. min^{-1} respectively.

CONCLUSIONS

(1) The proposed model predicts substantial TB deposition for fibres that conventionally are considered respirable. Thin fibres (< 1 μm in diameter) have significant TB deposition if they are longer than 5–10 μm.

(2) Length as well as diameter must be taken into account in models of fibre deposition in the human lung. The conventional practice of summarising fibre length and diameter using an aerodynamic equivalent diameter is not sufficient to evaluate the role of fibre length in regional lung deposition. This is especially important for fibres > 5 μm in length.

Model predictions

(3) As fibre length increases, interception becomes an important mode of deposition in the first five generations of the lung. Fibres having diameters > 1 μm and lengths > 10 μm, deposit quite efficiently (> 25%) in the first five generations of the TB region, the hypothesised target tissue for lung cancer. Fibres of these dimensions are common in many MMVF exposure settings.

(4) Interception predominates over impaction in the first five generations of the lung for fibres with lengths > 20 μm. As a result, these long fibres may deposit efficiently (> 20%) in the first five generations even with diameters as small as 0.2 μm. Conventionally, particles with $D_{ac} < 2$ μm are considered to penetrate to the alveolar region with greater than 90% efficiency.

(5) As the respiratory flow rate increases, the TB deposition efficiency increases with length, especially for fibres with lengths > 10 μm. This is primarily because airway turbulence increases, thus increasing the probability of long fibres intercepting with the airway walls.

REFERENCES

Kannerstein, M. and Churg, J. (1972) Pathology of carcinoma of the lung associated with asbestos exposure. *Cancer* **30**(1), 14–21.

Parkes, W. R. (1982) *Occupational Lung Disorders*. Butterworth, London.

Quinn, M. M. (1992) A biologically-based quantitative method for characterizing airborne fiber exposure. Doctoral dissertation, UMI Dissertation Information Service, Ann Arbor, MI, U.S.A.

Quinn, M. M., Smith, T. J., Ellenbecker, M. J., Schneider, T., Eisen, E. A. and Wegman, D. H. (1996) Biologically-based indices of fiber exposure for use in epidemiology. *Occ. Hyg.* **3**, 103–111.

Robbins, S. and Cotran, R. (1979) *Pathologic Basis of Disease*, pp. 866–872. Saunders, Philadelphia.

Sussman, R. (1988) The effect of fiber dimension on the deposition of asbestos fibers in casts of the human tracheobronchial tree. Doctoral dissertation, New York University, New York.

Sussman, R., Cohen, B. and Lippmann, M. (1991a) Asbestos fiber deposition in a human tracheobronchial cast. I. Experimental model. *Inhal. Toxicol.* **3**, 145–160.

Sussman, R., Cohen, B. and Lippmann, M. (1991b) Asbestos fiber deposition in a human tracheobronchial cast. II. Empirical model. *Inhal. Toxicol.* **3**, 161–179.

WHO/EURO (1985) Reference methods for measuring airborne man-made mineral fibres (MMMF). In WHO/EURO Reference Scheme, World Health Organization, Copenhagen.

 Pergamon

Ann. occup. Hyg., Vol. 41, Supplement 1, pp. 203–209, 1997
© 1997 Published by Elsevier Science Ltd on behalf of BOHS
Printed in Great Britain. All rights reserved
0003–4878/97 $17.00 + 0.00
Inhaled Particles VIII

PII: S0003–4878(96)00063–4

DEFINITION OF LOWER LIMITS FOR AIRBORNE PARTICLE ANALYSES BASED ON COUNTS AND RECOMMENDED REPORTING CONVENTIONS

D. P. Fowler

Fowler Associates, 643 Bair Island Road, Suite 305, Redwood City, CA 94063, U.S.A.

INTRODUCTION

In all of the standard methods for counting asbestos fibres or structures on filters by microscopy, a lower limit is established, below which a count obtained using the method is considered to be practically indistinguishable from background. In the phase contrast microscopy (PCM) method most widely used in the U.S. (NIOSH, 1994), that lower limit is stated as the "Limit of Detection" and is quantitatively given at 7 fibres mm^{-2} (roughly equivalent to 5.5 fibres/100 fields counted). In the most widely used (in the U.S.) transmission electron microscopy (TEM) method (the "AHERA" Method—EPA, 1987), the "background level" for asbestos structures is stated as 70 asbestos structures mm^{-2} filter surface, which is the approximate equivalent of four structures counted/10 electron microscope grid openings. In some TEM methods promulgated by the American Society for Testing and Materials (ASTM), the limit of detection has been explicitly stated as the counting of four structures (ASTM, 1996). Relatively little attention has been paid to this issue in the analysis of environmental asbestos samples and it is generally not recognised that such arbitrary detection limits across the universes of sample media and laboratory cleanliness and technical proficiencies of field and laboratory personnel are likely to misstate the "true" detection limits. In fact, the detection limit for any analytical method will necessarily be a function of both the method itself and the application of that method to obtain sample blank levels. Further, when faced with arbitrary detection limit language, some analysts do not report actual results for samples for which the detection limits are not attained, but simply report that the results are "less than the detection limit". In this paper both the recommended procedure to determine the detection limit, and the recommended reporting conventions are given.

GENERAL CONSIDERATIONS

In any environmental measurement process, it is usual to take samples from the environment of interest and to compare the results of those samples to some standard or to "blanks" and to attempt to come to an understanding of whether the analyte of interest is present and if so, at what level. In the case of asbestos measurement where the output is to be some form of counts per environmental unit (stuctures cm^{-3}), the analytical data take the form of discrete counts (of structures or fibres) from a defined portion of a specific sample medium (usually a filter) observed with a microscope. The count data for each sample (structures per 10 grid openings, or structures per 100 fields) may then be converted into counts per unit area of the filter (structures mm^{-2}) and thence into counts per environmental unit. Where several individual samples have been taken from one or more of the settings of interest, blank(s), or standard(s), then straightforward comparisons may be made, using statistical procedures appropriate to the distributions of the results (examples might be the "t-test" or various non-parametric tests).

However, where it is believed that there is some "background" level of the analyte, or that the method may yield false indications of the presence of the analyte it is sometimes believed appropriate to set a "detection limit"—an amount or concentration of analyte which if observed in a single measurement is believed to indicate the presence of the analyte with reasonable certainty and, which, if not observed leads to suspicion that the analyte is not truly present.

GENERAL CONCEPTS REGARDING DETECTION LIMITS

It is important to understand that our count of asbestos structures from a sample is intended to provide an estimate of the concentration of structures over the whole area of the filter to be analysed. If one structure is found on the (approximately) 1/7000 of the filter area that is analysed in the U.S. EPA AHERA method we should understand that this estimate of the true average structure count per 10 grid openings might be in error. If it is assumed that the distribution of fibres or structures across the filter is approximately Poisson, then the lower limit of the 90% confidence interval surrounding this estimate is 0.0513; the upper bound is 4.74 (Hahn and Shapiro, 1994). What this means is that, given a structure count of 1 structure observed in 10 grid openings, the true average concentration of structures per 10 grid openings will be between 0.0513 and 4.74 90% of the time. Ten per cent of the time, the true average concentration may be less than 0.0513 or greater than 4.74. [Note that this statement is true only if the distribution is truly Poisson. If, as is likely, the distribution of structure counts can be more appropriately described by the negative binomial distribution (Javitz and Fowler, 1981; Oehlert et al., 1995), the confidence interval will be wider.]

Similar kinds of statements can be made if the number of structures counted is 0, or 2, or 3, or 500, or 1000. That is, the number counted is an estimate (and our best estimate assuming we have not biased the results) of the true average number of structures that would be counted if an approximately infinite number of successive samples of 10 grid openings each were taken from that filter. Except by happenstance, that average is unlikely to be an integer. However, we count structures in discrete units, thus, we must express our detection limits or other important waypoints on the journey toward rational decision-making as units, as counts, unless we wish to define "minimum detectable concentrations" or other expressions that may involve the volume of air, or area of surface, or the like. The principal locus of efforts to define and understand "detection limits" for count data and their application has been the health physics community (see, as an example, ANSI, 1989). Fundamental references in this area are the writings of Lloyd Currie (Currie,

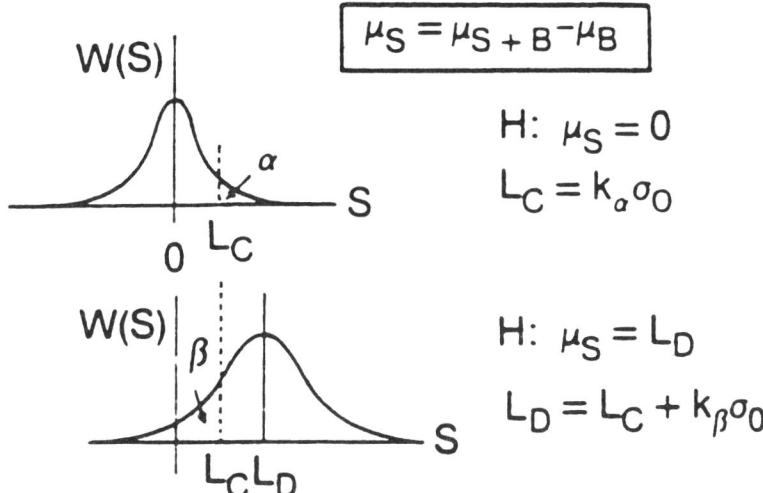

$$\mu_S = \mu_S + B - \mu_B$$

$$W(S)$$

$$0 \quad L_C$$

$$S$$

$$H: \mu_S = 0$$

$$L_C = k_a \sigma_0$$

$$W(S)$$

$$\beta$$

$$L_C L_D$$

$$S$$

$$H: \mu_S = L_D$$

$$L_D = L_C + k_\beta \sigma_0$$

Fig. 1. Bases for "decision level" and "detection limit" (adapted from Currie, 1984). W(S) = distribution of net count, L_C = decision level—risk of α error is acceptable, L_D = detection limit—risk of β error is acceptable.

1968, 1984), in which the basic concepts of decision level, detection limit and "determination limit" were carefully and clearly defined. [Currie made the simplifying assumption that the normal distribution could be used, an assumption that is often valid when the count of a Poisson-distributed variable rises above about 50–70. Brodsky (1992) extended the work of Currie to the cases where small numbers are counted, as in our work.]

The first two concepts are related to the common statistical concepts of Type I error (alpha error or false positive error) and Type II error (beta error or false negative error) and are defined pictorially in Fig. 1, adapted from Currie (1968). In that figure, if we assume alpha to be 5%, the decision level (L_C) is shown to be the level at which the upper 5% tail of the distribution of sampling results (signals) falls when the true mean of sampling results is zero. Another way of expressing this would be to say that measurements above L_C may be assumed to fall outside the distribution of the results for a true zero; our conclusion would be that we did detect a signal and we would be right 95% of the time. To put this concept into the context of structure counting, we count the number of structures in 10 grid openings and from that count wish to conclude whether the entire filter average count is "above background" or "greater than blank" with 95% confidence. If our sample count is above the decision level and the decision level is properly set, then our conclusion that the filter average count is positive will be wrong no more than 5% of the time. In the structure or fibre counting context, the decision level must be an integer, of course. [Note that the decision level is not the detection limit, but that useful information is found if counts exceed the decision level, even though the number counted is less than the detection limit. That information would be lost if one were not to report all of the data available.]

The example above illustrates the probability of committing a Type I error. A

Type II error arises when we observe a signal and falsely conclude that a true signal (above background) has not been demonstrated. Such errors can be avoided 1–β% of the time by establishing a detection limit (L_D in Fig. 1) such that the lower β% point of the distribution of results about the mean detection limit is set at the decision level. To put this into the context of structure counting, a detection limit for the number of structures in 10 grid openings must be set high enough that our risk of falsely concluding that some number at or greater than the detection limit is negative is less than (say) 5%. That is, if we are counting structures from a filter, we will not falsely reject a filter that truly contains more structures than a blank filter by incorrectly concluding that it is like a blank filter more than ~ 5% of the time. Finally, we come to the "determination limit" (sometimes called the "limit of quantification"), a number intended to give "reliable" results. A typical requirement for ordinary (non-counting) analyses might be that the relative standard deviation of a set of replicate samples should be less than 10% of the mean value. This is similar in concept to the recommendation in NIOSH 7400 that the desirable minimum fiber density is 100 fibres mm^{-2}, or a minimum fibre count of about 80, in order to reduce the uncertainty in the method to ~ +213%/−49%, or to the recommendation of Ogden (1982) that the fibre count should be ". . . at least 50 fibres . . . where an accurate answer is required." This number is a matter of choice. If one chooses to accept the hypothesis that fibre or structure counts are distributed as a Poisson variable, then one may refer to (as an example) Table 4-2 in Hahn and Shapiro (1994) to see the inherent, irreducible, error that is a consequence of counting small numbers of things, exclusive of any other random or determinate errors that may arise from the sampling and analytical process or method. Figure 1 in NIOSH 7400 shows the uncertainty associated with that overall method, including other sources.

CALCULATION OF LIMITS

Using the formulae in Brodsky (1992), I have calculated the "decision level" and "detection limit" approximately corresponding to α = 0.05 and β = 0.05 for a variety of different average blank levels ranging from 0 to 5 structures/unit area (which may be assumed to be 10 grid openings, or 100 fields, or any other counting area). In making these calculations, it has been assumed, with Brodsky, that a Poisson distribution is apt. The results of those calculations are shown in Table 1. Those calculations are based on analysis of a single blank sample for each set of field samples to be analysed. For this case, (a single blank) if the count of interest is structures/10 grid openings, then the most usual blank count will typically be either 0 or 1; the decision level will be either 1 or 3 and the detection limit will be either 3 or 8. [It may be noted that if the true "background level" (average blank count) is as stated in the AHERA method at ~ 70 s mm^{-2} = 4 structures/10 grid openings, then the detection limit for the method should be 13 structures/10 grid openings.]

The values in the first column of the table are means and that the values in the second column are standard deviations, calculated as the square roots of the individual means, in accordance with the assumption that the distributions are Poisson. The likelihood is low that the true standard deviations will be as low as this theoretical Poisson value, even if one assumes that the Poisson distribution is the

Table 1. Exact decision levels and detection limits for structure counts—blanks and samples

Mean of blank count (a)	SD of blank count (b)	Decision levels		Detection limits	
		Critical (decision) level (c)	Next integer	Detection limit (Brodsky) (d)	Next integer
0	0	0.0000	1	3.00	3
0.0001	0.0100	0.0233	1	3.05	4
0.0005	0.0224	0.0521	1	3.10	4
0.001	0.0316	0.0737	1	3.15	4
0.005	0.0707	0.1648	1	3.33	4
0.01	0.1000	0.2330	1	3.47	4
0.05	0.2236	0.5210	1	4.04	5
0.1	0.3162	0.7368	1	4.47	5
0.5	0.7071	1.6476	2	6.29	7
1	1.0000	2.3300	3	7.65	8
2	1.4142	3.2951	4	9.58	10
3	1.7321	4.0357	5	11.05	12
4	2.0000	4.6600	5	12.30	13
5	2.2361	5.2100	6	13.40	14

(a) Mean blank count—structures/10 grid openings.
(b) Standard deviation of blank counts = sq. root of mean (Poisson assumption).
(c) Critical or decision level = 2.33 (SD of blank counts).
(d) Detection limit = 4.65 (SD of blank counts) + 3.

most apt. The references previously cited, as well as NIOSH 7400, may be consulted on this issue.

With regard to a determination limit for the TEM methods for asbestos, it may be argued that it is unlikely that those methods are inherently more reliable than the PCM methods. The fraction of the filter surface examined in the AHERA method is approximately 1/7000, as opposed to the 1/500 for the NIOSH PCM method. Note also that the sample preparation required of a piece of filter prior to its submission to the TEM is more elaborate and likely to distort the true underlying particle distribution on that filter than is the preparation for PCM. If an indirect preparation method is used, then the resulting particle distribution is more likely to be Poisson than if the original filter was used. Accordingly, at a minimum, the 90% confidence interval about the sample structure count (within which it is 90% certain that the true sample mean must lie) cannot be much smaller (if at all) than the interval for NIOSH 7400. That is, assuming that we wish to have our confidence interval as small as +213%/−49%, then our limit of determination should be set at least at 80 structures. If, for reasons of cost, inconvenience or time, it is wished to count fewer structures, the confidence interval will be wider, and that confidence interval should be reported.

REPORTING PRACTICES

A continuing area of concern with environmental analysis methods, including those for asbestos, is the practice of not reporting the actual best estimate of concentration for any sample that falls below some "detection limit". Of course, self-censoring the results reported destroys any information that might be present below the "detection limit". The reasons why this should not be done have been eloquently expressed by Chambless *et al.* (1992) and by Helsel (1990). There is useful information available below the detection limit; the detection limit is not the point at which a sample fails to be distinguishable from a blank sample. In particular, it is noted that the decision level, as defined above, is the point at which useful information about the sample is first apparent. Finally, as noted above, the definition of detection limit appears to vary from one method to another and today's definition of detection limit may not be valid tomorrow. Substantial efforts are now needed in many environmental evaluations to attempt to reconstruct censored data taken in past years, where no report of some of the actual measurements made is available, except for the dreaded phrase "less than detection level" [see Helsel (1990) and Chambless *et al.* (1992) for examples]. It is further noted that specialized software for attempted reconstruction of the overall distributions of LDL data is available and needed (Newman, 1995). Of course, the actual sample specific numbers are lost forever.

CONCLUSIONS AND RECOMMENDATIONS

It is absolutely essential that all of the information obtained in the analytical process be recorded and reported for the client to evaluate. Statements as to the reliability of the data should, of course, be part of the report. The methods and values given here provide a useful initial approach. Olcerst (1995) has delineated a similar approach, although he depends upon a definition of the distribution of fibre counts as Poisson, which may not be valid. In order to reduce the complexity of the required calculations and to reduce the intrusion of sampling variability, it is recommended that any statement of detection limit in these methods should be stated as a minimum number of particles (structures or fibres) to be counted.

REFERENCES

American National Standards Institute (ANSI) (1989) *Draft American National Standard for Performance Criteria for Radiobioassay, N13.30*, ANSI, 11 West 42nd Street, New York, NY 10036 [see especially Appendix A].
American Society for Testing and Materials (ASTM) (1996) *Method D5755-95—Test Method for Microvacuum Sampling and Indirect Analysis of Dust by Transmission Electron Microscopy for Asbestos Structure Number Concentrations* and *Method D5756-95—Test Method for Microvacuum Samling and Indirect Analysis of Dust by Transmission Electron Microscopy for Asbestos Mass Concentration*, ASTM, 100 Barr Harbor Drive, West Conshohocken, PA 19428-2959, U.S.A.
Brodsky, A. (1992) Exact calculation of probabilities of false positives and false negatives for low background counting. *Health Phys.* **63**, 198–204.
Chambless, D. A., Dubose, S. S. and Sensintaffar, E. L. (1992) Detection limit concepts: foundations, myths and utilization. *Health Phys.* **63**, 338–340.
Currie, L. A. (1968) Limits for qualitative detection and quantitative determination: application to radiochemistry. *Analyt. Chem.* **40**, 586–593.
Currie, L. A. (1984) Lower Limit of Detection: Definition and Elaboration of a Proposed Position for Radiological Effluent and Environmental Measurements. National Bureau of Standards Report available from U.S. Government Printing Office, through National Technical Information Service, Springfield, VA, U.S.A.
Environmental Protection Agency (USEPA) Asbestos Hazard Emergency Response Act—Title 40, United States Code of Federal Regulations, Part 763, Appendix A to Subpart E, 1987. (Available from U.S. Government Printing Office, through National Technical Information Service, Springfield, Virginia, U.S.A.)

Hahn, G. J. and Shapiro, S. S. (1994) *Statistical Models in Engineering*. Wiley, New York.

Helsel, D. R. (1990) Less than obvious. Statistical treatment of data below the detection limit. *Environ. Sci. Technol.* **24**, 1766–1774.

Javitz, H. S. and Fowler, D. P. (1981) Statistical analysis of microscopical counting data. In *Electron Microscopy and X-Ray Applications to Environmental and Occupational Health Analysis*, Vol. 2. Ann Arbor Science Publishers, Inc., Ann Arbor, Michigan, U.S.A.

National Institute for Occupational Safety and Health (NIOSH) Method 7400, Asbestos and other fibres by PCM. *NIOSH Manual of Analytical Methods*, 4th ed. NIOSH, Cincinnati, OH, U.S.A. (Available from U.S. Government Printing Office, through National Technical Information Service, Springfield, Virginia, U.S.A.)

Newman, M. C. (1995) *UNCENSOR, v.4.0*. Available from Dr Michael C. Newman, University of Georgia, Savannah River Ecology Laboratory, P.O. Drawer E, Aiken SC 29801 (Tel. 803 725-2472).

Oehlert, G. W., Lee, R. J. and Van Orden, D. (1995) Statistical analysis of asbestos fibre counts. *Environmetrics*, **6**, 115–126.

Ogden, T.L. (1982) *The Reproducibility of Asbestos Counts*. Health and Safety Executive, U.K.

Olcerst, R. (1995) Analytical limits of asbestos fiber detection. *Appl. occup. Environ. Hyg.* **10**(9) 776–782.

 Pergamon

Ann. occup. Hyg., Vol. 41, Supplement 1, pp. 210–212, 1997
© 1997 British Occupational Hygiene Society
Published by Elsevier Science Ltd. All rights reserved
Printed in Great Britain
0003–4878/97 $17.00 + 0.00
Inhaled Particles VIII

PII: S0003–4878(96)00065–8

EFFECTS OF EXPOSURE PERIOD AND LUNG BURDEN ON CLEARANCE RATE OF INHALED ALUMINIUM–SILICATE CERAMIC FIBRE FROM RAT LUNG

T. Oyabu, I. Tanaka, S. Ishimatsu*, H. Yamato, Y. Morimoto, T. Tsuda, H. Hori* and T. Higashi

Institute of Industrial Ecological Sciences, *School of Health Sciences, University of Occupational and Environmental Health, Japan, 1–1 Iseigaoka, Yahatanishi, Kitakyushu 807, Japan

INTRODUCTION

It has been shown that occupational exposure to various types of asbestos may lead to asbestosis, bronchial cancer, pleural and peritoneal mesotheliomas. Recently the production of man-made fibres (MMFs) has increased as asbestos substitutes. Therefore, the effects of MMFs on health should be investigated because the health effects are not always clear for many MMFs.

In general, asbestosis and fibrosis occur after long retention of asbestos in the lung. Therefore, the clearance rate of deposited particles is considered to be a very important factor of these diseases. The lower the clearance rate, the higher the possibility of disease. In our previous study (Tanaka *et al.*, 1994) the clearance rate of glass fibre in rat lungs after a 12 month exposure experiment became slower than that after 1 month exposure.

In this study, the effect of exposure period on the clearance rate of aluminium–silicate ceramic fibre is examined by inhalation studies. In addition, the relationship between lung burden and clearance rate is discussed.

MATERIALS AND METHODS

The exposure system and the experimental procedure have been shown in a previous paper (Tanaka and Akiyama, 1990). Exposure concentration in the chamber was monitored continuously by a light scattering method (Dust Monitor AP-632, Shibata Sci. Tech. Japan). The mass concentration of RCF was measured gravimetrically each day by the isokinetic sampling of air through a glass fibre filter. The mass median aerodynamic diameter (MMAD) and the geometric standard deviation (GSD) of RCF in the exposure chamber were measured by using an Andersen cascade impactor (AN-200 Sibata Sci. Tech. Japan). The RCF in the chamber were collected by an asbestos-sampler, which conformed to the asbestos measurement method of NIOSH. The diameter and the length of the RCF were measured by a scanning electron microscope (S-700, Hitachi, Japan) and a digitizer (KD3030L, Graphtec, Japan).

Experimental conditions are summarised in Table 1. The Wistar male rats were randomly allocated to control and test groups. Eighty Wistar male rats in the test

Table 1. Experimental conditions of RCF inhalation

Exposure period (months)	1	3	6	12
Number of exposed rats	21	15	15	27
Number of control rats	22	15	16	27
Exposure concentration (mg m^{-3})	20 ± 3	2.6 ± 1.3	2.6 ± 1.1	2.8 ± 1.0
MMAD (GSD) (μm)	4.0 (1.9)	4.6 (1.8)	4.6 (1.8)	4.6 (1.8)
GMD (μm)	1.1	1.3	1.3	1.3
GML (μm)	8.2	9.2	9.2	9.2
Sacrifice time after exposure (months)	3 days, 3, 6, 9	3 days, 3, 6	3 days, 3, 6	3 days, 12

groups were divided into four groups depending on the exposure period and were exposed to aluminium–silicate ceramic fibre (RCF, content: SiO_2 52% and Al_2O_3 48%) for 6 h day^{-1}, 5 days week^{-1} by inhalation. The exposure period was 1, 3, 6 and 12 months. Control rats were exposed to clean air in identical, adjacent chambers under similar conditions of flow, temperature and humidity. The exposed rats were sacrificed at 3 days after each exposure period. The controls were also sacrificed at the same time.

At each sacrifice time the body and the wet lung were weighed and then the lungs were ashed by a low temperature asher (PR-503, Yamato Sci. Co., Japan). Ashed samples were digested with phosphoric acid in a teflon flask and the silicon and aluminium were measured by an inductively coupled plasma—atomic emission spectroscopy (SPS-1500R, Seiko Instruments Inc., Japan). The wavelengths of the detector were 251.611 nm for silicon and 396.152 nm for aluminium.

RESULTS AND DISCUSSION

Figure 1 shows the measured lung burden of the RCF with the clearance time. The solid lines in Fig. 1 are regression lines based on a single compartment model.

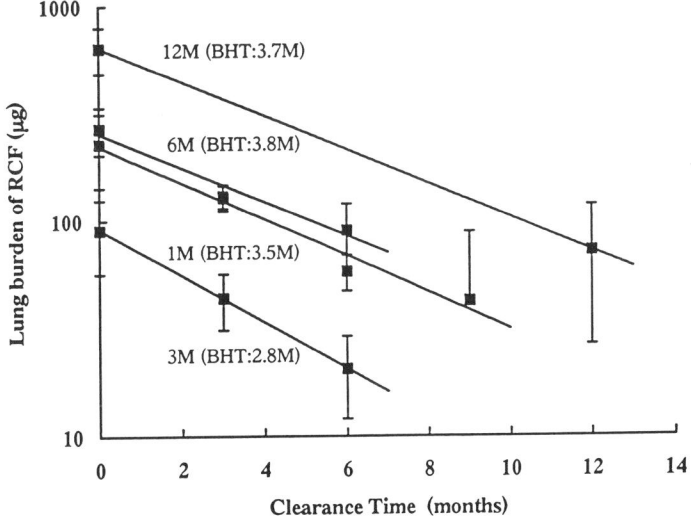

Fig. 1. Clearance of inhaled RCF from rat lungs.

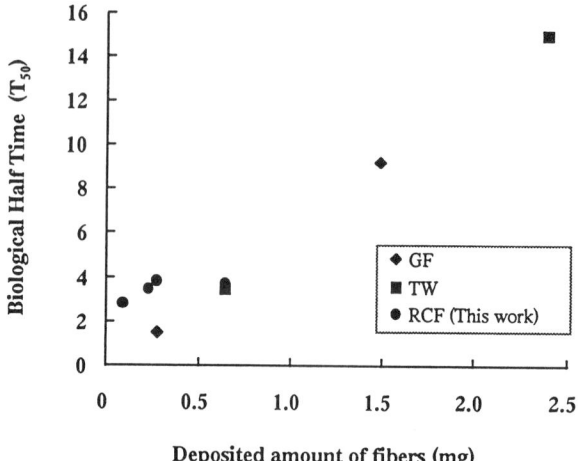

Fig. 2. Relationship between biological half time and deposited amount of fibres in rat lungs.

This figure shows that the biological half times (BHTs) of four groups have almost the same value in spite of the difference in the exposure period.

In the previous papers (Tanaka *et al.*, 1994; Fujino *et al.*, 1995), the BHTs in a 12 month exposure period [glass fibres (GF): 9.2 months and potassium titanate whisker (TW): 15 months] were much longer than those in a 1 month exposure period (GF: 1.5 months and TW: 3.5 months).

One possible reason for this discrepancy may be the different amounts of the deposited fibres. Figure 2 shows the relationship between the BHTs and the deposited amounts of fibres just after the termination of the exposure periods in our previous studies. The BHTs show almost the same level (1.5–4 months) when the lung burden is lower than 0.65 mg. On the other hand, the BHTs increase significantly with the lung burden which is over 1.5 mg.

Morrow (1992) indicated that when the volume of the particles phagocytized by alveolar macrophages is over the limitation, macrophage-medicated clearance is impaired and finally ceases. The limited amount is not exactly estimated but it is probable that when the deposited amount is over a threshold value, BHT will become longer.

We suggest that the BHTs show almost the same time in spite of the different exposure periods (1, 3, 6 and 12 months), as the deposited amount of fibres in rat lungs may be less than the threshold.

REFERENCES

Fujino, A., Hori, H., Higashi, T. *et al.* (1995) *In vitro* biological study to evaluate the toxic potentials of fibrous materials. *Int. J. occup. Environ. Health* **1** (1), 21–28.

Morrow, P. E. (1992) Dust overloading of the lungs: update and appraisal. *Toxicol. appl. Pharmacol.* **113**, 1–12.

Tanaka, I. and Akiyama, T. (1990) Pulmonary deposition fraction of a glass fibre in rats by inhalation. In *Aerosols: Sci., Ind., Health and Environ.* (Edited by S. Masuda and K. Takahashi) pp. 1242–1245. Pergamon Press, Oxford

Tanaka, I., Oyabu, T., Ishimatsu, S. *et al.* (1994) Pulmonary deposition and clearance of glass fibre in rat lungs after long-term inhalation. *Environ. Health Perspect.* **102** (S5), 215–216.

 Pergamon

Ann. occup. Hyg., Vol. 41, Supplement 1, pp. 213–218, 1997
© 1997 British Occupational Hygiene Society
Published by Elsevier Science Ltd. All rights reserved
Printed in Great Britain
0003–4878/97 $17.00 + 0.00
Inhaled Particles VIII

PII: S0003–4878(96)00066–X

BIOPERSISTENCE AND BRONCHOALVEOLAR LAVAGE INVESTIGATIONS IN RATS AFTER A SUBACUTE INHALATION OF VARIOUS MAN-MADE MINERAL FIBRES

O. Creutzenberg, B. Bellmann and H. Muhle

Fraunhofer Institute of Toxicology and Aerosol Research, Nikolai-Fuchs-Strasse 1,
D-30625 Hannover, Germany

INTRODUCTION

The production of man-made mineral fibres as a substitute for carcinogenic asbestos fibres has increased permanently in the last 15 years. A lot of different fibre types have been developed aiming at a low biopersistence in lungs. Significant factors for a potential carcinogenicity of mineral fibres are the length and diameter distribution and the biodurability. In an integrated project three different administration modes were compared in the rat: inhalation, intratracheal instillation and intraperitoneal injection. Further variables are the fibre type, the exposure time and the exposure dose. The final goal is to establish a standardized method to test fibrous particulates in animal experiments. Only the inhalation study is reported in this paper.

Biopersistence of fibres is dependent on three main items: dissolution, mechanical disintegration and dislocation of fibres. Contribution of these processes for removal of fibres out of rat lungs was analysed by determination of the number of retained fibres and their length and diameter. Additionally, the macrophage-mediated clearance capacity was measured using tracer particles ($^{46}Sc_2O_3$). The inflammatory potency of fibres was characterised by analysis of the bronchoalveolar lavage (BAL). Parts of the study are not totally completed. For fibre counts intermediate results are given.

MATERIALS AND METHODS

Origin and characterisation of test substances

The test materials were glass fibres B-01/0.9, stone wool MMVF 21 and ceramic fibres RCF 1. The latter two materials were used in chronic inhalation studies (McConnell *et al.*, 1994; Mast *et al.*, 1995). The type B-01/0.9 is an experimental fibre developed to achieve a low biopersistence. MMVF 21 and RCF 1 fibres were commercial samples which were sized to provide a respirable fraction for rats.

Exposure

Female Wistar rats of the strain Crl:(WI)BR were used. At the start of the inhalation the rats were 9 weeks old.

Table 1. Experimental design

Exposure groups	Exposure time (weeks)	Aerosol concentrations (mg m^{-3})		Number of particles (1 ml^{-1})	WHO fibres (1 ml^{-1})
		Nominal	Actual		
Clean air	3	—	—	—	—
Glass fibres B-01/0.9 low	1	30	30.1	1125	723
Glass fibres B-01/0.9 high	3	40	38.0	1324	902
Stone wool MMVF 21 low	1	30	38.7	1113	695
Stone wool MMVF 21 high	3	40	43.8	1423	879
Ceramic fibres RCF 1	3	40	51.2	1298	679
Glass spheres (negative control)	3	40	43.9	5032	—

Aerosol generation

For each nose-only exposure unit the fibre aerosol was generated by a high-pressure pneumatic disperser. The dispersers were fed with test material under computerized control, with feedback to the actual aerosol concentrations measured by an aerosol photometer.

Fibre characterisation and analysis

Test materials were characterised by analysis of the fibre size distribution in the aerosol of exposure chambers using a scanning electron microscope (SEM). The lengths and diameters of about 200 fibres per exposure day were measured. The distribution of fibre length, fibre diameter and aerodynamic diameter is given in Table 2.

Lungs of sacrificed rats were oxidized by low-temperature plasma ashing for at least 6 h. SEM analysis was done on 200 fibres per sample and the total number of fibres per lung was calculated. The mass was estimated assuming cylindrical geometry. Clearance kinetics was calculated using a regression analysis of logarithm of number or mass of fibres vs time after exposure end for individual rats. The resulting clearance rate constants k with their 95% confidence limit were transformed to the corresponding half-times $t_{1/2}$ by: $t_{1/2} = \ln2/k$.

Bronchoalveolar lavage (BAL)

Bronchoalveolar lavage was performed by the method of Henderson *et al.* (1987) with minor modifications. After preparation of lungs they were lavaged with 2×4 ml of saline without massage. From this lavagate a differential cell count was done. After centrifugation at 160 **g**, in the supernatant lactic dehydrogenase (LDH), total protein and γ-glutamyl transferase were analysed. This first lavage series was followed by a 4×5 ml series including massage to harvest 2×10^6 macrophages. After lysis of the macrophages, in the supernatant the titres of reduced and oxidized glutathione were determined (Griffith, 1985).

Tracer inhalation and clearance measurement

After end of fibre exposure rats inhaled a radioactive aerosol (Sc_2O_3) for about 1 h by nose-only exposure. Up to 90 days the thoracic γ-activity was measured twice a week using NaI scintillation detectors (Bellmann *et al.*, 1991). From the results of

Table 2. Fibre size distribution

Fibre sample	Number of fibres	Fibre length (µm)				Fibre diameter (µm)				Calculated aerodynamic diameter (µm)[a]			
		< 10%	< 50%	< 90%	< 99%	< 10%	< 50%	< 90%	< 99%	< 10%	< 50%	< 90%	< 99%
B-01/0.9 low	2982	3.5	8.6	27.4	75.6	0.60	0.95	1.55	2.19	3.3	6.0	8.5	10.6
B-01/0.9 high	1068	3.4	7.9	26.1	75.0	0.56	0.99	1.55	2.19	3.3	5.9	8.9	11.3
MMVF 21 low	1041	3.2	7.7	25.4	63.4	0.52	0.99	1.59	2.32	3.3	5.9	9.3	11.2
MMVF 21 high	2439	3.1	7.6	24.6	65.8	0.52	0.95	1.68	2.41	3.2	5.9	8.9	11.5
RCF 1	2514	2.4	6.3	22.2	68.3	0.34	0.82	1.81	2.67	3.7	6.8	10.7	14.5
Glass spheres	2819	0.5	0.9	2.2	4.3	0.49	0.90	2.17	4.24	2.2	5.0	10.0	11.3

[a] Calculated according to Harris and Fraser (*Am. Ind. Hyg. Ass. J.* **37**, 73, 1976), weighted by mass of fibres.

Table 3. Results of fibre retention and clearance measurements

Retention and clearance of fibres	Retained number of fibres after end of exposure[a] (WHO fibres—10^6/lung)					$t_{1/2}$ (days)[b] (95% C.L.) WHO fibres	$t_{1/2}$ (days)[c] (95% C.L.) tracer particles
	Day 3	Day 17	Day 31	Day 93	Day 365		
Control	—	—	—	—	—	—	55 (46–68)
Glass fibres B-01/0.9 low	6.58	1.91	0.94	0.47	0.02	47 (39–59)	66 (58–76)
Glass fibres B-01/0.9 high	12.59	4.47	1.93	1.61	0.02	37 (34–42)	55 (46–69)
Stone wool MMVF 21 low	10.92	6.67	4.86	2.95	0.26	75 (62–94)	57 (48–70)
Stone wool MMVF 21 high	28.67	23.49	15.44	8.82	1.36	92 (79–109)	68 (55–88)
Ceramic fibres RCF 1	29.69	24.57	21.00	13.08	2.17	113 (97–133)	102 (85–128)
Glass spheres[d]	328.3	133.8	166.0	99.6	17.2	101 (82–129)	1200 (573–10 000)

[a] Not yet all lungs evaluated.
[b] Clearance half-times calculated from fibre retention measurements of days 3–365.
[c] Clearance half-times calculated from measurements with tracer particles of days 3–90.
[d] Number of glass spheres per lung.

days 15–90 (excluding the initial fast bronchio-tracheal clearance) clearance half-times were calculated for individual animals.

Statistics

Differences of group means were considered to be statistically significant if $p <$ 0.05 (Dunnett's test).

RESULTS

Retention and clearance of fibres

The results of the fibre retention measurements are given together with the clearance half-times in Table 3. Three days after end of exposure, in the ceramic fibre group and the stone wool high dose group the retained numbers of WHO fibres per lung were 29.7×10^6 and 28.7×10^6. In the glass fibre high dose group the value was 12.6×10^6 (Table 3). On a mass basis the corresponding values were 0.69, 0.66 and 0.15 mg lung^{-1}. In the glass sphere group the value was 0.50 mg lung^{-1}. The clearance half-times of WHO fibres in the 1 year period after exposure were 113 (ceramic fibres), 92 (stone wool high) and 37 days (glass fibre high); for comparison, $t_{1/2}$ in the glass sphere group was 101 days. The corresponding data for tracer particles in a 3 month period after exposure were 1200, 102 and 57 days (glass spheres: 55 days).

Bronchoalveolar lavage

Significant increases of LDH, total protein and γ-glutamyl transferase were found in the stone wool and ceramic fibre groups. Significantly increased levels of reduced glutathione (GSH) were found in these groups only after 3 days of exposure. The glass fibre and glass sphere groups did not show any significant alteration compared to clean air controls.

The differential cell count showed a significant increase of PMNs in the ceramic fibre (18.9%) and in the stone wool groups (dose-dependent: 4.7 and 13.4%) 3 days after end of exposure (Table 4). Also lymphocyte percentages were increased in these groups: 14.5 and 4.9/6.5%, respectively. In the ceramic fibre group these inflammatory response persisted up to 1 year whereas in the stone wool groups a normalization occurred after 3 months of recovery. All other groups did not show significant alterations.

DISCUSSION

Retention and clearance of fibres

After end of exposure the retained numbers of fibres were similar in the stone wool and ceramic fibre groups. In the glass fibre groups retention was 2 to 3-fold smaller. This indicates the higher solubility of this experimental fibre type which has been developed to achieve this characteristic. The fibre retention measurements up to 1 year revealed a retardation of fibre clearance which can be ranked in this order: ceramic fibres > stone wool > glass fibres. In the stone wool groups a clear dose dependence was observed. Clearance of Sc_2O_3 tracer particles was

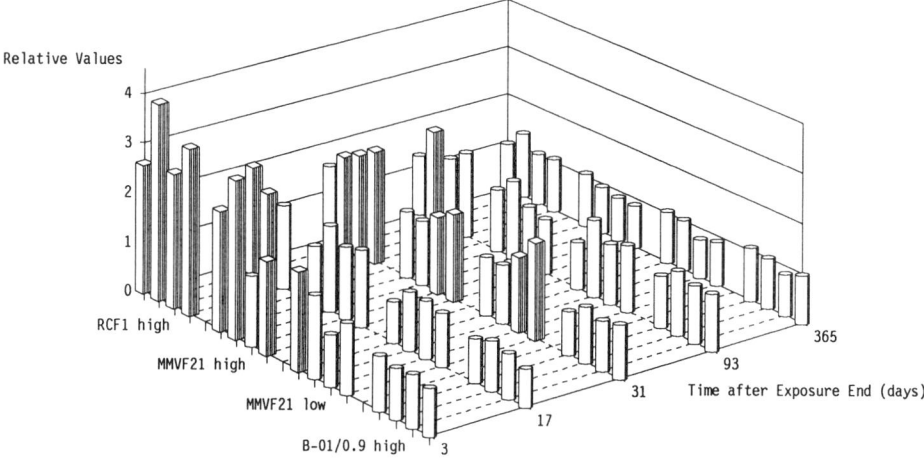

Fig. 1. Results of bronchoalveolar lavage (BAL) measurements (order of cylinders per group and date from front to the rear: LDH, total protein, γ-GT, GSH; cubes: values significantly different compared to sham controls, B-01/0.9 low and glass sphere groups at control levels).

strongly decreased in the ceramic fibre group. The ranking for clearance retardation of tracer particles is as follows:

ceramic fibres > stone wool > glass fibres = glass spheres = control.

Bronchoalveolar lavage

After end of exposure glass fibres (type B-01/0.9) did not show any significant inflammatory alterations in BAL, probably due to the low lung retention. In the stone wool fibre groups (type MMVF 21) a significant dose-dependent inflammatory reaction in lungs was observed 3 days after end of exposure. In the post-exposure period these effects normalized within a 3 month period. Ceramic fibres (RCF 1) showed in all parameters measured the most pronounced significant response with long-lasting effects. A partial normalization occurred after 1 year. A particular result in the ceramic fibre group was the high lymphocyte induction which reached the level of PMN concentration after end of fibre exposure. Up to 1 year this effect remained significant. Levels of reduced glutathione (GSH) were

Table 4. Differential cell count

Exposure groups	PMNs (% after 0.1, 0.5, 1, 3 and 12 months)	Lymphocytes
Clean air	1.1 – 0.3 – 0.2 – 0.3 – 3.0	2.5 – 1.5 – 0.2 – 0.5 – 0.3
Glass fibres B-01/0.9 low	0.2 – 0.2 – 0.1 – 0.6 – 0.9	0.8 – 1.5 – 0.3 – 0.8 – 0.1
Glass fibres B-01/0.9 high	1.0 – 0.4 – 0.1 – 0.6 – 1.0	1.5 – 2.1 – 0.5 – 0.5 – 0.1
Stone wool MMVF 21 low	**4.7** 2.9 – **2.1** – 0.5 – 1.1	4.9 – 2.3 – **1.5** – 0.3 – 2.0
Stone wool MMVF 21 high	**13.4 – 4.0 – 2.3** – 0.8 – 0.8	**6.5** – 2.3 – **2.0** – 0.5 – **2.9**
Ceramic fibres RCF 1	**18.9 – 11.1 – 15.5 – 13.5** – 3.4	**14.5 – 3.1 – 3.5** – 2.0 – **4.8**
Glass spheres (negative control)	0.4 – 0.2 – 0.3 – 0.2 – 0.7	1.8 – 2.0 – 0.2 – 0.1 – 0.0

Values given in bold font were significantly different to clean air controls.

significantly increased on day 3 after end of exposure in the stone wool and ceramic fibre groups. Already after 2 weeks the concentrations returned to normal. Under the given conditions mineral fibres do not deplete the pool of reduced glutathione (GSH). However, the increase of GSH shows the intense use of this system, particularly by reactive oxygen species (Gillisen and Wiethege, 1996).

The outstanding increase of tracer clearance half-time (1200 days) compared to fibre clearance (111 days) in the ceramic fibre group seems to be a combinatory effect. The mass median aerodynamic diameters (MMAD) of tracer particles (about 1 µm) and fibres (6–7 µm) will have implied a varying deposition behaviour, with the tracers more pronounced in the alveolar region. Additionally, RCF 1 fibres induced the strongest inflammatory reaction in the lung which persisted over the 3 month tracer clearance measurement period. Thus, the macrophage-mediated clearance will have been impaired. In the MMVF 21 high dose group, however, the PMN influx returned to almost normal already 2 weeks after end of exposure.

Acknowledgements—This investigation was supported by the German "Bundesministerium für Bildung, Wissenschaft, Forschung und Technologie (BMBF)" and the "Hauptverband der gewerblichen Berufs-genossenschaften e.V."

REFERENCES

Bellmann, B. *et al.* (1991) Lung clearance and retention of toner, utilizing a tracer technique, during chronic inhalation exposure in rats. *Fund. appl. Toxicol.* **17**, 300–313.
Gillissen, A. and Wiethege, T. (1996) Zelluläre Reaktionen der Lunge auf Mineralfasern. *Pneumonologie* **50**, 5–17.
Griffith, O. W. (1985) *Methods of Enzymatic Analysis*, 3rd edn, Vol. VIII (Edited by H. U. Bergmeyer), pp. 521–529. VCH-Verlagsgesellschaft, Weinheim.
Henderson, R. F. *et al.* (1987) Comparative study of bronchoalveolar lavage fluid: effect of species, age and method of lavage. *Exp. Lung Res.* **13**, 329–342.
McConnell, E. E. *et al.* (1994) Chronic inhalation study of size-separated rock and slag wool insulation fibers in Fischer 344/N rats. *Inhal. Toxicol.* **6**, 571–614.
Mast, R. W. *et al.* (1995) Multiple-dose chronic inhalation toxicity study of size-separated kaolin refractory ceramic fiber in male Fischer 344 rats. *Inhal. Toxicol.* **7**, 469–502.

Ann. occup. Hyg., Vol. 41, Supplement 1, pp. 219–223, 1997
© 1997 British Occupational Hygiene Society
Published by Elsevier Science Ltd. All rights reserved
Printed in Great Britain
0003–4878/97 $17.00 + 0.00
Inhaled Particles VIII

Pergamon

PII: S0003–4878(96)00165–2

FIBRE TOXICOLOGY—THE REGULATORY PERSPECTIVE

M. Meldrum

Toxicology and ESR Unit, Health & Safety Executive, Magdalen House, Stanley Precinct, Bootle, U.K.

INTRODUCTION

The scale and severity of the asbestos-related diseases has generated substantial concerns over the potential human health effects of man-made fibres. Despite the longstanding nature of these concerns, there is still a lack of an agreed scientific framework within which the toxicological hazards of fibrous materials can be investigated. Neither is there any clear agreement regarding which toxicity tests are appropriate or how these tests should be undertaken. Even with highly expensive and technically difficult animal inhalation studies there is often uncertainty surrounding the interpretation of results.

Until now, fibre manufacturers have not received any regulatory guidance in relation to fibre toxicity testing. The U.K. Health and Safety Executive (HSE) has attempted to address this by providing information and guidance in a *Review of Fibre Toxicology* (Meldrum, 1996). The primary objective of the document is to provide a structured scientific approach to the interpretation of fibre toxicity data. The guidance has particular relevance in the case of newly emerging fibres, which have only limited toxicological databases, often with no worthwhile human data. For such fibres, there is a need for a rational toxicity testing strategy to underpin hazard identification. Achievement of this aim has been limited by the lack of standardised test protocols. However, it is possible, based on current knowledge, to propose a testing strategy which should provide a rational basis for hazard assessment. Accordingly, the HSE document contains a proposal for a three-stage fibre toxicity testing strategy.

FIBRE TOXICITY TESTING STRATEGY

The following is a summarised version of the U.K. HSE proposal for a fibre toxicity testing strategy (see Fig. 1), supplemented where appropriate with information taken from other parts of the HSE document.

Stage 1. Physico–chemical evaluations and short-term toxicity tests

Solubility testing. Fibre durability is recognised to be an important determinant of toxicity. The potential to cause chronic lung damage will be less with a soluble fibre than with a more durable one. Although there is currently a lack of an agreed test protocol for measuring fibre solubility, methods are available in the literature (for example Potter and Mattson, 1991; Scholze and Conradt, 1987). Overall, it is

FIBRE TOXICITY TESTING STRATEGY

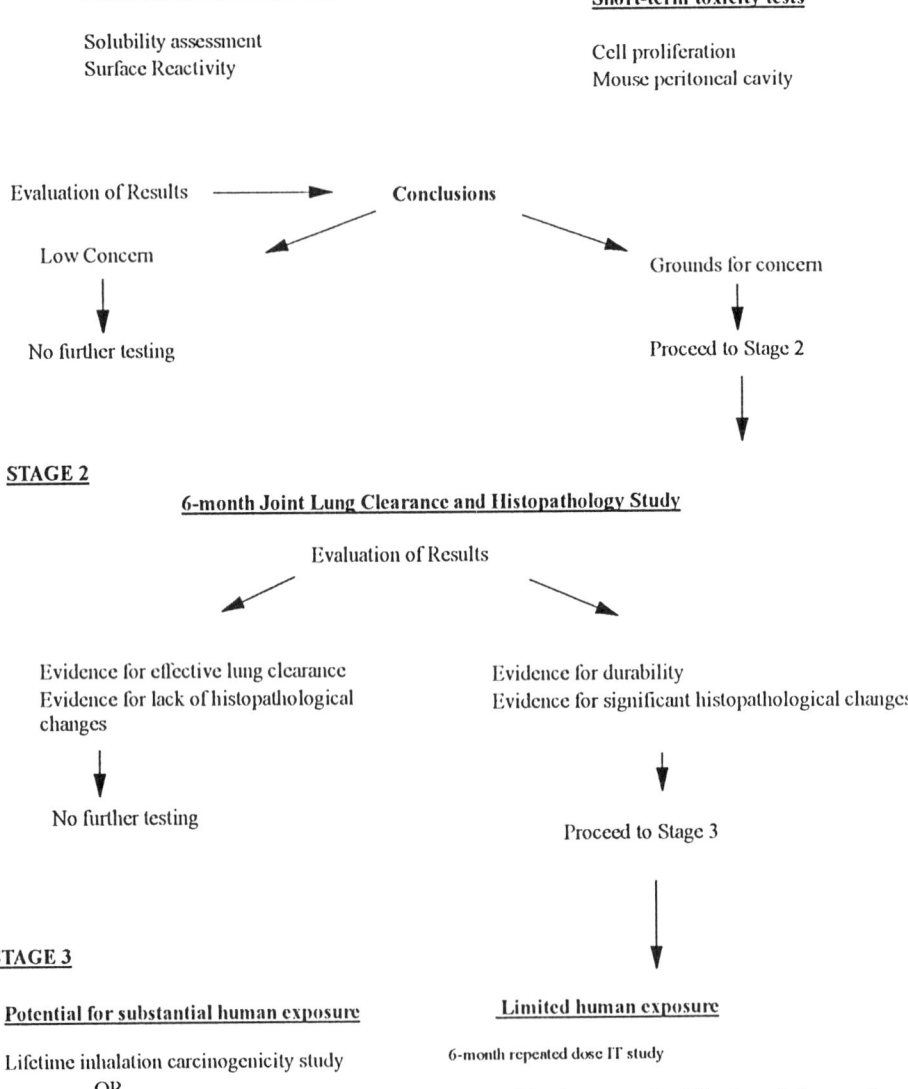

STAGE 1

Physico-chemical evaluation

Solubility assessment
Surface Reactivity

Short-term toxicity tests

Cell proliferation
Mouse peritoneal cavity

Evaluation of Results ⟶ **Conclusions**

Low Concern

↓

No further testing

Grounds for concern

↓

Proceed to Stage 2

↓

STAGE 2

6-month Joint Lung Clearance and Histopathology Study

Evaluation of Results

Evidence for effective lung clearance
Evidence for lack of histopathological
changes

↓

No further testing

Evidence for durability
Evidence for significant histopathological changes

↓

Proceed to Stage 3

↓

STAGE 3

Potential for substantial human exposure

Lifetime inhalation carcinogenicity study
OR
Lifetime IT carcinogenicity study

Limited human exposure

6-month repeated dose IT study

If irreversible pulmonary or pleural fibrosis results then regard
as potentially carcinogenic. Fibre must be regarded as carcinogenic
unless counteracted by results from a lifetime carcinogenicity study.

Fig. 1.

considered possible, based on currently available methods, to obtain useful *in vitro* measures of fibre solubility under simulated physiological conditions, which will provide a broadly reliable index of the likely *in vivo* durability of fibres.

Fibre reactivity. The cytotoxicity of asbestos fibres is reduced when scavengers of active oxygen species are present, and this may be related to surface reactivity and the ability to mobilise iron (Kane, 1993). Measures of the ability of fibres to generate oxygen-containing free radicals in cell free systems would therefore provide a further useful index of potential toxicity.

Short-term toxicity tests. As noted above, fibre durability is thought to be an important determinant of the potential to elicit chronic cell damage, but the influence of this particular parameter is unlikely to be discernible in short-term tests for toxicity. Nevertheless, fibre durability may not be the only property affecting cytotoxicity. Therefore, although short-term toxicity tests with fibres may have limitations, there is sufficient justification to anticipate that meaningful results can be derived from well designed short-term tests.

One group of workers has been particularly active in developing short-term toxicity *in vitro* and *in vivo* tests (Donaldson *et al.*, 1988, 1993). For example, they reported on the cellular inflammatory response obtained in mice following intra-peritoneal injection of mineral dusts. The response to titanium dioxide and coal dust was small, while aramid fibres, chrysotile and quartz induced a marked and sustained inflammatory response. This test appears to provide a rapid, simple and reliable *in vivo* test of the pathogenicity of particulate material.

The mechanisms of fibre-induced carcinogenicity would appear to involve repeated episodes of cytotoxicity followed by regenerative hyperplasia. Therefore, measures of the ability of fibres to provoke increased cellular proliferation in target tissues (for example bronchiolar and alveolar epithelium) following short-term *in vivo* exposures would be useful. Such studies have been undertaken with aramid fibres and the methodology is widely accessible (Warheit *et al.*, 1992).

From the *in vitro* assessments of fibre solubility and reactivity, and the results of perhaps three or more well designed short-term toxicity tests, conclusions could be drawn regarding the likely inhalation health hazards of the fibre and the need for further testing. A fibre which was judged to be highly soluble, of low surface reactivity, and of low biological activity in carefully chosen short-term toxicity tests might be judged to be of low concern, and may not require any further testing. However, if any of the above criteria are not met, then a move to the next stage in the testing strategy would be required.

Stage 2. Joint lung clearance and histopathology

Lung clearance. The clearance of fibres from the lungs is achieved by macrophage mechanisms and is also influenced by fibre dissolution. Although there is no standardised protocol, there have been many animal studies in which the pulmonary clearance rates of either asbestos or other fibre types have been measured (for example Bellman *et al.*, 1987; Warheit *et al.*, 1992). In the context of a fibre toxicity testing strategy it is essential to have a standard protocol for measuring lung clearance. Factors to be considered include the route of administration, dose, fibre size distribution, methods for recovering fibres from the lungs and counting and sizing the fibres, and also the duration of the study. For relatively durable fibres, 6 months might be insufficient to allow a meaningful evaluation of lung clearance.

Evidence for substantial lung clearance would suggest a relatively low degree of concern. In contrast, evidence for a lack of substantial lung clearance would raise concerns for the potential to cause long term damage. However, what is meant by 'substantial' lung clearance would need to be defined.

Histopathology evaluations. Lung tissue samples from the clearance studies could be taken for histopathological evaluation; this would allow conclusions to be drawn about the inflammatory/fibrogenic potential of the fibres. The absence of signs of fibrosis/chronic inflammation in the target tissues would suggest an absence of hazard, and together with evidence for substantial lung clearance and *in vitro* solubility would suggest that no further testing may be warranted. If these criteria were not met, then this would provide grounds for concern about potential human health effects, and a move to the next stage in the testing strategy would be justified.

Stage 3. Carcinogenicity testing

By this stage in the testing strategy, enough data should be available (*in vitro* solubility, fibre reactivity, inflammatory potential, lung clearance kinetics, cell proliferation activity, etc.) to enable a decision as to whether further toxicity testing is warranted. If there are grounds for concern remaining at this stage then one option would be to conduct a long-term carcinogenicity study. If this could not be justified (for example limited commercial use of the fibre) then another, less costly option, would be to conduct a 6-month repeated dose intra-tracheal (IT) study.

If there was evidence for the development of pulmonary or pleural fibrosis within the 6-month study, then the tested fibre might be presumed to be carcinogenic, and should be regarded as such. This conclusion is based on observations from long-term rat inhalation studies with asbestos (chrysotile and crocidolite) and with four type of refractory ceramic fibre, in which the early onset of fibrosis (within 3–6 months of exposure) in the target tissues consistently preceded tumour development (McConnell *et al.*, 1994; Mast *et al.*, 1995). In contrast, in rat inhalation studies with glass, rock and slag wool fibres involving similar fibre exposure conditions, fibrosis did not develop within 6 months of exposure and there were no treatment-related increases in tumours. Thus, fibrosis appears to act as a marker for a subsequent increased risk of tumour formation. Criteria for the severity of the fibrosis which would warrant this conclusion need to be developed. On the other hand, if only minor pulmonary or pleural changes were seen in the 6-month study, then the potential for carcinogenesis with more prolonged exposures might be viewed as negligible.

An alternative to performing the 6-month study would be to conduct a life-time inhalation or IT carcinogenicity study. Only an inhalation study would allow an investigation of dose–response characteristics. Dose–response relationships in fibre studies involving IT instillations appear to be highly erratic (Coffin *et al.*, 1992), which is presumably due to the uneven fibre deposition patterns which may occur with this method of administration.

CONCLUSIONS

There is a need from the regulatory perspective for an agreed fibre toxicity testing strategy which will allow a rational approach to hazard identification. This is needed to guide regulatory decision making in relation to classification and labelling and other occupational control measures. The proposed strategy is not presented as being definitive or rigidly prescriptive, and acknowledges the limitations imposed by the lack of agreed standardised test method protocols. It is presented as a starting point which can be modified both in the light of advances in the field of fibre toxicology and in test method development. An agreed testing strategy would benefit fibre manufacturers, as it should avoid unnecessary and costly testing. From the point of view of animal welfare, the strategy aims to reduce the overall burden on animal testing by promoting the use of *in vitro* testing.

REFERENCES

Bellman, B., Muhle, H., Pott, F., Konig, H., Kloppel, H. and Spurny, K. (1987) Persistence of man-made mineral fibres (MMMF) and asbestos in rat lungs. *Ann. occup. Hyg.* **31**, 693–709.

Coffin, D. L., Cook, P. M. and Creason, J. P. (1992) Relative mesothelioma induction in rats by mineral fibers: comparison with residual pulmonary mineral fiber number and epidemiology. *Inhal. Toxicol.* **4**, 273–300.

Donaldson, K., Miller, B. G., Sara, E., Slight, J. and Brown, R. C. (1993) Asbestos fibre length-dependent detachment injury to alveolar epithelial cells *in vitro*: role of a fibronectin-binding receptor. *Int. J. Path.* **74**, 254–250.

Donaldson, K., Bolton, R. E. and Brown, R. C. (1988) Inflammatory cell recruitment as a measure of mineral dust toxicity. *Ann. occup. Hyg.* **32**, 299–306.

Kane, A. B. and McDonald, J. L. (1993) Mechanisms of mesothelial cell injury, proliferation, and neoplasia induced by asbestos fibres. In *Fibre Toxicology*, (Edited by D. Warheit) Chap. 14, pp. 323–348. Academic Press, New York.

McConnell, E. E., Kamstrup, O., Musselman, R., Hesterberg, T. W., Chevalier, J., Miiller, W. C. and Thevanaz, P. (1994) Chronic inhalation study of size-separated rock and slag wool insulation fibres in Fischer 344/N rats. *Inhal. Toxicol.* **6**, 571–614.

Mast, R. W., McConnell, E. E., Hesterberg, T. W., Chevalier, J., Kotin, P., Thevanez, P., Bernstein, D., Glass, L. R., Miller, W. and Anderson, R. (1995) Multiple-dose chronic inhalation study of size-separated kaolin refractory ceramic fiber in male Fischer 344 rats. *Inhal. Toxicol.* **7**, 469–502.

Meldrum, M. (1996) *Review of Fibre Toxicology*. EH65/30 Health and Safety Executive. HSE Books.

Potter, R. M. and Mattson, S. M. (1991) Glass fibre dissolution in a physiological saline solution. *Glastech. Ber.* **64**, 16–28.

Scholze, H. and Conradt, R. (1987) An *in vitro* study of the chemical durability of siliceous fibres. *Ann. occup. Hyg.* **31**, 683–692.

Warheit, D. B., Kellar, K. A. and Hartsky, M. A. (1992) Pulmonary cellular effects in rats following aerosol exposures to ultrafine Kevlar Aramid fibrils: evidence for biodegradability of inhaled fibres. *Toxicol. Appl. Pharmacol.* **116**, 225–239.

 Pergamon

Ann. occup. Hyg., Vol. 41, Supplement 1, pp. 224–230, 1997
© 1997 British Occupational Hygiene Society
Published by Elsevier Science Ltd. All rights reserved
Printed in Great Britain
0003–4878/97 $17.00 + 0.00
Inhaled Particles VIII

PII: S0003–4878(96)00067–1

THE BIOPERSISTENCE OF FIBRES FOLLOWING INHALATION AND INTRATRACHEAL INSTILLATION EXPOSURE

D. M. Bernstein,* C. Morscheidt,† A. de Meringo,† M. Schumm,‡
H. G. Grimm,§ U. Teichert,¶ Ph. Thevenaz‖ and L. Mellon†

*Consultant, 40 Chemin de la Petite Boissiere, CH1208, Geneva, Switzerland; †St. Gobain, Paris, France;
‡Grünzweig & Hartmann, Germany; §Consultant, Germany; ¶GSA, Neuss, Germany;
‖RCC, Füllinsdorf, Switzerland

INTRODUCTION

The biopersistence of synthetic material fibres (SMF) is considered an important parameter in determining the potential of a fibre to cause a pathological response. At one extreme are the asbestos form fibres which are very persistent in lung and have been shown in both humans and in animal studies to cause a tumorigenic response. At the other extreme are the new generation of glass and stone wools which dissolve in the lung in a few days. Some soluble fibres have been evaluated in long-term animal studies with no pathological response. Although no experience with human exposure is yet available, the even higher solubility of this new generation of fibres would suggest that it would be very unlikely that an effect would ever occur.

Two methods are used for the evaluation of the biopersistence of SMF in animals, either exposure by inhalation or by intratracheal instillation (IT). Inhalation exposure follows the physiological route by which humans are exposed and could, therefore, be considered preferential (McClellan *et al.*, 1992; WHO, 1992). However, there are concerns whether the dose delivered allows prediction of biopersistence following chronic low dose exposure (as would likely be encountered in humans). As such, in parallel, the biopersistence of fibres has also been evaluated using administration by intratracheal instillation.

Fibre characteristics

The fibres used in this study were manufactured according to the known compositions of these fibres to be largely rat respirable (~ 1 μm diameter) for use in the animal studies. The chemical composition for each of these preparations is shown in Table 1. As shown in the table, the glasses are usually characterized as having higher Na_2O concentrations while the stone wools generally have higher CaO concentrations. Also shown in Table 1 is an index proposed in June 1994, by the Federal Ministry of Labour in Germany which distinguishes different groups of fibres based upon their chemical composition and which is the basis for the German classification scheme for SMF (TRGS 905, 1994). This index which is referred to as the K index is defined as = Σ Na, K, B, Ca, Mg, Ba–oxide − 2 × Al–oxide. This

Table 1. Fibre composition

Fibre components	C	G	M	O	P	R* (RCF 1a)
				Per cent		
SiO_2	61.7	60.1	57.9	50.3	65.4	47.7
Fe_2O_3	0.11	6.05	0.10	0.40	0.06	0.97
TiO_2	0.02	0.05	0.0	0.10	0.02	2.05
Al_2O_3	0.97	0.45	0.40	2.90	0.93	48.0
CaO	7.15	18.80	8.40	31.3	7.40	0.07
MgO	2.94	8.30	3.40	10.3	3.20	0.98
Na_2O	16.06	5.50	17.9	4.7	15.7	0.54
K_2O	0.59	0.15	0.30	0.20	0.37	0.16
B_2O_3	9.20	0.0	11.9	0.0	6.10	0.0
P_2O_5	1.05	0.08	0.0	0.0	1.0	0.0
SO_3	0.20	0.05	0.0	0.0	0.0	0.0
Cr_2O_3	0.0	0.0	0.0	0.0	0.0	0.0
MnO	0.01	0.0	0.0	0.0	0.0	0.0
ZrO_2	0.0	0.0	0.0	0.0	0.0	0.11
Total	100	99.53	100.3	100.2	100.18	100.58
Density	2.54	2.70	2.51	2.78	2.47	2.58
AGS K_{index}	34.0	31.9	40	40	30.1	−95.5

index has been used by the French and German authorities in proposals to the European Union for classification of SMF.

METHODS

Both the inhalation and intratracheal instillation (IT) studies reported here were performed at Research and Consulting Company Ltd (RCC) in Füllinsdorf, Switzerland.

INHALATION EXPOSURE

The experimental design has been presented in detail previously (Bernstein *et al.*, 1994, 1995 and 1996) and is summarised below. In particular, details of the counting and sizing procedures are reiterated as these are considered essential to the successful interpretation of these studies.

Groups of 56 weanling (4–8 weeks old) male Fischer 344 rats (SPF quality) were exposed by flow-past nose-only exposure to a target fibre aerosol concentration of 30 mg m^{-3} for 6 h day^{-1} for a period of five consecutive days. This concentration corresponded to the highest gravimetric concentration used for the MMVF 11 fibre in an inhalation oncogenicity study (Hesterberg *et al.*, 1993). The fibre number concentrations in the studies evaluated here were generally higher due to the differences in fibre diameters. In addition, a negative control group was exposed in a similar fashion to filtered air. To be comparable with current and previous fibre inhalation studies, Fischer 344 rats [CFD (F-344)/CrlBR] obtained from Charles River Laboratories (Kingston, NY, U.S.A.) were used.

---

* The inhalation data for Fibre R (RCF 1a), used in comparison to the IT data, is preliminary data provided by the North American Insulation Manufacturers Association (NAIMA) through Dr Mellon.

Groups of eight animals were sacrificed at 1 h, 1 day, 5 days and 4 weeks following the last (5th) day of exposure and at 13, 26 and 52 weeks following the start (1st day) of exposure, the lungs digested using low temperature ashing (LFE™ LTA 504 at 300 watts for at least 16 h) and the number and bivariate size distribution of the recovered fibres determined. The 52-week animals were not analyzed as the number of fibres longer than 20 µm was for most of the lung samples near or below the limit of detection at 26 weeks (25 000 fibres lung^{-1}). All animals were anesthetised with an intraperitoneal injection of sodium pentobarbital and sacrificed by exsanguination. The lungs and the lower half of the trachea were removed, weighed and immediately deep frozen at approximately −20°C to minimise dissolution of the fibres.

When sizing or counting fibres, an object was accepted as a fibre if the ratio of length to diameter was at least 3:1. All other objects were considered particles. Fibres lying with both ends in the field were counted as one fibre. There was no upper or lower limit of the size. All fibres and particles visible at an initial magnification of 2000 were measured using as high a magnification as necessary. The length and diameter of each particle or fibre measured was recorded. The recording of particles was stopped when a total of 30 particles was observed. The evaluation of fibres was stopped when 300 WHO fibres (1 > 5 µm, $d < 3$ µm) (WHO, 1985) or a total of 1000 fibres and particles were recorded, even if a total of 300 countable WHO fibres was not reached. For lung samples, five lungs/fibre/time point were analysed and the means reported.

INTRATRACHEAL INSTILLATION

Groups of male Wistar rats (SPF quality) were used in order to be comparable to the current series of inhalation biopersistence studies performed at RCC. The animals were instilled a total of four times, once on each of four successive days by injection of 0.5 mg of fibres suspended in 0.3 ml of physiological saline. The test material in suspension was removed from a glass vessel (under constant stirring using a magnetic stirrer ~ 750 RPM) using a 1 ml plastic syringe. The tracheal cannula was approximately 60 mm long with an internal diameter of 1.4 mm. The fibres were suspended in a vehicle of 0.9% NaCl in distilled water. For each test article and for control article, suspensions were prepared such that each dose stock suspension contained a fibre concentration of 0.5 mg per 0.3 ml suspension.

Samples were collected on GELMAN membrane filters (Metricel DM-450, pore size 0.45 µm) for gravimetry and on NUCLEPORE filters (PC membrane, pore size 0.2 µm) for size determination by SEM. Samples were collected from the top, middle and bottom of the fibre suspension flask both immediately prior to and following instillation. The counting and sizing rules used were the same as for the inhalation studies described above.

DATA ANALYSIS

Curve fitting procedures

The clearance of both the number of WHO fibres and the number of fibres remaining in the lung by length category (< 5 µm, 5–20 µm and > 20 µm) as a function of time following cessation of exposure was fitted using non-linear techniques to a double exponential function. The double exponential function:

Per cent fibre remaining = a1 * exp (− b1 * Time) + a2 * exp (− b2 * Time)

was fitted to the data using the quasi-Newton non-linear regression (Statistica, 1995), with the loss function weighted by the inverse of the variance (Neter *et al.*, 1990). For each curve two clearance half-times were obtained, one for the coefficient b1 and another for coefficient b2 as follows:

$$T_{1/2} - 1 = \ln 2 / b1 \quad \text{and } T_{1/2} - 2 = \ln 2 / b2.$$

These clearance half-times often correspond to a faster clearance phase followed by a slower clearance phase. In order to provide an index of the complete clearance which includes both the fast and slower clearance half-times ($T_{1/2} - 1$ and $T_{1/2} - 2$), the combined weighted clearance times ($W - T_{1/2}$) were determined by summing the product of each half-time weighted by its coefficient a_x as follows:

$$W - T_{1/2} = \left(\frac{a_1}{a_1 + a_2} \right) \times T_{1/2} - 1 + \left(\frac{a_2}{a_1 + a_2} \right) \times T_{1/2} - 2.$$

RESULTS

In evaluating the results, it should be kept in mind that the fibre lung burdens in this study were obtained by the digestion of the entire lung using the low temperature ashing techniques described above. While a complete quantification of the inorganic particle and fibre content of the lung is obtained, no differentiation can be made from these measurements alone as to where in the lung these particles or fibres were located. Thus, if a fibre is on the bronchial tree it is counted equally as a fibre in the alveolar region. The same is true for shorter fibres which may be phagocytised by macrophages and removed to the bronchial associated lymphoid tissue (BALT) or to lymph nodes.

Comparison of clearance following inhalation and intratracheal instillation exposure

The clearance of fibres by length fraction following inhalation and intratracheal instillation exposure is shown for fibres C, G, M, O, P and R in Figs 1–6, respectively. The weighted half-times for the WHO fibres and fibres with lengths greater than 20 μm are shown in Table 2. For the more soluble fibres (C, G, M, O, P), the long fibre fractions (> 20 μm) appear to clear faster than the shorter fibre fractions following both inhalation and IT exposure. However, for these fibres a greater difference in clearance rate is observed between the long and short fibre fraction following inhalation than following intratracheal instillation. For the relatively insoluble fibre R (RCF 1a), very little clearance was observed following IT exposure especially for longer fibres. SEM examination of the lungs is now planned in an effort to determine if this lack of clearance is associated with possible fibre agglomeration in the bronchi. Following inhalation, there occurs a fast clearance of 60–80% of the fibres depending upon length category followed by an extended slower clearance phase.

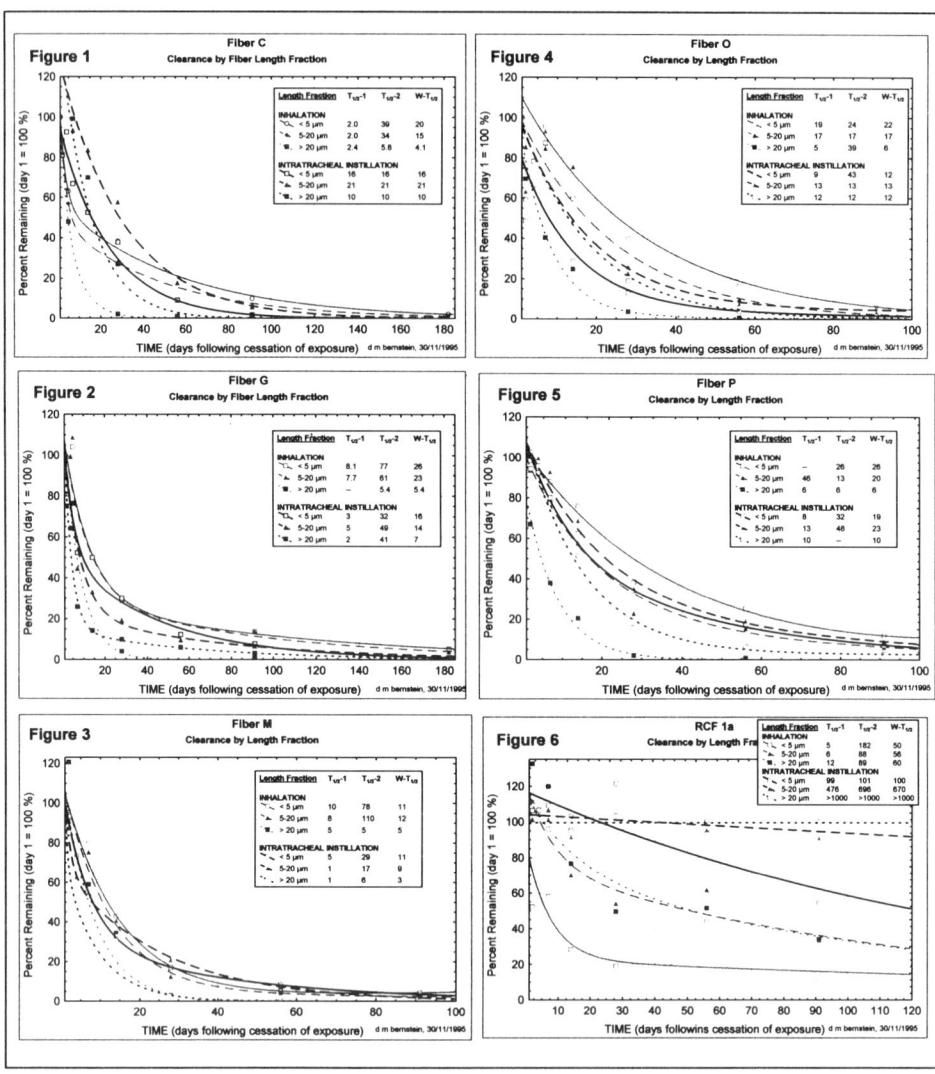

Table 2. Summary of WHO and long fibre clearance

| Fibre | KI | Inhalation | | Intratracheal instillation | |
		$W - T_{1/2}$ WHO	$W - T_{1/2}$ $L > 20\ \mu m$	$W - T_{1/2}$ WHO	$W - T_{1/2}$ $L > 20\ \mu m$
C	34.0	14	4	21	10
G	31.9	22	5	12	7
M	40	12	5	8	3
O	40	17	6	13	12
P	30.1	18	6	22	10
R (RCF 1a)	−95.5	57	64	> 1000	> 1000

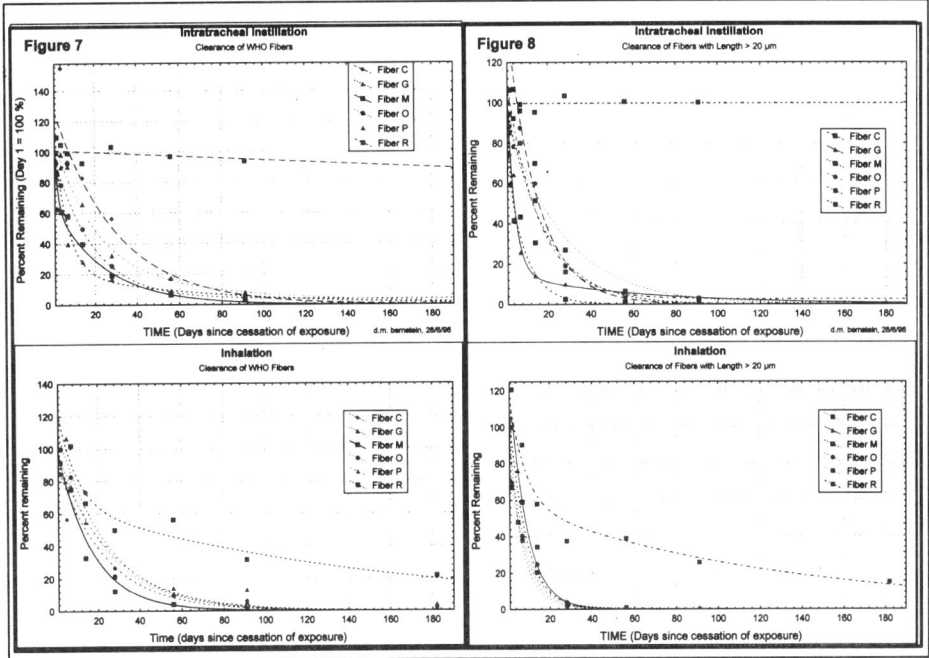

WHO fibre clearance

Figure 7 shows for each fibre the clearance of the WHO fibres following either inhalation or IT exposure. The weighted half-times are presented in Table 2. Roughly similar trends are seen although a wider dispersion is seen during the earlier time points following IT exposure.

Clearance of fibres longer than 20 μm

The clearance of fibres longer than 20 μm is shown in Fig. 8 for both inhalation and IT exposure. By inhalation, the clearance of the long soluble fibres (C, G, M, O, P) is as would be predicted by the Kdis (Bernstein *et al.*, 1996) nearly identical. By IT, however, the clearance is again more dispersed.

Inter-laboratory comparison

Bernstein *et al.*, (1996) have previously reported that by inhalation, the fibre MMVF11 was evaluated in two separate biopersistence studies at RCC. The weighted clearance half-times determined in the two studies were 38 d/47 d ($L < 5$ μm), 32 d/42 d (L 5–20 μm) and 13 d/9 d ($L > 20$ μm).

The studies reported here provide the opportunity to examine the inter-laboratory reproducibility following IT exposure performed on the same fibres at two different laboratories, RCC and the Fraunhofer-Institut für Toxikologie und Aerosolforschung (Fraunhofer) in Hannover, Germany. The results from the Fraunhofer studies are preliminary and will be published in detail subsequently. Two examples illustrate the range of half-times at RCC/Fraunhofer, respectively. For fibre G the $T_{1/2}$ were: 16 d/24 d ($L < 5$ μm), 14 d/30 d (L 5–20 μm) and 7 d/19 d

$(L > 20 \, \mu m)$. For fibre M the $T_{1/2}$ were: 11 d/19 d $(L < 5 \, \mu m)$, 9 d/14 d $(L \, 5\text{--}20 \, \mu m)$ and 3 d/6 d $(L > 20 \, \mu m)$. The fibre R (RCF 1a) presented above was also evaluated independently by Fraunhofer with much shorter clearance half-times than reported above. This may be due to differences in size distributions of the fibres instilled and/or possible agglomeration in the RCC studies which is currently being investigated.

SUMMARY

Both inhalation and IT exposure provide estimates of the biopersistence of soluble fibres of similar magnitude. Inhalation systematically shows the faster clearance of longer fibres $(L > 20 \, \mu m)$. For the one less soluble fibre evaluated in this study (RCF 1a), a large difference was observed between inhalation and IT exposure although as mentioned above this may be due to possible agglomeration in the bronchi following IT exposure.

The results from both inhalation and IT studies were reproducible. Both methods require extensive controls to be valid. Inhalation exposure requires the aerosol to be rat respirable while IT requires confirmation of no agglomeration of fibres in the fibre suspensions used for injection and in the bronchi of the lungs following injection.

REFERENCES

Bernstein, D. M., Mast, R., Anderson, R., Hesterberg, T. W., Musselman, R., Kamstrup, O. and Hadley, J. (1994) An experimental approach to the evaluation of the biopersistence of respirable synthetic fibres and minerals. *Environ. Health Perspect.* **102** (Supplement 5), 15–18.
Bernstein, D. M., Morscheidt, C., Grimm, H. G. and Teichert, U. (1996) The evaluation of soluble fibres using the inhalation biopersistence model, a nine fibre comparison. *Inhal. Toxicol.* **8**, 345–385.
Bernstein, D. M., Morscheidt, C., Tiesler, H., Grimm, H. G., Thevenaz, Ph. and Teichert, U. (1995) Evaluation of the biopersistence of commercial and experimental fibres following inhalation. *Inhal. Toxicol.* **7**(7), 1029–1056.
Hesterberg, T. W., Mast, R., McConnell, E. E., Chevalier, H. J., Hadley, J., Bernstein, D. M., Thevenaz, P. and Anderson, R. (1993) Chronic inhalation toxicity of size-separated glass fibres in Fischer 344 rats. *Fundam. appl. Toxicol.* **20**, 464–476.
McClellan, R. O., Miller, F. J., Heslerberg, T. W., Warheit, D. B., Bunn, W. B., Kane, A. B., Lippmann, M., Mast, R. W., McConnell, E. E. and Reinhert, C. F. (1992). Approaches to evaluating the toxicity and carcinogenicity of man-made fibers: summary of a workshop held 11–13 November, 1991, Durham, North Caroline. *Reg. Tox. Pharm.* **16**, 321–364.
Neter, J., Wasserman, W. and Kutner, M. H. (1990) *Applied Linear Statistical Models*, 3 edn. Irwin, Inc., Homewood, Il.
Statistica (1995) *STATISTICA for Windows*, Version 5, (Computer program manual). Tulsa, OK: StatSoft, Inc., 2325 East 13th Street, Tulsa, OK 74104, U.S.A. Tel.: (918) 583–4149, Fax: (918) 583–4376.
TRGS 905 (Technische Legeln für Gefahrstoffe) (1994) Justification for the classification of dusts from natural and man-made mineral fibres. Bundesarbeitsblatt, Nr. 6, p. 57. Bundesministerium für Arbeit und Sozialordnung, Verlag W. Kohlhämmer, ISSN 0007-5868.
WHO (1992) Validity of methods for assessing the carcinogenicity of man-made fibres. Executive Summary of a Consultation held 19–20 May 1992, prepared by the WHO Regional Office for Europe, Copenhagen.
WHO, World Health Organization (1985) Reference methods for measuring airborne man-made mineral fibres (MMMF), prepared by the WHO Regional Office for Europe, Copenhagen.

 Pergamon

Ann. occup. Hyg., Vol. 41, Supplement 1, pp. 231–236, 1997
© 1997 British Occupational Hygiene Society
Published by Elsevier Science Ltd. All rights reserved
Printed in Great Britain
0003–4878/97 $17.00 + 0.00
Inhaled Particles VIII

PII: S0003–4878(96)00068–3

LUNG FIBRE CONTENT FOR MESOTHELIOMA IN THE 1891–1920 BIRTH COHORT OF QUEBEC CHRYSOTILE WORKERS: A DESCRIPTIVE STUDY

B. W. Case*, A. Churg,† A. Dufresne*, P. Sébastien*, A. McDonald*
and J. C. McDonald*

*Departments of Pathology and Epidemiology and Occupational Health, Faculty of Medicine,
McGill University, 3775 University Street, Montreal, Canada H3Y 3B4; and
†Department of Pathology, Faculty of Medicine, University of British Columbia, Vancouver, Canada

INTRODUCTION

Since the discovery of the crocidolite/mesothelioma relationship by Wagner *et al.* (1960) human studies have indicated that the amphibole forms of asbestos, and not chrysotile, are responsible for asbestos-related mesothelioma with rare, if any, exceptions (Elmes, 1994; and the accompanying workshop discussion reported by Churg *et al.*, 1994). A recent paper by Stayner *et al.* (1996), reaffirms that ". . . the differences in mesothelioma response observed among chrysotile- and amphibole (primarily crocidolite)- exposed workers are so striking that alternative explanations for these differences appear unlikely." There remain observers (including Stayner and colleagues) who are nonetheless unconvinced that amphibole co-exposure (to tremolite and commercial amphiboles) explains mesotheliomas observed in the 1891–1920 birth cohort of Quebec chrysotile miners and millers. This is true even though they can otherwise document only 13 cases in 11 other cohorts of ". . . workers exposed to predominantly chrysotile asbestos" (Stayner *et al.*, 1996, Table 1, page 180). This, the chrysotile hypothesis for mesothelioma, has implications for risk assessment both for chrysotile and for tremolite (Case, 1991).

The 1891–1920 birth cohort consists of 11 000 men employed in the Quebec chrysotile industry for at least one month, including over 700 who were first employed in an asbestos products factory at the town of Asbestos, where commercial amphiboles were used. Miners and millers at the local Jeffrey Mine replaced sick factory workers, and visited the factory building on a regular basis. Lung-retained fibre analysis for workers with and without mesothelioma at Asbestos (Case and Sébastien, 1987, 1988; Dufresne *et al.*, 1995, 1996) showed excesses of crocidolite, and to a lesser extent amosite asbestos, in about 75% of lungs analysed, in addition to tremolite and chrysotile. In the separate region around Thetford and Black Lake, a number of mines were exploited, but there were no products factories and no known exposure to commercial amphiboles among men. Recently the first full analysis of mesotheliomas from these two areas has been performed (McDonald *et al.*, 1996). A maximum of 38 possible mesotheliomas among 8009 cohort deaths were identified, with highest proportional mortality (1.08%) in the factory and lowest (0.21%) in miners and millers at

Asbestos. Miners and millers at Thetford Mines had most mesothelioma deaths but an intermediate rate (25 of 4125 observed deaths; 0.62%; $P < 0.01$ vs. Asbestos, Chi-square). Internal analyses confirmed the previous observation by McDonald and McDonald (1995) that mesothelioma deaths in Thetford were largely confined to those who had worked in five central mines. These appeared both from historical geologic data and from lung fibre analysis to be mineralogically distinct, particularly in terms of an increased intrapulmonary tremolite burden. In the current study we analysed lung fibre content in both Asbestos ($N = 7$) and Thetford Mines ($N = 15$) workers with mesothelioma. Subgroups examined included (1) miners and millers in the "central mines" of Thetford, near the town site ($N = 9$); (2) those who worked more peripheral to the town of Thetford ($N = 1$); (3) Thetford miners and millers who had worked in the central area but also had some exposure in the peripheral mines and mills ($N = 5$); (4) men who worked in the asbestos products factory at Asbestos ($N = 2$); and (5) those who worked exclusively in the mine and mill at Asbestos ($N = 5$). Data on cumulative exposure allowed us for the first time to relate external exposure measurement to lung fibre content.

METHODS

We contacted scientists who had published lung fibre analyses for workers in the Quebec chrysotile industry and asked to obtain results from any cases of mesothelioma, together with pathology accession numbers. We then matched the latter to those obtained for our own list of 38 possible mesotheliomas among cohort members. Altogether, 17 male cohort members with mesothelioma and available results were identified. Lung analyses produced in the past by four investigators (AC, AD, BC and PS) were used. Five additional cases were obtained from pathologists and analysed in our laboratory. Detailed methods of fibre analyses are available in the references cited. All analyses but one included fibres of "all lengths"; fibre definition (aspect ratio) was greater than 3:1; and results were expressed as fibres/µg dry lung. Both "wet" (formalin-fixed) and "block" (paraffin-embedded) tissues were used. All analyses included filtered bleach digestion; low-temperature ashing had also been used in specimen preparation in the McGill laboratory.

Fibre identification was performed using energy dispersive X-ray spectrometry (EDS) coupled with transmission electron microscopy (TEM). For analysis, results from all laboratories were combined: where results were available from more than one laboratory we chose our own (McGill) laboratory first (AD, BC, PS): where results were available from more than one investigator within our own laboratory we chose the analysis performed by AD. Ultimately 17 of the 22 cases came from the McGill laboratory and 16 were analysed by AD; the remaining five cases were analysed by AC in his Vancouver laboratory.

Statistics

We limit our observatiouns in this paper to fibre concentration data and parameters of exposure. Fibre concentration data are expressed as geometric mean fibres per microgram dry lung in the groups and subgroups of interest: comparisons for statistical significance were made using two-sample *t*-tests having independent

Table 1. Subject characteristics and exposure profiles in five subgroups (median values)

Median demographic and exposure variables:	N	Year start	Age start	Net years worked	Latency (years)	Exposure (MPCFY)[1]	Cessation interval[2]
Asbestos miners/millers	5	1936	18	37	49	373	11
Asbestos factory	2	1944	25	16	37	42*	22
Thetford: both areas	5	1935	22	38	48	188	6
Thetford central	9	1935	18	32	50	319	16
Thetford peripheral	1	1947	31	34	38	325	4

[1] Median cumulative exposure in (million particles per cubic foot) X (years worked) (see text).
[2] Interval between end of work and time of death from mesothelioma.
* $P < 0.05$ vs. Thetford central area men but no other significant differences (Mann–Whitney).

variance. All values were log-transformed and where fibre concentration was "not detected" a value of 0.1 fibres μg^{-1} dry lung was used. Parameters of exposure were compared to lung fibre concentrations using Pearson and Spearman rank correlation coefficients. All comparisons were made using Minitab® version 11.12 for Windows 95.

RESULTS

(1) *Subject characteristics and exposure history*

Median values for the five subgroups are provided in Table 1. With the exception of the two factory workers in Asbestos, who began work in 1940 and 1948, most men started in the 1930s ($N = 11$) or prior to 1930 ($N = 7$). The factory workers performed 3 and 28 years of factory work (much less than the average total work performed by the other groups); the man with only 3 years also worked in a shipyard for 1 year. Their cumulative exposures derived from historical midget impinger counts were an order of magnitude lower than those in other groups. The median period of time between the end of their employment and their death from mesothelioma was twice that in other groups, but the latency period for the appearance of their tumors (37 and 38 years) was 10 years shorter, on average. Other subgroups were not distinguishable from one another by demographic, disease or exposure variables.

(2) *Fibre concentrations*

Geometric mean fibre type concentrations for all Thetford men combined and for all Asbestos men combined and for each of the five exposure subgroups, are presented in Table 2. Crocidolite concentrations were significantly higher in workers at Asbestos, and when individual values are examined (10.1 and 4.0 fibres μg^{-1} in the Asbestos factory workers; 14.7, 7.0, 2.0, 0.7 and 0.00 fibres μg^{-1} in the Asbestos miners and millers) six of seven men had levels we consider diagnostic of significant occupational exposure to commercial amphiboles. Three of the six also had elevated intrapulmonary amosite levels. Men in Thetford Mines had strikingly high levels of intrapulmonary tremolite. This excess of tremolite was not explained by cumulative exposure levels, which were not significantly different from those in Asbestos miners and millers. In addition, division of intrapulmonary concentra-

Table 2. Asbestos fibre concentrations in lungs of mesothelioma patients in the Québec cohort: comparison of two mining regions and five subgroups

Geometric mean fibre concentrations[1]	N	Tremolite	Chrysotile	Crocidolite	Amosite	MPCFY[2]
Asbestos miners/millers	5	7.50	4.35*	1.69	0.25	126
Asbestos factory workers	2	0.47[‡]	2.12	6.34	0.32	41[†]
All Asbestos men	7	3.40**	3.54*	2.47**	0.27	92
Thetford central area	9	119	10.5	n.d.[3]	n.d.	298
Thetford peripheral area	1	101	32.0	n.d.	n.d.	325
Thetford both areas	5	85	15.7	n.d.	n.d.	196
All Thetford men	15	105	12.9	n.d.	n.d.	260

[1] Geometric mean of concentrations for all fibres of given type in fibres of all lengths, aspect ratio greater than 3:1, per microgram dry lung.
[2] Geometric mean of cumulative lifetime exposure in (million particles per cubic foot) X (years worked); note difference between these values and median values in Table 1.
[3] Below limit of detection. For geometric mean calculations set at 0.1 fibres μg^{-1} dry lung.
* $P < 0.05$ vs. Thetford men who worked in central area or in both areas.
** $P < 0.01$ vs. all Thetford men combined.
[†] $P <$ vs. Thetford men in central area and $P < 0.05$ vs. Thetford men in both areas.
[‡] $P < 0.05$ vs. Thetford men in both areas.

tions for crocidolite and tremolite by total cumulative exposure left a 20-fold excess of tremolite fibres μg^{-1} dry lung/unit MPCFY in the 15 Thetford men ($P < 0.01$) and a 25-fold excess for crocidolite/unit MPCFY in lungs of the seven subjects from Asbestos ($P < 0.01$). Chrysotile content was higher in the lungs of miners and millers in Thetford, regardless of subgroup, and lowest in Asbestos factory workers. However, in this instance adjustment of intrapulmonary chrysotile fibre levels through division by total cumulative exposure left no residual differences between the two areas.

In Asbestos, it is notable that crocidolite not only formed 53% of all asbestos fibres in the lungs of factory workers, but also more than 20% of asbestos fibres in the five miners and millers studied. Thus, while the tremolite per cent fraction was significantly lower in Asbestos chrysotile workers, their total amphibole concentrations constituted about the same fraction of asbestos lung burden as those observed in workers in Thetford Mines.

(3) *Correlation of fibre concentrations with exposure measurements*

As in a previous study of chrysotile miners, millers and textile workers (Sébastien *et al.*, 1989), lung tremolite and chrysotile concentrations were both good indicators of past exposure. Pearson correlation coefficients across all 22 men between cumulative exposure in MPCFY and lung chrysotile and tremolite content were 0.64 ($P < 0.01$) and 0.61 ($P < 0.01$), respectively, or 0.57 and 0.71 for log-transformed values. Tremolite and chrysotile concentrations were closely correlated (0.94, $P < 0.001$), principally in the Thetford groups of men. Chrysotile concentration correlated with crocidolite in the five miners and millers from Asbestos ($r = 0.88$, $P < 0.05$). Duration of work did not correlate with any lung fibre concentration. Work was broken down in years spent in the mines, mills, factory or as tradesmen, but only the latter showed weak correlation with any aspect of fibre burden and that with crocidolite asbestos alone ($r = 0.45$, $P < 0.01$).

Spearman rank correlation coefficients generally showed the same pattern, with especially high rank correlation between tremolite concentration and cumulative exposure (MPCFY) in the seven Asbestos men ($R = 0.96$; $P < 0.05$). There was also an apparent "clearance effect" for crocidolite in this group (Cessation interval vs. crocidolite concentration $R = -0.92$; $P < 0.05$), possibly due to a greater available time for fibre clearance between cessation of work and death than that observed in the Thetford workers (Table 1). Interval between date of retirement and date of lung tissue acquisition (cessation interval) was not correlated with any other lung fibre concentration parameter in any group.

DISCUSSION

In the absence of comparable data on denominators or appropriate referents, questions of risk cannot be properly examined with this data alone. However, in this cohort of "chrysotile" workers, crocidolite, amosite and/or tremolite were present in substantial concentrations in the lungs of most mesothelioma cases. These known biopersistent fibres corresponded well with cumulative exposure, but so did chrysotile—the human pulmonary dynamics of which are perhaps more complex than has heretofore been acknowledged. Although the tremolite component of exposure in these groups is well described, the past commercial amphibole problem in the region which has lower tremolite (Asbestos) has only recently come to attention (Case and Sébastien, 1987, 1988; Dufresne et al., 1995, 1996). These results cannot be explained by the chrysotile hypothesis of mesothelioma genesis. They are consistent with variable amphibole exposure to workers in both areas as reported in epidemiological and geological observations in the mining region (McDonald et al., 1995, 1996). Recent observations indicate increased mesothelioma mortality among women in Thetford Mines (Camus, M. personal communication), as opposed to Asbestos. This finding, coupled with previous observations of pleural disease and both airborne and intrapulmonary tremolite excess near the mines of the Thetford/Black Lake area, support the idea of a biological role for this amphibole (Case and Sébastien, 1989; Case, 1991).

This extended abstract cannot examine all of the issues involved in the interpretation of lung-retained fibre. Every measure of exposure used in epidemiological studies is subject to errors of measurement, classification and interpretation. Careful use is nonetheless essential if we are to adequately explore past exposures. That proven carcinogens which accumulate in lung (such as crocidolite and tremolite) may be less harmful than fibres which do so to a lesser degree (such as glass fibre, or chrysotile) seems biologically untenable, even given our inadequate knowledge of mechanism. The idea that "pleural burden" may be more useful in assessing a pleural tumour appeals until we realise the inadequacies of any underlying biological model for the disease and most particularly the opportunities for error due to specimen contamination by short chrysotile fibres; opportunities which have been realised in some misleading published work (Case, 1994). To ignore the overall evidence (epidemiology, geology and exposure assessment using retained internal dose) could lead to overestimation of chrysotile risk for mesothelioma in humans while inadequately addressing the problem of fibrous tremolite.

REFERENCES

Case, B. W. (1991) Health effects of tremolite: now and in the future. *Ann. NY Acad. Sci.* **643**, 491–504.

Case, B. W. (1994) Biological indicators of chrysotile exposure. *Ann. occup. Hyg.* **38**, 503–518.

Case, B. W. and Sébastien, P. (1987) Environmental and occupational exposures to chrysotile asbestos: A comparative microanalytic study. *Arch. Environ. Health* **42**, 185–191.

Case, B. W. and Sébastien, P. (1988) Biological estimation of environmental and occupational exposure to asbestos. *Ann. occup. Hyg.* **32**, 181–186.

Case, B. W. and Sébastien, P. (1989) Fibre levels in lung and correlation with air samples. In *Non-occupational Exposure to Mineral Fibres* (Edited by J. Bignon, J. Peto and R. Saracci), International Agency for Research on Cancer Scientific Publication, no. 90, pp. 207–218. International Agency for Research on Cancer, Lyon.

Churg, A., Neuberger, N. and McDonald, J. C. (rapporteurs) (1994) Discussion of the paper by Elmes, P. (1994). *Ann. occup. Hyg.* **38**, 416–417.

Churg, A., Wright, J. L. and Vedal, S. (1993) Fiber burden and patterns of asbestos-related disease in chrysotile miners and millers. *Am. Rev. Resp. Dis.* **148**, 25–31.

Dufresne, A. D., Bégin, R., Churg, A. and Massé, S. (1996) Mineral fiber content of lungs in patients with mesothelioma seeking compensation in the province of Québec. *Am. Rev. Resp. Crit. Care Med.* **153**, 711–718.

Dufresne, A. D., Harrison, M., Massé, S. and Bégin, R. (1995) Fibers in lung tissues of mesothelioma cases among miners and millers of the township of Asbestos, Quebec. *Am. J. ind. Med.* **27**, 581–592.

Elmes, P. (1994) Mesotheliomas and chrysotile. *Ann. occup. Hyg.* **38**, 547–553.

McDonald, J. C., Liddell, F. D. K., Dufresne, A. and McDonald, A. D. (1993) The 1891–1920 birth cohort of Quebec chrysotile miners and millers: mortality 1976–1988. *Br. J. ind. Med.* **50**, 1073–1081.

McDonald, J. C. and McDonald, A. D. (1995) Chrysotile, tremolite and mesothelioma. *Science* **267**, 775–776.

McDonald, A. D., McDonald, J. C., Liddell, F. D. K., Case, B. W., Dufresne, A. D., Churg, A., Gibbs, G. W. and Sébastien, P. (1996) Mesothelioma in Québec miners and millers: epidemiology and etiology (in preparation).

Sébastien, P., McDonald, J. C., McDonald, A. D., Case, B. W. and Harley, R. (1989) Respiratory cancer in chrysotile textile and mining industries: exposure influences from lung analysis. *Br. J. ind. Med.* **46**, 180–187.

Stayner, L. T., Dankovic, D. A. and Lemen, R. A. (1996) Occupational exposure to chrysotile asbestos and cancer risk—a review of the amphibole hypothesis. *Am. J. Pub. Health* **86**, 179–186.

Wagner, J. C., Sleggs, C. A. and Marchand, P. (1960) Diffuse pleural mesothelioma and asbestos exposure in the North Western Cape Province. *Br. J. ind. Med.* **17**, 266–271.

 Pergamon

Ann. occup. Hyg., Vol. 41, Supplement 1, pp. 237–243, 1997
© 1997 British Occupational Hygiene Society
Published by Elsevier Science Ltd. All rights reserved
Printed in Great Britain
0003–4878/97 $17.00 + 0.00
Inhaled Particles VIII

PII: S0003–4878(96)00117–2

OVERLOADING OF CLEARANCE OF PARTICLES AND FIBRES

C. L. Tran, A. D. Jones, R. T. Cullen and K. Donaldson*

Institute of Occupational Medicine, 8 Roxburgh Place, Edinburgh EH8 9SU;
and *Napier University, Edinburgh, U.K.

INTRODUCTION

Overloading of the macrophage defence system can retard clearance of inert particles from the lung. The purpose of this paper is to compare the burden-dependence of clearance of particles and fibres using data from inhalation experiments at the IOM and at RCC, Geneva. Due to the different experimental protocols (in exposure concentrations and periods of exposure and recovery), the comparisons benefit from the use of a mathematical model of retention and clearance. The hypothesis to be investigated is that clearance of short fibres (length < 15 µm) is the same as for non-fibrous particulate dust of low toxicity.

METHODS, MATERIALS AND DATA

A mathematical model, describing the mechanisms of clearance for an inert dust (TiO$_2$) and a toxic dust (quartz) (Tran *et al.*, 1994), has recently been extended to include processes relevant to fibres such as the disintegration of long fibres ($l > 20$ µm). Predictions from the model are compared with data for rats exposed to TiO$_2$ at 1, 10, 50 mg m^{-3} or inert man made mineral fibres (glass wool, MMVF10) at 3, 16 and 30 mg m^{-3}. The TiO$_2$ data consist of alveolar lung burdens (and lymph node burdens) of titanium dioxide measured by atomic absorption. The MMVF10 data consist of the number and size distribution of fibres recovered from the accessory lobe. The accessory lobe is approximately 1/10th of the lung (by mass) and, therefore, the accessory lobe burden is scaled up proportionately to estimate the whole lung burden. The data and predictions from the same model for lung burden are compared below; the lymph node burden data for TiO$_2$ served to determine some of the parameters in the model.

For particles, the model's parameters include rates for: alveolar macrophages' phagocytosis and clearance, fibre disintegration, alveolar sequestration, interstitialisation, transfer to lymph nodes. Figure 1 shows the main translocation routes for particles (or short fibres). This part of the model comprises nine compartments (X_1–X_9) describing the fate of inhaled particles and short phagocytosable fibres. These are, at the alveolar level, X_1 free (un-phagocytosed) on the alveolar surface; X_2 successfully phagocytosed by active alveolar macrophages; X_3 inside inactive alveolar macrophages; X_4 sequestrated in overloaded alveolar macrophages. Then there are similar compartments for particles in the interstitium (X_5–X_8) and a final compartment for particles in the lymph nodes (X_9). Differential equations with

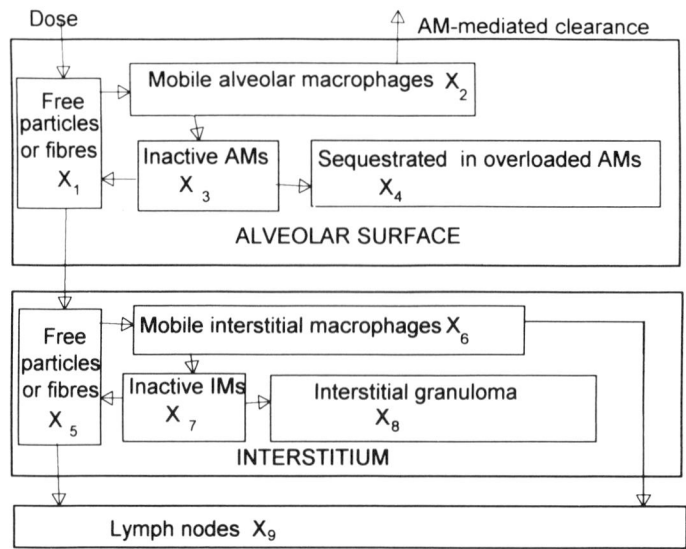

Fig. 1. Schema of the compartments and translocations of particles and short fibres in the lung.

appropriate rate constants describe the translocation of particles or short fibres from one compartment to another. The transfer to overloaded macrophages starts as the lung burden reaches a certain threshold level.

For fibres, the model consists of two parts; the first part, describing the kinetics of clearance of short fibres, is as for particles. The second part, describing the kinetics of fibre disintegration for intermediate length and long fibres, is shown in Fig. 2. For intermediate length fibres ($15 \, \mu m < l < 20 \, \mu m$), phagocytosis is assumed to be less successful and their removal from the alveolar region is treated as being by AM-mediated clearance or disintegration by dissolution in extra-cellular fluid. For long fibres, phagocytosis is not successful and the removal of long fibres is due to dissolution and disintegration in extra-cellular fluid.

THE MODEL'S PARAMETERS

The objective was to examine the extent to which the same model of clearance kinetics accounts for the lung burden data for the TiO_2 and the MMVF10 fibres.

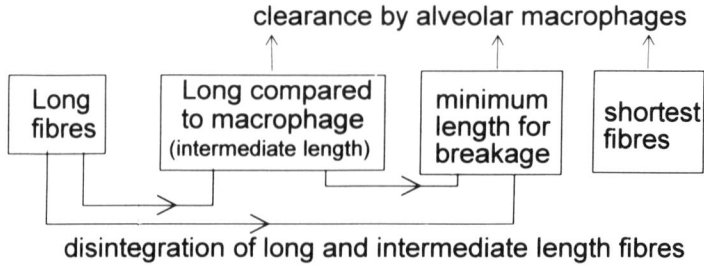

Fig. 2. Diagram of the disintegration of long fibres into shorter phagocytosable fragments.

Therefore, the model's parameters were assigned values which were either based on previous studies of particulate clearance (e.g. Stöber *et al.*, 1989; Katsnelson *et al.*, 1994) or derived from obtaining a "best-fit" to the TiO_2 data, or based on assumptions. For fibres, the clearance parameter values were retained, except those pertaining specially to the fibre experiments (e.g. deposition rates and disintegration rates) as shown in Table 1. For the TiO_2, the low-dose (1 mg.m^{-3}) data were used to calculate the deposition efficiency of TiO_2 particles in the alveolar region. The parameters pertaining to overload were subsequently estimated from the mid-dose (10 mg.m^{-3}) experimental data. This procedure evaluated all the parameter values. The data from the high-dose experiment (50 mg.m^{-3}) were then compared with the model simulation to check consistency between the model and this part of the dataset.

The strategy for fibres was similar to TiO_2 in that clearance parameters were established initially by comparison with the low-dose data; and this showed that the values as derived from particles were suitable.

Additionally, an estimation was made of the disintegration rate of long fibres initially from published dissolution rates and then by (relatively minor) adjustment to improve consistency between the MMVF10 data and the extended model. Because the fibre aerosols were perhaps not equally well dispersed at all concentrations, the deposition efficiency was estimated for each exposure concentration. However, the parameters of normal clearance and fibre disintegration remained the same for all concentrations. The parameters of overload were established using the data from the mid-dose experiment. As for TiO_2, the high-dose experiment was used to check the consistency between the model and data.

RESULTS AND DISCUSSION

(a) TiO_2

Figure 3(a) shows the data for the accumulating burden at the lowest exposure concentration (1 mg.m^{-3}) and the prediction for the model. Each data point is the mean burden for groups of four animals at 3 days post-exposure. Lung burdens were also measured for further groups at 10 and 38 days post-exposure, giving the data in Fig. 3(b) where the curves are the predictions from the same model. At this lowest concentration, the clearance rate of 1.5% per day (consistent with other reported clearance rates) was independent of lung burden; and lymph node burdens were below the detectable level (i.e. < 10 μg). Since TiO_2 has a density of 4.26 gm cc^{-1}, the critical volume V_{crit}, of 100 nl (Morrow, 1992), above which clearance starts to gradually become retarded, corresponds to a mass burden of 0.426 mg of TiO_2. At mid-dose (10 mg.m^{-3}), the accumulated lung burden exceeds this 0.4 mg. Lung clearance becomes progressively retarded when lung burden level reaches approximately 10 mg; this impairment of clearance is also evident post exposure. At 50 mg.m^{-3}, this lung burden (of 10 mg) is reached by about 5 days of exposure and the data show that clearance impairment occurs within 25 days of exposure and the lung burden accumulated almost linearly; there is also considerable impairment of clearance post-exposure (Fig. 5). Figure 5 shows that the model, with parameters estimated from the 1 and 10 mg.m^{-3} data, fits well with the 50 mg.m^{-3} data.

Table 1. Values of clearance parameters in the model. Note that subscript j means parameters refer to fibres in length category j for fibres which can be phagocytosed (these parameters also apply to particles), whereas subscript k is used for fibres which are too long to be successfully phagocytosed

Parameters	Symbol	TiO$_2$	MMVF10	Unit	Basis or source for value
Rate of successful phagocytosis by AMs/IMs for particles or fibres					
< 15 μm long	r_j	4	4	day^{-1}	Stöber *et al.* (1990)
Reduced r_j for fibres with length in 15–20 μm range	r_j	NA	0.4	day^{-1}	estimated, MMVF10 data
Reduced r for fibres with length in 20–25 μm range	r_k	NA	0.2	day^{-1}	estimated, MMVF10 data
Reduced r for fibres longer than 25 μm	r_k	NA	0	day^{-1}	estimated, MMVF10 data
The critical length beyond which phagocytosis is no longer successful	l_{crit}	NA	25	μm	estimated
AM-mediated clearance of particles and fibres	cl_j	0.015	0.015	day^{-1}	estimated from TiO$_2$ data
Transfer rate of particles and fibres from active to inactive macrophages	ρ_j	0.036	0.036	day^{-1}	Stöber *et al.* (1990)
Release rate of particles and fibres back to the alveolar surface for re-phagocytosis	δ_j	0.14	0.14	day^{-1}	Stöber *et al.* (1990)
Kinetics of particles and fibres					
Interstitialisation rate of free particles and fibres	i_j	0.03	0.03	day^{-1}	assumption
Removal rate of particles and fibres to the lymph nodes	e_j	0.005	0.001	day^{-1}	Stöber *et al.* + TiO$_2$ lymph node data
Disintegration rate of unphagocytosed fibres in length category k	b_k	NA	0.041	day^{-1}	dissolution data + adjustment
Disintegration rate of phagocytosed fibres in length category j (relevant when $l < l_{crit}$)	b_{LPj}	NA	0	day^{-1}	dissolution data
No. of fibres of length category j generated by one disintegrating fibre of length category k	n_{jk}	NA	2	day^{-1}	assumption + MMVF10 data
Overload and sequestration					
Critical volumetric lung burden of particles or fibres	V_{crit}	100	78	nl	from MMVF10 data
Motility decay coefficient	β	0.01	0.01	m^{-1}, day^{-1}	estimated from TiO$_2$, MMVF10 data
Alveolar sequestration rate	ϕ_j	0.14	0.14	day^{-1}	Stöber *et al.* (1990)

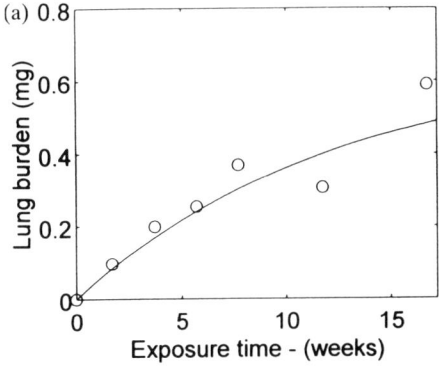
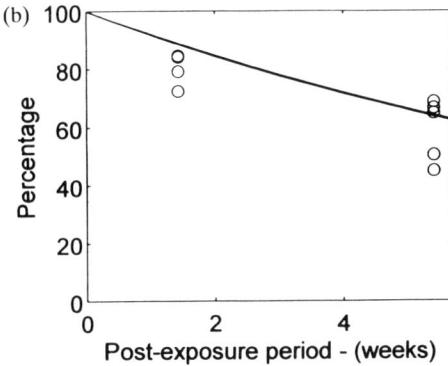

Fig. 3. Lung burden for rats exposed to TiO$_2$ at 1 mg m^{-3}: (a) accumulating during exposure up to 117 days and (b) with clearance post-exposure for up to 38 days. (All data points correspond to the same curve.)

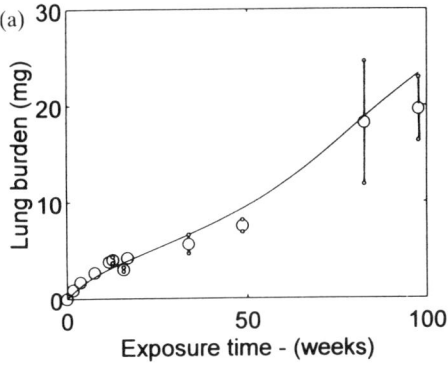
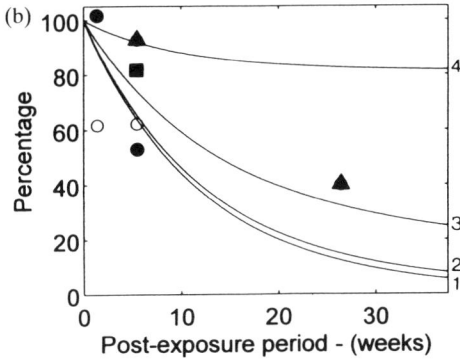

Fig. 4. Lung burden for rats exposed to TiO$_2$ at 10 mg m^{-3}: (a) accumulating during exposure for up to 684 days and (b) during clearance for up to 185 days post-exposure (● corresponds to curve 1; ○ curve 2; ▲ curve 3; ■ curve 4).

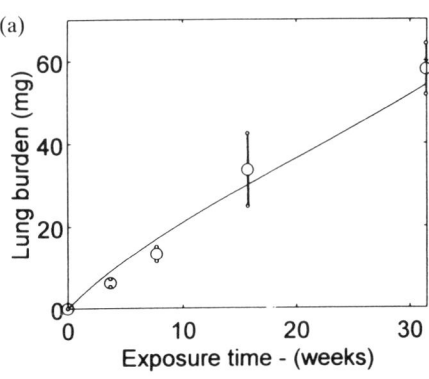
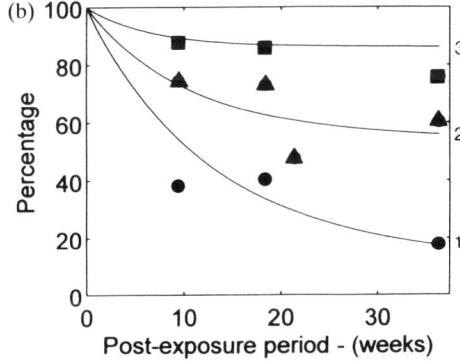

Fig. 5. Lung burden for rats exposed to TiO$_2$ at 50 mg m^{-3}: (a) accumulating during exposure and (b) during clearance post-exposure (● corresponds to curve 1; ▲ curve 2; ■ curve 3). The vertical bars in Figs 4 and 5 indicate ± 2 SE.

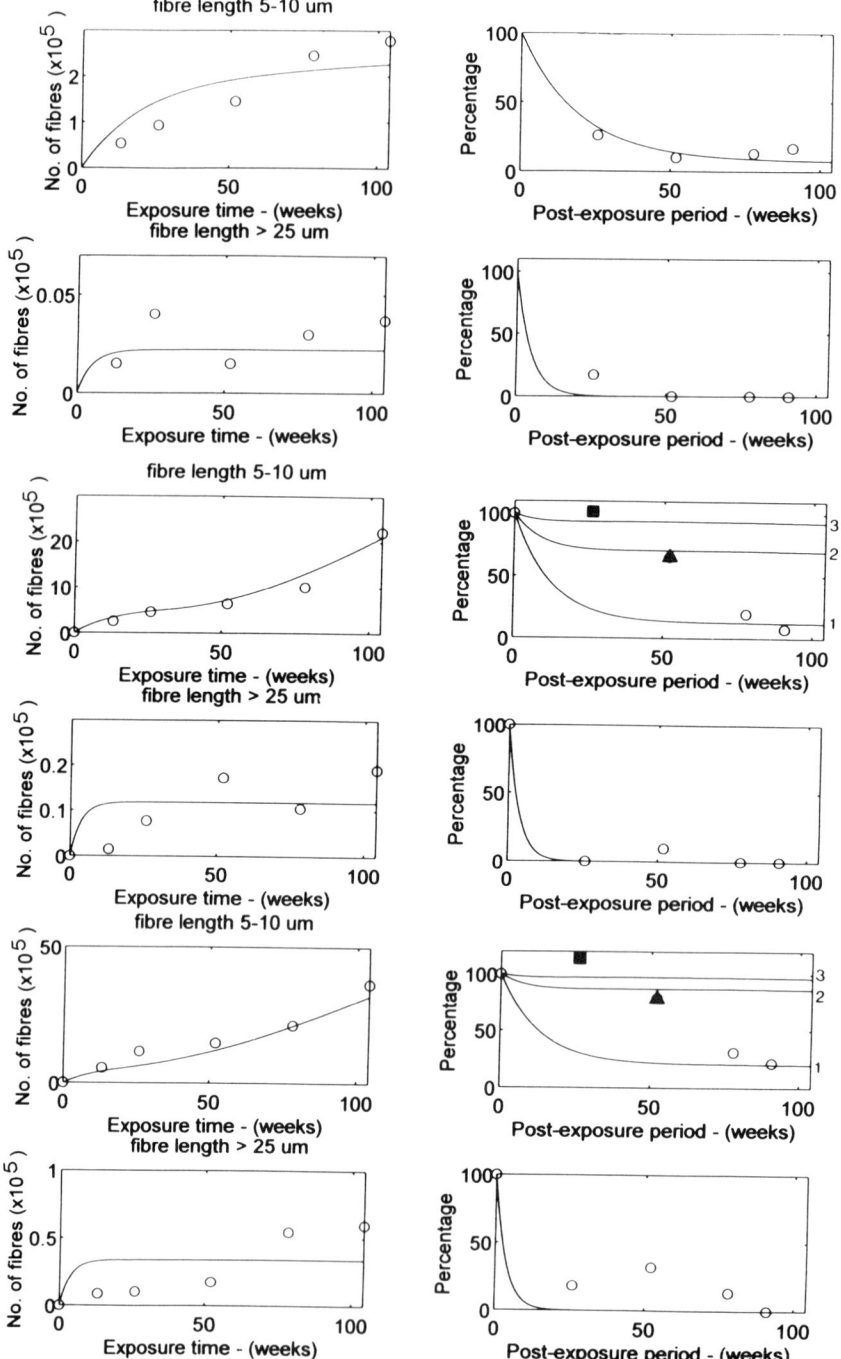

Fig. 6. Lung burden of fibres during exposure (left hand graphs) and post exposure (right hand graphs) for fibres 5–10 μm long and for fibres longer than 25 μm. Top four graphs are for the low dose experiment, middle four are for the mid-dose (16 mg.m^{-3}) experiment, and the bottom four are for the high-dose (30 mg.m^{-3}) experiment (O corresponds to curve 1; ▲ curve 2; ■ curve 3).

(b) MMVF10

For MMVF10 fibres, at the lowest concentration (3 mg.m^{-3}) short fibres ($l < 15$ μm) are cleared effectively with the same effective clearance rate as TiO_2; long fibres ($l > 25$ μm) appear to disappear substantially faster. At 16 mg.m^{-3}, for short fibres, clearance starts to become retarded after 30 weeks exposure when the total fibre volume in the lung is approximately 80 nl. However, the disappearance of long fibres occurs at the same rate as in the low dose experiment. At 30 mg.m^{-3}, there is the same pattern of retardation of clearance for short fibres and disappearance of long fibres.

CONCLUSION

The overloading of the lung by fibres or particles follows the same kinetics and, therefore, the hypotheses underlying overload, as quantified by the model, are the same for particles and short fibres ($l < 15$ μm). The influence of overload on the removal of fibres diminishes with increasing fibre length. Indeed, long fibres ($l >$ approximately 25 μm) disappear at a rate which is independent of lung burden. This implies that long fibre disappearance is probably due to *in vivo* dissolution leading to disintegration of long fibres into shorter fibres or fragments which can be cleared by macrophages.

Acknowledgements—The studies at the IOM were supported by JEMRB for the fibres and by the CEC and British Coal for the particles.

REFERENCES

Hesterberg, T. W., Miller, W. C., McConnell, E. E., Chevalier, J., Hadley, J. G., Bernstein, D. M., Thevanaz, P. and Anderson, R. (1993) Chronic inhalation toxicity of size-separated glass fibers in Fischer −344 rats. *Fundam. appl. Toxicol.* **20**, 464–476.

Katsnelson, B. A., Konysheva, L. K., Privalova, L. I. and Morosova, K. I. (1992) Development of a multicompartmental model of the kinetics of quartz dust in the pulmonary region of the lung during chronic inhalation exposure in rats. *Brit. J. ind. Med.* **49**, 172–181.

Morrow, P. E. (1992). Dust overloading of the lungs: update and appraisal. *Toxicol. appl. Pharmacol.* **113**, 1–12.

Stöber, W., Morrow, P. E. and Morawietz, G. (1990) Alveolar retention and clearance of insoluble particles in rats simulated by a new physiology-oriented compartmental kinetics model. *Fundam. appl. Toxicol.* **15**, 329–349.

Tran, C. L., Jones, A. D. and Donaldson, K. (1994) Development of a dosimetric model for assessing the health risk associated with inhaling coalmine dusts. Edinburgh Institute of Occupational Medicine (IOM Report TM/94/01), pp. 1–59.

Pergamon

Ann. occup. Hyg., Vol. 41, Supplement 1, pp. 244–250, 1997
© 1997 British Occupational Hygiene Society
Published by Elsevier Science Ltd. All rights reserved
Printed in Great Britain
0003–4878/97 $17.00 + 0.00
Inhaled Particles VIII

PII: S0003–4878(96)00069–5

THE COLT FIBRE RESEARCH PROGRAMME: ASPECTS OF TOXICOLOGICAL RISK ASSESSMENT

A. D. Jones, B. G. Miller, R. T. Cullen, A. Searl, J. M. G. Davis, D. Buchanan, K. Donaldson,* C. A. Soutar and R. E. Bolton

Institute of Occupational Medicine, 10 Roxburgh Place, Edinburgh EH8 9SU, U.K.

INTRODUCTION

The Colt Fibre Research programme was established to examine the toxicity of man-made mineral fibres, in order to understand what characteristics of fibre dimension, composition and durability affect toxicity and to what extent tests of inflammation, cell proliferation and intra-peritoneal mesothelioma production predict the results of chronic inhalation studies of carcinogenicity. A number of man-made fibres, including special purpose glass, glass wool, stone (or rock) wool, slagwool, refractory ceramic fibres and silicon carbide whiskers have now been studied (Davis *et al.*, 1996). This paper examines: (i) the extent to which the responses can be related to fibre characteristics, using statistical regression analysis; and (ii) the form of the estimated relationships and their implications.

METHODS

Rats injected intraperitoneally with high doses of 11 different fibres were followed to death and examined for mesothelioma. Rats inhaling three of these fibres were followed to death and assessed for lung tumours and mesothelioma. Results for inhalation experiments with the other eight fibres were available from studies at RCC (Rossiter and Chase, 1995; Hesterberg *et al.*, 1993). Fibre characteristics measured included concentrations of WHO fibres by PCOM, bivariate size distributions (for aerosol and lung burdens) by scanning electron microscopy (SEM), dissolution *in vitro* and biopersistence after instillation in the rat lung by measuring, by SEM, the lung burden in groups of four rats at 3 days and 12 months (and intermediate time points yet to be measured).

RESULTS

The tests of durability *in vitro* and persistence of long fibres *in vivo* produced a very similar ranking of the fibre types, as shown in Table 1. The *in vivo* biopersistence data showed variation between individual animals which arise from several causes, including uncertainties due to delivery of the dose by intratracheal injection rather than inhalation. Figure 1 illustrates, for two examples of the fibre

---

* Present address: Napier University, Edinburgh, U.K.

Table 1. Comparison of tests of dissolution in a closed cell *in vitro* test and retention of long fibres after instillation into rat lungs

Fibre type	*In vitro* durability % of silicon dissolved (at pH 7)	Persistence of long (> 15 μm) fibres % retained 12 months after instillation into the rat lung
SiC whiskers	0.1	42.8
RCF4	0.2	100
RCF1	0.4	74.5
RCF2	0.4	63.8
RCF3	0.4	—*
Long fibre amosite	1.7	22.8
MMVF21	2.8	21.1
Code 100/475	5.9	13.3
MMVF10	15.3	8.8
MMVF11	29.4	—*
MMVF22	52.7	5.7

* These fibre types (RCF3 and MMVF11) are included in the instillation test.

types, the relative retention of fibres in specified length classes ($l < 5$, 5–10, 10–15, 15–20 and $l > 20$ μm) with ± 2 SE. These tests indicated that for some fibre types (for example long fibre amosite) there was substantial clearance of the shorter fibres. The rates of change of numbers of longer fibres depended on fibre type, with the more durable fibres (such as the amosite or the RCFs) showing no apparent reduction, whereas the less durable fibres (MMVF10) showed major reductions in the numbers of long fibres.

The intraperitoneal injection tests produced mesotheliomas in a high proportion of the animals (between 33% for code 100/475 and approximately 90% for three

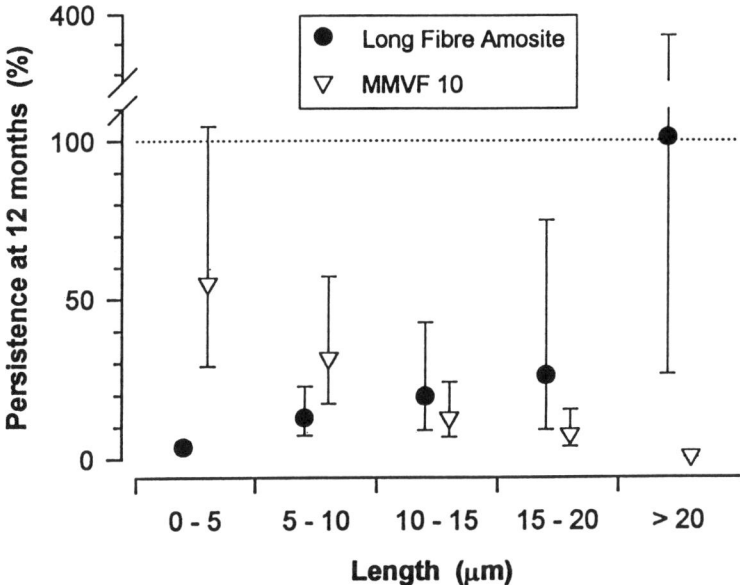

Fig. 1. The relative retention of fibres in specified length classes ($l < 5$, 10–15, 15–20 and $l > 20$ μm) with ± 2 SE for a durable fibre (long fibre amosite) and a fibre with a high dissolution index (MMVF10).

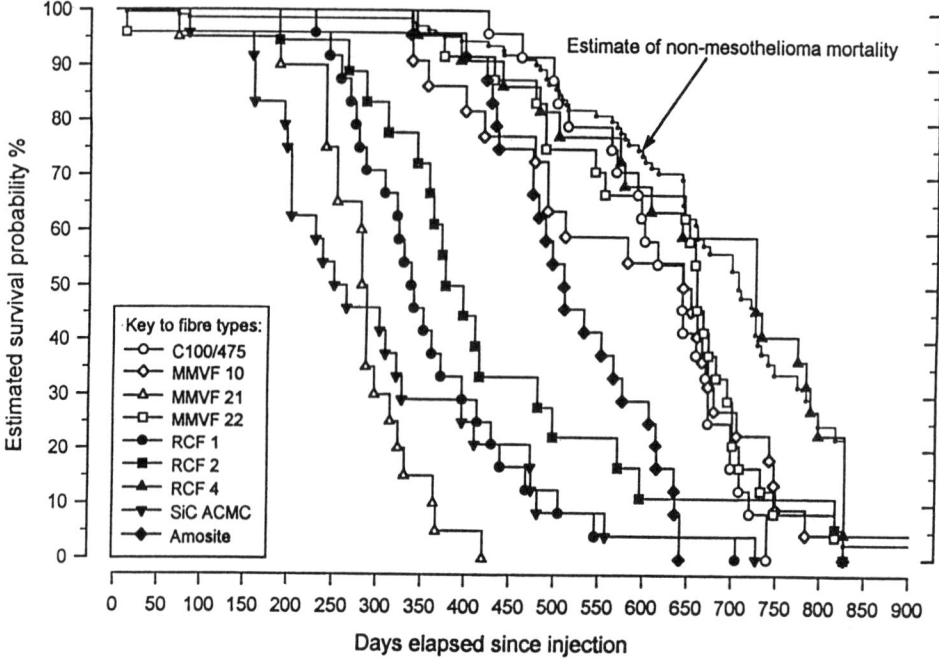

Fig. 2. Kaplan Meier Survival functions for deaths from all causes for groups of rats exposed to each fibre type.

fibre types); only RCF4 (material derived from RCF1 by heat treatment simulating typical use) produced no mesotheliomas. Figure 2 shows survival curves for the groups of rats exposed to each fibre type. The curves cross in some instances, but overall the median survival time is a good indicator of the severity of life shortening due to mesothelioma production; SiC whiskers reduced the median survival from approximately 700 to 300 days.

Multiple linear regression analysis of these data examined the dependence of the median survival time on several measures of injected dose (number of fibres with length > 5 μm, > 8 μm, > 15 μm, etc) and the two indices of fibre durability. The injected number of fibres longer than 15 μm was the first significant explanatory parameter. The indices of durability were statistically significant, with the log [% silicon dissolved *in vitro*] slightly ahead of clearance of long fibres in the instillation experiment. After either durability index was included in the regression, the other became redundant and after the number of fibres > 15 μm was included, the other length categories became redundant. Figure 3, with survival times plotted (on the vertical axis) in relation to the two parameters identified, shows that the longer median survival times correspond to high dissolution rates. The regression predictions are illustrated in Fig. 4 for three typical examples of fibre dissolution, ranging from the less soluble fibres to the most soluble (as measured by % silicon dissolved in this assay). The prediction curves illustrate, as would be expected, the median survival decreasing as fibre dose is increased. They also show the major effect of fibre durability on the response, with much shorter median survival for the durable fibres than for the low durability fibres.

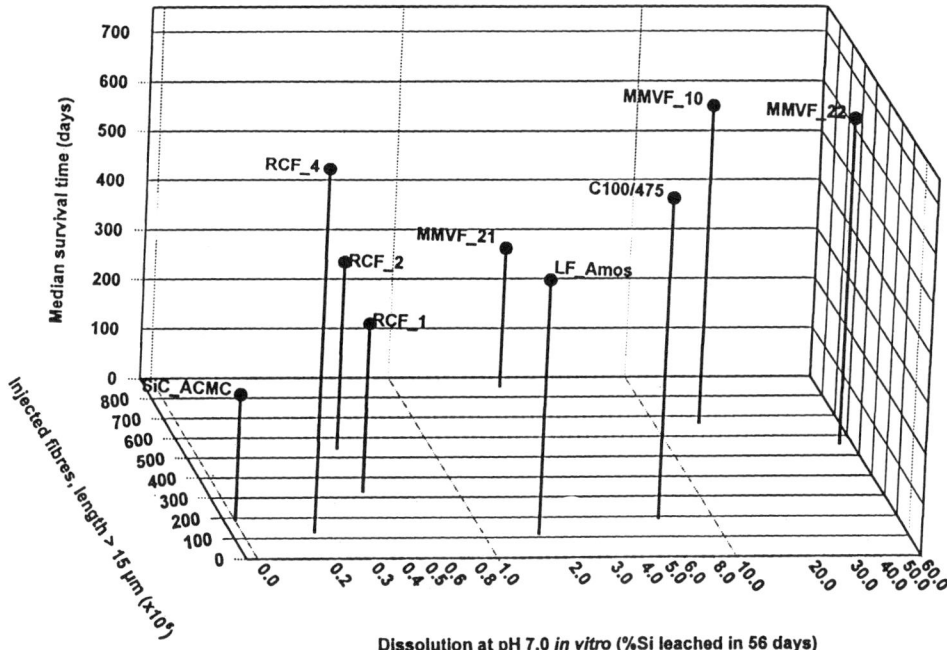

Fig. 3. The median survival in the intraperitoneal injection assay plotted against the number of long fibres injected, and the percentage dissolution in the *in vitro* assay.

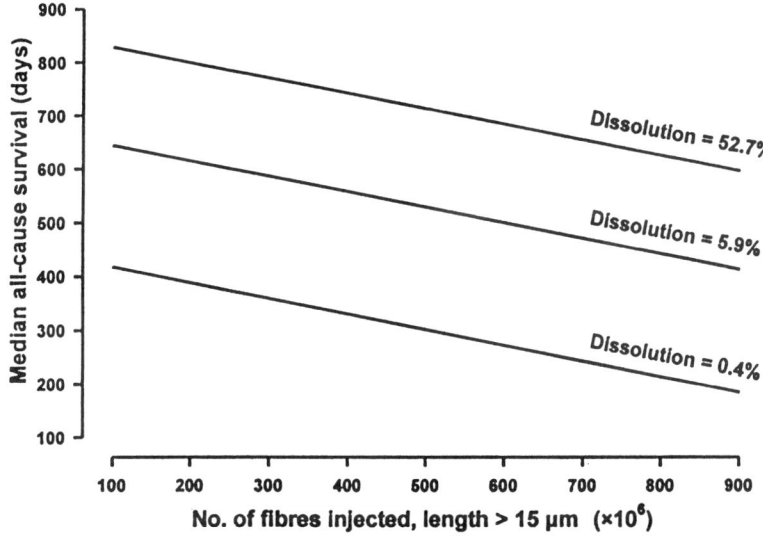

Fig. 4. The regression predictions illustrated for three examples of the dissolution index, ranging from the less soluble fibres to the most soluble (as measured by % silicon dissolved in the *in vitro* assay).

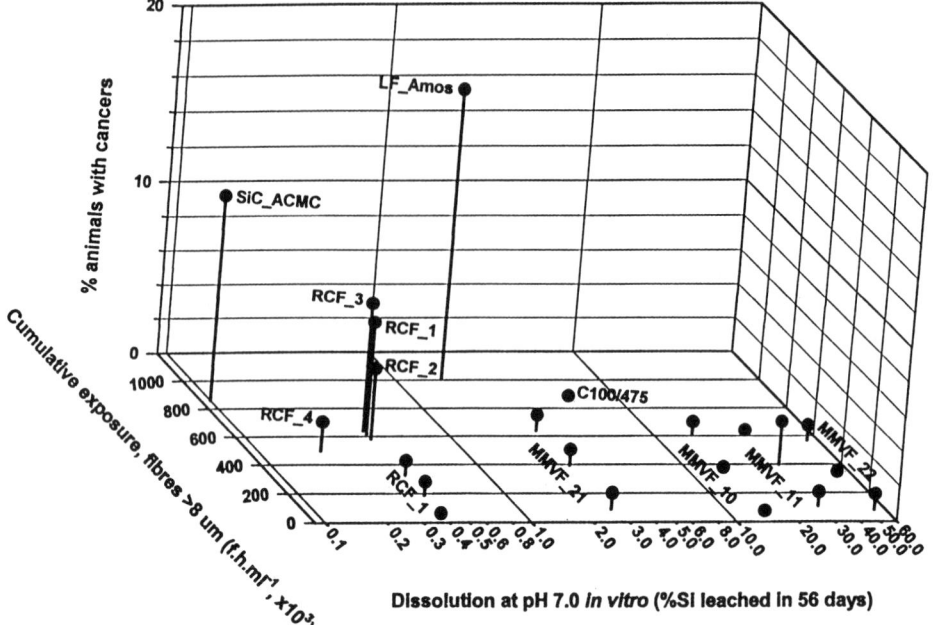

Fig. 5. The percentage incidence of cancer (vertical axis) plotted against the two significant explanatory parameters: cumulative fibre exposure (fibre hours/litre) and the index of dissolution.

Multiple logistic regression analysis of data from the inhalation studies examined the dependence of the percentage incidence of malignant lung cancers on various measures of the cumulative exposure to fibres (fibres longer than 5, 8, 10, 15 and 20 µm) and on the indices of durability. The most significant parameter was the cumulative fibre exposure to fibres longer than 8 µm, but the apparent distinction between 8 µm and the other measures of long fibres may merely reflect the fact that there was more data for the shorter lengths. The second most statistically significant parameter was the index of fibre durability from the *in vitro* test, closely followed by the persistence of fibres > 15 µm (for the fibres instilled into the lung).

Figure 5 shows the percentage incidence of malignant lung cancers (vertical axis) plotted against the two significant explanatory parameters: cumulative fibre exposure (fibre h l.$^{-1}$) on one horizontal axis and the index of dissolution on the other axis. This shows that low durability fibres produced a very low cancer incidence (virtually indistinguishable from background levels). The logistic regression produced prediction curves for cancer incidence as functions of the cumulative exposure to fibres (longer than 8 µm); three examples of these curves are shown in Fig. 6 for the same three values of the fibre durability index as used in Fig. 4. The upper end of the cumulative exposure axis corresponds to the concentrations attained in the experiments with the low durability fibres; for the high durability fibres, the experimental results included data for exposures at higher cumulative exposures. These curves indicate the range of cumulative exposures which are desirable for reliable discrimination in inhalation experiments between cancer incidence for low and high durability fibres.

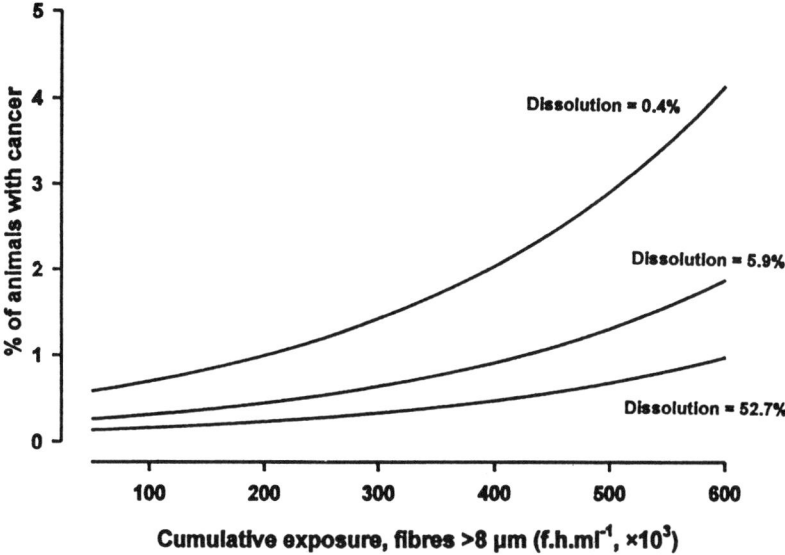

Fig. 6. Prediction curves, from the logistic regression, for cancer incidence as functions of the cumulative exposure to fibres (longer than 8 μm) for the same three values of the fibre durability index as used for Fig. 4.

CONCLUSIONS

The number of lung fibres is confirmed (by inhalation and injection tests) as the most significant factor affecting cancer incidence and latency for a wide range of fibres. The second significant factor, for both the inhalation and intraperitoneal injection tests, was an index of fibre durability, estimated from either dissolution *in vitro* or biopersistence after intratracheal instillation into the lung. The results of the intraperitoneal test at 10^9 WHO fibres, which produced a strong response at this sensitive site, showed that the median survival time can distinguish clearly between the carcinogenicity of durable and less durable fibres, as demonstrated by the dose–response relationships from linear regression. The finding that fibre durability affected production of mesotheliomas with fibres injected into the peritoneum, supports and emphasises the importance of durability.

The analysis of the inhalation data has shown a clear dependence on two parameters, cumulative exposure and a durability index. The predictions may be useful in determining, for future experiments on fibres, the cumulative exposures needed if such experiments are to distinguish the carcinogenicity of durable fibres from that of less-durable fibres. Extrapolation to longer time scales relevant to human exposure may require complex modelling of the biologically effective dose at the target organ [Tran *et al.*, this conference (1996)], for which the present results provide a sound foundation as the above two significant parameters largely determine the biologically effective dose.

Acknowledgements—This independent scientific programme has been supported by the Colt Foundation, by industrial sponsors and by the U.K. Health and Safety Executive.

REFERENCES

Davis, J. M. G., Donaldson, K., Brown, D. M., Cullen, R. T., Jones, A. D., Miller, B. G., McIntosh, C., Searl, A. and Whittington, M. (1996) A comparison of methods of determining and predicting the pathogenicity of mineral fibres. *Inhal. Toxicol.* **8**, 747–770.

Hesterberg, T. W., Miller, W. C., McConnell, E. E., Chevalier, J., Hadley, J. G., Bernstein, D. M., Thevanaz, P. and Anderson, R. (1993) Chronic inhalation toxicity of size-separated glass fibres in Fischer-344 rats. *Fundam. appl. Toxicol.* **20**, 464–476.

Rossiter, C. E. and Chase, J. R. (1995) Statistical analysis of results of carcinogenicity studies of synthetic vitreous fibres at Research and Consulting Company, Geneva. *Ann. occup. Hyg.* **39**(5), 759–769.

Pergamon

Ann. occup. Hyg., Vol. 41, Supplement 1, pp. 251–255, 1997
© 1997 Published by Elsevier Science Ltd on behalf of BOHS
Printed in Great Britain. All rights reserved
0003–4878/97 $17.00 + 0.00
Inhaled Particles VIII

PII: S0003–4878(96)00118–4

DETERMINANTS OF MAN-MADE MINERAL FIBRE HAZARD

R. C. Brown,* M. Moore,† L. D. Maxim‡ and C. G. Burley§

*4 Bramble Close, Uppingham, Rutland LE15 9PH; †Morgan Materials Technology, Stourport,
Worcs, and ‡Carborundum Company, Rainford, St Helens, Merseyside, U.K.;
§Everset Consulting, Cranbury, New Jersey, U.S.A.

INTRODUCTION

The term "hazard" is used by some as meaning the inherent properties of a chemical contributing to any toxic properties and is distinguished from "risk". While this might have some validity with simple compounds any distinction between risk and hazard with complex materials such as mineral fibres is difficult to justify. Toxicologists have long been quoted Paracellsus in saying that "all substances are poisons" and, therefore, concentrating on "dose" and thus presumably "risk". Whatever the requirements of regulators to toxicologists "hazard and risk are quite similar in meaning and include considerations of both intrinsic toxicity and the circumstances specific to exposure" (Klaasen and Eaton, 1991). In this paper we are, therefore, concerned with the routes available to reduce any properties of high temperature insulation glass wools (including refractory ceramic fibres) that could contribute to any real or perceived possibility of ill health arising from them during normal handling and use.

Most man-made vitreous fibres (MMVF) are manufactured and sold as coherent "wools" or in product forms containing fibres many centimetres long. In order to pose any danger of adverse health effect these products must be subjected to "work" to break fibres and then release and aerosolise fibrous dust. This must be inhaled and enter the deeper regions of the lung. The composition of the fibre and its physical properties determine how easily the lung fibres will break. The mass of the fragments will determine the energy needed to suspend them in air and their falling speed will determine how long they remain airborne and the probability of their deposition in the alveolae. Their composition, surface area and perhaps the presence of surface defects then determines how long fibres will persist in lung tissue.

Only two routes for hazard reduction seem practical: to change chemical composition so that fibres are less persistent or to alter the ability of the wools to release fibrous respirable dust. This paper will concentrate on the second route. Some efforts to reduce such dustiness are simply aimed at risk reduction by adding binders or dust suppressants, however the inherent properties of the fibre itself may be altered by, for example, changing diameter distribution. At one extreme products containing no respirable fibres, such as some continuous filament products, may be regarded as "hazard or risk free". Studies using experimental "dustiness" tests, measurements of fibres in the workplace and theoretical modell-

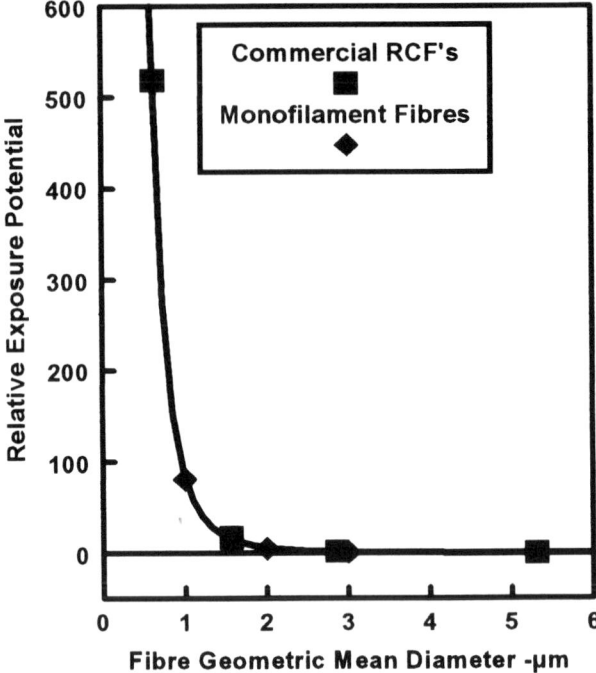

Fig. 1. The theoretical relationship between fibre diameter and exposure potential.

ing have been used to predict the effect of some product properties, particularly diameter, on fibre exposure.

METHODS AND RESULTS

Theoretical modelling

The diameter distribution of a number of refractory ceramic fibres was determined by electron microscopy using the length weighted approach recommended by the U.K. HSE (Burdett and Revell, 1994). The relative settling rate (S) for each distribution was calculated for each distribution normalised to a 3 μm diameter fibre having a settling rate of 1. Assuming a uniform fibre length of 30 μm the weight of a similar fibre number for each diameter distribution (W) was then calculated and the exposure potential for unit weight of each material (E) calculated by taking $E = S/W$. The exposure potential of hypothetical monodisperse fibres was also calculated. The relationship between length weighted mean geometric mean diameter and exposure potential is illustrated in Fig. 1.

Experimental "dustiness testing"

The release of fibres from a series of RCFs was studied using a testing chamber based on that described by Dogdson *et al.* (1987). Essentially a non-electrostatic PVC box was fitted with a stainless steel mesh box at the top of the chamber which could be shaken by a speed controlled reciprocating shaft. Fibre samples were placed in this mesh box so that dust would be generated by repeated collisions

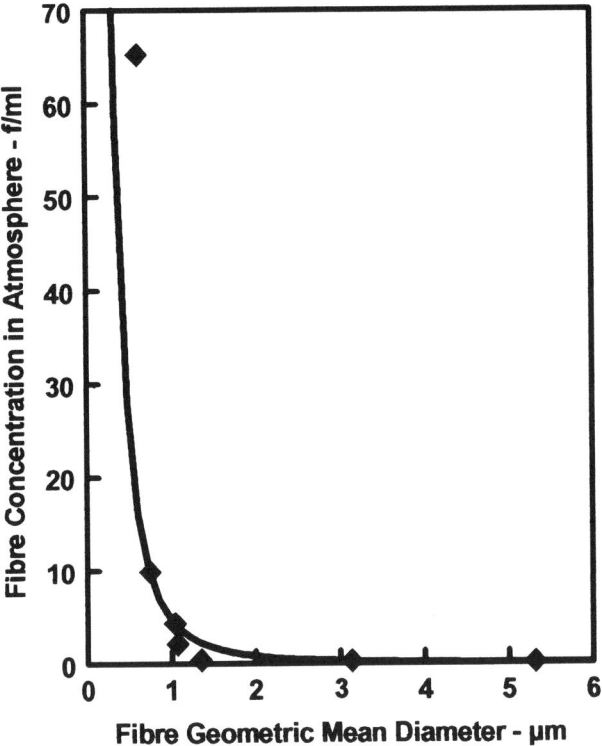

Fig. 2. The relationship between fibre diameter and airborne fibre level in an experimental dustiness test. In practice fibre products are not treated to this degree of mechanical abrasion.

between product and the mesh. Sampling heads were arranged below this and a mixing fan maintained uniformity of the dust cloud. Air to make up that lost through the sampling cassettes was admitted freely through an HEPA filter. Preliminary studies were needed to optimise filter loading and counting parameters and to ensure that the dust emission was stable over at least some minutes. The results of this study are shown in Fig. 2. Full details will be published in Class *et al.* (in press).

WORKPLACE FIBRE LEVELS

The records of fibre counts made by the RCF industry were examined for counts that also listed the product being used at the time of sampling. The nominal diameter of these products was then obtained from the manufacturers. Clearly this nominal diameter is not as meaningful or accurate as the length weighted diameters used in the theoretical and experimental treatments outlined above. In some cases it is also likely that the recorded product being used was not the only source of fibre. Despite these uncertainties statistically significant relationships existed between nominal diameter and average fibre concentration in both manufacturing and customer facilities (Fig. 3). The full results of this study may be seen in ECFIA (1994).

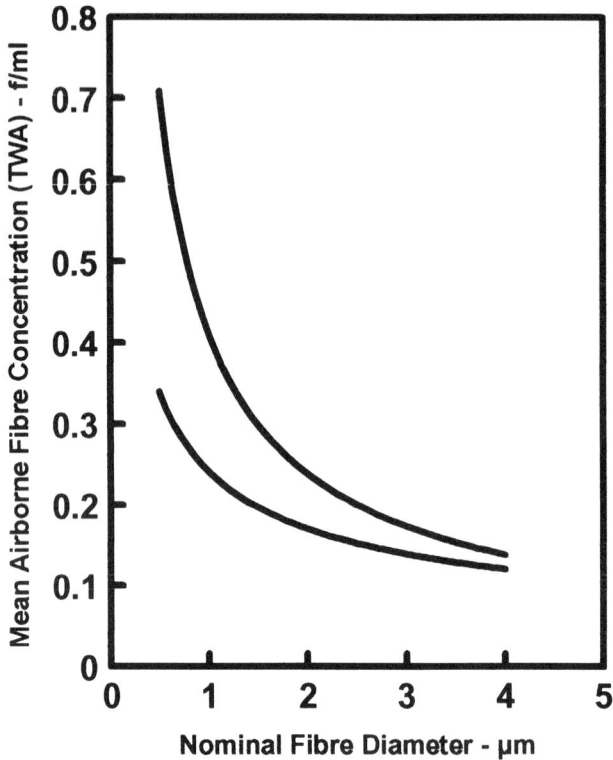

Fig. 3. Relationship between airborne fibre concentration and product nominal diameter in manufacturing plants (lower curve) and in customer premises (upper curve).

DISCUSSION

From Stokes' law, the falling speed or settling velocity of a fibre is proportional to the square of the diameter, suggesting that as the diameter of a fibre increases the airborne fibre concentration resulting from its use should decline with its diameter raised to the power of −2 (an elasticity of −2). Since there are a variety of diameter fibres in any commercial RCF integration for the number/diameter size distributions of actual RCF samples gave a theoretical elasticity of approximately −3.8. Several investigators have developed experimental means for determining the "dustiness" of fibre wools. For example Dodgson *et al.* (1987) determined that for some mineral wools fibre concentration was proportional to nominal product diameter with an elasticity of −1.52. Testing RCFs in a similar experimental apparatus resulted in a relationship in which fibre level the airborne fibre concentration was related to the length weighted geometric mean diameter raised to the power −2.5.

In a study of glass wool, rock wool, and slag wool plants by Esmen *et al.* (1979) the value for the elasticity was −1.54. The similar study of RCF production and processing facilities reported here showed a best-fit elasticity of approximately −0.5. There was a stronger dependency of fibre concentration on fibre diameter in customers' premises than in those of the manufacturer's. This, and the less pronounced relationship was that seen in the laboratory, is probably due to the better control of the highest exposures in RCF plants. Processes producing the most dust and using the finest fibres are naturally the subject of the most intense control efforts and this flattens the apparent concentration–diameter curve. The confounding effect of workplace controls means that laboratory data obtained under standardised conditions provide the most reliable and repeatable estimate of the intrinsic relationship between concentration and diameter.

In theory, in laboratory simulations and in the workplace there is clearly a relationship between diameter and airborne fibre level. Thus the use of a product with a length weighted diameter of 1 micron will result in lower fibre concentrations than one with a 0.5 micron diameter by a factor in the range of 1.4 to nearly 14 depending on workplace controls. The underlying relationship between diameter and dustiness is certainly as inherent a property as any other likely to be considered in hazard identification. Certainly finer fibres are to be avoided if workplace levels are to be controlled and such controls will be easier to implement and have greater effect as fibre diameter increases.

Thermal insulation at high temperature requires fine fibres however the propensity of a wool to break and release fibrous dust increases with reducing diameter. The use of ceramic fibres with length weighted geometric mean diameters of more than 1 μm reduces hazard and enables dust levels to be controlled while maintaining the utility of these materials. Changes in manufacture and composition could further reduce fibre fragmentation by changing diameter distribution so as to reduce the content of very fine fibres. Effects of composition and manufacturing method on fibre fragility and on other method of reducing dust emission should be undertaken.

REFERENCES

Burdett, G. and Revell, G. (1994) Development of a standard method to measure the length-weighted geometric mean fibre diameter. Results of the second inter laboratory exchange Health and Safety Executive Research and Laboratory Services Division. Report number IR/MF/94/07.

Class, P., Brown, R. C., Alexander, I., Jubb, G. and Deghilage, P. Evaluation of the relation between the nominal fibre diameter of bulk refractory ceramic fibres and airborne fibre concentration (dustiness) using a laboratory shaking test box. *Gefahrstoffe, Rheinhaltung der Luft* (in press).

Dogdson, J., Cherrie, J. and Groat, S. (1987) Estimates of past exposure to respirable man-made mineral fibres in the European insulation wool industry. *Ann. occup. Hyg.* **31**, 567–582.

ECFIA (1994) Refractory ceramic fibres: the relationship between product nominal diameter and airborne fibre concentration: a report of an ad hoc working party to the DGXI working group on dangerous substances, November, 1994. Available from ECFIA 3, rue du Colonel Moll, Paris, France.

Klaassen, C. D. and Eaton, D. L. (1991) *Casarett and Doulls Toxicology* (Edited by M. O. Amdur, J. Doull and C. D. Klaassen), p. 37. Pergamon Press, New York.

 Pergamon

Ann. occup. Hyg., Vol. 41, Supplement 1, pp. 256–260, 1997
British Occupational Hygiene Society
Crown Copyright © 1997 Published by Elsevier Science Ltd
Printed in Great Britain
0003–4878/97 $17.00 + 0.00
Inhaled Particles VIII

PII: S0003–4878(96)00119–6

THE USE OF A DUSTINESS TEST FOR MEASURING FIBRE EMISSIONS FROM MAN-MADE VITREOUS FIBRES (MMVFS)

G. J. Burdett, G. Revell and J. Brammer

Health and Safety Laboratory, Broad Lane, Sheffield S3 7HQ, U.K.

INTRODUCTION

The EC through a DG XI committee has been debating the hazard classification of man-made vitreous fibres (MMVFs) since the late 1980s. Most of the debate has focused on the differences between the carcinogenic hazard of the main groups of MMVF produced, e.g. continuous fibres, glass wool, slag wool rock and refractory ceramic fibres (RCFs), and whether the toxicological evidence is sufficient to classify these various subtances as a category 0, 3 or 2 carcinogen under the Dangerous Substances Directive EEC 548/67. The toxicological data has been vigorously debated at workshops and in publications, EHP (1994). It is, however, generally agreed that the carcinogenic hazard is related to the number of potentially respirable fibres and their durability in the lung. The risk posed to humans when handling the substance is the probability that hazardous fibres will be inhaled. This probability is a function of the relative ease which respirable fibres (fibres with widths < 3 μm) can become airborne or the "dustiness" of the material. The objective of this work was to use a bench scale test to investigate the relationship between the bulk fibre dimensions of MMVFs and their relative fibre release. Particular attention was given to RCF materials.

METHOD AND MATERIALS INVESTIGATED

The method for assessing the dimensions of the bulk fibre has been developed in co-operation with the European Ceramic Fibres industry Association (ECFIA) and involves first heating the sample to remove organic coatings and crushing the fibres at 10 KPa to reduce the fibre lengths. A representative sub-sample is then suspended in filtered distilled water and a known aliquot filtered onto a 0.2 μm pore size polycarbonate filter. The filter is then splutter coated with gold to give a conductive coating suitable for scanning electron microscope (SEM) evaluation. Some 300 fibre diameters are measured at 10 000 magnification in the SEM which is calibrated against a traceable standard. An unbiased estimate of the average fibre diameter is achieved by only measuring the diameter of fibres (particles with an aspect ratio of $> 3:1$) which cross a horizontal line drawn on the SEM monitor. This length weights the distribution and enables the length weighted geometric mean diameter (LWGMD) to be calculated. All materials used for the bench scale tests were analysed for their LWGMD by combining five core samples across the width of the blanket from which the test piece was taken.

The method for bench scale testing the bulk materials is based on a HSE/WSL Mk1 standard rotating drum dustiness tester, which has been used for about 10 years for comparing the dustiness of powders. The tester consists of a 0.3 m diameter stainless steel drum with six internal 2.5 cm vanes, which is rotated at 30 rpm to repeatedly lift and drop the material, to impart a tumbling motion on the test sample and to abrade any surfaces that come into contact. The vanes also mix the air within the drum to give an even fibre distribution. Two conical end pieces fitted with rubber gaskets are used to seal the drum, with a respirator filter cartridge on the inlet and a 22 mm i.d. tube on the outlet. A simple airtight seal was placed on the outlet, to allow rapid interchange between cowled sampling cassettes, containing a 0.8 μm pore size membrane filter, which were connected via a rotatable pipe coupling to a pump.

The test uses a 60×60 mm square sample of blanket material cut from within the roll which has the same nominal thickness as the blanket. Five consecutive 2 min samples were collected at a flow rate of $2 \, \text{l min}^{-1}$ by stopping the drum and pump and replacing the cowled cassette in a simple push fitting (over about 10–15 s) and restarting the drum and pump for a further 2 min. The low sampling rate and times were used as it allowed optimum loading of fibres to be collected for analysis by phase contrast light microscopy (PCM) using the current reference method (MDHS 59, 1988). The fibres counted were defined as (particles > 5 μm long, < 3 μm width and with an aspect ratio > 3:1) and have the same fibre size and visibility criteria which are used as the index of respirable fibres exposure in the workplace. Two samples, which gave very low fibre counts were also collected using a flow rate of $27 \, \text{l min}^{-1}$, (the normal flow rate used for gravimetric analysis of dusts) to obtain countable samples above background.

The products investigated are listed in Table 1. All the RCF materials were 25 mm thick blankets except for Saffil which was supplied both as a blanket and a bulk material. The commercial glasswool and rockwool blankets were 100 mm thick, except for one rockwool blanket (25 mm). The Schuller fibres were supplied as small pieces with sides 100–500 mm long and approximately 20–25 mm thick. The test pieces were cut from these sections and classed as blanket rather than a bulk material.

RESULTS AND DISCUSSION

Table 1 summarises the results from the LWGMD analysis carried out on the bulk materials used in the tests and the average result from the bench scale testing. The repeatability of the dustiness test has been investigated by carrying out at least two runs for most RCF materials.

There is usually reasonable consistency between repeat runs. One sample in the repeat analysis (Kerlane 45) has given very different values in the repeat fibre count analysis and this is currently under investigation.

The relationship between the bulk fibre LWGMD and the fibre concentration released by bench scale testing is given in Fig. 1. It can be seen that for RCF blankets alone the relationship is nearly horizontal, except that a few products appear to give much higher fibre levels. When the glass and rockwool materials are considered, a more obvious (inverse) relationship between bulk fibre diameter and

Table 1. Summary of bulk fibre diameter and dustiness test average airborne fibre concentrations

Material	Sampling rate (l min.$^{-1}$)	Average airborne fibre concentration (f ml^{-1})	LWGMD of bulk material (μm)	Geometric standard deviation
RCF Blanket materials				
Saffil	2	7.45	2.47	1.83
Cerachrome	2	54.9	2.22	1.89
Cerachem	2	18.62	1.46	2.53
Cerawool	2	10.62	3	2.48
Cerablanket	2	12.12	1.57	2.6
Kerlane 50	2	8.04	1.27	3.01
Kerlane 45	2	57.72	1.65	2.15
Kaowool S	2	8.09	1.98	2.8
Zirlane	2	24.56	1.56	2.37
Superwool	2	15.3	2.45	2.42
Insulfrax	2	5.33	2.93	2.04
Durablanket 1460a	2	12.16	4.13	2.12
Durablanket 1260	2	12.12	2.78	2.29
Durablanket 1460b	2	1.37	2.13	2.74
Other MMVF Blankets				
Glasswool	27	0.61	4.21	2.47
Rockwool	27	0.23	3.64	2.65
Rockwool Gmbh	2	0.92	2.63	2.31
Schuller 108A	2	10.88	1.05	1.84
Schuller 108B	2	49.82	1.15	1.91
Schuller 90	2	21.67	0.74	0.82
Bulk materials				
Saffil bulk	2	70+	2.47	1.83
Heated MMVFs				
Glasswool (heated)	2	91.6	4.21	2.47
Rockwool (heated)	2	186.2	3.64	2.65

Note: RCF has been used as a term for all high temperature performance fibres.

fibre concentration is seen. However, it should be noted that the low concentrations from the large diameter products are due mainly to oil suppression of the blanket fibres. When the oil was removed by heating to 400°C for several hours, glasswool and rockwool gave concentrations of 92 and 186 f ml^{-1} respectively, when tested in the rotating drum. The finest of the Schuller fibres appeared quite different in surface texture to the other two and may have had some suppressant or binder added, resulting in lower fibre emission.

The precision of the analysis is largely dependent on the number of fibres counted, as the flow rates were accurately measured using a calibrated bubble flow meter and timing errors were estimated to be < 5%. With the current stopping rules in MDHS 59 (200 ends or 200 fields) with a 4 l volume of air sampled, this is equivalent to a fibre concentration of 12 f ml^{-1}. This would give an upper and lower 95% confidence interval of approximately 8–20 f ml^{-1} (66–166%) for a repeat count based on the estimated precision from manual fibre counting. This means that any fibre concentrations higher than this are counted to the same precision because of the stopping rules but lower concentrations are less precise. The average count for the RCF blanket materials investigated, was 19.1 f ml^{-1} so most of the

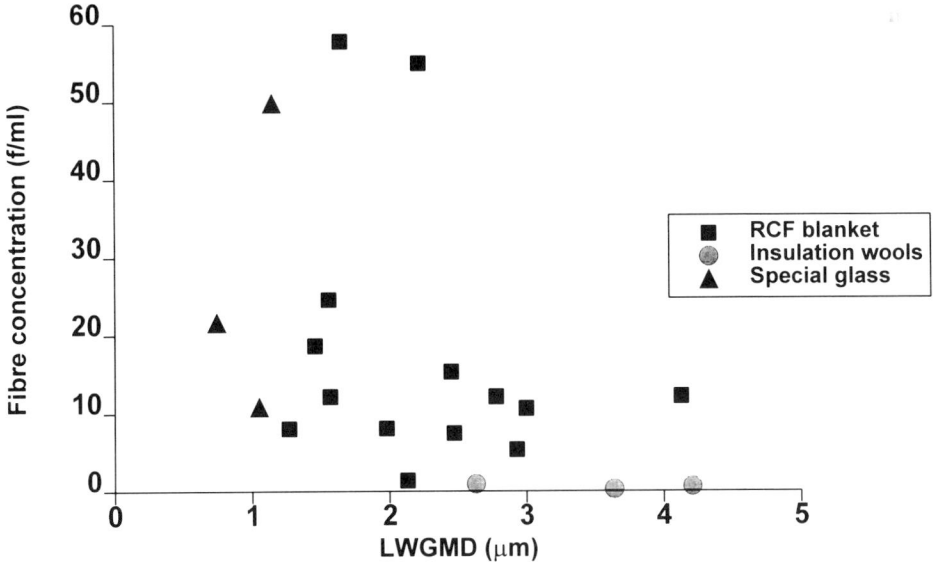

Fig. 1. Scatter plot to show relationship between the LWGMD of the RGF, glass and rockwool blanket and WHO fibre concentration.

samples were counted to this precision based on the 200 end (100 fibre) stopping rule being applied. The rockwool and glass wool however gave much lower counts.

The precision of the analysis was improved by taking additional samples for analysis. In these tests five samples were averaged for each run and a repeat run was performed, which should give the equivalent of 1000 fibres counted, and will reduce the counting error. However, the five consecutive 2 min samples takes no account of the output being time dependent and this will introduce a further variable. It is, therefore, estimated that the 95% confidence interval for a count of 12 f ml^{-1} is of the order of 80–130% (9.6–15.6 f ml^{-1}) of the average count based on 10 samples analysed.

The limited precision of the fibre counting explains why, even when a large number of fibres are counted, it may be difficult to see any underlying relationship between bulk fibre diameter and fibre concentration. The additional results confirm that although LWGMD is a factor which will determine the numbers of respirable fibres released it makes relatively little difference over the limited range of LWGMDs (1.3–4.1 um), which RCF material is currently produced. Some particular products (e.g. Cerachrome and Kerlane 45) appeared to release more fibres than would be expected. Kerlane 45 gave large variation in repeat tests. Durablanket 1460 had the largest LWGMD of the RCF fibres tested but had a relatively high count of 12.2 f ml^{-1}.

The other MMVF products were tested primarily to extend the range of LWGMDs to give an increased change of obtaining a relationship with the respirable fibre counts. While an underlying inverse trend is seen with increasing count for the finer bulk LWGMDs, many other factors may modify this relationship (e.g. the density of the fibre, the fibre strength and brittleness, the efficiency of needling, the area of cut surface, the friability of the material in the drum tester, the environmental conditions, amount of heating, the presence of dust suppressants) which will reduce the correlation between the bulk LWGMD and the fibre emission. This appears to defeat any meaningful attempt to fit a curve with a reasonable correlation coefficient (e.g. attempts to fit linear, exponential and power curves to the RCF data all gave correlation coefficients of less than 0.1).

It is, therefore, concluded that it is necessary to carry out direct measurements of the respirable fibre release from MMVF materials using bench scale tests and/or work simulation exercises to measure and compare the relative risks. This conclusion becomes even more pertinent as it was derived mainly from tests on new RCF blanket materials, but MMVF materials are produced in increasingly wider ranges of product types and forms, all of which will give different levels of release both when new and used.

REFERENCES

Burdett, G. J. and Revell, G. (1994) Development of a standard method to measure the length weighted geometric mean fibre diameter: results of the second inter-laboratory exchange. HSL internal report number IR/L/MF/94/02.

EHP (1994) Biopersistence of respirable synthetic fibres and minerals. *Environ. Health Perspect.* **102** (Supplement 5).

MDHS 59 (1988) Method for determining hazardous substances. Man-made mineral fibre: airborne number concentration by phase contrastlight microscopy. Health and Safety Executive, London.

 Pergamon

Ann. occup. Hyg., Vol. 41, Supplement 1, pp. 261–266, 1997
© 1997 British Occupational Hygiene Society
Published by Elsevier Science Ltd. All rights reserved
Printed in Great Britain
0003–4878/97 $17.00 + 0.00
Inhaled Particles VIII

PII: S0003–4878(96)00120–2

EXPOSURE AND RESPIRATORY HEALTH OF WORKERS IN THE AUSTRALIAN REFRACTORY CERAMIC FIBRE PRODUCTION INDUSTRY

A. Rogers,* P. Yeung,† G. Berry,‡ G. Conaty‡ and L. Apthorpe†

*Alan Rogers OH and S Pty Ltd, PO Box 2128, Clovelly, NSW 2031, Australia; †Occupational Hygiene Unit, Worksafe Australia, PO Box 58, Sydney, NSW 2000, Australia; and ‡Department of Public Health and Community Medicine, University of Sydney, NSW 2006, Australia

INTRODUCTION

Refractory ceramic fibre (RFC) is a subset of synthetic or vitreous mineral fibres (SMF) being composed of aluminium and silicon oxides. They have a high melting point and hence provide insulation application at temperatures up to 1400°C. To achieve the insulation properties, fibre is manufactured with nominal diameters of around 3–4 micrometers which is finer than other major commercial SMF types. In most instances the produce is manufactured without binder which with its finer fibre diameters results during handling in the generation of larger quantities of respirable fibres than does other commercial SMF. In addition the chemical composition of RCF makes it more durable in biological systems than most other commercial SMF. These factors of diameter, dose and biological durability have been recognised from experiences in asbestos and other natural and synthetic mineral fibres as major influences on fibre toxicity and carcinogenicity (Glass *et al.*, 1995).

In 1988 as a result of increasing concern within the community about the possible health effects from exposure to SMF the National Occupational Health and Safety Commission established a Technical Working Group to investigate typical levels of exposure in manufacturing and user industries and possible health effects and, based on available information, recommend exposure standards and safe working practices. In examining data derived from the Australian RCF industry, the Technical Group found upper respiratory tract irritation at concentrations above 1 fml^{-1} and for employees under medical surveillance (cross-sectional studies) no evidence of lung fibrosis. In terms of evidence for cancer, no suitable epidemiological studies were found anywhere for determining lung cancer or mesothelioma risk although data from the Australian Mesothelioma Surveillance Program indicated no cases associated with the manufacture of SMF (Worksafe, 1989). As part of its recommendations the Technical Group prepared a standardised membrane filter method for monitoring airborne SMF fibres and a gravimetric method for inspirable dust. Through the tripartite and public comment processes the Commission declared a national exposure standard of 0.5 fml^{-1} for all types of SMF and a secondary standard of 2 mgm^{-3} of inspirable dust to protect against upper respiratory tract irritation from large diameter non-respirable fibre (Work-safe, 1990).

Since 1990 representatives from the original Technical Working Group have updated the Commission on an annual basis with a review of the considerable published and unpublished research developments on *in vitro* testing, animal experiments and epidemiological and exposure investigations in the SMF industry most of which has now been recently formally reviewed (Glass *et al.*, 1995). This study provides an overview of a more detailed and up to date analysis which is currently being undertaken into the health status of the Australian RCF manufacturing workforce.

MATERIALS AND METHODS

Industry size and structure

Refractory ceramic fibre products are manufactured by two companies in Australia; Thermal Ceramics which commenced fibre production in 1976 at Alexandria, NSW and Saint Gobain (formerly Carborundum) at Thomastown, Victoria in 1977. Combined production is approximately 3000 tonnes of fibre per annum and all RCF material produced in Australia is synthetic, being blown from a melt of approximately 50:50 mix of refined alumina and high purity silica. Both plants have a fibre production line and specialty products areas making modules, vacuum formed shapes, insulation boards and wet and dry mixes. The current combined production workforce is approximately 40 men. Company personnel records which indicated commencement and duration of specific duties are available for almost all employees and occupational health investigations have been carried out on the workforce since the mid-1970s by external consultants.

Health studies

Plant 1 has conducted medical examinations on the production workforce at approximately 2 year intervals since 1985. Such examinations (except for one set of examinations the results of which were unfortunately lost in a fire) were conducted and reviewed by a respiratory physician who is a B reader and also member of the medical panel of the local workers compensation dust diseases board. The examination consisted of a full size chest radiograph, simple spirometry, ECG, full blood count and biochemical tests as well as an investigation of respiratory symptoms, occupational and smoking history.

Plant 2 has conducted medical examinations at first employment then at approximately 3 year intervals since commencement of the fibre production line in 1977. All medical data has become lost to the study due to a number of changes in ownership of the medical consultant's practice. One complete copy of the set of records included pulmonary function tests, chest radiograms, respiratory symptoms and smoking history associated with all the production workforce conducted in 1990 was available to this study. This resulted in a review of the health status of the production workers by the University of Cincinnati (Lockey *et al.*, 1991). For both plants lung function was adjusted for age, height and smoking using Crapo predictive values.

Exposure studies

Personal air monitoring records for Plant 1 had been obtained since 1983 by both government inspectors and two consultancy groups. Much of the records particularly in the later years relate to targeting specific operations so as to test the efficacy of various control systems. At Plant 2 both personal and static airborne fibre monitoring had been conducted by the same group of consultants at approximately 6 month intervals since 1989. A comprehensive external company survey was carried out in July 1996.

RESULTS AND DISCUSSION

Fibre production has been undertaken for nearly 20 years although small quantities of fibre were imported prior to this period. During local production some 152 men have been involved in the manufacture of bulk fibre and specialty products of which 43 are currently employed with some having moved on to management positions. There has been a reasonably stable workforce of 26% with more than 5 years, 15% with more than 8 years and 9% with more than 12 years experience in the industry. The current employees have a mean employment duration of 10.8 and 8.6 years, respectively. Within the last 10 years there have been ongoing information sessions to acquaint the workforce with an update on the developing knowledge of the toxicological and epidemiological studies as they occur. This has included labelling of products, the encouragement in the use of respiratory protection and the development of attitudes to handling practices which result in less dust generation.

Airborne fibre levels in both plants have gradually been reduced over time with geometric mean concentrations falling from 0.52 to 0.11 fml^{-1} at Plant 1 with a lesser reduction at Plant 2 of 0.29–0.27 fml^{-1}. This was due to the combination of a number of factors such as the introduction of an exposure standard (Yeung and Rogers, 1996), the use of various control and handling technologies and an increased awareness of dust suppression by the workforce. Results presented in Table 1 associated with specific processes represent daily time weighted averages which often include duties such short term tasks such as cutting, shaping, fibre and blend preparation for specialty products and cleaning duties which create the higher dust levels typically 1–3 fml^{-1} but in the past have ranged up to 10 fml^{-1}. Recent ventilation and product handling procedures have been targeted with a view to reducing these peak dust levels. Workforce exposures are less than the values listed above due to the use of approved disposable respirators which provide around a 10-fold protection factor. These respirators are used during operations when the fibre levels are likely to exceed the occupational exposure limit. In attributing these values to cumulative workplace exposures, it should be kept in mind that the results are based on a number of short term surveys many of which were designed to investigate specific high dust activities rather than a methodological occupational hygiene investigation of all exposures in the industry. Such a comprehensive survey is just about to commence. Australian airborne levels indicate different levels but a process trend similar to those reported from production plants in Japan, Europe and America (Hori *et al.*, 1993; Trethowan *et al.*, 1995; Rice *et al.*, 1994).

Table 1. Respirable fibre exposure pre- and post-NOHSC national standards

Process	Plant	n	Pre-NOHSC standard 1983–90			n	Post-NOHSC standard 1991–96		
			GM (fml^{-1})	GSD	Range (fml^{-1})		GM (fml^{-1})	GSD	Range (fml^{-1})
Fibre line	1	12	0.68	4.4	0.03–10.0	16	0.08	4.2	0.01–1.1
	2	19	0.30	2.6	0.03–2.3	63	0.22	3.4	0.01–2.0
Mixing/forming	1	9	0.23	3.1	0.05–1.3	16	0.08	4.3	0.01–1.1
	2	7	0.46	2.0	0.20–1.0	19	0.36	3.0	0.05–2.0
Finishing	1	14	0.86	2.5	0.20–6.0	13	0.30	3.5	0.02–2.8
	2	2	0.48	1.3	0.40–0.58	9	0.35	3.9	0.03–1.80
Assembly	1	1	1.5	—	1.5	6	0.11	2.4	0.06–0.60
	2	2	0.16	3.1	0.07–0.35	6	0.47	3.2	0.09–2.8
Auxiliary	1	4	0.19	7.3	0.01–0.70	2	0.19	2.9	0.09–0.40
	2	3	0.10	1.1	0.09–0.10	5	0.23	2.5	0.10–0.70
All activities	1	40	0.52	3.9	0.01–10.0	53	0.11	4.1	0.01–2.8
	2	33	0.29	2.5	0.03–2.3	102	0.27	3.3	0.01–2.8
All years	1	93	0.22	4.8	0.01–10.0				
(Pre + post)	2	135	0.27	3.1	0.01–2.8				

A summary of the findings of the last available set of medical examinations for 43 current production workers is presented in Tables 2 and 3. For Plant 1 all workers X-ray findings were classified as 0/0 with no pleural abnormalities detected. No longitudinal degradation in lung function had been detected in individuals from this plant over the many years of testing except due to factors of lifestyle such as smoking. In Plant 2 the radiographic findings were similar with the exception of one individual with 1/0. Medical history suggested that his radiographic change is not related to working in RCF manufacturing due to the short and recent duration of employment. Overall the reviewers of Plant 2 medical data indicated that there appears to be no pleural or parenchymal changes in the current workforce related to RCF manufacturing (Lockey *et al.*, 1991). No pleural plaques were found in either of the plants. Studies in U.S. production facilities have found pleural changes and pleural plaques which appear to be related to duration and time since first exposure, however no evidence of interstitial fibrosis was detected (Lemasters *et al.*, 1994). European morbidity studies from 7 production facilities indicate that 13% of chest radiographs had small opacities of greater than or equal to 0/1 but these were not related to cumulative exposure to RCF (Trethowan *et al.*, 1995).

Duration of employment based on a 5 year split (Table 3) appears to have a slight effect in decreasing lung function in non-smokers and current smokers although the trend is reversed in former smokers. Overall these trends are not significant. The small number of workers does not allow a finer categorisation to determine if a dose related effect is present. The small number of workers involved in the industry also does not allow further statistical determinations to be made on the data and the results are not distinguishable from the normal population nor do they have any clinical significance. Larger population studies conducted in U.S.A. and Europe have reported slight decreases in lung function associated with smoking and duration of exposure but these appear not to be clinically significant and mimicked findings of the interaction of smoking and dust in other dusty industries (Lemasters *et al.*, 1994; Trethowan *et al.*, 1995).

In conclusion the Australian data, whilst being small in numbers compared with

Table 2. Major findings of the health surveillance programs

	Plant 1	Plant 2
Duration of program	1985–96	1977–96
Pre-employment exam	Yes	Yes
Frequency	every 2 years	every 3 years
Program components	Chest radiography, spirometry, respiratory symptoms, occupational history, smoking history, ECG, full blood count and biochemistry tests	Chest radiography, spirometry, respiratory symptom questionnaire, occupational history and smoking history
Current medical status: Workforce		
• N	18[1]	25[2]
• Mean age (years)	44.8	38
• Mean years of service	10.8	3.7
• Non smokers	10 (56%)	5 (20%)
• Ex-smokers	4 (22%)	3 (17%)
• Current smokers	4 (22%)	15 (63%)
Respiratory symptoms	none	regular cough (64%), bronchitis (20%), chest pain (8%), asthma# (8%), chest tightness (4%)
Lung function	see Table 3	see Table 3
Radiological changes		
• X-ray films read by:	one "B" reader	3 "B" readers
• parenchymal changes (ILO)	0/0 (all)	(1/24): ≥ 1/0+ (2/3 readers) (2/24): = 0/1+ (1/3 readers)
• pleural abnormalities	none	(1/24) – pleural thickening+ (1/3 readers only)

(1) Current round of medical examination (1996).
(2) 1990 results (Lockey et al., 1990) presented.
one worker with asthma showed airway obstruction in lung function test.
+ No films had both pleural and parenchymal abnormalities.

Table 3. Mean spirometric percent predicted by smoking and years of RCF exposure

	Plant 1 (tests done in 1996)		Plant 2 (tests done in 1990)	
	< 5 years	≥ 5 years	< 5 years	≥ 5 years
Non-smokers				
n	4	6	4	1
FVC	106.6	99.1	113.9	104.7
FEV1	115.1	99.4	122.5	100.6
FEV1/FVC	107.8	100.5	108.7	96.3
Former smokers				
n	0	4	2	2
FVC	NA	95.2	100.9	113.9
FEV1	NA	101.3	103.7	117.3
FEV1/FVC	NA	106.4	103.0	103.1
Current smokers				
n	4	0	11	4
FVC	94.0	NA	104.1	99.1
FEV1	90.0	NA	107.3	98.0
FEV1/FVC	95.6	NA	103.5	99.1

other countries, indicate no clinically detectable respiratory health effects in the current manufacturing workforce who have been employed over a number of years or in past employees whose records are available. There has been a considerable reduction in exposures and awareness of the issues associated with RCF within the industry and this is being extended to end users. The Australian studies are being prepared to link directly with European morbidity studies and CARE (controlled and reduced exposure) programme.

REFERENCES

Glass, L., Brown, R. and Hoskins, J. (1995) Health effects of refractory ceramic fibres: scientific issues and policy considerations. *Occup. Environ. Med.* **52**, 433–440.

Hori, H., Higashi, T., Fujino, A., Yamato, H. *et al.* (1993) Measurement of airborne ceramic fibres in manufacturing and processing factories. *Ann. occup. Hyg.* **37**, 623–629.

Lemasters, G., Lockey, J., Rice, C., McKay, R. *et al.* (1994) Radiographic changes among workers manufacturing refractory ceramic fibre and products. *Ann. occup. Hyg.* **38** (Suppl. 1), 745–751.

Lockey, J., Lemasters, G., Grumski, K. and Giles, D. (1991) Australian workers of carborundum resistant materials. Report prepared by University of Cincinnati for Carborundum Resistant Materials, Thomastown, Vic 3074, Australia.

Rice, C., Lockey, J., Lemasters, G., Dimos, J. and Gartside, P. (1994) Assessment of current fibre and silica exposure in the U.S. refractory ceramic fibre manufacturing industry. *Ann. occup. Hyg.* **38** (Suppl. 1), 739–744.

Rossiter, C., Gilson, J., Sheers, G., Thomas, J., Trenthowan, W., Cherrie, J. and Harrington, J. (1994) Refractory ceramic fibre production workers. Analysis of radiograph readings. *Ann. occup. Hyg.* **38** (Suppl. 1), 731–738.

Trethowan, W., Burge, P., Rossiter, C., Harrington, J. and Calvert, I. (1995) Study of the respiratory health of employees in seven European plants that manufacture ceramic fibres. *Occ. Environ. Med.* **52**, 97–104.

Worksafe Australia (1989) Technical report on synthetic mineral fibres and guidance note on the membrane filter method for the estimation of airborne synthetic mineral fibres. Australian Government Publishing Service, GPO Box 84, Canberra ACT 2601, Australia.

Worksafe Australia (1990) Synthetic mineral fibres—national standard and national code of practice. Australian Government Publishing Service, GPO Box 84, Canberra ACT 2601, Australia.

Yeung, P. and Rogers, A. (1996) A comparison of synthetic mineral fibres exposures pre- and post- the NOHSC national exposure standard and code of practice. *J. occup. Health Safety—Aust NZ* **12**, 279–288.

 Pergamon

Ann. occup. Hyg., Vol. 41, Supplement 1, pp. 267–272, 1997
© 1997 Published by Elsevier Science Ltd on behalf of BOHS
Printed in Great Britain. All rights reserved
0003–4878/97 $17.00 + 0.00
Inhaled Particles VIII

PII: S0003–4878(96)00121–4

REFRACTORY CERAMIC FIBRES: THE MEASUREMENT AND CONTROL OF EXPOSURE

C. G. Burley*, R. C. Brown† and L. D. Maxim‡

*Carborundum Company, Rainford, St Helens, Merseyside, U.K.; †4 Bramble Close, Uppingham, Rutland; ‡Everest Consulting, Cranbury, New Jersey, U.S.A.

INTRODUCTION

Refractory ceramic fibres (RCF) are vitreous aluminosilicates manufactured by melting alumina, silica and other materials; a stream of the molten material is either blown by high pressure air or directed onto a series of spinning wheels to form fibres which have a random diameter and length distribution. The industrial applications for which these materials are used require resistance to temperatures between 1000 and 1460°C, precluding the use of organic binders. For such temperatures fibres with median diameters typically between 1.5 and 3.0 μm are needed for efficient insulation.

Many fibre types, including RCFs, injected or implanted into the pleural or peritoneal cavities of test animals can cause mesothelioma (e.g. Pott *et al.*, 1987) and despite the rejection of intracavity experiments by most authorities (reviewed McClellan *et al.*, 1992), the International Agency for Research on Cancer (IARC) classified man-made mineral fibres (MMMF), including RCF, as category 2B carcinogens: "possibly carcinogenic in humans" (IARC, 1988).

As animal inhalation experiments had given inconclusive results (Davis *et al.*, 1984; Smith *et al.*, 1987) the MMMF industry carried out additional experiments most of which used RCFs (the so called "RCC" experiments reviewed by Bunn *et al.*, 1993). However problems in interpretation remain (e.g. Brown *et al.*, 1995). While positive results can be obtained in animal experiments studies of human populations exposed to RCFs have revealed no ill health (reviewed Glass *et al.*, 1995). However the RCF industry remains determined to ensure the safest possible practices in the manufacture and use of its products and to do so in as public a way as possible; thus providing increasing reassurance to its workers, customers and regulators.

METHODS

The principal tool established by the manufacturers to ensure that RCF products will not present any unreasonable risk to human health or the environment is a comprehensive product stewardship programme (PSP) concerning the entire product life cycle from design, through manufacture, application to disposal of RCF materials.

This programme includes seven major areas; communication, health effects research, product research, special studies and three areas concerned with exposure. These three areas, exposure assessments, workplace monitoring and workplace controls, have been combined in Europe into the Ceramic Fibre Industries Association's (ECFIA's) controlled and reduced exposure (CARE) programme. Refractory ceramic fibres are generally used in industry, so that exposure to airborne fibre is more readily controllable than it would be if it were employed in consumer products. In North America and Europe only about 50 000 people could be exposed to RCFs and these three facets of the PSP aim is to identify, quantify and control exposure in both manufacturing and user industries.

The PSP arose from the industry's earlier work on measuring RCF exposure and recommending methods of control. While some countries had exposure limit values for RCFs ranging from 0.5 to 2 f cc^{-1} in many there was no regulatory exposure limit. Therefore the RCF industry took the initiative and established an exposure guideline of 1 f cc^{-1} (8 h time-weighted average). The industry then increased the collection of workplace monitoring samples, initially using various protocols and methods. During 1993 in the U.S.A. the industry and the EPA then constructed a voluntary consent in order to gather sufficient data to assess exposure in a statistically valid way. This data could then be used to monitor exposure reduction programmes and to enable the dissemination of the best available handling and use techniques.

A standardised, protocol has been developed to collect the necessary data, the number of fibres in the air is measured for all types of task where exposure to RCFs could occur both "internally", in manufacturing plants and "externally" in customers' premises. As the aim is to get a statistically valid assessment of airborne fibre levels a random sampling strategy has to be used so that samples do not over represent worst- or best-case conditions with the consequent underestimated or exaggerated measure of exposure. Clearly the concentration of airborne fibre varies with the process being used, therefore, the various tasks involved in making and using RCFs have been classified and samples are taken in proportion to the number of people involved in each task. While each manufacturing plant can be sampled this is not possible for the greater number of customer's premises. Therefore, customers are selected for sampling in a way proportional to the amount of RCF purchased in the previous year. The selection process has also been weighted to ensure that measurements are made in all European countries; an identical programme in Australasia is also being integrated with that in Europe. In all countries the choice of premises and work stations to be sampled is being made by one of the present authors who is provided with confidential sales and production data by all manufacturers. All samples are taken and fibres counted using approved occupational hygiene methods and quality control so that the measurements always comply with the requirements of local regulatory authorities. Some customers wish to volunteer for sampling and measurements in these are in addition to those taken "at random".

Some corrections for sampling and counting procedures will be made before all the data collected world-wide can be pooled. The results of this programme will be reported regularly to all interested parties including regulatory authorities and trades unions.

Table 1.

Job category	Number of samples per year in RCF manufacturing plants	Number of samples per year in industries using RCFs
Fibre production in manufacture only	70	Not applicable
Finishing Sawing, grinding, sanding, die cutting, milling or routering	65	115
Assembly Encapsulation/lamination, stapling, sewing, cutting, module manufacture, ball milling and dry mixing	65	50
Mixing/forming Wet mixing, vacuum forming	65	50
Auxiliary Maintenance, handling/shipping, cleanup, supervision, laboratory work	30	25
Other Textile automotive and others n.e.c.	25	60
Installation fitting, packing, wrapping, pounding, tamping and hardware installation	Not applicable	75
Removal mould knockout, cleanup disposal, furnace maintenance	Not applicable	25
Total	320	400

RESULTS

Tasks in the industry have been classified into 8 categories and an analysis of the RCF life-cycle and the number of people in each category has resulted in the type of statistical plan described in Table 1 together with some of the job titles in each category. In the U.S.A. over 2000 samples have been collected and analysed using the standard protocol and the results shared with the authorities. Over the next 3–5 years the U.S. PSP efforts will analyse 750–1000 samples/year and the CARE initiative in Europe will add another 720 samples annually. An analysis of the fibre counts has demonstrated that they are distributed log-normally and so are best described by geometric mean values. As the European programme only commenced in 1996 in this paper we report the historical data which was collected in Europe in a non-random way and compare this U.S. data in the same categories. In Figs 1 and 2 we show the geometric mean time-weighted average counts for various functional categories in both U.S. and Europe over the period 1991–1996.

The measurement of fibre levels cannot, in itself, lead to a reduction in airborne fibre levels, however levels have been dropping on both sides of the Atlantic, as is illustrated in Fig. 3.

DISCUSSION

Over 90% of all 8 h time-weighted averages so far measured have been below the

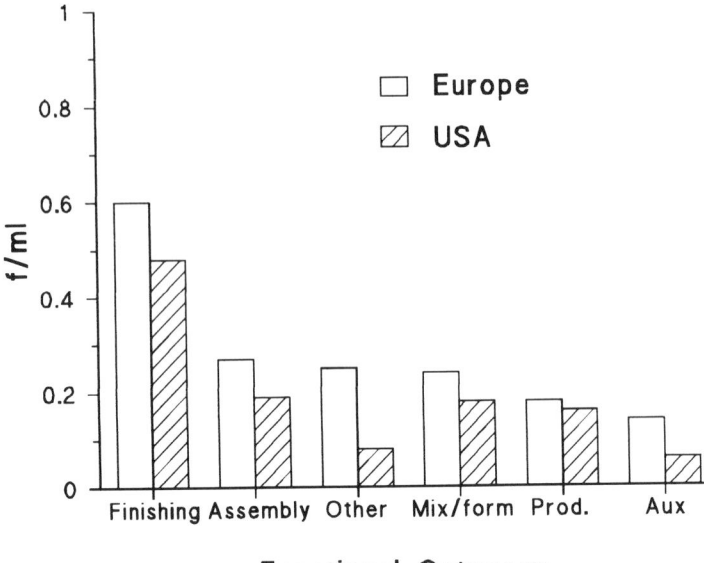

Fig. 1. The distribution of samples taken in the European ceramic fibre manufacturing and user industries as part of the CARE programme. The actual sites sampled are selected using a randomised sampling plan weighted by the quantity of fibre used. Thus accurate information on exposure to RCFs is obtained.

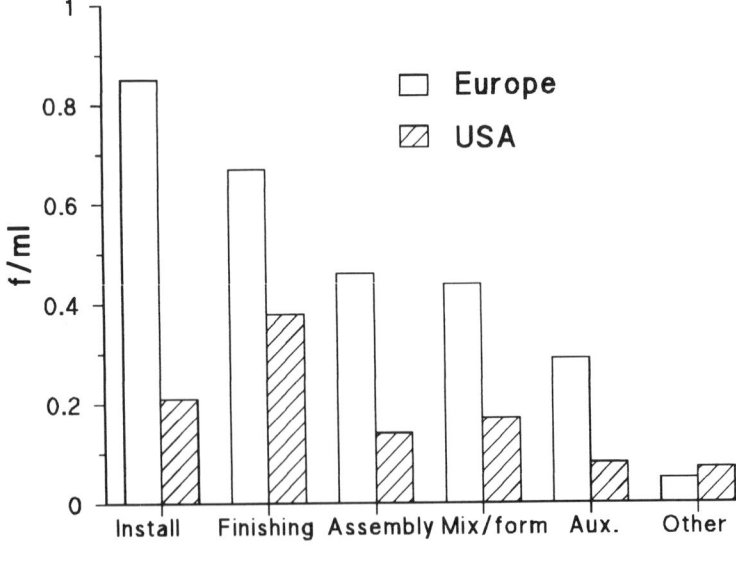

Fig. 2. Geometric mean fibre concentration by functional category at customer facilities, 1991 to 1996.

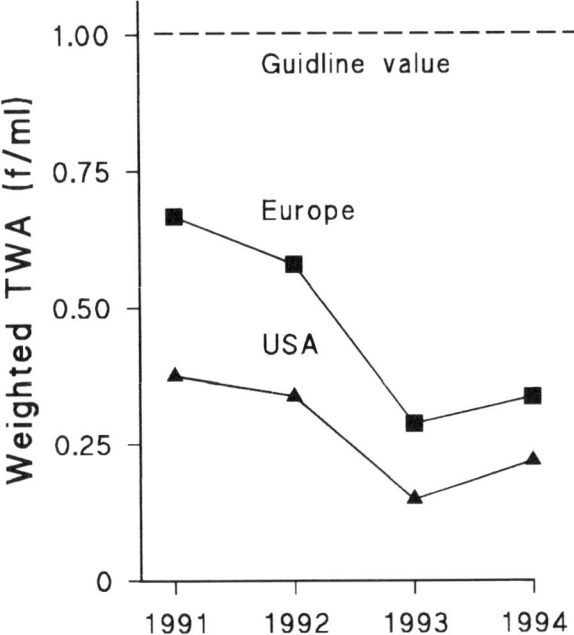

Fig. 3. Historical trend of average workplace RCF exposures at European and U.S. manufacturing facilities weighted by distribution of workers in functional categories.

guideline value of 1 f cc^{-1}. Exposures in the user industries are higher than those in manufacturing and in Europe higher than those in the U.S.A. Since the European counts have not been taken at random they may contain high values taken due to "problems" and thus may appear to drop when sufficient samples have been taken under the CARE programme. In nearly all areas exposures have been reduced substantially so that in the U.S.A. average levels in both manufacturing and user industries have fallen to about 0.2 f ml^{-1} and in Europe to below 0.4 f ml^{-1}. A small proportion of all counts have been above the guideline value of 1 f ml^{-1} but although taken with personal samplers these represent area values as workers under these conditions are protected through the use of personal protective equipment. Subsequently workplace controls have to be developed, this takes time and each situation has to be analysed and new equipment and technologies devised and implemented. Similarly the introduction of new work practices means that considerable time and effort must be expended in plant layout changes and/or employee training.

The operation of this programme represents a much more thorough study of airborne fibre concentrations than that required by law in any country. The aim is to continue to reduce exposure to the lowest practical levels as a minimum the full operation of the "CARE" programme will lead to a reduction of airborne fibre concentrations in Europe to similar values to those found in the U.S.A. and, for similar operations, the differences between the manufacturing and user industries should also reduce.

Due to the present level of concern and the uncertainty regarding any possible MMMF health risks there is a potential for considerable over-regulation. However, tougher regulation does not necessarily mean safer practices and the RCF industry has voluntarily undertaken many of the initiatives that regulation might impose. Success in reducing exposure and the continuing human health surveillance results suggest that any potential risks from RCF exposure, whilst not to be underestimated, are remote enough to permit sufficient time to investigate these issues thoroughly prior to any action. The pressure to take quick, and possibly damaging, action has been reduced so that in the long term any regulations that may prove necessary can be both objective and well founded.

REFERENCES

Brown, R. C., Hoskins, J. A. and Glass, L. R. (1995) The *in vivo* biological activity of ceramic fibres. *Ann. occup. Hyg.* **39**, 705–713.

Bunn, W. B., Bender, J. P., Hesterberg, T. W., Chase, G. R. and Konzen, J. L. (1993) Recent studies of man-made vitreous fibres. Chronic animal inhalation studies. *J. occ. Med.* **35**, 101–113.

Davis, J., Addison, J., Bolton, R., Donaldson, K., Jones, A. and Wright, A. (1984) The pathogenic effects of fibrous ceramic aluminium silicate glass administered to rats by inhalation and peritoneal injection. In *Biological Effects of Man-made Mineral Fibres* (Edited by J.C. Wagner). World Health Organisation, Copenhagen.

Glass, L. R., Brown, R. C. and Hoskins, J. A. (1995) Health effects of refractory ceramic fibres: scientific issues and policy considerations. *Occ. Environ. Med.* **52**, 433–440.

IARC *Monographs of the Evaluation of the Carcinogenic Risk of Chemicals to Humans; Man-made Mineral Fibres and Radon*, Vol. 43, pp. 15–169. International Agency for Research on Cancer, Lyon, France.

McClellan, R. O., Miller, F. J., Hesterberg, T. W., Warheit, D. B., Bunn, W. B., Kane, A. B., Lippmann, M., Mast, R. W., McConnell, E. E. and Reinhardt, C. F. (1992) Approaches to evaluating the toxicity and carcinogenicity of man-made fibers: summary of a workshop held November 11–13, 1991, Durham, North Carolina. *Reg. Tox. Pharmacol.* **16**, 321–364.

Pott, F., Ziem, U., Reiffer, F., Huth, F., Ernst, H. and Mohr, U. (1987) Carcinogenicity studies of fibres, metal compounds and some other dusts in rats. *Exptl. Path.* **32**, 129–152.

Smith, D., Oritz, L., Archuleta, R. and Johnson, N. (1987) Long term health effects in hamsters and rats exposed chronically to man-made vitreous fibres. *Ann. occup. Hyg.* **31**, 731–750.

Pergamon

Ann. occup. Hyg., Vol. 41, Supplement 1, pp. 273–278, 1997
© 1997 British Occupational Hygiene Society
Published by Elsevier Science Ltd. All rights reserved
Printed in Great Britain
0003–4878/97 $17.00 + 0.00
Inhaled Particles VIII

PII: S0003–4878(96)00071–3

SIMULATION TESTS OF MAN-MADE VITREOUS FIBRE EMISSIONS

G. Revell, G. J. Burdett and J. Brammer

HSE, Room 016, Robens Laboratory, HSL, Broad Lane, Sheffield S3 7HQ, U.K.

INTRODUCTION

This paper describes laboratory scale simulation tests, which have been undertaken using a standard work procedure, to measure the average release of fibre when installing MMVF insulation blanket. The objective of these tests was to measure the relative dustiness and airborne fibre concentration produced by various types of MMVFs, when being used for a relatively simple insulating task carried out under standard conditions. The fibre levels were monitored when installing the MMVF blanket, when the MMVF was removed and when the chamber was cleaned. These results were compared to the hazard as represented by LWGMFD measurements on the bulk materials and maximum emissions obtained from bench scale dustiness tests described by Burdett *et al.* (1996).

MATERIALS AND METHODS

Simulations were carried out on four refractory ceramic fibres (RCFs), with bulk length weighted geometric mean fibre diameters (LWGMFD) ranging from 1.56 to 4.13 μm and a soluble amorphous fibre, Insulfrax (LWGM FD = 2.93 μm). A rockwool (LWGMFD = 4.17 μm) and a glasswool (LWGMFD = 4.21 μm) were also tested to extend the range of "expected" dustiness levels. The RCFs and Insulfrax were purchased as 0.61 m wide by 7.3 m long, nominal thickness 25 mm standard rolls. The rockwool was purchased as 0.9 m × 0.44 m × 38 mm bats and the glasswool as 0.61 m wide by 10 m long, nominal thickness 50 mm rolls. The simulation tests were carried out inside a 9.2 m³ (2.1 × 2.1 × 2.1 m) test chamber in HSL's Sheffield laboratories. When the door was closed the chamber was fully sealed, except for a HEPA filtered inlet. A three stage airlock connected the chamber to the laboratory.

Experiments were carried out in still air (extraction off) but air extraction was used as part of the decontamination both between and after experiments. The chamber had two 0.65 × 0.65 m viewing ports one of which was used to video the installation and removal of MMVFs. All lights and electricity sockets were sealed so they could be washed down and decontaminated.

The simulations were carried out on three 2 m long, 0.1 m diameter pipes at 1.3, 1.6 and 1.9 m above the floor. These pipes were supported on a frame close to the back wall of the chamber. A table was placed in the centre of the chamber so that

0.3 × 0.4 m sections of MMVF could be cut ready for wrapping round the pipe. Testing was carried out using the procedure shown below.

(i)   The materials to be tested were stored in the laboratory and were unwrapped and conditioned overnight in the chamber before the start of the simulation tests.

(ii) The chamber was ventilated at > 1 m^3 s^{-1} overnight before the test.

(iii) The ventilation was switched off and background monitoring was carried out in still air for at least 1 h immediately before starting installation. Static samplers operating at 8 l. min^{-1} were used for background monitoring.

(iv) Installation: rectangular 0.3 by 0.4 m sections were cut from the RCF roll, wrapped round the pipe and fastened in position using two lengths of plastic coated wire (garden ties). The installation was carried out in still air with the door to the chamber closed.

(v)   Background monitoring was carried out in still air for at least 1 h immediately after installation was completed.

(vi) The chamber was vented at > 1 m^3 s^{-1} for 1 h after background sampling was complete and before clean up was started.

(vii) Clean-up: the ventilation was switched off and the insulation was removed and bagged. The chamber was thoroughly cleaned, initially by rigorous brushing of all walls and surfaces followed by vacuuming.

(viii) Background sampling was carried out for 1 h immediately after clean-up.

Between materials the chamber was left to ventilate at > 1 m^3 s^{-1} for a period of not less than 16 h.

The variables monitored were: (i) temperature using a mercury in glass thermometer and a T_1/T_2 thermocouple, (ii) relative humidity measured using a solid state detector and a hair hygrometer, and (iii) airborne fibre concentration in accordance to MDHS 59.

Sampling was carried out using 0.8 μm pore size membrane filters in conductive cowled sampling cassettes. The operative carrying out the installation, removal and clean-up wore two personal samplers, one sampling at 2 l. min^{-1} and the other at 3.5 l. min^{-1}. A series of static samples operating at flow rates of 2, 3.5 and 8 l. min^{-1} were sited at various locations close to the work area. The relative positions of sampling heads 1–6 around the pipes are shown in Fig. 1 but two other sampling heads (7 and 8) placed on the floor close the legs of the cutting table are not shown. The sampling heads for all the floor based equipment were about 0.3 m above the floor. Installation and clean-up (including removal) were monitored as separate activities.

Airborne fibre levels were determined by fibre counting using phase contrast light microscopy (PCM) fibre counts carried out in HSL's laboratories in accordance with MDHS 59, to count fibres > 5 μm long, < 3 μm wide and with an aspect ratio of $> 3:1$.

RESULTS

The temperature rose from the ambient of 20–24°C by up to 6°C whenever work activity was taking place in chamber (ventilation off). It quickly fell back to

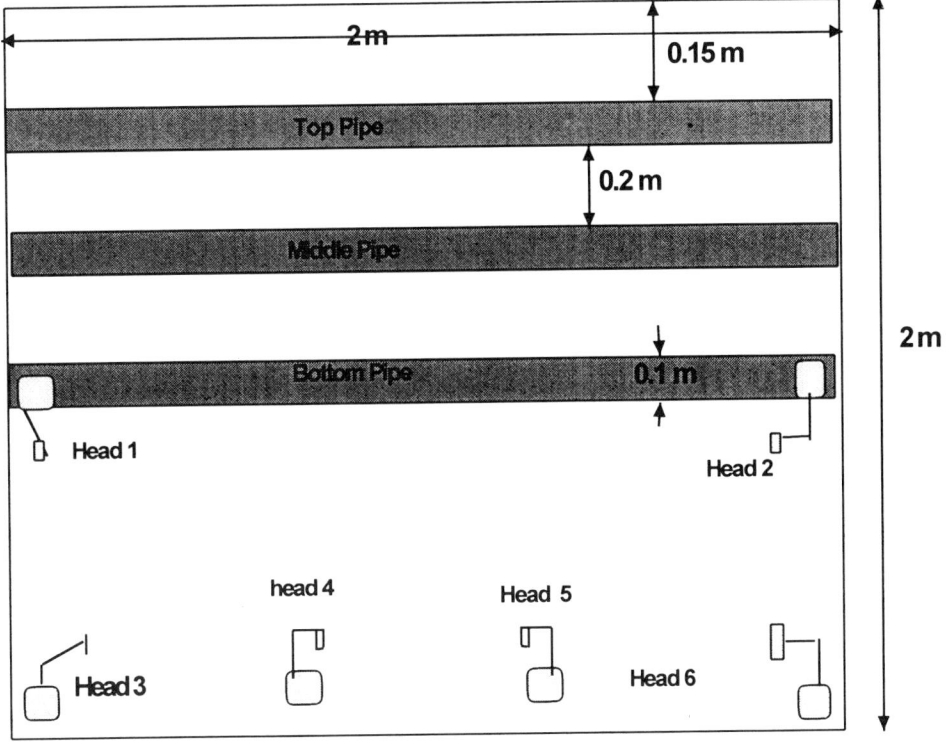

Fig. 1. Schematic layout of the back wall of the test chamber showing the positions of the pipes and the sampling heads.

ambient levels when the work activity ceased and the air extraction was switched on. Similarly the relative humidity rose from ambient levels of 50–60% to > 90% during work activity, in still air and fell back to ambient when work activity ceased and the air extraction was switched on.

Although this report concentrates on the results of airborne fibre concentration measurements from personal monitoring a small number of static monitoring values are included to show the background levels when no work activity was taking place. A more comprehensive report including all the results will be published later. A summary of the personal PCM airborne fibre levels are shown in Table 1. Generally the results are from single tests but those from Kaowool are the average from three repeat tests on the same material (two rolls purchased at the same time). The results from repeat runs from the Durablanket 1460 are reported as two individual tests as rolls purchased from different sources, at different times, gave substantially different bulk LWGMFDs.

DISCUSSION

The low fibre levels in the chamber before work commenced indicate that there was little or no cross contamination between experiments. Very low airborne fibre levels were recorded in the hour immediately after the clean-up indicating that this

Table 1. Summary data from MMMVF simulation tests showing PCM fibre concentrations in fibre ml^{-1}

	R'wool	Insulf.	C'chrom	Z'lane	Durablanket Run 1	Durablanket Run 2	P. G'wool	K'wool ave.
Pre-installation*	< 0.01	< 0.01	< 0.01	< 0.01	< 0.01	< 0.01	—	< 0.012
Installation[P]	0.08	1.49	2.36	1.92	0.46	0.92	0.11	0.99
Post-installation*	0.01	0.24	0.5	0.44	0.25	0.25	< 0.01	0.51
Removal and clean[P]	0.06	0.32	0.7	0.29	0.16	0.32	0.09	0.32
Post-removal*	< 0.01	0.02	0.1	0.03	0.01	0.02	< 0.01	—
Dustiness test	0.23	5.33	54.9	24.56	12.16	12.16	0.61	8.09
LWGMFD (μm)	4.17	2.93	2.22	1.56	4.13	2.13	4.21	1.98

* = static sampling only; [P] = Personal samples only; Insul. = Insulfrax; R'wool = Rockwool; P. G'wool = Pilkington Glasswool; Z'lane = Zirlane; C'chrom = Cerachrome; Durablanket = Durablanket 1460; K'wool ave = average of 3 Kaowool runs.

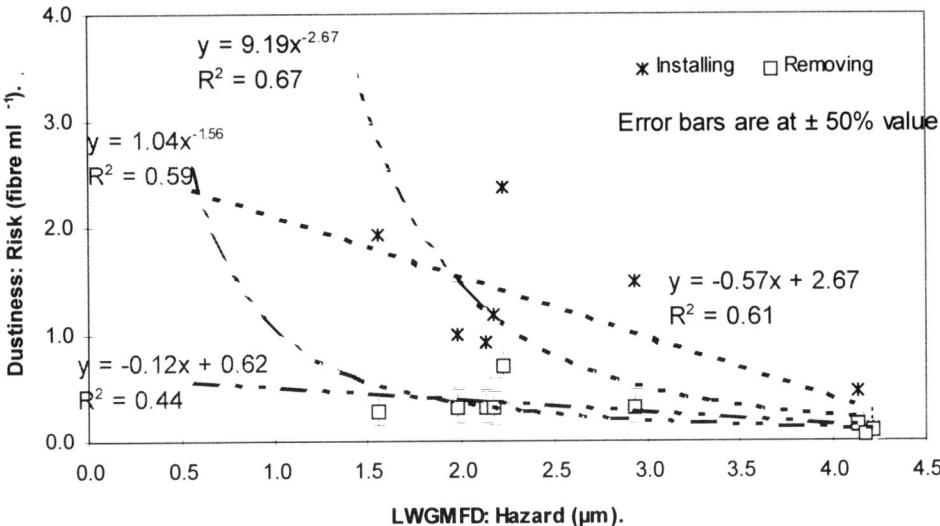

Fig. 2. Graph plotting simulation exercise dustiness values (risk) against LWGMFD (hazard): including best fit linear and power law lines.

was effective. A further check on the clean-up procedures carried out by monitoring between experiments using electric fans to disturb the air inside the sealed chamber (air extraction off) produced airborne fibre levels of < 0.01 fibre ml^{-1}.

Some tidying was carried out between installation and removal but no attempt was made to prevent fibres released during installation being disturbed during removal. PCM airborne fibre concentrations of up to 0.51 fibre ml^{-1} (Kaowool) were observed in the hour immediately after insulation was completed showing that fibres remain airborne for a considerable period after work activity has ceased. Running the ventilation for an hour between installation and removal removed the residual airborne fibres but loose fibres on the floor and other surfaces could be disturbed and contribute to the fibre levels recorded during clean-up. These fibre levels were much lower, being on average about one third those produced when installing the material.

One material, Kaowool when tested three times produced consistent results. When installing the average personal PCM airborne fibre level was 0.99 ± 0.19 fibre ml^{-1} and during removal and clean up it was 0.32 ± 0.12 fibre ml^{-1}. There did, however, appear to be wide differences if what was nominally the same material was purchased from different sources at different times. The two batches of Durablanket 1460 were found to have large variations in the length weighted geometric mean fibre diameter (batch 1 = 4.13 μm and batch 2 = 2.13 μm) and this was reflected in the airborne fibre levels which varied by a factor of two, both when installing (0.46 and 0.92 fibre ml^{-1}) and when removing (0.16 and 0.32 fibre ml^{-1}).

A comparison of hazard as measured by LWGMFD and risk as measured by the personal dustiness levels determined in these simulation tests is shown in Fig. 2. Although the dustiness (risk) decreases with increasing LWGMFD the best fit linear correlation is poor. Dodgson *et al.* (1987) suggested that there was an inverse power law relationship ($M^{-1.52}$) between dustiness and LWGMFD but best fit power law curves for this data indicate only a limited correlation with $R^2 = 0.67$ (installing) and 0.59 (clean-up). This indicates that bulk fibre diameter alone does not give a good prediction of the fibre concentration.

Although the airborne dust levels produced by the rotating drum (Burdett *et al.*, 1996) were an order of magnitude greater than those produced by the simulation exercise, a Spearman rank-correlation test showed that there was significant correlation between the two methods; a linear regression fit gave correlation (R^2) of 0.58 (installing) and 0.69 (clean-up). The soluble Insulfrax fibre appeared to be atypical and if this set was omitted then a marked increase in the correlation was observed, $R^2 = 0.79$ (installing) and 0.83 (clean-up).

CONCLUSIONS

- The simulation exercise reported here proved to be a relatively simple and practical method of measuring the fibre release of MMVF insulation blanket.
- The results showed that during simulated installation on pipes the different type of materials having bulk LWGMFDs ranging from 1.56 to 4.21 μm gave PCM airborne fibre concentrations of 0.46–2.36 fibre ml^{-1} for RCF, 1.49 fibre ml^{-1} for a soluble amorphous fibre and 0.08–0.11 fibre ml^{-1} for mineral wools.
- The airborne fibre concentrations for the RCFs tested either approached or exceeded the concentration in the maximum exposure limit (MEL) of 2 fibre ml^{-1}. This indicates that the MEL may be exceeded when RCF insulation is installed in greater amounts and/or under conditions of limited ventilation.
- Although there is an underlying inverse relationship between the LWGMFD of the bulk material and respirable fibre concentration, this relationship is secondary to other factors affecting the releasability and friability of the bulk fibres. This is particularly the case with the limited range of LWGMFDs in this study.

REFERENCES

Burdett, G. J., Revell, G. and Brammer, J. (1996) The use of a dustiness test for measuring fibre emissions from man made vitreous fibres (MMVFs) (this conference).

MDHS 59 (1988) Method of determining hazardous substances. Man made mineral fibre: airborne number concentration by phase contrast microscopy, Health and Safety Executive, London.

Dodgson, J., Cherrie, J. and Groat, S. (1987) Estimates of past exposure to respirable MMMFs in the European Insulation Wool Industry. *Ann. occ. Hyg.* **31**, 567–582.

 Pergamon

Ann. occup. Hyg., Vol. 41, Supplement 1, pp. 279–286, 1997
© 1997 British Occupational Hygiene Society
Published by Elsevier Science Ltd. All rights reserved
Printed in Great Britain
0003–4878/97 $17.00 + 0.00
Inhaled Particles VIII

PII: S0003–4878(96)00166–4

SURFACE SAMPLING FOR ASBESTOS RISK ASSESSMENT

D. P. Fowler* and E. J. Chatfield†

*Fowler Associates, 643 Bair Island Road, Suite 305, Redwood City, CA 94063, U.S.A.;
†Chatfield Technical Consulting, 2071 Dickson Road, Mississauga, Ontario, Canada L5B 1Y8

Surface sampling techniques, coupled with various analytical procedures, have been useful in the nuclear, food processing and pesticide industries to evaluate and control the potential influence of surface contamination on human health. In particular, surface sampling has been used in the nuclear industry for many years as a means of monitoring radioactive contamination which can occur during both laboratory and production operations (Eisenbud *et al.*, 1954; Fish, 1964). Although for alpha particle emitting radioactive materials such as plutonium, inhalation of particles is the primary hazard, measurements of loose surface contamination have become a useful and simple means of monitoring the spread of contamination in facilities where these materials are manipulated. The maximum routine working level for loose surface contamination in areas occupied by unprotected workers is derived using the maximum permissible airborne concentration and an appropriate "resuspension factor". A resuspension factor is the value or values such that, under specified conditions of disturbance of surface dust, the mass concentration of airborne dust can be related to the mass concentration of surface dust by the relationship:

$$\text{Airborne Concentration} = K \times \text{Surface Concentration}$$

where K is a resuspension factor with dimensions of length^{-1}. A number of studies (both published and unpublished) have established typical resuspension factors in the nuclear and other industries, and Sansone and colleagues (1978, 1987) and Caplan (1993) have independently reviewed these. The range of exponents for resuspension factors found was 10^{-2} to 10^{-8}. The mean exponent was $\sim 10^{-4.6}$, with a 95% confidence interval of $10^{-4.2}$–$10^{-4.9}$ for work directly disturbing the settled material. This is reasonably consistent with what has been thought of as the "typical" resuspension factor in the nuclear industry of 10^{-6}, and occasionally 10^{-5} (Brodsky, 1980).

In the U.S.A., surface sampling has been used to determine the concentrations of asbestos on surfaces in buildings where asbestos-containing building materials (ACBM) are present, and some consultants have used simple resuspension factor calculations in attempts to predict airborne asbestos fibre concentrations in general occupancy buildings. Usually, such attempts assume no direct disturbance of the ACBM, but still use resuspension factors such as those noted above, found during such activities as sweeping of deposited dust or other disturbances.

In this paper we consider the feasibility of predicting airborne asbestos concentrations on the basis of surface contamination measurements, using the methods that have been so used in the U.S.A. It is our contention that although surface-sampling data may be useful in tracing the source of asbestos-containing dust, modifications of the size distribution during analysis preclude the use of surface concentrations to predict air concentrations.

The resuspension factor for a given kind of dust will vary over a wide range, depending on a number of variables including the natures of the dust and the disturbance, the type of surface, and air movement in the space affected. For a specific situation in which the resuspension factor has been measured beforehand, the airborne concentrations associated with the same level of disturbance, but with different amounts of the same type of surface contamination, can be predicted. However, if the nature of the settled dust changes in any way, these predictions may be seriously in error. Clearly, if there is any change in the particle size distribution of the measured particle species, or if other types of dust are added which interact with, and become associated with, the measured particle species, the nature and amount of resuspended material will change, and the resuspension factor will be different. This is exemplified by the effectiveness of sweeping compounds in suppression of dust. In the U.S.A. and elsewhere, large numbers of measurements have been made of the airborne asbestos concentrations in general occupancy buildings. These concentrations have almost invariably been very low (Health Effects Institute—Asbestos Research, 1991). It has been argued that short duration, high peaks of airborne asbestos occur, and that these are unlikely to be observed in the relatively short period during which air samples are being collected. As evidence for this theory, large numbers of surface dust measurements have been made, showing that asbestos can be detected at various concentrations in dust found on surfaces in the buildings. It is asserted that this asbestos has settled out from a series of "episodic" releases from asbestos-containing building materials, and that the contribution of these releases to the occupants' exposure is not accounted for by the air sampling measurements alone. Attempts have also been made to predict the concentration of airborne asbestos which could arise by resuspension of the asbestos-containing dust, using resuspension factors obtained from a number of published sources, most of which are unrelated to asbestos. These predicted airborne concentrations of asbestos are generally orders of magnitude higher than any of the airborne concentrations actually observed during general occupancy air sampling, in the absence of direct disturbance of the dust deposit of interest. If the release of asbestos from the building materials, and the resuspension of surface dusts are both episodic, it would be expected that at least some of the measured airborne concentrations would indicate elevated levels of asbestos. Moreover, in occupied buildings in normal operation, the measured airborne concentrations should include any contribution from resuspension of dust caused by normal activities in the building.

Theoretically, TEM samples can be prepared from a surface dust sample by either a direct- or indirect-transfer procedure. In direct-transfer procedures, attempts are made, using an adhesive film, to transfer dust particles from a surface to a TEM grid without causing any changes in either the particles or the spatial relationships between the particles. Clearly, a direct-transfer technique can be used only if the loading and distribution of particles on the surface are suitable. If, as is most probably the case, the dust consists of widely separated, but relatively large aggregates, it is unlikely that the dust particles found on the small area normally examined on a TEM specimen grid correctly represent the full range of particles to be found in the dust deposit from the larger surface to be characterized (for example, the upper surface of the suspended ceiling in a building). In view of the aggregated nature of surface dust, it is also unlikely that either the number or the mass of any asbestos fibres present in the dust can be correctly determined by a direct-transfer preparation. Therefore, a direct-transfer procedure cannot be considered to provide quantitative results. However, since some of the particles can be examined in almost an unmodified condition, the method offers a means for determining the possible source of the dust. If quantitative results are desired, it is necessary to average the results over a much larger area of the sampled surface, and this can be achieved only by indirect-transfer analytical methods. As we shall show, there are also serious deficiencies in the indirect-transfer approach when an attempt is made to define resuspension factors.

Certain approaches to surface dust evaluation have been undertaken by the American Society for Testing and Materials (ASTM) Subcommittee D22.07. ASTM is a voluntary consensus standards organization which (among many other activities) has recently been involved in the development of standard analytical methods for determination of asbestos in bulk materials, air samples and dust samples. Two ASTM standard methods for determination of asbestos in dust samples have received final approval. These are: D5755-95—Test Method for Microvacuum Sampling and Indirect Analysis of Dust by Transmission Electron Microscopy for Asbestos Structure Number Concentrations; and, D5756-95—Test Method for Microvacuum Sampling and Indirect Analysis of Dust by Transmission Electron Microscopy for Asbestos Mass Concentration. In the present discussion, we will only consider the method D5755-95 for structure counting, as it is the use of this method as a precursor to speculation about resultant airborne asbestos concentration that has been the source of the misuse of the resuspension factor concept. In both of the ASTM indirect-transfer procedures, the dust sample is collected from the surface using an air sampling cassette fitted with a small diameter plastic tube. The device is used as a miniature vacuum cleaner (microvacuum) to collect dust from a known area of the surface to be evaluated by drawing air through the sampling train with a conventional personal sampling pump operated at two litres·minute^{-1}. The tip [¼″ (about 6 mm) inside diameter; ⅜″ (about 10 mm) outside diameter, cut at a 45° angle] is used to sample a defined area [usually either 100 cm² (10 cm × 10 cm) or one square foot (about 930 cm²)]. The ASTM definition of settleable particulate material was adopted for all ASTM asbestos dust sampling methods, being any particles which are capable of passing through a 1 mm screen. The dust collected in the cassette is washed out using a mixture of ethanol and water, and the suspension is passed through a 1 mm plastic screen to remove particles exceeding 1 mm in size. The pH of the suspension is adjusted to between 3 and 4, to assure optimum dispersion of chrysotile asbestos. The suspension is treated for a short time in an ultrasonic bath, and aliquots of the liquid suspension are then filtered to provide membrane filters (either polycarbonate or cellulose ester) of appropriate particle loading for TEM specimen preparation. TEM specimens are then prepared from the membrane filters using a direct-transfer procedure appropriate for the filter type used. Results are reported as asbestos structures · cm^{-2} of the original surface.

The reproducibility of the analysis (independent of the preparation) has been tested by analysis of 11 sets of duplicate aliquots of the final preparation. The results are about as would be expected due to chance counting variability alone, that is ± a factor of 2, but with a good overall correlation between the data sets. However, when 12 sets of side-by-side environmental samples were taken and analyzed by separate organizations, (that is, one organization took a microvacuum sample from a building surface, and the other organization took a sample next to the first, and the two samples were separately analyzed) the correspondence of the results was poor, with negative (but statistically insignificant) correlation between the two sets of data. Indeed, differences between two purportedly similar samples were often marked by substantial discrepancies, with six of the samples (two from one lab and four from the other) having no asbestos structures visible, whilst the corresponding samples analyzed by the other lab had counts of 16 700 to 399 000 structures·cm^{-2} (Van Orden, personal communication, 1993). This is not surprising, given that one can anticipate variable surface concentrations and variable efficiency in the collection of the particles found, since the collection system is likely to collect only large particles. Note that the velocity of air across the surface being microvacuumed will be variable, but in the best of circumstances, assuming a constant distance between the tip and the surface of 0.1 mm, will be approximately 9 ms^{-1}. At this velocity, only the relatively large—≥ ~ 10 μm AED—particles will be collected with some certainty, and only the very largest particles will be reliably collected ~ 90% of the time.

We now examine resuspension factors calculated on the basis of such indirect preparation sample results. In a limited and incompletely-described study (Kelman et al., 1994) a resuspension factor of ~ 10^{-5} cm^{-1} (for all structures seen by TEM) was found when comparing asbestos measurements in the breathing zone of a simulated cleaning worker performing vigorous cleaning work with measurements of surface asbestos deposits in a small apartment unit. The surface samples were analyzed with an indirect preparation Transmission Electron Microscopy (TEM) method similar to ASTM D5755-95, and the air samples by both a variant of the AHERA direct preparation TEM method (USEPA, 1987), and the conventional Phase Contrast Microscopy (PCM) method. It was noted by the authors that only one asbestos fibre > 5 μm length was seen by TEM in the air samples (that single fibre was noted to be too slender to be seen by PCM), and that the PCM concentrations (in only two air samples) were below the stated limit of detection of 0.04 f·ml. If we were to consider only the TEM-visible fibres longer than 5 μm, and the PCM results, none of the samples would have been above the method detection limits, and thus the resuspension factors should have been undefineable.

In preparation for this paper, we have considered some previously unpublished data. The first set of data relates to the measurement of asbestos in dust in occupied buildings using an earlier version of ASTM D5755-95 compared with a series of subsequent measurements of asbestos in air, made using conventional air sampling techniques, with subsequent analysis by the International Organization for Standardization (ISO) direct-transfer method ISO10312. Those results are given in Table 1, where resuspension factors are also given. As can be seen there, the

Table 1. Measurements of asbestos concentrations in buildings and resulting resuspension factors

| Building | Airborne asbetos concentrations* | | Surface asbestos concentrations** | | | Mean resuspension factor (cm^{-1}) |
	Mean concentration (structures·ml^{-1})	Number of samples	Mean concentration (structures·CM^{-2}·E06)	Range (structures·CM^{-2}·E06)	Number of samples	
Large office building	0.0017	53	2.2	0.04 – 6.2	11	7.7 E-10
University 1	0.00095	28	1.8	0.4 – 3.0	3	5.3 E-10
University 2	0.0018	25	24	7.8 –50	5	7.5 E-11
High School 1***	0.00085	17	0.26	0.015 – 0.68	4	3.3 E-09
			17	11 –28	6	5.0 E-11
High School 2	0.011	10	1.4	0.29 – 3.4	3	7.9 E-09
Elementary School	0.0011	11	0.94	0.12 – 2.4	3	1.2 E-09
Mean of Means (unweighted)						2 E-09

* Measurements of airborne asbestos made with direct transfer method.
** Measurements of surface asbestos made with indirect transfer method.
*** Two sets of dust measurements were made: the lower values were obtained in the occupied areas; and, the higher values were obtained above the suspended ceiling.

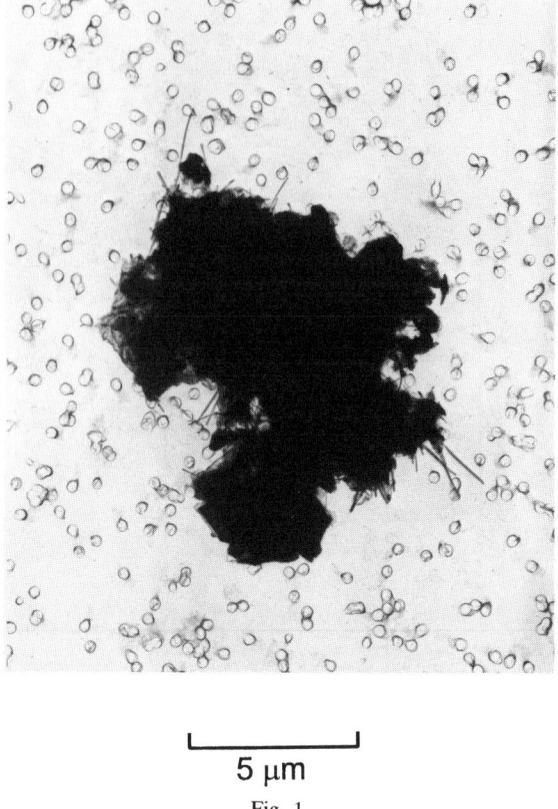

5 µm

Fig. 1.

resuspension factors are in the range of 10^{-9}–10^{-11}, which are substantially lower than the "typical" resuspension values given in the literature (for example Brodsky, Caplan and Sansone, previously cited), or calculated by Kelman *et al.* (1994) during cleaning. The predictions of airborne concentrations that would have been made if those published resuspension factors had been used would have been 3–5 orders of magnitude higher than they were in actuality. As discussed above, for quantitative determination of the asbestos present in surface dust, an indirect-transfer analytical procedure is required, and the size distribution of asbestos-containing dust is substantially modified by those procedures. Fibres are released from matrix materials, and in the case of chrysotile, very large numbers of short fibres are generated which, in the original sample, were loosely adherent to fibre bundles. Clusters are also dispersed into their fibre or fibre bundle components. In most indirect-transfer dust analyses, the data consist almost completely of single fibrils of chrysotile. The volume of a 20 µm diameter sphere is $\sim 4200 \ \mu m^3$, whilst the volume of a 10 µm diameter sphere is only $\sim 520 \ \mu m^3$. If the volume of a "typical" asbestos fibril may be taken as $\sim 0.008 \ \mu m^3$ a difference in structure counts of 4–$5 \cdot 10^3$ is easily foreseeable if in one sample a single 10 µm particle—consisting entirely of asbestos fibrils—is collected, disaggregated, and counted whilst in another taken nearby a single 20 µm particle similarly constituted is similarly

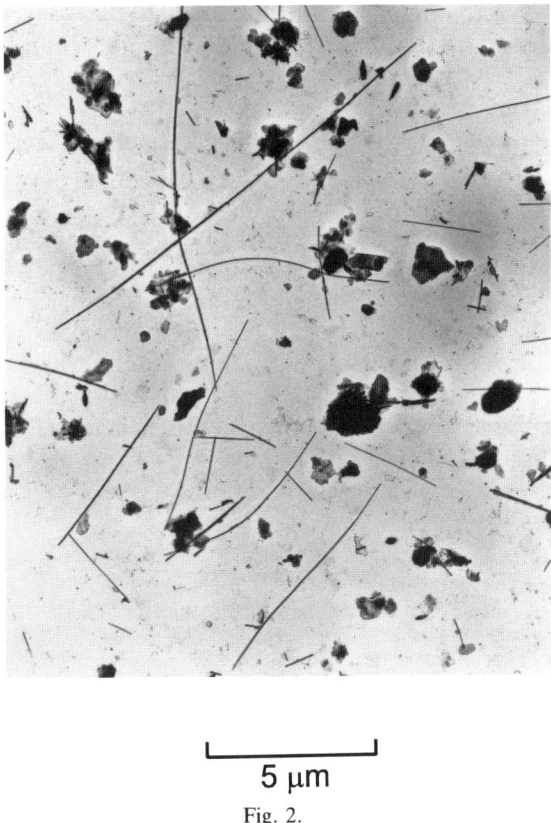

5 μm

Fig. 2.

collected, disaggregated, and counted. Of course, a single 1 mm particle might contain roughly $5 \cdot 10^8$ such hypothetical fibrils. It is therefore self-evident that the surface dust measurement does not represent the nature of the dust as it existed on the surface, because the settling velocities of single fibrils of chrysotile are too low for such material to have arrived on the surface by sedimentation.

An experiment was performed in which the edge of an asbestos-containing ceiling tile was abraded, as it might be when inserted into or removed from the supporting framework. The airborne particles liberated by this procedure were collected on a membrane filter after passage through a vertical elutriator with a vertical air velocity of 0.32 cm·s^{-1}. TEM specimens were prepared from the air filters by a direct-transfer method. Figure 1 shows a TEM micrograph of a typical aggregate found on the air filters. Examination of the TEM specimens showed that all of the asbestos-containing particles consisted of structures similar to that of Fig. 1, and no isolated individual fibres were found. Moreover, the majority of the particles were in a narrow size range close to the size of the particle in Fig. 1. The settled dust from the abrasion experiment (that is those dust particles *too large* to be carried up the vertical elutriator) was prepared according to the ASTM D5755-95 procedure. Figure 2 shows a TEM micrograph of the particles which were observed. Clearly, a numerical asbestos structure concentration derived from this specimen has little value in predicting the potential of the dust, in its original form, to become airborne as a result of a disturbance other than one which would produce complete disintegration of the dust particles and could cause the disintegrated dust to become airborne.

Based on the results shown here, as well as on our experience, we have concluded that: (1) The only practical *quantitative* measurement method for asbestos in surface dust is an indirect-transfer method such as ASTM 5755-95 or ASTM 5756-95; (2) Neither of the ASTM methods, or similar indirect-transfer methods, preserves the size distribution of the particles as they existed on the sampled surface; (3) Without detailed knowledge of the size distribution of the original dust on the sampled surface it is not possible to predict how much of the dust can become airborne under stated conditions of disturbance; (4) Calculations of airborne asbestos concentrations made using the results of ASTM 5755-95 and published values for resuspension factors for other materials and circumstances predict airborne asbestos concentrations much higher than those actually observed in buildings under normal occupancy conditions.

REFERENCES

Brodsky, A. (1980) Resuspension factors and probabilities of intake of material in process (or 'is 10^{-6} a magic number in health physics?'). *Health Physics* **39**(6), 992–1000.

Caplan, K. J. (1993) The significance of wipe samples. *Am. ind. Hyg. Assoc. J.* **54**, 70–75.

Eisenbud, M., Blatz, H. and Barry, E. V. (1954) How important is surface contamination? *Nucleonics* 12–15, August.

Fish, B. R., Ed. (1964) *Symposium on Surface Contamination, Gatlinburg, TN, 1964*. Pergamon Press, Oxford.

Health Effects Institute—Asbestos Research (1991) *Asbestos in Public and Commercial Buildings: A Literature Review and Synthesis of Current Knowledge*. National Technical Information Service, 5285 Port Royal Road, Springfield, VA 22161, U.S.A.

Kelman, B. J., Millette, J. R. and Bell, J. U. (1994) Resuspension of asbestos in settled dust in an apartment cleaning situation. *Appl. occup. Environ. Hyg.* **9**, 876–878.

Sansone, E. B. (1987) Redispersion of indoor surface contamination and its implications. In *Treatise on Clean Surface Technology*, K. L. Mittal, Ed., Plenum Press, New York.

Sansone, E. B. and Stein, M. W. (1978) Redispersion of indoor surface contamination: a review. *J. Hazardous Materials*, **2** 347–361.

USEPA (1987) 40 CFR Part 763 Appendix A: asbestos containing materials in schools: final rule and notice. *Federal Register* **52**, 41870–41893.

 Pergamon

Ann. occup. Hyg., Vol. 41, Supplement 1, pp. 287–292, 1997
© 1997 British Occupational Hygiene Society
Published by Elsevier Science Ltd. All rights reserved
Printed in Great Britain
0003–4878/97 $17.00 + 0.00
Inhaled Particles VIII

PII: S0003–4878(96)00070–1

LACK OF TREMOLITE IN UICC REFERENCE CHRYSOTILE AND THE IMPLICATIONS FOR CARCINOGENICITY

A. L. Frank, R. F. Dodson and M. G. Williams

Department of Cell Biology and Environmental Sciences, University of Texas Health Center at Tyler,
P.O. Box 2003, Tyler, TX 75710, U.S.A.

INTRODUCTION

In 1935, Lynch and Smith reported that in addition to causing asbestosis, asbestos might also cause lung cancer and soon after Heuper (1942) felt sufficient evidence existed to call asbestos a carcinogen. There is continuing controversy about aspects of asbestos-related disease. Few argue about the carcinogenicity of asbestos; disagreements exist regarding other issues, such as if asbestosis is needed before a lung cancer can be attributed to asbestos (Churg *et al.*, 1984). Some of these controversial issues have been addressed elsewhere by Frank (1994).

One area of controversy has been the ability of chrysotile to produce lung cancer and mesothelioma. In spite of much evidence to the contrary, some still hold to the view that chrysotile, as chrysotile, does not cause these. Central to this belief has been what has been called the "amphibole hypothesis" (Wagner, 1986; McDonald *et al.*, 1989). Certain policy questions may be influenced by the claimed lack of chrysotile's ability to cause disease. One should recall that some have advocated that only certain fibre types, such as tremolite in chrysotile, produce disease, especially cancer (Churg *et al.*, 1984; Wagner, 1986), but other data have shown this to be untenable (Begin *et al.*, 1992; Stayner *et al.*, 1996). There is now evidence by Karjalainen *et al.* (1994) of the ability even of anthophyllite to produce mesothelioma.

Few populations have ever been identified where only chrysotile exposure had taken place, but Mancuso (1988) documented mesotheliomas among such workers in the U.S.

Despite evidence to the contrary, some authors (Case, 1991; Dunnigan, 1988; Mossman and Gee, 1989; Mossman *et al.*, 1990) have persisted in putting forth information that is not in keeping with most other scientific evidence.

One reason for the erroneous conclusion that chrysotile does not cause mesothelioma may be that tremolite is more readily found in lung tissue than chrysotile (Churg *et al.*, 1984). Although tissue analysis has been useful for some issues, given the differences between amphibole and chrysotile persistence *in vivo*, concluding that fibres remaining caused disease is fraught with difficulty. A detailed analysis of the issue of the amphibole hypothesis has recently been published by Stayner *et al.* (1996), with the conclusion that dismissing chrysotile as a cause of mesothelioma and other asbestos-related diseases is inappropriate.

The study reported here was undertaken to address whether a statement could be made regarding the ability of chrysotile, considered solely as chrysotile, to produce mesotheliomas.

The most widely used standardized preparations of asbestos are those of the International Union Against Cancer. Timbrell and Rendall (1971/2) reviewed their preparation and Timbrell (1970) has published an assessment of their characterisation. There are five samples, one each of crocidolite, amosite and anthophyllite, and two of chrysotile, one from Zimbabwe (then Rhodesia) called chrysotile A and the other a mixture from eight Canadian mines, called chrysotile B.

These reference samples have long been used for *in vivo* and *in vitro* experiments. Among the relevant *in vivo* experiments that are central to this paper is an inhalation experiment of Wagner *et al.* (1974) which produced lung cancers and mesotheliomas and one reporting mesotheliomas following intrapleural inoculation (Wagner *et al.*, 1973).

It is this finding of mesothelioma after UICC Chrysotile B inhalation or inoculation that is of relevance to this paper. We have not been able to locate any specific report of tremolite contamination of this material and we are not aware of a systematic search for such tremolite. The present report documents the findings of such a detailed evaluation of this specific specimen.

MATERIALS AND METHODS

Aliquots of UICC Chrysotile B obtained from the PRU Johannesburg were used. A representative sample was obtained by combining ten separate subsamples. The combined sample was reduced in size by chopping in filtered (0.1 μm pore) deionized water. The homogeneous suspension was collected on a polycarbonate filter with 0.1 μm pores. The filter was dried, carbon coated and extraction replicas were prepared on 200 mesh grids for electron microscopy.

The replicas were examined at 330 000× (33 000× direct magnification × 10× optical magnification) in a JOEL 100CX analytical electron microscope. The analysis consisted of three phases.

First, 500 randomly selected fibres were identified as chrysotile by morphology and electron diffraction (ED). X-ray energy dispersive spectroscopy (XEDS) was used for additional confirmation. Next, 10 072 additional consecutive fibres were examined for morphology during a linear scan across 81 total grid squares on three grids. The electron beam intensity and diameter were limited to permit observation of the typical tubular morphology and of any subsequent characteristic beam sensitivity. All fibres were determined to be chrysotile. ED and XEDS were used to resolve any questionable morphology.

During the third phase of electron microscopic analysis 50 fibres, purposefully selected at lower magnification because of their "amphibole-like" shape, were sought out; all gave chrysotile electron diffraction patterns.

Polarized light microscopy techniques were used to examine an additional 10 mg aliquot. A Leitz Laborator Lux 12 POL microscope with a 100 watt light source was used. Dispersion staining was performed at 100 and 150×, while extinction angle, sign of elongation and bright field microscopy were done at 100–400×.

Extinction angles were measured with a rotary stage scale and an ocular graticule

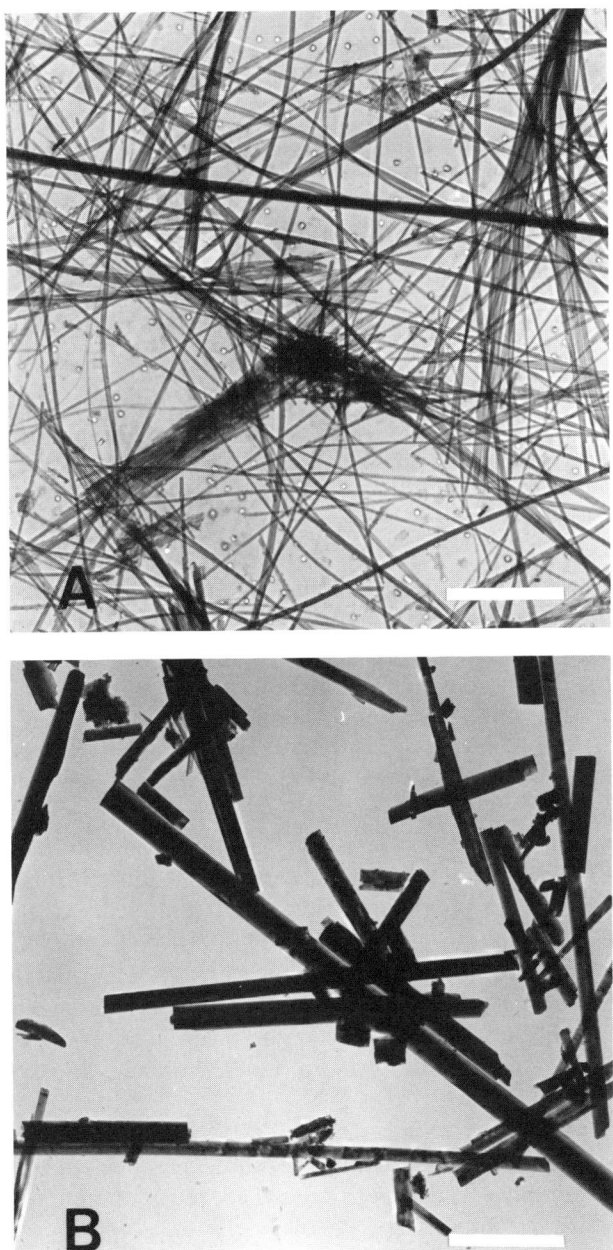

Fig. 1. Transmission electron photomicrographs of (A) UICC Chrysotile B and (B) tremolite asbestos (McCrone). (Marker equals 2 μm; instrument magnification 5000×).

calibrated with nylon (bright) and anthophyllite asbestos fibres. The stated refractive indices of the oils (1.550 and 1.605, Cargile) were confirmed with a Fisher refractometer. Room temperature was maintained at 25°C ± 1°C.

NIST SRM 1867 Tremolite was used as a standard material. Fibres that had a high possibility of being tremolite based upon one or another of the battery of observations were individually eliminated based upon the complete array of tests run with polarizing light microscopy. The tests included refractive index, type of extinction (undulose), sign of elongation, morphology and colour.

Because of the known presence of trace amounts (0.01–0.1%) of calcium (Timbrell, 1970) further testing was done. A solution of 10% hydrochloric acid was placed on an aliquot of the standard and off-gassing was noted. This would lead to the conclusion that calcium carbonate might be present.

RESULTS

No tremolite fibres were identified in the UICC Chrysotile B standard among more than 10 000 fibres evaluated by transmission electron microscopy or in a 10 mg aliquot evaluated by polarizing light microscopy techniques.

DISCUSSION

These results, using the same material that has produced mesotheliomas, lead to the conclusion that chrysotile was uncontaminated by tremolite and chrysotile is capable of producing mesotheliomas.

As noted above, no systematic examination of this material related to possible tremolite content was found in the published literature. Wagner (1986) alluded to work to be undertaken by Pooley in a paragraph that began with the words "Our view that chrysotile, if uncontaminated, is probably a material causing little disease, . . ." has to our knowledge never been reported in print.

Other chrysotile laboratory test materials have not been reported as containing tremolite. Specifically, tests of two NIEHS chrysotile materials from California and Quebec did not report the presence of tremolite (Campbell *et al.*, 1980).

With the use made of the "amphibole hypothesis" to minimize the health implications of exposure to chrysotile, it is curious that this finding, to our knowledge, has not been previously reported. The confounding caused by such, now unsubstantiated, statements have been significant in terms of public health and public policy. The health of workers and those environmentally exposed has been potentially diminished by these views.

There are two policy implications regarding this finding. The first is public policy regarding asbestos in public buildings. Chrysotile has been the major form of asbestos used in buildings and should be considered as a potentially significant public health hazard in that setting (Nicholson, 1991).

Similarly, the second issue also depends on an understanding of the hazards of chrysotile. There has been an effort to move asbestos use from the developed world to developing countries (Frank, 1993). In addition, one should note the concomitant issue of the increased efforts to sell tobacco products in these same settings and its health implications.

The lack of tremolite in one of the best characterised and most widely used chrysotile standards does not support the amphibole hypothesis. The hazards of chrysotile must be fully recognised, and it should not be considered as "a material

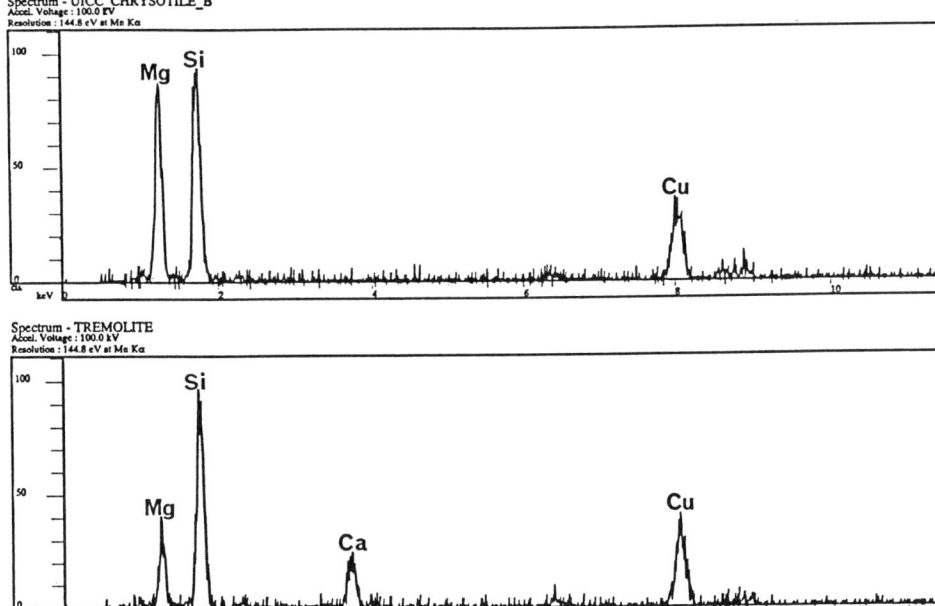

Fig. 2. X-ray energy dispersive spectra of UICC Chrysotile B and tremolite asbestos (McCrone).

causing little disease" (Wagner, 1986). Until other data to the contrary are offered, this work might even be thought by some to have relevance in the continuing medical and legal matters related to asbestos.

REFERENCES

Begin, R. *et al.* (1992) Work related mesothelioma in Quebec, 1967–1990. *Am. J. ind. Med.* **22**, 531–542.

Campbell, W. J., Higgins, C. W. and Wylie, A. G. (1980) Chemical and physical characterization of amosite, chrysotile, crocidolite and nonfibrous tremolite for oral ingestion studies by the National Institute of Environmental Health Sciences. U.S. Department of the Interior, Bureau of Mines Report of Investigations, RI 8452, Washington, D.C.

Case, B. W. (1991) Health effects of tremolite. *Ann. N.Y. Acad. Sci.* **643**, 491–504.

Churg, A. *et al.* (1984) Lung asbestos content in chrysotile workers with mesothelioma. *Am. Rev. Respir. Dis.* **130**, 1042–1045.

Dunnigan, J. (1988) Linking chrysotile asbestos with mesothelioma. *Am. J. ind. Med.* **14**, 205–209.

Frank, A. L. (1993) Global problems from exposure to asbestos. *Env. Health Persp.* **101** (Suppl. 3), 165–167.

Frank, A. L. (1994) Failure of prevention in the workplace and the added outrage of "revisionist" science. In *The Identification and Control of Environmental and Occupational Diseases; Asbestos and Cancers* (Edited by M. Mehlman and A. Upton), Vol. XXIII, pp. 549–557. Advances in Modern Environmental Toxicology. Princeton Scientific Publishing Co. Inc., Princeton.

Heuper, W. (1942) *Occupational Tumors and Allied Diseases*. C. C. Thomas, Springfield.

Karjalainen, A., Meurman, L. O. and Pukkala, E. (1994). Four cases of mesothelioma among Finnish anthophyllite miners. *Occup. Environ. Med.* **51**, 212–215.

Lynch, K. M. and Smith, W. A. (1935) Pulmonary asbestosis: Carcinoma of lung in asbestosilicosis. *Am. J. Cancer* **24**, 56–64.

McDonald, J. C. *et al.* (1989) Mesothelioma and asbestos fiber type. *Cancer* **63**, 1544–1547.

Mancuso, T. F. (1988) Relative risk of mesothelioma among railroad machinists exposed to chrysotile. *Am. J. ind. Med.* **13**, 639–657.

Mossman, B. T. and Gee, J. B. L. (1989) Asbestos-related diseases. *New Engl. J. Med.* **26**, 1721–1730.

Mossman, B. T. *et al.* (1990) Asbestos:scientific developments and implications for public policy. *Science* **247**, 294–301.

Nicholson, W. J. (1991) Comparative dose–response relationship of asbestos fiber type: magnitudes and uncertainties. *Ann. N.Y. Acad. Sci.* **643**, 74–84.

Stayner, L. T., Dankovic, D. A. and Lemen, R. A. (1996) Occupational exposure to chrysotile asbestos and cancer risk: a review of the amphibole hypothesis. *Am. J. Pub. Health* **86**, 179–186.

Timbrell, V. (1970) Characteristics of the international union against cancer standard reference samples of asbestos. In *Pneumoconiosis. Proceedings of the International Conference* (Edited by H. A. Shapiro), Johannesburg, 1969, pp. 28·36. Oxford University Press, Capetown.

Timbrell, V. and Rendall, R. E. G. (1971/2) Preparation of the UICC standard reference samples of asbestos. *Powder Tech.* **5**, 279–287.

Wagner, J. C., Berry, G. and Timbrell, V. (1973) Mesotheliomata in rats after inoculation with asbestos and other materials. *Br. J. Cancer* **28**, 173–175.

Wagner, J. C. *et al.* (1974) The effects of inhalation of asbestos in rats. *Br. J. Cancer* **29**, 252–269.

Wagner, J. C. (1986) Mesothelioma and mineral fibers. *Cancer* **57**, 1905–1911.

 Pergamon

Ann. occup. Hyg., Vol. 41, Supplement 1, pp. 293–297, 1997
© 1997 British Occupational Hygiene Society
Published by Elsevier Science Ltd. All rights reserved
Printed in Great Britain
0003–4878/97 $17.00 + 0.00
Inhaled Particles VIII

PII: S0003–4878(96)00122–6

PULMONARY HYPERPLASTIC AND NEOPLASTIC CHANGES IN RATS TREATED CONCURRENTLY WITH CHRYSOTILE ASBESTOS AND *N*-NITROSOHEPTAMETHYLENEIMINE (NHMI)

P. T. C. Harrison,* J. A. Hoskins,† R. C. Brown,‡ P. M. Hext§
and G. H. Pigott¶

*MRC Institute for Environment and Health, University of Leicester, 94 Regent Road, Leicester LE1 7DD, U.K.; †MRC Toxicology Unit, University of Leicester, P.O. Box 138, Leicester LE1 9HN, U.K.; ‡4 Bramble Close, Uppingham, Rutland LE15 9PH, U.K.; §Zeneca CTL, Alderley Park, Macclesfield, Cheshire SK10 4TJ, U.K.; and ¶27 Windermere Drive, Alderley Edge, Cheshire SK9 7UP, U.K.

INTRODUCTION

Asbestos and smoking

Exposure to asbestos can result in pulmonary fibrosis, mesothelioma and lung cancer. The amphiboles are generally considered the most dangerous forms of asbestos, especially with respect to mesothelioma induction. For mesothelioma, asbestos acts as if it were a complete carcinogen, whereas for lung cancer it is more generally regarded as a tumour promoter, resulting in enhancement of the tumorigenicity of other agents. Thus, it has convincingly been demonstrated that smoking and asbestos exposure act synergistically in the production of lung cancer (e.g. Huilan and Zhiming, 1993). Since nearly half the working population are still smokers any interaction of tobacco smoke with fibres will be the cause of considerable pathology. While various studies have provided information regarding possible mechanisms, experimental corroboration of the smoking–asbestos interaction, and of the promoting activity of asbestos more generally, has proved somewhat elusive.

Experimental models

In the absence of a widely accepted animal model for the effects of smoking, many studies (e.g. Miller *et al.*, 1965; Topping and Nettesheim, 1980) have used a pure chemical (usually a proven lung carcinogen) in place of smoke or smoke condensate, and this has been combined with treatment with asbestos, usually chrysotile. Because such combination experiments are complex and lengthy (and can be prohibitively expensive), intratracheal instillation (ITI) is often chosen as the method of delivery of the asbestos to the lungs. In a previous experiment (Harrison and Heath, 1988) this technique was used in a study of chrysotile and NHMI interactions. However, a major disadvantage of ITI is the propensity for uneven distribution of material in the lung, resulting in local "overload" and parenchymal damage. Experiments using ITI are thus open to the criticism that reported effects may be at least partly due to the non-physiological mode of fibre delivery.

A number of *in vitro* studies have demonstrated phenomena that may be indicative of synergistic interactions between asbestos and other carcinogens. For example, asbestos has been shown in several laboratories to augment the activities of carcinogens such as benzo(a)pyrene (Brown *et al.*, 1983; Eastman *et al.*, 1983), and synergism has also been reported between asbestos and ionising radiation (Hei *et al.*, 1984, 1985), using end points such as transformation of cultured cell lines. A relationship between fibre dimensions and synergism with chemical carcinogens has been shown by Brown *et al.* (1983) who found that milled crocidolite does not augment the activity of benzo(a)pyrene in causing the *in vitro* transformation of C3H10T1/2 cells.

Objectives

The primary aim of this study was to investigate whether early hyperplastic changes induced in the lungs of rats by NHMI [a specific lung carcinogen (Lijinsky *et al.*, 1969) particularly when delivered by the sub-cutaneous route (Taylor and Nettesheim, 1975)], were enhanced by concomitant inhalation of chrysotile asbestos. A previous study using a single dose level of NHMI and delivering the asbestos by ITI did produce an apparent synergistic interaction in the induction of tumours (Harrison and Heath, 1988, 1991). It was anticipated that this present experiment would contribute towards a clearer understanding of the findings of the previous study and would, more importantly, demonstrate a useful experimental model for testing the ability of chrysotile asbestos (and perhaps other inhaled particles) to augment or otherwise modify the tumorigenicity of known lung carcinogens.

MATERIALS AND METHODS

N-nitrosoheptamethyleneimine (NHMI) was a generous gift from Dr W. Lijinsky (The Eppley Institute for Research in Cancer, University of Nebraska College of Medicine, Omaha, NB 68105, U.S.A.). Confirmation of its purity was by gc and of its structure by ms. Solutions of NHMI in 8% ethanol in isotonic saline suitable for injection were prepared (0.12 and 0.4%). UICC Rhodesian chrysotile asbestos (Timbrell *et al.*, 1968) was used for inhalation.

Animals

Male, specific pathogen-free, Alpk:APfSD (Wistar derived), albino rats (Alderley Park, Cheshire, U.K.) were used. After allocation to groups, they were housed in cages in long-term exposure chambers during the study, i.e. prior to and during the exposure period. After the treatment phase of the study had been completed they were removed to conventional animal accommodation. Food and water *ad libitum* were allowed except during the exposure periods. Animal room temperatures were nominally $21 \pm 2°C$, the relative humidity was 40–60% and there was a 12 h light/dark cycle.

Experimental design

Male rats (from approximately 6 weeks old) were given 3 or 10 mg kg^{-1} NHMI, s.c., or vehicle only, weekly for 10 weeks and exposed for 4 weeks (6 h/day, 5 days/week), either to 50 mg m^{-3} (respirable) of chrysotile asbestos, or clean air as control, by nose-only inhalation. Animals were killed for pathological examination after 15 months.

Two days after the first injection of NHMI those rats which were also to be exposed to asbestos were restrained in polycarbonate tubes (Battelle, Geneva) and placed in ICI-designed nose only exposure chambers and exposed to an aerosol of UICC Rhodesian chrysotile asbestos generated using the Timbrell Dust Dispenser (Rank Bros, Cambs, U.K.).

Atmosphere analysis

The test atmospheres were analysed at least three times per exposure period for gravimetric concentration using 25 mm open-faced filter holders and VM-1 filters (both from Gelman Sciences). The aerodynamic size distribution of each atmosphere was determined using a Marple 296 Cascade Impactor (Schaeffer Instruments, Wantage, Oxon, U.K.) at least once per exposure period.

The mean aerodynamic equivalent diameter (by mass), the D_{50}, was calculated as was that percentage of the aerosol which is considered respirable (i.e. capable of entering the non-ciliated regions of the respiratory tract). At least one sample/week was taken from each atmosphere for fibre counting.

Pathology

At the end of the post-exposure period animals were anaesthetised by exposure to halothane BP vapour (FLUOTHANE, ICI Pharmaceuticals, Macclesfield, U.K.) and killed by exsanguination using cardiac puncture. A full post-mortem examination was carried out on each animal. Larynx, trachea, lungs and any visible abnormalities were preserved in ethanol–acetic acid–formol saline (Harrison, 1984) for subsequent histological examination. The head, for nasal cavity examination, was stored in 10% buffered formol saline.

The larynx, trachea, lungs (all lobes) and any abnormalities were embedded in paraffin wax. Five micron sections were cut, stained with haematoxylin and eosin and Masson's Trichrome stain and examined by light microscopy.

RESULTS

Originally the study was planned to continue until the rats were at least 18 months old but an outbreak of Kilham virus in the animal house 15 months after the start of the experiment prevented this. Although some of the animals in the experiment were infected this was considered not to compromise the objectives of this study. However, because of the infective nature of the virus veterinary advice was that the experiment should be terminated.

Scoring the lung sections for hyperplastic/metaplastic and neoplastic changes showed there were clear trends in incidences of hyperplastic lesions in the lung, with a dose–response relationship for NHMI treatment and, in each case, an augmentation by chrysotile exposure. Type II cell hyperplasic, alveolar epithelialisation, focal bronchiolo-alveolar cell hyperplasia and adenomatoid hyperplasia were present to a greater extent in the groups treated with both NHMI and chrysotile than in the controls or animals treated with NHMI alone. Furthermore, apart from malignant lymphoma, the only lung tumours detected (adenoma and

Table 1. Epithelial changes and neoplasms in the lungs of male rats

Group no.	1	3	4	5	7	8
Lesion						
Type II cell hyperplasia	1	5	9	16	18	19
Alveolar epithelialisation	1	3	4	7	10	15
Focal bronchiolo-alveolar cell hyperplasia	0	9	10	9	15	15
Adenomatoid hyperplasia	0	2	3	1	4	7
Adenoma	0	0	0	0	1	4
Carcinoma	0	0	0	0	1	0
Malignant lymphoma	0	1	0	1	1	0
$N =$	15	20	20	20	20	20

1 Vehicle only	No chrysotile.
3 3 mg kg^{-1} NHMI	No chrysotile.
4 10 mg kg^{-1} NHMI	No chrysotile.
5 Vehicle only	Chrysotile.
7 3 mg kg^{-1} NHMI	Chrysotile.
8 10 mg kg^{-1} NHMI	Chrysotile.

carcinoma) were in animals treated with both NHMI and asbestos. For adenoma there was a clear dose response to NHMI treatment (Table 1).

DISCUSSION

This study has demonstrated that hyperplastic changes induced by NHMI are enhanced by chrysotile exposure, and confirms a "promoting" effect for chrysotile asbestos in the induction of lung tumours. The wider implications of the study relate to the methodology developed. A short period of asbestos inhalation exposure was sufficient to produce characteristic pathological lesions and to provide the opportunity for the demonstration of chrysotile's tumour promoting activities, whilst NHMI produced hyperplastic changes which were clearly susceptible to enhancement by combined treatment with asbestos. This was achieved despite the enforced curtailment of the study at 15 months. It is suggested that this method could be utilised to investigate, over a relatively short experimental period, either: (i) the ability of other inhaled particles to augment NHMI carcinogenicity; or (ii) the propensity of other chemical carcinogens (including tobacco smoke or smoke condensate) to interact with asbestos in the production of pulmonary neoplasms.

REFERENCES

Brown, R. C., Poole, A. and Gleming, G. T. A. (1983) The influence of asbestos dust on the oncogenic transformation of C3H10T1/2 cells. *Cancer Lett.* **18**, 221–227.
Eastman, A., Mossman, B. T. and Bresnick, E. (1982) Influence of asbestos on the uptake of benzo(a)pyrene and DNA alkylation in hamster tracheal epithelial cells. *Cancer Res.* **43**, 1251–1254.
Harrison, P. T. C. (1984) An ethanol–acetic acid–formol saline fixative for routine use with special application to the fixation of non-perfused rat lung. *Lab. Animals* **18**, 325–331.
Harrison, P. T. C. and Heath, J. C. (1988) Apparent synergy between chrysotile asbestos and N-nitrosoheptamethyleneimine in the induction of pulmonary tumours in rats. *Carcinogenesis* **9**, 2165–2171.
Harrison, P. T. C. and Heath, J. C. (1991) Apparent promotion by chrysotile asbestos of NHMI-initiated lung tumours in the rat. In *Mechanisms in Fibre Carcinogenesis* (Edited by R. C. Brown, J. A. Hoskins and N. F. Johnson), pp. 469–479. Plenum Press, New York.

Hei, T. K., Geard, C. R., Osmak, R. S. and Travisano, M. (1985) Correlation of *in vitro* genotoxicity and oncogenicity induced by radiation and asbestos fibres. *Br. J. Cancer* **52**, 591–597.

Hei, T. K., Hall, E. J. and Osmak, R. S. (1984) Asbestos radiation and oncogenic transformation. *Br. J. Cancer* **50**, 717–721.

Huilan, Z. and Zhiming, W. (1993) Study of occupational lung cancer in asbestos factories in China. *Br. J. Indust. Med.* **50**, 1039–1042.

Lijinsky, W., Tomatis, L. and Weynon, C. E. M. (1969) Lung tumours in rats treated with N-nitroso-heptamethyleneimine and N-nitrosooctamethyleneimine. *Proc. Soc. Exp. Biol. Med. (NY)* **130**, 945–949.

Miller, L., Smith, W. E. and Berliner, S. W. (1965) Tests for the effects of asbestos on benzo(a)pyrene carcinogenesis of the respiratory tract. *Ann. NY Acad. Sci.* **132**, 489–500.

Taylor, H. W. and Nettesheim, P. (1975) Influence of administration route and dosage schedule on tumour response to nitrosoheptamethyleneimine in rats. *Int. J. Cancer* **15**, 301–307.

Timbrell, V., Gilson, J. C., Webster, I. (1968) UICC standard reference samples of asbestos. *Int. J. Cancer* **3**, 406–408.

Topping, D. C. and Nettesheim, P. (1980) Two stage carcinogenesis studies with asbestos in Fischer 344 rats. *J. Nat. Cancer Inst.* **65**, 627–630.

 Pergamon

Ann. occup. Hyg., Vol. 41, Supplement 1, pp. 298–303, 1997
© 1997 British Occupational Hygiene Society
Published by Elsevier Science Ltd. All rights reserved
Printed in Great Britain
0003–4878/97 $17.00 + 0.00
Inhaled Particles VIII

PII: S0003–4878(96)00123–8

BIOPERSISTENCE OF VARIOUS TYPES OF MINERAL FIBRES IN THE RAT LUNG AFTER INTRATRACHEAL APPLICATION

B. Bellmann and H. Muhle

Fraunhofer Institute of Toxicology and Aerosol Research, Nikolai-Fuchs-Str. 1, D-30625, Hannover, Germany

INTRODUCTION

From present knowledge the biopersistence of fibres is a critical factor for the possible cancerogenic potency of these materials. To test the relationships between chemical composition, biopersistence and cancerogenicity, in the present study fibre materials were used which preferably had been already tested for carcinogenic potency in an intraperitoneal test or an inhalation study.

MATERIALS AND METHODS

These types of fibres were tested: three glass wools MMVF11, TL wool, X607 (experimental wool); M-Stone wool, M-Slag wool, ceramic fibre RCF1, glass fibre B-01 and polypropylene fibres. The chemical compositions are given in Table 1. Respirable fractions of these materials were used. The bivariate size distribution of about 400 fibres per sample was analysed by scanning electron microscopy (SEM). For this the length and the diameter of the fibres was measured on photos or videoprints using a computer with a digitizing tablet. Different magnifications were chosen to measure the thin and short fibres and the long and thick fibres. To avoid double counting, fibre length limits were set for the count at each magnification. The results are shown in Table 2.

Female Wistar rats strain Han:WIST were exposed by intratracheal instillation of up to 2 mg of the test materials (see Table 2) suspended in 0.3 ml saline. After serial sacrifice of rats up to 18 months after treatment, the fibres were analysed in the lungs after removal of the lung tissue by drying and plasma ashing. For the polypropylene-treated rats a combination of an alkaline and an acid hydrolysis was used to dissolve the lung tissue (Muhle *et al.*, 1990a). By SEM about 200 fibres per rat were counted and the size distribution was analysed.

RESULTS

The calculated number of fibres retained in the lungs is shown in Table 3. Half-times were calculated as characteristic parameter for the kinetics of the elimination of fibres. In Table 4 the half-times calculated for the number of WHO fibres, of fibres with length > 20 μm, and for the mass of retained fibres is shown.

Table 1. Chemical composition of test materials

Material	K_I	CaO	MgO	Al_2O_3	TiO_2	FeO_x	B_2O_3	SiO_2	Na_2O	K_2O	ZnO	BaO
M-475[a]	20.1	3.0	—	5.8	—	0.1	10.7	57.9	10.1	2.9	3.9	5.0
M-E[a]	2.3	17.4	4.7	14.1	0.5	0.3	8.0	54.4	0.4	—	—	—
MMVF11[b]	23.5	7.4	2.8	3.9	0.1	0.2	4.4	63.4	15.4	1.3	—	0.0
TL[c]	25.3	7.0	2.9	3.1	0.1	0.3	4.7	64.9	15.3	1.5	—	0.1
X607[b]	38.5	38.3	0.6	0.2	—	0.1	—	59.6	0.0	—	—	—
B-01[a]	37.4	16.5	3.2	—	—	—	3.3	61.0	15.4	—	—	—
M-Stone[a]	18.8	24.7	11.4	9.9	2.0	8.0	—	38.4	1.6	0.8	—	0.1
M-Slag[a]	27.6	37.4	10.0	10.3	0.5	0.3	—	38.2	0.4	0.4	—	—
RCF1[b]	~95.1	0.1	0.1	48.0	2.0	1.0	—	47.7	0.5	0.2	—	—

Chemical composition (%)

[a] From Professor Pott, chemical analysis and sizing by Manville Technical Centre and by Bayer (B-01).
[b] From TIMA, sizing and chemical analysis by Manville Technical Centre.
[c] From market, chemical analysis by Grünzweig und Hartmann, sizing at ITA.

Table 2. Size distribution of test materials [percentiles of cumulative frequencies of fibres (L/D ≥ 3)]

Material	Dose [mg]	Fibre length [μm]				Fibre diameter [μm]			
		10%<	50%<	90%<	99%<	10%<	50%<	90%<	99%<
M-475/104	2.0	1.3	2.6	8.1	27.5	0.09	0.14	0.32	0.86
M-E/104	2.0	0.5	2.3	8.8	27.8	0.12	0.17	0.51	0.98
MMVF11	2.0	4.6	13.5	38.2	99.0	0.26	0.94	2.05	4.51
TL-Wool	2.0	1.8	6.5	27.0	76.4	0.21	0.51	1.44	3.00
X607	2.0	5.6	21.8	63.8	153.8	0.38	1.50	3.75	6.25
M-Stonewool	1.0	3.6	8.8	25.1	75.1	0.39	0.84	1.68	2.52
M-Slagwool	1.0	3.3	7.8	24.3	68.7	0.38	0.77	1.35	2.31
RCF 1	2.0	3.7	8.8	34.3	104.6	0.26	0.85	1.87	3.32
B-01/0,9	0.35	2.8	8.2	32.7	104.3	0.32	0.70	1.18	2.35
Polypropylene	0.2	5.0	12.4	40.3	115.6	0.26	0.51	1.36	2.64

The half-times for the elimination of fibres with length > 20 μm were significantly lower for all fibre samples besides the ceramic fibres RCF1 and polypropylene fibres.

DISCUSSION

In Table 5 half-times for the clearance of WHO fibres calculated from the results of various inhalation studies are compared to intratracheal instillation. Relatively good agreement was found between chronic inhalation and intratracheal instillation for similar fibre types, whereas the half-times in studies using short term inhalation are usually significantly shorter. One important point for this different behaviour is the lower lung burden after short term inhalation of 5 days. After chronic inhalation and intratracheal instillation the lung burden was above 1 mg fibres per lung. For this lung burden a retardation of the macrophage-mediated clearance can be assumed from studies with insoluble particles (Muhle et al., 1990b). At low lung burden in the rat for the macrophage-mediated clearance of spherical particles half-times of about 50–70 were detected. Therefore, half-times of 50–70 days after short-term inhalation of fibres can be explained by macrophage-mediated clearance without dissolution of fibres.

In humans the half-time of the macrophage-mediated clearance is in the range of 300–500 days. Therefore the retardation of the macrophage-mediated clearance by a lung burden of about 2 mg in the rat simulates the behaviour of fibres in the human lung where the dissolution of fibres is the most important clearance process.

For the analysed fibre materials the order of decreasing biopersistence is identical to the order of decreasing carcinogenic potency determined by intraperitoneal injection tests (Roller et al., 1996): ceramic fibres (RCF) > glass wool (MMVF11, TL) > M-Stone wool > M-Slag wool > special glass wool (X607, B-01/09).

Acknowledgement—This study was funded by the German "Bundesantalt für Arbeitsschutz", Dortmund.

Table 3. Analysis of fibres in the lung ash (Means and standard deviation of five rats per date)

Material	Sacrifice	Anal. fibres Mean	Anal. fibres S.D.	WHO-fibres [10^6/lung] Mean	WHO-fibres [10^6/lung] S.D.	Fibres (L>20μm) [10^6/lung] Mean	Fibres (L>20μm) [10^6/lung] S.D.	Mass of fibres [mg] Mean	Mass of fibres [mg] S.D.
M-475/104	2 Days	170	7	291.6	32.9	22.14	6.29	3.71	0.63
	1 Month	128	12	161.9	35.2	11.95	4.41	1.75	1.00
	3 Months	113	16	132.7	55.7	7.32	2.77	1.21	0.43
	6 Months	130	20	62.3	15.8	4.07	0.68	0.72	0.23
	12 Months	113	15	54.0	24.6	4.01	1.94	0.71	0.36
	18 Months	105	28	28.9	9.0	2.86	1.42	0.37	0.11
M-E/104	2 Days	150	11	290.4	41.4	23.14	6.83	1.73	0.71
	1 Month	179	15	248.4	26.0	15.18	1.79	1.18	0.14
	3 Months	199	12	126.1	26.5	7.99	1.36	0.80	0.16
	6 Months	152	17	112.3	32.0	7.33	0.73	0.54	0.08
	12 Months	192	7	67.4	26.0	4.97	0.33	0.33	0.06
	18 Months	110	20	52.0	22.8	3.48	1.35	0.27	0.09
MMVF11	2 Days	204	6	22.8	2.1	4.99	1.14	2.05	0.51
	1 Month	199	4	15.0	3.2	2.82	1.09	1.19	0.33
	3 Months	204	3	12.1	1.3	1.52	0.24	0.73	0.15
	6 Months	196	4	7.0	1.8	0.53	0.18	0.36	0.05
	12 Months	195	5	5.5	0.9	0.28	0.16	0.23	0.03
	18 Months	179	29	2.9	1.2	0.14	0.06	0.08	0.04
TL-Wool	2 Days	200	12	18.9	4.7	3.93	1.02	1.98	0.49
	1 Month	192	27	17.3	2.6	2.53	0.62	1.53	0.28
	3 Months	194	9	15.2	1.6	0.73	0.10	1.03	0.27
	6 Months	140	26	7.2	0.6	0.16	0.03	0.37	0.04
	12 Months	89	21	6.1	2.2	0.11	0.06	0.21	0.06
	18 Months	101	35	2.6	1.0	0.00	0.00	0.07	0.02
X607	2 Days	209	10	5.7	1.0	1.41	0.14	1.58	0.32
	14 Days	188	16	2.7	0.3	0.50	0.09	0.66	0.22
	1 Month	189	19	2.8	0.2	0.48	0.16	0.49	0.10
	2 Months	204	4	1.5	0.3	0.20	0.03	0.22	0.05
	3 Months	134	36	0.9	0.2	0.10	0.03	0.19	0.03
	6 Months	50	35	0.3	0.2	0.04	0.03	0.09	0.06
M-Stonewool	2 Days	201	5	8.1	0.9	2.12	0.20	1.04	0.18
	1 Month	204	5	6.4	0.6	1.47	0.21	0.78	0.18
	3 Months	205	4	5.2	0.4	0.88	0.19	0.48	0.13
	6 Months	203	4	3.0	0.5	0.53	0.23	0.26	0.07
	12 Months	115	42	0.8	0.2	0.06	0.04	0.06	0.01
	18 Months	65	10	0.4	0.0	0.03	0.02	0.04	0.01
M-Slagwool	2 Days	204	12	16.8	1.1	3.37	0.76	0.91	0.09
	1 Month	205	11	13.1	2.3	1.10	0.60	0.40	0.07
	3 Months	205	5	8.2	1.2	0.74	0.48	0.17	0.02
	6 Months	192	34	2.3	0.5	0.18	0.06	0.03	0.00
	12 Months	25	9	0.4	0.2	0.01	0.01	0.00	0.00
	18 Months	26	8	0.2	0.0	0.01	0.01	0.00	0.00
RCF 1	2 Days	200	11	14.9	1.4	4.47	0.61	1.70	0.24
	1 Month	196	5	11.6	1.8	3.13	0.45	1.09	0.20
	3 Months	195	3	11.9	0.9	3.53	0.45	1.11	0.15
	6 Months	197	4	9.2	1.2	2.94	0.56	0.91	0.13
	12 Months	198	6	7.4	0.6	1.89	0.27	0.56	0.10
	18 Months	196	2	4.5	1.7	1.15	0.40	0.36	0.08
B-01/0,9	2 Days	208	7	4.8	0.8	0.27	0.04	0.088	0.033
	1 Month	85	40	1.0	0.5	0.01	0.01	0.012	0.006
	3 Months	15	4	0.2	0.1	0.00	0.00	0.003	0.001
	6 Months	5	2	0.1	0.1	0.00	0.00	0.001	0.001
	12 Months	1	1	0.0	0.0	0.00	0.00	0.000	0.000
	18 Months	2	1	0.0	0.0	0.00	0.00	0.001	0.002

Table 4. Half-time and confidence limit (C.L.) of the elimination of fibres (L/D ≥ 3)

	Half-time in days calculated from		
Material	Number of WHO-Fibres Mean (95% C.L.)	Number of fibres (L > 20 μm) Mean (95% C.L.)	Mass of fibres Mean (95% C.L.)
M-475/104	183 (151–230)	206 (156–301)	197 (151–281)
M-E/104	218 (180–277)	226 (188–281)	216 (183–264)
MMVF11	199 (172–235)	107 (93–126)	126 (113–144)
TL-Wool	188 (163–220)	66 (55–81)	113 (104–125)
X607	46 (40–54)	39 (33–46)	49 (40–63)
B-01/0,9[a]	32 (26–45)	9 (7–11)	29 (24–37)
M-Stonewool	116 (108–126)	82 (75–92)	106 (96–118)
M-Slagwool	81 (75–89)	66 (56–81)	60 (54–69)
RCF 1	343 (291–416)	300 (256–363)	270 (235–317)
Polypropylene	229 (142–590)	362 (90–∞)	∞ (1858–∞)

[a] Calculation only for sacrifice dates up to 6 months (number of fibres for later dates was too low).

Table 5. Comparison of retention kinetics of fibres for different types of application

Type	Material	Half-time in days (95% confidence limit) calculated from		
		Short-term inhalation	Chronic inhalation	Intratracheal instillation
Asbestos	Crocidolite	155[a2] 608[a1]		695 (354–∞)[b]
Glasswool	MMVF 10	77[a2] 148[a1]	323[c]	
	MMVF 11	50[a2] 127[a1] 42 (37–47)[d]	249[c]	199 (172–235)
	TL		182 (56–∞)[e]	188 (163–220)
Stonewool	MMVF 21	77[a2] 107[a1] 86[g]		291 (221–406)[b]
	SG		203 (82–∞)[e]	
	M			116 (108–126)
Slagwool	MMVF 22	55[a2] 73[a1]		
	M			81 (75–89)
Ceramicwool	RCF 1	111[g]	141[c] 316 (24–∞)[f] 177 (93–1984)[f]	343 (291–416)

[a1] Musselman et al., 1994 (calculation day 1–545).
[a2] Musselman et al., 1994 (calculation day 1–180).
[b] Bellmann et al., 1994, 1995.
[c] Hesterberg et al., 1993 (365 days exposure, 365 days clearance).
[d] Bernstein, 1995 (calculation day 1–180).
[e] LeBouffant et al., 1987 (365 days exposure, 490 days clearance).
[f] Yamata et al., 1992, 1994.
[g] Creutzenberg et al., 1997 (3 weeks inhalation, calculation day 3–365).

REFERENCES

Bellmann, B., Muhle, H., Kamstrup, O. and Draeger, U. F. (1994) Investigation of the biodurability of man-made vitreous fibers in rat lungs. Environ. Health Perspect. 102 (Suppl. 5), 185–189.
Bellmann, B., Muhle, H., Kamstrup, O. and Draeger, U. F. (1995) Investigation on the biodurability of chemically different stone wool fibres. Exp. Toxic. Pathol. 47, 195–201.
Bernstein, D. M., Morscheidt, C., Tiesler, H., Grimm, H.-G., Thévenaz, P. and Teichert, U. (1995)

Evaluation of the biopersistence of commercial and experimental fibers following inhalation. *Inhal. Toxicol.* **7**, 1031–158.

Creutzenberg, O., Bellmann, B. and Muhle, H. (1997) Biopersistence and bronchoalveolar lavage fluid analysis after a subacute inhalation of rats to various man-made mineral fibres (this meeting).

Hesterberg, T. W., Miiller, W. C., McConnell, E. E., Chevalier, J., Hadley, J. G., Bernstein, D. M., Thevenaz, P. and Anderson, R. (1993) Chronic inhalation toxicity of size-separated glass fibers in Fischer 344 rats. *Fundam. appl. Toxicol.* **20**, 464–476.

Le Bouffant, L., Daniel, H., Henin, J. P., Martin, J. C., Normond, C., Tichoux, G. and Trolard, F. (1987) Experimental study on long-term effects of MMMF on the lung of rats. *Ann. occup. Hyg.* **31**, 765–790.

Muhle, H., Bellmann, B., Creutzenberg, O., Fuhst, R., Mohr, U., Takenaka, S., Morrow, P., Kilpper, R., MacKenzie, J. and Mermelstein, R. (1990a) Subchronic inhalation study of toner in rats. *Inhal. Toxicol.* **2**, 341–360.

Muhle, H., Bellmann, B., Creutzenberg, O., Heinrich, U., Ketkar, M. and Mermelstein, R. (1990b) Dust overloading of lungs after exposure of rats to particles of low solubility. Comparative studies. *J. Aerosol Sci.* **21**, 374–377.

Musselman, R. P., Miiller, W., Easters, W., Hadley, J., Kamstrup, O., Thevenaz, P. and Hesterberg, T. (1994) Biopersistence of crocidolite versus man-made vitreous fibers in rat lungs after brief exposures. In *Toxic and Carcinogenic Effects of Solid Particles in the Respiratory Tract*, pp. 451–454. ILSI Press.

Roller, M., Pott, F., Kamino, K., Althoff, G.-H. and Bellmann, B. (1996) Results of current intraperitoneal carcinogenicity studies with mineral and vitreous fibres. *Exp. Toxic. Pathol.* **48**, 3–12.

Yamato, H., Tanaka, I., Higashi, T. and Kido, M. (1992) Determinant factor for clearance of ceramic fibres from rat lungs. *Br. J. ind. Med.* **49**, 182–185.

Yamato, H., Tanaka, I., Higashi, T. and Kido, M. (1994) Clearance of inhaled ceramic fibres from rat lungs. *Environ. Health Perspect.* **102** (Suppl. 5), 169–171.

 Pergamon

Ann. occup. Hyg., Vol. 41, Supplement 1, pp. 304–311, 1997
© 1997 British Occupational Hygiene Society
Published by Elsevier Science Ltd. All rights reserved
Printed in Great Britain
0003–4878/97 $17.00 + 0.00
Inhaled Particles VIII

PII: S0003–4878(96)00167–6

THE RELATIONSHIP BETWEEN FOUR METHODS OF ASSESSING CLEARANCE OR TOXICITY OF SYNTHETIC MINERAL FIBRES WITH CHEMICAL COMPOSITION

J. Bignon,* D. M. Bernstein,† P. Brochard‡ and C. Morscheidt§

*Unité INSERM 139, Faculté de Médecine, Créteil, France; †CH 1208 Geneva, Switzerland;
‡Médecine du Travail, Université II, Bordeaux, France; and §St Gobain Branche Isolation,
92096, Paris la Défense, France

INTRODUCTION

Risk assessment of exposure to manmade vitreous fibres is a worldwide major concern, and has caused particular problems in the European Union, where consensus must be obtained between 15 member states. Two expert workshops have been held in Paris (Bignon, 1994; Bignon et al., 1995) and from these two primary issues have emerged: (1) the relationship between solubility and biopersistence; and (2) carcinogenic potential of MMVF in rodents. The German authorities (TRGS 905, 1994) disagree with most experts, because German experts believe that long-term inhalation tests in rodents are not sufficiently sensitive as a test of carcinogenic potential and that intraperitoneal injection (IP) tests are a safer basis for classification.

The German authorities propose that MMVF should be classified using a formula reflecting biopersistence in terms of chemical composition:

$$KI = \Sigma \ (Na,K,B,Ca,Mg,Ba\text{–oxide—}2 \times Al\text{–oxide}).$$

A fibre with $KI < 40$ will be classified as a possible carcinogen, unless an IP test at high doses (5×10^9 fibres) gives $< 10\%$ tumours.

A French proposal has been provided to DGXI of the European Commission (French Proposal, 1995) taking into account results of the IP test and weight of evidence from all available data on toxicity and biopersistence. This paper presents analysis of the data on *in vivo* biopersistence, *in vitro* dissolution, IP results and the KI index which have been presented in these two proposals.

MATERIALS AND METHOD

The data which are analysed here appeared in the French Proposal (1995). It has been updated according to the most recent reports. The *in vitro* accelular dissolution coefficients (Kdis) (Table 1) are now based on Bernstein et al. (1996). The half-time ($T_{1/2}$ in days) for WHO fibres and long fibres after inhalation (Table 2) are also based on Bernstein et al. (1996). The half time ($T_{1/2}$ in days) for WHO fibres after intratracheal instillation (Table 3) are based on TRGS 906 (1995). The dose (TD25) in critical fibres to induce 25% tumours in an intraperitoneal injection

Table 1. Kdis based on dissolution of the network former (SiO2), fine fibres, measured at pH 7.4, at SA/V of 175–190 hcm^{-1} (Table 5 French Proposal, 1995 and Bernstein et al, 1996)

Fibre type	Description	Kdis ng (cm2h)	KI
B	exp. glasswool	320	39
C	exp. glasswool	309	34
J	exp. stonewool	104	37
H	exp. stonewool	169	27.5
G	exp. stonewool	129	31.8
A	exp. glasswool	129	27.6
MMVF 11	glasswool	71	25
F	exp. stonewool	96	30
L	stonewool	23	0

Table 2. Half-time ($T_{1/2}$ in days) for WHO fibres and long fibres after inhalation (Table 4, French Proposal, 1995 completed with $T_{1/2}$ $L > 20$ and Bernstein et al, 1996)

Fibre type	Description	Median fibre size L (µm)	D (µm)	$T_{1/2}$ WHO fibres	$L > 20$	KI
B	exp. glasswool	12.0	0.5	11	2.4	39
C	exp. glasswool	12.0	0.5	14	4.1	34
J	exp. stonewool	10.0	0.5	23	9.8	37
H	exp. stonewool	13.0	0.6	26	13	27.5
G	exp. stonewool	10.5	0.6	22	5.4	31.8
A	exp. glasswool	8.8	0.5	15	3.5	27.6
MMVF 11	glasswool	13.5	0.9	28	13	25
F	exp. stonewool	9.9	0.5	28	8.5	30
L	stonewool	17.5	0.6	54	45	0

Table 3. Half-time ($T_{1/2}$ in days) for WHO fibres after intratracheal instillation (Table 3, French Proposal, 1995 and TRGS 906, 1995)

Fibre type	Median fibre size L (µm)	D (µm)	Dose mg	$T_{1/2}$ WHO	KI
B-01/0.9	7.5	0.7	0.35	32	39.1
X-607	21.8	1.5	2.0	46	38.5
M-Schlacke	7.8	0.8	1.0	81	27.6
TL-WOLLE	6.5	0.51	2.0	188	25.2
MMVF11	13.5	0.94	2.0	199	23.8
M-753	3.2	0.21	0.1	45	23.4
M-475	2.3	0.14	2.0	183	20.1
M-Stein	8.8	0.84	1.0	116	15.1
M-E	2.3	0.17	2.0	218	2.3
RCF-1	8.8	0.85	2.0	343	−95.1

test (Table 4) are also based on TRGS 906 (1995). The data was analysed by non linear regression (Statistica, 1995).

Characteristics of fibres comparatively analysed

The sizing and the KI of the fibres are given in Tables 2–4 of the French Proposal. When not indicated, the KI is calculated on the basis of the chemical composition provided in the references.

Table 4. Dose (TD25) in critical fibres necessary to induce 25% tumours in an intraperitoneal injection test (Table 2, French Proposal, 1995 and TRGS 906, 1995)

Fibre type	Description	Median fibre size L (µm)	D (µm)	TD 25 (1OE9 fibres, $L > 5$ µm) $L/D > 3$, $D < 3$ µm Linear	KI
R-stone-E3	Experimental Rockwool	15.4	1.10	2.60	45.20
B-01-0.9	Experimental Bayer	7.5	~0.7	12.50	39.00
B-09-2.0	Experimental Bayer	8.6	1.30	1.60	32.00
M-slag	Slagwool	7.8	0.80	1.00	28.00
M-stone	Stonewool Manville	8.8	0.84	0.30	15.10
MMVF-11	Glasswool Saint Gobain	13.5	0.90	0.31	24.00
B-20-2.0	Experimental Bayer	6.6	0.80	0.27	14.00
MMVF-21	Stonewool Rockwool	14.6	1.00	0.04	4.00
M-753-104	Glasswool Manville	3.2	0.22	0.21	23.80
B-20-0.6	Experimental Bayer	3.1	0.30	0.33	14.00
AI-Silicat	Ceramic fibre	5.5	0.50	0.02	−90.00
Basalt	Stone wool	17.00	1.10	0.02	0.00
M475/104	Glass wool Manville	2.30	0.10	0.45	20.00
B-02	Experimental Bayer	5.60	0.30	0.13	20.00

In vitro *acellular solubility assays (Table 1)*

The *in vitro* acellular assays are based on a comprehensive physiological approach which attempts to mimic what happens *in vivo*, with the assumption that chemical processes will run in the same manner *in vitro* as *in vivo* (TRGS 906, 1995). These *in vitro* acellular assays aim to evaluate the dissolution rates based on the kinetics of Si removal from the fibres in a fluid (Gamble's or derived solutions), at neutral pH (pH = 7.4–7.6) which is similar to the pH of the alveolar and interstitial fluids (Thélohan, 1994). These studies are conducted using a flow-through dynamic procedure.

Although a standardised protocol has yet to be agreed upon by the fibre industry, the results published by Bernstein *et al.* (1996) showed a satisfactory link at pH 7.4 with *in vivo* biopersistence studies in the rat by inhalation (Bignon, 1994). Some authors have proposed to measure *in vitro* acellular solubility at pH 4.5, in order to mimic the pH existing inside the phagolysosome of the alveolar macrophage. It was not possible however to show a correlation between these measurements and the *in vivo* biopersistence of either short or long fibres (Bernstein *et al.*, 1994, 1995). For this reason measurements at pH 4.5 were not included in the French Proposal and therefore are not analysed here.

In vivo *biopersistence studies (Tables 2 and 3)*

These studies have been conducted according to two different methods, as described in the report on the 2nd MMF Workshop of September 94 in Paris (Bignon, 1994):

- by inhalation (IH) after a short-term (5 days) exposure (Bernstein *et al.*, 1995, 1996; Bignon, 1994). During the Paris Workshop there was an international consensus regarding the protocol for conducting this test.
- by intratracheal instillation (IT) carried out according to the "Muhle" protocol (Bellmann and Muhle, 1994). This test consisted of one or several instillations, with different doses (from 0.1 to 2 mg).

Long-term animal experiments by intraperitoneal injection (IP test) (Table 4)

The intraperitoneal injection test is also described in the report on the Workshop of September 1994 in Paris, although there was no consensus on the protocol to be used for this test. The results reported in Table 4 are from IP tests carried out by Pott *et al.* (Bellman *et al.*, 1994). The dose, in number of critical fibres ($D > 2$, $L > 5$, $L/D > 5$), producing 25% tumours in this test (TD25) has been calculated by Pott and is presented in TRGS 906 (1995).

Intercomparison of data by means of a multiregression analysis

In order to assess the reality of possible interactions of well characterised mineral fibres with the biological milieu, we compared and, when possible, correlated the different biological experimental results with the physical (D, L) and chemical index (KI) characteristics of several samples of chemically well-defined MMVFs. The data was analysed using non-linear regression (Statistica, 1995).

Dependant variables were: Kdis, $T_{1/2}$ WHO IH, $T_{1/2}$ $L > 20$ IH, $T_{1/2}$ WHO IT, T D25. Independent variables were: KI, median diameter ($D50$), median length ($L50$), mass injected. The estimated percent of variance explained is given for each calculation. For the mathematical extrapolation coefficients were used only when they could be considered significant ($P < 0,05$).

RESULTS OF THE INTER COMPARISON OF DATA BY A MULTIREGRESSION ANALYSIS

Correlation between KI index and in vitro *solubility acellular index (Kdis)*

The *in vitro* solubility (Kdis in mg cm^{2}h) measured at pH 7.4 for 9 MMVFs is given in Table 1. From these data, it appears that two fibres (B and C) with a KI $\cong$ 34 and 39 have also a high Kdis (> 300 ng cm^{-2}h). These results suggest that a KI in the range of 40 might correspond to an *in vitro* dissolution rate in the range of 500 ng cm^{-2}h; conversely, lower KI might correspond to less soluble fibres (Table 1). However, exceptions are possible, which need further investigation.

Such exceptions could be due either to a specific chemical composition or to an inadequate choice of the different parameters selected for the acellular *in vitro* assay, such as flow rate, pH or chemical composition of the test fluid. If we take the data obtained according to the Scholze and Conradt method (Thélohan, 1994) it appears that, for the nine MMVFs investigated, there was generally an excellent correspondence between KI and Kdis (Fig. 1).

Correlation between KI and in vivo *biopersistence after inhalation (IH)*

Based on animal experiments carried out by Bernstein (1996) (Table 2), the correlation between *in vivo* biopersistence ($T_{1/2}$) for WHO fibres and the AGS chemical index (KI) was acceptable (variance explained: 83%). At present, with the MMVFs investigated, the chemical index KI can be used as a predictor of the approximate outcome of *in vivo* biopersistence measurements (Fig. 2). When using $T_{1/2}$ calculated on fibres longer than 20 μm which are also given in Bernstein *et al.* (1996) the correlation is slightly improved (variance explained: 94%) (Fig. 3). Preliminary results of new biopersistence measurements after inhalation show that fibres with KI $\geq$ 40 are indeed very "biosoluble". Preliminary results with fibres of

K_{dis} as a function of KI

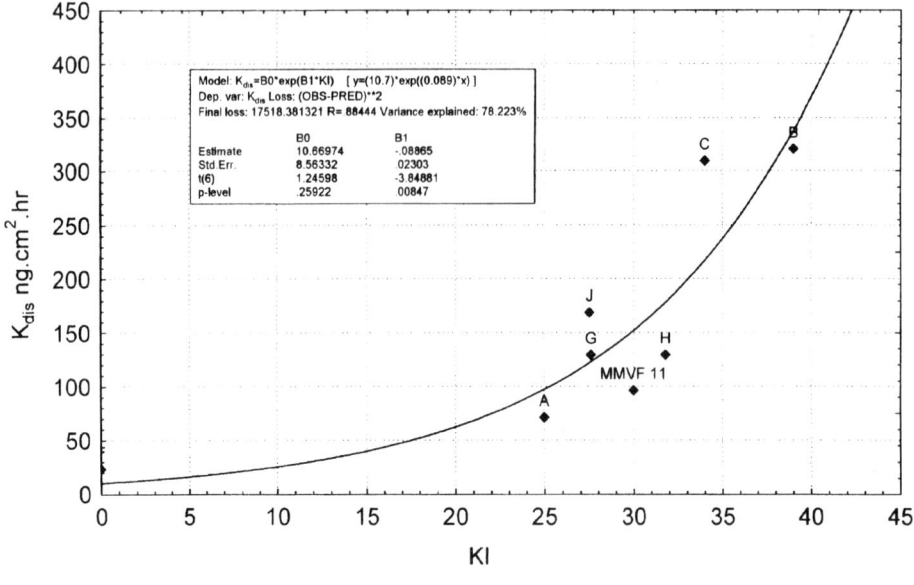

Fig. 1.

low solubility or a low KI are coherent with the results presented by Bernstein (1996) and the numbers calculated here. Presently, there is one exception to the calculated correlation between KI and *in vivo* biopersistence after inhalation: preliminary results of *in vivo* biosolubility of the fibre (HT) with a negative KI of −40 showed a solubility equivalent to a fibre with a KI = +30. Comparison of these

T_{1/2} WHO Fibers versus KI

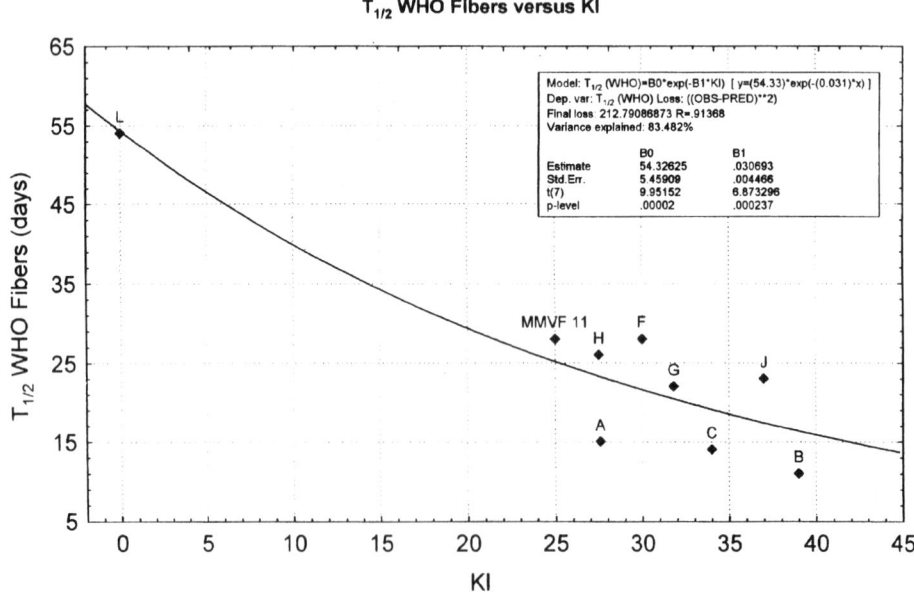

Fig. 2.

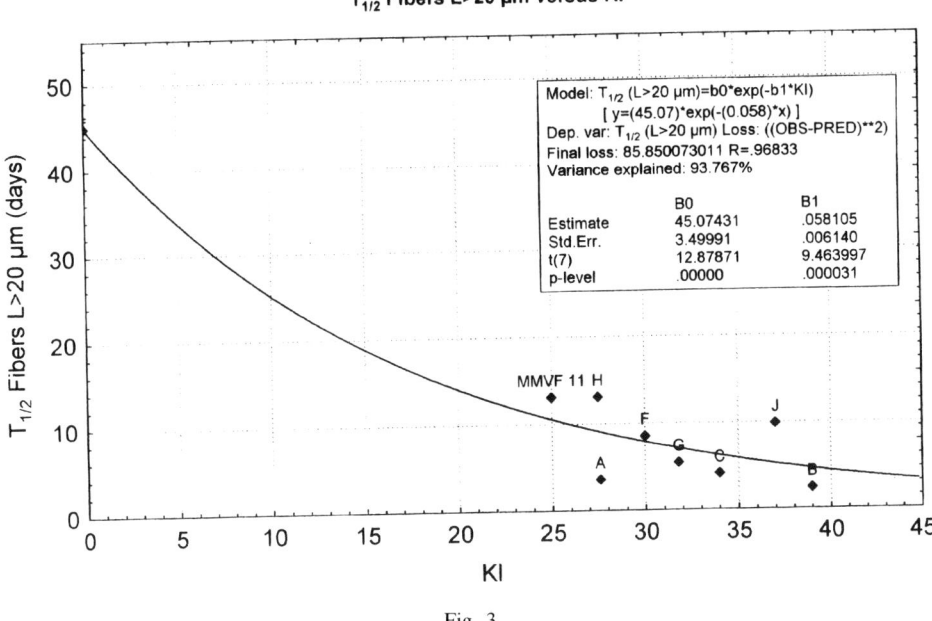

Fig. 3.

results with the TD25 however is not possible due to the lack of an IP study on this fibre.

Correlation between KI and in vivo *biopersistence after intratracheal instillation*

The tests presented in Table 3 have been conducted with fibres of different dimensions and with the intratracheal instillation (IT) of different masses. The correlation is not very good (variance explained: 60%) when only KI is used (Fig. 4); by contrast it improves when mass is added (variance explained: 83%), but it does not improve when diameter and length are involved. The instilled mass is a very important parameter in this type of test and as mentioned by Bellmann (1995), the value of $T_{1/2}$ of 45 days for fibre M753 (KI = 23) cannot be compared to the other results because the mass instilled was only 0.1 mg. New experiments are in progress using the Muhle IT protocol in two laboratories. The preliminary results of these two IT studies show that the $T_{1/2}$ WHO of KI $\geq$ 40 fibres is in the order of 30 days. The analysis of these new results is in progress and will be reported when complete. If the IT studies follow what would be expected, the biopersistence measurements after inhalation the correlation with KI would also be better for $T_{1/2}$ long fibres than for $T_{1/2}$ WHO fibres. This could not be confirmed as the $T_{1/2}$ for long fibres has not been published in TRGS 906 (1995).

Correlations between KI and the IP test in the rat

The aforementioned studies (*in vitro* acellular assays and *in vivo* biopersistence experiments), even if they correlate well to each other, need to be compared with the carcinogenic potential assessed by long term animal experiments. In order to be able to assess the statistical coherence of Pott's published data, similar experiments by intraperitoneal injection in the rat are needed.

T$_{1/2}$ WHO Fibers versus KI

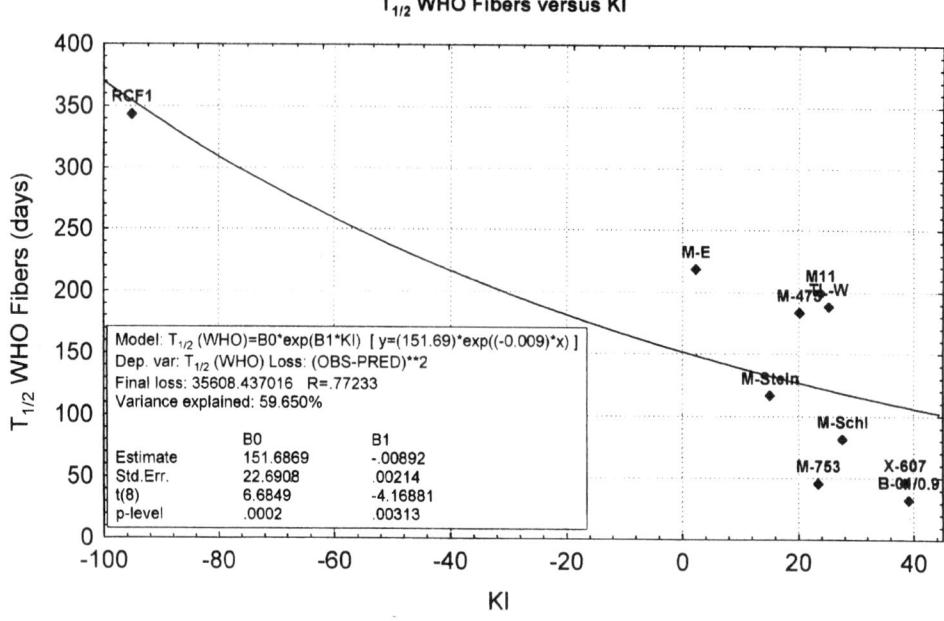

Fig. 4.

At present the statistical analysis is limited to the data published in Germany for the validation of the chemical index (TRGS 905, 1994) (Table 4). A more complete analysis with experiments conducted in different laboratories is under way, the preliminary results are not in contradiction with the conclusions:

- TD25 increases with the diameter for durable fibres.
- TD25 decreases with the length for all fibres.
- TD25 increases with KI.

In the French Proposal, as suggested by Pott, TD25, which is the dose necessary to produce 25% tumours in experimental animals, is presented as published in TRGS 906 (1995). A multi regression analysis has been carried out taking into account length, diameter and KI. Even if length and diameter are taken into account, the correlation between observed and calculated TD25 is not very good (variance explained: 49%). This is probably due to experimental conditions which were not comparable or to a poor assessment of the fibre geometry, length and diameter, but the KI has a statistically significant influence on the TD25 ($P = 0.005$). Presently, it does not seem acceptable to base a classification as "probable" carcinogen of a fibre on the value of this index. Unless a better relationship can be developed, the animal experiments cannot be avoided for a definitive risk assessment. However, we still have to decide what should be the "Gold Standard" long-term animal experiment: inhalation or IP test?

CONCLUSIONS

On the basis of the statistical analysis conducted, the increasing values of the German chemical index appear to be associated with greater biosolubility. These are consistent with the results of the four major "Biotests" presented in the French Proposal for classification of MMVF (French Proposal, 1995):

- *in vitro* solubility measured at pH 7.4.
- *in vivo* biopersistence $T_{1/2}$ after inhalation.
- *in vivo* biopersistence $T_{1/2}$ after intratracheal instillation.
- intraperitoneal injection.

With the analysis of the new data from the Biotests under way it appears that it is possible to predict in sufficient confidence the outcome of a Biotest on the basis of the chemical index or eventually to indicate for which ranges of composition this is not possible. These results confirm that the decision not to classify a fibre as possible carcinogen when KI $\geq$ 40 represents an acceptable solution and could be the basis of a consensus in the European Union. However, the data indicate that some fibres with KI < 40 are also highly soluble.

REFERENCES

Bellmann, B. and Muhle, H. (1994) Untersuchung über die Verweildauer verschiedener Mineral-fasertypen in der Rattenlunge als Basis für die Beurteilung, ob eine "hinreichende Beständigkeit" für die Krebserzeugung vorliegt. Fraunhofer (ITA) Report 28/03/94.

Bellmann, B., Muhle, H., Kamstrup, O. and Draeger, U. (1994) Investigation on the durability of man-made vitreous fibers in rat lungs. *Environ. Health Persp.* **102** (Suppl. 5) 185–189.

Bellmann, B. and Muhle, H. (1993) Biobeständigkeit verschiedener Mineralfastertypen. BAU Fb 711.

Bellman. B. and Muhle, M. (1994) Investigation of the biodurability of woolastonite and xonotlite. *Environ. Health Persp.* **102**, 191–195.

Bernstein, D. M., Mast, R., Anderson, R., Hesterberg, T. W., Musselman, R., Kamstrup, O. and Hadley, J. (1994) An experimental approach to the evaluation of the biopersistence of respirable synthetic fibers and minerals. *Environ. Health Perspect.* **102** (Suppl. 5) 15–18.

Bernstein, D. M., Morscheidt, C., Grimm, H., Thevenaz, Ph. and Teichert, U. (1996) Evaluation of soluble fibers using the inhalation biopersistence model, a nine-fiber comparison. *Inhal. Toxicol.* **8**, 345–385.

Bernstein, D. M., Morscheidt, C., Tiesler, H., Grimm, H. G., Thevenaz, Ph. and Teichert, U. (1995) Evaluation of the biopersistence of commercial and experimental fibers following inhalation. *Inhal. Toxicol.* **7**, 1029–1056.

Bignon, J. (1994) Summary Report of 2nd Workshop, International Cooperative Research Programme of Assessment of MMF's Toxicity, prepared by Margaux Orange Organization, 46 rue de la Clef, 75005 Paris, France.

Bignon, J., Bernstein, D. M., Pairon, C. and Brochard, P. (1995) Linkages between *in vitro* findings and quantitative risk assessment of mineral fibres based on animal studies. *5th Int. Symp.*, Hanover, Germany.

Bignon, J., Brochard, P., Brown, R., Davis, J. M. G., Vu, V., Gibbs, G., Greim, M., Oberdörster, G. and Sebastian, P. (1995) Assessment of the toxicity of man-made fibers. A final report of a workshop held in Paris. *Ann. occ. Hyg.* **39**, 29–106.

Bignon, J., Saracci, R. and Touray, J. C. (1994) Introduction: workshop on biopersistence of respirable synthetic fibres and minerals. *Environ. Health Persp.* **102** (Suppl. 5), 3–5.

French Proposal (1995) Assessment of MMF's toxicity submission by French authorities to DGXI of the European union. Prepared by Margaux Orange Organisation, 46, rue de la Clef, 75005 Paris France.

Statistica (1995) *Statistica for Windows*, Version 5. Stat.Soft, Inc., 2325 East 13th Street, Tulsa, OK 74104, U.S.A.

Thélohan, S. and De Meringo, A. (1994) *In vitro* dynamic solubility test—influence of various parameters. *Environ. Health Persp.* **102** (Suppl. 5), 91–96.

TRGS 905 (Technische Regeln für Gefahrstoffe) (1994) Classification of dusts from natural and man-made mineral fibers. *Bundesarbeitsblatt* Nr.6, p.57. Bundesministerium für Arbeit und Sozialordnung, Verlag W. Kohlhämmer, ISSN 0007–5868 (1994).

TRGS 906 (Technische Regeln für Gefahrstoffe) (1995) Justification for the classification of types of inorganic fibre dust (excluding asbestos). *Bundesministerium für Arbeit und Sozialverordnung.* Verlag W. Köhlhammer, Bundesarbeitsblatt 10.

 Pergamon

Ann. occup. Hyg., Vol. 41, Supplement 1, pp. 312–319, 1997
© 1997 British Occupational Hygiene Society
Published by Elsevier Science Ltd. All rights reserved
Printed in Great Britain
0003–4878/97 $17.00 + 0.00
Inhaled Particles VIII

PII: S0003–4878(96)00049–X

BIOPERSISTENCE OF INSULATION GLASS FIBRES

M. A. Moore,* L. M. Hanna,† D. M. Grumm,† P. Turnham,†
C. P. Yu‡ and G. A. Jubb§

*Morgan Crucible Company plc, Windsor, U.K.; †Sciences International Inc, Alexandria,
Virginia, U.S.A.; ‡State University of New York, Buffalo, New York, U.S.A.; and
§Morgan Materials Technology Ltd, Bewdley Road, Stourport-on-Severn DY13 8QR, U.K.

INTRODUCTION

Insulation fibres, often referred to as man-made vitreous fibres (MMVFs), are silicate glasses, widely employed in buildings, land, marine and air transport, consumer goods and industrial process equipment. The potential hazard to human health posed by respirable fibrous dust from these materials has been the subject of extensive scientific investigation and debate. Although the scientific issues are complex, toxicologists agree that the potential hazard depends on the lung burden of long, thin fibres (WHO, 1994).

The lung burden (that is dose) of such fibres is determined by exposure time and airborne respirable fibre concentration and by the biopersistence of fibres in the lung. A direct measure of biopersistence is obtained from fibre clearance rates following a single or short-term administration of the fibre. Fibre solubility measured *in vitro* has been used as a surrogate for fibre biopersistence, but solubility is an incomplete descriptor of the relevant fibre properties determining biopersistence.

Studies following the administration of fibres by inhalation (Bernstein, 1996), or intratracheal (IT) instillation (Bellmann and Muhle, 1995), have shown that, in addition to generally faster clearance rates for higher solubility fibres, fibres longer than 20 μm may clear faster than shorter fibres (Fig. 1). Hypotheses to explain this disparity invoke either preferential breakage or preferential dissolution of long fibres in contrast to the macrophage-mediated clearance rate of shorter fibres (Morris *et al.*, 1995; Bernstein *et al.*, 1995; Eastes and Hadley, 1995, 1996). Preferential dissolution alone of long fibres is unable to account for observations in animals that the mean diameter of fibres recovered from the lungs may change little with residence time, whereas there is very rapid clearance of long fibres immediately post-administration accompanied by a short-term increase in the number of fibres < 20 μm in length (Bernstein *et al.*, 1995; Bellmann and Muhle, 1995), (Fig. 2). Additionally, the observed clearance rates for long fibres are much faster than the times required for complete dissolution, estimated from *in vitro* dissolution rates (Christensen *et al.*, 1994; Eastes and Hadley, 1995).

Evident from *in vivo* and *in vitro* studies is that MMVFs are susceptible to local chemical attack and to preferential leaching of certain phases and constituents, which are detrimental to the strength and toughness of silicate glasses (Paul, 1990).

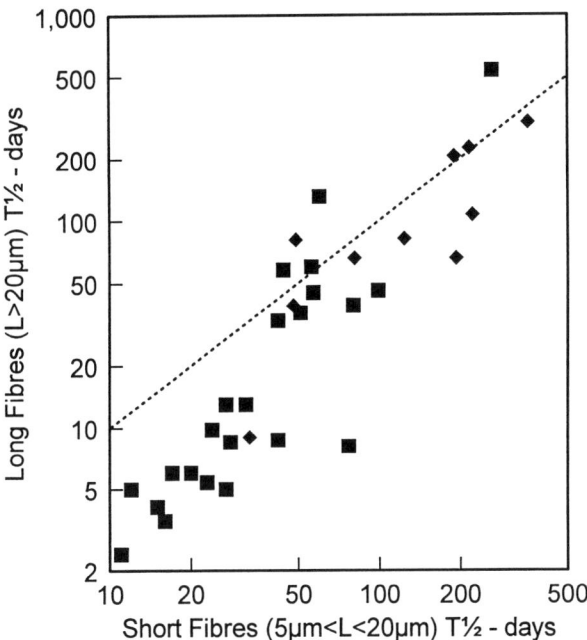

Fig. 1. Comparison of *in vivo* clearance half-lives from rat lungs of long ($>$ 20 μm) and short (5–20 μm) mineral fibres, ■ inhalation; ◆ IT instillation. Data from Bernstein (1996), for inhalation and Bellmann and Muhle (1995), for IT instillation.

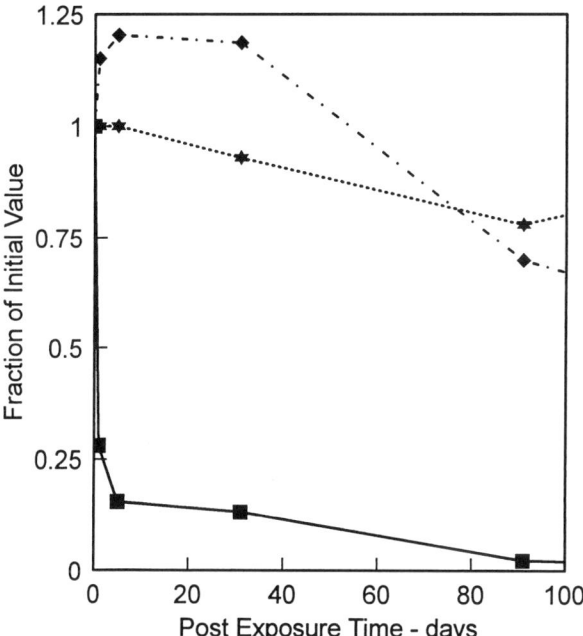

Fig. 2. Early post exposure changes in number of fibres of length $>$ and $<$ 20 μm and mean fibre diameter for MMVF10 recovered from rat lungs after inhalation, –■– number of fibres 720 μm, ·–◆–· number of fibres $<$ 20 μm, ---★--- mean fibre diameter. Data from RCC/TIMA study.

Glasses are brittle materials, having very limited plasticity, which fail (break) at stresses much lower than their theoretical strength because of the presence of inherent flaws. They fail catastrophically when the stress intensity at a flaw, which is proportional to $\sigma.a^{1/2}$, where σ is the applied stress and a is the flaw size, reaches a critical value. Additionally, in aqueous environments silicate glasses demonstrate time-dependent failure at stresses below their catastrophic fracture strength; a phenomenon known as stress–corrosion cracking and illustrated in Fig. 3 (Wiederhorn and Bolz, 1970), for three glasses having very similar catastrophic fracture strengths, but for which the stress–corrosion effect is most pronounced in the more chemically active soda-lime glass. The mechanism of stress–corrosion failure in glasses involves ion exchange and, therefore, depends on the availability of exchangeable cations present as silicate network modifiers (typically the alkali and alkaline earth metals), the pH and the kinetics of ion exchange. The effect on failure time of pH depends on glass composition as shown in Fig. 4 (Wiederhorn and Johnson, 1973).

The time to failure for silicate glasses, at a specific sub-critical stress, is described by:

$$t_f \propto a_i^{1/2}.(v_i.\beta)^{-1},$$

where v_i is the initial crack velocity, a_i is the initial flaw size (usually the depth of a surface flaw) and β depends on glass chemistry and the environment. The initial crack velocity

$$v_i \propto \exp(\beta.\sigma.a_i^{1/2}),$$

where σ is the applied stress. Values of v_i in glasses typically range from less than 10^{-10} m.s.$^{-1}$ to greater than 10^{-4} m.s^{-1}, depending on the applied stress, initial flaw size, specific glass and the environment. As the flaw enlarges by sub-critical growth the crack velocity increases, until the flaw size is large enough to satisfy the criterion for catastrophic failure, which is:

$$K_{IC} = \sigma(\pi.a)^{1/2},$$

where K_{IC} (the materials-dependent critical stress intensity factor) for silicate glasses has a value of ~ 1 MN.m$^{-3/2}$.

The cumulative probability of catastrophic failure, POF, as a function of applied stress, is determined by the statistical distribution of flaw sizes, which is a feature of all glases, such that:

$$POF = 1 - \exp-(\sigma/\sigma_n)^m,$$

where m is known as the Weibull modulus, with a value which increases as the flaw size distribution decreases (typically in the range 5–15 for glasses) and σ_n is a normalising stress (usually taken to be the mean of the failure stress distribution and, therefore, a material property).

Thus, if breakage of long fibres is a significant contributory mechanism, the clearance rate should be a function of the stress on the fibre and of time. Assuming that long fibres in the lung are subject to bending forces, F, the stress on the fibre is:

$$\sigma \propto F.l.d^{-3},$$

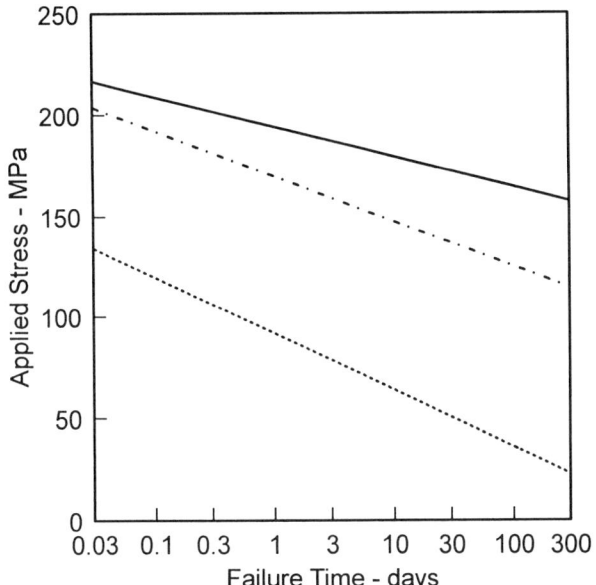

Fig. 3. Time dependent failure of glasses in distilled water, —— silica, –·–·– aluminosilicate, ----
soda-lime. After Wiederhorn and Bolz (1970). Silica glass—99.8%; aluminosilicate—57% silica, 20%
alumina; soda-lime—72% silica, 14% soda, 7% calcia.

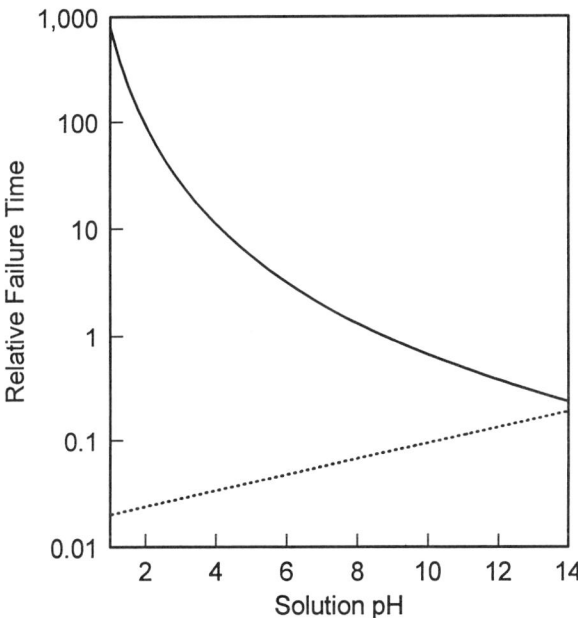

Fig. 4. Time dependent failure of glasses in aqueous solutions of different pH, —— silica, ---- soda-lime.
After Wiederhorn and Johnson (1973). Silica glass—99.8%; soda-lime, 72% silica, 14% soda, 7% silica.

M. A. Moore *et al.*

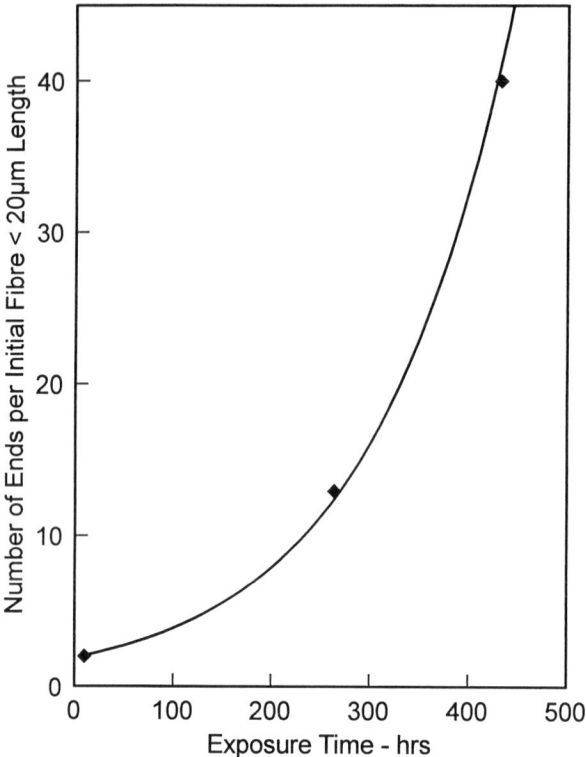

Fig. 5. Time dependent breakage of a glass fibre exposed to simulated body fluid. After Bauer (1995).

where l and d are the fibre length and diameter, respectively.

Bauer (1995) has demonstrated *in vitro* time-dependent fracture of a silicate glass fibre exposed to simulated body fluid, (Fig. 5). The cumulative probability of failure, the number of ends per initial fibre in his data, may be represented as an exponential function of time for the early stages of exposure (up to 400 h). This is consistent with the general equations for glass failure if the applied stress were less than the critical stress to cause catastrophic failure at the largest flaw present at time zero, but sufficient to provide failure of many of the fibres within a few hundred hours by sub-critical crack growth.

Combining the principles of the sub-critical time to failure equation with the cumulative probability of failure equation and subsuming fibre diameter distribution with flaw size distribution, we have curve fit $> 20 \ \mu m$ long fibre clearance data for MMVFs 10, 11, 21 and 22 (RCC/TIMA durability studies post inhalation by rats) to expressions of the form:

$$POF = 1 - N_t/N_0 = 1 - \exp{-b(t).l,}$$

where N_t and N_0 are the number of fibres of length l at time t and time 0, respectively (Fig. 6). We have found that the time-dependent coefficients $b(t)$, determined from these fits, are best described by the kinetic expression commonly known as the Michaelis–Menten equation:

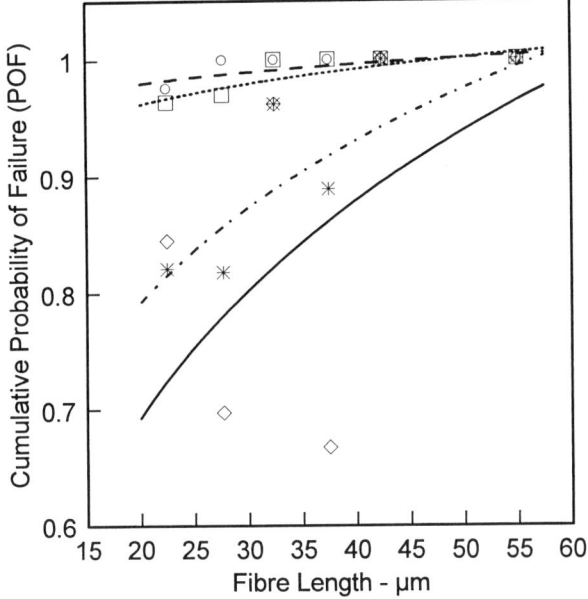

Fig. 6. Cumulative probability of failure against fibre length curve fits at various post exposure times for MMVF10. ——— 5 days curve fit, ◇ 5 days data; —·—·— 31 days curve fit, * 31 days data; ---- 91 days curve fit, □ 91 days data; – – – – 546 days curve fit, ○ 546 days data.

$$b(t) = C_0.t/(C_1 + t),$$

where C_1 is the kinetic half-life to reach one half the maximum C_0, as shown in Fig. 7. Values of C_0 and C_1 for each MMVF are given in Table 1.

The small differences in the C_0 values for the four fibres are consistent with fibre failure rates after long periods being both low and relatively independent of the initial fibre properties when most of the glass network modifiers have been depleted, whereas the differences in the C_1 values are consistent with early post-administration fibre failure rates being both high and dependent on fibre properties.

Specifically, C_1 values rank with measured *in vitro* fibre dissolution rates at near-neutral pH (Fig. 8); this correlation is not because dissolution itself is causing fibres to fail, but because dissolution rate is a measure of ion exchange kinetics. Thus, the most soluble and chemically active fibres, MMVF 10 and 11, have the lowest C_1 values and highest early fibre breakage rates, or shortest time to failure, consistent with the general stress–corrosion failure of glasses discussed earlier. It is reasonable to assume that long fibres start to break by this mechanism immediately post-deposition, providing an "apparent" early and rapid clearance, although the fragments will remain to be cleared by other mechanisms, such as complete dissolution or phagocytosis, and may include further breakage. At long post-deposition times the analysis indicates that some long fibres may remain; the cumulative probability of failure is less than 1, determined by the C_0 coefficient, but approaches 1 more closely for the longest fibres. This suggests that a few fibres neither have a large enough sub-critical flaw to fail, nor do they disappear by complete dissolution, within several hundreds of days post deposition.

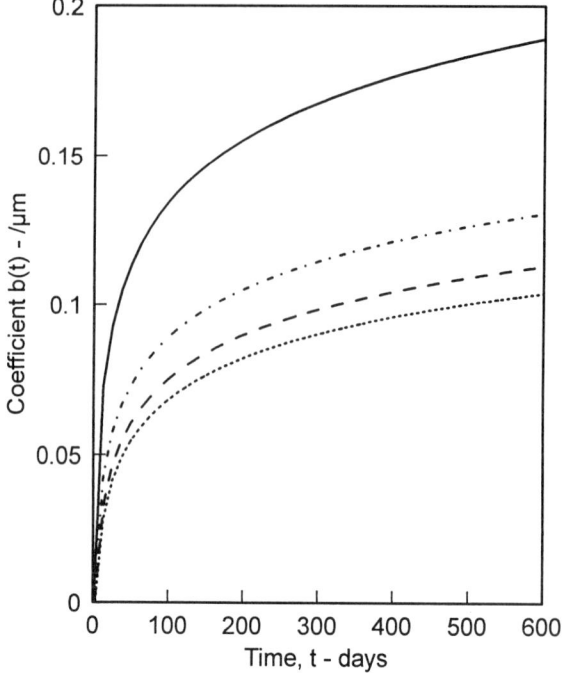

Fig. 7. The time dependence of the cumulative probability of failure (POF) of fibres of length l, $b(t)$ in the equation POF $= 1 - \exp-b(t).l$. —— MMVF10, –·–·– MMVF11, ······ MMVF21, ---- MMVF22.

Many chemically active fibres do not dissolve in the lungs in a uniform and congruent manner, but rather the silicate network modifiers are leached from the structure leaving a residual skeletal form which may be similar in dimensions to the original fibre, Bauer (1995). If such fibres do not fail during depletion of the glass network modifiers, the remaining, relatively insoluble silicate skeletal structure will be resistant to both stress–corrosion failure and to complete dissolution.

The analysis suggests that the biopersistence of long fibres depends on general or local dissolution providing the ion exchange for stress–corrosion cracking, thereby increasing the probability of failure. Hence, clearance rates are not alternately described by either dissolution or simple mechanical breakage, but rather by a process integrating these two mechanisms, both of which are time-dependent.

In the case of the MMVFs studied, the relatively small change in diameter suggests that stress–corrosion failure of the fibres dominates, whereas other fibres may be more resistant to stress–corrosion failure and may dissolve, resulting in diameter reduction leading to complete disappearance of the fibre or to increased

Table 1. Comparison of C_0 and C_1 coefficients

Fibre type	C_0	C_1
MMVF10	0.188	31.4
MMVF11	0.161	92.2
MMVF21	0.225	371
MMVF22	0.189	219

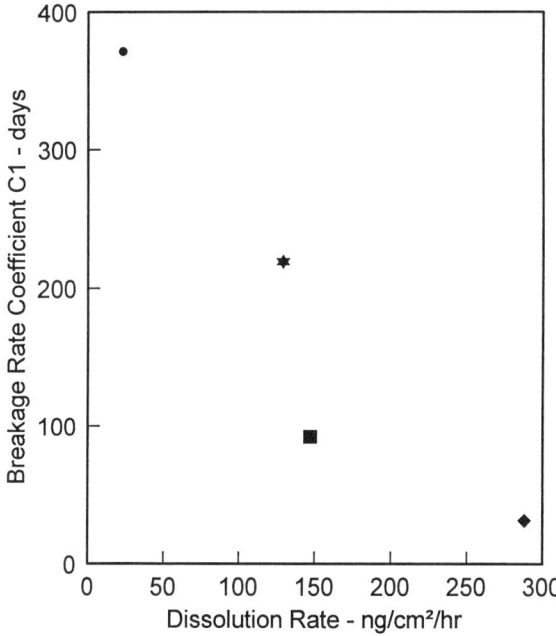

Fig. 8. The half-life for time-dependent long fibre breakage (coefficient C1) correlated to *in vitro* dissolution rate for MMVFs 10, ◆; 11, ■; 21, ●; and 22, ⋆. Dissolution rates are NAIMA values.

probability of breakage. We are currently undertaking analysis of other fibre types to understand these phenomena in terms of specific fibre properties.

REFERENCES

Bauer, J. F. (1995) *Glass Fiber in the Lung—A Materials Science Perspective.* Meeting of the American Ceramic Society, New Orleans.
Bellmann, B. and Muhle, H. (1995) *Biopersistence of Various Types of Mineral Fibres in the Rat Lung after Intratracheal Application.* Bundesanstalt für Arbeitsschutz, Dortmund.
Bernstein, D. M. (1996) RCC study results presented at various European regulatory meetings. Research and Consulting Co. Ltd, Geneva.
Bernstein, D. M., Morscheidt, C., Tiesler, H., Grimm, H-G., Thévenaz, P. and Teichert, U. (1995) Evaluation of the biopersistence of commercial and experimental fibers following inhalation. *Inhal. Toxicol.* **7**, 1031–1058.
Christensen, V. R., Lund Jensen, S., Guldberg, M. and Kamstrup, O. (1994) Effect of chemical composition of man-made vitreous fibers on the rate of dissolution *in vitro* at different pHs. *Environ. Health Perspect.* **102** (suppl. 5), 83–86.
Eastes, W. and Hadley, J. G. (1995) Dissolution of fibers inhaled by rats. *Inhal. Toxicol.* **7**, 179–196.
Eastes, W. and Hadley, J. G. (1996) A mathematical model of fiber carcinogenicity and fibrosis in inhalation and intraperitoneal experiments in rats. *Inhal. Toxicol.* **8**, 323–343.
Morris, K. J., Launder, K. A., Collier, C. G. and Morgan, A. (1995) *In Vivo and In Vitro Comparisons of the Dissolution of Calcium Magnesium Silicate Fibres.* 5th Int. Inhalation Symp., Hanover.
Paul, A. (1990) *Chemistry of Glasses*, 2nd edn. Chapman and Hall, London.
WHO (1994) *Validity of Methods for Assessment of Carcinogenicity of Fibres.* WHO European Regional Office, Copenhagen.
Wiederhorn, S. M. and Bolz, L. H. (1970) Stress corrosion and static fatigue of glass. *J. Am. Ceram. Soc.* **53**, 543–548.
Wiederhorn, S. M. and Johnson, H. (1973) Effect of electrolyte pH on crack propagation in glass. *J. Amer. Ceram. Soc.* **56**, 192–197.

 Pergamon

Ann. occup. Hyg., Vol. 41, Supplement 1, pp. 320–326, 1997
British Occupational Hygiene Society
© 1997 AEA Technology plc. Published by Elsevier Science Ltd
Printed in Great Britain
0003–4878/97 $17.00 + 0.00
Inhaled Particles VIII

PII: S0003–4878(96)00050–6

LUNG CLEARANCE OF EXPERIMENTAL MAN-MADE MINERAL FIBRES, PRELIMINARY DATA ON THE EFFECT OF FIBRE LENGTH

C. G. Collier*, O. Kamstrup†, K. J. Morris*, R. A. Applin* and I. A. Vatter*

*AEA Technology Plc, Biomedical Research Department, 551 Harwell, Didcot, Oxfordshire OX1 0RA, U.K.; †Rockwool International A/S, Hovedgaden 854, DK-2640 Hedehusene, Denmark

INTRODUCTION

The hazard from inhaled fibres is believed to be related to their dimensions, with long, thin fibres being potentially more carcinogenic than short, thick ones (Stanton *et al.*, 1981; Pott, 1978) and to their biopersistence, as fibres which dissolve rapidly in the lung are unlikely to induce long-term pathological changes (Morgan and Holmes, 1986). Following deposition in the lung, short fibres (< 10–20 μm in length) are ingested by alveolar macrophages, exposing them to the acidic environment of the phagolysome (pH 4.5). Longer fibres remain free in the lung tissue and are exposed to the neutral environment of the lung fluid. These longer fibres may be partially engulfed by macrophages, with macrophages gathering along the fibre like pearls on a string, exposing different parts of the fibre to different environments.

These pH differences can explain the length dependent biopersistence observed for experimental glass fibres (Eastes *et al.*, 1995). For these fibre types, short fibres (< 20 μm) showed only limited decline in diameter with time, indicating low dissolution. For longer fibres, diameter reduction was faster, indicating higher dissolution, similar to that seen in *in vitro* studies. This is consistent with the fact that glasses are more soluble at neutral pH than acidic pHs. For stonewool fibres, which are more soluble at low pHs, the situation should be reversed, with short stonewool fibres showing lower lung biopersistence than longer fibres. If pH is the determining factor, adjustment of the chemical composition of fibres may result in higher dissolution of longer fibres within the lung, reducing the potential hazard from these fibres.

Studies on lung clearance of fibres (and hence biopersistence) present problems as thick fibres (respirable by humans) can not be inhaled by rats. With intratracheal instillation, longer and thicker fibres may be administered to the lungs of animals and measurements of biopersistence made. In this study, initial lung biopersistence results for two experimental MMVFs are compared and the effect of fibre length on clearance demonstrated. Corresponding data for biopersistence of both fibres in the peritoneal cavity and for HTN lung retention measured by SEM is currently being collated.

METHODS

Two stonewool fibre types were studied, MMVF21 and HTN. The MMVF21 was obtained from the NAIMA (North American Insulation Manufacturers Association) repository and was the same material as that used in inhalation studies conducted at RCC (Research and Consulting Centre, Geneva, Switzerland) (McConnell *et al.*, 1994). HTN was an experimental stonewool fibre developed and supplied by Rockwool International A/S, Denmark. Prior to administration, approximately 2 g of each fibre type was irradiated in the Imperial College reactor at Silwood Park, for 8 h in a neutron flux of 2.5×10^{12}, to activate some of the Na present to ^{24}Na, detectable by external γ-counting. A preliminary test of the effects of hypochlorite digestion on the morphometry of neutron activated fibres showed that the methods of sample preparation did not significantly affect the fibres to be used.

Suspensions of each fibre type in sterile saline (0.2 ml; 6 mg fibres ml^{-1}) were administered by intratracheal instillation, to groups of 40 female Fischer F344 rats (Harlan Olac, U.K.) which were 12 weeks old, under halothane anaesthesia. Each rat received approximately 1.2 mg of fibres. Samples of the instillation suspension were taken at regular intervals during the administrations to act as γ-counting standards for the *in vivo* whole body counts. A group of 10 control animals received instillations of 0.2 ml of sterile saline only. Following recovery animals were returned to their cages and housed in standard conditions with food and water available *ad libitum*.

The whole body γ-activity of ^{24}Na in each animal was measured using NaI detectors at 2 days. From this the original fibre burden in each animal was estimated by comparison with aliquot activity, normalised by the number of fibres which had been measured in the aliquot samples. Fibre exposed animals were killed at 2 days ($n = 5$) and at 7, 14, 30, 90, 141 and 180 days ($n = 3$) and control animals ($n = 2$) at 2, 30 and 360 days after administration. The lung lobes were removed and digested in 60 ml of hypochlorite at 4°C for ~ 4 h. Measured volumes of each digest were filtered onto cellulose filters. Four, 0.45 µm pore size mixed cellulose ester filters were prepared for phase contrast optical microscopy (PCOM). These were dried, mounted, and cleared on glass slides using the acetic acid/formamide method of Le Guen and Galvin (1981). Two, 0.22 µm pore size mixed cellulose ester filters were mounted on copper stubs and sputter coated with gold for scanning electron microscopy (SEM) examination.

At each time point, bivariate size measurements were made on at least 600 (or as many as are found in a maximum of 300 fields) fibres sampled from all the animals killed, by SEM (Leica, S440 SEM, Leica, U.K.). At each time point, the median fibre length and diameter, the fraction of WHO fibres, and the fraction of fibres < 5 µm, 5–10 µm, 10–20 µm and > 20 µm in length were calculated. The number of fibres in each sample was estimated by PCOM. At least 200 fibres, up to a maximum of 200 fields were scored per filter, using modified WHO (1985) and NIOSH (1989) counting rules. From the known area of each field and the total filter area, the number of fibres on the filter was estimated. Two filters per lung digest sample were randomly selected and scanned. The number of fibres present in each lung was calculated from the average counts per animal corrected by appropriate scaling factors for sample volume and dilution.

Table 1. Median length and diameter of fibres recovered from lungs
at various times after intratracheal instillation (μm)

	MMVF21	MMVF21	HTN	HTN
Days	Length	Diameter	Length	Diameter
0	15.00	1.22	20.21	1.09
2	15.71	1.19	24.22	1.14
7	18.17	1.25	23.82	1.26
15	15.44	1.25	17.43	1.29
30	17.55	1.17	14.83	1.13
90	15.78	1.15	12.66	1.13
141	17.94	1.28	15.50	1.84
181	17.27	1.10	16.39	1.88

RESULTS

The median length and diameter of the fibres recovered from the lungs at the different time points are given in Table 1. For MMVF21, there was little change in diameter over the course of the study, with a trend towards a slight increase in fibre length, indicating some preferential removal of shorter fibres with time. For HTN, there was a decrease in length with time, with the minimum length being found 90 days after administration. The diameter of the fibres remained relatively constant up to 90 days and then showed a marked increase.

The fraction of the fibres falling into each size category at each time point is given in Table 2. Changes in the distribution of fibres between size categories with time (2–181 days) were not seen for MMVF21, but were noticeable for HTN fibres. For HTN, there was a decrease in the proportion of WHO fibres with time, increases in the proportion of fibres in fractions 5–20 μm and decreases in the proportions < 5 and > 20 μm.

Initial fibre burdens, measured at 2 days after administration, were similar for

Table 2. Percentage of fibres found in different size categories following recovery from lung after intra-tracheal instillation of fibres

Time	N fibres	% WHO	% < 5 μm	% 5–10 μm	% 10–20 μm	% > 20 μm
MMVF21						
0	1206	88.14	9.62	22.55	28.36	39.47
2	606	90.43	7.76	23.93	28.88	39.44
7	603	89.05	6.30	21.06	27.36	45.27
15	603	89.05	7.46	22.89	30.85	38.81
30	602	90.03	8.31	19.60	28.90	43.19
90	606	90.92	8.42	21.78	31.85	37.95
141	607	91.10	6.75	21.42	28.01	43.82
181	603	93.20	5.80	19.24	30.68	44.28
HTN						
0	1200	91.83	5.00	18.58	25.50	50.92
2	602	95.18	1.66	10.96	27.24	60.13
7	601	91.51	3.99	14.14	23.63	58.24
15	600	91.00	2.67	17.67	34.50	45.17
30	605	90.25	5.45	23.14	36.36	35.04
90	331	90.33	4.83	30.21	40.18	24.77
141	386	86.01	2.59	22.54	34.72	40.16
181	336	84.82	0.89	24.40	33.93	40.77

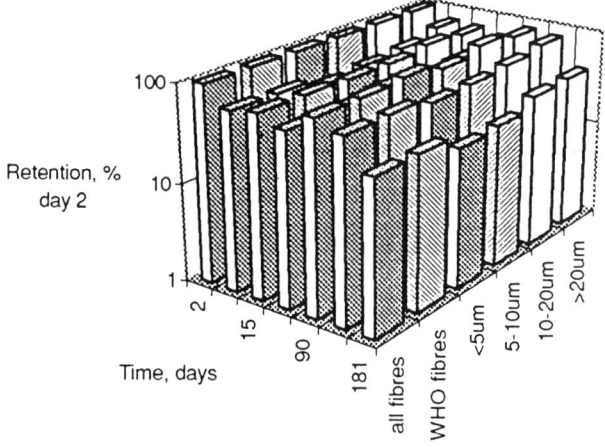

MMVF 21

Fig. 1. Retention of MMVF21 fibres in the lung of rats following intratracheal instillation. (% of burden at 2 days post-administration, mean $n = 3$–5.)

both groups ($6.2 \pm 0.3 \times 10^6$ and $4.8 \pm 0.4 \times 10^6$ fibres for MMVF21 and HTN respectively, mean $\pm$ SE, $n = 5$) and very few fibres were found in the control animals (350 ± 200, $n = 3$). From the total number of fibres recovered from each animal and the fraction of fibres in each size category (Table 2), the number of fibres in each size category remaining in the lungs of each animal was calculated. For each animal, the retention of each size fraction of fibres was calculated by comparing the number of fibres in that category at death with the number of that category present originally (2 days). Retention curves for each size fraction and fibre type are given in Figs 1 and 2. For MMVF21, there was some clearance of all

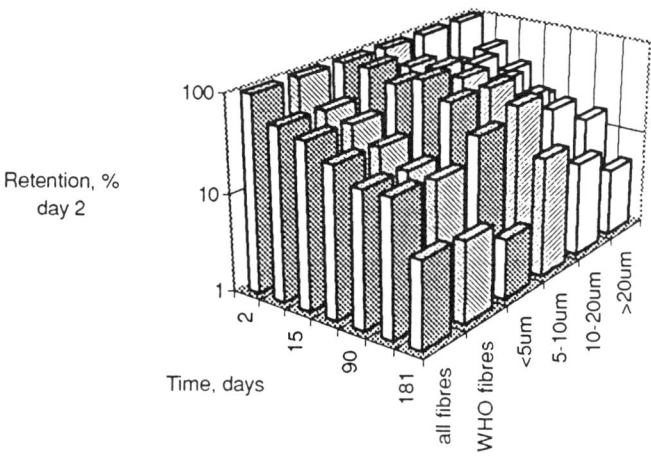

HTN

Fig. 2. Retention of HTN fibres in the lung of rats following intratracheal instillation. (% of burden at 2 days post-administration, mean $n = 3$–5.)

Table 3. Clearance half times and 95% confidence intervals for fibres and size fractions of fibres following intratracheal instillation (days)

	All fibres	WHO fibres	Fibres < 5 μm long	Fibres 5–10 μm long	Fibres 10–20 μm long	Fibres > 20 μm long
MMVF21	231 (167–375)	246 (174–415)	174 (130–261)	188 (139–290)	250 (175–433)	261 (186–437)
HTN	64 (58–71)	61 (55–68)	49 (43–56)	96 (85–110)	73 (65–81)	53 (47–61)

size fractions, with slightly faster clearance of shorter fibres than longer ones (retention at 181 days of < 5 μm fibres = 26.6 ± 6.4 and for > 20 μm fibres = 39.9 ± 9.6, mean ± SE). For the HTN fibres, clearance of all size fractions was faster than for MMVF21 and the shortest and longest fractions (< 5 μm and > 20 μm) cleared faster than fibres in the intermediate size ranges (5–20 μm).

Clearance half-times ($t_{1/2}$) for each size fraction were estimated by fitting a single exponential function to the data. A linear regression was performed on the natural logarithm of the retention data for individual animals. This method was chosen over fitting a single exponential to the basic retention as the standard deviation on $\log_e$ retention is relatively constant. The linear regression analysis provided a best estimate of the gradient and the 95% confidence limits on this estimate, from which the $t_{1/2}$ and its 95% confidence limits were derived (Table 3). The two fibre types showed marked differences in their clearance characteristics, with $t_{1/2\ \text{all fibres}}$ of 231 and 64 days for MMVF21 and HTN, respectively. For MMVF21, shorter fibres cleared faster than longer fibres. For HTN, length related differences in clearance were more marked. The longest and shortest fibres (< 5 and > 20 μm) had $t_{1/2}$ values which were considerably lower than the fibres in the intermediate size ranges. SEM of the fibres indicated that whilst MMVF21 fibres were relatively unchanged in appearance with time, HTN fibres were eroded and fragmented.

DISCUSSION

Instillation has been criticised as a route of administration as mechanical clearance from the lung may be impaired or even halted by overloading of the lungs. In this study, the lungs of animals exposed to both fibre types were loaded to a similar extent both in terms of mass and fibre number, so any effects on mechanical clearance would have been similar for both groups. The differences in clearance between the groups therefore directly reflect differences in lung biopersistence of the fibres.

For MMVF21, clearance was relatively slow and there was a slightly faster clearance of shorter fibres than longer fibres. This is consistent with the fact that stonewool fibres are more soluble in the macrophages, than in the neutral conditions of the lung tissue. The longer the fibre, the more likely it is to be in a neutral environment, therefore the slower its dissolution in the lung as indicated by the results of this study. The HTN fibre has a relatively low dissolution rate *in vitro* at pH 7.5, similar to that of normal stonewool, but a high dissolution rate at pH 4.5,

Knudsen *et al.*, (in press). Chemically compared to the traditional stonewool (MMVF21), the HTN fibre type is relatively low in silicon oxide and higher in aluminium oxide, Knudsen *et al.* (in press). For HTN, all fractions cleared considerably faster than MMVF21. The longest and shortest fibres appeared to be clearing faster than intermediate sized fibres. The > 20 μm fraction is too long for complete ingestion by alveolar macrophages (Morgan *et al.*, 1982) and following lavage of lungs, these fibres have been observed with macrophages clustered along their length like pearls on a chain. As HTN fibres are more soluble *in vivo* than MMVF21, the longer fibres would be more likely to dissolve at the sites of macrophage contact and hence more likely to fragment into shorter fibres. Evidence for this was seen in the appearance of the fibres under the SEM, with the longer HTN fibres appearing eroded and fragmented. Fractionation of the longer fibres will "enrich" the pool of fibres in the lower size ranges, reducing the overall clearance of these fractions. This phenomena has also been observed following inhalation studies with MMVF fibres (Mussleman *et al.*, 1994). In the present study, this enrichment appears to be occurring up to about 90 days after administration. Beyond 90 days far fewer longer fibres were available to enrich the shorter fractions and so clearance from the shorter fractions appeared to increase. Over the interval 141–181 days, clearance of the shortest fibre fraction is the fastest, with the other fractions clearing at similar rates.

This study has demonstrated that fibre length, fibre chemistry and fibre fragmentation all affect biopersistence. All length fractions of the HTN fibres were cleared much faster from the lungs than the traditional stonewool fibres (MMVF21). Shorter stonewool fibres were found to dissolve faster in the lung than longer fibres, confirming the hypothesis that the pH within the macrophages and lung environments determines fibre dissolution within the lung. Alterations in fibre chemistry can produce fibres with significantly shorter clearance half times which may prove to be less hazardous if inhaled. When fibre solubility at low pHs is high, digestion by macrophages at points along the length of the fibre, may result in significant fragmentation into shorter fibres which will enrich the shorter fibre populations. This study has shown the low biopersistence of the HTN fibre type which has a high *in vitro* solubility at pH 4.5, but low at pH 7.5.

Acknowledgement—This work was sponsored by Rockwool A/S Denmark.

REFERENCES

Eastes, W., Morris, K. J., Launder, K. A., Morgan, A., Collier, C. G., Davis, J. A., Mattson, S. M. and Hadley, J. G. (1995) Dissolution of glass fibres in rat lungs after intratracheal instillation. *Inhal. Toxicol.* **7**(2), 197–214.

Knudsen, T., Guldberg, M., Christensen, V. and Jensen, S. L. A new type of stonewool (HT fibre) with a high dissolution rate at pH 4.5. *Glass Sci. Technol.—Glastechnische Berichte* (in press).

Le Guen, J. M. M. and Galvin, S. (1981) Clearing and mounting techniques for the evaluation of asbestos fibres by the membrane filter method. *Ann. occup. Hyg.* **24**, 273.

McConnell, E. E., Kamstrup, O., Mussleman, R., Hesterberg, T., Chevalier, J., Miller, W. C. and Thevenaz, P. (1994) Chronic inhalation study of size separated rock and slag wool insulation fibers in Fischer 344/N rats. *Inhal. Toxicol.* **6**(6), 571–614.

Morgan, A. and Holmes, A. (1986) Solubility of asbestos and man-made mineral fibres *in vitro* and *in vivo*: its significance in lung disease. *Environ. Res.* **39**, 475–484.

Morgan, A., Holmes, A. and Davison, W. (1982) Clearance of sized glass fibres from the rat lung and their solubility *in vivo. Ann. occup. Hyg.* **25**, 317–331.

Mussleman, R., Miller, W., Eastes, W., Hadley, J., Kamstrup, O., Thevenaz, P. and Hesterberg, T. (1994) Biopersistence of MMVF and crocidolite in rat lungs. *Environ. Health Perspect.* **102**(5), 139–143.

NIOSH (1989) Manual of analytical methods: method 7400, Revision 3. NIOSH, U.S.A.

Pott, F. (1978) Some aspects on the dosimetry of the carcinogenic potency of asbestos and other fibrous dusts. *Staub Reinhalt Luft* **38**, 486–490.

Stanton, M., Layard, M., Tegaris, A., Miller, E., May, M., Organ, E. and Smith, A. (1981) Relationship of particle dimension to the carcinogenicity in amphibole asbestos and other fibrous minerals. *JNCL* **67**, 965–975.

WHO (1985) World Health Organisation. Reference methods for measuring airborne man-made mineral fibre (MMMF). WHO, Copenhagen.

 Pergamon

Ann. occup. Hyg., Vol. 41, Supplement 1, pp. 327–333, 1997
© 1997 British Occupational Hygiene Society
Published by Elsevier Science Ltd. All rights reserved
Printed in Great Britain
0003–4878/97 $17.00 + 0.00
Inhaled Particles VIII

PII: S0003–4878(96)00153–6

PULMONARY RESPONSES TO INHALED PARA-ARAMID FIBRILS IN EXPOSED RATS AND HAMSTERS

D. B. Warheit, S. I. Snajdr, M. A. Hartsky and S. R. Frame

DuPont Haskell Laboratory, PO Box 50, Elkton Rd, Newark, DE 19714–0050, U.S.A.

INTRODUCTION

This multifunctional study was designed to (1) compare the pulmonary effects of size-separated p-aramid vs. chrysotile asbestos inhalation exposure in rats; and (2) to compare the effects of p-aramid inhalation in rats and in hamsters following exposures to aerosolised, size-separated fibrils. A new term to characterise para-fibrils is also introduced herein. Respirable-sized, fibre-shaped particulates (RFP) have recently been described as the respirable component of para-aramid fibres (ECETOC). Para-aramid fibres are nonrespirable, having a diameter of 12–15 μm. Henceforth, the term p-aramid RFP will be used interchangeably with the term "p-aramid fibrils" and these denote the respirable-sized fraction of para-aramid fibres. Rats were exposed nose-only for 2 weeks to design fibre concentrations of 400 and 750 f ml^{-1} p-aramid RFP or chrysotile asbestos fibres, while hamsters were exposed whole-body to p-aramid for 2 weeks to design fibre concentrations of 350 and 700 f ml^{-1}. Following completion of exposures, the lungs of sham and fibre-exposed rats were evaluated immediately after (time 0), as well as 5 days, 1, 3, 6 and 12 months postexposure. The hamster study is currently ongoing; thus, sham and p-aramid fibril-exposed animals have been evaluated immediately after 2 week exposures as well as 10 days, 1 and 3 months postexposure. The major endpoints of this study are assessments of (1) fibre deposition and clearance; (2) biopersistence of inhaled fibrils; and (3) cellular proliferation of terminal bronchiolar, pulmonary parenchymal, subpleural and mesothelial cells.

The mean fibre concentrations over the 2 week period were 458 and 782 f ml^{-1} for chrysotile asbestos; mean RFP concentrations were 419 and 772 f ml^{-1} for nose-only p-aramid exposed rats and 358 and 659 f ml^{-1} for p-aramid exposed hamsters. p-Aramid fibrils were recovered from the lungs of exposed animals using a diluted Clorox digestion technique. Fibre lung burdens for the two groups of exposed rats immediately after p-aramid exposure were 4.8×10^7 and 7.6×10^7 fibrils/lung, respectively; similar numbers of chrysotile fibres > 5 μm were recovered from the lungs of asbestos exposed rats. The lung burden for the high dose hamster group was approximately 8×10^6 fibrils/lung. Biopersistence studies in rats indicated that there was an initial decrease (relative to asbestos) in the numbers of cleared p-aramid fibrils and this corresponded to a reduction in the mean lengths of inhaled fibrils during this time period, signifying biodegradability of inhaled fibrils. Clearance of fibrils was more rapid after the first month postexposure. In contrast, clearance of *short chrysotile asbestos fibres* was rapid, but there appeared to be slow or insignificant clearance of the *long chrysotile asbestos fibres* and this was evidenced by a progressive increase over time in the mean lengths of fibres recovered from the lungs of exposed rats. Biopersistence/pulmonary clearance results in the p-aramid hamsters are preliminary but there are indications of a reduction in mean fibril lengths at 1 and 3 months postexposure relative to dimensions of retained fibrils immediately after exposure. This finding demonstrates biodegradability of inhaled p-aramid fibrils in two rodent species. Two week, high-dose exposures to p-aramid in both rats and hamsters produced transient increases in pulmonary inflammatory responses, which were evident immediately after exposure but were not significantly different from controls at later time points. Similar transient effects were measured in pulmonary cell labeling indices of p-aramid exposed rats and hamsters, wherein increases relative to controls were measured on terminal bronchiolar surfaces immediately after exposure but were not different from controls following a 5 day postexposure period. In contrast, inhalation of size-separated chrysotile asbestos fibres in rats produced sustained cell proliferative responses in terminal bronchiolar, pulmonary parenchymal, subpleural and pleural regions. These results suggest that high-dose p-aramid exposures in rats and hamsters produce a transient inflammatory response and that inhaled fibrils are biodegradable in the lungs of exposed animals.

METHODS

Groups of male Crl:CDBR rats (7–8 weeks old, Charles River Breeding Laboratories, Kingston, New York) or male Syrian Golden hamsters were exposed 6 h day^{-1}, 5 days week^{-1} for 2 weeks. For this study, Kevlar® was utilised as a representative para-aramid RFP (respirable-sized, fibre-shaped particulates; see ECETOC reference). After completion of exposures, the lungs of p-aramid or chrysotile-exposed animals and aged-matched sham controls were subsequently evaluated for airway, parenchymal, subpleural and mesothelial cell proliferation and clearance at 0 h, 5 days 1, 3, 6 and 12 months postexposure. The lungs of p-aramid-exposed hamsters were evaluated immediately after exposure, as well as 10 days, 1 and 3 months postexposure.

The general experimental design, RFP preparation, inhalation exposure techniques, and methods for pulmonary lavage, biochemical assays, cell proliferation and lung digestion/biopersistence studies were similar to those previously carried out in p-aramid and chrysotile asbestos-exposed rats (with the exception that this was a whole-body exposure vs a nose-only exposure for the rats). The methods for the studies in rats are described elsewhere (Warheit *et al.*, 1995, 1996). In this study, four rats or hamsters/exposure group/time period each were utilised in cell proliferation and lung tissue studies. Statistics were carried out using a two-tailed Student's *t*-test on a Microsoft Excel software program.

RESULTS AND DISCUSSION

Clearance data from rats and hamsters exposed to p-aramid RFP as well as rats exposed to chrysotile asbestos fibres are presented in Fig. 1. Clearance of chrysotile

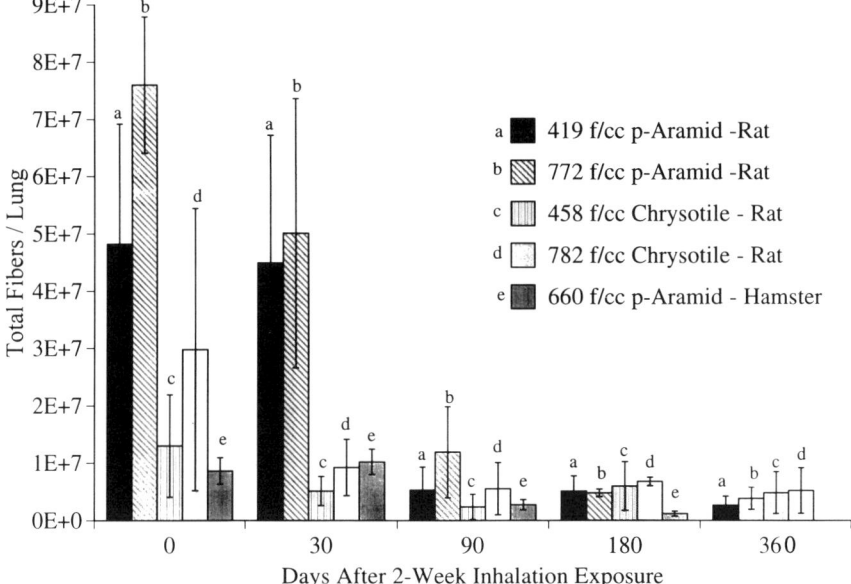

Fig. 1. Lung retention and clearance of inhaled p-aramid RFP in exposed rats and hamsters, and inhaled chrysotile asbestos fibres in exposed rats. The numbers of retained RFP recovered from the lungs of exposed hamsters was increased relative to the fibril burden of animals immediately after exposure. This was accounted for by a decrease in mean fibril lengths (see Fig. 2), indicating breakage or shortening of retained RFP in the lungs of exposed hamsters. Mean values of 4 hamsters or rats/dose at the 0, 1, 3, 6 and 12 month postexposure time periods.

asbestos fibres was rapid, however, a subpopulation of longer fibres was cleared at a slow rate. In this regard, compared to p-aramid fibrils which appeared to be biodegradable in the lungs of exposed rats, the median and mean lengths of chrysotile asbestos fibres recovered from digested lung tissue were increased over time, suggesting that the subpopulation of short asbestos fibres were selectively cleared from the lungs, with apparent insignificant pulmonary clearance of the subpopulation of long chrysotile asbestos fibres (Fig. 2). Inhalation exposures to p-aramid fibrils in rats and hamsters produced initially a slower clearance pattern, concomitant with a shortening of retained fibrils over time (Fig. 2), indicating a biodegradation of inhaled fibrils. Subsequently (at the 3 month postexposure time period), there was rapid clearance of inhaled p-aramid RFP (Fig. 1). In p-aramid exposed rats, increased pulmonary cell labelling effects relative to controls were measured on terminal bronchiolar and subpleural surfaces immediately following exposure to fibrils but were not different from controls following a 5 day postexposure period (Figs 3 and 4). In contrast, significant increases in asbestos-exposed rats compared to controls in pulmonary cell labelling indices were measured on terminal bronchiolar and subpleural surfaces immediately after exposure, persisting through a period of 3 months postexposure and this correlated with slow or significant clearance of long chrysotile asbestos fibres (Figs 2–4). The dimensional changes of asbestos fibres as well as the pulmonary cell labelling data indicate that chrysotile asbestos fibres may produce greater long-term pulmonary effects when compared with p-aramid fibrils.

D. B. Warheit *et al.*

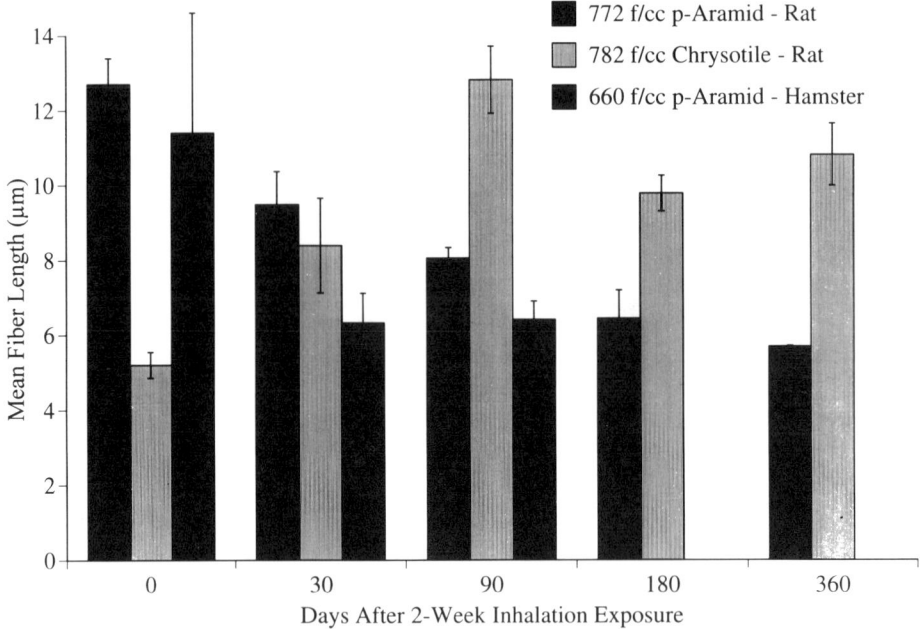

Fig. 2. Mean lengths of chrysotile or para-aramid fibrils recovered from the lungs of exposed rats or hamsters following a two-week inhalation exposure. Values given are mean lengths ± SD immediately after, as well as 1 and 3 months postexposure. These results indicate a shortening of the inhaled p-aramid fibrils over time suggesting biodegradability of inhaled fibrils. In contrast, a progressive increase in mean lengths of chrysotile-exposed animals signifies clearance of short asbestos fibres concomitant with insignificant clearance of long asbestos fibres.

For p-aramid exposed hamsters, the mean aerosol RFP concentrations over the 2-week exposure period were 358 and 659 f cc^{-1}. Mean lung burden for the high dose hamster group was 8.6×10^6 RFP/lung. The numbers of retained p-aramid RFP were increased to 1.02×10^7 fibres lung^{-1} at 1 month postexposure, before decreasing to 2.75×10^6 fibrils at 3 months postexposure (Fig. 1). Mean lengths of fibrils recovered from hamster lungs (time 0) were 10.4 μm and fibril lengths were progressively decreased to 6.3 and 6.1 μm, respectively, by 1 and 3 months postexposure (Fig. 2). This progressive reduction in the length dimensions of retained fibrils over time signifies a reduction in length or shortening of retained RFP and is likely to account for the enhanced numbers of fibrils (relative to the retained lung burden immediately after the end of the 2 week exposure) measured at 1 month postexposure. These data are consistent with the results of earlier studies carried out in p-aramid exposed rats, wherein the mean and median lengths of retained fibrils were progressively reduced with increasing residence time in the lung (Warheit *et al.*, 1992, 1995, 1996; Kelly *et al.*, 1993; Searl, 1997).

The pulmonary cell proliferative data demonstrating persistent cellular labelling effects in chrysotile asbestos-exposed rats presented here correlate with findings reported by several other investigators (Brody and Overby, 1989; McGavran *et al.*, 1990; Coin *et al.*, 1992a,b). In previous studies in rats and mice, acute inhalation exposures to chrysotile asbestos fibres resulted in dramatic increases in epithelial

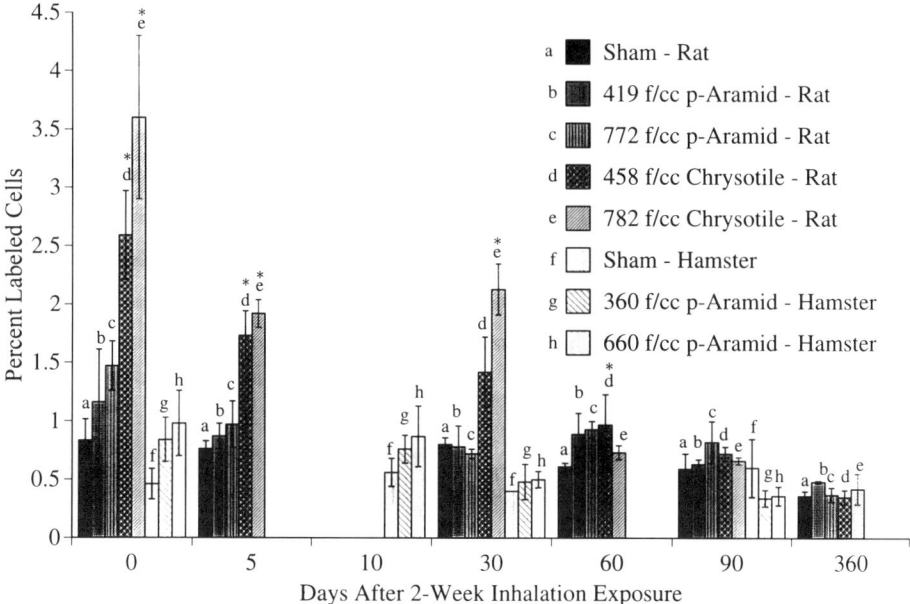

Fig. 3. BrdU labelling index of terminal bronchiolar cells in rats exposed to chrysotile asbestos fibres or p-aramid fibrils or hamsters exposed to p-aramid fibrils for two weeks. Values are expressed as mean percentages of labeled cell ± SD. Substantial increases in BrdU immunostaining compared to controls were measured in airway regions of asbestos-exposed rats. Significant increases in labeling were measured in high dose groups immediately after, 5 days, 1 and 3 months postexposure before returning to control levels at 6 months postexposure. In contrast, transient increases in cell labeling indices were in p-aramid-exposed rats or hamsters relative to controls immediately after exposure but were not measured 5 days later ($P < 0.05$).

cell DNA synthesis, followed several days later by enhanced labelling of interstitial cells (Brody and Overby, 1989). This time-course was extended in follow-up studies, resulting in a prolonged duration of increased cell labelling (Coin et al., 1992a). In another study, Coin and colleagues (1992b) reported that a 5 h exposure to chrysotile fibres in mice produced substantial increases in mesothelial and subpleural cell labelling indices at 2 and 8 days postexposure.

In summary, pulmonary cell proliferation results demonstrated substantial increases in lung parenchymal, airway, pleural/subpleural, and mesothelial cell proliferation effects following chrysotile exposures, suggesting that chrysotile produces a potent proliferative response in the airways, lung parenchyma, and subpleural/pleural regions. In contrast, p-aramid exposures produced only transient effects in airway and subpleural regions.

Fibre biopersistence/durability results demonstrate that short chrysotile fibres are cleared rapidly, however the longer chrysotile fibres were retained in the lung or cleared at a slow rate. In contrast, para-aramid fibrils have low biodurability in the lungs of exposed animals. Based on these comparisons, we conclude that the proliferative effects and enhanced biodurability of chrysotile that are associated with the induction of chronic disease do not occur with para-aramid fibrils. In addition, the finding that inhaled p-aramid fibrils are biodegradable in the lungs of

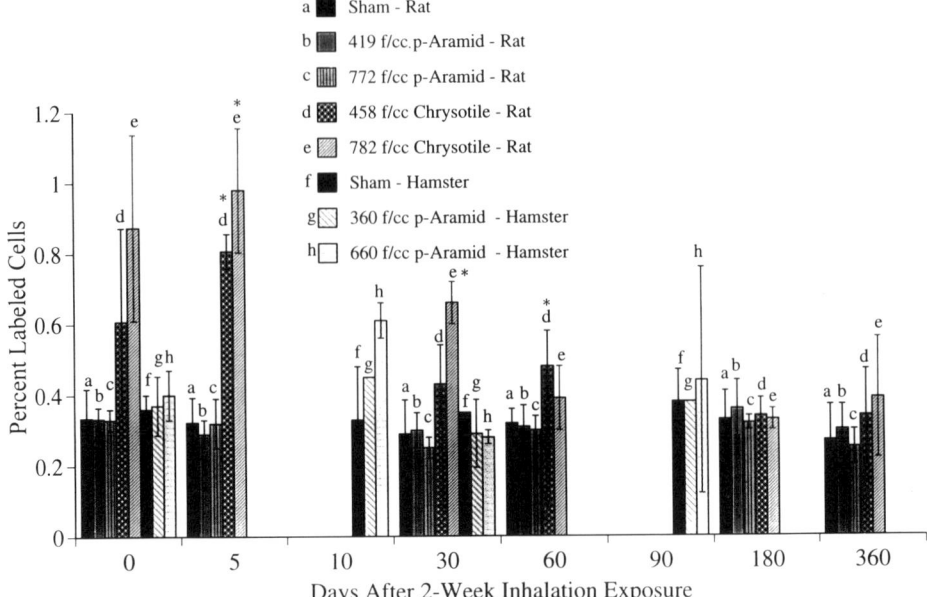

Fig. 4. BrdU labelling index of subpleural and pleural cells in rats exposed to chrysotile asbestos fibres or p-aramid fibrils or hamsters exposed to p-aramid fibrils for two weeks. Values are expressed as mean percentages of labeled cell ± SD. Substantial increases in BrdU immunostaining compared to controls were measured in subpleural regions of asbestos-exposed rats. Significant increases in labeling were measured in high dose groups immediately after, 5 days, 1 and 3 months postexposure before returning to control levels at 6 months postexposure. In contrast, transient increases in cell labeling indices were in p-aramid-exposed rats or hamsters relative to controls immediately after exposure ($P < 0.05$).

two rodent species strongly suggest that para-aramid RFP are likely to be biodegradable in the lungs of exposed humans.

Acknowledgements—This study was sponsored by the DuPont Co. and Akzo Nobel Corp.

REFERENCES

Brody, A. R. and Overby, L. H. (1989) Incorporation of tritiated thymidine by epithelial and interstitial cells in bronchiolar–alveolar regions of asbestos-exposed rats. *Am. J. Pathol.* **134**, 133–140.

Coin, P. G., Roggli, V. L. and Brody, A. R. (1992a) Deposition, clearance and translocation of chrysotile asbestos from peripheral and central regions of the rat lung. *Environ. Res.* **58**, 97–116.

Coin, P. G., Roggli, V. L. and Brody, A. R. (1992b) Pulmonary fibrogenesis and BRDU incorporation after three consecutive inhalation exposures to chrysotile asbestos. *Amer. Rev. Respir. Dis.* **145**, A328.

ECETOC (1996) (European Centre for Ecotoxicology and Toxicology of Chemicals) Technical Report no. 69. Toxicology of man-made organic Fibres.

Kelly, D. P., Merriman, E. A., Kennedy, G. L. Jr and Lee, K. P. (1993) Deposition, clearance and shortening of Kevlar para-aramid fibrils in acute, subchronic and chronic inhalation studies in rats. *Fundam. appl. Toxicol.* **21**, 345–354.

McGavran, P. D., Butterick, C. J., Brody, A. R. (1990) Tritiated thymidine incorporation and the development of an interstitial lesion in the bronchiolar alveolar regions of the lungs of normal and complement deficient mice after inhalation of chrysotile asbestos. *J. Environ. Pathol. Toxicol. Oncol.* **9**, 377–92.

Searl, A. (1997) A comparative study of the clearance of respirable para-aramid, chrysotile and glass fibres from rat lungs. *Ann. occup. Hyg.* **41**, 217–233.

Warheit, D. B., Kellar, K. A. and Hartsky, M. A. (1992) Pulmonary cellular effects in rats following aerosol exposures to ultrafine Kevlar aramid fibrils: Evidence for biodegradability of inhaled fibrils. *Toxicol. appl. Pharmacol.* **116**, 225–239.

Warheit, D. B., Hartsky, M. A., Butterick, C. J. and Frame, S. R. (1995) Pulmonary toxicity studies with man-made organic fibres: Preparation and comparisons of size-separated para-aramid with chrysotile asbestos fibres. In *Toxicology of Industrial Compounds*, (Edited by H. Thomas), Chapter 8, pp. 119–130.

Warheit, D. B., Hartsky, M. A. and Frame, S. R. (1996) Pulmonary effects in rats inhaling size-separated chrysotile asbestos fibres or p-aramid fibrils: differences in cellular proliferative responses. *Toxicol. Lett.* **88**, 287–292.

 Pergamon

Ann. occup. Hyg., Vol. 41, Supplement 1, pp. 334–339, 1997
© 1997 British Occupational Hygiene Society
Published by Elsevier Science Ltd. All rights reserved
Printed in Great Britain
0003–4878/97 $17.00 + 0.00
Inhaled Particles VIII

PII: S0003–4878(96)00154–8

EFFECT OF THE pH AND BIOCHEMICAL MILIEU OF THE LUNG ON RELEASE OF IRON FROM THE SURFACE OF RESPIRABLE INDUSTRIAL FIBRES AND ABILITY TO CAUSE A RESPIRATORY BURST IN ALVEOLAR MACROPHAGES

K. Donaldson, P. Gilmour, P. H. Beswick, C. Fisher
and D. M. Brown

Biomedicine Research Group, Department of Biological Sciences, Napier University, 10 Colinton
Road, Edinburgh EH10 5DT, U.K.

INTRODUCTION

Free radicals are considered to play an important role in the pathogenicity of a number of lung conditions including asbestos-related lung diseases (Donaldson *et al.*, 1994). We have been investigating pro-oxidant characteristics of asbestos and other respirable fibres which cause cancer and fibrosis in humans or experimental animals and have demonstrated that asbestos has strong hydroxyl radical activity that is mediated by iron (Gilmour *et al.*, 1996). An important role for iron has been demonstrated in the TNF response of macrophages exposed to asbestos and, indeed, the chelating of surface iron effectively abolished the TNF-stimulating activity of amphibole asbestos (Simeonova and Luster, 1995). Any particle, including fibres, that deposit in the lung first encounter the lung lining fluid which is a complex mixture of proteins and phospholipids. We have previously demonstrated that the biological activity of fibres can be modulated by coating them with immunoglobulin (Hill *et al.*, 1996), but little is known of the modifying effect of lung lining fluid. The pH of the alveolar surface is close to neutrality, whilst the pH of the phagolysosome is 4–5 (Nyberg *et al.*, 1992). Therefore the ability of any particle depositing in the lung to release iron will be an important factor in dictating its pathogenicity. Furthermore, the release of reductants such as H_2O_2 or superoxide from the macrophage respiratory burst could have important effects in causing redox cycling of iron with the generation of hydroxyl radical. These putative factors that contribute to oxidative stress in fibre-exposed lung are summarised in Fig. 1.

As part of an HSE-funded study we are investigating a panel of respirable fibres of differing pathogenicity to determine which *in vitro* endpoints, related to the above sequence, correlate with pathogenicity, with a view to refining the short-term assay approach for the prediction of fibre pathogenicity. We report here on:

(1) the hydroxyl radical activity of the fibre panel;
(2) the loss of Fe III from the panel;
(3) the effect of lung lining fluid on the loss of Fe III;
(4) the ability of H_2O_2 to modulate the hydroxyl radical activity;

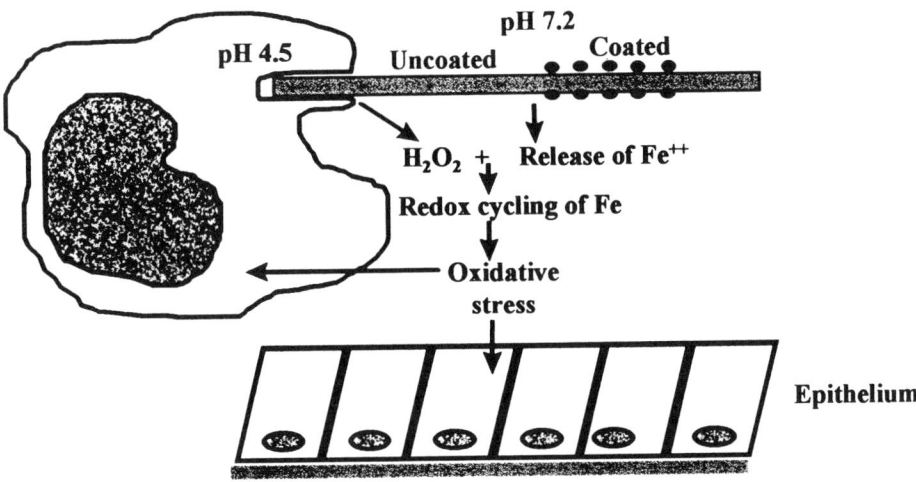

Fig. 1. Possible events that could lead to oxidative stress in macrophages and epithelial cells following deposition of fibre.

(5) the ability of lung lining fluid to modulate the oxidative burst activity of the fibres.

METHODS

The fibres used in the study are shown in the table:

Fibre	Composition	Qualitative estimate of pathogenicity
Amosite asbestos	Amphibole asbestos	Pathogenic
RCF1	Kaolin-based refractory ceramic fibre	Pathogenic
Silicon carbide	Refractory fibre	Pathogenic
RCF4	High temperature-treated sample of RCF1	Non-pathogenic
MMVF10	Fibre glass	Non-pathogenic
JM Code 100/475	Special purpose glass fibre	Non-pathogenic

The ability to break strands of supercoiled plasmid DNA was the sensitive assay used here for detecting particle surface-associated free radicals (Gilmour $et\ al.$, 1995). Briefly φ X174 RF plasmid DNA was incubated with the differing quantities of particles. Each treatment or control sample was incubated at 37°C in a water bath for 8 h. The proportion of the plasmid forms, which provide a measure of the free radical damage to the plasmid, were quantified by scanning laser densitometry of bands in agarose gels. Free radical damage to DNA was expressed as depletion of the supercoiled DNA band. The free radical activity of fibres was also assessed in the presence of H_2O_2. To confirm the role of hydroxyl radicals in the DNA damage, mannitol was used as a blocking agent. Desferal was used to measure the release of Fe II in leachates from fibres at pH 4.5 and pH 7.0. In an additional study the leaching of iron into saline was compared with leaching into lung surfactant solution. The ability of fibres to cause stimulation of the respiratory burst in rat

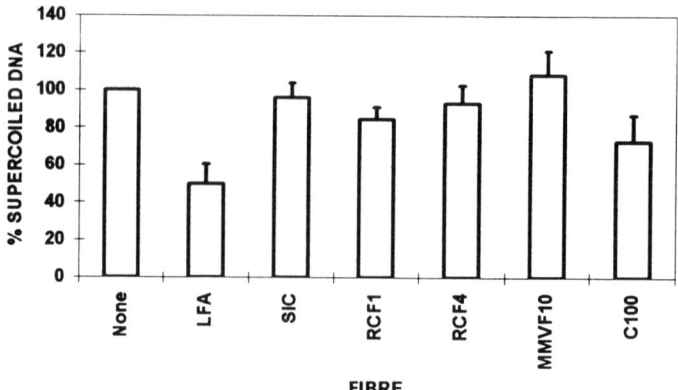

Fig. 2. Hydroxyl radical activity of the fibre panel measured as ability to cause depletion of supercoiled plasmid DNA.

alveolar macrophages was obtained by assessing the release of superoxide anion by the cytochrome c reduction method for either uncoated fibres or fibres incubated in lung lining fluid obtained from the lungs of sheep and rats. All data are given as mean + SEM from three or more separate experiments.

RESULTS

Only long amosite had significant ability to cause free radical damage to plasmid DNA (Fig. 2), this was blockable by mannitol showing it to be mediated by hydroxyl radical (Gilmour *et al.*, 1996).

Different types of fibre showed considerable differences in the ability to release iron (Fig. 3) and in general there was two to three times more release of iron at pH 4.5 than at pH 7.2 (Fig. 3).

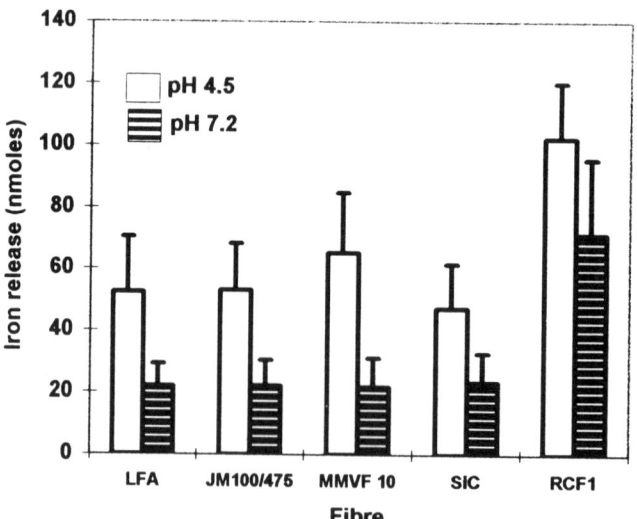

Fig. 3. Release of Fe III from the fibre panel at pH 4.5 and pH 7.2; LFA = long fibre amosite.

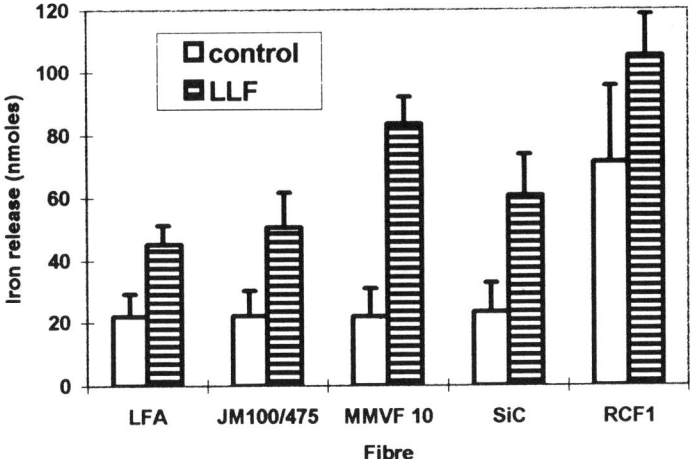

Fig. 4. Release of iron from different fibre types at pH 7.2 in the presence or absence of lung lining fluid (LLF); LFA = long amosite.

Furthermore, when the fibres were placed in lung lining fluid there was much more release of iron than when they were placed in saline (Fig. 4).

When fibres were incubated in lung lining fluid, washed and then tested for their ability to stimulate an oxidative burst, there was substantial inhibition of the oxidative burst with some fibres but little effect with others and there were slight differences between rat and sheep lung lining fluid (Fig. 5).

Hydrogen peroxide on its own was capable of causing moderate free radical

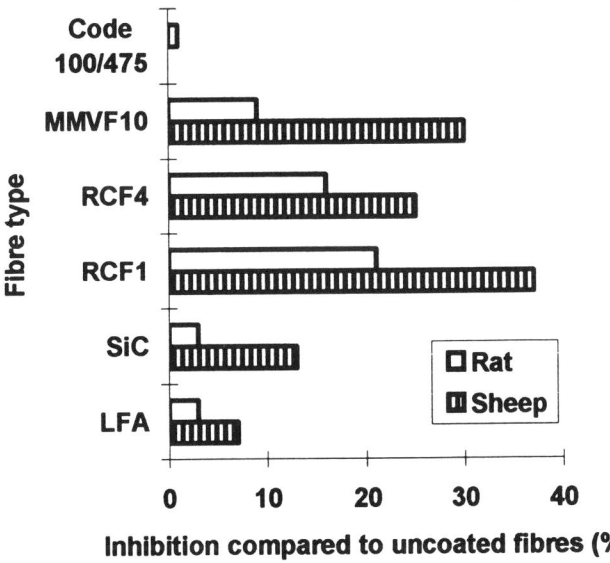

Fig. 5. Effect of coating with rat or sheep lung lining fluid on the oxidative burst caused by different fibre types.

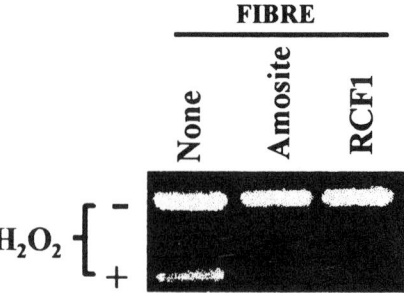

Fig. 6. Image of supercoiled band following incubation with fibres in the presence or absence of H_2O_2.

damage to supercoiled plasmid DNA but in the presence of some fibres, notably amosite and RCF1, there was increased generation of free radicals (Fig. 6). Work is continuing investigating other fibre types in this regard.

Using a range of fibres shown to have different pathogenicities in animal studies we demonstrate that none of the endpoints measured here showed a pattern of response with the different fibres that was consistent with the pathogenicity of the fibres. The data do however confirm the complexity of the events that occur when fibres deposit in the lung and this very likely explains the lack of a correlation between any single endpoint and *in vivo* activity.

With reference to Fig. 1, we confirm that all of the factors shown are likely to dictate the response of the lung to deposited fibre and that these very likely differ between fibre types. For example, only asbestos had substantial intrinsic free radical activity at the surface of the fibres whilst SiC, which is a very pathogenic fibre, had no real activity in this regard. It was clear that the pH milieu affects iron release and that an acid pH will lead to increased leaching of iron. The presence of endogenous "chelators" such as lung lining fluid serve to increase the mobilisation of iron from the fibre and so are major determinants of how much iron is released from fibres. Iron mobilised from fibres could be a factor in leading to oxidative stress but the release of iron under various conditions did not reflect the pathogenicity of the fibre types. Furthermore H_2O_2 from the macrophage respiratory burst can lead to generation of more free radicals by interacting with the surface of some fibre types and this was shown to occur with long amosite and RCF1. On the other hand, a coating of lung lining fluid on a fibre is shown here to inhibit the extent of the oxidative burst for some fibre types, principally RCF1, RCF4 and MMVF10, so decreasing this particular avenue for the generation of oxidative stress.

These preliminary studies form part of a large HSE-funded programme that should allow us to verify the efficacy or otherwise of a number of short-term assays that have been ostensibly used for the determination and prediction of pathogenicity of fibres. Our studies indicate that, on their own, neither intrinsic free radical activity nor ability to release iron in the short term discriminate the pathogenic from the non-pathogenic fibres. Hopefully further studies such as these will allow us to develop rational approaches to short-term testing for fibre toxicity as well as elucidating some aspects of pathogenic mechanism.

Acknowledgements—We acknowledge the financial assistance of the Health and Safety Executive in funding this project and we are grateful to the British Occupational Hygiene Research Foundation for an equipment grant.

REFERENCES

Donaldson, K., Brown, R. C. and Brown, G. M. (1993) Respirable industrial fibres: mechanisms of pathogenicity. *Thorax* **48**, 390–395.

Gilmour, P., Beswick, P. H. and Donaldson, K. (1995) Detection of surface free radical activity of respirable industrial fibres using supercoiled ϕX 174 plasmid DNA. *Carcinogenesis* **16**, 2973–2979.

Hill, I. M., Beswick, P. H. and Donaldson, K. (1996) Enhancement of the macrophage oxidative burst by immunoglobulin coating of respirable fibres: fibre-specific differences between asbestos and man-made fibres. *Exp. Lung Res.* **22**, 133–148.

Nyberg, K., Johansson, V., Johansson, A. and Cramer, P. (1992) Phagolysosomal pH in alveolar macrophages. *Environ. Health Perspect.* **97**, 149–152.

Simeonova, P. P. and Luster, M. I. (1995) Iron and reactive oxygen species in the asbestos-induced tumor necrosis factor-alpha response from alveolar macrophages. *Am. J. Resp. Cell Mol. Biol.* **12**, 676–683.

SECTION 4

COAL, SILICA AND OTHER DUSTS

 Pergamon

Ann. occup. Hyg., Vol. 41, Supplement 1, pp. 341–345, 1997
British Occupational Hygiene Society
Published by Elsevier Science Ltd
Printed in Great Britain
0003–4878/97 $17.00 + 0.00
Inhaled Particles VIII

PII: S0003–4878(96)00168–8

EXPOSURE–RESPONSE FOR COAL WORKERS' PNEUMOCONIOSIS IN UNDERGROUND COAL MINERS: A DISCUSSION OF ISSUES AND FINDINGS

M. Attfield, E. Kuempel and G. Wagner

DRDS & EID, NIOSH, U.S.A.

INTRODUCTION

Over a dozen exposure–response curves for coal workers' pneumoconiosis (CWP) and dust exposure currently exist. Although there is a great deal of similarity in the general approaches adopted to assessing exposure–response in these analyses, there are many points of difference. In particular, some of the factors are: the healthy worker survivor effect, threshold effects and reader effects and their implications for inter-study comparison. This paper takes a preliminary look at these methodological differences and explores them through specific analyses of data on a large group of underground coal miners.

METHODS

The new findings presented in this paper were derived from analyses undertaken on a subset of U.S. coal miners examined in 1969–71 from Round 1 of the National Study of Coal Workers' Pneumoconiosis (NSCWP) (Attfield and Morring, 1992a). The original dataset consisted of X-ray readings on over 9000 underground coal miners, together with cumulative dust exposure and age and coal rank information. For this analysis, only miners for the western Appalachian region (eastern Kentucky, West Virginia, Virginia, Ohio, western Pennsylvania and Alabama) were included, in order to obtain a group that was relatively homogeneous in terms of geography, geology and social aspects ($N = 4500$). The X-rays had been read by one reader using the UICC/Cincinnati scheme (Bohlig *et al.*, 1970). Derivation of the dust exposures is explained in detail in Attfield and Morring (1992b). Briefly, it involved the merger of work history information with job-specific dust concentrations derived from measurements made during 1970–72 after factoring to pre-1970 conditions (before the reduced levels mandated by the 1969 Coal Mine Health and Safety Act) using data from some intensive dust measurement surveys undertaken in 1968.

In all analyses logistic models were used, with cumulative exposure and age as predictor variables and three response dichotomies: category 1+ (category 1/0 or greater), category 2+ (category 2/1 or greater) and presence of PMF. Goodness of fit of the model was ascertained using the coefficient χ^2 value and also by examination of the reduction in the log-likelihood value.

The healthy worker survivor effect was studied in two ways: first, by progressively censoring the miners with the greatest dust exposures, 1% at a time and second, after examination of the data, by censoring those age 55 or greater with exposures 200 ghm^{-3} or greater. Models were fit with and without censoring and the model fit used to assess the evidence for worker selection.

The question of a threshold was examined by employment of a method suggested by Morfeld (Morfeld *et al.*, 1994) in which cumulative exposure is set to zero if less than the threshold value and otherwise set to the difference between the true level and the threshold value. Logistic models were fit as the threshold value was progressively increased from 0–2 mgm^{-3}, and the model goodness of fit assessed. This was done after censoring miners aged 55 or greater with exposure of 200 ghm^{-3} or greater.

In a separate study, concerns about relying on results from only one reader led to the decision to undertake a re-reading exercise of a stratified sample from the original dataset (Attfield and Morring, 1992a). Accordingly, a random sample of 2380 cases was taken and re-read by three readers using the ILO 1980 scheme (International Labour Office, 1980). Particular care was taken to select readers who were representative and to do this, readings from reader participants in the NIOSH Coal Workers' X-ray Surveillance Program (CWXSP) were tabulated and ranked. From these, a set of three were chosen whose readings appeared to straddle the median prevalence level. In the current paper the median determinations based on the results of the re-reading are compared with the original determinations, both in terms of absolute prevalence, and in terms of exposure–response. This leads on to a comparison of findings from some additional NIOSH analyses (Attfield and Seixas, 1995; Attfield and Morring, 1992a; Attfield, 1996a, 1996b) and from one British study (Hurley and Maclaren, 1987).

RESULTS

Exploration of the healthy worker survivor effect by progressive censoring of the extreme cumulative exposures in 1% increments showed that the logistic coefficient for category 1+ on exposures increased by 25% initially (from 0.008 for all data to 0.011 after 1% censored) and then stabilized. This increase was accompanied by a rise in the χ^2 value from 79 to 107, which occurred despite the fewer observations and reduction in exposure range (two factors which would normally tend to lower statistical significance). Tabulation of the observed prevalences by age and exposure together with predicted exposures from a logistic fit to the complete dataset showed that the observed prevalences were much greater than predicted for high cumulative exposures (150 ghm^{-3} or above) in younger miners, but much lower in those 55 or older with cumulative exposures 200 ghm^{-3} or greater, suggesting healthy worker selection. After censoring cases aged 55 or over with exposure 200 ghm^{-3} or above, on the assumption that they were affected by healthy worker selection, the observed and predicted prevalences throughout showed good agreement. As with the previous approach, the exposure coefficient showed a 25% rise and increased χ^2 value. The effect was apparent for category 1+, category 2+ and PMF, although the χ^2 did not increase for the category 2+ response. Predictions for a working life were reduced at zero exposure and

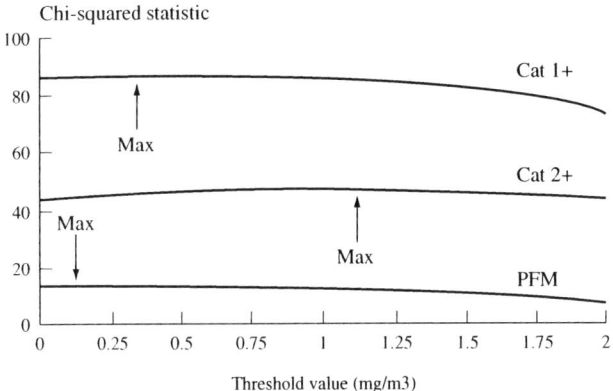

Fig. 1. Examination of threshold. Plot of χ^2 statistics against threshold value for category 1+, category 2+ and PME.

increased at 2 mgm^{-3} censoring, consistent with a rotation of the exposure–response curve around the overall mean exposure and response.

Determination of the existence of a threshold effect, through use of the transformation CE–(CE-TLV)· H (CE-TLV), where CE is cumulative exposures and H (CE-TLV) = 0 if CE < TLV, and 1 otherwise, was examined using the $\chi 2$ value for the coefficient for transformed exposure variable. Figure 1 shows a plot of this statistic for three outcomes: category 1+, category 2+ and PMF for a range of TLV from 0–2 mgm^{-3}. It is clear from this figure that there was little convincing indication of a threshold. For category 1+ and PMF, χ^2 peaked close to 0 mgm^{-3}, while for category 2+ the peak was near to 1 mgm^{-3} but the curve was virtually flat, suggesting great uncertainty in the location of any threshold. Use of the log-likelihood value in place of χ^2 suggested even less evidence for a threshold. In other analyses (Attfield and Morring, 1992a; Attfield and Seixas, 1995), rather than a threshold, there was evidence of a non-zero baseline of response at zero dust exposure.

Figure 2 shows the working-life predicted prevalences of category 1+ obtained from identical logistic models applied to the original study (Attfield and Morring, 1992a) and the median of the re-readings (Attfield, 1996b) for 0, 1 and 2 mgm^{-3}. The new classifications gave rise to much higher predicted prevalences than did those from the earlier study by a factor of about four. What is interesting, however, is that the logistic coefficients from the two studies for cumulative exposure were almost identical (0.008 for the original study and 0.010 for the re-readings). The age coefficients were identical at 0.034, but the intercepts were quite different, at −5.03 and −3.66. From this it is apparent that the estimated odds ratio for different exposure levels is virtually the same from both studies, although the predicted absolute prevalence levels were drastically different.

This finding prompted a comparison of findings from five different studies. These included: (R1X), the original investigation by Attfield and Morring (1992a] upon which the new analyses presented here are based; (R1XRR), the re-readings study (Attfield, 1996b); (R4X), a cross-sectional study from Round 4 of the NSCWP (Attfield and Seixas, 1995); (R4L), a longitudinal study from Rounds 1 and 2 to

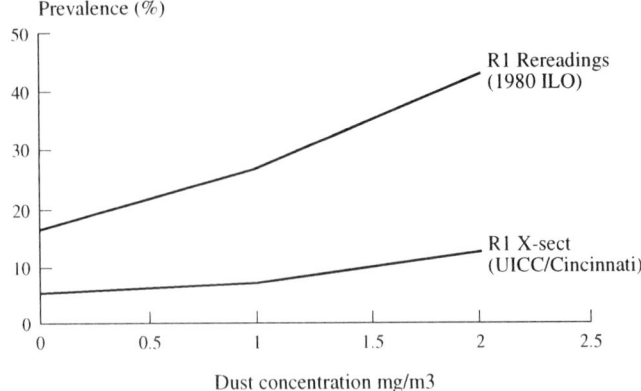

Fig. 2. Prevalence of category 1+ for original study and re-readings.

Round 4 of the NSCWP (Attfield, 1996a); and (GBL), a longitudinal study of British miners (Hurley and Maclaren, 1987). In all cases the results presented pertain to predicted prevalences for a 40-year working life at age 58 for high volatile bituminous coal at dust concentrations of 1 and 2 mgm.$^{-3}$

Table 1 gives the predicted prevalences for the five studies for three outcomes. For category 1+, there was a wide range in predicted prevalence, from 27 to 4% for 1 mgm^{-3} and 43 to 9% for 2 mgm^{-3}. In contrast, the ranges in predictions for the other two outcomes were much less, particularly for category 2+. The predictions from the British study were consistently lower than for those from the U.S. studies. What is noteworthy about the results in Table 1 is that despite the disagreement in absolute predicted prevalences, when the relative risks for 1 v. 2 mgm^{-3} are considered, there is much greater uniformity across studies. This uniformity occurs for all three outcomes. Moreover, the OR is similar across the three outcomes and averages to about 0.55 overall.

Had the exposure–response curves been such that they forced there to be zero predicted prevalences at zero dust exposure, the finding of similar ORs across the different outcomes and studies would have been a trivial outcome resulting from interpolating an approximately linear relationship halfway between zero and 2

Table 1. Predicted working life prevalences at age 58 of category 1+, category 2+, PMF for five different studies and relative risks and associated relative risks for 1 v. 2 mgm^{-3}

| | | Predicted prevalences (%) | | | RR for 1 v. 2 mgm^{-3} | | |
		1+	2+	PMF	1+	2+	PMF
1 mgm^{-3}	R1XRR	27	3.3	1.6	0.63	0.71	0.63
	R4L	9	2.3	1.2	0.45	0.42	0.36
	R4X	8	1.7	0.9	0.56	0.56	0.63
	R1X	4	2.3	1.7	0.61	0.56	0.77
	GBL	4	1.0	0.3	0.43	0.40	0.43
		Geometric mean RR			0.54	0.53	0.56
2 mgm^{-3}	R1XRR	43	4.7	2.5	R1 Re-readings		
	R4L	20	5.3	3.3	R4 Longitudinal		
	R4X	14	3.1	1.4	R4 Cross-sectional		
	R1X	12	4.0	2.2	R1 Cross-sectional		
	GBL	9	2.5	0.7	British longitudinal study		

mgm^{-3}. However, as can be seen from Fig. 1, for the re-readings data (R1XRR), the predicted prevalence at zero dust exposure was much greater than zero. Hence, the observed result in this case is not so simply explained. Instead, it appears to point to the possibility that relative contrasts in exposure–response relationships of CWP can be ascertained much more consistently than can absolute prevalences levels be estimated. Moreover, the similarity in RR across studies provides consistent evidence of the health benefits of reducing dust exposures.

CONCLUSIONS

The findings from this study indicate: (1) the presence of a healthy worker survivor effect, leading to over- and under-estimation of predicted prevalences at the extremes of dust exposure; (2) the lack of convincing evidence for a threshold, at least above 1 mgm^{-3}; (3) that reader effects can drastically affect estimates of absolute prevalence levels of CWP, but that estimated relative effects are remarkably consistent. Confirmation for these findings is being examined in other cohorts of coal miners. So far, the results appear to support those presented here.

REFERENCES

Attfield, M. D. (1996a) Working-life predictions of risk of coal workers' pneumoconiosis in U.S. underground coal miners based on incidence data (submitted).
Attfield, M. D. (1996b) Exposure–response for coal workers' pneumoconiosis in U.S. underground coal miners: results from re-readings of radiographs taken 1969–71 (submitted).
Attfield, M. D. and Morring, K. (1992a) An investigation into the relationship between coal workers' pneumoconiosis and dust exposure in U.S. coal miners. *Am. ind. Hyg. Assoc. J.* **53**, 486–492.
Attfield, M. D. and Morring, K. (1992b) The derivation of estimated dust exposures for U.S. coal miners working before 1970. *Am. ind. Hyg. Assoc. J.* **53**, 248–255.
Attfield, M. D. and Seixas, N. S. (1995) Prevalence of pneumoconiosis and its relationship to dust exposure in a cohort of U.S. bituminous coal miners and ex-miners. *Am. J. ind. Med.* **27**, 137–151.
Bohlig, H., Bristol, L. J., Cartier, P. H., Felson, B., Gilson, J. C., Grainger, T. R., Jacobson, G., Kiviluoto, R., Lainhart, W. S., McDonald, J. C., Pendergrass, E. P., Rossiter, C. E., Selikoff, I. J., Sluis-Cremer, G. K. and Wright, G. W. (1970) UICC/Cincinnati classification of the radiographic appearance of pneumoconiosis. *Chest* **58**, 57–67.
Hurley, J. F. and Maclaren, W. M. (1987) Dust-related risks of radiological changes in coalminers over a 40-year working life: report on work commissioned by NIOSH. Institute of Occupational Medicine, Edinburgh, Scotland.
International Labour Office (1980) International classification of radiographs of pneumoconiosis (revised edition 1980) (Occupational Safety and Health Series no. 22 (Rev. 80)), International Labour Office, Geneva, pp. 1–48.
Morfeld, P., Rohleder, F., Vautrin, H.-J., Kampmann, B. and Piekarski, C. (1994) An epidemiological approach in estimating a threshold limit value for respirable dust in German hard coal mining. *Ann. occup. Hyg.* **38**, 799–803.

 Pergamon

Ann. occup. Hyg., Vol. 41, Supplement 1, pp. 346–351, 1997
© 1997 British Occupational Hygiene Society
Published by Elsevier Science Ltd. All rights reserved
Printed in Great Britain
0003–4878/97 $17.00 + 0.00
Inhaled Particles VIII

PII: S0003–4878(96)00155–X

OVERALL MORTALITY AND CANCER MORTALITY OF COAL MINERS: ATTEMPTS TO ADJUST FOR HEALTHY WORKER SELECTION EFFECTS

P. Morfeld, K. Lampert, H. Ziegler, C. Stegmaier, G. Dhom and C. Piekarski

Institut für Arbeitswissenschaften der Ruhrkohle AG, Wengeplatz 1, 44369 Dortmund, Germany

INTRODUCTION

The presumption of a possibly elevated cancer risk in coalminers is based mainly on two hypotheses. Firstly most coalmine dusts include a relevant quartz dust fraction. With a view to the possibly causal relationship between pure quartz dust exposure/silicosis and lung cancer excess risk it is of interest to investigate whether there exists an elevated lung cancer risk in coalminers. Secondly a fraction of the inhaled coalmine dust particles is cleared from the lung. It is probable that some of the cleared dust is swallowed, which then interacts with the acidic environment in the stomach and finally may cause an elevated gastric cancer risk in coalminers. To explore whether such elevated cancer risks exist, we organised a mortality follow up study on German hard coal miners.

MATERIAL AND DESCRIPTIVE RESULTS

We focused on the cancer risk of coalminers of the Saar area (the second biggest hard coal mining area of Germany) because firstly there is a rather high quartz content in the respirable coalmine dust of that area (on average about 12%) and secondly because the only reliable population based cancer registry in West Germany is working there. Because we had no means to enroll a classical inception cohort we had recourse to a study group of Saar coalminers used for a pneumoconiosis study. The main features of the inclusion criteria at that time were firstly that the miners were known to the personnel departments in 1977 or 1979 and secondly that they had worked at least 5 years underground.

All in all, 4578 coalminers alive at 1st January 1980 were enrolled in the study for a 12 year mortality follow-up to the end of 1991, constructing 52 967 person–years at risk. Vital status could be determined for all coalminers (100%). 317 deaths were documented. For five cases, 1.6%, the cause of death remained unknown.

BASIC ANALYSIS

Methods

Mortality was standardised indirectly by 5 year intervals of age and calendar time on the male population of the Saar area by calculating observed and expected

Table 1. Estimates of relative standardised mortality ratios RSMR and 0.95-confidence intervals for overall mortality, cancer mortality, lung cancer mortality and stomach cancer mortality (4578 German coalminers, follow up from 01/01/1980 through 12/31/1991)

Mortality end point	Observed number	RSMR	0.95-Conf.int.
Cancer	104	1.03	0.84–1.25
Lung cancer	41	1.11	0.80–1.51
Stomach cancer	6	0.98	0.36–2.11
[15 year follow-up	13	1.46	0.78–2.50]

numbers of deaths with the help of the person–years program. We derived SMRs for all causes, all cancers, lung cancer and stomach cancer as well as relative SMRs ($RSMR = SMR_{specific}/SMR_{overall}$) for the cancer causes of death to adjust for the expected "healthy worker"-selection. Confidence limits for the SMR were estimated via the Poisson distribution and for the RSMR by multiplying the expected number with $SMR_{overall}$.

Results

Among the 317 deaths 104 cancer deaths, 41 lung cancer deaths and 6 stomach cancer deaths were recorded: SMRs for all causes (0.63), cancer (0.65), lung cancer (0.70) and stomach cancer (0.62) are below one; with the exception of stomach cancer even significantly below one. The low relative overall mortality of 63% points to a pronounced healthy worker selection.

Based on RSMRs (Table 1) slight non-significant increases in cancer and lung cancer risk are described, but they can easily be explained by smoking habits or even as a chance finding. It should be remarked that a 15 year follow-up through 1994 is nearly completed now and the risk estimates for cancer and lung cancer seem to be stabilised near the results of the 12 year follow-up, but for stomach cancer the longer follow up yields an elevated RSMR of about 1.5. Therefore, this basic analysis gives no hint at a cancer or lung cancer excess risk in coalminers but leaves some doubt about stomach cancer risk.

META-ANALYSIS

Material and methods

12 SMR-follow up studies on coalminers (Enterline, 1964; Ortmeyer et al., 1973, 1974; Costello et al., 1974; Rockette, 1977; Armstrong et al., 1979; Atuhaire et al., 1985; Maclaren, 1992; Kuempel et al., 1995; Starzynski et al., 1995; Swaen et al., 1995; Morfeld et al., 1996) and five OR-case control studies on coalminers with external controls (Meijers et al., 1990; Morabia et al., 1992; Muller et al., 1995; Swaen et al., 1987; Gonzales et al., 1991) were pooled according to Greenland, 1987. Because the study of Enterline (1964) gives strikingly high risk estimates and this probably due to an underestimate of the number of exposed, we performed two analyses including and excluding the study of Enterline (1964).

Results

The study designs are clearly different, inclusion criteria vary a lot and it is not surprising that the χ^2-statistics prove a significant heterogeneity between the studies

Table 2. Pooling of relative risk estimates from 17 studies on coalminers. RR_{pool}:point estimate, $0.95-CI_{pool}$:interval estimate, CHI^2:statistics of heterogeneity between studies, 0.95–FRACTILE of the corresponding χ^2-distribution

	RR_{pool}	$0.95-CI_{pool}$	χ^2	0.95–Fractile
Overall mortality (11 studies)				
with Enterline, 1964	1.19	1.18–1.21	2206	18.31
without Enterline, 1964	1.08	1.06–1.09	507	16.92
Cancer risk (7 studies)				
with Enterline, 1964	1.11	1.07–1.14	253	12.59
without Enterline, 1964	0.98	0.94–1.01	24	11.07
Lung cancer risk (13 studies)				
with Enterline, 1964	1.02	0.98–1.07	116	21.03
without Enterline, 1964	0.96	0.92–1.01	47	19.68
Stomach cancer risk (10 studies)				
with Enterline, 1964	1.56	1.45–1.69	73	16.92
without Enterline, 1964	1.34	1.23–1.46	16.23	15.51

on the 5%-level for each endpoint considered. As expected heterogeneity is increased by including Enterline (1964). Therefore the interpretation of the pooled point and interval estimates is strictly limited. However, the results on relative risks of cancer and lung cancer demonstrate no excess risk in coalminers. Moreover we find it remarkable, that the pooling of nine studies on stomach cancer risk in coalminers yields a significantly elevated relative risk estimate of 1.34, together with a just borderline significant statistic of heterogeneity.

Thus to summarise, from Table 2 (Table 1), one might get the impression, that there exists no cancer and no lung cancer excess risk in coalminers, but an elevated gastric cancer risk is indicated, which of course will have to be discussed on the background of confounding in detail.

EXPLORING HEALTHY-WORKER–SURVIVOR EFFECTS

Being aware of a growing epidemiological evidence of an association of quartz dust exposure and lung cancer excess risk the negative results on coalminers' lung cancer risk are somewhat surprising. Therefore we checked whether a healthy worker survivor effect (HWSE) is at work that may bias downward the estimates of relative risk of lung cancer death. A survivor effect working on morbidity endpoints has already been demonstrated for coalminers. A plausible pathway of exposure effects taking a HWSE for mortality into account explicitly can therefore be outlined as follows.

Long term exposure to coal mine dust leads to (lung-)diseases like pneumoconiosis or chronic bronchitis. After developing a severe degree of disease exposure is stopped (= change in work status). Therefore higher mortality is expected as an indirect outcome of long term exposure mediated by a selection out of the job. Simultaneously this longitudinal selection produces a lower mortality as a direct effect of continued exposure. This selection effect may be amplified by an intervention of occupational physicians stopping exposure for diseased coalminers as it is enforced by law in Germany.

Table 3. Coefficients in SMR–Poisson models by cohort restriction and by inclusion of work status as a covariable. Variables: TSFE = time since first exposure/year, status = 0, during exposure; 1, after cessation of exposure

ICD	TSFE	Model I b_{TSFE}	Model II b_{STATUS}	Model II b_{TSFE}
001–999	≥ 0 year	+ 0.0053	+ 0.634	− 0.0089
	≥ 10 years	+ 0.0050	+ 0.640	− 0.0094
	≥ 20 years	+ 0.0106	+ 0.578	− 0.0022
140–208	≥ 0 year	+ 0.0098	+ 1.385	− 0.0134
	≥ 10 years	+ 0.0149	+ 1.493	− 0.0089
	≥ 20 years	+ 0.0174	+ 1.362	− 0.0042
162	≥ 0 year	+ 0.0077	+ 1.529	− 0.0136
	≥ 10 years	+ 0.0075	+ 1.529	− 0.0139
	≥ 20 years	+ 0.0149	+ 1.386	− 0.0044

Methods

One simple method to check for the HWSE is to split the total follow up period by work status, into the time during work underground and the time after cessation of work underground and to compare SMR/RSMR-estimates. A death is counted to occur during exposure when cessation of exposure and death happened in the same month.

Time since first exposure (= TSFE) and work status are logically and causally independent. Thus it makes sense to fit SMR/RSMR-Poisson models including the general mean and TSFE (model I) and models extended by work status (model II). Under the HWSE-assumption that nearly all detrimental effects of exposure are indirect, approximately all adverse effects of TSFE are absorbed by the status variable. Therefore the coefficient for TSFE in model II is approximating the direct effect of exposure, the HWSE. Additionally a left censoring of TSFE up to 10 years and 20 years was tried, a time dependent restriction of cohort.

Cox models with time dependent covariables (time underground − duration of exposure, calendar time) are constructed to analyse mortality by age. Specific risk windows are explored (left censoring of the risk period from 0 to 30 years, left censoring in the range from 5 years before up to 5 years after cessation of exposure) together with a lagging of exposure (0–20 years). Risk window specification is an extension of the restriction method proposed by Fox and Collier (1976); lagging to control for the HWSE is proposed by Gilbert (1982).

Results

(4578 German coalminers, follow up from 01/01/1980 through 12/31/1991).

SMRs during/after cessation of work underground for overall (0.40/0.70), cancer (0.21/0.76) and lung cancer mortality (0.19/0.81) clearly demonstrate a longitudinal healthy worker selection (even when relying on RSMRs).

The restriction of cohort produced the expected outcome: the TSFE-coefficient is rising in general (model I, Table 3). The approximation of the HWSE is empirically reflected by the change in sign of the TSFE-effect (model I vs model II, Table 3).

Adjusted coefficients (probably still biased downward) can simply be derived by calculating the difference between the coefficients in model I and model II. These

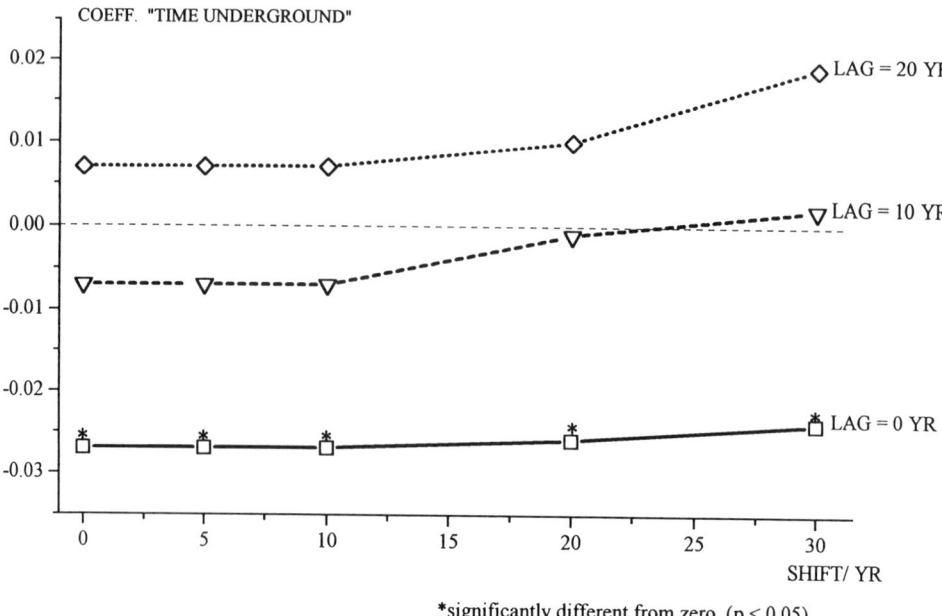

Fig. 1. Coefficients in Cox-models with time dependent covariables analysing overall mortality by age. Risk window: 01/01/1980–12/31/1991 and after start of work underground with shift; covariables: time underground with lag, calendar time.

adjusted coefficients are quite independent from cohort restriction and remarkably greater than the coefficients from standard analyses (model I, TSFE ≥ 0 year; Table 3). So all the evidence points to the conclusion that a standard analysis suffers from a severe underestimate of the exposure effect on overall mortality, cancer mortality and lung cancer mortality.

Thus, it is not surprising that traditional Cox-models show negative coefficients for the effect of time underground on overall ($P = 0.009$), cancer ($P = 0.12$) and lung cancer mortality ($P = 0.20$).

The first square of the bottom line in Fig. 1 represents the coefficient of the traditional model. Again the results illustrate the possibility of a severe misinterpretation of the effect of time underground on overall mortality in standard models without considering healthy worker survivor effects.

Because death cannot occur before cessation of exposure it is of interest to explore the dependence of risk coefficients on differently specified risk windows starting around the end of work underground. The most interesting result was a change of the significantly negative coefficient of the traditional model into a significantly positive one by choosing a left censoring up to, but excluding the end of work underground (no lag). Basically the same relationship was found for cancer mortality too. This may point at an additional direct adverse effect of exposure on mortality that a standard analysis is unable to detect.

SUMMARY AND CONCLUSION

We think that, based on standard analyses, no cancer, no lung cancer, but a stomach cancer excess risk is demonstrable among coalminers. These findings of our study are broadly in accordance with the results from other investigations. Regarding stomach cancer risk the possibility of confounding has to be considered in detail.

A more sophisticated analysis trying to explore the degree of healthy worker survivor effects points at probably masked excesses in overall, cancer and lung cancer mortality among coalminers. Therefore, it has to be discussed whether models, beyond those traditionally based on cumulative exposure and duration of exposure, are acceptable or even necessary to describe the exposure–response relationship for lung cancer in coalminers adequately.

Acknowledgements—The work was organised and funded by the Arbeitsgemeinschaft des Saarlands zur Erforschung und Förderung des Gesundheitsschutzes im Bergbau e. V. (AGiB).

REFERENCES

Armstrong, B. K., McNulty, J. C. and Levitt, L. J. (1979) Mortality in gold and coal miners in Western Australia with special reference to lung cancer. *Brit. J. ind. Med.* **36**, 199–205.

Atuhaire, L. K., Campbell, M. J., Cochrane, A. L. *et al.* (1985) Mortality of men in the Rhondda Fach 1950–80. *Brit. J. ind. Med.* **42**, 741–745.

Costello, J., Ortmeyer, C. E. and Morgan, W. K. C. (1974) Mortality from lung cancer in U.S. coal miners. *Am. J. Publ. Health* **64**, 222–224.

Enterline, P. E. (1964) Mortality rates among coal miners. *Am. J. Publ. Health*, 758–768.

Fox, A. J. and Collier, P. F. (1976) Low mortality rates in industrial cohort studies due to selection for work and survival in the industry. *Br. J. Prev. Soc. Med.* **30**, 225–230.

Gilbert, E. S. (1982) Some confounding factors in the study of mortality and occupational exposures. *Am. J. Epidemiol.* **116**, 177–188.

González, C. A., Sanz, M., Marcos, G., *et al.* (1991) Occupation and gastric cancer in Spain. *Scand. J. Work. Environ. Health* **17**, 240–247.

Greenland, S. (1987) Quantitative methods in the review of epidemiologic literature. *Epidem. Rev.* **9**, 1–30.

Kuempel, E. D., Stayner, L. T., Attfield, M. D., *et al.* (1995) Exposure response analysis of mortality among coal miners in the United States. *Am. J. ind. Med.* **28**, 167–184.

Maclaren, W. M. (1992) Coalminers' mortality in relation to low-level exposures to radon and thoron daughters. Institute of Occupational Medicine, Edinburgh IOM Report TM/92/06.

Meijers, J. M. M., Swaen, G. M. H., van Vliet, K. and Borm, P. J. A. (1990) Epidemiologic studies on inorganic dust-related lung diseases in the Netherland. *Exp. Lung Res.* **16**, 15–23.

Morabia, A., Markowitz, S., Garibaldi, K. and Wynder, E. L. (1992) Lung cancer and occupation: results of a multicentre case-control study. *Brit. J. ind. Med.* **49**, 721–727.

Morfeld, P., Lampert, K., Ziegler, H, *et al.* (1996) Coal mine dust exposure and cancer mortality in German coal miners. 2nd International Conference on the Health of Miners, Pittsburgh 1995. *J. appl. occ. Environ. Hyg.* (in press).

Muller, M., Bartsch, P., Albert, A., Huybens, D. and Rondia, D. (1995) Epidemiological study about occupational risk factors of lung cancer in the province of Liege, Belgium. *5th Int. Conf. Environmental and Occupational Lung Diseases*, Orlando, Florida.

Ortmeyer, C. E., Baier, E. J. and Crawford, G. M. (1973) Life expectancy of Pennsylvania coal miners compensated for disability. *Arch. Environ. Health* **27**, 227–230.

Ortmeyer, C. E., Costello, J. and Morgan, W. K. C. (1974) The mortality of Appalachian coal miners, 1963–1971. *Arch. Environ. Health* **29**, 67–72.

Rockette, H. E. (1977) Cause specific mortality of coal miners. *J. occ. Med.* **19**, 795–801.

Starzynski, Z., Marek, K., Kujawska, A., *et al.* (1995) Mortality among silicotics in Poland. A cohort study. 11th International Symposium on Epidemiology in Occupational Health, Noordwijkerhout. The Netherlands, September 1995. Abstracts of papers presented at the meeting. *Epidemiol.* **6**, S.118.

Swaen, G. M., Aerdts, C. W. H. M. and Slangen, J. J. M. (1987) Gastric cancer in coalminers: final report. *Brit. J. ind. Med.* **44**, 777–779.

Swaen, G. M., Meijers, J. M. M. and Slangen, J. J. M. (1995) Risk of gastric cancer in pneumoconiotic coal miners and the effect of respiratory impairment. *Occ. Environ. Med.* **52**, 606–610.

Pergamon

Ann. occup. Hyg., Vol. 41, Supplement 1, pp. 352–357, 1997
British Occupational Hygiene Society
Published by Elsevier Science Ltd
Printed in Great Britain
0003–4878/97 $17.00 + 0.00
Inhaled Particles VIII

PII: S0003–4878(96)00169–X

THE ROLE OF COAL MINE DUST EXPOSURE
IN THE DEVELOPMENT OF PULMONARY EMPHYSEMA

V. Vallyathan,* F. H. Y. Green,‡, P. Brower† and M. Attfield†

*Division of Health Effects Laboratory, NIOSH, Morgantown, WV, U.S.A.;
†Division of Respiratory Disease Studies, NIOSH, Morgantown, WV, U.S.A.; and ‡Department of
Pathology, University of Calgary, Calgary, Alberta, Canada T2N 4N1

INTRODUCTION

Coal miners have long been recognised to be at increased risk for several forms of pneumoconioses, including macules, nodules, progressive massive fibrosis (PMF) and silicosis, as well as for other chronic lung diseases (Kleinerman et al., 1979). Although focal emphysema is recognised as an integral part of the lesion of simple CWP (Kleinerman et al., 1979), the relationship between coal mining and disabling emphysema is an issue that still sparks considerable scientific debate and controversy (Hurley and Soutar, 1986; Gee and Morgan, 1979; Seaton, 1990).

In this study we examine the relationship between severity of emphysema and years of employment in underground coal mining, years smoked, retained dust in the lung, and CWP. Our findings show a positive association between coal mining, and emphysema and are similar to previously reported autopsy studies from Europe and Australia (Ruckley et al., 1984; Leigh et al., 1994; Lyons et al., 1981; Cockcroft et al., 1982).

METHODS

Population

The study group consisted of 266 underground coal miners autopsied at the Beckley Southern Appalachian Regional Hospital, Beckley, West Virginia from 1957–1971. All miners had at least 1 year of underground bituminous coal mining experience and had worked in various mines within a 100 mile radius of Beckley. Available information consisted of the following: age at death, smoker (never, ever), years smoked, underground coal mining tenure and cause of death. A comparison population of 75 non-miners comprised 25 (13 male, 12 female) autopsies from the same hospital during the same period of time and a series of 50 autopsies (all male) collected and processed in a similar manner at the University of Vermont from 1972–1978. The latter were derived from a population of Medical Examiner deaths. At autopsy, whole left lungs (except when trauma or neoplasm precluded their use) were removed and cannulated through the bronchus and infused with a 4% buffered formaldehyde solution at a constant pressure of 30 cm of water for 1 h. A sagittal slice of the whole lung was embedded in a gelatin solution according to the procedure of Gough et al. (1952) and semi-thin sections

($\sim 200\ \mu m$) were cut on a sledge microtome. Tissue blocks were taken for histologic examination from representative areas of the right and left lungs.

Pneumoconiosis classification and grading
The whole lung sections were evaluated by two investigators (VV and FHYG) according to the diagnostic criteria established by the joint committee of the College of American Pathologists and the National Institute for Occupational Safety and Health (Kleinerman et al., 1979). Macules, micronodules, macronodules and PMF were each graded on a four-point scale of increasing severity within the whole lung section by both readers independently, in random order without knowledge of historical information, radiographic classification, or autopsy diagnosis. The grades were based on a series of reference standards established and scored for CWP prior to the grading process (Hu et al., 1990). Items of disagreement were subsequently resolved by consensus with reference to the standard lung sections and the consensus scores were used in the analyses.

Emphysema grading
The severity of emphysema (emphysema index) in whole lung sections was graded by two pathologists (VV and FHYG) using a set of photographic standards (Thurlbeck et al., 1974) in conjunction with a 10 segment grid (Saito and Thurlbeck, 1995). Each zone was graded as though it were the whole lung using the photographic standard panel to give a maximum score of 100 for each zone and a total maximum score of 1000 for each lung.

Analysis of the lungs for total dust, total mineral, coal and silica
For a subgroup of 63 miners, a 1.5 cm slice of lung adjacent to the whole lung section had been analysed for total dust, coal dust, total mineral and free silica using standardized protocols. Data were expressed as $\mathrm{mg(dust)g^{-1}}$ of dry lung.

Statistical analysis
Statistical analyses including Wilcoxon rank score test, t-test, Pearson's correlation coefficient and graphical techniques were used to explore the relationships between the primary variable emphysema and the following: years smoked, years worked underground, and retained dust for miners (never-smokers and ever-smokers) and non-miners (never-smokers and ever-smokers).

RESULTS

Population characteristics
Tables 1 and 2 show the breakdown of the cases used in this study by age at death, underground mining tenure and smoking history. The overall average age of the non-miner (51.7) group was less than that of the coal miner (66.6) group (Table 1). However this difference was not statistically significant. There was, however, a statistically significant difference in age between the 50 medical examiner cases from Vermont (49 ± 16 years) and the 25 West Virginia non-miners (57 ± 14 years) and the 266 miners (66.1 ± 10.1 years).

V. Vallyathan *et al.*

Table 1. Demographic characteristics of study groups

	Coal miners	Non-miners
Number of cases	266	75
Age at death	66.6 ± 10.1*	51.7 ± 15.8
Years of underground mining	32.1 ± 10.6	—
Years smoked	36.9 ± 14.5	28.2 ± 15.1

* Means ± standard deviation. Differences between groups were not statistically significant.

Table 2. Study groups: distribution by smoking status

Group	Never smokers N	Ever smokers N	Total
Non-miners	14	61	75
Miners	41	225	266
Total	55	286	341

Table 3. Distribution of emphysema by mining and smoking

Group	Never smokers	Ever smokers	P-Value
Non-miner	46 ± 15.5	152 ± 18.4	0.0038*
Miner	295 ± 36.4	392 ± 16.3	0.0015†
P-Value	0.0001*	0.0001*	—

* Wilcoxon rank score test. † t-test.

Relationships between emphysema index, mining tenure and smoking

The distribution of emphysema scores for the miners and non-miners by smoking status is shown in Figs 1 and 2. The distribution of emphysema index appears to be influenced independently by smoking and years underground (figure not shown). Centriacinar (including focal) emphysema was the most prevalent type of emphysema in both never-smokers and ever-smokers.

The mean emphysema index and its standard error for non-miners and miners by smoking status are shown in Table 3. The emphysema index was significantly greater for ever-smokers than for never-smokers within the mining group (P-values < 0.005) and for miners compared to non-miners within the smoking group (P-value $= 0.0001$). The effect of smoking and mining on the emphysema index appears additive in this basic model.

Table 4 shows the correlation between emphysema and years worked underground and years smoked for miners. For miners the relationship between emphysema and years of underground mining was stronger for ever-smokers ($r = 0.45$) than for never-smokers ($r = 0.18$).

Relationship between pneumoconiosis and emphysema index

A majority of miners had macules (97%), 67% had micronodules, 41% had macronodules, 27% had PMF and 13% had silicosis. There was a good overall correlation between emphysema index and severity of CWP by pathologic grading ($r = 0.51$, $P = 0.0001$).

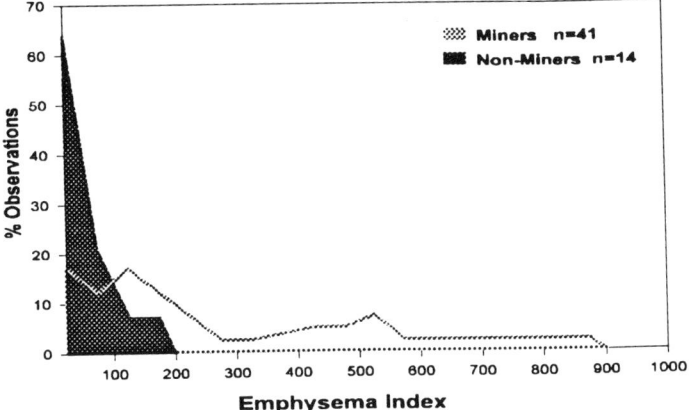

Frequency distribution of Emphysema Index for Never-smokers

Fig. 1.

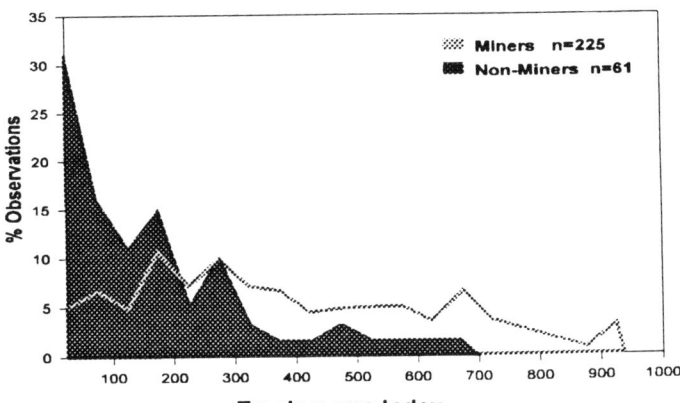

Frequency distribution of Emphysema Index for Ever-smokers

Fig. 2.

Table 4. Correlation between emphysema index and years worked underground and years smoked for miners

		Years worked underground	P-Value		Years smoked	P-Value
Emphysema	All miners	0.21	0.0006	All ever-smokers	0.27	0.0003
	Never-smoke miners	0.45	0.0028	Non-miner smokers	0.28	0.0692
	Ever-smoke miners	0.18	0.0080	Miner ever-smokers	0.18	0.0411

Table 5. Relationship between emphysema index and lung dust burden for 63 miners

Lung dust characteristics	mg gm^{-1} dry lung (mean $\pm$ SE) Ever-smokers	Never-smokers	Overall correlation	*P*-Value
Total	7.1 $\pm$ 0.58	6.9 $\pm$ 1.5	0.34	0.006
Coal	4.4 $\pm$ 0.50	5.1 $\pm$ 1.2	0.46	0.0002
Total mineral	2.8 $\pm$ 0.31	1.8 $\pm$ 0.46	-0.09	0.49
Silica	0.20 $\pm$ 0.02	0.18 $\pm$ 0.03	0.09	0.47

Relationship between emphysema index and retained dust for 63 miners

Overall, emphysema score was significantly correlated with total dust ($P = 0.006$) and coal dust ($P = 0.0002$) but not with total minerals and silica (Table 5).

DISCUSSION AND CONCLUSION

In this study we show that the severity of emphysema in coal workers is associated with mining tenure, retained coal dust, severity of pneumoconiosis and years of smoking. Our findings are consistent with the results of autopsy studies on coal miners from the United Kingdom (Cockcroft *et al.*, 1982; Lyons *et al.*, 1981; Ruckley *et al.*, 1984) and Australia (Leigh *et al.*, 1994). Also they are consistent with epidemiologic studies of dusts other than coal dust in showing that an association with airways obstruction and emphysema (Oxman *et al.*, 1993). Plausible mechanisms for mineral dust induced emphysema have been advanced; these implicate increased protease activity and oxidative inactivation of α-1 antitrypsin in its pathogenesis (Rom *et al.*, 1990).

REFERENCES

Cockcroft, A. E., Wagner, J. C., Ryder, R., *et al.*, (1982) Post mortem study of emphysema in coal workers and non-coal workers. *Lancet* **2**, 600–603.

Gee, J. B. L. and Morgan, W. K. C. (1979) Coal mining, emphysema and compensation. Letter to the Editor. *Br. J. ind. Med.* **103**, 70–72.

Gough, J. (1952) The pathological diagnosis of emphysema. *Proc. Roy. Soc. Med.* **45**, 576.

Hu, S. N., Vallyathan, V., Green, F. H. Y. and Weber, K. C. (1990) Pulmonary arteriolar muscularization in coal workers' pneumoconiosis and its correlation with right ventricular hypertrophy. *Arch. Path. Lab. Med.* **114**, 1063–1070.

Hurley, J. F. and Soutar, D. A. (1986) Can exposure to coal mine dust cause severe impairment of lung function? *Brit. J. ind. Med.* **43**, 150–157.

Kleinerman, J., Green, F. H. Y., Harley, R., *et al.* (1979) Pathology standards for coal workers pneumoconiosis: report of the pneumoconiosis Committee of the College of American Pathologists to the National Institute for Occupational Safety and Health. *Arch. Pathol. Lab. Med.* **103**, 375–431.

Leigh, J., Driscoll, T. R., Cole, B. D., Beck, R. W., Hull, B. P. and Yang, J. (1994) Quantitative relation between emphysema and lung mineral content in coal workers. *Occup. Environ. Med.* **51**, 400–407.

Lyons, J. P., Ryder, R. C., Seal, R. M. E. and Wagner, J. C. (1981) Emphysema in smoking and non-smoking coal workers with pneumoconiosis. *Bull. Euro. Physiopath. Resp.* **17**, 75–85.

Oxman, A. D., Muir, D. C. F., Shannon, H. S., Stock, S. R., Hnizdo, E. and Lange, H. O. (1993) Occupational dust exposure and chronic obstructive pulmonary disease. A systematic overview of the evidence. *Am. Rev. Respir. Dis.* **148**, 38–48.

Rom, W. N., (1990) Basic mechanisms leading to focal emphysema and coal workers' pneumoconiosis. *Environ. Res.* **53**, 16–28.

Ruckley, V. A., Gauld, S. J., Chapman, J. S., Davis, J. M. G., Douglas, A. N., Furney, J. M.,

Jacobsen, M. and Lamb, D. (1984) Emphysema and dust exposure in a group of coal workers. *Am. Rev. Respir. Dis.* **125**, 528–532.

Seaton, A. (1990) Editorial: coal mining, emphysema and compensation. *Br. J. ind. Med.* **47**, 433–435.

Saito, K. and Thurlbeck, W. M. (1995) Measurement of emphysema in autopsy lungs with emphasis on interlobar differences. *Am. J. Respir. Crit. Care Med.* **151**, 1373–1376.

Thurlbeck, W. M., Ryder, R. C. and Stornby, N. (1974) A comparative study of the severity of emphysema in necropsy populations in three different countries. *Am. Rev. Respir. Dis.* **109**, 239–248.

 Pergamon

Ann. occup. Hyg., Vol. 41, Supplement 1, pp. 358–362, 1997
© 1997 British Occupational Hygiene Society
Published by Elsevier Science Ltd. All rights reserved
Printed in Great Britain
0003–4878/97 $17.00 + 0.00
Inhaled Particles VIII

PII: S0003–4878(96)00156–1

BIOLOGICAL EFFECTS OF BUILDING MATERIALS IN COAL MINING

A. Brammertz

Institute of Occupational Health, Department of Hygiene and Environmental Health,
Rheinisch-Westfälische Technische Hochschule Aachen, Pauwelsstr. 30, D 52 057 Aachen, Germany

INTRODUCTION

Fine coal dusts are complicated mixtures. Analyses of dust samples of underground building materials showed that the coal dust of the soam is mixed with natural anhydrite, cements, or REA gypsum (Rosmanith and Weller, 1991). Until now the experiments with mixed dust samples required the investigation of the biological effect of separate samples of fine coal dust in combination with building material dust in animal experiments and correlation of the results with the physical characteristics (Schyma and Rosmanith, 1988; Schyma et al., 1991).

As a general rule we conclude from our experiments that every dust reacts in a different way. Furthermore the effects of mixed samples cannot be predicted by the reaction of its components. An animal experimental study was separately and simultaneously run with samples of REA gypsum and Dörentruper quartz DQ 12 in increasing amounts.

MATERIAL AND METHOD

The experimental procedure was a non-invasive intratracheal instillation of rats (Brammertz, 1992). 70 female Sprague Dawley rats (weight approx. 200 g), both bred and kept under SPF conditions (Department of Experimental Zoology and central Laboratory for Experimental Animals in the Medical Faculty of the RWTH Aachen) were divided into seven experimental groups, each of 10 animals.

The animals in the first group were instilled intratracheally once non-invasively under direct control during brief inhalation anaesthesia (ether) with 5 mg of the REA gypsum in 0.5 ml of saline solution, those of the second with 10 mg, the third with 5 mg DQ 12, the fourth with 10 mg DQ 12, the fifth with a combination of 5 mg REA gypsum and 5 mg DQ 12, the sixth with a combination of 10 mg REA gypsum and 10 mg DQ 12 quartz in a manner analogous to the first. 10 nontreated rats were used as control group.

After treatment the rats were placed in cages (five animals per cage) and were given the standard diet (Atromin) and water ad libitum. The experimental period was 6 months. The animals were sacrificed, half of the lungs with the regional lymph nodes of the animals were fixed in a 4% formalin solution dehydrated (in the normal manner) in an alcohol series, then embedded in paraffin and the prepared

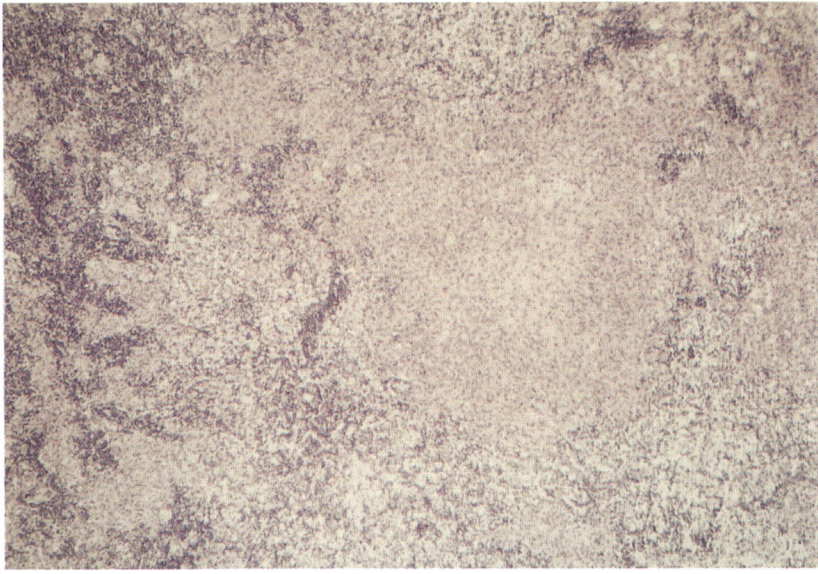

Fig. 1. Lymph nodes with quartz-typical areas (QTA), 10 mg quartz DQ 12 i.t., HE stain, magnification 50×.

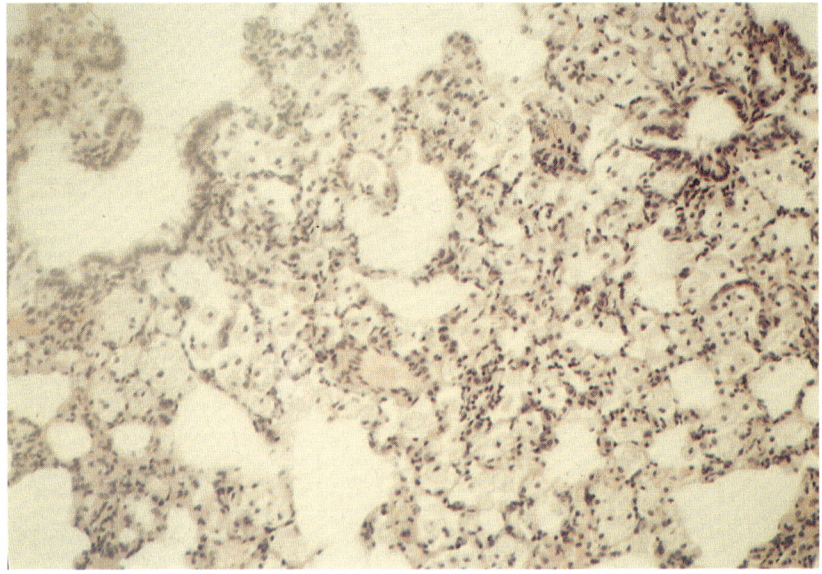

Fig. 2. Lung tissue, 6 months after i.t. instillation of 10 mg REA gypsum, HE stain, magnification 50×.

histological section stained with HE and Azan (Figs 1–5). The other lung halves were assessed for the total lipid content (determined according to Folch, 1957) and total hydroxiproline content (according to Stegemann, 1958) were compared with that of the control group.

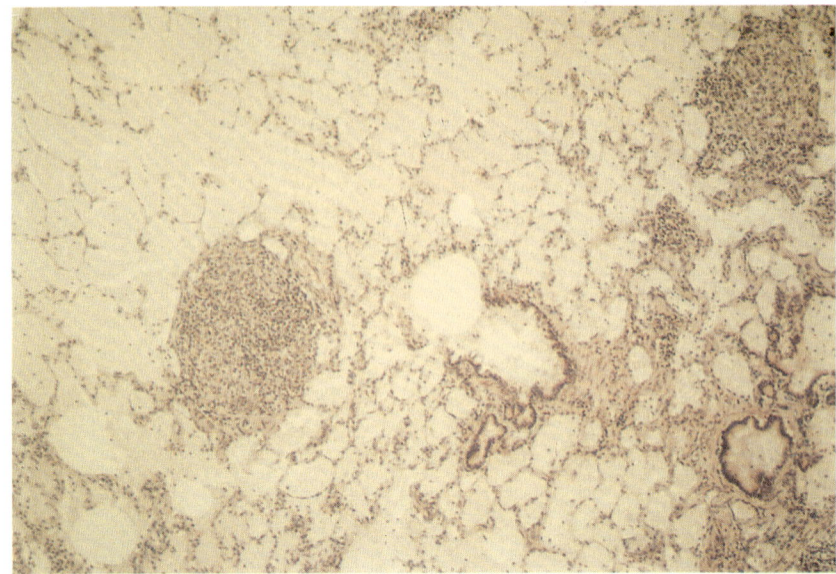

Fig. 3. Lung tissue, 6 months after 10 mg quartz DQ 12 i.t., bundles of collagene fibres, HE stain, magnification 25×.

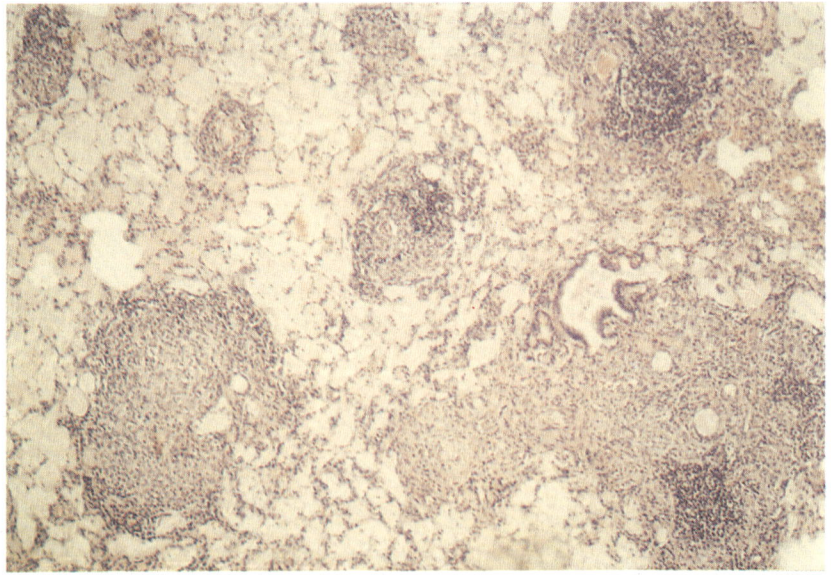

Fig. 4. Lung tissue, 6 months after intratracheal instillation of 5 mg REA gypsum and 5 mg quartz DQ 12, HE stain, magnification 25×.

RESULTS

The result of the biochemical investigation showed a small increase in the moist weight of the lungs compared with the control group in a non-invasive intratracheal

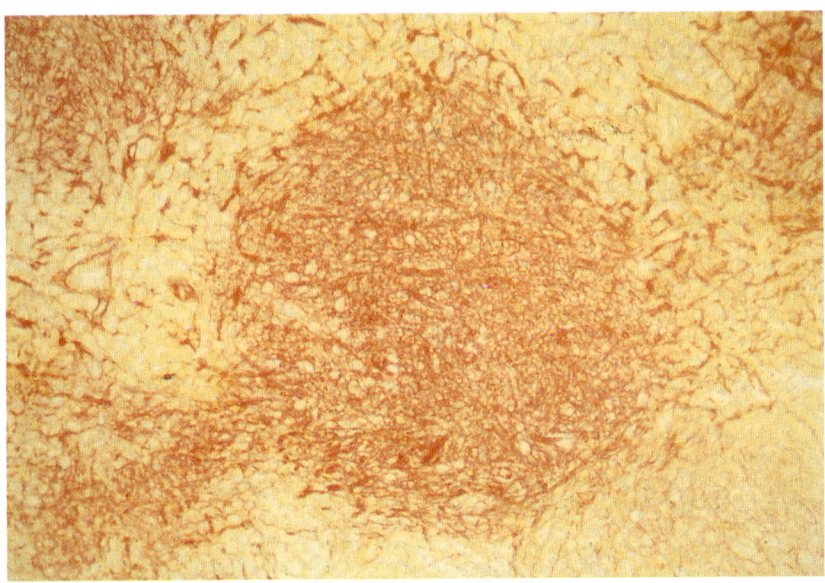

Fig. 5. Lymph node 6 months after intratracheal instillation of 10 mg REA gypsum and 10 mg quartz DQ 12, Saturn Red stain for collagen fibres, magnification 50×.

instillation on rats of REA gypsum, quartz and mixtures of both (Table 1). The hydroxiproline content and total lipid content of the lungs indicated that the sample of REA gypsum dust in different amounts (5, 10 mg per animal) did not act fibrogenically compared with the quartz dust (5, 10 mg per animal). Nevertheless, the results of the combination of REA gypsum and quartz indicated that REA gypsum did not prevent the fibrogenic effect, on the contrary, an intensification of the biological effect of quartz could be observed (Fig. 4).

DISCUSSION

The biochemical analyses indicated a small hydroxiproline increase as well as a discrete increase of the total lipid content compared with the control group. The histological examination showed clear signs of fibrosis. Recently there are inten-

Table 1. Median value ($\bar{x}$) and standard deviation (s) of the moist weight of the lungs, total hydroxi-proline content and total lipid content of the experimental animals (REA, quartz DQ 12)

Group	Moist weight of the lungs (mg)		Hydroxyproline (mg lung^{-1})		Total lipid (mg lung^{-1})	
	$\bar{x}$	s	$\bar{x}$	s	$\bar{x}$	s
5 mg DQ 12	2.00	1.01	5.71	2.05	89.58	56.73
5 mg REA	1.30	0.15	4.35	0.42	61.64	34.66
10 mg DQ 12	2.42	0.99	4.11	1.76	97.62	93.84
10 mg REA	1.33	0.08	3.60	0.30	52.40	8.80
5 mg REA + 5 mg DQ 12	2.51	1.01	5.94	2.18	130.10	88.95
10 mg REA + 10 mg DQ 12	2.76	1.37	6.61	3.53	184.82	136.67
control group	1.31	0.10	5.67	0.95	51.63	11.76

tions to mix the building materials with filtration plant dust containing a lot of as yet unknown risk factors. Therefore, further investigations should be performed.

REFERENCES

Brammertz, A. (1992) Intratracheale instillation von staüben im experiment an ratten (vergleich der methoden und beitrag zur wertung hinsichtlich der stellung in der pneumokonioseforschung). Thesis, Medizinische Fakultät der Rheinisch-Westfälischen Technischen Hochschule Aachen.

Folch, J. M., Less, M. and Sloane, S. (1957) A simple method for the isolation and purification of total lipids from animal tissues. *J. Biol. Chem.* **226**, 497–509.

Rosmanith, J. and Weller, W. (1991) Effects of new dam building materials in animal experiments on rats. In *Results of Studies on Dust Suppression and Silicosis Prevention in Hard Coal Mining*, Vol. 18, pp. 243–249. Steinkohlenbergbauverein, Essen.

Schyma, S. B., Schyma, U. J. and Buchholz, H. (1991) The physical characteristics of dust mixtures in a suspended and disperse state. In *Results of Studies on Dust Suppression and Silicosis Prevention in Hard Coal Mining*, Vol. 18, pp. 109–116. Steinkohlenbergbauverein, Essen.

Schyma, S. B., and Rosmanith, J. (1988) Physikalische charakteristik der in tierversuchen intratracheal instillierten Stäuben in dispersem und suspensem Zustand. *Wissenschaft und Umwelt* **1**, 39–52.

Stegemann, H. (1958) Mikrobestimmung von hydroxiprolin mit chloramin T und p-dimethyamino-benzaldehyd. *Hoppe Seylers Z. Physiol. Chem.* **312**, 41–42.

Ann. occup. Hyg., Vol. 41, Supplement 1, pp. 363–367, 1997
© 1997 British Occupational Hygiene Society
Published by Elsevier Science Ltd. All rights reserved
Printed in Great Britain
0003–4878/97 $17.00 + 0.00
Inhaled Particles VIII

PII: S0003–4878(96)00157–3

LUNG FUNCTION AND RESPIRATORY SYMPTOMS IN FOUNDRY WORKERS

P. F. Woerfel, R. Mc. L. Niven and N. M. Cherry

University of Manchester, Centre for Occupational Health, Stopford Building, Oxford Road,
Manchester M13 9PT, U.K.

INTRODUCTION

Foundry workers are more likely to die from chronic bronchitis and emphysema than men in other occupations (OPCS, 1995) but a report published by the U.K. Department of Social Security in 1992 (DSS, 1992) concluded that it was "unable on the present evidence to recommend prescription of chronic bronchitis and emphysema in metal workers". Although studies cited in the report found an excess of non-malignant respiratory disease, the reviewers felt that inadequate allowance had been made for potential confounders and, specifically, for age, smoking history and previous exposure in other dusty industries.

The present study was set up, in collaboration with the Foundry Section of the Amalgamated Engineering and Electrical Union, to determine whether foundry workers had an excess risk, compared to unexposed men in similar occupations, of either chronic bronchitis or decreased lung function (FEV_1) and whether any such excess persisted after allowance for confounders. Further, it was of interest to determine any relationship between exposure, within the foundry workers, and symptoms or decrease in FEV_1. The present paper addresses particularly this latter question and attempts to determine which, if any, index of cumulative exposure to foundry dust best explains the reduction in lung function or an increase in chronic bronchitis.

METHODS

Foundry workers and comparison groups (from assembly, sheet metal and machine shops) were identified at three ferrous foundries in the north west of England, all working with cast iron and mild steel. The cohorts selected were those who had started work in the foundries since 1 January 1970. All those currently working in foundry or comparison jobs were taken, together with a random sample of ex-workers in the cohort who had been employed for at least 12 months. Questionnaires were developed to collect detailed information on work since joining the foundry, as well as previous occupational history, symptoms indicative of chronic bronchitis (MRC, 1966) and smoking history. These were distributed individually to current employees at work. The completed forms were checked through by the research team on the following day, immediately prior to lung

Table 1. Job exposure matrix (JEM) concentrations of inhalable dust
(TWA—mg m^{-3})

	Core/mould	Melt	Finishing	Other
Foundry 1				
1970–74	6.0	12.0	21.0	9.0
1975–94	2.0	4.0	7.0	3.0
Foundry 2				
1970–77	4.0	3.0	6.0	4.0
1978–94	3.0	3.0	6.0	4.0
Foundry 3				
1970–94	4.5	8.0	18.0	8.0

function testing. FEV_1 was measured using a vitalograph series 3 dry wedge bellows spirometer, the best of three forced expiratory procedures being used for the study. Questionnaires and lung function for ex-workers were completed, where possible, at the foundry at which the worker had last been employed. Where an ex-worker was unable or unwilling to attend for lung function testing, a questionnaire was mailed to the home.

Estimates of cumulative exposure were made in three ways. First, duration in foundry work was calculated from the work history, including work in other foundries. Second, intensity of exposure since 1970 was estimated from rather limited records of results of past sampling (of inhalable dust) modified by information on process change and on reports of "comparative dustiness" from long term workers. The resulting matrix of exposure by time and task (core, melt, finish, other) at each foundry (Table 1) was used to derive a cumulative exposure for each worker, relating task and period to the matrix. An adjusted estimate, having allowed for reported use of respiratory protection, was also calculated, intensity being weighted by a factor of 0.5 for those reporting that they wore protection "always" in a particular job, of 0.8 for those wearing it "sometimes" and of 0.95 for those wearing it occasionally.

Because of the limited number of hygiene survey results and uncertainty about their representativeness, a further approach was adopted to estimate cumulative exposure. Personal sampling was performed in accordance with standard methodology in each factory to estimate inhalable, respirable and thoracic dust in each of the four main tasks (core, melt, finish, other) (Table 2). These estimates were then applied to the job history such that a cumulative exposure estimate (not reflecting change over time) was obtained for each dust fraction. Indices based on current measures were calculated only for the 135 workers whose foundry exposure had been entirely with their present employer.

The potential confounders of particular interest were age, self reported smoking history (total years smoking × average packs day) and exposure to dust before entering foundry work. The index reflecting past exposure was derived from years in dusty work and an intensity measure, the median rating on a scale (1 = low, 2 = medium, 3 = high) used by 3 independent judges (occupational hygienists or physicians) to assess each non foundry job reported.

Table 2. Mean current dust exposure levels (inhalable, respirable and thoracic TWA—mg m^{-3})

	Core/mould	Melt	Finishing	Other
Foundry 1				
Inhalable dust	1.6	5.2	4.0	3.6
Respirable dust	0.3	1.3	1.5	1.0
Thoracic dust	1.2	2.2	2.4	1.9
Foundry 2				
Inhalable dust	2.0	1.9	11.1	5.0
Respirable dust	0.2	0.3	0.3	0.3
Thoracic dust	0.8	0.7	2.0	1.5
Foundry 3				
Inhalable dust	5.4	5.1	25.3	9.0
Respirable dust	0.4	0.7	1.5	1.0
Thoracic dust	1.4	1.8	7.3	5.9

RESULTS

Amongst 206 current foundry workers 200 completed the questionnaire and 196 the lung function test. Average duration in foundry work was 15.2 years. Response from the comparison group was somewhat lower, 226 of 251 (90%), completing each part of the study. Amongst the ex-workers, response was less satisfactory. Of 163 ex-foundry workers identified from company records only 88 were traced alive and only half of these were willing or able to attend the factory for lung function measurement. Almost all (91%) however filled in the respiratory questionnaire. Response figures for ex-workers from the comparison group were very similar (50% traced; of these 57% with lung function, 94% with symptom questionnaire). Because of the low response rate for lung function testing in ex-workers, comparison of current and ex-workers has been restricted to respiratory symptoms.

Men currently in foundry and comparison jobs were similar in age (41.6 years foundry; 41.8 years comparison) and in time with the present company (12.6 years foundry; 14.4 years comparison). Their lung function results were also very similar (Table 3). Coefficients for the relation between FEV_1 and each of the exposure indices are shown in Table 4 in which column 1 gives the correlation unadjusted, column 2 the regression coefficient (beta) after allowance for height and age and column 3 the coefficient after further adjustment for smoking and previous work in dusty jobs. It is evident, from column 1, that whilst all measures were, as expected, negatively related to lung function, duration had the largest (negative) correlation coefficient; weighting by intensity of exposure (as in the exposure matrices) did not improve this. After allowance for age and height no relation of any importance remained with only three of the six indices having a negative regression coefficient consistent with a decrement in function. Allowance for smoking and previous exposure did not change this pattern.

Current and ex-foundry workers had very similar rates of chronic bronchitis (10% current, 11% ex-workers) and these were only slightly (and insignificantly) higher than those for the comparison group (current 6.7%, ex-workers 10.3%). Having allowed, in a logistic regression analysis, for age, pack years and previous

Table 3. Mean forced expiratory volume (litres) (FEV₁) by exposure group and foundry

| | Current Workers | | |
	Foundry	Comparison	P
Foundry 1			
Mean	3.80	3.81	0.92
SD	0.72	0.70	
N	72	66	
Foundry 2			
Mean	3.71	3.75	0.73
SD	0.73	0.77	
N	61	91	
Foundry 3			
Mean	3.43	3.61	0.22
SD	0.72	0.56	
N	63	62	
Overall			
Mean	3.67	3.73	0.36
SD	0.73	0.70	
N	196	219	

dust exposure, no relation was found between symptoms of chronic bronchitis and any of the exposure indices developed for this study.

DISCUSSION

In this group of foundry workers no relation was found between indices of exposure and either lung function (FEV₁) or chronic bronchitis. At least three reasons may be put forward to explain the apparent discrepancy between the negative results in the present study and the previously observed (OPCS, 1995) high PMR for foundry workers in the Manchester area of the North West of England. First it might be that workers in poor health had left foundry employment disproportionately; however the lack of any apparent difference in rates of chronic bronchitis in current and ex-workers suggests that this may not have been an important factor. Second, it may be that the information available on exposure was simply too poor to arrive at adequate estimates of past exposure. Were any marked

Table 4. Relation of exposure indices to FEV₁ with and without adjustment

Exposure indices	Column 1 Correlation unadjusted	Column 2 Adjusted for age, height	Column 3 Adjusted for age, height, pack years and previous exposure
Duration	−0.47*	0.01	0.13
JEM unadjusted	−0.38*	−0.06	−0.03
JEM adjusted	−0.43*	−0.08	−0.03
Inhalable dust.years	−0.31*	−0.07	−0.04
Thoracic dust.years	−0.25*	0.03	0.06
Respirable dust.years	−0.26*	0.01	0.04

* $p < 0.05$.

effect of exposure to be present, however, it would be reflected in duration of exposure; this was not the case. Third, it may be that either duration or intensity of exposure was too low for an important effect to be apparent. The average length of exposure was only 15 years and the mean cumulative exposure to respirable dust only 6.7 mg m $^{-3}$.years (range 0.1–34.5). Studies in the coal industry, in which FEV_1 decline has been demonstrated with cumulative dust exposure, have generally had rather higher levels (Seixas et al., 1992 had a mean cumulative exposure to respirable dust of 15.6 mg m^{-3}.years and a range of 0.6–40.8).

Finally, it is important to consider the power of the study to detect a difference if one were present. With the sample size eventually achieved, the study had 90% power (assuming $\alpha = 0.05$ (one sided)) to detect a difference of 5% between the exposed and comparison groups. If an effect does exist at these levels of exposure, but is undetected by this study, it seems unlikely that it is one of major significance for health.

Acknowledgements—This work was supported by a grant from the Health and Safety Executive, London. The views expressed are not necessarily those of the HSE or any other government department.

REFERENCES

DSS (1992) Department of Social Security. Chronic bronchitis and emphysema. HMSO, London.
MRC (1966) Committee on research into chronic bronchitis. Questionnaire on respiratory symptoms and instructions for its use. Medical Research Council, London.
OPCS (1995) Occupational Health Decennial Supplement, Series DS no 10. (Edited by F. Drever) Office of Population Censuses and Surveys, HMSO, London.
Seixas, N., Robins, T., Attfield, M. D. and Moulton, L. H. (1992) Exposure–response relationships for coal mine dust and obstructive lung disease following enactment of the Federal Coal Mine Health and Safety Act of 1969. *Amer. J. ind. Med.* **21**, 715–734.

Pergamon

Ann. occup. Hyg., Vol. 41, Supplement 1, pp. 368–373, 1997
© 1997 British Occupational Hygiene Society
Published by Elsevier Science Ltd. All rights reserved
Printed in Great Britain
0003–4878/97 $17.00 + 0.00
Inhaled Particles VIII

PII: S0003–4878(96)00158–5

ANGULAR AND FIBROUS PARTICLES IN LUNG IN RELATION TO JOB CATEGORY

A. Dufresne,* R. Bégin,† C. Dion,‡ J. Jagirdar,§ W. N. Rom,§
P. Loosereewanich,* D. C. F. Muir,¶ A. C. Ritchie¶ and G. Perrault‡

*McGill University, Department of Occupational Health, Faculty of Medicine, 3450 University Street, suite 22, Montréal (Québec), Canada, H3A 2A7; †Service de pneumologie, Centre Hospitalier de l'Université de Sherbrooke, Sherbrooke (Québec), Canada J1H 5N4; ‡Institut de recherche en santé et sécurité du travail, Direction des laboratoires, 505 Ouest, de Maisonneuve, Montréal (Québec), Canada H3A 3C3; §Bellevue NYU Medical Centre, 27th Street-1st Avenue, New York, NY 10016, U.S.A.; and ¶McMaster University, 1200 Main Street West, Hamilton, Ontario, Canada L8N 3Z5

INTRODUCTION

The objective of the research work was to evaluate the differences in lung mineral burdens in relation to job categories. The lung dust data will reveal information on unknown past exposure to mineral dusts including asbestos and heavy metals. The lung burden data may, thus, provide a supplementary information for an accurate identification of the cause of disease of interest.

MATERIAL AND METHODS

Case selection

All information related to lung tissue sample location, occupational histories, pathological diagnoses and other nominative information were available through the files of the Commission de la Santé et Sécurité du Travail du Québec (CSST—Workers Compensation Board). We wrote to pathologists at each hospital (more than 50) where the CSST's files indicated that lung parenchyma and biopsy samples were likely to be held. When cases were stratified into their job categories we obtained the following distribution (Table 1): 21 miners (metallic mines, gold and copper mines), 18 iron foundry workers, 22 non-iron foundry workers, four welders, three sand blast workers, 4 construction workers, three technicians and professionals, seven workers in other trades excluding welding, 13 asbestos miners and finally the 20 people representing the background population.

Transmission electron microscopy and phase contrast microscopy analyses

Detailed analytical procedures for analytical transmission electron microscopy (ATEM) and phase contrast microscopy (PCM) have been described previously (Dufresne, 1993, 1994 and 1995). The detection limit for fibrous particles ≥ 5 mm was 70 fibres mg^{-1} dry lung tissue (based on 15 mg of lung tissue) and for non-fibrous particles it was around 35 000 particles mg^{-1} dry lung tissue (based on

10 mg of lung tissue). No asbestos fibers were detected in the blank that was run with the samples. The limit of detection is 0.04 ferruginous bodies per milligram of dry lung tissue.

X-ray diffraction analysis

X-ray diffraction (XRD) was employed to characterise the crystalline silica of the extracted lung particles, which were deposited on a silver membrane filter, according to a method described elsewhere (Dufresne *et al.*, 1987).

RESULTS

The highest concentrations (Table 2) of quartz were found in miners (metallic mines), iron foundry workers and sand blast workers. Notable amounts of quartz were also found in welders and professionals when their concentrations were compared to the background population. Also, in Table 2, is shown that the highest concentrations of metallic particles were found in welders and iron foundry workers. These particles were rich in Al, N i, Mn and Cr. The highest concentrations of short fibres (Table 3) were found in non-iron foundry workers, asbestos miners and construction workers. The highest concentrations of long fibres (Table 4) were found in non-iron foundry workers and asbestos miners. Similarly, the highest concentrations of ferruginous bodies (Table 4) were found in non-iron foundry workers and asbestos miners. The non-iron foundry workers were exposed to ceramic fibres and asbestos fibres.

DISCUSSION

In the present study, we undertook a descriptive analysis of the lung concentration of particles in workers who were exposed in different occupational settings. From an analytical point of view, the most interesting observation is that analysing the element silicon by energy dispersive spectroscopy does not imply that the analyst analysed a crystalline form of silica. Indeed, it can be seen in Table 2 that although the concentration of silicon particles in the lungs on the non-iron foundry workers was comparable to the miners or the iron foundry workers the quartz concentation was much lower. In fact, in many of these workers the element Si corresponded to silicon carbide. Still, the relative comparison between the different job categories in which the workers were likely to have been exposed to quartz (metallic mine miners, iron foundry workers and sand blasting workers) showed, on average, the highest concentrations of crystalline silica. Cristobalite was not found in any of the exposed groups.

Again, one should be cautious in interpreting the present results since the fibrotic workers in this study had been likely exposed to heterogeneous dust intensities in different calendar years and occupational settings. Workers in the early years were likely exposed to higher intensities of dust owing to poorly organised work practices and the lack of dust control measures. Since the lung particle burden study has to rely on the availability of lung tissue samples, the investigators had no control on selection but had to make use of the available lung samples. It should be stressed

Table 1. Occupational history of workers

	Miners (metals)	Foundrymen (iron)	Foundrymen (non-iron)	Welders	Sand blasting workers	Construction workers	Technician and professional	Other trades	Miners (asbestos)	Background population (Sherbrooke)
n	24	18	22	4	3	4	3	7	13	20
DOB	1923	1917	1922	1931	1920	1913	1922	1924	1910	1946
DOD	1986	1987	1986	1990	1985	1980	1986	1980	1985	1990
AGE	63	69	64	59	62	66	63	56	70	43
YOE	24	30	30	31	10	36	40	41	34	—
FYE	1947	1945	1948	1948	1954	1936	1946	1939	1936	—
LYE	1974	1973	1978	1989	1976	1972	1982	1979	1973	—
YSLE	11.0	11.0	7.5	1.0	9.2	6.2	3.0	4.0	11.0	—

n = number of cases = 118. YOE = years of exposure. DOB = average date of birth. DOD = average date of death. AGE = average age. FYE = average first year of exposure. LYE = average last year of exposure. YSLE = average year since last exposure.

Table 2. Median concentration of angular particles (million of particles $\geq$ 0.25 µm mg^{-1} lung tissue dry weight)

Types	Miners (metals)	Foundrymen (iron)	Foundrymen (non-iron)	Welders	Sand blasting workers	Construction workers	Technician and professional	Other trades	Miners (asbestos)	Background (Sherbrooke)
Silicon	1.41	1.85	1.74	0.65	1.13	1.21	5.70	0.55	0.74	0.34
Feldspar	0.49	0.49	0.37	1.02	0.11	1.70	0.12	0.45	1308	0.28
Clay	0.12	0.18	0.26	12.2	0.08	0.43	0.83	0.83	0.48	0.37
Mica	0.18	0.07	<0.01	0.15	0.02	0.57	0.11	0.23	0.37	0.09
Other silicates	0.08	0.16	0.06	0.12	0.06	0.13	0.11	0.13	0.11	<0.01
Metal rich	0.15	1.06	0.17	5.61	0.46	0.51	0.30	0.24	0.33	0.15
Iron rich	0.09	0.58	0.15	0.60	0.27	0.08	0.11	0.18	0.22	0.30
Phosphorous calcium rich	<0.01	<0.01	<0.01	<0.01	<0.01	<0.01	<0.01	<0.01	0.10	0.15
Other non-silicates	<0.01	0.07	0.09	0.17	0.04	0.02	0.06	0.20	0.04	0.01
All	4.63	7.91	5.02	10.1	2.12	5.57	14.2	3.41	3.99	2.03
Quartz	2.1	2.1	0.3	0.8	2.0	0.2	0.7	<0.1	0.4	0.3

Table 3. Median concentration of fibrous particles (thousand of particles < 5 μm mg^{-1} lung tissue dry weight)

Types	Miners (metals)	Foundrymen (iron)	Foundrymen (non-iron)	Welders	Sand blasting workers	Construction workers	Technician and professional	Other trades	Miners (asbestos)	Background (Sherbrooke)
Silicon	2.51	1.15	32.5	1.37	0.21	1.38	5.39	0.70	< 0.04	0.23
Feldspar	0.47	0.33	< 0.04	1.42	< 0.04	2.67	1.80	< 0.04	< 0.04	0.30
Clay	2.63	3.98	0.42	0.58	4.94	5.66	6.98	1.89	0.63	0.61
Mica	2.62	0.68	< 0.04	0.58	< 0.04	6.60	2.09	0.52	2.09	0.14
Other silicates	1.34	6.20	2.84	5.57	1.35	7.07	9.77	1.85	4.19	0.47
Metal rich	0.27	12.6	0.80	6.77	0.63	3.14	0.90	0.63	3.14	0.99
Iron rich	< 0.04	1.83	< 0.04	< 0.04	0.84	< 0.04	< 0.04	0.79	< 0.04	0.07
Phosphorous calcium rich	< 0.04	< 0.04	< 0.04	< 0.04	< 0.04	< 0.04	< 0.04	< 0.04	< 0.04	
Other non-silicates	< 0.04	< 0.04	< 0.04	< 0.04	< 0.04	< 0.04	< 0.04	0.31	< 0.04	< 0.04
Asbestos	0.07	0.38	< 0.04	3.59	< 0.04	0.14	< 0.04	0.37	16.8	0.52
All	26.6	21.2	56.6	21.8	13.5	31.7	26.0	11.8	45.6	6.11

Table 4. Median concentration of fibrous particles (thousand of particles ≥ 5 μm mg^{-1} lung tissue dry weight)

Types	Miners (metals)	Foundrymen (iron)	Foundrymen (non-iron)	Welders	Sand blasting workers	Construction workers	Technician and professional	Other trades	Miners (asbestos)	Background (Sherbrooke)
Silicon	0.21	< 0.04	6.55	0.12	< 0.04	0.07	< 0.04	< 0.04	< 0.04	< 0.04
Feldspar	< 0.04	< 0.04	< 0.04	0.23	< 0.04	0.07	< 0.04	< 0.04	< 0.04	< 0.04
Clay	< 0.04	< 0.04	< 0.04	< 0.04	< 0.04	< 0.04	< 0.04	< 0.04	< 0.04	0.09
Mica	< 0.04	< 0.04	< 0.04	0.23	< 0.04	0.16	0.21	< 0.04	< 0.04	< 0.04
Other silicates	< 0.04	< 0.04	< 0.04	0.12	< 0.04	0.12	0.42	0.21	0.21	< 0.04
Metal rich	< 0.04	< 0.04	< 0.04	0.65	< 0.04	< 0.04	< 0.04	< 0.04	< 0.04	0.07
Other non-silicates	< 0.04	< 0.04	< 0.04	< 0.04	< 0.04	< 0.04	< 0.04	< 0.04	< 0.04	< 0.04
Asbestos	< 0.04	< 0.04	< 0.04	0.23	< 0.04	0.07	0.84	0.21	2.25	0.18
All	0.63	0.73	8.82	1.72	< 0.04	0.73	1.26	0.42	3.99	0.45
Ferruginous bodies (f.b./gram dry lung)	186	2100	21900	300	500	800	500	141	8600	50

that we have used the terms fibrosis instead of silicosis because in many autopsy reports the diagnosis of silicosis was not always clear.

The higher concentrations of short and long silicon and other silicates fibres in non-iron foundry workers is explained by their exposure to silicon carbide whiskers (for silicon carbide workers) and ceramic fibres (for workers involved in aluminum, copper, tin and other smelters) as we have shown in previous manuscripts (Dufresne et al., 1995, 1996).

Although few individual cases from the different occupational settings had long asbestos fibres comparable to asbestos miners, on average, each group had a concentration of fibres lower than the asbestos miners. Occupational exposure to asbestos or ceramic fibres is also supported by the ferruginous bodies concentrations. One can see in Table 4 that the concentration of ferruginous bodies in the non-iron foundry workers is high when it is compared to the asbestos exposed workers.

CONCLUSION

Results obtained from lung tissue must always be interpreted cautiously. However, these results are consistent with the hypothesis that workers in some trades (not all of them) were exposed not only to quartz but also to asbestos, ceramic fibres, metal-rich non-fibrous particles and other likely carcinogenic chemicals. The wide range of particles types identified in the lungs of these workers illustrates the complexity of trying to determine disease origins in these work environments.

REFERENCES

Dufresne, A., Lesage, J. and Perrault, G. (1987) Evaluation of occupational exposure to mixed dusts and polycyclic aromatic hydrocarbons in silicon carbide plants. Am. ind. Hyg. Assoc. J. 48, 160–166.

Dufresne, A., Krier, G., Muller, J. F. et al. (1993) Measurement by TEM and LAMMA of metallic particles extracted from lung parenchyma of three electricians. Am. ind. Hyg. Assoc. J. 54, 564–568.

Dufresne, A., Case, B. W., Fraser, R. et al. (1994) Protocol of lung particulate analysis by electron transmission microscopy for decoding occupational history from lung retention. Ann. occup. Hyg. 38 (supplement 1), 503–517.

Dufresne, A., Loosereewanich, P., Armstrong, B. et al. (1995) Pulmonary retention of ceramic fibres in silicon carbide (SIC) workers. Am. ind. Hyg. Assoc. J. 56, 490–498.

Dufresne, A., Loosereewanich, P., Armstrong, B. et al. (1996) Inorganic particles in the lungs of aluminum smelter workers with pleuro pulmonary cancer. Am. ind. Hyg. Assoc. J. 57, 370–375.

Pergamon

Ann. occup. Hyg., Vol. 41, Supplement 1, pp. 374–378, 1997
© 1997 British Occupational Hygiene Society
Published by Elsevier Science Ltd. All rights reserved
Printed in Great Britain
0003–4878/97 $17.00 + 0.00
Inhaled Particles VIII

PII: S0003–4878(96)00159–7

HISTOLOGICAL CHANGES IN HILAR LYMPH GLANDS IN RELATION TO COAL WORKERS' PROGRESSIVE MASSIVE FIBROSIS AND LUNG COAL AND QUARTZ CONTENTS. EVIDENCE IN SUPPORT OF THE "CENTRAL" HYPOTHESIS

J. Leigh, M. Todorovic and T. Driscoll

National Institute of Occupational Health and Safety, GPO Box 58, Sydney 2001, NSW, Australia

INTRODUCTION

The pathogenesis of progressive massive fibrosis (PMF) in coal workers is still not fully understood. Principal mechanisms include total dust load, quartz and rheumatoid diathesis. Recently, Seal *et al.* (1986) proposed a further possible mechanism whereby dust and necrotic material in central (hilar) lymph nodes is released into bronchi or vessels (or both) by rupture of the node capsule and erosion through bronchial or vessel walls and then transported to the lung parenchyma where it may act as a "seed" for PMF development. Seal *et al.* found a statistically significant association between severity of central lymph node pathology grade and presence of PMF lesions > 5 cm diameter and that lungs having smaller areas of PMF also had capsular rupture of central lymph nodes. In an earlier report, where we studied only lungs with PMF, we could find no association between the magnitude of PMF and the severity of central lymph node changes (Leigh and Wiles, 1987). We have now extended this study to a larger series of post mortem coal workers' lungs, with and without PMF, and have examined lymph node pathology, PMF lesions and, in a subset, have also analysed lung coal and quartz contents.

METHODS

The data analysed in this study were obtained from the Joint Coal Board which has maintained a continuous series of post mortem studies on pulmonary tissue (pulmonary parenchyma and hilar lymph glands) from deceased New South Wales coal workers since 1949. 1002 cases were analysed for central lymph node histopathology grade in relation to PMF. They represented 92.3% of the total 1086 deceased coalworkers who underwent post mortem examinations from 1949 to 1987. For 274 cases (27.3%) of this sample, measurements for lung coal and silica (quartz) content were available. The quartz and coal contents were then analysed in relation to lymph node (LN) histopathology grade. The study measured internal relationships between the variables of interest without attempting in any way to measure prevalence of PMF or lymph node changes in the general mining population. Moreover, the group of miners studied were exposed to much higher dust levels than today, so the present study should be seen as an example of "historical pathology". There is good evidence that the deceased miners who were subject to post mortem examinations had a higher prevalence of lung disease than the overall mining population in the study period.

The main reasons for exclusion of subjects were inadequacy of lung specimens and missing data. However, it is very unlikely that significant selection bias was introduced, as there is no reason to believe that the relationship should differ between those in this sample studied and the full group. At the second stage of selection, when choosing the subset of 274 subjects for mineral analysis from the full group, selection bias was unlikely because of a similar lifetime exposure to coal and mineral dust in the subset and the full sample.

Histopathological assessment of the central lymph nodes was done without knowledge of the severity of PMF in the lung parenchyma. For each case both right and left lymph nodes were examined. The scoring system used for LN grading was the same as that used by Seal et al. (1986). Criteria for the recognition of the four grades of LN changes were defined as follows:

- Grade one—packing of the medulla with macrophages laden with coal dust.
- Grade two—small areas of necrosis and fibrosis with collagen deposition. Necrosis and fibrosis can progress further until the node is completely replaced.
- Grade three—the fibrotic and necrotic material together with dust laden macrophages breaches the capsule of the lymph node. Usually, the space between the node and surrounding structures (bronchi and vessels) is filled by collagenous and necrotic material, making the node appear enlarged and irregularly shaped.
- Grade four—the necrotic and sclerotic material accompanied with the macrophages erodes the walls of large central bronchi (the cartilaginous plates and the epithelial surface) as well as the branches of the pulmonary arteries (their muscular and elastic tissues can be destroyed).

Assessments were made using a light microscope at a magnification of × 10– × 40. All lymph node histopathological assessments were made by MT. Analysis was limited by the historical tissue sections available. However under the Joint Coal Board pathology program protocol, the most severely affected nodes would have been selected for sectioning for histology by the examining pathologist. Post mortem assessment of PMF in the lung parenchyma had been obtained from the original data. PMF was defined as any dust lesion more than 2 cm in diameter. The method used for analysis of lungs for quartz and coal content is described in full by Leigh et al. (1994). The method was, in brief, an acid digestion, ashing and gravimetric determination of coal with Fourier Transform IR Spectroscopy for quartz in the ash.

A total of 274 pairs of lungs were analysed. One observer made all measurements. Results were reported as total quartz and coal content in both lungs. Quartz and coal content measurements in the lymph nodes were not determined because lymphatic tissue had been previously dissected.

Table 1. Right central lymph node histopathology grades and the presence or absence of progressive massive fibrosis (PMF) in the right lung from 1002 coal workers

| | PMF | | | | |
| | Present | | Absent | | |
	No	%	No	%	Total
1	18	3.2	535	96.7	553
2	22	7.3	281	92.7	303
3	28	20.9	106	79.0	134
4	4	33.3	8	66.6	12
Total	72		930		1002

Table 2. Left central lymph node histopathology grades and the presence or absence of progressive massive fibrosis (PMF) in the left lung from 1002 coal workers

| | PMF | | | | |
| | Present | | Absent | | |
	No	%	No	%	Total
1	18	3.5	485	96.4	503
2	22	6.4	324	93.6	346
3	28	20.0	112	80.0	140
4	4	30.7	9	69.2	13
Total	72		930		1002

Statistical analysis

The significance of the relationship between the extent of the central lymph node histopathology grades and the presence or absence of PMF in the lung parenchyma was tested with the extended Mantel–Haenszel χ^2 test using the computer package programme Excel 4.0 for Windows. The overall (heterogeneity) chi square χ^2_H was partitioned into two component chi-squares, one of which measured trend χ^2_T in the proportions across the four groups while the other measured the scatter about the trend or the departure from linearity χ^2_{NL}.

The significance of the association between the predictor variables (quartz and coal) and the outcome variable (lymph node histopathology grade) was examined by simple linear regression analysis using the statistical package MINITAB.

RESULTS

Lymph node histopathology grade and PMF

Central lymph node histopathology grades in relation to the presence or absence of PMF in the right and left lungs from 1002 coal workers are shown in Tables 1 and 2. It can be seen from Tables 1 and 2 that PMF was associated with increasing LN histopathology grade. To test the Seal hypothesis, an extended Mantel–Haenszel chi-square test was undertaken separately for the right and left lungs.

The association between increasing right central node changes and the presence of PMF in the right lung (Table 1) was highly significant ($\chi^2_T = 55.68$, $P < 0.001$).

The overall (heterogeneity) chi-square was highly significant ($\chi^2_H = 62.83$, $P < 0.001$). χ^2_{NL} was also significant ($\chi^2_{NL} = 7.19$, $P < 0.05$). Therefore, considering χ^2_T significant and χ^2_{NL} significant together, it can be said that there was evidence for a statistically significant non-linear trend in the proportions across the four groups.

A statistically significant non-linear trend was also obtained from the data for the left central lymph node and lung (Table 2). The results showed a highly significant trend ($\chi^2_T = 45.94$, $P < 0.001$) in the proportions across the four groups, significant heterogeneity chi-square ($\chi^2_H = 55.48$, $P < 0.001$) and significant χ^2 non-linear trend ($\chi^2_{NL} = 9.53$, $P < 0.01$).

The results justify the conclusion that there was a significant association between the extent of a lymph node histopathology grade and presence of PMF in the lung parenchyma. The findings are also consistent with a pure high dose effect. In other words, high dust levels cause PMF and lymph node changes independently in parallel. However, the demonstrated non-linearity (significant non-linear trend) suggests capsular damage to be a triggering factor for PMF in the lung parenchyma, more consistent with the central hypothesis.

Quartz and coal content in the lungs and LN histopathology grade

Linear regression analysis was undertaken with the maximum of right and left LN score as the dependent (outcome) variable and total quartz and coal content as the independent (predictor) variables. Simple univariate regression analysis showed that lymph node pathology grade (LNPG) was related to coal content (gm in both lungs):

$$LNPG = 0.0714 \text{ coal} + 1.38$$
$$t = 6.08 \quad R^2 = 12.0\%$$

and to quartz content (gm in both lungs)

$$LNPG = 2.01 \text{ quartz} + 1.30$$
$$t = 8.45 \quad R^2 = 20.5\%.$$

A variety of other univariate polynomial and multivariable regression models were fitted but the overall conclusion was that LNPG was more strongly related to quartz content than coal content. Similar findings were obtained when coal and quartz were expressed as concentrations per weight of dry lung.

DISCUSSION

It has been demonstrated clearly in the present study that high dust levels are associated with PMF and LN changes separately in parallel. However, a statistically significant non-linear trend in the proportions across the four histopathological groups of LN changes (LN 1–4) supports the "central" hypothesis of Seal *et al.* in that rupture of the nodal capsule is an initiating factor for PMF development in the pulmonary parenchyma. Statistical examination of data from 274 cases indicated that the lung quartz content contributed more to LN changes than coal content. This may be due to quartz predilection for lymph glands (Churg and Green, 1988; Ruckley *et al.*, 1981) but certainly does not diminish the influence of coal content on pneumoconiosis development.

The hilar lymph nodes were assessed histopathologically without knowledge of PMF in the lung parenchyma. Dust analysis (quartz, coal) was performed using well established techniques from the literature and there was good agreement between duplicate samples (Leigh *et al.*, 1994). Unfortunately, in the present study it was not possible to make measurements of quartz and coal content in the lymph glands due to previously dissected lymphatic tissue.

REFERENCES

Churg, A. and Green, F. H. Y. (1988) Coal workers' pneumoconiosis and pneumoconiosis due to other carbonaceous dust. In *Pathology of Occupational Lung Disease*, pp. 89–139. Igaku-Shoin, New York.

Leigh, J. and Wiles, A. N. (1987) Central lymph node changes and progressive massive fibrosis in coalworkers. (Letter) *Thorax* **42**, 559.

Leigh, J., Driscoll, T. R., Cole, B. D., Beck, R. W., Hull, B. P. and Yang, J. (1994) Quantitative relation between emphysema and lung mineral content in coalworkers. *Occ. Environ. Med.* **51**, 400–407.

Ruckley, V. A., Chapman, J. S., Collings, P. L., Douglas, A. N., Fernie, J. M., Lamb, D. and Davis, J. M. G. (1981) Autopsy studies of coalminers' lungs. Phase II Report TM/81/18, pp. 5–23. Institute of Occupational Medicine, Edinburgh.

Seal, R. M. E., Cockcroft, A. and Wagner, J. C. (1987) Central lymph node changes and progressive massive fibrosis in coalworkers. (Letter) *Thorax* **42**, 559–560.

Seal, R. M. E., Cockcroft, A., Kung, I. and Wagner, J. C. (1986) Central lymph node changes and progressive massive fibrosis in coalworkers. *Thorax* **41**, 531–537.

 Pergamon

Ann. occup. Hyg., Vol. 41, Supplement 1, pp. 379–383, 1997
© 1997 British Occupational Hygiene Society
Published by Elsevier Science Ltd. All rights reserved
Printed in Great Britain
0003–4878/97 $17.00 + 0.00
Inhaled Particles VIII

PII: S0003–4878(96)00160–3

MINERAL DUSTS OXIDIZE METHIONINE RESIDUES: PROBABLE MECHANISM OF INACTIVATION OF ALPHA-1-ANTITRYPSIN

K. Li, K. Zay and A. Churg*

Department of Pathology, University of British Columbia, 2211 Westbrook,
Vancouver, BC, Canada V6T 2B5

INTRODUCTION

There is now extensive evidence that mineral dusts induce airflow obstruction, and that exposure to some types of mineral dust, especially coal and silica, is associated with the presence of radiologic and pathologic emphysema (Becklake, 1989; Oxman *et al.*, 1993; Leigh *et al.*, 1983; Cockroft *et al.*, 1982; Ruckley *et al.*, 1984). However, little is known of the mechanism of emphysema induction by mineral particles. Because many mineral dusts can catalyse the formation of active oxygen species (AOS) in aqueous solution (Pezerat *et al.*, 1989; Fubini *et al.*, 1995) and because all mineral dusts evoke an inflammatory reaction in the lung, we postulated that mineral particles may induce emphysema by mechanisms similar to those thought to apply cigarette smoke; namely, excess release of proteolytic enzymes from dust-evoked inflammatory cells with subsequent breakdown of alveolar wall connective tissue and oxidative inactivation of α1AT, the major anti-proteolytic protein in the lung, by dust-catalysed active ozygen species (Evans and Pryor, 1992), as well as by AOS released from dust-evoked inflammatory cells.

We have previously shown that intratracheal instillation of mineral dusts does result in increased levels of desmosine, a measure of elastin breakdown and hydroxyproline, a measure of collagen breakdown, in lavage fluid (Li *et al.*, 1996) and also that exposure of purified α1AT to crystalline silica results in loss of antiproteolytic activity through a hydrogen peroxide mediated mechanism (Zay *et al.*, 1995). It is believed that cigarette smoke inactivates α1AT by oxidizing methionine residues (Carp *et al.*, 1982), although few reports of this phenomenon have been published. We now extend our observations by testing whether mineral dusts can similarly oxidize methionine, either as a pure amino acid, or in whole proteins.

MATERIALS AND METHODS

Samples of coal, amosite asbestos, silica, or titanium dioxide (rutile) were mixed with either 0.5 mg ml^{-1} methionine in phosphate buffered saline pH 7.4, or whole

---

* Address for correspondence: Andrew Churg, MD, Department of Pathology, University of British Columbia, 2211 Westbrook Mall, Vancouver, BC, Canada V6T 2B5.

Table 1. Characterisation of dusts used

Dust	Source/identification	Geometric mean size
Silica	Minusil, U.S. Silica Corp	0.76 μ
Amosite asbestos	UICC sample	3.8 × 0.20 μ
Titanium dioxide	Sigma (rutile)	0.51 μ
Coal 1520 (sub-bituminous)	Penn State coal bank	1.8 μ

protein in phosphate buffered saline. The specific dusts used are shown in Table 1. Human α1AT was used at a concentration of 0.2 mg ml^{-1}, but because of the cost of purified α1AT, most experiments with whole protein were done with human albumin at a concentration of 5 mg ml^{-1}. The mixtures were incubated for 24 h at 20°C. In some experiments scavengers of active oxygen species were added to the mixture. A variety of dust concentrations were tried initially; because high concentrations of titanium dioxide or amosite precipitated protein and adsorbed amino acids, only relatively low (10 or 20 mg ml^{-1}) concentrations of these dusts were employed. Reagent hydrogen peroxide was used as a control.

The amino acid preparations were analysed directly as described below. The whole proteins were first hydrolysed in 4 M potassium hydroxide at 105°C for 16 h. The hydrolysate was neutralized with 6 M hydrochloric acid and its amino acid composition determined as follows:

50 μl samples were dried using a Waters Pico-Tag Vaccum Station and redried in 50 μl of 2:1:1 ethanol:water:triethylamine (TEA). Dried samples were derivated with 50 μl solutions of 7:1:1:1 ethanol:water:TEA:phenylisothiocyanate (PITC) for 20 min at room temperature. The derivatised samples were dried again and dissolved in 700 μl phosphate buffer for analysis.

Analysis was performed on a Waters HPLC system. Amino acid separation was achieved on a Partisil ODS-2 C18, 10 μm, 4.6 × 250 mm Whatman column. The mobile phase was programmed at a flow rate of 1.6 ml per min starting with 100% solvent A (60 ml acetonitrile and 940 ml of 138 mM acetate buffer pH 6.4 with 0.05% TEA), followed by a linear gradient to 50% solvent B (60% acetonitrile in water) for an additional 15 min. The column was then cleaned with 100% solvent B for 12 min and equilibrated with 100% solvent A for another 12 min before injection of the next sample. Absorbance was read at 254 nm. The detection limit of methionine was 0.1 ng with a 30 μl injection with a signal to noise ratio of five. All operations were performed at room temperature. Results were expressed as the percent of total methionine represented by methionine sulphoxide. Statistical comparisons to the control value (amino acid or protein) were performed by analysis of variance.

RESULTS

Tables 2 and 3 show the percentage of methionine oxidized to methionine sulphoxide after exposure of methionine (Table 2) or whole protein (Table 3) to various mineral dusts. All dusts examined caused methionine oxidation when the

Table 2. Percent of methionine oxidized to methionine sulphoxide
using pure methionine in saline

Dust/treatment	Methionine sulphoxide (% ± SD)
Methionine alone	0
H_2O_2 10 μM	0.29 ± 0.27
H_2O_2 50 μM	1.26 ± 0.26
H_2O_2 100 μM	2.48 ± 0.42
TiO_2 10 mg ml^{-1}	0.19 ± 0.15
TiO_2 20 mg ml^{-1}	0.91 ± 0.06
TiO_2 10 mg ml^{-1} + catalase	0.07 ± 0.01*
TiO_2 10 mg ml^{-1} + deferoxamine	0.07 ± 0.04*
TiO_2 10 mg ml^{-1} + mannitol	0.20 ± 0.01
TiO_2 10 mg ml^{-1} + DMSO	0.16 ± 0.07
Silica 10 mg ml^{-1}	0.17 ± 0.07
Silica 100 mg ml^{-1}	0.38 ± 0.02
Silica 200 mg ml^{-1}	0.68 ± 0.26
Silica 10 mg ml^{-1} + catalase	0
Silica 10 mg ml^{-1} + deferoxamine	0.01 ± 0.02*
Silica 10 mg ml^{-1} + mannitol	0.08 ± 0.09*
Silica 10 mg ml^{-1} + DMSO	0.15 ± 0.12
Coal 10 mg ml^{-1}	1.84 ± 0.27
Coal 10 mg ml^{-1} + catalase	1.39 ± 0.05*
Coal 10 mg ml^{-1} + deferoxamine	1.44 ± 0.04*
Coal 10 mg ml^{-1} + mannitol	1.44 ± 0.20
Coal 10 mg ml^{-1} + DMSO	1.61 ± 0.09
Amosite 10 mg ml^{-1}	1.00 ± 0.51
Amosite 10 mg ml^{-1} + catalase	0.09 ± 0.02*
Amosite 10 mg ml^{-1} + deferoxamine	0.88 ± 0.26
Amosite 10 mg ml^{-1} + mannitol	0.15 ± 0.02*
Amosite 10 mg ml^{-1} + DMSO	0.56 ± 0.23*

* Indicates significantly less than dust alone.

pure amino acid was used and all but titanium dioxide was able to oxidize methionine in whole proteins.

DISCUSSION

Because oxidation of methionine residues lowers the affinity of α1AT for neutrophil elastase, it has been proposed that oxidative inactivation of α1AT by cigarette smoke may play an important role in the pathogenesis of emphysema (Carp et al., 1982; Evans and Pryor, 1992). However, there is little information about the amount or even presence of oxidized α1AT in the lungs of human smokers. Carp et al. (1982) reported oxidized methionine in α1AT recovered from smoker's lavage fluid. Evans and Pryor (1992) exposed methionine or α1AT to aqueous extracts of cigarette tar and observed oxidation of methionine to methionine sulphoxide. Their results are, in a general sense, very similar to ours and they also found, by use of scavengers or pure hydrogen peroxide, that the oxidative process appeared to be driven by hydrogen peroxide, and less certainly by hydroxyl radical, since hydroxyl radical scavengers such as DMSO or mannitol and

Table 3. Percent of methionine oxidized to methionine
sulphoxide using whole proteins

Dust/treatment	Methionine sulphoxide (% ± SD)
α-1-Antitrypsin	
α1AT alone	0
H_2O_2 1 μM	6.1 ± 2.6*
H_2O_2 10 μM	31.6 ± 4.9*
Coal 100 mg ml^{-1}	51.6 ± 9.9*
Coal 200 mg ml^{-1}	68.3 ± 21.3*
Albumin	
Albumin alone	17.6 ± 4.1*
H_2O_2 1 μM	21.7 ± 2.4*
H_2O_2 10 μM	32.8 ± 1.9*
H_2O_2 20 μM	36.3 ± 2.4*
Coal 10 mg ml^{-1}	30.5 ± 2.4*
Coal 100 mg ml^{-1}	35.4 ± 2.0*
Silica 10 mg ml^{-1}	15.9 ± 1.6
Silica 100 mg ml^{-1}	14.3 ± 4.4
Silica 200 mg ml^{-1}	25.4 ± 3.8*
Amosite 10 mg ml^{-1}	19.9 ± 1.7
Amosite 20 mg ml^{-1}	26.6 ± 4.7*
TiO_2 10 mg ml^{-1}	17.4 ± 2.9

* Indicates significantly greater than pure protein.

iron chelators such as deferoxamine, provided partial and inconsistent protection. With cigarette smoke the hydrogen peroxide is derived from the tar (Evans and Pryor, 1992); with mineral partiles, it is derived from particle catalyzed reduction of molecular oxygen (Pezerat *et al.*, 1989; Fubini *et al.*, 1995), as well as from the oxidative burst of inflammatory cells.

It is of interest that there were quite marked differences in our data in terms of relative potency of the different dusts. Titanium dioxide, which was able to oxidize pure methionine, could not attack whole proteins. In terms of compact particles, on a weight for weight basis, coal was more potent than silica, and silica more potent than titanium dioxide. If one adjusts, using the mean size values shown in Table 1, for numbers of particles and compares the amount of oxidation observed in Table 2 or 3, then the order of potency remains the same, but it is clear that coal is much more potent on a particle for particle basis than is silica and silica is similarly much more potent than titanium dioxide. These results are generally similar to the impressions gained from review of epidemiologic and pathologic studies of human lungs regarding the potency of these various dusts as agents of emphysema.

REFERENCES

Becklake, M. R. (1989) Occupational exposures: evidence for a causal association with chronic obstructive pulmonary disease. *Am. Rev. Respir. Dis.* **140**, S85–S91.
Carp, H., Miller, F., Hoidal, J. R. and Janoff, A. (1982) Potential mechanism of emphysema:

a-1-proteinase inhibitor recovered from lungs of cigarette smokers contains oxidized methionine and has decrease elastase inhibitory capacity. *PNAS* **79**, 2041–2045.

Cockcroft, A., Wagner, J. C., Ryder, R., Seal, R. M. E., Lyons, J. P. and Anderson, N. (1982) Post mortem study of emphysema in coalworkers and non-coalworkers. *Lancet* **2**, 600–603.

Evans, M. D. and Pryor, W. A. (1992) Damagen to human α-1-proteinase inhibitor by aqueous cigarette tar extracts and the formation of methionine sulphoxide. *Chem. res. Toxicol.* **5**, 655–660.

Fubini, B., Mollo, L. and Giamello, E. (1995) Free radical generation at the solid/liquid interface in iron containing minerals. *Free Rad. Res.* **23**, 593–614.

Leigh, J., Outhred, K. G., McKenzie, H. I., Glick, M. and Wiles, A. N. (1983) Quantified pathology of emphysema, pneumoconiosis and chronic bronchitis in coal workers. *Br. J. ind. Med.* **40**, 258–263.

Li, K., Keeling, B. and Churg, A. (1996) Mineral dusts cause elastin and collagen breakdown in the rat lung: A potential mechanism of dust-induced emphysema. *Amer. J. Respir. Crit. Care Med.* **153**, 644–649.

Oxman, A. D., Muir, D. C. F. and Shannon, H. S. (1993) Occupational dust exposure and chronic obstructive pulmonary disease: a systematic overview of the evidence. *Amer. Rev. Respir. Dis.* **148**, 38–48.

Pezerat, H., Zalma, R., Guignard, J. and Jaurand, M. C. (1989) Production of oxygen radicals by the reduction of oxygen arising from the surface activity of mineral fibers. In *Biological Effects of Mineral Fibers in the Nonoccupational Environment* (Edited by J. Bignon *et al.*), pp. 100–111. IARC, Lyon, France.

Ruckley, V. A., Gauld, S. J., Chapman, J. S., Davis, J. M. G., Douglas, A. N., Fernie, J. M., Jacobsen, M. and Lamb, D. (1984) Emphysema and dust exposure in a group of coal workers. *Amer. Rev. Respir. Dis.* **129**, 528–532.

Zay, K., Devine, D. and Churg, A. (1995) Quartz inactivates alpha-1-antiproteinase: possible role in mineral dust induced emphysema. *J. appl. Physiol.* **78**, 53–58.

 Pergamon

Ann. occup. Hyg., Vol. 41, Supplement 1, pp. 384–389, 1997
© 1997 Published by Elsevier Science Ltd on behalf of BOHS
Printed in Great Britain. All rights reserved
0003–4878/97 $17.00 + 0.00
Inhaled Particles VIII

PII: S0003–4878(96)00092–0

RELATIONSHIPS BETWEEN LUNG DUST BURDEN, PATHOLOGY AND LIFETIME EXPOSURE IN AN AUTOPSY STUDY OF U.S. COAL MINERS

E. D. Kuempel,*† E. J. O'Flaherty,† L. T. Stayner,* M. D. Attfield,‡ F. H. Y. Green§ and V. Vallyathan‡

*National Institute for Occupational Safety and Health (NIOSH), Education and Information Division, 4676 Columbia Parkway, MS C32, Cincinnati, OH 45226; †University of Cincinnati, Department of Environmental Health, Cincinnati, OH; ‡NIOSH, Division of Respiratory Disease Studies, Morgantown, West Virginia, U.S.A.; and §University of Calgary, Department of Pathology, Calgary, Alberta, Canada

INTRODUCTION

Previous pathological studies of U.S. miners have described correlations between radiographical and pathological evidence of coal workers' pneumoconiosis and lung dust burden (Vallyathan *et al.*, 1996; Attfield *et al.*, 1994). The present study includes estimates of miners' lifetime cumulative exposures to respirable coal mine dust and investigation of the relationships between duration and cumulative exposure, lung dust burden and pathological response. Also investigated is the estimated mean lung dust clearance during the retirement period.

METHODS

Cases were collected systematically from 1957 to 1971 by the late Dr Laqueur in the Beckley, West Virginia, area from consecutive autopsies of approximately 700 former coal miners (Attfield *et al.*, 1944). The coal mined in that area is primarily high volatile bituminous. Whole lung sections were evaluated for macules, micronodules, macronodules and progressive massive fibrosis (PMF), using a four-point scale (absent, slight, moderate, severe) described in Vallyathan *et al.* (1996) and the guidelines of Kleinerman *et al.* (1979). For 131 miners, the mass and composition of dust (coal, noncoal, silica and total dust) in the lungs were determined using gravimetric or spectrophotometric analysis (Carlberg *et al.*, 1971). Occupational and smoking histories were determined from the next-of-kin. Cumulative exposure was estimated from tenure in each job and job-specific measurements of mean concentration of respirable coal mine dust, using the approach of Attfield and Morring (1992). Table 1 describes characteristics of the cohort.

Lung dust burden was used as either a response variable (in exposure–dose relationships) or a predictor variable (in dose–response relationships) Lung dust burdens are expressed in mg dust g^{-1} dry lung. Initial linear and logistic regression analyses were performed using all available measures of lung dust burden—coal, silica, noncoal and total dust. Further analyses focused on coal lung dust because it

Table 1. Characteristics of autopsy study population of U.S. miners*

Variable (units)	Smokers ($n = 91$)		Nonsmokers ($n = 24$)		Whole cohort[†] ($n = 131$)	
	Mean (SD)					
Age (years)						
Start of mining	20	(7.1)	20	(8.3)	21	(7.8)
Retirement	56	(7.4)	61	(6.9)	57	(7.3)
Death	66	(9.4)	74	(9.0)	67	(9.8)
Exposure[‡]						
Cumulative (mg-year m^{-3})	108.4	(42.7)	122.1	(49.0)	107.8	(43.4)
Duration (years)	36.2	(9.6)	40.8	(10.6)	36.0	(10.0)
Intensity (mg m^{-3})	3.0	(0.9)	2.9	(0.6)	3.0	(0.8)
Post-exposure[‡] duration (years)	9.8	(6.5)	12.8	(6.7)	10.3	(6.7)
Lung dust burdens (mg g^{-1} dry lung)						
Total dust	67.3	(40.0)	82.3	(42.7)	69.1	(40.3)
Coal dust	42.0	(34.0)	61.9	(35.4)	45.5	(34.7)
Noncoal dust	26.1	(21.1)	20.4	(13.0)	24.1	(19.1)
Silica dust	1.8	(1.1)	2.2	(1.3)	1.9	(1.1)
	Percentage (count)					
Pathological responses[§] (%)						
No disease	1	(1)	8	(2)	3	(4)
Macules	87	(79)	83	(20)	84	(110)
Micronodules	79	(72)	75	(18)	77	(101)
Macronodules	42	(38)	58	(14)	47	(61)
PMF	24	(22)	42	(10)	30	(39)
Race (%)						
White	73	(66)	46	(11)	66	(86)
Black	25	(23)	54	(13)	31	(41)
Other	2	(2)	0	(0)	3	(4)

* Subset of Laqueur autopsy study (Attfield *et al.*, 1994) for which lung dust burden data and occupational histories are available; all miners in the study are male.
[†] Smoking status is unknown for 16 miners.
[‡] Respirable coal mine dust exposure.
[§] Macules: moderate severity or greater; micronodules, macronodules, and PMF: slight severity or greater.

was a significant predictor of all pathological responses evaluated in the study. Pathological response was defined in logistic regression analyses as the presence or absence of a given response and all responses of higher severity.

Analyses were performed separately for smokers and nonsmokers because of possible differences in the deposition and retention of dust in the lungs of smokers and nonsmokers and because of observed differences in factors that may influence pathological response (age of retirement, age of death, cumulative exposure, lung dust burden) (Table 1). Departures from linearity were evaluated using quadratic and cubic terms of duration of exposure, cumulative exposure and lung dust burdens; two-way interactions between covariates were evaluated.

Mean clearance rates of dust from the lungs were estimated from the slope of the relationship between the lung dust burden (normalized on cumulative exposure) and the post-exposure duration (years between retirement and death). These values were computed in linear regression analyses of the whole cohort, smokers and nonsmokers, and smokers stratified by lung dust burden (tertiles) for each of the measured lung dusts.

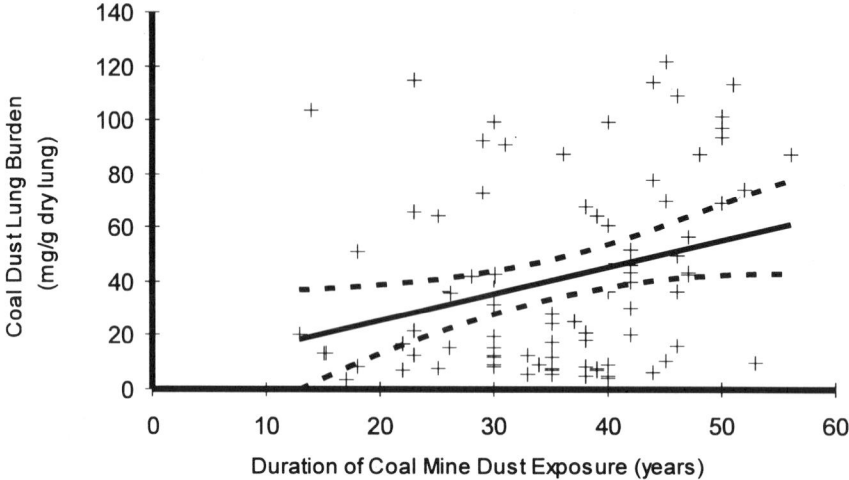

Fig. 1. Actual and predicted coal dust lung burden by duration of exposure to respirable coal mine dust in smokers (95% confidence intervals).

RESULTS

Exposure–dose

Coal lung dust burden increased in a linear relationship with duration of exposure ($P < 0.006$, $r^2 = 0.08$) (Fig. 1). Coal lung dust burden also increased with cumulative exposure ($P < 0.0001$; $r^2 = 0.17$) (Fig. 2). Departure from linearity was suggested by marginally significant quadratic and cubic terms for cumulative exposure ($P < 0.05$; skewness observed in residuals). The model fit shown in Fig. 2 excludes the two highest cumulative exposures; exposure–dose was also significant with these cases included ($P < 0.007$). Smoking status was a significant, negative

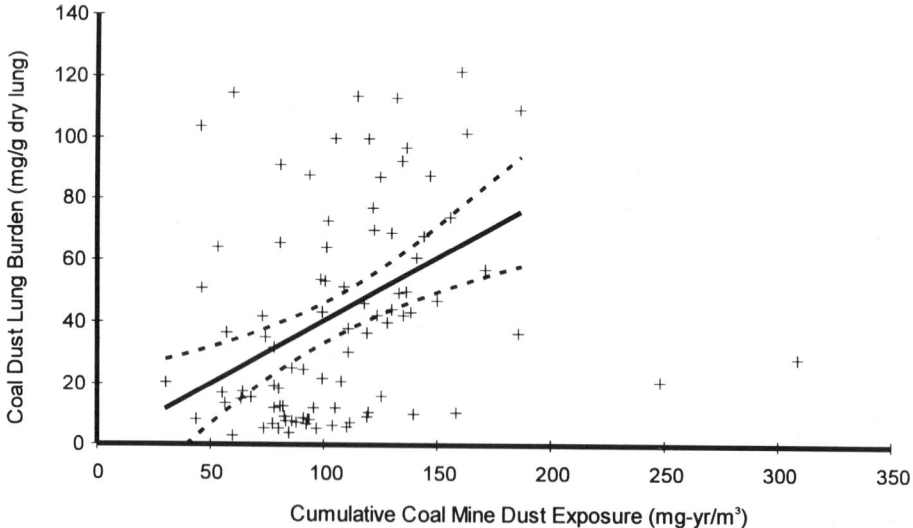

Fig. 2. Actual and predicted coal dust lung burden by cumulative exposure to respirable coal mine dust in smokers (95% confidence intervals).

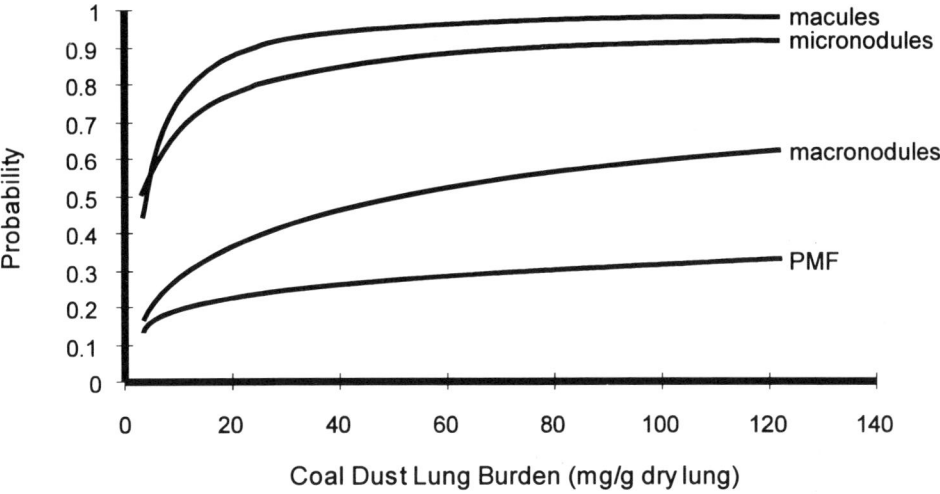

Fig. 3. Probability of pathological response associated with (log) coal dust lung burden in smokers.

predictor of coal dust lung burden ($P < 0.02$), apart from the positive effect of cumulative exposure (effect of smoking: -17 mg g^{-1} dry lung).

Dose–response

The presence and severity of pathological lesions (macules of moderate or greater severity, nodules and PMF) were significantly associated with the coal, silica and total lung dust burdens in the whole cohort (P values = 0.02–0.0001) and in smokers (P values = 0.04–0.003) (Fig. 3). The best predictor of macules was coal lung dust burden, while the best predictor of nodules and PMF was silica lung dust burden (based on comparison of model likelihoods). Smoking status was not a significant predictor of macules, nodules, or PMF (P values = 0.2–0.6). Age at death was a significant and positive predictor of all pathological responses evaluated except moderate or severe PMF.

Exposure–response

Lung dust burden was a better predictor of pathological response in the whole cohort than was duration of exposure or cumulative exposure to respirable coal mine dust (based on comparison of model likelihoods). Separate analysis by smoking status showed improved model fit for exposure–response in smokers (Fig. 4).

Clearance

No evidence of lung dust clearance during the post-exposure period was observed in the full cohort, smokers, or nonsmokers for any of the dusts evaluated—total, coal, noncoal, silica lung dust burdens (P values = 0.15–0.9). Post-exposure lung dust clearance was not detected even in the lowest coal lung dust burden stratum (9.8 mg g^{-1} dry lung, SD = 5.2), thus no dose-dependent clearance was observed.

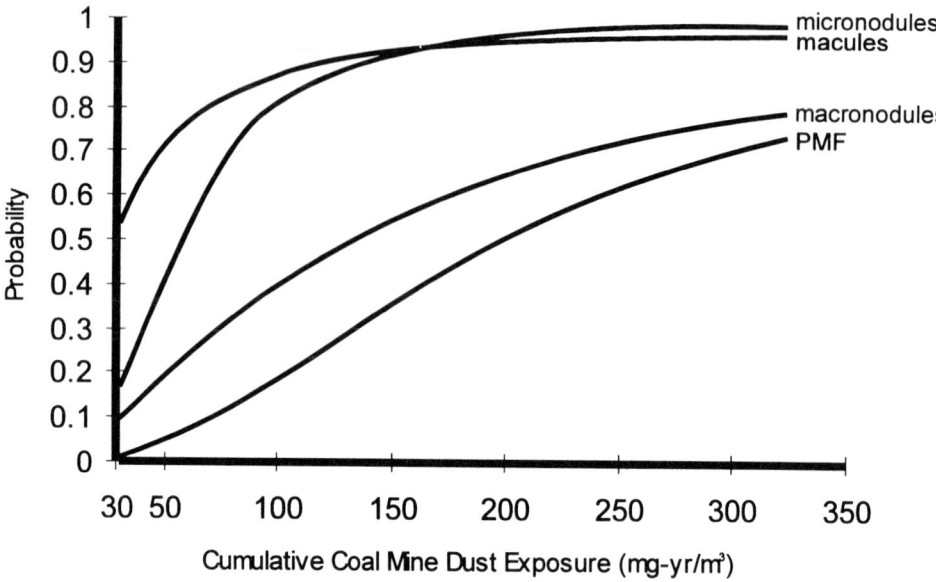

Fig. 4. Probability of pathological response associated with (log) cumulative exposure to respirable coal mine dust in smokers.

DISCUSSION AND CONCLUSIONS

The linear relationship between duration of exposure and lung dust burden suggests the absence of a time-dependent change in lung clearance during the study period, although a nonlinear relationship cannot be ruled out given the variability in the data. Lung dust burden at death increased with increasing lifetime exposure (a linear model may not be sufficient). The negative influence of smoking status on lung dust burden, after accounting for the positive effect of cumulative exposure, may suggest lower deposition of dust in the alveolar region in smokers (perhaps due to mucous hypersecretion and dust trapping in the proximal airways combined with enhanced cough clearance). No post-exposure clearance of dust from miners' lungs was observed in this study, although clearance of a relatively small portion of the total lung dust burden may have been undetectable (e.g. the alveolar dust burden relative to the interstitial dust burden).

Lung dust burden (coal, silica or total dust) was a better predictor of the presence and severity of pathological responses than were duration or cumulative exposure to respirable coal mine dust, although this distinction was less apparent when the analyses were stratified by smoking status. The significant association between increased age at death and increasing probability of having macules, nodules, or PMF (slight severity) may reflect the increased time for development and progression of these pathological responses.

The lung dust burdens of most miners in this study exceed those observed in animal studies in which overloading of lung clearance has been observed (Morrow, 1992). Further analyses of these data will include toxicokinetic modeling to evaluate hypotheses about relationships between lifetime exposures, retained lung dust, overloading of lung clearance and disease development.

REFERENCES

Attfield, M. D. and Morring, K. (1992) The derivation of estimated dust exposures for U.S. coal miners working before 1970. *Am. ind. Hyg. Assoc. J.* **53**(4), 248–255.

Attfield, M. D., Vallyathan, V. and Green, F. H. Y. (1994) Radiographic appearances of small opacities and their correlation with pathology grading of macules, nodules, and dust burden in the lungs. *Ann. occup. Hyg.* **38** (suppl. 1), 783–789.

Carlberg, J. R., Crable, J. V., Limtiaca, L. P., Norris, H. B., Holtz, J. L., Mauer, P. and Wolowicz, F. R. (1971) Total dust, coal, free silica and trace metal concentrations in bituminous coal miners' lungs. *Am. ind. Hyg. Assoc. J.* **32**(7), 432–440.

Kleinerman, J., Green, F., Harley, R. A., Lapp, N. L., Laqueur, W., Naeye, R. L., Pratt, P., Taylor, G., Wiot, J. and Wyatt, J. (1979). Pathology standards for coal workers' pneumoconiosis. *Archs. Path. Lab. Med.* **103**, 375–432.

Morrow, P. E. (1992) Dust overloading of the lungs: update and appraisal. *Toxicol. appl. Pharmacol.* **113**, 1–12.

Vallyathan, V., Brower, P. S., Green, F. H. Y. and Attfield, M. D. (1996) Radiographic and pathologic correlation of coal workers' pneumoconiosis. *Am. J. crit. Resp. Med.* **154**, 741–748.

 Pergamon

Ann. occup. Hyg., Vol. 41, Supplement 1, pp. 390–396, 1997
© 1997 British Occupational Hygiene Society
Published by Elsevier Science Ltd. All rights reserved
Printed in Great Britain
0003–4878/97 $17.00 + 0.00
Inhaled Particles VIII

PII: S0003–4878(96)00093–2

BACTERIAL MODULATION OF THE LUNG RESPONSE TO COAL MINE DUST

R. T. Cullen, J. Slight and W. M. Maclaren

Institute of Occupational Medicine, Roxburgh Place, Edinburgh EH8 9SU, U.K.

INTRODUCTION

Occupational exposure to coal mine dust is correlated with the development of a number of respiratory diseases. The mechanisms through which dust causes these conditions are incompletely understood but probably involve the inflammatory, growth regulatory and immune functions of leukocytes, particularly those of pulmonary macrophages. A number of studies have also examined the role of infection as a co-factor in dust-induced lung disease (Seaton, 1984).

Numerous reports in the literature describe alterations in immune parameters following occupational inhalation exposure to mineral dusts, including coal. It is also well known that infection or exposure to the products of some bacteria can lead to a series of changes in macrophage structure, metabolism and function; a process known as activation (Adams and Hamilton, 1984).

Infection can alter the lung response to mineral dusts (Chiappino and Vigliani, 1982). It is likely that the toxic properties of dust components, particularly silica, on pulmonary macrophages potentiate mycobacterial and other infections (Snider, 1978). It has also been shown that smoke inhalation caused lung injury in rats only when there was a concurrent, remote inflammatory stress induced by intraperitoneal injection of glycogen (Thom *et al.*, 1994). Co-exposure or sequential exposure of workers to infection and dust could result in potentiation of lung damage.

AIMS OF THE STUDY

The objective of this study was to examine, in an animal model, the effects of bacteria (killed *Corynebacterium parvum*) and a bacterial product, lipopolysaccharide (endotoxin), on the pulmonary response to instilled coal mine dust.

MATERIALS AND METHODS

Coal mine dust was collected from the air of a British coal mine producing a low rank coal. Analysis of the dust showed that it contained 6.7% quartz, 18.1% kaolin, 53.2% ash and no mica. Dust was pasteurised by exposure to 150 kV radiation and stored under nitrogen gas until use.

Coal mine dust instillation

One microgram quantities of coal mine dust were suspended in phosphate-buffered saline (PBS) and instilled into the lungs of SPF male Wistar rats by surgical excision of the trachea under halothane anaesthesia. Lungs were lavaged 18 h later and differential counts were made on the recovered cells.

Tumour necrosis factor (TNF) assay

In some experiments, alveolar macrophages recovered in lavage were cultured overnight for the production of the pro-inflammatory cytokine TNF. TNF was measured using the L929 cell bioassay (Flick and Gifford, 1984).

Corynebacterium parvum *(CP) and lipopolysaccharide (LPS)*

Rats were treated with lyophilised, heat-killed whole cells of the bacterium *C. parvum* (RIBI ImmunoChem Research Inc., Hamilton, MT, U.S.A.) either as an intraperitoneal injection of 2.8 mg in 2 ml PBS or as an intra-tracheal instillation of 1.4 mg in 0.2 ml PBS. LPS (from *E. coli* strain 0.127:B8), also known as endotoxin, was used either to stimulate alveolar macrophages *in vitro* to produce TNF or as an intraperitoneal injection of 100 μg in 2 ml PBS.

Statistical analyses

Data were analysed using the analysis of variance directives of the Genstat Statistical Analysis Program (Genstat 5, Reference Manual, 1989). The various response variables (i.e. numbers of cells of various types and TNF production) were transformed to the log scale before analysis. Adjusted treatment means obtained from analysis of variance were re-transformed for presentation purposes and are interpretable as estimated geometric means.

Treatment schedule

Day −1, −3 or −7	Pre-treatment with LPS or *C. parvum*.
Day 0	Lung instillation of 1 mg coal mine dust.
Day +1	Bronchoalveolar lavage of rat.

RESULTS

The pulmonary inflammatory response to instilled coal mine dust

The effect of an instillation of coal mine dust on the numbers of neutrophils recovered in lavage is shown in Fig. 1. Neutrophil numbers peak within the first 18 h post treatment and return to normal levels by day 7. Note that neutrophils are absent or in very low numbers in lavage from untreated lungs. It is the effect of LPS and CP on this initial, 1 day, inflammatory response which we studied.

The effect of treatment with C. parvum *prior to instillation of dust*

CP was injected i.p. 1, 3 or 7 days before dust instillation into the lung. There was no effect of CP on cell recruitment or on TNF production (data not shown). CP was also instilled into rat lungs 3 or 7 days before dust. CP caused the greatest leukocyte infiltrate at day 3 but this was significantly reduced when dust was also instilled. The effect on the macrophage response is shown in Fig. 2(a). This effect

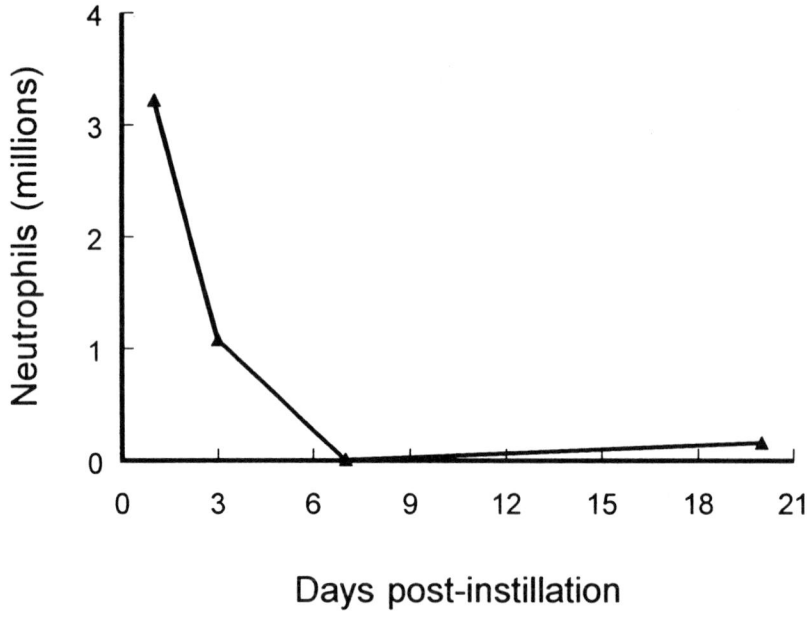

Fig. 1.

was no longer present when there was a gap of 7 days between lung instillations. TNF production by alveolar macrophages [Fig. 2(b)] was lower for the dust alone group than for the CP group ($P < 0.01$) or for the CP + dust combined treatment ($P < 0.05$). TNF in the CP + dust group macrophages increased with the length of the interval between CP and dust treatments.

The effect of treatment with lipopolysaccharide prior to instillation of dust

There was no difference in the neutrophil (inflammatory) response between the dust alone and the dust + LPS groups [Fig 3(a)]. However, neutrophils in the LPS + dust group were significantly raised for the day 7 interval compared to the 1 day interval. Total cell and macrophage numbers were unaffected by the LPS pre-treatment (data not shown).

Alveolar macrophages from the i.p. LPS experiment were cultured for TNF production with and without the presence of LPS. Where LPS was absent from cultures, LPS pre-treatment of the peritoneal cavity did not affect TNF concentrations (data not shown). When LPS was added to the cultures, TNF was enhanced for the day 7 dust + LPS group compared to the dust alone group ($P < 0.01$) [Fig. 3(b)]. This effect was not seen with the day 1 group.

The effect of opsonising coal mine dust for instillation on subsequent in vitro *TNF production*

Rat lungs were lavaged 1 day following the intratracheal instillation of coal mine dust pre-treated with either saline or 10% normal rat serum. There was a significant effect of serum treatment ($P < 0.05$) (Fig. 4).

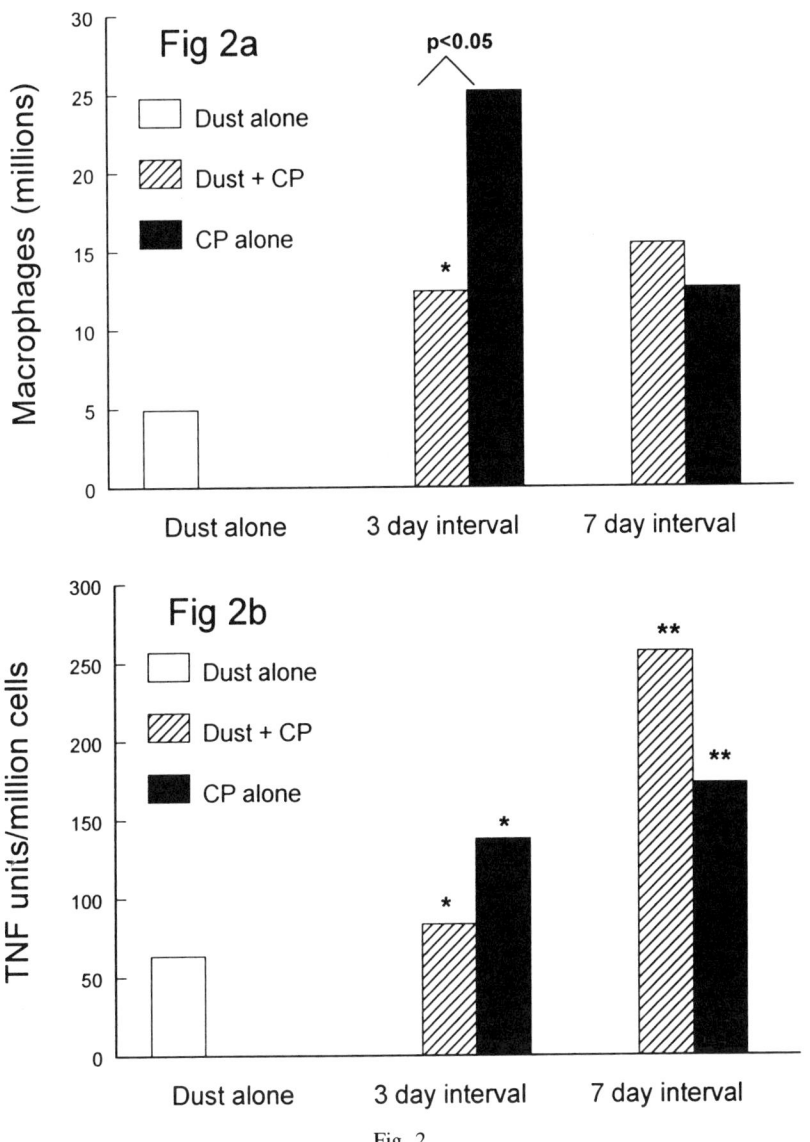

Fig. 2.

DISCUSSION

Repeated dust exposure and infection can both separately cause modulation of the immune and phagocytic defence systems and can lead to permanent lung damage. We examined the interactive effects of C. parvum or LPS on the lung inflammatory response to coal mine dust and alveolar macrophage production of TNF. We found that pre-treatment with intratracheal C. parvum or intraperitoneal LPS could affect cellular and TNF responses but that the effects depended on the time interval between bacterial and LPS treatments and dust instillation.

In the experiments where C. parvum was instilled into lungs a few days before

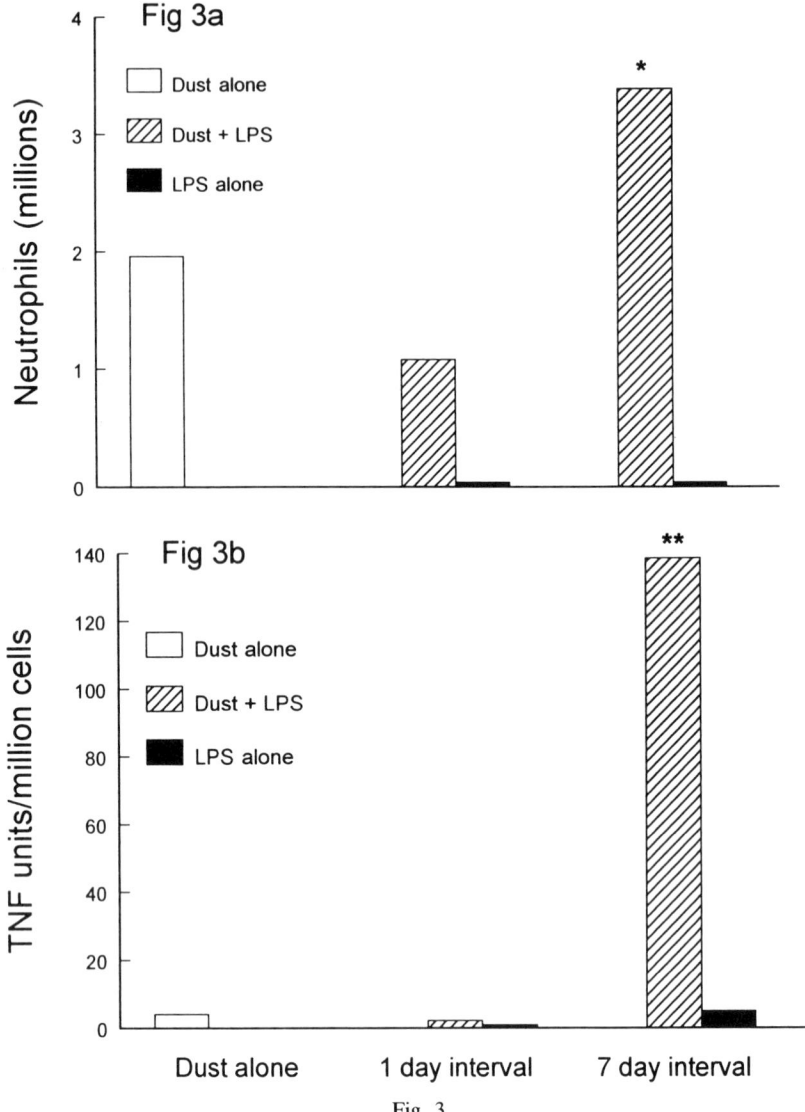

Fig. 3.

dust, there was a reduction in the numbers of cells recovered by lavage in the dust + CP group. This would appear to be a manifestation of the "macrophage (or leukocyte) disappearance reaction" well known in peritoneal and pleural studies involving LPS, *C. parvum* and other acute inflammatory stimulants (reviewed in Barth *et al.*, 1995). In this phenomenon macrophages are thought to aggregate or adhere to surfaces more firmly and thus resist lavage.

In contrast to our previous studies using quartz (e.g. Cullen and Li, 1994), we did not find that intratracheal instillation of untreated coal mine dusts led to an enhancement of TNF production *in vitro* by lavaged macrophages. However, opsonin (serum) treatment of dusts before instillation did enhance the subsequent

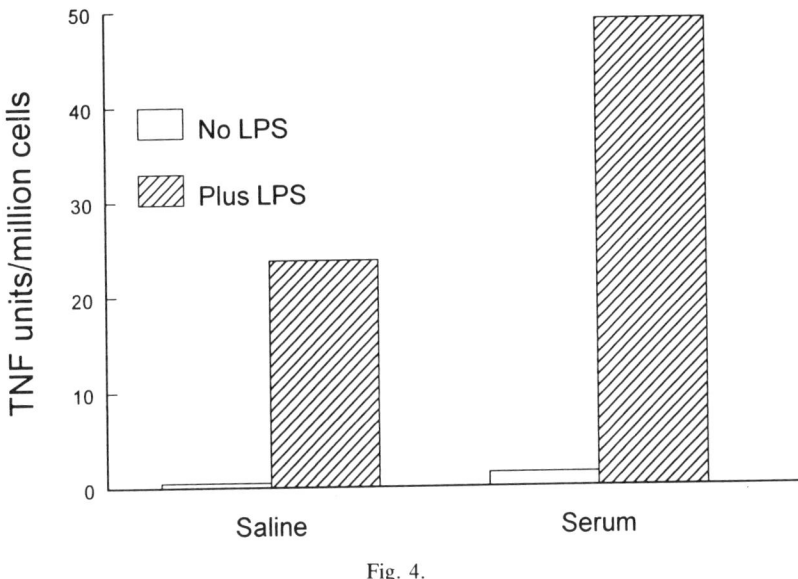

Fig. 4.

in vitro production of TNF. Neither *C. parvum* nor LPS peritoneal pre-treatment affected spontaneous *in vitro* production of TNF. However, when macrophages were stimulated *in vitro* with LPS, the LPS (but not *C. parvum*) pre-treatment was associated with enhanced TNF release when there was a gap of 7 days, but not of 1 day, between LPS injection and dust instillation. These LPS experiments suggest that at remote inflammation sites (i.e. the peritoneal cavity in this case) time is required for the development of systemic signals, possibly involving lymphocytes, which subsequently activate alveolar macrophages to produce cytokines such as TNF and the neutrophil chemokines CINC or MIP-2 (Wolpe and Cerami, 1989; Blackwell *et al.*, 1994). One of the many cellular products produced in response to infection or microbial products is interferon gamma, a powerful activator of macrophages (Adams and Hamilton, 1984). Macrophages themselves can be primed to produce IFN-gamma on subsequent exposure to LPS which in turn leads to enhanced production of TNF, IL-1 and IL-6 (Smith *et al.*, 1993).

In a study in mice, Ayala *et al.* (1992) showed that peritoneal infection did not affect spontaneous production of TNF by alveolar macrophages but that in the presence of LPS *in vitro*, TNF was reduced at 1 h post-infection but not at 24 h. Several studies have reported similar effects of time where the body's response to a second agent is affected by its response to a previous agent. For example, in a paper by Ghaffar and Sigel (1978) intraperitoneal (i.p.) injection of *C. parvum* 1 to 16 days before i.p. injection of antigen resulted in marked depression of antibody responses whereas intravenous administration of *C. parvum* enhanced antibody production.

Evidence is now emerging that LPS could be exerting its many effects by mimicking the action of the intracellular second messenger molecule ceramide, with which it shares structural features (Wright and Kolesnick, 1995). Ceramide, as part of the sphingomyelin cycle of signal transduction, is now thought to be important in mediating the effects of cytokines such as TNF and IL-1 (Pushkareva *et al.*, 1995). Therefore, there is now a mechanism to explain many of the effects of LPS on the host defence systems of the body.

In conclusion, our study suggests that infection, both in the lung and at remote sites, could modulate the activity of the alveolar macrophage, a key cell in dust pathogenesis because of its ability to initiate and regulate inflammation and fibrogenesis.

Acknowledgements—This study was supported by the Commission of the European Communities.

REFERENCES

Adams, D. O. and Hamilton, T. A. (1984) The cell biology of macrophage activation. *Ann. Rev. Immunol.* **2**, 283–318.

Ayala, A., Perrin, M. M., Kisala, J. M., Ertel, W. and Chaudry, I. H. (1992) Polymicrobial sepsis selectively activates peritoneal but not alveolar macrophages to release inflammatory mediators (interleukins-1 and -6 and tumor necrosis factor). *Circulatory Shock* **36**, 191–199.

Barth, M. W., Hendrzak, J. A., Melnicoff, M. J. and Morahan, P. S. (1995) Review of the macrophage disappearance reaction. *J. Leuk. Biol.* **57**, 361–367.

Blackwell, T. S., Holden, E. P., Blackwell, T. R., DeLarco, J. E. and Christman, J. W. (1994) Cytokine-induced neutrophil chemoattractant mediates neutrophilic alveolitis in rats: association with nuclear factor k B activation. *Am. J. Resp. Cell Molec. Biol.* **11**, 464–472.

Chiappino, G. and Vigliani, E. C. (1982) Role of infective, immunological, and chronic irritative factors in the development of silicosis. *Br. J. ind. Med.* **39**, 253–258.

Cullen, R. T. and Li, X. Y. (1994) Inflammatory cell recruitment and cytokine production in a rat model of acute silicosis: modifying effects of endotoxin. Inhaled particles VII: Proceedings of an international symposium on inhaled particles organised by the British Occupational Hygiene Society. *Ann. occ. Hyg.* **38** (1), 383–388.

Flick, D. A. and Gifford, G. E. (1984) Comparison of *in vitro* cell cytotoxicity assays for tumor necrosis factor. *J. Immunol. Meth.* **68**, 167–175.

Genstat 5 Committee (1989, 1993) *Genstat 5 Reference Manual.* Clarendon Press, Oxford.

Ghaffar, A. and Sigel, M. M. (1978) Immunomodulation by *Corynebacterium parvum.* I. Variable effects on anti-sheep erythrocyte antibody responses. *Immunol.* **35**, 685–693.

Ohmori, Y. and Hamilton, T. A. (1994). Regulation of macrophage gene expression by T-cell-derived lymphokines. *Pharmacol. Therapeut.* **63**, 235–264.

Pushkareva, M., Obeid, L. M. and Hannun, Y. A. (1995) Ceramide: an endogenous regulator of apoptosis and growth suppression. *Immunol. Today* **16**, 294–297.

Seaton, A. (1984) Infectious diseases. In *Occupational Lung Diseases* (Edited by W. K. C. Morgan and A. Seaton), 2nd edn. pp. 643–656. WB Saunders Co, London.

Smith, S. R., Calzetta, A., Bankowski, J., Kenworthy-Bott and Terminelli, C. (1993) Lipopolysaccharide-induced cytokine production and mortality in mice treated with *Corynebacterium parvum. J. Leuk. Biol.* **54**, 23–29.

Snider, R. E. (1978) The relationship between tuberculosis and silicosis. *Am. Rev. Resp. Dis.* **118**, 455–460.

Thom, S. R., Mendiguren, I., Van Winkle, T., Fisher, D. and Fisher, A. B. (1994) Smoke inhalation with a concurrent systemic stress results in lung alveolar injury. *Am. J. Resp. crit. Care Med.* **149**, 220–226.

Wolpe, S. D. and Cerami, A. (1989) Macrophage inflammatory proteins 1 and 2: members of a novel superfamily of cytokines. *FASEB J.* **3**, 2565–2573.

Wright, S. D. and Kolesnick, R. N. (1995) Does endotoxin stimulate cells by mimicking ceramide? *Immunol. Today* **16**, 297–302.

 Pergamon

Ann. occup. Hyg., Vol. 41, Supplement 1, pp. 397–402, 1997
© 1997 Published by Elsevier Science Ltd on behalf of BOHS
Printed in Great Britain. All rights reserved
0003–4878/97 $17.00 + 0.00
Inhaled Particles VIII

PII: S0003–4878(96)00161–5

SILICOSIS AND LUNG CANCER—A STUDY OF A COMPENSATION REGISTER IN AUSTRALIA

A. Rogers,* P. Yeung,† G. Berry,‡ J. Lee,§ I. Gardiner§ and L. Apthorpe†

*Alan Rogers OH&S Pty Ltd, PO Box 2128, Clovelly, NSW 2031, Australia; †Worksafe Australia,
GPO Box 58, Sydney, NSW 2001, Australia; ‡University of Sydney, NSW 2006, Australia; and
§Workers' Compensation (Dust Diseases) Board, NSW, Australia

INTRODUCTION

A study is undertaken to review the silicosis register of the New South Wales Workers' Compensation (Dust Diseases) Board. Its principal purpose is to determine if there is evidence of increased lung cancer mortality in workers compensated for silicosis in NSW. Preliminary results from a random sample of 863 (55%) subjects in this cohort were reported (Yeung *et al.*, 1995). This paper is an update of the 1995 findings.

The association of silica exposure and lung cancer continues to be a subject of debate at scientific forums (San Francisco, 1993; Baltimore, 1994). The International Agency for Research on Cancer (IARC, 1987) classified crystalline silica as a probable human carcinogen (category 2A). This category is used when there is limited evidence of carcinogenicity in humans and sufficient evidence in experimental animals. Sufficient evidence in animals consisted of rats exposed intrapleurally, intraperitonally, intratracheally and by inhalation.

Although the present collective epidemiological evidence is not completely conclusive, a number of studies of cohorts and compensation registries of silicotics have shown consistent excesses of lung cancer (see reviews by Berry, 1996; McDonald, 1995; Weill and McDonald, 1996). In most earlier studies of silicotics, the standardised mortality ratios (SMRs) for lung cancer were raised by some two to four-fold. However, interpretation of these studies is complicated by a number of factors, including confounding due to cigarette smoking and exposures to known carcinogens such as PAHs, radon and asbestos, and diagnostic bias towards those subjects who received compensation.

MATERIALS AND METHODS

The Dust Diseases Board (DDB) is a statutory authority established by the NSW Workers' Compensation (Dust Diseases) Act 1942. It has exclusive jurisdiction under the Act to determine claim for compensation from any NSW worker (except coal miners) whose disability is attributable to one of the 14 specified dust diseases (which include silicosis and silicotuberculosis). Applicants for compensation are usually examined on DDB's behalf by respiratory physicians. The examinations include a detailed pulmonary function test, PA chest radiograph and questionnaire on work history, medical symptoms and since 1972, cigarette-smoking habits. Determinations of pneumoconiosis are made by DDB's Medical Panel which consists of three experienced respiratory physicians. All films are evaluated for pneumoconiosis according to the ILO Classification (ILO, 1981). Diagnosis of silicosis is based on film reading and the history of silica exposure. Regular medical surveillance is also provided to those workers who are occupationally exposed to silica, as well as to those who have been assessed as having a disability due to silicosis.

The register consisted of 1575 silicotics who claimed for compensation for silicosis and silicotuberculosis and who had survived until 1968 if first diagnosed earlier or who had been first diagnosed in or after 1968. Of these, 108 were not compensated because clinical disability had not developed and 52 of the 1467 successful claimants had incomplete records. This paper is concerned with the 1415 successful claimants for whom complete records have been identified.

Each of the 1415 claimants has on file a series of medical records, chest radiographs and questionnaires in chronological sequence. Since 1972, detailed and high-quality information has been collected on standardised forms. This information comprised occupation and industry history, smoking habit, exercise capacity, respiratory symptoms, lung function, silicosis classification (ILO system) and causes of death if applicable. The data on record have been transferred to a computerised database and stratified and analysed statistically. The underlying causes of death were abstracted from the death certificates and are to be coded according to the International Classification of Diseases—9th revision.

Age-adjusted cause-specific SMRs are computed as a ratio of observed to expected number of deaths for comparison of the silicotics to the NSW male population. 95% confidence levels were estimated assuming that the observed number of deaths follow a Poisson distribution. The lung cancer SMRs are adjusted for the cigarette smoking habits of the silicotics, which are compared to the reported patterns of tobacco smoking among the Australian male population in different periods of time (Hill *et al.*, 1991).

RESULTS

Industry and occupation profile

These 1575 subjects were foundry workers (25%), labourers (21%), machine operators (20%), miners or quarry workers (14%), brick workers (9%), construction workers (5%), or others (6%). They worked in different industries: building and construction (26%), foundries (19%), brickworks (17%), engineering work (16%), mining and quarrying (14%), glass making (2%) or others (6%).

A draft technical report produced by Worksafe Australia (1993) considers the risk of lung cancer with the current distribution of exposure levels. About 136 000 male workers in Australia (or 2% of the workforce) are likely to be occupationally exposed to silica. Of these, 70% are in construction, 19% in mining and 11% in manufacturing. The highest exposures (≥ 0.2 mg m^{-3}) are experienced by workers from the following industries: earthmoving and dredging, metallic and non-metallic mineral work and clay brick manufacture. The distribution of occupations and

Table 1. Distribution of smoking habits (in percent) of NSW silicotics in comparison with the Australian population

| Smoking status | NSW silicotics | | Australian men of same age distribution as silicotics in 1989 |
	last medical (N = 982)	1989 (N = 378)	
Current	34.3	26.7	22.4
Ex-	48.1	52.4	46.4
Non-	17.6	20.9	31.2

industries of the NSW silicotics are generally consistent with the national exposure profile.

Smoking habits

Among the 1415 silicotics, appropriate cigarette smoking habits are kept for 982 cases (69%). Table 1 summarises the pattern of their smoking habits at the last medical examination and in 1989, in comparison with the age-adjusted results of a market research survey conducted in Australia in 1989 (Hill *et al.*, 1991). The overall proportions of smokers and ex-smokers in this cohort are higher than in the general population of Australian men in 1989. Similar smoking data of the Australian population are available for the period 1974–92.

Date of diagnosis

The 1415 "compensable silicosis" cases (with disability due to silicosis) were diagnosed between 1937–1994. Table 2 shows the distribution of these claims in relation to calendar years. About two thirds of the cases were diagnosed before 1970 and, as some of those diagnosed in 1967 or earlier will have died before 1968, the incidence in the three decades 1937–1969 was higher than given in Table 2. It is clear that the incidence has declined markedly since the 1970s. The high incidence before the 1970s can be attributed to the increased building and water work activities in Sydney in the postwar years (Sydney sandstone contains practically 100% crystalline silica) and a large-scale project carried out in the Snowy Mountains area between 1948–60, which involved extensive tunnelling and underground work.

Table 2. Distribution of silicosis cases in relation to the date of initial diagnosis

Date of initial diagnosis	N	Percent
1937–49	109	7.7
1950–59	257	18.2
1960–69	527	37.2
1970–79	398	28.1
1980–89	99	7.0
1990–94	25	1.8
All cases	1415*	100.0

* Note: non-compensable silicosis cases (without disability) are not included.

Table 3. Distribution of deceased silicosis cases in
relation to the age at death

Age at death	N	Percent
−49	28	2.3
50–59	84	7.0
60–69	328	27.2
70–79	465	38.6
80–89	280	23.2
> 89	20	1.7
All cases	1205	100.0

Table 3 summarises the age at death of 1205 deceased silicotics. These are from
the 1457 silicotics as some with incomplete records had death information.
However, the cause of death has yet to be established for 63 (5%) of these deaths.

Mortality rate

The mortality study is of the 1415 compensated silicotics with complete records,
except that nine of them were not included because they were 85 years old either
before 1968 or when compensated. So this is a study of 1406 men. The follow-up
period is to the end of 1994. Age- and period-adjusted mortality ratios up to age 85
from all causes and from lung cancer are presented in Tables 4 and 5. The overall
SMR for all causes was 1.7 (95% CI 1.6–1.8) and for lung cancer was 2.2 (95% CI
1.7–2.7). A trend between lung cancer mortality and the degree of disability on the
first diagnosis of silicosis is observed, however, it is not statistically significant ($P >$
0.5). In both tables, the deaths in the first year were discarded in the calculation of
the overall SMRs. This is necessary because the inclusion of early deaths (including
a few who were compensated at death) would introduce some selection bias.

Adjustment for smoking habits

The mortality due to lung cancer is influenced by smoking and the data given in

Table 4. All causes age-standardised mortality in relation to the degree of disability and the length of
survival following diagnosis of silicosis

	N	Observed	Expected[a]	SMR	95% CI
Disability on silicosis diagnosis					
1– 25%	312	156	122.3	1.3	1.1–1.5
25– 49%	782	611	354.4	1.7	1.6–1.9
50– 74%	98	79	41.1	1.9	1.5–2.4
75–100%	214	178	80.5	2.2	1.9–2.6
Total	1406	1024	598.3	1.7	1.6–1.8
Years of survival after silicosis diagnosis					
< 1	652	35	13.2	2.7	1.8–3.7
1– 4	869	127	68.2	1.9	1.6–2.2
5– 9	972	200	122.6	1.6	1.4–1.9
10–19	913	407	249.1	1.6	1.5–1.8
20–59	446	255	145.2	1.8	1.6–2.0
Total (1–59)		989	585.1	1.7	1.6–1.8

(a) Expected values based on NSW male death rates.

Table 5. Lung cancer age-standardised mortality in relation to the degree of disability and the length of survival following diagnosis of silicosis

	N	Observed	Expected[a]	SMR	95% CI
Disability on silicosis diagnosis					
1– 25%	312	17	9.4	1.8	1.1–2.9
25– 49%	782	55	23.9	2.3	1.7–3.0
50– 74%	98	6	2.6	2.3	0.8–5.0
75–100%	214	11	4.3	2.6	1.3–4.6
Total	1406	89	40.2	2.2	1.8–2.7
Years of survival after silicosis diagnosis					
< 1	652	4	1.0	4.1	1.1–10.4
1– 4	869	12	5.0	2.4	1.2–4.2
5– 9	972	9	8.6	1.1	0.5–2.0
10–19	913	42	16.2	2.6	1.9–3.5
20–59	446	22	9.4	2.3	1.5–3.6
Total (1–59)		85	39.2	2.2	1.7–2.7

[a] Expected values based on NSW male death rates.

Table 5 will only give an unbiased assessment if the silicotics had similar smoking habits to the NSW male population which has been used to calculate the expected mortality. It was noted from Table 1 that a higher proportion of the silicotics had smoked at some time than for Australian men, in 1989 and after adjusting for age. The expected mortality due to lung cancer should be increased to allow for the excess mortality due to smoking. A factor for adjustment has been calculated as follows. Doll *et al.* (1994) found that for British doctors the relative death rates for lung cancer, relative to non-smokers, were 15 and 4.1 for current smokers and ex-smokers, respectively. These relative rates have been applied to the observed and expected smoking distributions of silicotics in 1989 (Table 1). The relative death rate for the observed smoking distribution was 6.36 and for the age-adjusted Australian population it was 5.57. The ratio of these death rates is 1.14 and represents the increased death rate expected in the silicotics as a result of their increased smoking. Applying this factor to lung cancer mortality more than 1 year after compensation gives an expected number of 39.2 × 1.14 = 44.7 and an adjusted SMR of 1.9 (95% CI 1.5–2.4).

DISCUSSION

The cohort of the NSW silicotics shows an excess of lung cancer mortality (SMR = 2.2, 95% CI 1.7–2.7) prior to any adjustment for smoking. An initial cross-sectional analysis of the smoking data available in 1989 has shown that the silicotics had smoked more than the age-adjusted average rate in Australia and for this reason their excess lung cancer risk was estimated at 1.14; applying this factor reduces the SMR in the silicotics to 1.9 (95% CI 1.5–2.4). However, applying the 1989 smoking data uniformly across the whole cohort, particularly to the earlier periods, might have underestimated the overall lung cancer risk. Stratified analyses of smoking data are necessary and being undertaken.

Mortality due to all causes shows an elevated risk (SMR = 1.7; 95% CI 1.6–1.8) and this is consistent with the initial observation that excess deaths were due to tuberculosis and non-malignant respiratory diseases.

The result has shown a trend between lung cancer SMR and the degree of initial disability (as an indicator of the initial severity of silicosis), however, it was not statistically significant ($P > 0.5$). In future analysis, radiographic progression of silicosis (simple silicosis vs progressive massive fibrosis) will be evaluated in relation to lung cancer mortality. This will help clarify the possible relationship between the severity of silicosis and lung cancer.

The results from this study have reinforced the collective evidence from the numerous overseas studies that silicotics have shown consistent excess of lung cancer. While the totality of epidemiological evidence linking silica exposure in humans to lung cancer may still be inconclusive, patients suffering from silicosis are likely to be at an increased risk of lung cancer, even if exposure to silica may not be the only aetiological factor. Thus control measures for the prevention of silicosis will be effective in reducing any excess in lung cancer.

Acknowledgements—The authors thank the New South Wales Workers' Compensation (Dust Diseases) Board for its support to this study and its staff for providing assistance and access to the silicosis register. The authors also thank Mr John Tay and Ms Julia Davies of Worksafe Australia for their assistance in data management.

Disclaimer—The conclusions reached and scientific views expressed in this paper are those of the authors and do not necessarily reflect those of the organisations in which they work.

REFERENCES

Baltimore Conference (1995) Proceeding of the international conference on crystalline silica health effects: current state of the art. *Appl. occup. Environ. Hyg.* **10**(12), 981–1156.
Berry, G. (1996) Crystalline silica: health impacts and possible lung cancer risks. *J. occ. Health Safety—Aust NZ*, **12**(2), 157–167.
Doll, R., Peto, R., Wheatley, K., Gray, R. and Sutherland, I. (1994) Mortaltiy in relation to smoking: 40 years' observations on male British doctors. *Br. Med. J.* **309**, 901–910.
Hill, D. J., White, V. M. and Gray, N. J. (1991) Australian patterns of tobacco smoking in 1989. *The Medical J. Australia* **154**, 797–801.
ILO Committee on Pneumoconiosis (1981) Classification of radiographs of the pneumoconiosis. *Med. Radiogr. Photogr.* **57**, 2–17.
International Agency for Research on Cancer (1987) Silica and some silicates. *Monographs on the Evaluation of the Carcinogenic Risks of Chemicals to Humans*, Vol. 42, pp. 104–108. Lyon.
McDonald, C. (1995) Silica, silicosis, and lung cancer: an epidemiological update. *Appl. occup. Environ. Hyg.* **10**(12), 1056–1063.
San Francisco Conference (1995) Proceedings of the second international symposium on silica, silicosis and lung cancer, San Francisco, October 28–30. *Scand. J. Work Environ. Health* **21** (suppl 2), 1–120.
Weill, H. and McDonald, J. C. (1996) Crystalline silica exposure and lung cancer risk: the epidemiological evidence. *Thorax* **51**, 97–102.
Worksafe Australia (1993) Draft technical report on crystalline silica completed in July 1992, Sydney.
Yeung, P., Rogers, A., Berry, G. and Davies, J. (1995) Silicosis and lung cancer—preliminary results from a study of a compensation register in New South Wales. *Proc. 14th Annual Conf. of Australian Inst. Occupational Hygienists* (Edited by D. Pisaniello) pp. 166–169, December 9–13, Adelaide.

 Pergamon

Ann. occup. Hyg., Vol. 41, Supplement 1, pp. 403–407, 1997
© 1997 British Occupational Hygiene Society
Published by Elsevier Science Ltd. All rights reserved
Printed in Great Britain
0003–4878/97 $17.00 + 0.00
Inhaled Particles VIII

PII: S0003–4878(96)00162–7

COHORT MORTALITY STUDY OF STAFFORDSHIRE POTTERY WORKERS: (I) RADIOGRAPHIC VALIDATION OF AN EXPOSURE MATRIX FOR RESPIRABLE CRYSTALLINE SILICA

G. L. Burgess,* S. Turner,* J. C. McDonald† and N. M. Cherry*

*Centre for Occupational Health, University of Manchester, Stopford Building, Oxford Road, Manchester M13 9PT, U.K.; and †Department of Occupational and Environmental Medicine, National Heart and Lung Institute, Imperial College, London, U.K.

CONSTRUCTION OF THE MATRIX

The population for study was defined (Cherry *et al.*, 1996) as male pottery workers in North Staffordshire born 1916–1945 and employed in the "dustier" trades, as specified by the Pottery (Health and Welfare) Special Regulations, 1950. These regulations required workers performing certain activities to be examined every 2 years, with full-size chest radiographs every 4 years. The geographical catchment area covered by the Stoke centre included most, but not all British pottery workers.

To derive the exposure estimates, it was necessary to review information on air sampling specific to this population, to determine a method for categorising workers into similar exposure groups and to find a way of estimating exposure where quantitative information was not available.

All available sampling data were gathered from a variety of published references as well as unpublished surveys. Most of the 1390 results obtained were from surveys performed by government enforcement agencies and industrial groups. In all instances the reported sampling strategy had been to determine representative exposures.

Methods designed to collect static (breathing area) samples for particle counting were widely used in the U.K. until the 1960s, when a cyclone was introduced to collect personal samples of respirable-sized dust followed by analysis of gravimetric silica mass. After a review of the issues surrounding conversions of results from one method to another, it was decided to make use of that used by Rice *et al.* (1984) for the dusty trades of North Carolina, U.S.A. This conversion equates 1 million particles per cubic foot of silica to $0.09 \ \mathrm{mg \ m^{-3}}$. The use of respiratory protective devices might have complicated exposure estimates but this does not appear to have been a control measure used in the pottery industry (Hopkinson, 1993).

Job titles were used to indicate similar exposure unless information was available to the contrary. By this means the industry was divided into 11 major process groups comprising, in all, 569 job titles.

Air sample results were averaged, particle counts converted to gravitational mass, categorised by job and decade, and tabulated into a preliminary exposure matrix. Published literature and unpublished reports of dust control measures or

Table 1. Matrix of estimated exposures, reported as daily 8 h time weighted average airborne concentrations of respirable crystalline silica in micrograms per cubic meter of air ($\mu g\ m^{-3}$)

Process	1930–39	1940–49	1950–59	1960–69	1970–79	1980–89	1990–
Material preparation	600	370	250	175	100	60	45
Body preparation	400	300	225	170	115	60	50
Primary shaping (production)	400	325	250	175	70	60	50
Primary shaping (non-production)	200	160	125	90	35	30	25
Secondary shaping	500	425	340	180	40	50	40
Firing	800	650	500	220	24	20	20
Glazing	250	200	150	100	80	29	25
Tiles	550	450	370	95	80	60	50
Maintenance	230	190	150	90	50	40	30
Mould making	150	110	70	40	30	20	15
Pottery support activities	20	16	12	9	3	3	2
Pottery activity, no job details known	200	160	125	90	35	30	25

changes to the process or workrate were then used to refine the matrix, particularly for the periods where no sample results were available. Judgement was inevitably needed in taking account of engineering changes.

The final version of the matrix is presented in Table 1. In general, it depicts an overall trend toward reduction in exposure during the 60 year span, with considerable variations with process and decade.

VALIDATION

A sub-cohort of pottery workers was selected from the main mortality cohort test to test whether exposure estimates based on the matrix were related to radiographic changes in a reasonable way. To ensure that members of the sub-cohort had sufficient exposure to produce a health effect and to eliminate confounding exposures, it was restricted to men with at least 10 years in the potteries, beginning before 1960 and without employment in other occupations which might have contributed to the development of pneumoconiosis. A total of 1080 workers met these criteria.

The routine medical examinations performed by the Stoke Medical Boarding Centre (MBC) included details of employer, job title and chest radiographic reading. The readings were made and recorded by medical officers of the Department of Social Security trained and experienced in the ILO classifications in current use. A reading of 1/0 or more was recorded in 64 members of the sub-cohort. For these 64 cases and for the remaining 1016 men without recorded opacities work histories were extracted without knowledge of the radiographic readings and cumulative and average lifetime exposures to crystalline silica were calculated from the matrix up to the time of the first positive reading. Basic information presented in Table 2 shows that workers exhibiting radiographic changes were born earlier, started work earlier and had an earlier date of last radiograph than those without changes.

Table 2. Descriptive characteristics of the study cohort

	Changes on radiograph ($\geq$ 1/0)		
	No	Yes	Overall
Year of birth			
Mean	33.1	24.6	32.6
SD	6.3	6.5	6.6
Year started in pottery			
Mean	49.7	40.3	49.1
SD	6.6	7.1	7.0
Year of last radiograph			
Mean	80.3	71.1	79.8
SD	9.8	11.2	9.9
N	1016	64	1080

Table 3. Prevalence of radiographic changes ($\geq$ 1/0) by cumulative exposure

Cumulative exposure (μg m^{-3}.year)	At risk (*N*)	Prevalence (%)
< 2000	109	0.0
2000 < 4000	449	2.0
4000 < 6000	297	6.4
$\geq$ 6000	225	16.0
Overall	1080	5.9

VALIDATION RESULTS

Table 3 shows that the prevalence of small radiographic opacities was related systematically to cumulative exposures, with no case observed at less than 2000 μg m^{-3}.year^{-1}. It is evident from Table 4, however, that while prevalence was related to time-weighted average exposure intensity there was no apparent relation to years of pottery employment. Nevertheless within duration categories, prevalence tended to increase with intensity.

DISCUSSION

An exposure matrix such as the one developed here provides only an approxima-

Table 4. Prevalence of radiographic changes $\geq$ (1/0) by time-weighted average intensity and duration of pottery employment

Average intensity (μg m^{-3})	Duration of pottery employment (years)									
	10 < 20		20 < 30		30 < 40		$\geq$ 40		Overall	
	N	%	N	%	N	%	N	%	N	%
< 100	35	0.0	33	3.0	81	0.0	16	0.0	165	0.6
100 < 150	57	3.5	88	1.1	175	2.9	40	0.0	360	2.2
150 < 200	97	1.0	66	3.0	74	8.1	73	4.1	310	3.9
$\geq$ 200	117	9.4	69	30.4	33	24.2	26	11.5	245	17.6
Overall	306	4.6	256	9.8	363	5.2	155	3.9	1080	5.9

tion to the true exposure of an individual worker. Raw materials vary in the pottery industry as do temperatures to which they are subjected; both of which may significantly affect the concentration and nature of respirable silica in the workroom air. Many additional factors will affect the quantity of silica dust that reaches a worker's lower respiratory tract.

The essential validity of the matrix is supported by the good correlations observed in the sub-cohort between prevalence of small radiographic opacities and estimates of both cumulative exposure and of time-weighted average intensity at varying durations of employment. The evidence is further strengthened by the absence of an association with duration of employment *per se*. In other words, the matrix was not merely a dilute surrogate for the measure of exposure time but rather the predicted concentrations within the matrix were necessary to achieve validation.

Given this evidence, it is reasonable to expect that the matrix will prove equally reliable in assessing exposures for the 5000 or so subjects in the main mortality cohort. In the meantime the observed exposure–response relationships for radiographic opacities add one more set of data to the few widely varying studies currently available. It is not inconceivable, of course, that the indices of cumulative exposure and average intensity simply act as surrogates for very high short term exposures.

SUMMARY AND CONCLUSIONS

A job exposure matrix for respirable crystalline silica was constructed from and for application to male workers employed in the Staffordshire potteries from 1930 to the present. Information was derived from more than 1300 air samples, published literature and unpublished reports of dust control innovations and process/workrate changes and interviews with current and former occupational hygienists in the industry.

To validate the matrix, cumulative and time-weighted average exposures were calculated for a sub-cohort of 1080 men who had at least 10 years of work in the potteries, had begun employment before 1960 and had no recorded occupational exposure to other known causes of pneumoconiosis. Indices of cumulative and average intensity of exposure both demonstrated a strong association with the prevalence of small radiographic opacities ($\geq$ 1/0). No association was detected between radiographic changes and duration of work which did not take account of exposure intensity. It is concluded that the matrix can be expected to provide reliable estimates of crystalline silica exposure for subjects in the main mortality cohort.

Acknowledgements—This work was supported in part by a grant from the Department of Social Security, London. The views expressed are not necessarily those of the DSS or any other government department.

REFERENCES

Cherry, N. M., Burgess, G. L., McNamee, R., Turner, S. and McDonald, J. C. (1996) A cohort mortality study of British pottery workers. *App. occup. Environ. Hyg.* **10**, 1042–5.

Hopkinson, D. (1993) The use of respiratory protection amongst workers in the UK pottery industry, April 8, 1993. (Private conversation), Mr D. Hopkinson, Health and Safety Executive, The Marches, Newcastle under Lyme, U.K.

Rice, C. H., Harris, R. L., Lumsden, J. C. and Symons, M. J. (1984) Reconstruction of silica exposure in the North Carolina dusty trades. *Am. ind. Hyg. Ass. J.* **45**, 689–696.

Pergamon

Ann. occup. Hyg., Vol. 41, Supplement 1, pp. 408–411, 1997
© 1997 British Occupational Hygiene Society
Published by Elsevier Science Ltd. All rights reserved
Printed in Great Britain
0003–4878/97 $17.00 + 0.00
Inhaled Particles VIII

PII: S0003–4878(96)00143–3

COHORT STUDY OF STAFFORDSHIRE POTTERY WORKERS: (II) NESTED CASE REFERENT ANALYSIS OF LUNG CANCER

N. M. Cherry,* G. L. Burgess,* S. Turner* and J. C. McDonald†

*Centre for Occupational Health, Stopford Building, University of Manchester, Oxford Road, Manchester M13 9PT, U.K.; and †Department of Occupational and Environmental Medicine, National Heart and Lung Institute, Imperial College, London SW3 6LY, U.K.

INTRODUCTION

A mortality study of Staffordshire pottery workers was undertaken to determine whether, in a cohort of workers unselected for silicosis or compensation, lung cancer was more common than in comparable populations without exposure to silica. Excess rates of lung cancer have been consistently observed in compensated or registered silicotics, but limited evidence for cohorts unselected in this way has been acquired, with excess in cohorts in pottery and with refactory brick work being more securely attributed to silica exposure than that in mines or quarries (McDonald, 1995). In the Staffordshire pottery cohort (Cherry et al., 1995), increased SMRs for both lung cancer (1.91, 90% CI 1.62–2.55) and non-malignant respiratory disease (2.87, 90% CI 2.28–3.58) were observed against rates for England and Wales. The rates for lung cancer (1.28, 90% CI 1.04–1.57) and, to a lesser degree, for non-malignant respiratory disease (2.04 90% CI 1.62–2.55) were much reduced when compared to rates for Stoke-on-Trent. Information on exposures within the cohort was seen to be very important in establishing the nature of the risk (if any) of lung cancer. A nested case referent study was carried out, with exposure being estimated from a matrix demonstrated (Burgess et al. this conference) to be useful in identifying risk of radiographic change in these workers.

METHODS

The cohort identified comprised 5115 men born 1916–1945, first registered as pottery workers in dusty trades in Stoke-on-Trent and with no known prior exposure to asbestos or foundry work and not more than one year's exposure to other dusts, including coal. All but 58 men (1.1%) were traced through Department of Social Security records. Of this total, 4352 were reported alive and 705 (13.9%) dead. Of these, 88 had a death certificate with a diagnosis of lung cancer (ICD9; 162). Fifty-two were included in the case referent analysis. Records for 32 had been destroyed, following a policy (inconsistently applied) of shredding forms and recycling radiographs when a worker reached his 70th birthday, had not worked in dusty pottery trade for 10 years or whose death had come to the attention of the local Department of Social Security office. The possibility that non-random application of these rules had introduced bias to the study (for

Table 1. Odds ratios for exposure indices (assuming a 10 year latency) before and after adjustment for smoking and radiographic change

	Unadjusted		Adjusted	
	OR	90% CI	OR	90% CI
Cumulative exposure				
$\geq 4000\ \mu g\ m^{-3}.y$	0.51	0.24–1.11	0.60	0.26–1.41
Duration				
$\geq 20\ y$	0.38	0.18–0.78	0.48	0.21–1.09
Mean intensity				
$\geq 200\ \mu g\ m^{-3}$	1.88	1.06–3.34	1.68	0.93–3.03
Maximum exposure				
$\geq 400\ \mu g\ m^{-3}$	2.16	1.11–4.18	2.07	1.04–4.14

example through the destruction of records for all those with compensated disease) was examined through information on the death certificates. For those with extant records, certificates reported pottery work for 50% and for those destroyed, 47%. Only three workers with lung cancer whose records had been destroyed had death certificates with mention of silicosis. This did not suggest important bias in destruction. Of those whose records were available, four cases were eliminated because of exposure to asbestos in work outside the pottery industry and two because they had developed lung cancer within 10 years of first pottery exposure.

Each of the remaining cases was matched (± 3 years) with three or four referents (depending on availability), on date of birth and date of first exposure. Referents were considered ineligible if they were known never to have smoked (as no case fell into this category), if they had been exposed to asbestos or if they had pre-deceased the case. Smoking information, recorded every 2 years since 1950, was missing for 12 cases and 33 referents and there was no record of a chest radiograph (taken every 4 years) for eight cases and 25 referents.

RESULTS

Cases and referents were matched successfully with mean year of birth 1930 and mean age at starting pottery work 19 years in each group. Amongst those with radiographs only three cases and 10 referents had an ILO reading of $\geq 1/0$. Many more cases (70% of those with data) than referents (47%) were heavy smokers (20 cigarettes per day or more). In both cases and referents, however, heavy smokers had a significantly shorter duration of exposure (5–6 years) than ex-smokers or those smoking less than 20 cigarettes a day.

Odd ratios (from conditional logistic regression) for each of four exposure indices are shown in Table 1, where column 1 gives ratios unadjusted for confounders and column 2 ratios after adjustment for smoking and radiographic change. Cumulative exposure was unrelated to lung cancer either with or without adjustment. Duration of exposure was significantly shorter in cases before adjustment but after allowance for confounders this no longer reached conventional levels of statistical significance. Mean intensity, as a function of cumulative exposure and duration, was also less strongly related to lung cancer after this adjustment.

The final exposure index shown in the table was included as a measure independent of duration; it indicates whether a worker had at any time been in a job with estimated exposure of more than 400 μg m^{-3}. Cases were more likely to have had such an exposure and the odds ratio remained significant after allowance for confounders.

COMMENTS AND CONCLUSIONS

In this nested case referent analysis cumulative exposure to silica was apparently unrelated to lung cancer risk and neither duration nor mean intensity showed a clearly positive relation after allowance for smoking and radiographic change. However, work in a job with high exposure (> 400 μg m^{-3}) appeared to increase risk of death from lung cancer. It may be that heavy exposure for some limited period rather than cumulative exposure over many years was responsible for the risk. Alternatively, such high exposures might have been characteristic of a job carrying a particular lung cancer risk because of the nature of the process or type of exposure.

Within this cohort exposure at such high levels was concentrated in jobs in firing, secondary shaping and, to a lesser degree, work with tiles. Firing and secondary shaping are of particular interest in that these jobs, together with glazing, are processes that incur exposure to silica that has been heated to 1000–1400°C with, particularly in the earlier years of this study, such high temperatures continuing for a number of days. Under these conditions conversion to cristobalite may occur and, although environmental samples in which cristobalite has been measured have not been identified specific to the firing and post-firing process, pottery samples in Stoke-on-Trent have been recorded with 8% cristobalite (HM Factory Inspectorate, 1959). In view of the suspected greater fibrogenicity of cristobalite than quartz and the apparent excess of lung cancer in cristobalite-exposed workers in the diatomaceous earth industry (Checkway *et al.*, 1993) and in the manufacture of refactory brick (Merlo *et al.*, 1991), cases and referents were classified by whether or not they had ever worked in firing or post firing occupations. When this was done, cases were in clear excess (31% cases compared with 18% of referents) with an odds ratio of 2.17 (90% CI 1.16–4.07) after adjustment for smoking and radiographic change. Having taken account of tasks in these areas, work at more than 400 μg m^{-3} no longer differentiated cases from referents. It thus appears that the only risk factor so far identified in this study was work in firing or post firing occupations.

Acknowledgements—This work was supported in part by a grant from the Department of Social Security, London. The views expressed are not necessarily those of the DSS or any other government department.

REFERENCES

Burgess *et al.*, This conference.
Checkoway, H., Heyer, N. J., Demers, P. A. and Breslow, N. E. (1993) Mortality among workers in the diatomaceous earth industry. *Br. J. ind. Med.* **50**, 586–597.

Cherry, N., Burgess, G., McNamee, R., Turner, S. and McDonald, J. C. (1995) Initial findings from a cohort mortality study of British pottery workers. *Appl. occup. Environ. Hyg.* **10**, 1042–1045.

HM Factory Inspectorate (1959) A survey of the pottery industry in Stoke-on-Trent, HMSO, London.

McDonald, J. C. (1995) Silica, silicosis and lung cancer: an epidemiological update. *Appl. occup. Environ. Hyg.* **10**, 1056–1063.

Merlo, F., Constantini, M., Reggiardo, G., Ceppi, M. and Puntoni, R. (1991) Lung cancer risk among refractory brick workers exposed to crystalline silica: a retrospective cohort study. *Epidemiol.* **2**, 299–305.

 Pergamon

Ann. occup. Hyg., Vol. 41, Supplement 1, pp. 412–414, 1997
© 1997 British Occupational Hygiene Society
Published by Elsevier Science Ltd. All rights reserved
Printed in Great Britain
0003–4878/97 $17.00 + 0.00
Inhaled Particles VIII

PII: S0003–4878(96)00144–5

COHORT STUDY OF STAFFORDSHIRE POTTERY WORKERS: (III) LUNG CANCER, RADIOGRAPHIC CHANGES, SILICA EXPOSURE AND SMOKING HABIT

J. C. McDonald,*, G. L. Burgess,† S. Turner† and N. M. Cherry†

*Department of Occupational and Environmental Medicine, National Heart and Lung Institute, Imperial College, Dovehouse Street, London SW3 6LY, U.K.; and †Centre for Occupational Health, Stopford Building, University of Manchester, Oxford Road, Manchester M13 9PT, U.K.

INTRODUCTION

The cohort of Staffordshire pottery workers, on which this and the accompanying papers by Burgess *et al.* and Cherry *et al.* (this conference) are based, is one of the very few designed to address the risk of lung cancer in workers exposed to crystalline silica where work histories, smoking habits, chest radiographic readings and prior exposure to other hazardous dusts were all recorded. Preliminary analyses of proportional mortality (McDonald *et al.*, 1995) and standardised mortality ratios (Cherry *et al.*, 1995) showed substantial evidence of excess lung cancer and non-malignant respiratory disease as compared with national (U.K.) mortality statistics but much less clearly when local (Stoke-on-Trent) rates were used. Full interpretation thus depended on the internal case-referent analyses of Cherry *et al.* (this conference), which took account of exposure intensity and duration, amount smoked and, so far as possible, chest radiography. For these analyses the use of exposure estimates derived from the validated matrix of Burgess *et al.* (this conference) were clearly important.

AIMS AND METHODS

The present paper explores the inter-relationships of silica exposure, smoking habit, small radiographic opacities and lung cancer in the sub-cohort of 1080 men used for validation of the exposure matrix. In this group 64 men (5.9%) had small opacities ($\geq 1/0$) on the chest X-ray and 16 (1.5%) died of lung cancer. Smoking habit was available for all but 14.

RESULTS

The main findings are presented in Tables 1–3. Table 1 shows that 14 of the 16 lung cancer deaths were in smokers and it is not unlikely that the few men without a recorded history were also smokers. Lung cancer mortality in this sub-cohort (1.5%) was similar to that in the main cohort (1.7%).

Table 2 relates exposure, radiographic opacities and lung cancer. There is little evidence here of an association between lung cancer and either cumulative

Table 1. Lung cancer deaths and smoking

Smoking habit	n	Deaths
Ever	811	14 (1.7%)
Never	255	0 (0.0%)
Not known	14	2 (14.3%)
All	1080	16 (1.5%)

Table 2. Lung cancer deaths, cumulative exposure and small opacities ($\geq$ 1/0)

Exposure (μg m^{-3}.year)	At risk	Small opacities	Deaths
< 2000	109	0 (0.0%)	7 (1.3%)
2000–	449	9 (2.0%)	
4000–	297	19 (6.4%)	9 (1.7%)
6000	225	36 (16.0%)	
All	1080	64 (5.9%)	16 (1.5%)

Table 3. Prevalence of small opacities ($\geq$ 1/0), cumulative exposure and smoking habit

Exposure (μg m^{-3}.year)	Never smoked at risk	prevalence	Ever smoked* at risk	prevalence
< 2000	30	0 (0.0%)	79	0 (0.0%)
2000–	112	1 (0.9%)	337	8 (2.4%)
4000–	70	2 (2.9%)	227	17 (7.5%)
6000	43	4 (9.3%)	182	33 (18.1%)
All	255	7 (2.7%)	825	58 (7.0%)

* Incl. 14 not known.

exposure or prevalence of small opacities, but the numbers are too small for confident interpretation. Of the 64 men with small opacities 3 (4.7%) died from lung cancer, compared with 13 (1.3%) of 1016 whose radiographs were normal; all three cases were in the higher exposure category (i.e. $\geq$ 4000 μg m^{-3}.y). These differences could readily be due to chance and take no account of age.

Table 3 presents an analysis of exposure response for radiographic small opacities in ever and never smokers. There is a clear gradient in prevalence in both categories, with rates about twice as high in smokers as non-smokers at each level of cumulative exposure above 2000 μg m^{-3}.y.

CONCLUSIONS

In most respects the results from these further analyses of this small sub-cohort, while entirely compatible with those from the main parent cohort, do not add much to them. The most important finding, however, concerns the evidence that the prevalence of small radiographic opacities was systematically related to cumulative exposure and about twice as high in smokers as non-smokers at each exposure level. A higher rate of small opacities in smokers was also reported by Finkelstein (1995) in his study of miners and dust-exposed surface workers in Ontario and by Weiss (1984) in several cohort and prevalence studies of asbestos workers. These observations are relevant to the silica-lung cancer question in that they may help to explain why men diagnosed as having silicosis are at increased risk of lung cancer. This is in addition, of course, to sources of selection bias which may affect decisions on compensation, and the virtual impossibility of correcting for smoking in persons who have respiratory symptoms, as would most applicants seeking compensation for silicosis.

Acknowledgements—This work was supported in part by a grant from the Department of Social Security, London. The views expressed are not necessarily those of the DSS or any other government department.

REFERENCES

Burgess *et al.* This conference.

Cherry, N., Burgess, G., McNamee, R., Turner, S. and McDonald, J. C. (1995) Initial findings from a cohort mortality study of British pottery workers. *Appl. occup. Environ. Hyg.* **10**, 1042–1045.

Cherry *et al.* This conference.

Finkelstein, M. M. (1995) Silicosis surveillance in Ontario from 1979 to 1992. *Scand. J. Work Environ. Health* **21** (2), 55–57.

McDonald, J. C., Cherry, N., McNamee, R., Burgess, G. and Turner, S. (1995) Preliminary analysis of proportional mortality in a cohort of British pottery workers exposed to crystalline silica. *Scand. J. Work Environ. Health* **21** (2), 63–65.

Weiss, W. (1984) Cigarette smoke, asbestos, and small irregular opacities. *Amer. Rev. Resp. Dis.* **130**, 293–301.

 Pergamon

Ann. occup. Hyg., Vol. 41, Supplement 1, pp. 415–419, 1997
British Occupational Hygiene Society
Published by Elsevier Science Ltd
Printed in Great Britain
0003–4878/97 $17.00 + 0.00
Inhaled Particles VIII

PII: S0003–4878(96)00145–7

RESPIRABLE QUARTZ LOSS OF AN ADSORBED PULMONARY SURFACTANT *IN VITRO* AND EXPRESSION OF CYTOTOXICITY OR GENOTOXICITY

X. Liu,* M. J. Keane,* T. Ong,* J. M. Antonini† and W. E. Wallace*

*National Institute for Occupational Safety and Health, 1196 Willowdale Road, Morgantown, WV 26505, U.S.A.; and †Harvard School of Public Health, Boston, MA, U.S.A.

INTRODUCTION

Inhaled particles will contact a surfactant-rich surface hypophase upon deposition in a pulmonary alveolus. Interactions with surfactant may affect the expression of particulate toxicity and may be a factor in distinguishing the pathogenic potentials of different dusts. Research has found that dipalmitoyl phosphatidylcholine (DPPC), a primary component of that surfactant, in aqueous dispersion will absorb onto the surface of respirable sized quartz particles and promptly suppress membranolytic activity (Wallace *et al.*, 1985, 1988). Cell-free systems have been used to investigate phospholipase enzymatic digestion of quartz-adsorbed DPPC, and the resultant surfactant removal and restoration of dust surface cytotoxicity (Wallace *et al.*, 1992; Keane and Wallace, 1995). Research reported here examines the expression of *in vitro* toxicity with time after challenge of pulmonary macrophage by DPPC-coated quartz dust, of the expression of micronucleus activity by a similarly challenged V79 cell line, and of the use of fluorescence-label substituted DPPC with dioleoyl-phosphatidyl choline (DOPC) adsorbed on quartz particles to track the removal of surfactant from quartz particles within cells.

MATERIALS AND METHODS

Quartz used was commercially obtained "Min-U-Sil 5" crystalline quartz. For toxicity studies, DPPC was ultrasonically dispersed into 0.165 M NaCl physiologic concentration salt solution (PSS). Quartz was mixed with this dispersion at a ratio of 0.1 g surfactant per g of dust to assure saturated coverage of quartz particle surfaces by surfactant (Wallace *et al.*, 1992). Cells for *in vitro* cytotoxicity assay were pulmonary macrophages obtained by lung lavage from male Sprague–Dawley rats. Cell suspensions of 2×10^6 cells in 4 ml complete medium were incubated for 2 h at 37°C and 5% CO_2 to permit cell adherence. Change to fresh medium containing quartz resulted in exposures of 0.4–3.2 mg quartz. After incubation of cells with dust for selected times of 1–7 days, cell viability was determined by the trypan blue dye exclusion method. Micronucleus induction by native and DPPC-treated quartz was measured using the V79 cell line, derived from pulmonary fibroblasts. Dishes (100 mm) were seeded with 2×10^5 cells in complete medium and were incubated 24 h at 37°C and 5% CO_2, the medium removed and the cells challenged with quartz suspensions in a total volume of 5 ml. A one-day incubation with native and DPPC-treated quartz used doses of 0, 20, 40, 80 and 160 µg quartz cm^{-2}. Challenge with DPPC-treated quartz over a 5 day period used a dose of 80 µg cm^{-2}. Cells were incubated for 1–5 days, released by trypsinisation, fixed with absolute methanol, and stained in DiffQuik #1 (Baxter, McGaw Park, IL) and Diff-Quik #2. One thousand mononucleated cells per slide were scored. Only slides that had a visible nucleus were scored and micronuclei that had diameters larger than 1/20th and smaller than 1/5th the nuclear diameter were counted. Three slides were scored for each treatment, a total of 3000 cells. Observation of the loss of surfactant from dusts used fluorescent-labelled surfactant consisting of a dipyrrometheneboron difluoride fluorescent group substituted into one of the acyl chains of DPPC (trademark name "Bodipy," Molecular Probes, Inc.), mixed with DOPC at 1:20 labeled to unlabeled surfactant. Cell-free studies have shown that the fluorescent-label substituted DPPC in DOPC can be used to track the fate of dust-adsorbed DOPC under phospholipase A2 (PLA2) digestion. This was shown by fluorescence assay of DPPC/DOPC-coated particles under cell-free PLA2 digestion, with parallel assay of particle-bound organics by elution, chromatographic separation and wet phosphate quantitation of products of digestion (Das, 1993). Quartz treated with 0.1 g of this surfactant per g was used for *in vitro* challenge of lavaged rat pulmonary macrophage. Suspensions in complete medium of 4×10^5 cells ml^{-1} were challenged with 0.1 mg surfactant-treated quartz ml^{-1} suspension. Following incubation at 37°C under 5% CO_2 for desired times between 1 and 7 days, cells were viewed by confocal laser scanning microscopy (CLSM) for fluorescence ($\lambda_{em} > 520$ nm), reflectance, and transmitted polarized light to observe the time-course of fluorescent surfactant associated with particles phagocytosed by cells. Qualitative observation indicated a major fraction of the fluorescence was lost over the 7 day period. Quantitative average fluorescence values were obtained by using image analysis tools to estimate the fluorescence profile across a rectangle inside the boundaries of 10 individual cells and computing the average value over all cells for a given treatment.

RESULTS

Cytotoxicity to lavaged macrophage of DPPC-coated quartz is shown in a representative replicate experiment in Fig. 1. As assayed by trypan blue dye exclusion, percent macrophage viability vs time is shown for 1, 3, 5 and 7 days following cell challenge. The control cells, not challenged with quartz, show a slow decline in measured viability over the 7 day period from approximately 93–85%. At day 1 the native quartz-challenged cells show on the order of a 50% attrition. The DPPC-treated quartz challenge shows little toxicity on day 1; however, restoration of cytotoxic activity is seen to begin by day 3 and by day 7 the DPPC-quartz challenge has resulted in 50% losses in cell viability.

There was a statistically significant positive dose–response for micronucleus induction during a one day challenge of V79 cells by untreated quartz: as concentrations increased from 0 to 160 µg quartz cm^{-2}, the frequency of micronucleus induction increases from 6.67 ± 0.33 per 1000 cells scored ($\pm$ SE, *N*

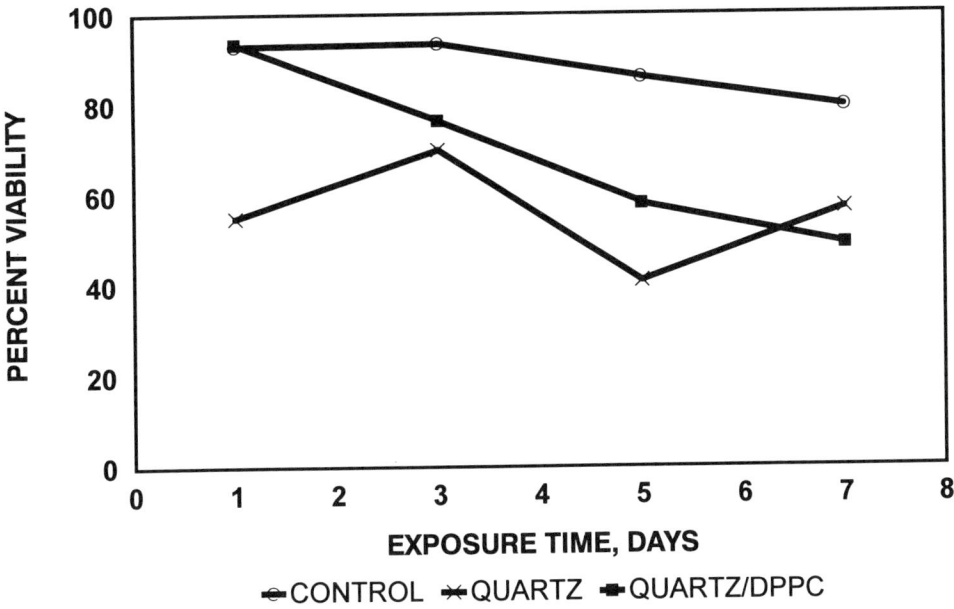

Fig. 1. Macrophage viability *in vitro* vs time after exposure to surfactant-coated quartz.

= 3), to 25.67 ± 0.33. Figure 2 shows the results of challenging with DPPC-treated quartz for 1, 3, and 5 days. Values are 5.67 ± 1.20, 6.00 ± 2.00 and 6.67 ± 0.88 micronuclei/1000 cells, respectively. DPPC dispersion controls gave comparable values of 6.33 ± 0.88, 6.00 ± 1.00, and 6.00 ± 0.00 at those times. In these

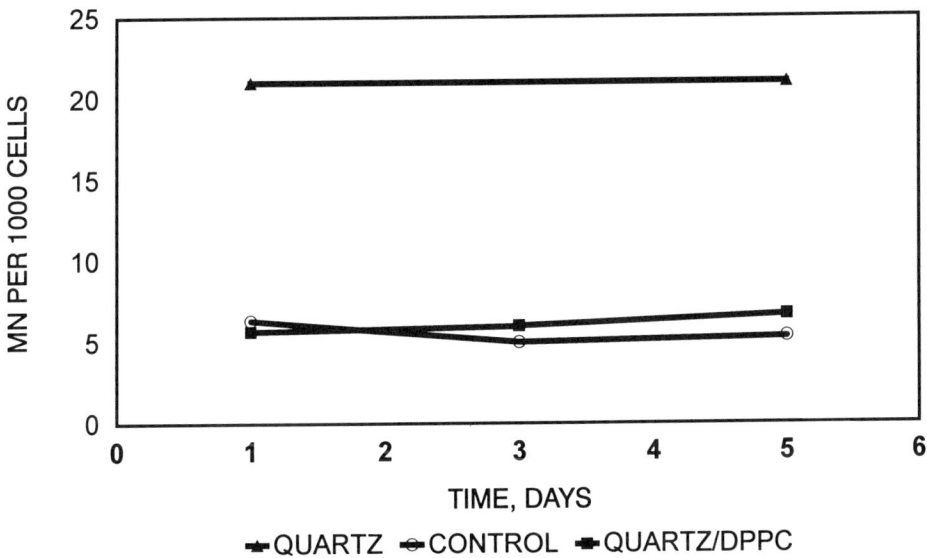

Fig. 2. Micronuclei (MN) induced per 1000 V79 cells *in vitro* vs time after exposure to surfactant-treated quartz.

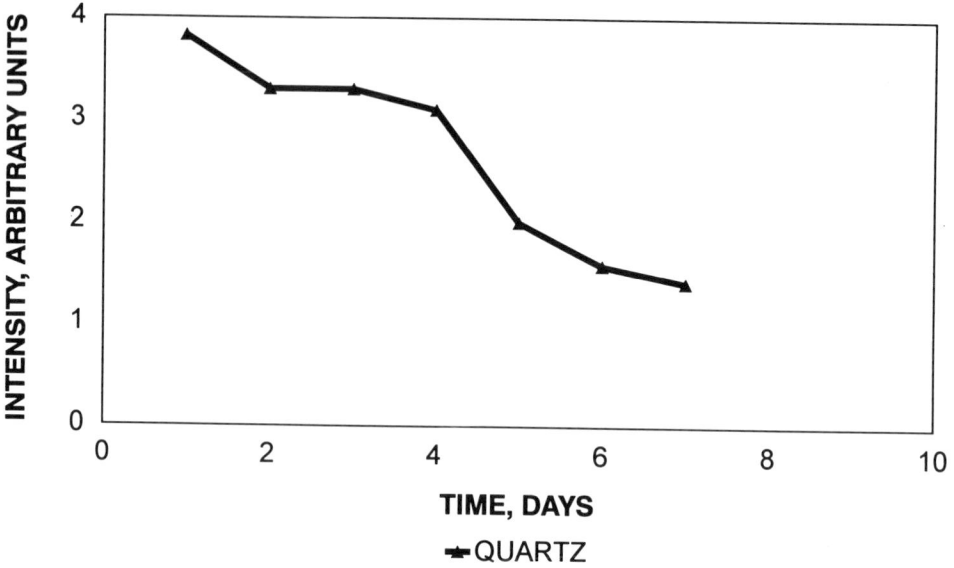

Fig. 3. Fluorescence intensity averaged over macrophages *in vitro* vs time after exposure to labelled-surfactant coated quartz.

extended-time micronucleus assays, the medium was changed on the second and fourth days after cell challenge; when not changed, the values for both the controls and the DPPC-dust sets increased with time without a statistically significant difference between the two.

Qualitative observation of the time-course of labeled surfactant fluorescence associated with particles phagocytosed by lavaged macrophage cells indicated a major fraction of the fluorescence was lost over the 7 day period. Quantitative relative fluorescence values are shown in Fig. 3. Fluorescence is seen to decrease on day 7 to approximately 25% of the intensity at day 1.

DISCUSSION

These studies move from the reported cell-free enzymatic system examination of lipase digestion of quartz-adsorbed surfactant (1, 2, 3, 4) to *in vitro* cellular digestive reactions and parallel examination of dust cytotoxicity and genotoxicity. Native quartz dust expressed prompt *in vitro* cytotoxicity as measured by trypan blue dye exclusion for lavaged rat pulmonary macrophage and genotoxic activity for V79 cells as measured by micronucleus induction. Coating the particles with diacyl phosphatidylcholine suppressed these activities through one day incubation with cells *in vitro*. Cytotoxic activity *in vitro* was restored to positive control values over a one week period. This is consistent with a study of the *in vitro* toxicity of Survanta (Ross Laboratories, Columbus, OH) pulmonary surfactant-treated quartz to lavaged rat macrophage, which found initial suppression of toxicity and beginning restoration over a 3 day period as measured by the trypan blue viability assay (Antonini *et al.*, 1994). Genotoxic activity was not restored in that period.

Fluorescence intensity of fluorophore-labelled phospholipid surfactant on quartz particles in rat alveolar macrophage *in vitro* decayed to near background levels in a 1 week period. This time-course is not inconsistent with measures of digestion with time of radio labelled DPPC from quartz particles by the P388D1 cell line (Hill *et al.*, 1995). Expression in both cell-free and *in vitro* systems of delayed cytotoxicity coincident with surfactant digestion suggests a prophylactic effect of adsorbed surfactant that diminishes with digestion. It must be noted that significantly different dose levels of quartz were used between the cytotoxicity and the digestion-tracking experiments reported here due to the exegencies of the assay methods. Untreated quartz expressed a statistically significant positive dose–response for micronucleus induction in the V79 cell line. We have not yet investigated whether this was due to clastogenic or aneuploidogenic activity, or to both. This was suppressed by dust incubation with DPPC and that activity was not restored over a time commensurate with surfactant removal (Liu *et al.*, 1996). It must be noted that the surfactant removal assay used primary macrophages while the micronucleus assay used a cell line; thus digestive kinetics might differ between the two cell types. An assumption of comparable digestion kinetics would suggest that the short-term expression of micronucleus activity seen with native quartz is, under these experimental conditions, associated with the native dust expression of cytotoxicity. Preliminary tests of fluorescent-labelled surfactant to track digestion kinetics *in vivo* are underway.

REFERENCES

Antonini, J. M., McCloud, C. M. and Reasor, M. J. (1994) Acute silica toxicity: attenuation by amiodarone-induced pulmonary phospholipidosis. *Environ. Health Perspect.* **102**, 372–378.

Das, A. R. (1993) Visualization of particle-macrophage interactions during phagocytosis *in vitro*. Dissertation, College of Engineering, West Virginia University.

Hill, C. A., Wallace, W. E., Keane, M. J. and Mike, P. S. (1995) The enzymatic removal of a surfactant coating from quartz and kaolin by P388D1 cells. *Cell Biol. Toxicol.* **11**, 119–128.

Liu, X., Keane, M. J., Zhong, B. Z., Ong, T. and Wallace, W. E. (1996) Micronucleus formation in V79 cells treated with respirable silica dispersed in medium and in simulated pulmonary surfactant. *Mutation Res.* **361**, 89–94.

Keane, M. J. and Wallace, W. E. (1995) Pulmonary surfactant adsorption and the expression of silica toxicity. In *Silica and Silica-induced Lung Diseases* (Edited by V. Castranova, V. Vallyathan and W. E. Wallace), pp. 271–281. CRC Press, FL.

Wallace, W. E., Keane, M. J., Vallythan, V., Hathaway, P., Regad, E. D., Castranova, V. and Green, F. H. Y. (1988) Suppression of inhaled particle cytotoxicity by pulmonary surfactant and re-toxification by phospholipase. *Ann. occ. Hyg.* **32**, 291–298 (Supplement 1).

Wallace, W. E., Keane, M. J., Mike, P. S., Hill, C. A., Vallyathan, V. and Regad, E. D. (1992) Contrasting respirable quartz and kaolin retention of lecithin surfactant and expression of membranolytic activity following phospholipase A2 digestion. *J. Toxicol. Environ. Health* **37**, 391–409.

Wallace, W. E., Vallyathan, V., Keane, M. J. and Robinson, V. (1985) *In vitro* biological toxicity of native and surface modified silica and kaolin. *J. Toxicol. Environ. Health* **16**, 415–424.

 Pergamon

Ann. occup. Hyg., Vol. 41, Supplement 1, pp. 420–425, 1997
British Occupational Hygiene Society
Published by Elsevier Science Ltd
Printed in Great Britain
0003–4878/97 $17.00 + 0.00
Inhaled Particles VIII

PII: S0003–4878(96)00146–9

PROTECTION BY IRON AGAINST THE TOXIC EFFECTS OF QUARTZ

R. T. Cullen,* V. Vallyathan,† S. Hagen* and K. Donaldson‡

*Institute of Occupational Medicine, 8 Roxburgh Place, Edinburgh EH8 9SU, U.K.; †NIOSH,
Morgantown, U.S.A.; and ‡Napier University, Edinburgh, U.K.

INTRODUCTION

Inhalation of quartz (crystalline silica) can lead to the fibrosing lung disease, silicosis (Morgan and Seaton, 1984). Exposure to quartz alone is rare and most exposures occur through the presence of quartz in other dusts produced by activities such as mining, quarrying and sandblasting. Consequently, silicosis is often combined with mixed dust pneumoconiosis. A number of studies in iron-ore miners (Reichel et al., 1977; Moore et al., 1987) and in animals (Gross et al., 1960) have indicated that iron can protect against the effects of quartz.

It is now believed that the characteristics of the surface of quartz particles determine their interaction with biological molecules and hence their toxicity. For example, the surface can generate a range of reactive oxygen species such as hydrogen peroxide (H_2O_2) and the hydroxyl radical ($\cdot OH$), agents known to be toxic to cells and tissues (Fantone and Ward, 1984). The production of reactive oxygen species can be measured using electron spin resonance (ESR) techniques. The two main chemical activities of the silica surface are those involving siloxane bridges (Si–O–Si) and silanols (SiOH) (Iler, 1979). Silanol groups can form hydrogen bonds with cell membranes, thought to be the main site of quartz toxicity. These groups can also dissociate at physiological pH giving a net negative charge on the silicate surface which can lead to adsorption, or coordination, of organic and inorganic cations, such as iron and aluminium.

This binding of metals has been shown to reduce the toxicity of quartz (Nolan et al., 1981). Paradoxically, the interaction of iron with H_2O_2 can also produce $\cdot OH$ radicals, extremely potent oxidants, through the Fenton reaction:

$$Fe^{2+} + H^2O^2 = Fe^{3+} + \cdot OH + OH^-$$

The H_2O_2 for this reaction can be produced at the quartz surface or be released from macrophages and neutrophils.

The objective of our study was to examine whether treating quartz particles with iron, using ferrous or ferric chloride solutions, or mixing quartz with carbonyl iron particles could alter the surface activity of quartz and the toxic and pulmonary inflammatory effects of quartz.

Table 1. Total cell and neutrophil numbers lavaged from lungs 7 days and 32 days post-instillation of quartz, carbonyl iron, or quartz/iron mixtures (geometric means) and tumour necrosis factor production in the presence of LPS by lavaged macrophages (geometric means and ranges)

Treatment	Total cells (millions) Day 7	Day 32	Neutrophils (millions) Day 7	Day 32	Tumour necrosis factor (units/10^6 AM)
Iron 1 mg	8.07	9.65	0.11	0.06	92 (55–156)
Quartz 0.08 mg	8.18	8.70	0.35	0.08	140 (58–340)
Quartz/Iron 0.08/0.92 mg	7.88	9.20	0.28	0.14	191 (113–322)
Iron 10 mg	6.37	7.60	0.61	0.04	166 (94–294)
Quartz 0.8 mg	24.60	44.66	9.01	19.06	976 (402–2373)
Quartz/Iron 0.8/9.2 mg	16.11	31.13	4.94	8.10	447 (253–792)
Iron 50 mg	12.58	10.68	1.77	0.72	384 (167–883)
Quartz 3.9 mg	47.12	123.36	30.19	65.76	562 (231–1366)
Quartz/Iron 3.9/46.1 mg	21.54	26.65	2.89	5.30	1331 (543–3261)
Control	3.82	4.42	0.01	0.02	61 (50–75)

MATERIALS AND METHODS

Samples of DQ12 quartz were treated with iron using ferrous or ferric chloride dissolved in distilled water to 1M concentrations. The presence of iron on quartz particles treated with iron chlorides was confirmed using an iron staining kit (Sigma, Poole, U.K.). For some experiments quartz particles were mixed with particles of elemental iron (carbonyl iron; Sigma). Treated and untreated quartz and quartz/iron mixtures were instilled at several doses into the lungs of specific pathogen-free (SPF) male Wistar rats which were lavaged 7 or 32 days later. The recovered cell types were enumerated and alveolar macrophages (AM) were cultured for the production of the inflammatory cytokine tumour necrosis factor alpha (TNFα) which was measured using the L929 cell bioassay (Flick and Gifford, 1984). *In vitro* experiments examined the effects of iron coating on (a) toxicity to macrophages, as lactate dehydrogenase (LDH) release, and their release of H_2O_2 and superoxide dismutase and (b) on radical activity on the surface of quartz particles and ·OH generation by quartz particles. Additionally, some comparisons between iron coating and aluminium lactate coating were made. Results were analysed using analysis of variance and regression methods using the statistical software of Genstat 5.1 and Minitab, release 8.2. Data from the instillation experiments were log-transformed prior to analysis.

RESULTS

The effect of iron particles on the pulmonary response to quartz

In the series of experiments with rat lungs instilled with various doses of quartz, the relatively non-pathogenic dust carbonyl iron, or quartz mixed with various proportions of particulate iron, inflammation (as total cell numbers and numbers of neutrophil leukocytes) increased with dose of quartz in both the quartz alone and quartz/iron mixtures (Table 1). At the lower doses (e.g. 1 mg of iron alone or iron mixed with 8% quartz) there was little difference in response between the treatments. With higher doses and the same percentage quartz mixture, it was evident that iron was reducing the strong response to quartz. Particulate iron alone produced very little response. Additional, experiments with lower ratios of iron to quartz and a range of doses showed no evidence of an iron-protective effect. It was clear that the ratio of iron to quartz (e.g. 12 to 1 weight for weight in Table 1) had to be sufficiently high for the effect to occur.

Histological examination of lungs from additional rats, 60 days following treatment with the same dusts, confirmed the lavage cell results, with quartz on its own proving to be the most pathogenic dust with evidence of lipoproteinosis and fibrosis.

Production of tumour necrosis factor (TNFα) by cells recovered in lavage

Alveolar macrophages lavaged from dust-treated lungs were cultured overnight with or without 10 µg ml^{-1} LPS (endotoxin) and their production of TNFα assayed. In the absence of LPS, there were no significant differences between the treatments for any dose at any time point (data not shown). In the presence of LPS there was considerably greater production of TNFα. Although results from the 32 day experiments tended to be higher than those of the other time points, there was no significant effect of time and results have been pooled across the time points for display in Table 1 as estimated geometric means. TNFα production increased with dose for iron and quartz/iron, but not for quartz alone where the middle dose gave the highest results. TNFα values for macrophages from the iron treatments were significantly lower than those from the quartz/iron treated group.

The effect of coating quartz with iron or aluminium on the pulmonary cellular response

The effect on the pulmonary leukocyte response of coating quartz particles with ferrous or ferric iron was studied. Aluminium lactate-treated quartz was also included as aluminium has previously been shown by us to reduce quartz toxicity in the lung (Brown *et al.*, 1989). An additional sample of quartz was subjected to the same coating procedure using distilled water in place of the metal salts. Rat lungs were lavaged 7 days or 32 days following intratracheal instillation of 1 mg amounts of the quartz dusts. Geometric means of total neutrophil counts, pooled from all experiments, are shown in Fig. 1. The strong neutrophil response to quartz was significantly reduced only by the aluminium lactate treatment, pre-treatment with iron had no effect. Similar statistically significant reductions were seen with aluminium treatment for total, macrophage and lymphocyte numbers (data not shown). No effect of iron coating was seen in additional experiments using quartz at lower doses.

Hydrogen peroxide production and superoxide dismutase activity by alveolar macrophages exposed in vitro to untreated quartz or to quartz coated with iron or aluminium

H$_2$O$_2$ production by alveolar macrophages incubated for 20 min with 0.5 mg ml^{-1} untreated quartz or quartz particles coated with iron or aluminium is summarised in Table 2. H$_2$O$_2$ levels were greatest with the water (control) and aluminium lactate treatments and lowest with Fe^{2+}-treated quartz.

As part of the body's defence system against oxidant injury, cells can produce the

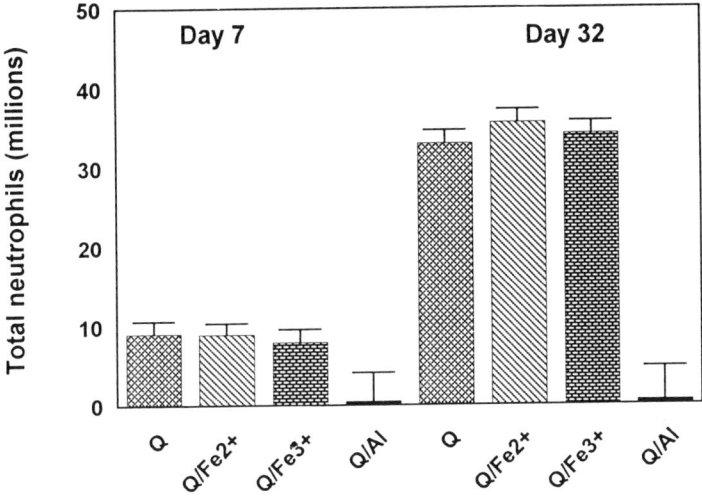

Fig. 1. Neutrophil numbers in lavage 7 and 32 days post-instillation of 1 mg metal-treated quartz (geometric mean and s.d.).

enzyme superoxide dismutase (SOD) which converts superoxide to oxygen and hydrogen peroxide. Accordingly, we assessed the levels of SOD in alveolar macrophages incubated for 1 h with 0.5 mg untreated quartz or quartz pre-treated with iron or aluminium. Both the intracellular and extracellular (supernatant) concentrations of SOD were assayed. Compared to untreated (no dust) macrophages, only untreated quartz caused a significant ($P < 0.05$) drop in the intracellular level of SOD (Table 2). All the quartz dusts caused some release of SOD extracellularly into the culture supernatant (Table 2), but only aluminium treated quartz produced a significantly lower concentration than untreated quartz ($P < 0.05$).

Toxicity of coated and uncoated quartz to alveolar macrophages

Lactate dehydrogenase (LDH) is a cytosol enzyme whose release from cells provides a reliable measure of cytotoxicity. We used this assay to identify any differences in cell injury between the treated and untreated quartz samples. Alveolar macrophages were incubated for 1 h with 1 mg ml^{-1} untreated quartz or quartz pre-treated with ferrous or ferric iron or aluminium lactate. All quartz

Table 2. *In vitro* treatment of alveolar macrophages with metal-coated quartz: superoxide dismutase levels, hydrogen peroxide production and lactate dehydrogenase release (mean and standard error)

Treatment	Superoxide dismutase Intracellular	Extracellular	H_2O_2 nmol/ 10^6 AM	LDH units/ 10^6 AM
None	111.5 (4.6)	1.5 (0.9)	0.33 (0.02)	59.8 (1.4)
Quartz	92.7 (6.8)	39.7 (4.2)	0.38 (0.22)	113.8 (1.7)
Quartz/water	96.8 (7.0)	24.1 (7.4)	1.09 (0.05)	101.4 (2.3)
Quartz/Fe2$^+$	103.5 (7.2)	19.6 (8.2)	0.01 (0.00)	113.0 (2.6)
Quartz/Fe3$^+$	101.4 (4.8)	31.1 (4.8)	0.42 (0.17)	110.0 (1.1)
Quartz/aluminium	106.6 (4.6)	16.9 (8.6)	1.44 (0.15)	99.4 (3.2)

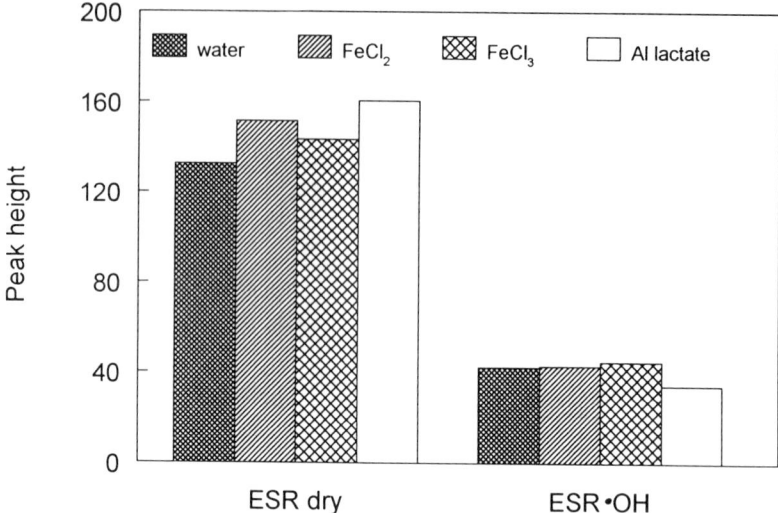

Fig. 2. Electron spin resonance (ESR) measurements as peak heights from 50 mg dry quartz samples and ·OH radical activity from 10 mg quartz samples in the presence of DMPO and H_2O_2 (mean of 3 experiments).

samples, treated or untreated, caused LDH release from macrophages (Table 2). There was no difference between untreated quartz and ferrous or ferric-treated quartz.

Measurement of radicals on the surface of quartz particles treated with iron or aluminium

Electron spin resonance (ESR) measurements were made on three batches of quartz treated with water, ferrous or ferric chloride, or aluminium lactate. All the treated quartz samples produced surface radical activity but there were no significant differences between the treatments (Fig. 2; data pooled from the three batches). To determine generation of ·OH, measurements were also carried out in liquid in the presence of the spin trap DMPO and hydrogen peroxide but there were no significant differences between the treatments (Fig. 2).

DISCUSSION

We have shown that particulate iron can reduce the inflammatory effects of quartz in the rat lung and modulate TNF production by alveolar macrophages provided that the ratio of iron to quartz is sufficiently high, 11.8–1 in our experiments. Coating quartz with ferrous or ferric iron did not affect the cellular response in marked contrast to the protective effects of aluminium in rat lungs, previously shown by our laboratory (Brown *et al.*, 1989) and in the lungs of sheep (Begin *et al.*, 1986). Nolan *et al.* (1981) reported the ameliorating effects of treating the quartz surface with ferric chloride, aluminium chloride, zinc chloride, or polyvinyl-pyridine-N-oxide (PVPNO, a hydrogen-bonding polymer) on the lysis by quartz of red blood cells. The reason for the lack of effect of iron coating, as opposed to aluminium treatment, in our *in vivo* assay is not known. Aluminium may be bound more strongly than iron and be better able to resist leaching from the quartz surface within the body. Geometrical considerations would indicate that ions with a similar charge to radius ratio to that of silicon, and a similar preference for tetrahedral sites, would fit more readily into the quartz surface. Aluminium^{3+} and, to a lesser extent, Fe^{3+} can both replace silicon in silicate minerals. Fe^{2+} being larger and with less charge would not fill tetrahedral sites or be held as strongly by electrostatic forces (Iler, 1979). We found metal binding had no effect on radical production at the quartz surface. Radical activity is normally greatest with freshly ground quartz surfaces (Shi *et al.*, 1988) and we did not investigate the effects of grinding in our assay.

The reason why particulate, elemental iron was able to protect against the inflammatory effects of quartz is not clear. It is unlikely that the iron is interacting directly with the quartz surface and thus it may be acting indirectly by reacting with and neutralising reactive oxygen species.

In conclusion, this study confirms previous work showing (a) the protective effect of modifying the quartz surface with aluminium (Brown *et al.*, 1989) and (b) that particulate iron can also reduce the pulmonary toxicity of quartz (Gross *et al.*, 1960). Accordingly, in industries with significant mineral dust exposures that include quartz and iron, inhaled dust may be less harmful than would be predicted merely from its quartz content.

Acknowledgement—This study was funded in part by London Underground Ltd.

REFERENCES

Begin, R., Masse, S., Rola-Pleszczynski, M., Martel, M., Desmarais, Y., Geoffroy, M., Le Bouffant, L., Daniel, H. and Martin, J. (1986) Aluminium lactate treatment alters the lung biological activity of quartz. *Exp. Lung Res.* **10**, 385–399.

Brown, G. M., Donaldson, K. and Brown, D. M. (1989) Bronchoalveolar leukocyte response in experimental silicosis: modulation by a soluble aluminium compound. *Toxicol. appl. Pharmacol.* **101**, 95–105.

Fantone, J. C. and Ward, P. A. (1984) Mechanisms of lung parenchymal injury. *Am. Rev. Resp. Dis.* **130**, 484–491.

Flick, D. A. and Gifford, G. E. (1984) Comparison of *in vitro* cell cytotoxicity assays for tumor necrosis factor. *J. Immunol. Meth.* **68**, 167–175.

Gross, P., Westrick, M. L. and McNerney, J. M. (1960) Experimental silicosis: the inhibitory effect of iron. *Diseases of the Chest* **37**, 35–41.

Iler, R. K. (1979) *The Chemistry of Silica*. Wiley, New York.

Morgan, W. K. C. and Seaton, A. (1984) *Occupational Lung Diseases*, 2nd edn. WB Saunders Co, Philadelphia.

Moore, E., Martin, J. R., Edwards, A. C. and Muir, D. C. F. (1987) A case-control study to investigate the association between indices of dust exposure and the development of radiologic pneumoconiosis. *Arch. Environ. Health* **42**, 351–355.

Nolan, R. P., Langer, A. M., Harington, J. S., Oster, G. and Selikoff, I. J. (1981) Quartz hemolysis as related to its surface functionalities. *Environ. Res.* **26**, 503–520.

Reichel, G., Bauer, H-D. and Bruckmann, E. (1977) The action of quartz in the presence of iron hydroxides in the human lung. In *Inhaled Particles IV* (Edited by W. H. Walton), pp. 403–411. Pergamon Press, Oxford.

Shi, X., Dalal, N. S. and Vallyathan, V. (1988) ESR evidence for the hydroxyl radical formation in aqueous suspension of quartz particles and its possible significance to lipid peroxidation in silicosis. *J. Toxicol. Environ. Health* **25**, 237–245.

 Pergamon

Ann. occup. Hyg., Vol. 41, Supplement 1, pp. 426–433, 1997
© 1997 British Occupational Hygiene Society
Published by Elsevier Science Ltd. All rights reserved
Printed in Great Britain
0003–4878/97 $17.00 + 0.00
Inhaled Particles VIII

PII: S0003–4878(96)00051–8

BIOLOGICAL AND BIOCHEMICAL CHARACTERISATION OF A "PROLIFERATION FACTOR" FROM QUARTZ DUST-TREATED HUMAN MACROPHAGES

N. H. Seemayer, H. Olbrück and Ursula Griwatz

Medical Institute of Environmental Hygiene, Gurlittstr. 53, D-40223 Düsseldorf, Germany

INTRODUCTION

Silicosis is a chronic fibrotic lung disease caused by inhalation of quartz-containing dusts in coal mine workers, quarriers and stone masons. The disease is characterised by an increase in collagen content of the lungs, leading to destruction of lung parenchyma and to impairment of lung function. Alveolar macrophages are the primary target for the noxious effect of quartz and coal mine dust. In the alveoli dust particles are phagocytised by alveolar macrophages and removed from the lung via the "mucociliary transportation mechanism". In the case of "overload", dust-laden macrophages migrate into the regional lymph node and the interstitium of the lung.

Macrophages produce more than 100 "biofactors" or "mediators" that are involved in various processes of inflammation and immunological defence mechanisms (Nathan *et al.*, 1980; Nathan, 1987). Various authors have reported that human monocytes and macrophages exposed to diverse soluble and particulate agents *in vitro*, release cytokines which stimulate human fibroblasts to cell multiplication (Austgulen *et al.*, 1987; Bitterman *et al.*, 1982; Glenn and Ross, 1981; Leslie *et al.*, 1984; Schmidt *et al.*, 1984; Seemayer *et al.*, 1987, 1988).

In this paper we present data on the biological and biochemical characterisation of a "proliferation factor" which is produced by human monocyte-derived macrophages in culture following incubation with quartz dust DQ12. The "proliferation factor" (PF) stimulates under serum-free conditions cell replication of "quiescent" human lung and dermal fibroblasts and of human pneumocyte type II (line A-549) as well as collagen synthesis of human and dermal fibroblasts.

MATERIALS AND METHODS

Cell cultures

Isolation of monocytes from the peripheral blood in a Ficoll–Hypaque gradient and cultivation of cells to maturation with characteristics of macrophages has been described in detail (Seemayer and Braumann, 1985, 1988). The cell line WI38 (human embryonal lung fibroblasts) was purchased from Flow Laboratories, Meckenheim, Germany, the cell line MRHF (human foreskin fibroblasts) from Api-BioMerieux, Nurtlingen, Germany, the cell line FH3 (human embryonal skin fibroblasts) was obtained by Biochrom, Germany. The cell line A-549 (human pneumocyte type II cells) was kindly provided by Dr G. M. Alink, Department of Toxicology, Agricultural University, Wageningen, The Netherlands. The cell lines WI38, MRHF and A-549 were cultivated routinely in Dulbecco's modified Minimum essential medium with 10% fetal calf serum and antibiotics.

Preparation of dust sample

Fibrogenic quartz dust DQ12 was used as a stimulus for cultures of human macrophages. This is Dörentruper crystal quartz flour (grinding no. 12) with a particle size less than 5 µm. Dust samples were suspended in serum-free RPMI 1640 medium. The samples were subjected to ultrasonic treatment (Sonifier B-12 from Branson Sonic Power Company, U.S.A.) in order to achieve a uniform distribution of the particles and to destroy germs.

Preparation of supernatants from human macrophages in culture

Suspended dust samples were added in a final concentration of 30 µg ml^{-1} in serum-free RPMI 1640 medium to cultures of human macrophages. After an incubation period of 24 and 48 h at 37°C the culture supernatants were centrifuged for 15 min at 1.000 rpm, filtered through Millipore filters (pore size 0.45 µm) and treated in a water bath by 56°C for 1 h. The supernatants were then deep frozen at −20°C until used.

Ultrafiltration

Supernatants of quartz-treated human macrophages were ultrafiltered using devices from Millipore and Amicon (Centriprep) with different nominal molecular weight limits (NMWL).

Determination of proliferation of lung-fibroblasts and pneumocytes type II

The proliferation assay has been described in detail elsewhere (Griwatz *et al.*, 1994).

Analysis of collagen synthesis

Whole cell-lysates were hydrolysed in 6N HCl at 110°C for 24 h and analysed for Hydroxyprolin (OH-Prolin) according to the method of Stegemann and Stadler (1967), using Chloramin-T and Dimethylaminobenzaldehyd.

RESULTS

In Fig. 1 we present the phagocytosis of quartz dust DQ12 by macrophages in culture. We could demonstrate that supernatants of these quartz dust DQ12 treated human macrophages release a soluble Proliferation Factor which stimulates "quiescent" human lung (FH-27) and dermal fibroblasts (FH-3) to a considerable cell multiplication (Seemayer *et al.*, 1987, 1988).

The process of stimulation of fibroblast proliferation by the PF was also visualised by morphological criteria, such as an increased rate of mitosis and DNA synthesis and by reaches high cell density of fibroblast cultures (Hübner and Seemayer, 1989).

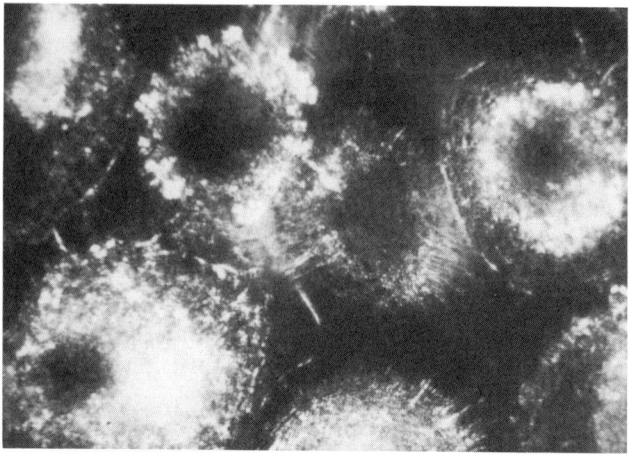

Fig. 1. Phagocytosis of quartz dust by human macrophages viewed with polarised light microscopy (1 000×, hematoxylin-eosin-staining).

A similar "proliferation stimulating activity" was observed with human lung fibroblasts (WI38) and dermal fibroblasts (FH-3) treated with supernatants of human macrophages incubated with coal mine dust TF-1 (Seemayer and Maly, 1990; Hübner and Seemayer, 1992).

We made further attempts to characterise the PF produced by quartz and coal mine dust treated human macrophages. We found that the PF is still active after incubation for 60 min at 56°C. For estimation of the approximate molecular weight of the PF we utilised ultrafiltration with devices with 10 000 and 30 000 NMWL. Results based on induced cell multiplication of "quiescent" dermal fibroblasts FH-3 and MRHF and on stimulation of DNA synthesis of lung fibroblasts WI38 by the supernatant of quartz dust DQ12 exposed human macrophages larger and smaller than 10 000 and 30 000 NMWl indicated a molecular weight of the factor of more than 30 kDa (Seemayer *et al.*, 1988; Hübner and Seemayer, 1989). In further studies we could demonstrate, that only a continuous presence of the PF induces a high number of DNA synthesising cells of WI38 cells followed by a considerable cell replication. DNA synthesis and cell replication of WI38 cells ceased very rapidly after removal of the PF from the medium (Seemayer *et al.*, 1988; Hübner and Seemayer, 1989).

To elucidate further the characterisation of the PF we performed complementation tests according to Stiles *et al.* (1979) and Bittermann *et al.* (1982). By addition of a competence factor such as fibroblast growth factor (FGF) or platelet derived growth factor (PDGF) cell growth of WI38 or MRHF human fibroblasts incubated with supernatant of quartz dust DQ12-treated macrophages was significantly enhanced. Similar results were observed with WI38 cells in presence of supernatant of coal mine dust TF-1 treated human macrophages. Addition of FGF or PDGF remarkable enhanced cell multiplication.

In further studies we were able to demonstrate that beside human fibroblasts, epithelial cells of the alveolar unit, such as pneumocyte type II cells (line A-549), also respond with strong proliferation activity in presence of supernatant of quartz

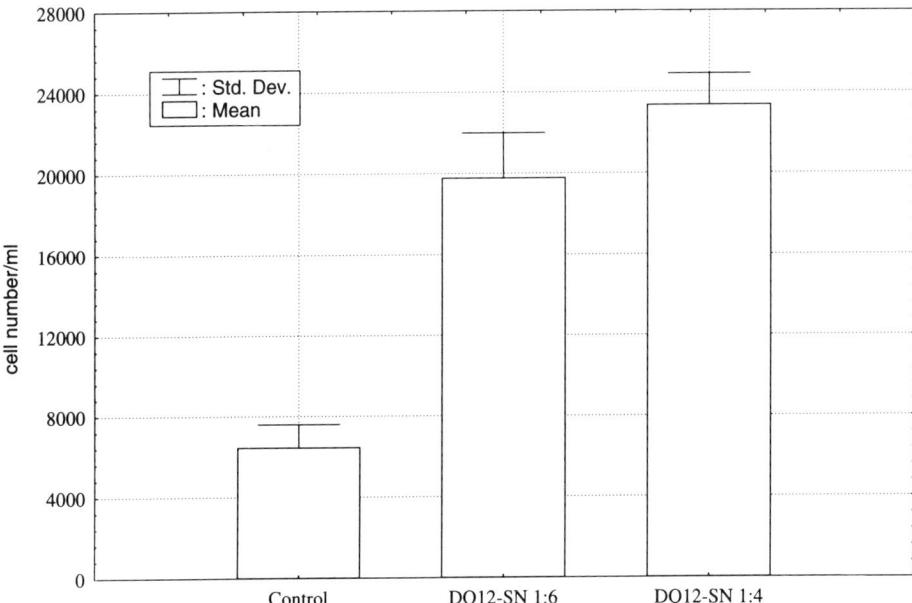

Fig. 2. Stimulation of cell proliferation of A-549 cells by a supernatant from quartz dust exposed human macrophages at various dilutions.

dust DQ12-treated human macrophages (Griwatz *et al.*, 1993, 1994, 1995). The stimulating effect of the supernatant led to a markable increase of cell growth of A-549 cells in comparison to control (Fig. 2).

To characterise further the proliferation factor, we incubated supernatants of quartz dust DQ12-treated human macrophages with mercaptoethanol according to Pledger *et al.* (1977). A strong loss of activity was caused by mercaptoethanol-treatment of the supernatant. Results indicate then presence of disulphide bridges in the molecule of the cytokine which get destroyed by mercaptoethanol (Griwatz *et al.*, 1995). A comparable loss of activity was observed after treatment of the supernatant with trypsin. For determination of the molecular mass of the proliferation factor, supernatants of quartz dust-exposed macrophages were concentrated by ultrafiltration and fractionated by gel-filtration on a column of Sephadex G 150. By gel filtration the proliferation activity eluted in the range of a molecular mass between 75 and 102 kDa (Griwatz *et al.*, 1994, 1995).

Using anion-exchange chromatography, the Proliferation Factor eluted in a linear NaCl-gradient between 210 and 250 mM. Figure 3 shows two of the active fractions in the A-549 proliferation assay, leading to an enhanced cell multiplication. In SDS, polyacrylamide gel electrophoresis of biological active fractions of the anion exchange chromatography revealed two bands with molecular masses of 76 and 79 kDa (Griwatz *et al.*, 1994, 1995).

Preliminary results demonstrate that supernatants of quartz-dust exposed human macrophages not only enhance the cell multiplication but also stimulate the collagen synthesis of human fibroblasts (Fig. 4). Importance has also been attributed to tumour necrosis factor alpha (TNF-α) in fibrosis, especially in

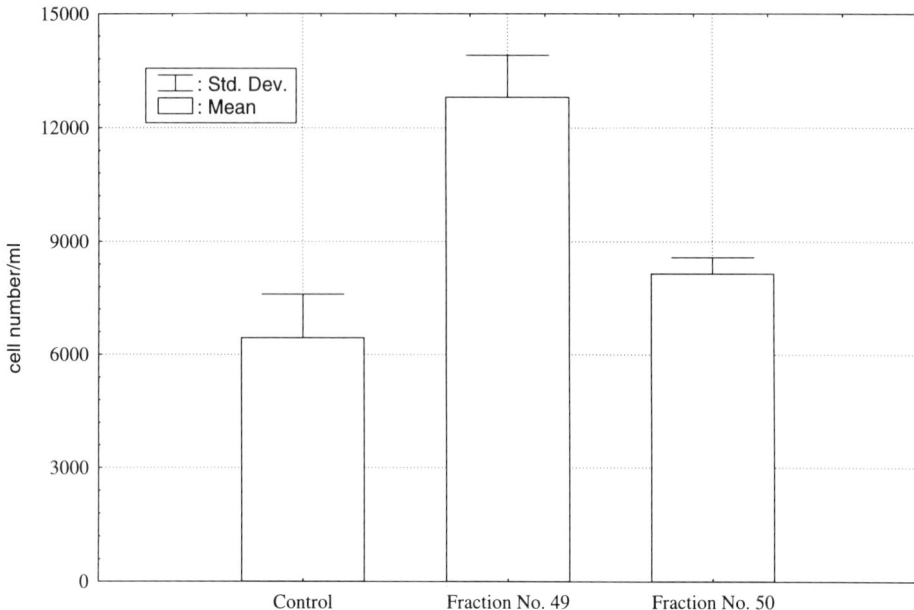

Fig. 3. Proliferation activity on A-549 cells by two active fractions from DQ12 supernatant after elution from anion exchange chromatography (Q-Sepharose) at 210–250 mM NaCl (20 mM Tris–Cl, pH 8.5).

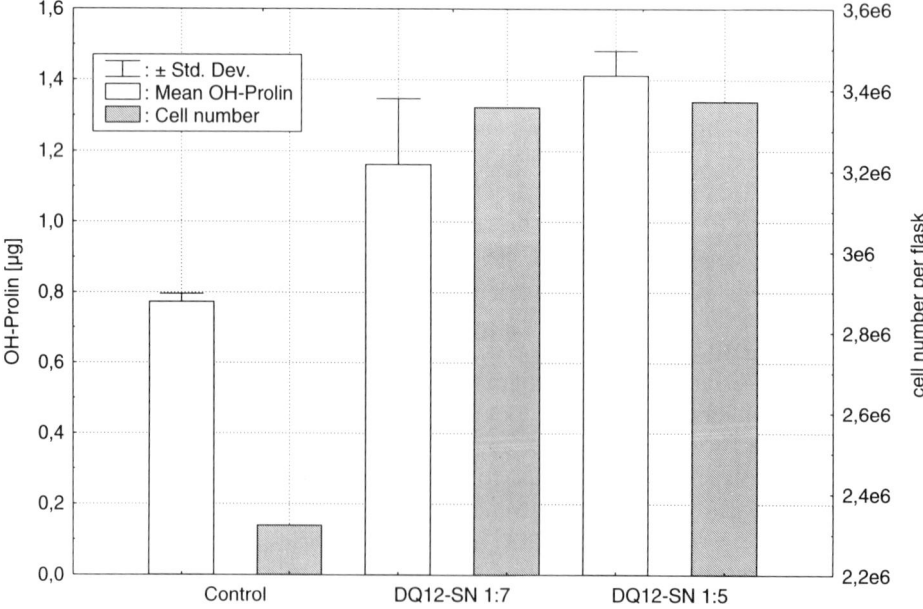

Fig. 4. Stimulation of collagen-synthesis (measured as OH–Prolin) and cell multiplication by a supernatant of DQ12-treated macrophages (DQ12-SN).

pathogenesis of silicosis and pneumoconiosis (Piguet *et al.*, 1990; Driscoll and Maurer, 1991). An important factor for TNF-α production and release by monocytes is endotoxin, a major lipopolysaccharide constituent of the cell wall Gram-negative bacteria (Carswell *et al.*, 1975). Because endotoxin contamination by bacteria of mineral dusts cannot be ruled out, we examined suspensions of coal mine dusts and quartz dust DQ12 for the presence of endotoxin utilising the very sensitive *Limulus* amoebocytes lysate test (LAL). Furthermore, we analysed supernatants of human macrophages treated with quartz dust DQ12 and with coal mine dusts for the presence of TNF-α by the cytotoxicity bioassay with L-929 cells (mouse fibroblasts). In addition, we investigated the effect of pure TNF-α on human pneumocyte type II cells (line A-549) alone and in combination with the Proliferation Factor from quartz dust DQ12 treated with human macrophages. Our results revealed that just a few samples of coal mine dusts from the Ruhr Valley contained endotoxin and only endotoxin-containing dusts stimulated human macrophages to produce TNF-α (Griwatz and Seemayer, 1994, 1995). While TNF-α led to a slight stimulation of cell proliferation of A-549 cells at low concentrations (1 ng ml^{-1}), a strong inhibition of cell proliferation of A-549 cells in presence of TNF-α was accompanied by morphological transformation of the epithelial cells to more spindle shaped cells. It is remarkable that strong stimulation of cell proliferation of A-549 cells by supernatants of quartz dust treated macrophages was completely inhibited in the presence of TNF-α (50 ng ml^{-1}) (Griwatz and Seemayer, 1994, 1995).

DISCUSSION

Results presented demonstrate that human macrophages in culture obtained by cultivation and differentiation of blood monocytes release a soluble factor following incubation with quartz dust DQ12 or coal mine dusts from the Ruhr Valley.

This factor stimulates "quiescent" or only moderately replicating human lung and dermal fibroblasts as well as pneumocyte type II cells (line A-549) to a considerable cell multiplication (Seemayer, 1989; Seemayer and Maly, 1990; Griwatz *et al.*, 1993, 1994, 1995). Therefore, we designated the factor as a Proliferation Factor. According to the dual control model of growth regulation (Stiles *et al.*, 1979) the Proliferation Factor was classified as a "progression factor" because a continuous presence was required for cell multiplication and an enhanced growth was observed by addition of competence factors such as FGF and PDGF (Seemayer and Maly, 1990; Seemayer, 1989; Hübner and Seemayer, 1989, 1992).

Bittermann *et al.* (1982) reported that human alveolar macrophages incubated *in vitro* with soluble and particulate agents, release an alveolar macrophage derived growth factor (AMDGF), exhibiting activity as a progression factor for human fibroblasts. The molecular weight of AMDGF of 18 kDa differs from the MW of the proliferation factor in the range of 75–102 kDa. Our described cytokine with proliferation activity differs by its molecular mass from known factors (Wahl, 1988). An explanation could be that activated macrophages produce a binding protein which forms a biological active complex, as described by Blum *et al.* (1989) for Somatomedin C. Leslie *et al.* (1989) described a cytokine in bronchioalveolar lavage fluid (BALF) of normal rats which stimulated cell proliferation of rat primary pneumocytes type II cells. When BALF was fractionated by gel filtration on a column Sephadex G150 a broad peak of activity eluted with an apparent molecular mass of approximately 100 kDa. Results presented suggest a complex interaction of cells of the lung, particularly of the alveolar unit in fibrotic lung diseases, especially silicosis.

REFERENCES

Austgulen, R., Hammerstrom, J. and Nissen-Meyer, J. (1987) *In vitro* cultured human monocytes release fibroblast proliferation factor(s) different from interleukin-1. *J. Leukocyte Biol.* **42**, 1–8.

Bittermann, P. B., Rennard, S. I., Hunninghake, G. W. and Crystal, R. G. (1982) Human alveolar macrophage growth factor for fibroblasts. Regulation and partial characterization. *J. Clin. Invest.* **70**, 806–822.

Blum, W. F., Jenne, E. W., Reppin, F., Kietzmann, K., Ranke, M. B. and Bierich, J. R. (1989) Insulin-like growth factor 1 (IGF-1) binding protein complex is a better mitogen than free IGF-1. *Endocrinology* **125**, 766–771.

Carswell, E. A., Old, L. J., Kassel, R. L., Green, S., Fiore, N. and Williamson, B. (1975) An endotoxin-induced serum factor that causes necrosis of tumors. *Proc. Natl. Acad. Sci. U.S.A.* **72**, 3666–3670.

Driscoll, K. E. and Maurer, J. K. (1991) Cytokine growth factor release by alveolar macrophages: potential biomarkers of pulmonary toxicity. *Toxicol. Pathol.* **19**, 398–405.

Glenn, K. C. and Ross, R. (1981) Human monocyte-derived growth factor(s) for mesenchymal cells. Activation of secretion by endotoxin and concavalin A. *Cell* **25**, 603–615.

Griwatz, U. and Seemayer, N. H. (1994) Release of cytokines by quartz and coal mine dust exposed cytokines. *J. Aerosol. Sci.* **25** (Suppl.) S495–S496.

Griwatz, U. and Seemayer, N. H. (1995) Tumour necrosis factor-alpha induction by endotoxin-containing coal mine dusts in cultures of human macrophages and its effects on pneumocyte type II cells. *Toxic. in Vitro* **9**, 403–409.

Griwatz, U., Jung, B., Seemayer, N. H. and Dehnen, W. (1993) Effect of cytokines produced by quartz dust treated human macrophages on human pneumocytes type II cells (A-549). *J. Aerosol. Sci.* **24** (Suppl. 1) 463–464.

Griwatz, U., Jung, B., Seemayer, N. H. and Dehnen, W. (1994) Biology and biochemical characterization of cytokines from quartz dust (DQ12)-treated macrophages. In *Cytokines in Hemopoieseis, Oncology and Immunology III* (Edited by M. Freund, H. Link, R. E. Schmidt and K. Welte), pp. 515–523. Springer, Berlin.

Griwatz, U., Jung, B., Seemayer, N. H. and Dehnen, W. (1995) The role of cytokines in the pathogenesis of silicosis: Biological and biochemical characterization of a "proliferation factor" from quartz dust exposed human macrophages. *Silicosis Report North-Rhine Westfalia* **19**, 259–273.

Hübner, K. and Seemayer, N. H. (1989) The role of a "fibroblast proliferation factor" in the pathogenesis of anthracosilicose. I. Stimulation of DNA synthesis of human lung fibroblasts (cell line WI38). *Silicosis Report North-Rhine Westphalia* **17**, 181–197.

Hübner, K. and Seemayer, N. H. (1992) Quartz and coal mine dust treated human monocyte-derived macrophages release a proliferation factor for human lung fibroblasts. In *Environmental Hygiene* (Edited by N. H. Seemayer and W. Hadnagy), pp. 227–230. Springer, Berlin.

Leslie, C. C., McCormick-Shannon, K., Mason, R. J. (1989) Bronchoalveolar lavage fluid from normal rats stimulates DNA synthesis in rat alveolar type II cells. *Am. Rev. Respir. Dis.* **139**, 360–366.

Leslie, C. C., Musson, R. A. and Menson, P. M. (1984) Production of growth factor activity for fibroblasts from human monocyte-derived macrophages. *J. Leucocyte Biol.* 143–159.

Nathan, C. E. (1987) Secondary products of macrophages. *J. Clin. Invest.* **79**, 319–326.

Nathan, C. E., Murray, H. W. and Cohn, Z. A. (1980) The macrophage as an effecto cell. *New England J. Med.* **303**, 622–626.

Piguet, P. F., Collart, M. A., Grau, G. E., Sappino, A. P. and Vassalli, P. (1990) Requirement of tumour necrosis factor for development of silica-induced pulmonary fibrosis. *Nature* **344**, 245–247.

Pledger, W., Stiles, C. D., Antoniades, H. N. and Scher, C. D. (1977) Induction of DNA synthesis in BALB/c 3T3 cells by serum components: Reevaluation of the commitment process. *Proc. Natl. Acad. Sci.* **74**, 4481–4485.

Schmidt, J. A., Oliver, C. N., Lepe-Zuiga, J. L., Green, I. and Gery, I. (1984) Silica-stimulated monocytes release fibroblast proliferation factors identical to interleukin 1. A potential role for interleukin 1 in the pathogenesis of silicosis. *J. Clin. Invest.* **73**, 1462–1472.

Seemayer, N. H. (1989) The role of a fibroblast proliferation factor in the pathogenesis of anthracosili-
    cosis. II. Biological characterisation of a fibroblast proliferation factor formed by human mac-
    rophage cultures following exposure to quartz dust and coal mine dust. *Silicosis Report North-Rhine
    Westphalia* **17**, 199–208.
Seemayer, N. H. and Braumann, A. (1985) Investigations on cytotoxic effects of quartz DQ12 and coal
    mine dusts on human macrophages *in vitro*. *Silicosis Report North-Rhine Westfalia* **15**, 301–320.
Seemayer, N. H. and Braumann, A. (1986) Effects of particle size of coal mine dusts in experimental
    anthracosilicosis. *In vitro* studies on human macrophages. *Ann. occup. Hyg.* **32**, 1178–1180.
Seemayer, N. H., Braumann, A. and Maly, E. (1987) Development of an *in vitro* test system with
    human macrophages and fibroblasts for analysis of the effect of quartz dusts and coal mine dusts. I.
    Formation of a fibroblast proliferation factor. *Silicosis Report North-Rhine-Westfalia* **16**, 143–155.
Seemayer, N. H., Behrend, H., Maly, E. and Hübner, K. (1988) The role of quartz and coal mine dust
    induced mediators from human macrophages in pathogenesis of silicosis. *J. Aerosol. Sci.* **19**,
    1129–1132.
Seemayer, N. H. and Maly, E. (1990) Release of a fibroblast proliferation factor from human
    macrophages *in vitro* treated with quartz dust DQ12 or coal mine dust. *Proc. VIIth Int.
    Pneumoconioses Conf. DHHS (NIOSH 9 Publ. No. 90–108, Part II.* US Department of Health and
    Human Services, pp. 926–929.
Stegemann, H. and Stadler, K. (1967) Determination of Hydroxyprolin. *Clin. Chim. Acta.* **18**, 267–273.
Stiles, C. D., Capone, G. T., Scher, C. D., Antoniades, H. N., van Wyk, J. J. and Pledger, W. J.
    (1979) Dual control of cell growth by somatomedins and platelet-derived growth factor. *Proc. Natl.
    Acad. Sci. U.S.A.* **76**, 1279–1283.
Wahl, S. M. (1988) Lymphocyte- and macrophage-derived growth factors. *Methods in Enzymol.* **163**,
    715–731.

Pergamon

Ann. occup. Hyg., Vol. 41, Supplement 1, pp. 434–439, 1997
© 1997 British Occupational Hygiene Society
Published by Elsevier Science Ltd. All rights reserved
Printed in Great Britain
0003-4878/97 $17.00 + 0.00
Inhaled Particles VIII

PII: S0003-4878(96)00052-X

SILICA INDUCED MICRONUCLEI IN PULMONARY ALVEOLAR MACROPHAGES *IN VIVO*

H. Wang,* J. Leigh,† A. Bonin† and M. Peters*

*University of Sydney, Sydney, Australia; and †NIOHS, GPO Box 58, Worksafe Australia, Sydney 2001, Australia

INTRODUCTION

In addition to the well known fibrotic effect, the potential lung carcinogenic effect of silica is now attracting much attention. Animal inhalation studies, at even quite low dosage, have yielded positive results in relation to silica-induced malignant tumours (Muhle *et al.*, 1989; Spiethoff *et al.*, 1992). Epidemiological studies show a consistently elevated risk of lung cancer in silicotic subjects (Smith *et al.*, 1995) and increasing but less consistent evidence for excess cancer in silica exposed workers (Berry, 1996; Goldsmith *et al.*, 1995).

Genotoxicity studies may help elucidate the carcinogenic effect in humans. In *in vitro* studies, silica-induced micronuclei (Hesterberg, 1986; Nagalakshmi, 1995), but failed to induce sister-chromatid exchanges in cultured hamster cells (Price-Jones, 1980). In *in vivo* studies, intraperitoneal injection of silica failed to induce micronuclei in the bone marrow of mice (Vanchugova, 1985) while intratracheally instilled silica did induce gene mutation in type II alveolar epithelial cells of rats (Driscoll, 1995).

From the results of the above studies, it is hypothesised that micronucleus formation of cells in the lungs of silica exposed rats may also be induced. This hypothesis is based on the following considerations: (a) bone marrow may not be a suitable tissue to test micronucleus formation in the intraperitoneal injection study because silica is not soluble in body fluid and transportation of it to bone marrow may be seriously limited; (b) silica-induced reactants such as various free radicals may be sufficiently diluted or inactivated before they reach bone marrow; (c) silica has been shown to induce micronuclei *in vitro*; (d) silica has been demonstrated to be able to induce DNA damage *in vivo*. In the present study, we modelled human inhalation exposure by intratracheal instillation of silica in rats and obtained pulmonary alveolar macrophages (PAM) by bronchoalveolar lavage.

MATERIALS AND METHODS

Animals

Specific pathogen-free male Wistar rats (200–230 g) were obtained from The University of New South Wales (Little Bay, NSW) breeding facility. The rats were in cages (five rats per cage) and food and water could be freely accessed. The lighting in the animal house was set to a 12/12 h on/off cycle.

Experimental design

Twenty rats were randomly distributed into four intratracheal instillation treatment groups (five in each group): (1) control, 0.5 ml saline; (2) 2.5 mg silica (Min-U-Sil 5 crystalline silica; Silica Corp., Berkeley Springs, West Virginia) suspended in 0.5 ml saline; (3) 7.5 mg silica in 0.5 ml saline; (4) 22.5 mg silica in 0.5 ml saline.

Intratracheal instillation

Rats were anaesthetized with intraperitoneal injection of a mixture of ketamine (100 mg kg^{-1}) and xylazine (3.3 mg kg^{-1}), (Sigma Chemical Company, Sydney NSW, Australia) dissolved in saline. Intratracheal instillation was performed by a procedure of tracheal exposure. After shaving, the skin in the ventral aspect of the neck was incised in the midline. Through blunt dissection, the trachea was exposed. Using a 1 ml disposable syringe with a #26 needle, 0.5 ml of a dust suspension or saline was injected. The incision was sutured carefully with interrupted silk sutures immediately after the injection and the rat was kept in a 30°C incubator until it regained consciousness.

Bronchoalveolar lavage

Ten days after the instillation, the rats were anaesthetized with 75 mg kg^{-1} of pentobarbital by intraperitoneal injection. A laparotomy was performed and the abdominal aorta was exposed. The rats were killed by transection of the aorta. Bronchoalveolar lavage was performed with two aliquots of 5 ml phosphate buffered saline (PBS) after *in situ* perfusion of the lung. The recovered fluid volume was recorded.

Histology

One hundred microlitres of the lavage fluid was placed on a slide by Cytospincentrifugation immediately after the lavage. At least two slides were made for each rat. The slides were stained with Diff-Quik (Lab Aids Pty Ltd, Sydney Australia).

Total cell number and LDH activity

The total cell number was counted with a haemocytometer and the remaining fluid was centrifuged (3000 rpm, 10 min). The supernatant was collected to measure LDH activity by a kit method (Trace, Sydney Australia).

Differential and micronucleus count

Slides were examined under oil immersion ($\times$ 1000). Five hundred leukocytes were counted to determine the frequency of differential types of cells by their morphology. Pulmonary alveolar macrophages were identified by their large size, vacuolated cytoplasm and non-ciliated margin. A minimum of 1000 PAM were scored to determine the proportion of micronucleated cells. The following criteria were used for identifying micronuclei: colour the same as main nucleus; diameter less than half of main nucleus; location within the cytoplasm as a round body completely separated from the main nucleus. Binucleated and multinucleated (three nuclei or more) PAM were also counted.

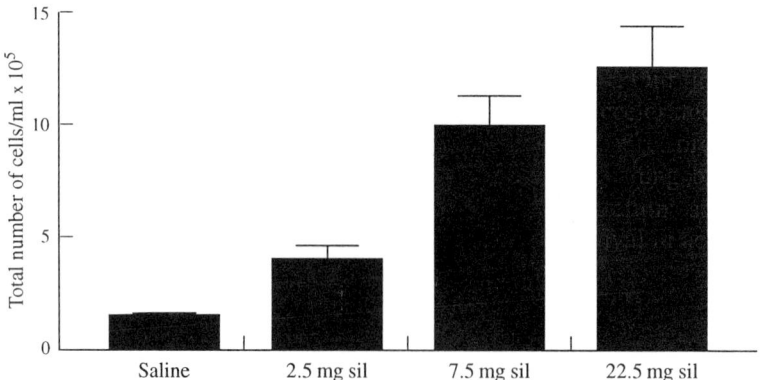

Fig. 1. Total cell number in 1 ml lavage fluid (means ± SEM., × 10⁵).

Statistical analysis

An unpaired 2-sided *t*-test (Welch modification if applicable) was used to compare the difference in various indices.

RESULTS

Inflammatory reaction

In this experiment, pulmonary inflammatory reaction was measured by total cell number and LDH activity in lavage fluid as well as the cell types of inflammatory cells. No accidental deaths of the animals occurred during the experimental period. The recovered volume of lavage fluid was between 9.0 and 9.7 ml and no significant difference could be detected among the four groups in the recovered volume.

Total cell number

Total cell number in lavage fluid was elevated dramatically in the three silica-instilled groups (Fig. 1). Compared with the control group, all silica-instilled groups had significantly higher cell numbers in lavage fluid ($P < 0.01$). Among the silica-instilled groups, total cell number of the 2.5 mg group was significantly lower than that of the other two groups ($P < 0.01$). There was no significant difference between the 7.5 and the 22.5 mg groups in total cell number.

LDH activity

LDH activity in lavage fluid of the silica-instilled groups was significantly higher than that of control. There was also a significant difference between the 2.5 mg silica group and the 22.5 mg silica group. There was no significant difference between the 7.5 mg silica group and the 22.5 mg silica group (Fig. 2).

Cell type of alveolar leukocytes

In the control group, the main cell type was macrophage whereas in the silica-instilled groups, the main cell type was neutrophil. Monocyte and lymphocyte proportions also increased with silica dosage (see Table 1).

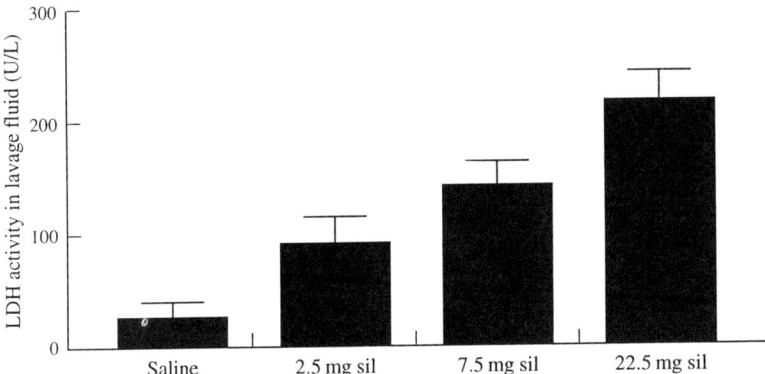

Fig. 2. Increased LDH activity in silica-instilled groups with dosage increase (mean ± SEM., U/L, International Unit per litre).

Micronucleus formation in PAM

Higher proportions of micronucleated PAM were identified in the silica-instilled groups compared with the control group. The proportion of micronucleated PAM was not related to silica dosage between 2.5 and 22.5 mg and there was no significant difference in the proportion of micronucleated PAM among the silica-instilled groups. The proportion of binucleated PAM in the 22.5 mg group was significantly higher than that of the control group, but no significant difference could be identified among the silica-instilled groups. Multinucleated PAM proportions in the silica-instilled groups were also significantly higher than that of control.

DISCUSSION

In the present study, inflammatory reaction in the silica-instilled rats was identified by an obvious infiltration of neutrophils into the alveolar space and an increase in total cell number as well as elevation of LDH activity in lavage fluid. The increase in total cell number and elevation of LDH activity in lavage fluid were

Table 1. Differential counting of leukocytes in lavage fluid (%)

Group	Macrophage	Monocyte	Neutrophil	Lymphocyte	Eosinophil
Saline	93.7 ± 0.7	3.4 ± 0.2	1.1 ± 0.4	1.8 ± 0.6	0
2.5 mg silica	15.2 ± 3.6	4.6 ± 0.3	78.1 ± 3.6	2.0 ± 0.2	0.2 ± 0.2
7.5 mg silica	8.8 ± 2.0	4.9 ± 0.3	83.4 ± 2.1	3.0 ± 0.5	0
22.5 mg silica	17.2 ± 1.3	9.9 ± 1.5	68.0 ± 1.5	4.5 ± 1.0	0.08 ± 0.05

Table 2. Proportion of different macrophages in lavage fluid (per 1000 cells)

Group	Micronucleated	Binucleated	Multinucleated
Saline	2.4 ± 0.68	18.80 ± 1.07	0.25 ± 0.25
2.5 mg silica	15.2 ± 1.56	25.80 ± 2.78	2.0 ± 0.84
7.5 mg silica	15.0 ± 2.30	22.00 ± 0.95	2.2 ± 1.32
22.5 mg silica	18.6 ± 1.33	25.80 ± 2.03	13.4 ± 5.30

dose-related. Therefore, it is considered that a valid animal model of silica-induced inflammation was established in this study.

In silica-induced inflammation, various substances such as free radicals and cytokines are elaborated. Because silica is insoluble in body fluid and silica-induced inflammation can be a chronic reaction, the cells physically existing in silica-exposed lungs will be exposed to not only silica particles but also silica-induced reactants. Since some silica-induced free radicals are genotoxic (Saffiotti, 1994; Castranova, 1994) and silica itself has a mutagenic effect (Driscoll, 1995), concern of a carcinogenic effect of silica is reasonably raised. This study is the first demonstration of increased micronucleus formation in silica-exposed animals *in vivo*. It was inconsistent with a previous *in vivo* study which administered silica by intraperitoneal injection and failed to detect increased micronucleus formation in bone marrow cells (Vanchugova, 1985). This failure may be because silica could not be transported to bone marrow or silica-induced substances were diluted/inactivated before they reached bone marrow. The type of animal may also be a possible explanation since silica failed to induce tumours in mice. Although the silica-induced inflammation was dose-related and silica-induced micronuclei *in vitro* are also dose-related (Nagalakshmi, 1995), the incidence of micronucleated PAM in silica-instilled rats was not dose-related in this study. A possible explanation for this is that 2.5 mg silica may saturate the reaction of micronucleus formation. Experiments with lower doses of silica are being conducted in our laboratory to verify this.

In an *in vitro* study (Nagalakshmi, 1995), it was proposed that silica induces micronuclei by damaging the spindle of dividing cells while in a recent study (Driscoll, 1995), it has been demonstrated that silica administration *in vivo* can induce gene mutation. We believe that the mechanism for micronucleus formation in macrophages in this study might be a result of the combined effects of silica particles and its induced reactants.

In summary, we have demonstrated that silica could induce micronucleus formation in PAM *in vivo*. Because the exposed organ was lung and the obtained PAM were from the silica-instilled lungs, we regard our study as more reliable than a previous *in vivo* study where intraperitoneal injection of silica was used and bone marrow cells were examined. We note that this positive result of micronucleus formation in rat cells was consistent with demonstrated carcinogenic effects of silica on rat lungs and consider that this might be an indication that silica-induced micronucleus formation is related to its carcinogenic effects. Examination of human cells exposed to silica to detect micronucleus formation *in vivo* may help to better assess the potential carcinogenic effects of silica in humans.

REFERENCES

Berry, G. (1996) Crystalline silica: health impacts and possible lung cancer risks. *J. occu. Health Safety—Aust. N.Z.* **12**, 157–167.

Castranova, V. (1994) Generation of oxygen radicals and mechanisms of injury prevention. *Environ. Health Perspect.* **102** (suppl. 10), 65–68.

Driscoll, K. E., Deyo, L. C., Howard, B. W., Poynter, J. and Carter, J. M. (1995) Characterizing mutagenesis in the hprt gene of rat alveolar epithelial cells. *Expl. Lung Res.* **21**, 941–956.

Goldsmith, D. F., Wagner, G. R., Saffiotti, U., Rabovsky, J. and Leigh, J. (1995) Future research needs in the silica, silicosis and cancer field. *Scand. J. Work Environ. Health* **21** (suppl. 2), 115–117.

Hesterberg, T. W., Oshimura, M., Brody, A. R. and Barrett, J. C. (1986) Asbestos and silica induce morphological transformation of mammalian cells in culture: a possible mechanism. In *Silica, Silicosis and Cancer. Controversy in Occupational Medicine*, pp. 177–190. Praeger, New York.

Muhle, H., Takenaka, S., Mohr, U., Dasenbrock, C. and Marmelstein, R. (1989) Lung tumor induction upon long-term low-level inhalation of crystalline silica. *Am. J. ind. Med.* **15**, 343–346.

Nagalakshmi, R., Nath, J., Ong, T. and Whong, W. Z. (1995) Silica-induced micronuclei and chromosomal aberrations in Chinese hamster lung (V79) and human lung (Hel 299) cells. *Mutation Res.* **335**, 27–33.

Price-Jones, M. J., Gubbings, G. and Chamberlain, M. (1980) The genetic effects of crocidolite asbestos: comparison of chromosome abnormalities and sister-chromatid exchanges. *Mutation Res.* **79**, 331–336.

Saffiotti, U., Daniel, L. N., Mao, Y., Shi, X., Williams, A. O. and Kaighn, M. E. (1994) Mechanisms of carcinogenesis by crystalline silica in relation to oxygen radicals. *Environ. Health Perspect.* **102** (suppl. 10), 159–63.

Smith, A., Lopipero, P. and Barroga, V. (1995) Meta-analysis of studies of lung cancer among silicotics. *Epidemiol.* **6**, 617–624.

Spiethoff, A., Wesch, H., Wegener, K. and Klimisch, H. (1992) The effects of thoratrast and quartz on the induction of lung tumors in rats. *Health Phys.* **63**, 101–110.

Vanchugova, N. N., Frash, V. N. and Kogan, F. M. (1985) The use of a micronucleus test as a short term method in detecting potential blastomogenicity of asbestos-containing and other mineral fibers. *Gig. Tr. Prof. Zabol.* **6**, 45–48.

 Pergamon

Ann. occup. Hyg., Vol. 41, Supplement 1, pp. 440–447, 1997
© 1997 British Occupational Hygiene Society
Published by Elsevier Science Ltd. All rights reserved
Printed in Great Britain
0003–4878/97 $17.00 + 0.00
Inhaled Particles VIII

PII: S0003–4878(96)00094–4

TWO CASES OF FATAL PMF IN AN ONGOING EPIDEMIC OF ACCELERATED SILICOSIS IN OILFIELD SANDBLASTERS: LUNG PATHOLOGY AND MINERALOGY

J. L. Abraham* and S. L. Wiesenfeld†

Departments of *Pathology, State University of New York, Health Science Center, 750 E Adams St, Syracuse, NY 13210; and †Medicine, Texas Tech University Health Sciences Center, Odessa, Texas, U.S.A.

INTRODUCTION

Epidemics of silicosis have occurred repeatedly in the U.S. at intervals of a few decades (1890s, 1930s, 1970s), during which collective memory appears to fall below a critical point of awareness (Rosner and Markowitz, 1991). Sandblasting with crystalline silica has long been recognised as one of the most potentially hazardous occupations for silicosis risk and in fact, silica sandblasting has been banned in the U.K. since the 1940s. Nevertheless, we have been able to study an ongoing epidemic of silicosis in West Texas Oilfield sandblasters (CDC, 1990; Wiesenfeld and Abraham, 1995) and to describe the clinical findings, lung pathology and mineralogy of the previously rarely described accelerated silicosis. In 1988, we independently began evaluating patients (SLW) or lung biopsies (JLA) from these workers. An early fatality from acute silico–proteinosis was reported (CDC, 1990). Here we present two recent fatalities from PMF in the same epidemic.

The setting for the epidemic

In the early 1980s the price of oil rose to $34 a barrel and the Permian Basin in Texas became a magnet for oil production. Working to capacity, sandblasting provided a means for preparing metal surfaces—pipes, tanks and manifolds. Working conditions were extremely dusty, little or no respiratory protection was used and in some operations, blasting sand was recirculated, becoming finer and finer until no longer useful. Workers worked in the midst of an aerosol so dense they could not see. Personal breathing zone air samples in 1988 showed respirable free silica from 400–700 µg m^{-3}. The work was arduous and lengthy, often 6–7 days/week, 10–12 h shifts. There were no cleaning facilities, pre-employment or annual medical exams. Summer temperatures routinely exceeded 100°F. A small group of Mexicans worked in Odessa foundries in the late 1970s and early 1980s. As work expanded, they recruited friends and relatives to do similar work.

MATERIALS AND METHODS

Clinical evaluation included: history, physical examination, chest X-ray (CXR),

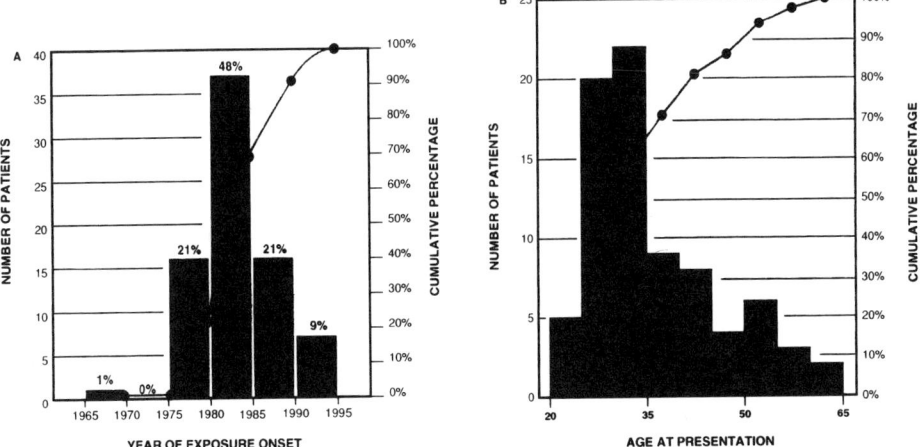

Fig. 1. (A) Histogram showing year of onset of sandblasting for all individuals studied. The ascending line represents the cumulative frequency and the dot represents the cumulative frequency at the end of each 2 year period; (B) histogram of age at presentation to physician for all individuals studied, by 5 year grouping. The ascending line indicates the cumulative frequency and the dot represents the cumulative frequency at the end of each 5 year period.

pulmonary function tests (PFT), CT scans, Ga^{67} scans and, in some cases, transbronchial or thoracoscopic lung biopsy. CXRs were scored using a 0–4 grading system as well as by ILO-1980 B-readings. Biopsies were analyzed by brightfield and polarized light microscopy and scanning electron microscopy/energy dispersive X-ray analysis for quantification of lung inorganic particulate burden (Abraham and Burnett, 1983). In 1995 and 1996, two men, with 36 and 48 months exposure, ceasing work at the time of diagnosis in 1988 and 1990, had progressed to PMF and died of respiratory failure, 11 and 10 years after initial exposure, respectively. Both had severe disease clinically and radiologically at the time of diagnosis. Autopsies were limited to the lungs.

RESULTS

A brief summary of the larger group is needed to put this report in perspective; 84 patients were initially studied (Wiesenfeld and Abraham, 1995). Figure 1(A) shows the year of first exposure, it correlates with the economic collapse in 1984, when the oil price plummeted. Age at presentation ranged from 20–64 years (mean 36.6 ± 10.6, median 33) [Fig. 1(B)]. Approximately 37% were smokers. Progressive dyspnea on exertion was insidious in onset and chest physical findings were initially mostly normal. Nearly all men had BCG vaccination as children in Mexico; no cases of M.TB were documented by sputum or biopsy cultures, although a few men years later developed atypical mycobacterial disease.

Chest X-ray, gallium scans and DLCO
Initially, only 35.7% of CXRs (30 of 84) showed any micronodularity or more

advanced changes. At 5 and 7 years of study, 9 and 18 cases or 18.7% and 33.3%, respectively, converted from normal to abnormal CXR. At 5 years of study the mean time from initial silica exposure to CXR abnormality was 8.6 ± 4.1 years (range 2.3–15.5 years). Biopsies confirmed accelerated silicosis and high silica burdens prior to CXR conversion. All but one had positive Ga scan prior to CXR conversion. The DLCO was reduced in these 18 cases prior to CXR conversion. The more positive CXRs and Ga scans are both associated with earlier age at presentation and shorter duration of exposure. Owing to differing work practices, patients could be divided into two cohorts: (1) one company in which sand was recirculated, and (2) all other companies in which sand was not recirculated. This factor of intensity of exposure apparently explains most of the at first seemingly paradoxical inverse relationship between severity of disease and age and duration of exposure. The worst initial DLCO values are seen in cohort 1. Even with a negative CXR, the mean initial % predicted DLCO is 75.7%.

Pathology

Commonly, subpleural dust and occasional nodules are seen at thoracoscopy. The most common finding is an interstitial infiltrate of dust-laden macrophages. There is a mixture of barely resolvable tiny opaque and birefringent particles with variable fibrosis. Silicotic nodules are rare and small (0.05–0.5 mm). Many alveoli contain dusty macrophages. Caseating granulomas are not found. Polymorphonuclear leukocytes are very rare. Increased interstitial lymphocytes and many Fe-stain positive macrophages are noted. Proteinosis is not observed in any of the biopsies, although focally noted in one of the PMF autopsy cases, as has been noted by Honma and Chiyotani (1991) and Shida *et al.* (1996). In the two autopsy cases the entire lungs were greatly reduced in volume, with the tiny upper lobes being grey and stony hard (Fig. 2). The lower lobes were firm and brown, with scattered palpable nodules up to 0.4 cm. Histologically, nearly all alveoli showed interstitial inflammation and fibrosis. The transbronchial biopsy obtained from case 1 in 1988 showed severe fibrosis and high dust concentrations (see Table 1).

Mineralogic findings and correlations

The silica concentrations in these sandblasters are among the highest seen in our analytical experience (Abraham *et al.*, 1991). The results show that pure silica exposure is rare, the more common pattern being mixed dust exposure (Table 1). The geometric mean concentration of silica is higher for cohort 1 than for cohort 2 ($p < 0.0004$; T-test). The geometric mean diameter of silica particles is 0.57 micrometers. The concentration of silica, but not that of other types of mineral particles, correlates with radiologic severity (Spearman Rank correlation = 0.686, $p < 0.0001$ by the Mann–Whitney U-test, for the correlation of the HRCT score with the log10 of lung silica particle concentration). Discriminant analysis (Hunt, 1986) was used to assess whether the microanalysis, performed without any knowledge of the work history or histopathology, facilitated distinguishing between cohorts. Major contributions to the analysis were made by the log10 of silica, aluminum silicates, chromium and titanium, in that order. Clear separation of the workers by their cohorts is evident based on the lung burden data (Fig. 3). Analytical results in the two PMF cases demonstrate extremely high concentrations

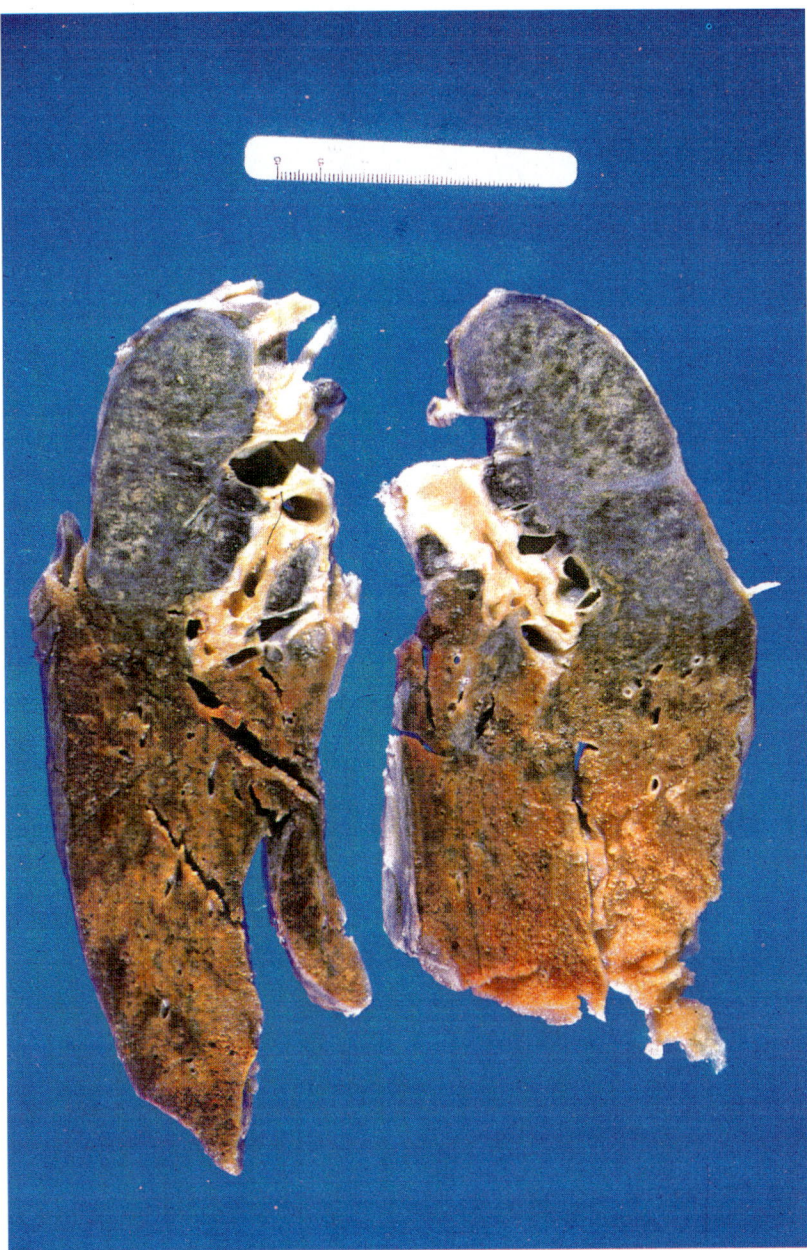

Fig. 2. Photograph of lungs from PMF case 1, showing stony grey remnant left upper and right upper and middle lobes and brown small lower lobes.

Table 1. Concentrations of inorganic particulates in lungs of two sandblasters with PMF*

	Total	Silica	Aluminium silicates	Metals (totals)	Cr	Ti
PMF Case 1						
Biopsy	3202	1827	597	669	334	485
upper	24 686	9111	10 433	4324	3719	2356
upper	26 636	14 570	8392	3652	2747	2133
middle	9995	6147	804	2741	1876	1698
lower	656	407	92	128	111	18
PMF Case 2						
upper	3953	2296	354	1303	1136	342
hilar	1090	394	245	451	340	243
lower	371	187	11	173	170	29
lower	269	190	13	66	36	32

* All concentrations in millions of particles cm^{-3} lung; site of lung sampling from biopsy unknown; others from autopsy. Cr and Ti concentrations = metal particles which contain that element.

of particles in the PMF lesions, and the lobar differences in dust concentration (Table 1).

DISCUSSION

Silicosis in its accelerated form occurs in sandblasters in this outbreak, with the exception of one earlier reported case of alveolar proteinosis (CDC, 1990). Two cases at least have progressed to death from PMF and respiratory failure. In accelerated silicosis the CXR is initially normal and there is little restriction on PFTs. An abnormal CXR is a late, global finding and is associated with more intense silica exposure and lung silica burden. Abnormal DLCO and Ga scan appear to be the only objective indicators of disease progression to positive CXR, while spirometry changes minimally and CXR may remain normal up to 8–12 years. In the absence of abnormal CXR, lung biopsies were of special value in demonstrating that suspect cases, based on dyspnea and impaired DLCO with or without a positive Ga scan, indeed had accelerated silicosis.

The lung pathology in accelerated silicosis has not been clearly defined (Craighead et al., 1988) or described in detail (Seaton, 1995). Silicotic nodules in classic chronic silicosis usually measure 2–3 mm diameter. In 1939, Gardner described the lung pathology in 19 cases of what the ILO in 1938 termed "rapidly developing silicosis", which would now be termed acute silicosis. Nearly all these cases had complicating TB. "Microscopic examination of the lungs revealed great numbers of minute silicotic nodules of ⅕th to half the size found in ordinary chronic cases."

The ability of the lung particulate burden data to allow separation of workers by their employment cohorts using discriminant analysis suggests differences in workplace materials and work practices may be investigated using lung burden microanalysis to reveal a type of fingerprint of the total and frequently complex lung burden retained in a persons pre-biopsy lifetime. The finding that, of all the types of retained particles measured in the lung, only the lung silica burden related

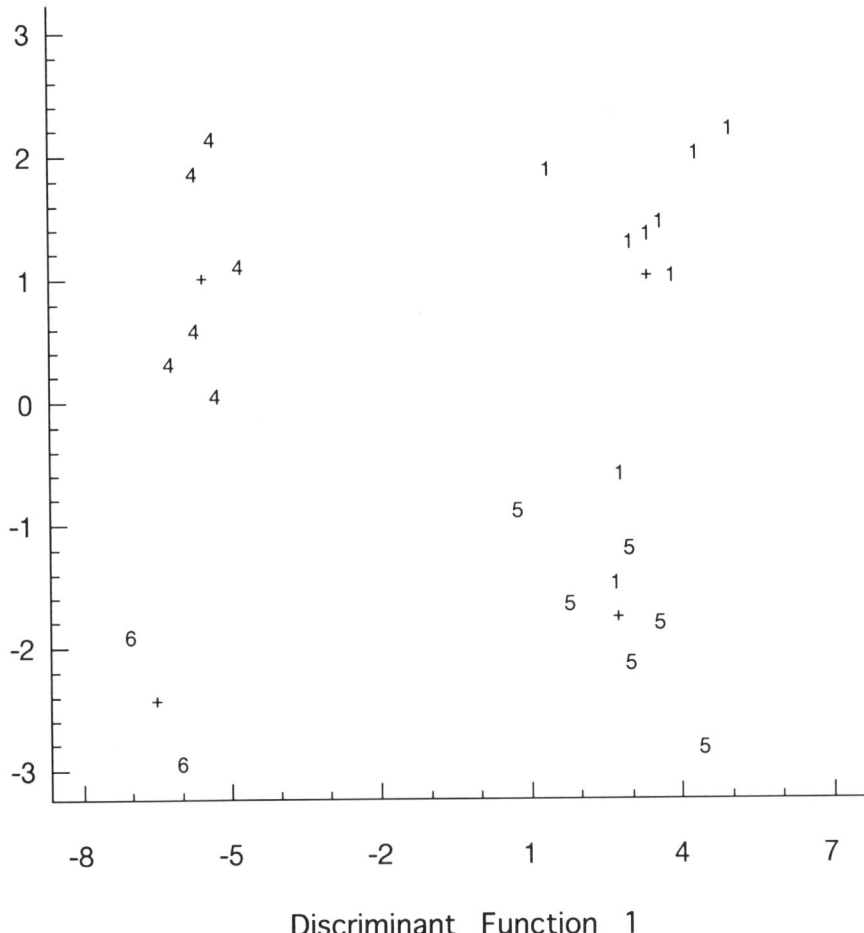

Fig. 3. Discriminant analysis plot showing the separation of cases according to employer, based on lung concentrations of silica, aluminium silicate, chromium and titanium particulates. Groups indicated as follows: 1 = cohort 1; 5 = workers at a second company; 6 = workers at a third company; and 4 = sandblasters from other states, not from this epidemic.

to most of the measured clinical and histopathologic measures of disease severity serves to further illustrate the utility of such analyses in investigating mixed exposures and their relationship to disease outcomes. It should not be surprising, however, to demonstrate that silica dust particles are the major pathogenic particle in silicosis. As Donald Cummings pointed out at the Fourth Saranac Silicosis Symposium (1939): "Among all the recognized phenomena in the realm of medicine none may be perceived with greater clarity than the true cause of silicosis."

Disease paradigms in silicosis and the nature of screening
    The diagnosis of silicosis for clinical epidemiological and compensation purposes has been inflexibly contingent on radiological demonstration of fibrotic lesions in a

person with a consistent occupational history. Contemporary dust standards are based on the premise that significant disease does not exist in absence of a positive CXR. Our findings in this study do not support this requirement for a positive CXR in accelerated silicosis. Currently, if the CXR is negative, even with a history of exposure, progressive dyspnea, a positive Ga scan, a deteriorating DLCO, plus a confirming biopsy, the diagnosis is contested and benefits denied! Lung transplantation, stated by Seaton (1995) to be the only possible hope for prolonged survival in accelerated silicosis, is unlikely to be available to the affected workers.

CONCLUSIONS

Almost a century ago, Betts (1990) described an outbreak of what would now be called accelerated or acute silicosis. He examined 30 of 200 deaths at one mill where the average age at death was 30, employment 1.2 years and time to death 2.4 years. For every epidemic reported, Betts suspected, as we do, hundreds of silent epidemics go unreported because of industry pressures, associated litigation, reluctance of medical colleagues to get involved and difficulties in studying migrant and immigrant populations. We have shown that fatalities from PMF with brief uncontrolled silica exposure in abrasive blasting continue to occur in industrialised nations. This epidemic seems to be unique in that tuberculosis is absent, most likely related to the high prevalence of childhood BCG vaccination in these workers. The zonal distribution of silica correlates with the severity of the disease. Tissue microanalysis delineates mixed exposures and facilitates analysis of specific dose–response relationships.

There are a number of effective substitute materials for silica; issues of price, demand and safety need to be kept in mind prior to mandatory use of any specific substitutes. NIOSH in 1974 proposed legislation to ban sand as an abrasive, but this was blocked by sand and related companies (Silica Safety Association, 1976), mirroring similar events decades ago (Rosner and Markowitz, 1993). Should similar legislation be proposed and not acted into law, the U.S.A. will be ensuring that it will continue to rediscover silicosis in the next millennium. Our cases unfortunately demonstrate the outcome of unregulated use of silica in the modern era.

REFERENCES

Abraham, J. L. and Burnett, B. R. (1983) Quantitative analysis of inorganic particulate burden *in situ* in tissue sections. *Scanning Electron Microscopy/1983* 2, 681–696.

Abraham, J. L. Burnett, B. R. and Hunt, A. (1991) Development and use of a pneumoconiosis database of human inorganic particulate burden in over 400 lungs. *Scanning Microsc.* 5, 95–108.

Betts, W. W. (1900) Chalicosis pulmonum, or chronic interstitial pneumonia induced by stone dust. *JAMA*, 34, 70–74.

CDC (1990) Silicosis: cluster in sandblasters—Texas, and occupational surveillance for silicosis. *MMWR* 39, 433–437.

Craighead, J. E., Kleinerman, J., Abraham, J. L., Gibbs, A. R., Green, F. H. Y., Harley, R. A., Ruettner, J. R., Vallyathan, N. V. and Juliano, B. (1988) Diseases associated with exposure to silica and nonfibrous silicate minerals. *Arch. path. Lab. Med.* 112, 673–720.

Cummings, D. E. (1939) The etiology of silicosis. In *Fourth Saranac Laboratory Symposium on Silicosis* (Edited by B. Kuechle) p. 20. Trudeau School of Tuberculosis, Saranac Lake, NY.

Gardner, L. (1939) Rapidly developing silicosis. In *Fourth Saranac Laboratory Symposium on Silicosis*.

(Edited by B. Kuechle), pp. 236–240. Trudeau School of Tuberculosis, Saranac Lake, NY.

Honma, K. and Chiyotani, K. (1991) Pulmonary alveolar proteinosis as a component of massive fibrosis in case of chronic pneumoconiosis: an autopsied study of 79 cases. *Zentralbl. Pathol.* **137**, 414–417.

Hunt, A. (1986) The application of mineral magnetic methods to atmospheric aerosol discrimination. *Phys. Earth Planet. Int.* **42**, 10–21.

Rosner, D. and Markowitz, G. (1991) *Deadly Dust: Silicosis and the Politics of Occupational Disease in Twentieth-Century America.* Princeton University Press.

Seaton, A. (1995) Accelerated silicosis, In *Occupational Lung Diseases*, (Edited by Morgan and Seaton), 3rd edn. Saunders, p. 237, 259.

Shida, H., Chiyotani, K., Honma, K., Hosoda, Y., Nobechi, T., Morikubo, J. and Wiot, J. F. (1996) Radiologic and pathologic characteristics of mixed dust pneumoconiosis. *RadioGraphics* **16**, 483–498.

Silica Safety Association (1976) Silica safety association comments concerning questions proposed by the department of labor. In *Federal Register, 1976*, 23721, Vol. 41.

Stern, R. M., Pigott, G. H. and Abraham, J. L. (1983) Fibrogenic potential of welding fumes. *J. appl. Toxicol.* **3**, 18–30.

Wiesenfeld, S. L. and Abraham, J. L. (1995) Epidemic of accelerated silicosis in Texas sandblasters: clinical, pathologic and microanalytic observations. *Am. J. Resp. crit. Care Med.* **151**, A710.

 Pergamon

Ann. occup. Hyg., Vol. 41, Supplement 1, pp. 448–453, 1997
© 1997 British Occupational Hygiene Society
Published by Elsevier Science Ltd. All rights reserved
Printed in Great Britain
0003–4878/97 $17.00 + 0.00
Inhaled Particles VIII

PII: S0003–4878(96)00095–6

RESPIRATORY HEALTH EFFECTS FROM AMBIENT SILICA EXPOSURE: A BENCHMARK DOSE ANALYSIS

J. S. Gift* and D. F. Goldsmith†

*United States Environmental Protection Agency, Research Triangle Park, North Carolina; and
†California Public Health Institute, 2001 Addison St., Suite 210, Berkeley, CA 94704–1103, U.S.A.

INTRODUCTION

The U.S. Environmental Protection Agency (EPA) has established National Ambient Air Quality Standards (NAAQS) of 150 µg m^{-3} (24 h average) and 50 µg m^{-3} (annual arithmetic mean) for particles < 10 µm in aerodynamic diameter (PM$_{10}$).* State health officials and some members of the public have expressed concern over whether this standard is adequate with respect to silica, a major component in some locations of airborne particulate matter. Quartz is the primary crystalline polymorph of silica and the predominant form found in the environment. This paper summarises what is known about ambient quartz levels in the United States (U.S.) and describes a benchmark concentration (BMC) analysis of the exposure–response relationship for fibrotic lung disease associated with chronic, low-level silica exposures.

One study was found that measured ambient silica levels in the U.S. Davis *et al.* (1984) report average and upper-bound quartz fractions of 5 and 10%, respectively, in particulate samples of less than 15 µm d$_{ae}$ taken from 22 United States metropolitan areas. In a mining environment, the silica fraction within airborne particle samples of less than 10 µm d$_{ae}$ did not differ markedly from those reported by Davis *et al.* (1984) for the larger particle size range (Verma *et al.*, 1994). Thus, 10% is a considered reasonable upper-bound estimate of the quartz fraction within PM$_{10}$ samplers. Davis *et al.* (1984) reported average and upper-bound quartz levels of 3 and 8 µg m^{-3}, respectively. As shown in Table 1, quartz levels estimated from 7 year average PM$_{10}$ levels (1987–1993) and quartz fractions reported by Davis *et al.* (1984) for 17 cities are slightly lower, but in general agreement with measured levels. Due to uncertainties in estimating quartz fractions in the 1987–1993 PM$_{10}$ samples, the direct measurements are preferred as realistic and conservative estimates of ambient U.S. silica levels.

METHODS

A BMC (or a benchmark dose) analysis is defined as the use of a mathematical model to determine an inhaled concentration (or oral dose) and its lower

---

* EPA has also recently proposed the addition of two new PH$_{2.5}$ standards at 15 µgm^{-3}, annual mean, and 50 µgm^{-3}, 24-h avegate (U.S. Federal Register, 1996).

Table 1. Comparison of estimates using 7 year average PM_{10} levels and measured quartz levels

Site	Quartz percentage of TDM[a] (weight %)	7 year avg. PM_{10}[b] ($\mu g\ m^{-3}$)	Estimated 7 year avg. quartz[c] ($\mu g\ m^{-3}$)	1980 Measured quartz levels[d] ($\mu g\ m^{-3}$)
Akron, OH	6.0	29	1.7	4.2
Boston, MA	5.7	32	1.8	8.0
Buffalo, NY	2.9	28	0.8	2.4
Cincinnati, OH	4.1	33	1.4	2.6
Dallas, TX	4.7	32	1.5	2.9
El Paso, TX	3.0	52	1.6	2.3
Hartford, CT	5.5	28	1.5	3.0
Honolulu, HI	5.2	21	1.1	2.4
Kansas City, KS	7.4	49	3.6	5.1
Kansas City, MO	7.4	49	3.6	4.3
Minneapolis, MN	8.2	33	2.7	3.8
Portland, OR	1.0	34	0.3	1.4
RTP, NC	2.5	29	0.7	1.3
Riverside, CA	2.8	84	2.4	3.0
St. Louis, MO	7.9	63	5.0	4.5
San Jose, CA	3.0	31	0.9	2.0
Seattle, WA	3.1	39	1.2	1.1
Averages	4.7	39	1.9	3.2

[a]Davis et al. (1984); quartz as percent of combined coarse and fine dust ($< 15\ \mu m\ d_{ae}$).
[b]U.S. Environmental Protection Agency (1996).
[c]Estimate of 7 year annual quartz level = 7 year annual PM_{10} level $\times$ quartz weight percent.
[d]Davis et al. (1984); combined coarse and fine quartz captured in a dichotomous sampler designed to eliminate particles $> 15\ \mu m$ aerodynamic diameter.

confidence bound (e.g. 95% confidence limit) that is associated with a predefined effect level. General methods for the performance of a BMC analyses are described in U.S.EPA (1994). The log-logistic model used for the purposes of this analysis is described in Appendix B of U.S.EPA (1996). The following is a discussion of how data for use in the BMC analysis were chosen.

Selection of noncancer endpoint

Health effects that have been associated with silica exposure include silicosis, silico-tuberculosis, cor pulmonale, auto-immune diseases, nephritis, emphysema and airflow abnormalities. Silico-tuberculosis and cor pulmonale are secondary effects of silicosis. Emphysema and airflow abnormalities are generally considered nonspecific pulmonary effects caused by dust exposures. Auto-immune diseases and nephritis from silica exposures have been documented (Sanchez-Roman et al., 1993; Steenland and Goldsmith, 1995), but the mechanisms involved and the relationship between exposure and effect are not well understood. Pulmonary fibrosis (as silicosis) determined by radiography or autopsy is the endpoint of choice for the benchmark analysis. It characterises a specific, primary health effect from silica exposure, and good occupational exposure–response data exist at exposure levels not far from estimated ambient concentrations (see Table 2).

Table 2. Key occupational studies

Reference	Study type[a]	Study population	Definition of silicosis	% Silica (quartz)
Hnizdo and Sluis-Cremer (1993)	LRC	2235 South African miners; started after 1938 and worked $\geq$ 10 years; followed to 1991.	313 cases of silicosis (ILO category $\geq$ 1/1)	30
Muir et al. (1989a,b), Muir (1991), Verma et al. (1989)	LRC	2109 Canadian miners; started 1940–1959; followed to 1982 or end of exposure.	32 cases of silicosis (ILO category $\geq$ 1/1)	6.0–8.4
Ng and Chan (1994)	XRC	338 Hong Kong granite workers; 132 past workers (1967–1985) and 206 current workers (1985); only most recent X-rays examined.	36 radiographical abnormalities, rounded opacities (ILO category $\geq$ 1/1)	27
Rice et al. (1986)	CC	U.S. (North Carolina) dusty trades workers diagnosed with silicosis, 1935–1980.	216 cases of silicosis; 672 controls	1–50
Steenland and Brown (1995)	LRC	3330 South Dakota gold miners who worked at least 1 year underground between 1940 and 1965; followed to 1990.	170 cases of silicosis (ILO category $\geq$ 1/1)	13

[a] CC = case control, L = longitudinal, RC = retrospective cohort, X = cross-sectional, Q = quartz.

Evaluation of key studies

Only three of the studies described in Table 2 attempted to define exposure–response relationships among silica-exposed miners at environmentally relevant levels: Hnizdo and Sluis-Cremer (1993), Muir et al. (1989b) and Steenland and Brown (1995). As can be seen from Fig. 1, the results of Muir et al. (1989b) are not consistent with data from other studies, possibly due to the lack of follow-up beyond retirement. While a recent comparison of X-ray and autopsy diagnosis (Hnizdo et al., 1993) suggests that X-rays diagnosis used by both Hnizdo and Sluis-Cremer (1993) and Muir et al. (1989b) may not be the most sensitive diagnostic method for determining a pulmonary fibrotic effect, Hnizdo and Sluis-Cremer's research was continuous and longitudinal, involving multiple X-ray examinations both before and beyond retirement. Further, autopsy results for the nearly complete cohort were available to Hnizdo and Sluis-Cremer (1993) to allow for more accurate interpretation of radiographic results, while death certificates were available for only half of the cohort from the Steenland and Brown (1995) study. Thus, Hnizdo and Sluis-Cremer (1993) study is considered the most appropriate for use in derivation of a BMC.

RESULTS AND DISCUSSION

Results of our log–logistic regression analysis are presented in Table 3. Several factors which limit our confidence in this analysis are discussed here. First,

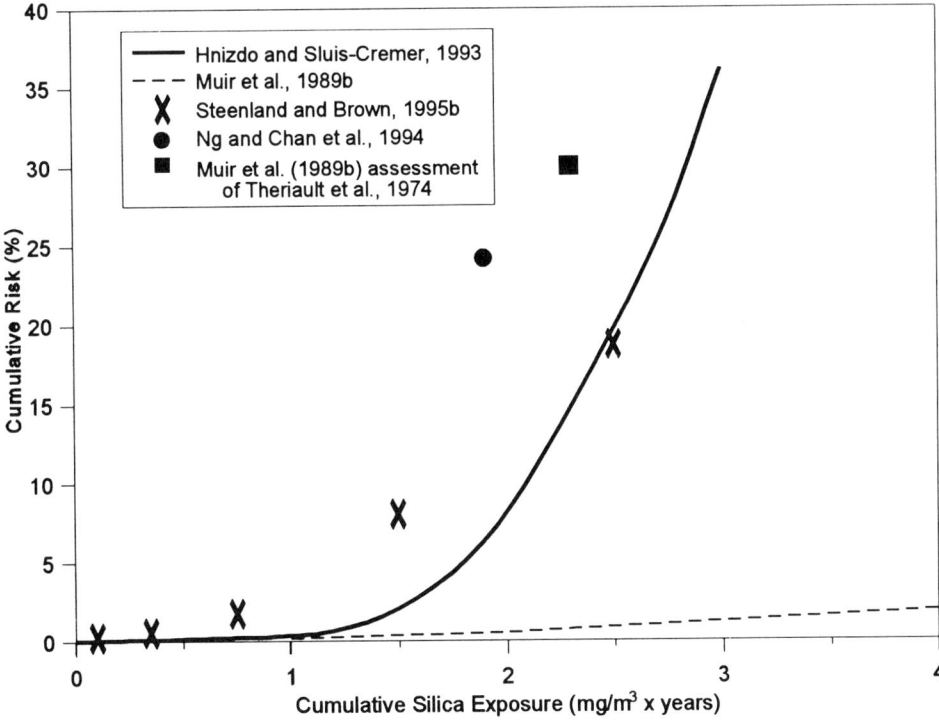

Fig. 1. Cumulative risk curves estimated for South African gold miners (—) (Hnizdo and Sluis-Cremer, 1993) and Canadian hardrock miners (– –) (Muir *et al.*, 1989b); cumulative silica risk points estimated for South Dakota gold miners (X) (Steenland and Brown, 1995), Hong Kong granite workers (●) (Ng and Chan, 1994) and Vermont granite miners (■) (Muir *et al.*, 1989b; Theriault *et al.*, 1974).

radiographs are not a particularly reliable or sensitive diagnostic tool. Hnizdo and Sluis-Cremer (1993) found that 57% of cases were diagnosed an average of 7.4 years after work in mines. While they attempted to follow workers beyond employment, participation by retired workers was voluntary. Also, the best of radiographers did not diagnose silicosis in 61% of 326 cases categorized at autopsy as slight to marked silicosis (Hnizdo *et al.*, 1993). Available exposure data are limited. All three studies had to estimate early exposures without high quality monitoring information. Further, peak exposures may be better predictor than cumulative silica exposure (Checkoway and Rice, 1992). In addition, differences

Table 3. Risk estimates from application of log–logistic model to data of Hnizdo and Sluis-Cremer (1993)[a,b]

	CSE for Risk of:			Risk for CSE of:		
	1%	5%	10%	0.6	1.0	1.6
Best-fit estimates	1.39	1.97	2.30	0.019%	0.21%	1.9%
Lower bounds on CSEs	1.31	1.90	2.24	—	—	—
Upper bounds on risks	—	—	—	0.032%	0.30%	2.4%

[a] See USEPA 1996, Appendix B for a discussion of methods used for estimating these risk levels.
[b] CSE = Cumulative silica exposure.
CSEs are in mg m^{-3}-years. Bounds are 95% lower bounds or 95% upper bounds, as appropriate.

between occupational and ambient exposure scenarios generally suggest that the mining environment may over estimate risk from ambient silica exposure. Silica in mining environment tends to be of smaller, more respirable size (Verma et al., 1994; Davis et al., 1984). More toxic, freshly fractured silica is more prevalent in occupational setting (Shoemaker et al., 1995). A higher fraction of silica has been reported in mining dust (e.g. 30% by Hnizdo et al., 1993) vs ambient upper-bound estimate of 10% (Davis et al., 1984). Further, miners may have used preventative measures such as masks and aluminium treatment (Muir et al., 1989b) not readily available to the general population. Finally, workers may be less susceptible to silicosis than the general population, including children, elderly and people with existing respiratory disease such as tuberculosis and pneumonia.

CONCLUSIONS

Using ambient silica measurements from Davis et al. (1984), average and high cumulative silica exposures (CSEs) of 0.6 and 1.6 mg silica m^{-3} years were estimated for 22 U.S. cities (U.S.EPA, 1996). Using a high estimate of 10% for silica fraction in PM_{10}, 1 mg silica m^{-3} years is the highest expected from continuous lifetime exposure at or below the annual PM_{10} NAAQS of 50 μm^{-3}. The 95% upperbound confidence limits on the risk estimated for CSEs of 0.6, 1 and 1.6 mg silica m^{-3} years were 0.032, 0.3 and 2.4%, respectively. Thus, even using conservative assumptions for example ambient environment not markedly different from mining environment with respect to duration of exposures, peak exposures, particle character and size distribution silica content in PM_{10} at 10%; cumulative risk when PM_{10} levels are maintained at or below the 50 are estimated to be close to 0%. This assessment may not be relevant for people with other respiratory ailments and for dusty environments containing more than 10% silica.

Disclaimer:—The views expressed in this paper are those of the author(s) and do not necessarily reflect the views or policies of the U.S. Environmental Protection Agency. The U.S. Government has the right to retain a non exclusive royalty-free license in and to any copyright covering this article.

REFERENCES

Checkoway, H. and Rice, C. H. (1992) Time-weighted averages, peaks, and other indices of exposure in occupational epidemiology. Am. J. ind. Med. 21, 25–33.

Davis, B. L., Johnson, L. R., Stevens, R. K., Courtney, W. J. and Safriet, D. W. (1984) The quartz content and elemental composition of aerosols from selected sites of the EPA inhalable particulate network. Atmos. Environ. 18, 771–782.

Hnizdo, E., Murray, J., Sluis-Cremer, G. K. and Thomas, R. G. (1993) Correlation between radiological & pathological diagnosis of silicosos: an autopsy population based study. Am. J. ind. Med. 24, 427–445.

Hnizdo, E. and Sluis-Cremer, G. K. (1993) Risk of silicosis in a cohort of white South African gold miners. Am. J. ind. Med. 24, 447–457.

Muir, D. C. F. (1991) Correction in cumulative risk in silicosis exposure assessment. Am. J. ind. Med. 19, 555.

Muir, D. C. F., Shannon, H. S., Julian, J. A., Verma, D. K., Sebestyen, A. and Bernholz, C. D. (1989a) Silica exposure and silicosis among Ontario hardrock miners: I. methodology. Am. J. ind. Med. 16, 5–11.

Muir, D. C. F., Julian, J. A., Shannon, H. S., Verma, D. K., Sebestyen, A. and Bernholz, C. D.

(1989b) Silica exposure and silicosis among Ontario hardrock miners: III. analysis and risk estimates. *Am. J. ind. Med.* **16**, 29–43.

Ng, T. P. and Chan, S. L. (1994) Quantitative relations between silica exposure and development of radiological small opacities in granite workers. In *Inhaled particles VII: Proceedings of an International Symposium* (Edited by J. Dodgson and R. I. McCallum), September 1991, Edinburgh, U.K. *Ann. occup. Hyg.* **38** (Suppl. 1), 857–863.

Rice, C. H., Harris, R. L., Jr, Checkoway, H. and Symons, M. J. (1986) Dose-response relationships for silicosis from a case-control study of N. Carolina dusty trades workers. In *Silica, Silicosis and Cancer: Controversy in Occupational Medicine* (Edited by D. F. Goldsmith, D. M. Winn and C. M. Shy), pp. 77–86. Praeger Publishers, New York.

Sanchez-Roman, J., Wichmann, I., Salaberri, J., Varela, J. M. and Nunez-Roldan, A. (1993) Multiple clinical and biological autoimmune manifestations in 50 workers after occupational exposure to silica. *Ann. Rheum. Dis.* **52**, 534–538.

Shoemaker, D. A., Petty, J. R., Ramsey, D. M., McLaurin, J. L., Khan, A., Teass, A. W., Castranova, V., Pailes, W. H., Dalal, N. S., Miles, P. R., Bowman, L., Leonard, S., Shumaker, J., Vallyathan, V. and Pack, D. (1995) Particle activity and *in vivo* pulmonary response to freshly milled and aged alpha-quartz. *Scan J. Work Environ. Health* **21** (suppl. 2), 15–18.

Steenland, K. and Brown, D. P. (1995) Mortality study of goldminers exposed to silica and nonasbesti-form amphibole minerals: an update. *Am. J. ind. Med.* **27**, 217–229.

Steenland, K. and Goldsmith, D. F. (1995) Silica exposure and autoimmune diseases. *Am. J. ind. Med.* **28**, 603–8.

Theriault, G. P., Peters, J. M. and Jonson, W. M. (1974) Pulmonary function and roentgenographic changes in granite dust exposure. *Arch. Environ. Health* **28**, 23–27.

U.S. EPA. (1994) Methods for derivation of inhalation reference concentrations and application of inhalation dosimetry [draft final]. Research Triangle Park, NC, Office of Health and Environmental Assessment, Environmental Criteria and Assessment Office, EPA/600/8–88/066F.

U.S. EPA. (1996) Ambient levels and noncancer health effects of inhaled crystalline and amorphous silica [draft final]. Research Triangle Park, NC, Office of Health and Environmental Assessment, Environmental Criteria and Assessment Office, report no. EPA/600/R-95/115.

U.S. Federal Register (1996) National Ambient Air Quality Standards for Particulate Matter; Proposed Role. F.R. (December 13) **61 (241)**, 65638–65714.

Verma, D. K., Sebestyen, A., Julian, J. A., Muir, D. C. F., Schmidt, H., Bernholz, C. D. and Shannon, H. S. (1989) Silica exposure and silicosis among Ontario hardrock miners: II. exposure estimates. *Am. J. ind. Med.* **16**, 13–28.

Verma, D. K., Sebestyen, A., Julian, J. A., Muir, D. C. F., Shaw, D. S. and MacDougall, R. (1994) Particle size distribution of an aerosol and its sub-fractions. *Ann. occup. Hyg.* **38**, 45–58.

 Pergamon

Ann. occup. Hyg., Vol. 41, Supplement 1, pp. 454–458, 1997
© 1997 British Occupational Hygiene Society
Published by Elsevier Science Ltd. All rights reserved
Printed in Great Britain
0003–4878/97 $17.00 + 0.00
Inhaled Particles VIII

PII: S0003–4878(96)00096–8

EVALUATION OF SOLUBLE TUMOUR NECROSIS FACTOR RECEPTOR IN RESPONSE TO SILICA CONTAINING DUST AND ASBESTOS FIBRES EXPOSURE

W. Hadnagy and H. Idel

Institute of Hygiene, Heinrich-Heine-University Düsseldorf, P.O. Box 10 1007,
D-40001 Düsseldorf, Germany

INTRODUCTION

Pneumoconiosis is a disease caused by chronic exposure to mineral dust such as coal mine dust, silica and asbestos. Current concepts in the pathogenesis of pneumoconiosis suggest that the alveolar macrophage plays a central role in the outcome of this disease. Upon stimulation with mineral dust the macrophages secrete a number of fibrogenic cytokines among them tumor necrosis factor α (TNFα) which is considered to play a key role in mineral dust induced pneumoconiosis (Dubois *et al.*, 1989; Donaldson, 1992; Donaldson *et al.*, 1992; Vanhée *et al.*, 1995).

Recently, naturally occurring inhibitors of TNFα activity have been identified showing high affinity binding to TNFα (Brockhaus *et al.*, 1990). These inhibitors have been characterised as extracellular domains of the TNFα receptors types I and II, respectively. In response to inflammatory stimuli that cause enhanced TNFα levels, they are shed from the cell surface and appear as extracellular receptors in body fluids (Porteu and Nathan, 1990). Since such extracellular soluble receptors are suggested to intervene in the regulatory mechanism for modulation of excessive TNFα activity (van Zee *et al.*, 1992; Spinas *et al.*, 1992), enhanced serum TNF-receptor levels may reflect overexpression of TNFα in response to chronic exposure of mineral dust.

In order to study the role of soluble TNF receptor as an individual factor in pneumoconiosis development we investigated serum levels of sTNFRII in groups exposed to silica containing dust or asbestos with normal chest radiographics and in groups of patients with silicosis and asbestosis.

MATERIALS AND METHODS

As a part of a preliminary orientation study on the relevance of soluble tumour necrosis factor receptor in pneumoconiosis development caused by mineral dust exposure, the following groups had been considered in this investigation: 51 nonexposed controls (randomly selected male volunteers), 33 miners with long-term exposure to coal dust (> 30 years) with normal chest radiographics, 22 asbestos exposed workers without radiological signs of asbestosis, 55 patients with silicosis occupationally exposed to silica containing dust other than coal mine dust

Table 1. Characteristics of study groups and serum levels of soluble tumor necrosis factor receptor (sTNFRII)

Study groups	n	Age (years)	Exposure duration (years)	sTNFRII (pgml^{-1})
Nonexposed controls	51	45.6 ± 11.5 (25–75)	—	1900 ± 114 (956–5000)
Coal dust exposed miners	33	> 45	> 30	2655 ± 156*** (1288–5000)
Patients with silicosis	55	64.5 ± 8.8 (46–83)	18.0 ± 11.0 (3–44)	3105 ± 174*** (424–5000)
Asbestos exposed workers	22	52.6 ± 12.2 (24–77)	17.6 ± 10.9 (3–37)	2261 ± 154* (1333–4014)
Patients with asbestosis	20	58.0 ± 7.2 (41–71)	17.2 ± 10.9 (3–35)	2365 ± 132** (1352–3381)

Values for age and exposure duration are given as means ± SD (range); values for sTNFRII represent means ± SE (range). Statistically increased as compared to nonexposed controls: *$P <$ 0.05; **$P < 0.001$; ***$P < 0.0001$ (Wilcoxon test).

(miners, quarrymen, stone masons, sandblaster, ceramic workers, glass-cutters, casters), 20 patients with asbestosis.

For determination of soluble tumour necrosis factor receptor (sTNFRII), blood samples were taken by venipuncture. Sera were obtained by centrifugation followed by immediate storage at $-20°C$. Quantification of sTNFRII in serum samples was carried out by a commercially available sandwich ELISA (Quantikine, R&D Systems) with a minimum detectable limit of 5 pgml^{-1}.

RESULTS

Serum levels of sTNFRII were significantly increased for all investigated study groups in comparison to the nonexposed controls (Table 1). The highest mean concentration was found for the group of patients with silicosis. According to the frequency distribution (Fig. 1), 66.7% of the nonexposed controls show sTNFRII values between 1000 and 2000 pgml^{-1} serum. This is in good agreement with the normal levels of healthy donors as described elsewhere (Aderka et al., 1992). In contrast, coal dust exposed miners, asbestos exposed workers as well as patients with silicosis or asbestosis show a distinct shift to higher values. These are predominately concentrated in the range of 2000–3000 pgml^{-1} serum with the exception of silicotic patients showing in more than 40% of cases sTNFRII values higher than 3000 pgml^{-1} serum. According to individual serum levels of sTNFRII, 23 out of 55 silicotic patients showed abnormally high values for sTNFRII above the serum normal level reaching in about the half of cases values of 5000 pgml^{-1} serum (Fig. 2). In general, individual sTNFRII values above the upper serum normal level of asbestos exposed workers and patients with asbestosis are lower than those of coal dust exposed miners. Odds ratios (Table 2) related to sTNFRII values above the upper serum normal levels revealed an increased prevalence for elevated sTNFRII values for all study groups in comparison to nonexposed

sTNFRII values in pg/ml serum: 1)<1000   2)1000 - 2000   3)2001 - 3000   4) >3000

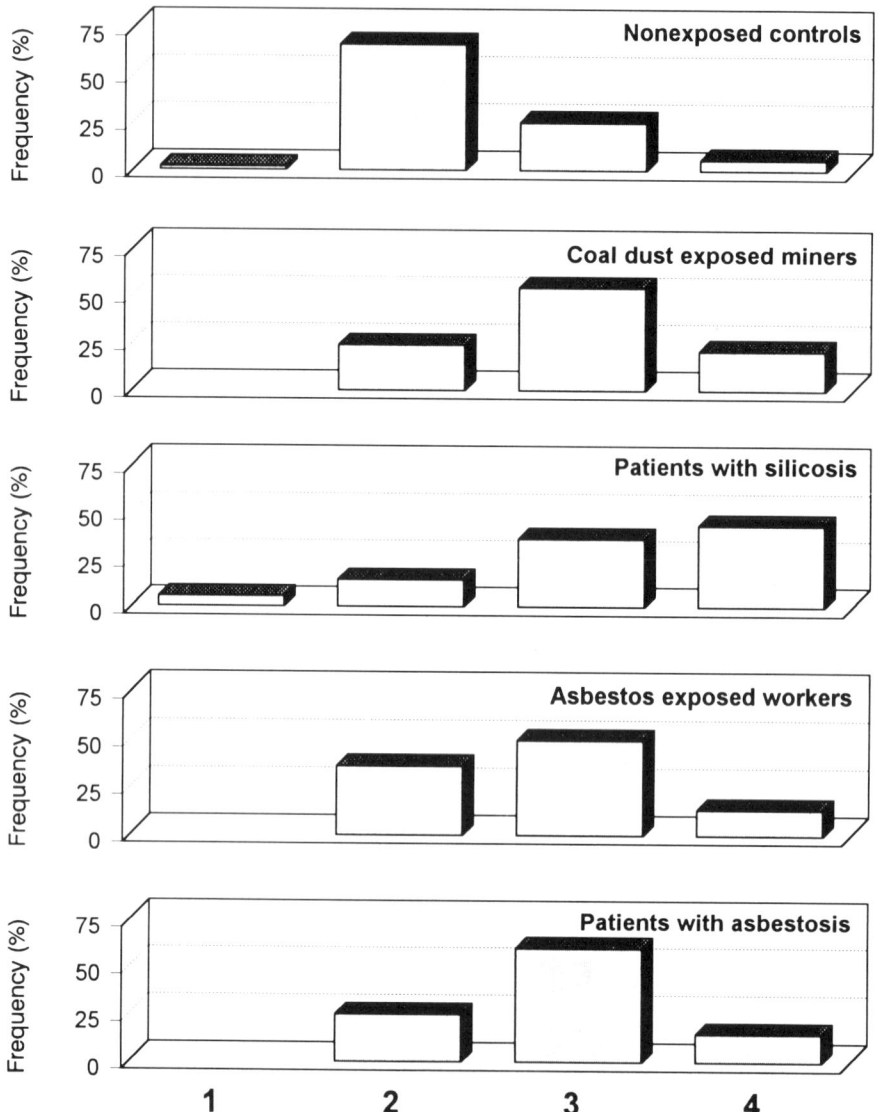

Fig. 1. Frequency distributions of sTNFRII values detected in serum of study groups.

controls. The highest discrimination was found for the group of silicotic patients giving a statistically significant increased odds ratio of 11.5.

DISCUSSION

As already described by Hadnagy and Idel (1995) sTNFRII serum levels of patients with silicosis occupationally exposed to silica containing dust are signifi-

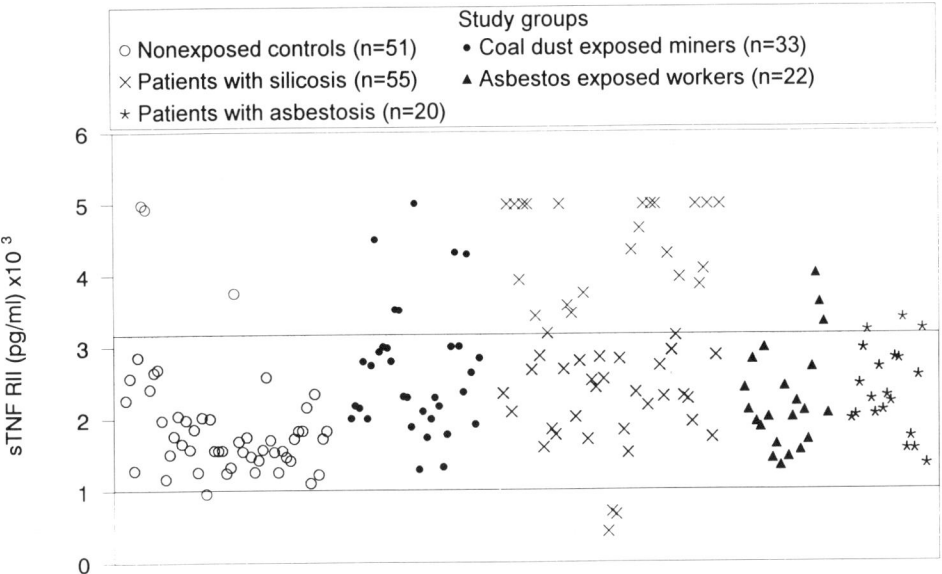

Fig. 2. Individual sTNFRII values detected in serum of study groups.

cantly increased as compared to nonexposed controls. In addition it could be shown that miners with long-term exposure to coal dust give rise to significantly elevated sTNFRII serum levels, but to a lower extent than found for silicotic patients. Based on these findings it is suggested that also in coal workers' pneumoconiosis elevated sTNFRII levels should be detectable. This assumption is supported by a recent study describing significantly increased plasma levels of sTNFRII in simple coal workers pneumoconiosis (Schins and Borm, 1995). Since there was no evidence that age, smoking, exposure duration or medication is related to sTNFRII (Hadnagy and Idel, 1995; Schins and Borm, 1995), it is considered that sTNFRII could be an important factor in TNF-mediated pneumoconiosis development. Furthermore, based on the findings by Hadnagy and Idel (1991) that silicotic patients with large opacities tend to higher sTNFRII levels, the implication of sTNFRII in pneumoconiosis progression could be suggested. Although asbestos exposed workers showed also increased sTNFRII serum levels comparable with those of coal dust exposed miners, no further elevation was found for patients with

Table 2. Prevalence of elevated levels of sTNFRII in study groups based on cases showing values above the upper serum normal level (> 3170 pgml^{-1}) and calculated odds ratios

Study groups	Prevalence (%)	Odds ratio	95% CI[a]	P
Nonexposed controls	5.9	1.0	—	—
Coal dust exposed miners	18.2	3.6	0.8–15.4	0.09
Patients with silicosis	41.8	11.5	3.2–41.5	0.0002*
Asbestos exposed workers	13.6	2.5	0.5–13.6	0.28
Patients with asbestosis	15.0	2.8	0.5–15.4	0.23

[a] Confidence interval; *Significantly different.

asbestosis. This difference of sTNFRII levels between silicosis and asbestosis could probably be attributed to coexisting inflammatory processes in silicosis such as chronic bronchitis contributing to enhanced sTNFRII levels. So far, it is considered that sTNFRII is an important factor indicating TNFα overexpression in response to chronic inhalation of inorganic dust. In view of pneumoconiosis development, however, the individual regulation of TNFα by the soluble TNF receptor could be of substantial significance. Therefore, the evaluation of TNFα in parallel to the soluble TNFα receptor seems to be of considerable value in the understanding of pneumoconiosis.

REFERENCES

Aderka, D., Engelmann, H., Shemer-Avni, Y., Hornik, V., Galil, A., Sarov, B. and Wallach, D. (1992) Variation in serum levels of the soluble receptors among healthy individuals. *Lymph. Cytokine Res.* **11**, 157–159.

Brockhaus, M., Schoenfeld, H. J., Schlaeger, E. J., Hunziker, W., Lesslauer, W. and Loetscher, H. (1990) Identification of two kinds of TNF receptors on human cell lines by monoclonal antibodies. *Proc. Natl. Acad. Sci. U.S.A.* **87**, 3127–3131.

Donaldson, K. (1992) *In vitro* correlates of the pathogenicity of a long fibre sample of amosite asbestos compared to a short fibre sample. In *Environmental Hygiene III* (Edited by N. H. Seemayer and W. Hadnagy), p. 223. Springer-Verlag, Heidelberg.

Donaldson, K., Brown, G. M., Brown, D. M., Slight, J. and Yang Li, X. (1992) Epithelial and extracellular matrix injury in quartz inflammed lung: role of the alveolar macrophage. *Environ. Health Perspect.* **97**, 221–224.

Dubois, C. M., Bisonette, E. and Rola-Pleszczynski, M. (1989) Asbestos fibers and silica particles stimulate rat alveolar macrophages to release tumor necrosis factor. *Am. Rev. resp. Dis.* **139**, 1257–1264.

Hadnagy, W. and Idel, H. (1995) Biologische Marker als Indikatoren für die durch quarzhaltige Stäube induzierte Lungenfibrose: Löslicher Tumornekrosefaktorrezeptor (sTNF RII). In *Dokumentationsband über die Verhandlungen der Deutschen Gesellschaft für Arbeitsmedizin und Umweltmedizin e.V.* (Edited by R. Schiele, B. Beyer and A. Petrovich), p. 301. Druckerei Rindt, Fulda.

Porteu, F. and Nathan, C. (1990) Shedding of tumor necrosis factor receptors by actived human neutrophils. *J. exp. Med.* **172**, 599–607.

Schins, R. P. F. and Borm, P. J. A. (1995) Plasma levels of soluble tumor necrosis factor receptor are increased in coal miners with pneumoconiosis. *Eur. Resp. J.* **8**, 1658–1663.

Spinas, G. A., Keller, U. and Brockhaus, M. (1992) Release of soluble receptors for tumor necrosis factor (TNF) in relation to circulating TNF during experimental endotoxinemia. *J. clin. Invest.* **90**, 533–536.

Vanhée, D., Gosset, P., Boitelle, A., Wallaert, B. and Tonnel, A. B. (1995) Cytokines and cytokine network in silicosis and coal workers pneumoconiosis. *Eur. resp. J.* **8**, 834–842.

Van Zee, K. J., Kohno, T., Fischer, E., Rock, C. S., Moldawer, L. L. and Lowry, S. F. (1992) Tumor necrosis factor soluble receptors circulate during experimental and clinical inflammation and can protect against excessive tumor necrosis factor α *in vitro* and *in vivo*. *Proc. Natl. Acad. Sci. U.S.A.* **89**, 4845–4849.

Pergamon

Ann. occup. Hyg., Vol. 41, Supplement 1, pp. 459–464, 1997
© 1997 British Occupational Hygiene Society
Published by Elsevier Science Ltd. All rights reserved
Printed in Great Britain
0003–4878/97 $17.00 + 0.00
Inhaled Particles VIII

PII: S0003–4878(96)00097–X

SILICA-INDUCED MORPHOLOGICAL CHANGE SIMILAR TO APOPTOSIS IN BRONCHOALVEOLAR LAVAGE CELLS AND GRANULOMATOUS CELLS

H. Wang,* J. Leigh,† A. Bonin,† and M. Peters*

*University of Sydney; and †NIOHS, GPO Box 58, Sydney 2001, Australia

INTRODUCTION

In silica-induced pathogenesis, it has been demonstrated that cell number in lavage fluid significantly increases and granulomas gradually develop in silica-exposed animals. The cell number increase in lavage fluid of silica-exposed animals is related to neutrophil infiltration into the alveolar space and the granuloma formation may be related to a hyperplasia reaction. Since there is no evidence that alveolar neutrophils can return to the circulation and granulomas eventually evolute into noncellular structure, we surmise that there should be a mechanism to delete the infiltrated and hyperplastic cells. Although the life span of these cells may play a role in the cell deletion, the exact pathway of cell death needs to be clarified. Silica is a toxicant and can interact with cells to induce toxicants such as free radicals. These toxicants may injure or induce cells to die and provide an alternative cell death pathway. We hypothesised that apoptosis (programmed cell death) might play a role in the silica induced inflammatory response, both acutely and chronically. We used a bronchoalveolar lavage method to obtain broncho-alveolar cells from silica-instilled rats to examine the existence of apoptosis in acute and chronic silica-induced effects.

MATERIALS AND METHODS

Animals

Specific pathogen-free male Wistar rats (200–230 g) were obtained from The University of New South Wales (Little Bay, NSW) breeding facility. The rats were held in cages (five rats per cage) and food and water could be freely accessed. The lighting in the animal house was set to a 12/12 h on/off cycle.

Experimental design

Twenty rats were randomly distributed into four intratracheal instillation treatment groups (five in each group): (1) control 0.5 ml saline; (2) 2.5 mg silica (Min-U-Sil 5 crystalline silica; Silica Corp., Berkeley Springs, West Virginia) suspended in 0.5 ml saline; (3) 7.5 mg silica in 0.5 ml saline; (4) 22.5 mg silica in 0.5 ml saline.

Intratracheal instillation

Rats were anesthetized with intraperitoneal injection of a mixture of ketamine

(100 mg kg^{-1}) and xylazine (3.3 mg kg^{-1}, Sigma Chemical Company, Sydney NSW, Australia) dissolved in saline. Intratracheal instillation was performed by a procedure of tracheal exposure. After shaving, the skin in the ventral aspect of the neck was incised in the midline. Through blunt dissection, the trachea was exposed. Using a 1 ml disposable syringe with a #26 needle, 0.5 ml of a dust suspension or saline was injected. The incision was sutured carefully with interrupted silk sutures immediately after the injection and the rat was kept in a 30°C incubator until it regained consciousness.

Bronchoalveolar lavage

Ten days after the instillation, the rats were anesthetized with 75 mg kg^{-1} of pentobarbital by intraperitoneal injection. A laparotomy was performed and the abdominal aorta was exposed. The rats were killed by transection of the aorta. Bronchoalveolar lavage was performed with two aliquots of 5 ml phosphate buffered saline (PBS) after *in situ* perfusion of the lung. The recovered fluid volume was recorded.

Histology

One hundred microlitres of the lavage fluid was placed on a slide by Cytospincentrifugation immediately after the lavage. At least two slides were made for each rat. The slides were stained with Diff-Quik (Lab Aids Pty Ltd, Sydney Australia).

Total cell number and LDH activity

The total cell number was counted with a haemocytometer and the remaining fluid was centrifuged (3000 rpm, 10 min). The supernatant was collected to measure LDH activity by a kit method (Trace, Sydney Australia).

Differential counting

The slides were read under oil immersion (× 1000). Five hundred leukocytes were counted to determine the frequency of different types of cells by their morphology.

Counting of apoptotic cells

A minimum of 1000 leukocytes were counted for the occurrence of cells with apoptotic features. Apoptotic features included formation of condensed chromatin bodies with sharp edges and convolution of the cell surface.

Extraction of genomic DNA from BAL cells and electrophoresis

Extraction of genomic DNA of the cells in lavage fluid was conducted by the classical procedure. Briefly, the obtained alveolar cell pellet was fixed by 70% alcohol immediately after lavage and centrifugation. The fixed cells were treated by lysis buffer and nuclear lysis buffer to obtain DNA. Agarose gel electrophoresis was used to examine if there was any ladder development during the electrophoresis.

Histology of the lung

A separate group of rats were intratracheally instilled with 12.5 mg of Min-U-Sil 5 silica to obtain tissues for histopathological observation. The rats were sacrificed

Table 1. LDH activity and total cell number in lavage fluid of rats (mean ± SEM)

Group	Number of rats	LDH(U/L)	Total cell ($\times 10^5$ ml^{-1})
saline	5	28.4 ± 12.3	1.528 ± 0.083
2.5 mg	5	94.0 ± 21.2	4.176 ± 0.543
7.5 mg	5	147.6 ± 19.1	10.176 ± 1.289
22.5 mg	5	218.6 ± 25.4	12.880 ± 1.756

at 56 days by intraperitoneal injection of pentobarbitol (75 mg kg^{-1}) and transection of the abdominal aorta. The lung tissue was fixed by inflation with 10% formalin and sectioned after embedding in paraffin. The sections were stained with hematoxylin and eosin. Under oil immersion, the lung sections were examined to identify if any apoptotic cells existed in silica-induced granulomas in lung tissue.

Statistical analysis

An unpaired 2 sided *t*-test (Welch *t*-test if applicable) was used to compare the means of various indices measured.

RESULTS

Inflammatory reaction

In this experiment, pulmonary inflammatory reaction was measured by the total cell number and LDH activity in lavage fluid as well as the cell types of inflammatory cells. The recovered volume of lavage fluid was between 9.0 and 9.7 ml and no significant difference could be detected among the four groups in the recovered volume. LDH activity and total cell number in lavage fluid of the three silica-instilled groups were significantly higher than that of the control group (Table 1).

Leukocyte apoptosis

Apoptotic leukocytes were extremely few in the saline-instilled group. In silica-instilled groups, however, typical nuclear changes of apoptosis in BAL cells were identified. The number of apoptotic cells in BAL fluid increased with dose (Fig. 1). There were a small number of apoptotic cells which could be identified to be macrophages by their morphological characteristics, but the great majority of apoptotic cells appeared to be neutrophils. Macrophage engulfment of apoptotic cells could also be identified in this experiment and all of these findings could still be identified in the rats sacrificed 56 days after silica instillation.

Changes in neutrophil number, macrophage number and apoptotic proportion with dose

The number of neutrophils increased dramatically in silica-instilled groups compared with the saline group. Among silica-instilled groups, the number of neutrophils increased with dosage initially and the response could be saturated when the dose reached 7.5–22.5 mg [Fig. 2(a)]. The number of macrophages, in contrast to the neutrophils, decreased in the 2.5 and 7.5 mg groups but significantly increased in the 22.5 mg group compared to the saline control group [Fig. 2(b)].

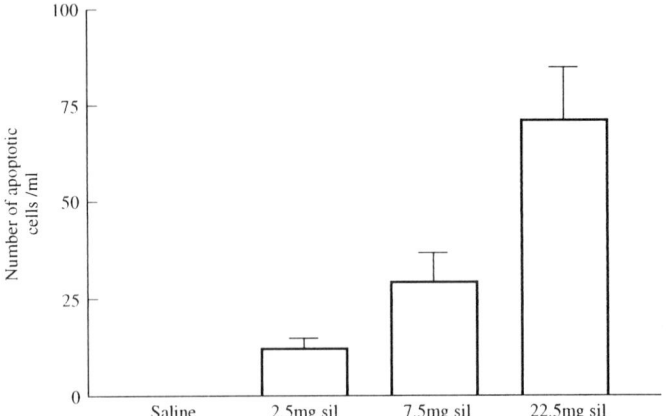

Fig. 1. Apoptotic cell number in bronchoalveolar lavage fluid (means $\pm$ SEM., $\times$ 10^3 ml^{-1}).

Interestingly, the proportion of apoptotic cells changed with dose in the same pattern as the number of macrophages [Fig. 2(c)].

Electrophoresis of DNA from bronchoalveolar lavage cells

A typical ladder was obtained in the electrophoresis of the extracted genomic DNA from silica-instilled rats. In saline-instilled rats, however, only one band could be seen parallelling the main bands of silica-instilled rats and no ladder could be detected.

Apoptosis in lung tissue

Fifty-six days after silica instillation, granulomas could be identified in lung tissue. The cell composition of the granuloma was mainly macrophage-like cells. Apoptotic cells could be identified in the granulomas.

DISCUSSION

We studied an animal model of silica-induced inflammation that was evidenced by influx of neutrophils into the alveolar space as well as by the elevation of LDH activity and increase of total cell numbers in BAL fluid. We have demonstrated that apoptosis occurs in BAL cells of silica-instilled rats both morphologically and biochemically. The number of the apoptotic cells in BAL fluid was clearly related to silica dosage. Since silica-induced increase in cell number and LDH activity in BAL fluid was also dose-related, the apoptosis might also be an aspect of silica-induced inflammation.

To our knowledge, this is the first documentation of apoptosis in silica-induced effects. Because cell death frequently occurs in silica-induced effects and is considered to be the outcome of silica toxicity, it is significant to note that the cell death can proceed via not only necrosis but also apoptosis. In addition, the apoptosis in our experiment could be detected in 10 and even 56 days after silica administration, indicating that apoptosis can also occur in chronic inflammation. Since silica-induced chronic inflammation poorly resolves and evolutes into fibrosis, the role of apoptosis in the poor resolution and gradual evolution is worthy of

(a)

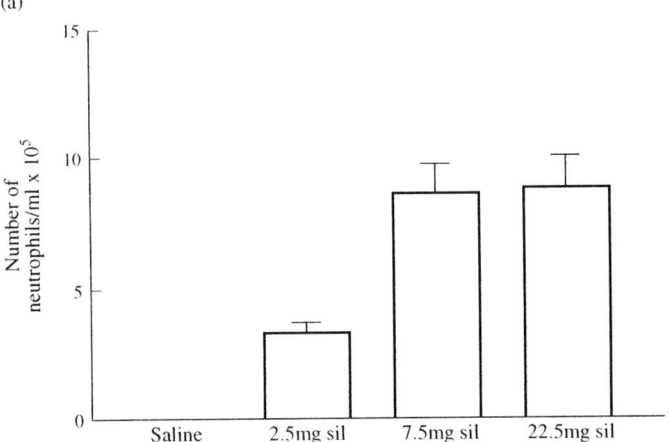

(b)

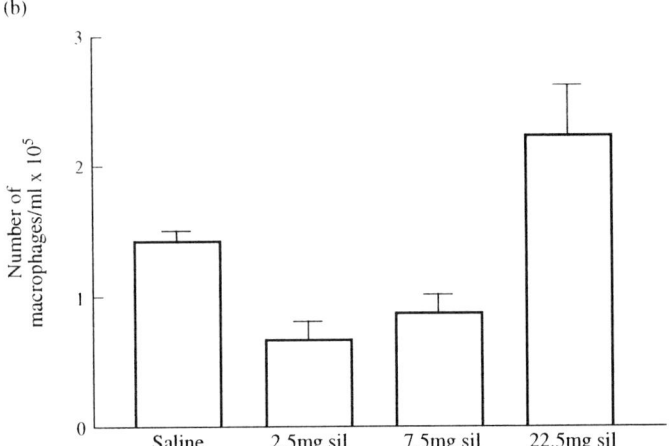

(c)

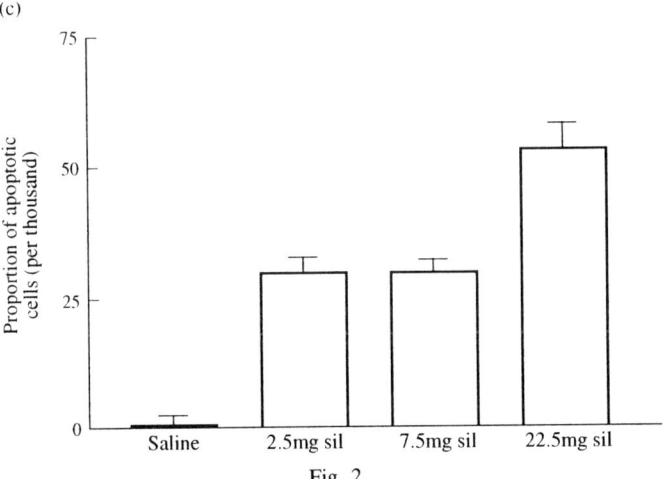

Fig. 2.

clarification. Notably, the two cytokines, tumour necrosis factor-α and transforming growth factor-β, which play important roles in silica-induced fibrosis (Piguet *et al.*, 1990; Williams *et al.*, 1993; Absher *et al.*, 1993), can also induce apoptosis in different type of cells (Pierce *et al.*, 1991; Sarin *et al.*, 1995; Bursch *et al.*, 1993). TNF can also induce apoptosis in neutrophils (Tsuchida *et al.*, 1995). TNF and TGF may regulate the silica-induced response via induction of apoptosis.

Engulfment of apoptotic cells was also noted in our study, consistent with the findings of Cox *et al.* (1995). The apoptotic cells, most of them neutrophils, may be deleted through this mechanism, maintaining a balance against the influx of circulating neutrophils. Indeed, in the present study, we noted interesting quantitative relationships between neutrophil number, apoptotic cell proportion and macrophage number. The neutrophil numbers, initially increasing with the increase of silica dose, can be saturated with the further increase of silica dose, whereas the macrophage number decreased initially with the silica administration but increased with dose increase, paralleling the proportion of apoptotic cells. This finding suggests that the increase in proportion of apoptotic cells may regulate the macrophage recruitment to engulf and delete the apoptotic neutrophils, maintaining a certain level of intra-alveolar neutrophils which is the main component of alveolar cells in inflammatory conditions. The finding also indicates that apoptosis and subsequent engulfment by macrophages is consistent with a homeostatic function.

In summary, we demonstrated nuclear damage morphologically and biochemically similar to apoptosis in silica-induced inflammation in the present study. This apoptotic change was related to silica dose. There was also evidence that the apoptotic cells were engulfed by macrophages. The apoptotic change of inflammatory cells and subsequent engulfment by macrophages may play an important role in the evolution of silica-induced effects and are consistent with a homeostatic function.

REFERENCES

Absher, M., Sjostrand, M., Baldor, L. C., Hemenway, D. R. and Kelley, J. (1993) Patterns of secretion of transforming growth factor-α (TGF-α) in experimental silicosis. Acute and subacute effects of cristobalite exposure in the rat. *Reg. Immunol.* **5**, 225–231.

Bursch, W., Oberhammer, F., Jirtle, R. L., Askari, M., Sedivy, R., Grasl, K. B., Purchio, A. F. and Schulte, H. R. (1993) Transforming growth factor-beta 1 as a signal for induction of cell death by apoptosis. *Br. J. Cancer* **67**, 531–536.

Cox, G., Crossley, J. and Xing, Z. (1995) Macrophage engulfment of apoptotic neutrophils contributes to the resolution of acute pulmonary inflammation in vivo. *Am. J. respir. Cell Mol. Biol.* **12**, 232–237.

Pierce, G. B., Parchment, R. E. and Lewellyn, A. L. (1991) Hydrogen peroxide as a mediator of programmed cell death in the blastocyst. *Differentiation* **46**, 181–186.

Piguet, P. F., Collart, M. A., Grau, G. E., Sappino, A. P. and Vassalli, P. (1990) Requirement of tumour necrosis factor for development of silica-induced pulmonary fibrosis. *Nature* **344**, 245–247.

Sarin, A., Conan, C. M. and Henkart, P. A. (1995) Cytotoxic effect of TNF and lymphotoxin on T lymphoblasts. *J. Immunol.* **155**, 3716–3718.

Tsuchida, H., Takeda, Y., Takei, H., Shinzawa, H., Takahashi, T. and Sendo, F. (1995) In vitro regulation of rat neutrophil apoptosis occurring spontaneously or induced with TNF-alpha or cycloheximide. *J. Immunol.* **154**, 2403–2412.

Williams, A. O., Flanders, K. C. and Saffiotti, U. (1993) Immunohistochemical localization of transforming growth factor-beta 1 in rats with experimental silicosis, alveolar type II hyperplasia, and lung cancer. *Am. J. Pathol.* **142**, 1831–1840.

 Pergamon

Ann. occup. Hyg., Vol. 41, Supplement 1, pp. 465–470, 1997
© 1997 Published by Elsevier Science Ltd on behalf of BOHS
Printed in Great Britain. All rights reserved
0003–4878/97 $17.00 + 0.00
Inhaled Particles VIII

PII: S0003–4878(96)00098–1

STRAIN AND SPECIES VARIATIONS IN PULMONARY RESPONSES TO INHALED MIN-U-SIL CRYSTALLINE SILICA PARTICLES IN MICE

D. B. Warheit, K. Kellar, T. McHugh, S. H. Gavett and M. A. Hartsky

DuPont Haskell Laboratory, P.O. Box 50, Elkton Rd, Newark, DE 19714–0050, U.S.A.

INTRODUCTION

Numerous mouse and rat strains are used for inhalation testing of pulmonary toxicants. Whether the pulmonary responses of various strains from a particular species to an inhaled toxicant are similar has not been systematically investigated. This acute inhalation study was designed to compare the acute pulmonary responses to inhaled silica in one common and one uncommon strain of mouse, namely the C57 black and the beige (neutrophil-defective) mouse; and to compare, in general, the pulmonary responses to acute silica exposure in mice to those recently studied in rats (Gavett *et al.*, 1992). In the current study, black and beige mice were exposed nose-only to Min-U-Sil crystalline silica in the same inhalation chambers for 5 days at 100 or 250 mg m^{-3}. Following the exposure period, cells and fluids from sham and silica-exposed animals were recovered by bronchoalveolar lavage (BAL). Granulocytes, as well as lactate dehydrogenase (LDH), N-acetyl-glucomsaminidase (NAG) and protein values were measured in BAL fluids at several postexposure time periods. In addition, the lungs of exposed mice were processed for BrdU pulmonary cell labeling, histopathological and morphometric analysis. The results showed that a 5 day exposure to silica produced a sustained, but low-level inflammatory response, measured up through 6 months postexposure in the lungs of black and beige mice. Cell labelling of terminal airway and lung parenchymal cells in beige and black mice were enhanced over corresponding unexposed controls. The results demonstrated minor differences in the pulmonary inflammatory or cell proliferative responses between the two strains of silica-exposed mice. Preliminary morphometric analysis of pulmonary lesions among silica-exposed beige and black mice did not show a dose response relationship, however, the lesions were similar in nature and in distribution at 6 months postexposure.

In general, the mice demonstrated a mild response to silica exposure when compared to the two strains of rats. We had expected to observe a reduced pulmonary response to inhaled silica particles in the neutrophil-defective beige mouse relative to the normal C57 black mouse. However, the species effect (i.e. reduced responsiveness in mice) seemed to take precedence over the differences in response among the two strains. Although it has been reported by others that exposure to silica may cause pulmonary fibrosis in mice, our results following acute exposures indicate that the two strains of mice tested here are much less responsive to the toxic effects of silica when compared to exposed rats.

METHODS

Male C57 black (C57BL/6J–bg^J/+) or beige (neutrophil-defective C57 BL/6J–bg^J/bg^J killer cell deficient) mice (Jackson Labs, Bar Harbor, Maine, U.S.A.) were exposed to Min-U-Sil silica particles for 5 days (6 h d^{-1}) at concentrations of 100 or 250 mg m^{-3}. The techniques for nose-only aerosol generation of particles have been previously reported (Warheit *et al.*, 1991). Following the completion of exposures, the lungs of sham and dust-exposed animals were perfused or lavaged immediately after, as well as 30, 90 or 180 days postexposure. Measures of cytotoxicity, and pulmonary inflammatory responses, cell proliferation, lung burden and morphometric parameters were assessed over an interval spanning 6 months postexposure using methods described elsewhere (Warheit *et al.*, 1984; 1992). BrdU pulmonary and airway immunostaining studies were implemented 3 days after the end of the 5 day exposure period. Groups of sham and dust-exposed mice were pulsed with an intraperitoneal injection of 5-bromo-2'-deoxyuridine (BrdU) dissolved in a 0.5 N sodium bicarbonate buffer solution at a dose of 100 mg kg^{-1} body weight as previously described (Warheit *et al.*, 1992). The animals were euthanised 2 h later by pentobarbital injection. Statistics were carried out using a Students t test on a Microsoft Excel software program.

RESULTS AND DISCUSSION

Five day exposures to Min-U-Sil at 100 or 250 mg m^{-3} produced a low-level, but sustained pulmonary inflammatory effects through a 6 month postexposure time period in black mice and this was greater than the response in beige mice (Fig. 1). Similarly, as evidenced by the levels of BAL fluid biomarkers, the biochemical response in black mice was substantially greater than in similarly-exposed beige mice. In this regard, BAL fluid values of lactate dehydrogenase (Fig. 2), protein (Fig. 3) and N-acetyl-glucosaminidase (Fig. 4) were consistently greater in the silica-exposed C57 black mice. Cell proliferation studies of terminal bronchiolar and pulmonary parenchymal surfaces 3 days after 5 day silica exposure demonstrated increased cell labeling indices in silica-exposed black and beige mice. However, the cell proliferative response appeared to be slightly increased in the silica-exposed beige mice when compared to the C57 black mice (Fig. 5). Preliminary morphometric analysis of parenchymal regions of silica-exposed black and beige mice at 6 months post 5-day exposure have thus far demonstrated increased thickened alveolar walls and foamy macrophages in silica-exposed beige and black mice (Fig. 6). These data do not demonstrate a dose response relationship and are preliminary.

The results of this study indicate that the pulmonary response to silica exposure in mice is mild when compared to rats. In earlier studies, we reported that short-term inhalation exposure to crystalline silica in rats resulted in a persistent pulmonary inflammatory response (Warheit *et al.*, 1991, 1995). Min-U-Sil exposure in mice produced a sustained, but low-level pulmonary inflammatory response, as evidenced by the low numbers of neutrophils recovered in BAL fluids. However, biomarker indices of pulmonary inflammation in silica-exposed black mice were greater than in similarly exposed beige mice. Cell proliferation studies of terminal bronchiolar and pulmonary parenchymal surfaces demonstrated that silica expo-

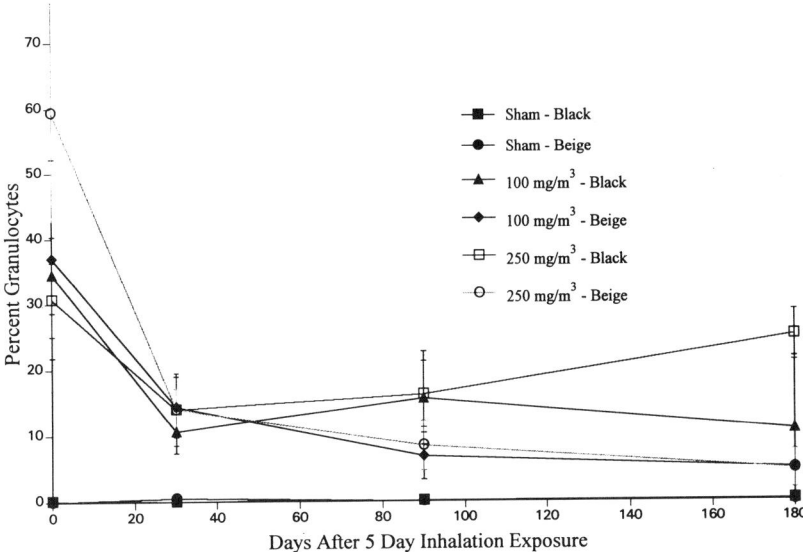

Fig. 1. Differential percentages of granulocytes recovered from bronchoalveolar (BAL) fluids of sham and black and beige mice. Exposure to silica produced a sustained but low-level pulmonary inflammatory response in both strains of mice as evidenced by a significant increase in neutrophils immediately after exposure. This increase persisted through 6 months postexposure and was substantially higher in the silica-exposed C57 black mice.

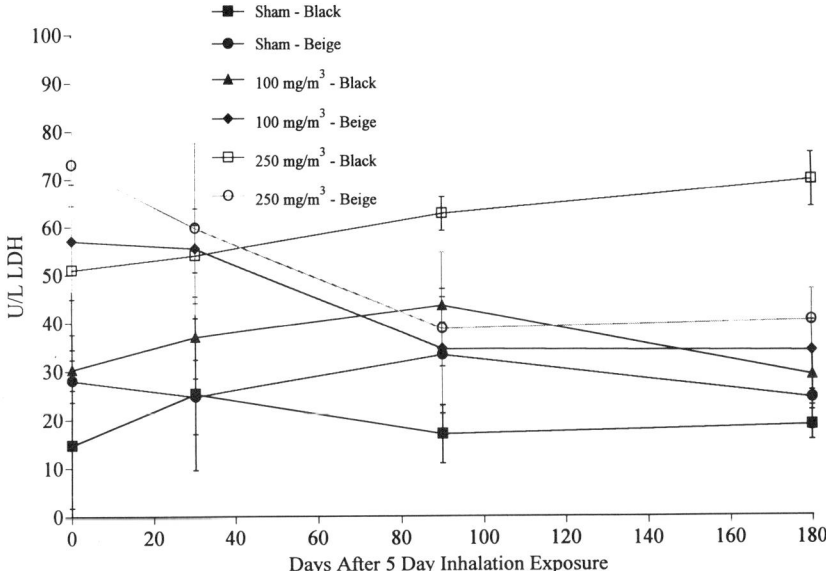

Fig. 2. Lactate dehydrogenase values in BAL fluids of silica-exposed black and beige mice. Values given are mean values ± SD. Five day exposures to silica produced an increased inflammatory response in the high dose group of the black mouse strain and seemed to persist through 6 months postexposure. The responses of the other silica-exposed black and beige groups were mild.

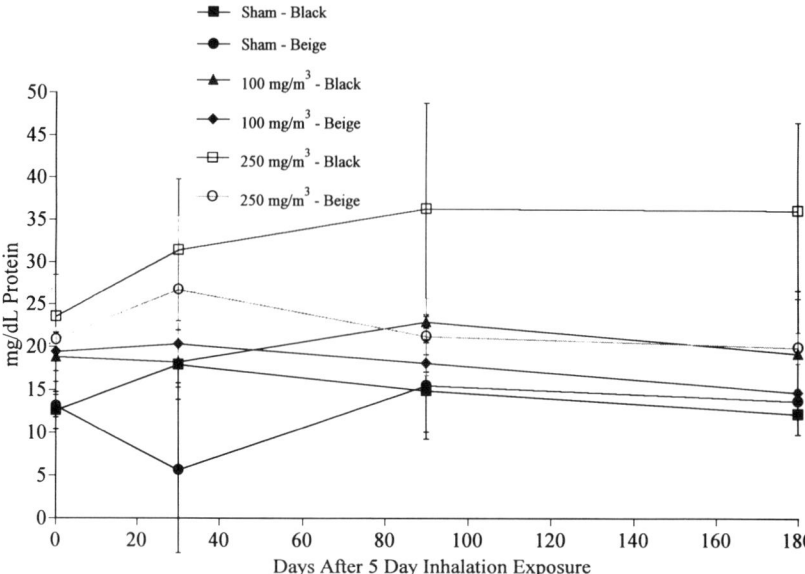

Fig. 3. Protein values in BAL fluids of silica-exposed, black and beige mice. Values given are mean values ± SD. Five day exposures to silica produced an increased inflammatory response in the high dose group of the black mouse strain and seemed to persist through 6 months postexposure. The responses of the other silica-exposed black and beige groups were mild.

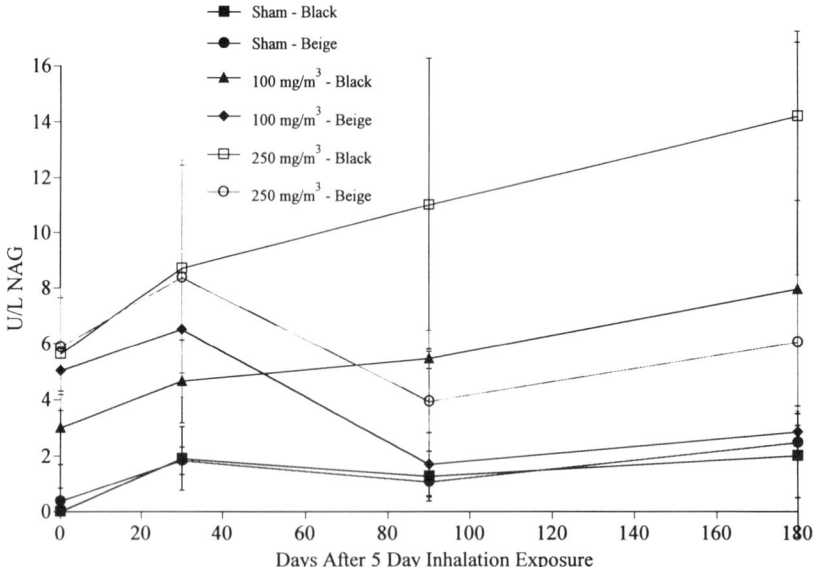

Fig. 4. N-acetyl-glucosaminidase (NAG) values in BAL fluids of silica-exposed, black and beige mice. Values given are mean values ± SD. Five day exposures to silica produced an increased inflammatory response in the high and low dose groups of the black mouse strain and seemed to persist through 6 months postexposure. The responses of the silica-exposed beige groups were not different from corresponding controls.

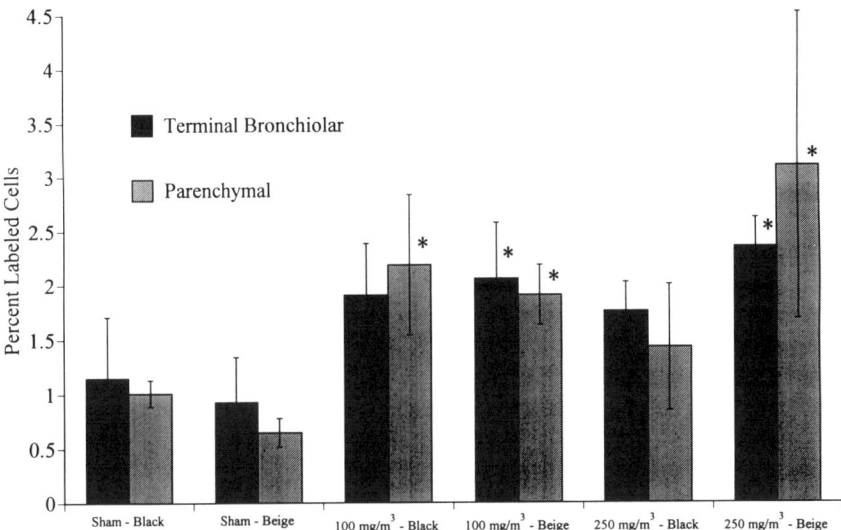

Fig. 5. BrdU labeling index of terminal bronchiolar and proximal lung parenchymal cells 3 days after a 5 day silica exposure in beige and black mice. Significant increases in BrdU immunostaining compared to controls were measured in parenchymal regions of silica-exposed low dose black and low and high dose beige mice. Significant increases were measured in labelling indices of terminal bronchiolar regions of silica-exposed low and high dose beige mice, while no significant airway effects were measured in silica-exposed, black mice ($P < 0.05$).

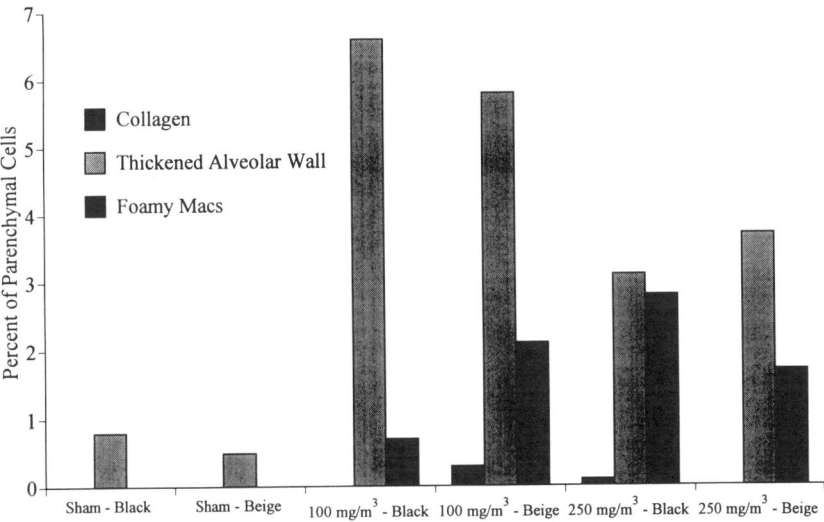

Fig. 6. Preliminary morphometric determinations of the lung parenchymal regions in silica-exposed beige and black mice 6 months after a 5 day inhalation exposure. Using a point counting method, light microscopy morphometry revealed increases relative to corresponding controls in thickened alveolar walls and foamy alveolar macrophages in both strains of silica-exposed mice. The results were not dose related. Although silica inhalation produced increased thickened alveolar wall and foamy macrophage endpoints in both strains, the histopathological responses were weak compared to those observed in similarly-exposed rats.

sure induced significant increases in cell labelling three days after exposure, with the beige mice demonstrating a greater cell proliferative response relative to black mice. In addition, preliminary morphometric evaluations thus far have demonstrated a substantial increase in thickened alveolar walls and foamy macrophages in silica-exposed mice relative to corresponding controls. However, the amounts of collagen were not significantly greater than controls and the histopathological analysis suggested that the response to silica in mice was significantly reduced when compared to silica-exposed rats.

These studies were undertaken (Gavett *et al.*, 1992) to determine the role of neutrophils in the pulmonary response to silica and to compare the silica-induced pulmonary effects in exposed rats and mice. In an earlier study, we reported that neutrophil-depleted rats exposed to silica showed no decrement in lung injury when compared to normal, silica-exposed animals. In this regard, in both groups of silica-exposed mice, numerous biomarkers of lung injury were increased significantly compared to BAL fluid measurements from corresponding sham-exposed groups of rats. Moreover, there were no differences between PMN-depleted and normal silica-exposed groups. As a consequence, our results suggest that recruitment of PMNs into the alveolar regions was not a necessary prerequisite for the observed increases in biochemical indicators of silica-induced acute lung injury (Gavett *et al.*, 1992). The current studies with silica were designed to extend upon our results reported in rats wherein we showed that neutrophils had little effect on biochemical indices of acute lung injury caused by silica. In this mouse study, we employed a neutrophil-defective mouse model (i.e. the beige mouse) and tested a similar hypothesis. The results reported here indicate that silica-exposed black mice demonstrated a greater biochemical pulmonary inflammatory response relative to the beige mouse, however, cell proliferative and morphological effects were similar for the two mouse strains. Perhaps more importantly, it appears that the two strains of mice tested here are much less responsive to the toxic effects of silica when compared to exposed rats.

REFERENCES

Gavett, S. H., Carakostas, M. C., Belcher, L. A. and Warheit, D. B. (1992) Effect of circulating neutrophil depletion on lung injury induced by inhaled silica particles. *J. Leuckocyte Biol.* **51**, 455–461.

Gavett, S. H., McHugh, T., Duespobl, K., Hartsky, M. A. and Warheit, D.B. (1992) Strain variations in pulmonary responses to inhaled crystalline silica particles in rats. *Toxicologist* 12, 225A.

Warheit, D. B., Carakostas, M. C., Hartsky, M. A. and Hansen, J. F. (1991) Development of a short-term inhalation bioassay to assess pulmonary toxicity of inhaled particles: comparisons of pulmonary responses to carbonyl iron and silica. *Toxicol. appl. Pharmacol.* **107**, 350–368.

Warheit, D. B., Chang, L. Y., Hill, L. H., Hook, G. E. R., Crapo, J. D. and Brody, A. R. (1984) Pulmonary macrophage accumulation and asbestos-induced lesions at sites of fiber deposition. *Am. Rev. respir. Dis.* **129**, 301–310.

Warheit, D. B., Kellar, K. A. and Hartsky, M. A. (1992) Pulmonary cellular effects in rats following aerosol exposures to ultrafine Kevlar aramid fibrils: evidence for biodegradability of inhaled fibrils. *Toxicol. appl. Pharmacol.* **116**, 225–239.

Warheit, D. B., McHugh, T. A. and Hartsky, M. A. (1995) Differential pulmonary responses in rats inhaling crystalline, colloidal or amorphous silica dusts. *Scan. J. Work, Environ. Health* **21** (suppl. 2), 19–21.

Pergamon

Ann. occup. Hyg., Vol. 41, Supplement 1, pp. 471–474, 1997
British Occupational Hygiene Society
Published by Elsevier Science Ltd
Printed in Great Britain
0003–4878/97 $17.00 + 0.00
Inhaled Particles VIII

PII: S0003–4878(96)00099–3

SILICA (CRYSTALLINE) AND ITS SELECTION AS AN OSHA PRIORITY SUBSTANCE FOR A COMPREHENSIVE STANDARD

L. D. Schuman

United States Occupational Safety and Health Administration, 200 Constitution Ave. NW,
RM N-3718, Washington, DC 20210, U.S.A.

INTRODUCTION

The mission of the Occupational Safety and Health Administration (OSHA), according to the Occupational Safety and Health Act of 1970, is "to assure as far as possible every working man and woman in the nation safe and healthful working conditions." OSHA has had substantial success over the past 25 years in fulfilling this mission. However, we recognise that many injuries, illnesses and deaths still occur as a result of hazards which are widely recognised and/or covered by existing standards, but which are nonetheless inadequately controlled. In order to deal with this problem, OSHA inaugurated the Priorities Planning Process in 1994. The objective of the Priorities Planning Process is to identify unmet needs and establish plans to address them.

METHODS

The Priorities Planning Process began with the selection of a committee composed of staff members from OSHA, the Mine Safety and Health Administration, the Environmental Protection Agency and the National Institute for Occupational Safety and Health (which is our sister research agency). The Committee reviewed available information on occupational fatalities, injuries and illnesses. The Committee also received input from representatives of labour, industry, professional and academic organisations and the public. These groups and agency staff nominated more than 125 hazards.

In order to select the priorities for action, four major criteria were applied to this list of 125 hazards: (1) the seriousness of the hazard; (2) the number of workers exposed or the magnitude of the risk; (3) the quality of available risk information; (4) the potential for risk reduction. Applying these criteria, the Committee selected 18 hazards as priorities for action. Five of these priorities were designated for rulemaking, i.e. comprehensive standards will be promulgated. The remaining priorities were designated for non-regulatory actions such as voluntary cooperative agreements between OSHA, labour and industry.

RESULTS

One of the 125 hazards nominated was crystalline silica. The following section will describe the results that were obtained during the process in which the selection criteria were applied to silica.

Silica exposure, an ancient hazard, remains a serious threat to nearly 2 million U.S. workers including more than 100 000 in high risk jobs, including sandblasters, foundry workers, stonecutters, rock drillers, quarry workers and tunnelers. Diseases associated with inhalation of silica-containing dusts include silicosis, chronic airways obstruction, bronchitis, tuberculosis and lung cancer.

Silicosis is a disabling, progressive and sometimes fatal disease involving scarring of the lung with resulting cough and shortness of breath. There are three types of silicosis. Acute silicosis, which occurs where exposure concentrations are the highest, can develop within a few weeks to 4 or 5 years after the initial exposure. Accelerated silicosis, which occurs when concentrations are high, develops 5–10 years after the first exposure. Chronic silicosis, which occurs as a result of exposure to relatively low silica concentrations, takes 10 or more years to develop (NIOSH, 1996).

The serious health hazard is indicated by continuing deaths from accelerated silicosis and silicotuberculosis in sandblasters, rock drillers and workers in other dusty trades. There are currently about 300 deaths reported per year from silicosis. However, the actual number of cases and the true risk is unknown due to inadequate case detection.

It is now clearly established that occupational exposure to silica causes lung cancer. Studies have demonstrated a statistically significant, dose-related increase in lung cancer in several occupational groups. Winter *et al.* (1990) observed that the lung cancer risk for pottery workers increased with estimated cumulative exposure at the low levels of silica found in potteries. Another study also found that the risk of lung cancer among pottery workers was related to exposure to silica, although the dose–response gradient was not significant (McLaughlin *et al.*, 1992). This study also analysed lung cancer risk in tin miners in China and found a significant trend of increasing risk of lung cancer with increasing cumulative respirable silica exposure. A significant dose–response relationship between death from lung cancer and silica dust particle–years has also been demonstrated for South African gold miners (Hnizdo and Sluis-Cremer, 1991). In this study a synergistic effect on lung cancer risk was found for silica exposure and smoking. Lung cancer risk among workers in the diatomaceous earth industry has been studied by Checkoway *et al.* (1993). Results showed increasing risk gradients for lung cancer with cumulative exposure to crystalline silica. The authors felt that this finding indicated a causal relationship. Several studies have demonstrated a relationship between the degree of silicosis disability and risk for lung cancer (Goldsmith, 1994). Since severity of silicosis reflects silica exposure, this may also indicate a dose–response relationship for silica exposure and lung cancer (Checkoway *et al.*, 1993).

Exposure studies indicate that some workers are still exposed to very high silica levels. OSHA currently has a permissible exposure limit for crystalline silica of (10 mg m^{-3}) ÷ (the percent of silica in the dust + 2)[respirable] and (30 mg m^{-3}) ÷ (the percent of silica in the dust + 2)[total dust]. A study by Freeman and Grossman (1995) examined data from OSHA-collected silica samples from 1982 to 1991. Respirable quartz was measured in 1655 inspections in 255 industries. In 52% of those industries respirable silica levels were below the OSHA PEL. In 48% of the industries the PEL for silica was exceeded. Among industries where more than 10 facilities were inspected, the most severe respirable silica exposures were found in fabricated structual metal; painting and paper hanging; nonresidential construction; shipbuilding and repair; masonry and other stone work; bridge, tunnel and elevated highway construction; metal coating; engraving and allied services; and special trades contractors.

Recent studies suggest that the current OSHA standard is insufficient to protect against silicosis. In one such study, Steenland and Brown (1995) estimated the risk of silicosis by cumulative exposure years in a cohort of 3330 gold miners. The miners worked at least 1 year underground from 1940 to 1965 (average 9 years) and were exposed to a median silica level of 0.05 mg m^{-3}. It was estimated that the risk of silicosis with a cumulative exposure of less than 0.5 mg m^{-3} years was under 1%. For those with a cumulative exposure of more than 4.0 mg m^{-3} years the risk of silicosis was 68–84%. It was estimated that, after adjustment for competing risks of death, a working life-time, 45 year exposure under the current OSHA PEL would lead to a lifetime risk of silicosis of 35–47%. The authors concluded that the current OSHA silica permissible exposure level is unacceptably high.

At the present time the OSHA standard for silica consists of only a permissible exposure limit. Many interested parties have urged OSHA to produce a comprehensive standard for crystalline silica. Comprehensive OSHA standards include many additional provisions in addition to the PEL. Some examples of comprehensive standards with which you may be familiar are asbestos, lead and cadmium. A comprehensive standard for silica would include provisions for product substitution, engineering controls, respiratory protection, medical screening and surveillance, and training and education. Under a comprehensive rulemaking OSHA would also do a new risk assessment and possibly propose a new PEL. The lack of a comprehensive silica standard containing these provisions has contributed to inadequate protection of workers.

DISCUSSION

OSHA used the information and data above to determine whether silica met the criteria to be selected as a priority for a comprehensive rulemaking. The first criterion was "seriousness of the hazard." It is clear that silica is a very serious hazard as indicated by continuing deaths from silicosis and the statistically significant increased risk of lung cancer in workers exposed to silica. The second criterion, "number of workers exposed or magnitude of risk" is certainly met by the fact that over two million workers are exposed with over 100 000 in high exposure/high risk occupations. The "quality of available risk information" (the third criterion) is very good with many studies providing excellent dose–response data for silica exposure and both silicosis and lung cancer. Finally, the "potential for risk reduction" is great as is indicated by studies demonstrating that, at current exposure levels and without the protection of the ancillary provisions of a comprehensive standard, the risk of silicosis and other silica-induced diseases is extremely high.

OSHA has concluded that there will be no significant progress in the prevention of silica-related diseases without the adoption of a full and comprehensive silica standard. A full standard will improve worker protection, ensure adequate prevention programs, and reduce silica-related disease. Silica meets the criteria and has been designated as a priority for a comprehensive standard.

Silica is not currently on OSHA's regulatory agenda. It will be added as other standards are completed and resources become available.

REFERENCES

Checkoway, H., Heyer, N. J., Demers, P. A. and Breslow, N. E. (1993) Mortality among workers in the diatomaceous earth industry. *Br. J. ind. Med.* **50**, 586–597.

Freeman, C. S. and Grossman, E. A. (1995) Silica exposures in workplaces in the United States between 1980 and 1992. *Scand. J. Work Environ. Health* **21** (Suppl. 2), 47–49.

Goldsmith, D. F. (1994) Silica exposure and pulmonary cancer. In *Epidemiology of Lung Cancer* (Edited by J. M. Samet), pp. 245–298. Marcel Dekker Inc., New York.

Hnizdo, E. and Sluis-Cremer, G. K. (1991) Silica exposure, silicosis, and lung cancer: a mortality study of South African gold miners. *Br. J. ind. Med.* **48**, 53–60.

McLaughlin, J. K., Chen, J-Q., Dosemeci, M., Chen, R-A., Rexing, S. H., Wu, Z., Hearl, F. J., McCawley, M. A. and Blot, W. J. (1992) A nested case-control study of silica exposed workers in China. *Br. J. ind. Med.* **49**, 167–171.

NIOSH (1996) Preventing silicosis and deaths in construction workers. *NIOSH Alert*, DHHS (NIOSH) Publication no. 96–112. NIOSH, Cincinnati, OH.

Steenland, K. and Brown, D. (1995) Silicosis among gold miners: exposure–response analyses and risk assessment. *Am. J. public Health* **85**, 1372–1377.

Winter, P. D., Gardner, M. J., Fletcher, A. C. and Jones, R. D. (1990) A mortality follow-up study of pottery workers: preliminary findings on lung cancer. In *Occupational Exposure to Silica and Cancer Risk (IARC Scientific Publications no. 97)* (Edited by L. Simonato, A. C. Fletcher, R. Saracci and T. L. Thomas), pp. 83–94. International Agency for Research on Cancer, Lyon, France.

Pergamon

Ann. occup. Hyg., Vol. 41, Supplement 1, pp. 475–479, 1997
© 1997 British Occupational Hygiene Society
Published by Elsevier Science Ltd. All rights reserved
Printed in Great Britain
0003–4878/97 $17.00 + 0.00
Inhaled Particles VIII

PII: S0003–4878(96)00100–7

DOES OCCUPATIONAL SILICA EXPOSURE OR SILICOSIS CAUSE LUNG CANCER?

D. F. Goldsmith

Public Health Institute, 2001 Addison Street, Suite 210, Berkeley, CA, 94704–1103 U.S.A.

INTRODUCTION

The International Agency for Research on Cancer's (IARC) assessment of silica as a probable human carcinogen rested on three evidentiary bases: (a) positive experimental rat model of pulmonary carcinogenesis via inhalation and injection; (b) elevated cancer (mostly lung) among silica-exposed workers; and (c) elevated pulmonary cancer risk among workers with extant silicosis, often diagnosed and recorded in conjunction with compensation registers (IARC, 1987). IARC indicated that the following epidemiology issues were of concern to the scientists on the Monograph writing committee 10 years ago. (1) Absence of dose–response findings; (2) Lack of adjustment for smoking; (3) Lack of adjustment for other possible confounding factors (i.e. radon, polycyclic aromatic hydrocarbons, arsenic); and (4) Diagnostic bias of lung cancer among compensated silicotics. For these reasons, the evidence among humans was judged to be "limited".

RATIONALE

If the scientific community has not addressed these issues in the 10 years since IARC's report, then there would be no justification for changing the Agency's assessment. However, if research has been undertaken to evaluate these limitations, then it may be justified to change assessment for silica exposure [see Goldsmith (1996) for an expanded review of the underlying issues]. Thus, the rationale for this paper is to critically evaluate the evidence over the past 10 years and offer a review of the causation issues for silica's carcinogenicity status. Each of the four weaknesses will be listed and evidence in response to it will be described. In conclusion, the criteria for causation will be examined for both occupational silica exposure and for silicosis.

Absence of dose–response findings

Fu *et al.* (1992) in a study of 1113 underground Chinese miners from 1960 to 1988 showed that cumulative dust (mg–y) relative risk (RR) for lung cancer rose from 1.0 for < 500 mg–y to 4.2 for 500–1449 mg–y, to 5.2 for 1500–2999 mg–y, to 13.0 for 3000–4449 mg–y, to 16.1 > 4500 mg–y. Using a case–control analysis to adjust for age and smoking McLaughlin *et al.* (1992) examined the risk among Chinese male miners by degree of cumulative exposure to respirable silica. There was a significant positive trend for cumulative dust exposure ($P = 0.02$) and for respirable silica ($P = 0.004$) among tin miners (87 cases, 371 controls). For cumulative respirable silica the lung cancer odds ratio (OR) rose from 1.5 for low cumulative exposure to silica (0.1–8.69 µg m^{-3} y^{-1}; 15 cases; 67 controls), to 1.9 for medium cumulative exposure to silica (8.7–26.2 µg m^{-3} y^{-1}; 22 cases; 82 controls), to 3.1 for high cumulative exposure to silica (> 26.3 µg m^{-3} y^{-1}; 35 cases, 108 controls). The authors examined cumulative exposure to arsenic, polycyclic aromatic hydrocarbons and radon decay products, and only arsenic produced a significant trend among tin miners, suggesting that arsenic and silica dusts may interact.

Muller and colleagues (1986) conducted a retrospective cohort study of 6972 underground gold miners from the Province of Ontario, Canada who were employed at least 60 months in mining. These men were followed from 1955 to 1977 using tracing information from the Ontario Workers' Compensation Board/Mining Master File, death records from Statistics Canada and Social Insurance files. When the years worked was defined, a dose–response emerged with the risk of < 5 years having a standardized mortality rate (SMR) of 139; 5–9 years a SMR of 143; 10–14 years SMR of 134; 15–19 years SMR of 163; and ≥ 20 years SMR was 202.

Lack of adjustment for smoking

A cohort of 3971 white South African gold miners (exposed to an environment containing ~ 30% airborne silica) born between 1916 and 1930 and actively employed in 1970, was followed for 9 years. Wyndham et al. (1986) demonstrated that overall, there was an elevated risk for lung cancer (SMR = 161; 39 observed; 24.2 expected; 95% CI = 115, 220). The authors showed that cumulative dust exposure had a smoking-adjusted rate ratio (RR) of 1.77 per 10 particle years of underground exposure [95% CI = 0.94–3.31 ($P = 0.06$)].

Hnizdo and Sluis-Cremer (1991) presented a study of 2209 South African white male gold miners who began mining during the years 1936–1943 and whose mortality was traced from 1968 to 1986. The authors classified the gold mine jobs into 11 groups based on the average respirable silica dust concentrations developed by Beadle (1971). Cumulative dust measures (in respirable particles–years/1000) were calculated for each miner. Overall, the authors showed that the 77 cases of lung cancer had greater levels of silica particle–years (SPY) exposure, smoked more, had greater pack–years, had a greater percentage of ever smokers (96% vs 88%) than control miners. Using a Cox proportional hazards model that adjusted for age and pack–years, the RR rose from 1.00 for ≤ 15 SPY/1000 to 1.54 for 16–30 SPY/1000, to 2.07 for 31–40 SPY/1000, to a RR of 2.92 for ≥ 41 SPY/1000 (95% CI = 1.02–8.4). According to the model, the RR for lung cancer per 1000 SPY (adjusted for smoking, year of birth and age) was 1.023 (95% CI = 1.005–1.042). This means that the most veteran gold miners (all employed ≥ 10 years) exposed to 50 000 SPY have 3.2 times greater likelihood of dying from lung cancer than those exposed to only 15 000 SPY. The data indicated the silica and smoking appeared to act synergistically and the synergy was greatest for miners with the most smoking and mining experience (Hnizdo and Sluis-Cremer, 1991). Hnizdo raised the possibility that radon exposure may have confounded the associations reported with silica dust (Hnizdo, 1994), though more recent research does not confirm a radon association (Hnizdo, personal communication).

Checkoway *et al.* (1993) have conducted an extensive study of the health of 2570 white male diatomaceous earth (DE) workers who were exposed to cristobalite, a silica polymorph. The men were employed at least 1 year in the industry and followed from 1942 to 1987. There was a SMR of 1.43 for lung cancer (59 observed; 41.4 expected; 95% CI = 1.09, 1.84). The authors found dose-related gradients for duration of employment and a semiquantitative silica exposure index for lung cancer with lagged exposure for 15 years. The risks were adjusted for age, calendar year, duration of follow-up and ethnicity. The lung cancer risk rose from 1.00 for < 50 exposure intensity–years, to 1.19 for 50 to 99 exposure intensity–years, to 1.37 for 100–199 exposure intensity–years, to 2.74 for ≥ 200 intensity–years. Having limited smoking histories on the DE workers, Checkoway *et al.* reported that smoking could not account for all the link between silica exposure and lung cancer because the exposure–response was the same for the cohort as it was for smokers only. They concluded that high cristobalite exposure prior to the 1950s was the most etiologically significant contributor to the lung cancer risk.

Lack of adjustment for other possible confounders (i.e. radon, PAH, arsenic)

There have been positive lung cancer findings among workers in the ceramic industries (Forastiere *et al.*, 1986; Winter *et al.*, 1990); in granite industry (Koskela *et al.*, 1987, 1990); in stone industry (Guenel *et al.*, 1989); and in diatomaceous earth processing (Checkoway *et al.*, 1993). All are industries characterized by little or no confounding from radon or arsenic (seen in mining) or polycyclic aromatic hydrocarbons (in foundries). Some residual confounding in uranium miners could indicate that silica and radon decay products may interact biologically in jointly producing excess of lung cancer (Goldsmith and Goldsmith, 1996).

Diagnostic bias among compensated silicotics

Rosenman *et al.* (1995) demonstrated the lung cancer risk was actually greater among hospitalized silicotics in Michigan and New Jersey who had not applied for Workers' compensation compared to those who had filed, PMR = 1.13 vs 1.60. Similarly, Merlo *et al.* (1995) found a SMR of 3.14 (37 observed; 11.8 expected; 95% CI = 2.21, 4.33) among hospitalized silicotics in Genoa, Italy without regard to compensation status. There have been five studies demonstrating lung cancer dose–response by severity of silicosis from chest X-rays (Chia *et al.*, 1991; Miller *et al.*, 1987; Ng *et al.*, 1990; Hnizdo and Sluis-Cremer, 1991; [for silicosis of hilar lymph glands only]; and Goldsmith *et al.*, 1995 [based on degree of compensation awarded]). Almost every study of silicotics (over 30 in the peer-reviewed literature since 1980) showed at least > 2 times lung cancer risk (see Smith *et al.*, 1995), including Chiyotani *et al.*'s (1990) RR of 2.22 among never-smoking Japanese silicotics.

SUMMARY OF EVIDENCE OF HUMAN CANCERS RELATED TO SILICA EXPOSURE

Table 1 shows that the modified criteria (after Hill, 1995) have been met for the most part for both silicotics and for silica-exposed workers (Goldsmith, 1996). Many studies have examined the cancer risk in silicotics and non-silicotics separately and found greater elevated lung cancer risks among silicotics. The

Table 1. Criteria for cancer causation for silica exposure and for silicosis

Point of evidence	Silica exposed workers	Workers with silicosis
• Strong relative risk	√	√√√
• Dose-response gradient	√√	√
• Consistent findings	√√	√√√
• Controlled confounding	√	√
• Biological plausibility	√√	±
• Temporal cogency	√	√
• Specificity	√√	√√
• Overall coherence	Yes	Yes

√ = criteria met. ± = incomplete evidence.

presence of silicosis itself may be an independent risk factor for pulmonary neoplasia, an indicator variable for duration and/or intensity of exposure, or a marker of genetic susceptibility to pulmonary damage by respirable size crystalline silica. The general trend is that silicotics show increased risk of pulmonary neoplasia as compared to both those exposed to silica without silicosis and the general population.

Winter *et al.* (1990) among U.K. ceramic workers, Fu *et al.* (1992) and McLaughlin *et al.* (1992) among Chinese tin miners, Hnizdo and Sluis-Cremer (1991) among South African gold miners and Checkoway *et al.* (1993) among U.S. diatomaceous earth workers demonstrated excess lung cancer risk associated with occupational exposure to crystalline silica and dose–response relationships. Some studies also demonstrated greater than additive interaction between crystalline silica and tobacco smoke. Other epidemiologic studies demonstrated dose–response gradients for the association between silicosis and lung cancer.

CONCLUSIONS

With two new positive animal studies since 1987 (Muhle *et al.*, 1989; Spietoff *et al.*, 1992) and silica's ability to bind with DNA (Daniel *et al.*, 1995), IARC's prior assessment does not appear to be wrong. Furthermore, epidemiology evidence in favour of silica's carcinogenic role appears to be sufficient for both silica-exposed workers and for silicotics (Goldsmith, 1996). The shortcomings in human studies noted by members of the last IARC Working Group on silica have been addressed and there is now a large body of scientific evidence to conclude that occupational silica dust exposure is carcinogenic to humans.

REFERENCES

Beadle, D. (1971) The relationship between the amount of dust breathed and the development of radiological signs of silicosis: an epidemiological study of South African gold miners. In *Inhaled Particles III*. (Edited by W. H. Walton) pp.953–964. Pergamon Press, Oxford.

Checkoway, H., Heyer, N. J., Demers, P. A. and Breslow, N. E. (1993) Mortality among workers in the diatomaceous earth industry. *Br. J. ind. Med.*, **50**, 586–597.

Chia, S-E., Chia, K-S., Phoon, W-H. and Lee, H-P (1991) Silicosis and lung cancer among Chinese granite workers. *Scand. J. Work Environ. Health* **17**, 170–174.

Chiyotani, K., Saito, K., Okubo, T. and Takahashi, K. (1990) Lung cancer risk among pneumoconiosis patients in Japan, with special reference to silicotics. In *Occupational Exposure to Silica and Cancer*

Risk (Edited by L. Simonato, A. C. Fletcher, R. Saracci and T. L. Thomas), pp. 95–104. International Agency for Research on Cancer (IARC): Lyon, France.

Daniel, L. N., Mao, Y., Williams, A. O. and Saffiotti, U. (1995) Direct interaction between crystalline silica and DNA—a proposed model for silica carcinogenesis. *Scand. J. Work Environ. Health* **21**(2): 22–26.

Forastiere, F., Lagorio, S., Michelozzi, P., Cavariani, F., Arca, M., Borgia, P., Perucci, C. and Axelson, O. (1986) Silica, silicosis and lung cancer among ceramic workers: a case-referent study. *Am. J. ind. Med.* **10**, 363–370.

Fu, H., Jing, X., Yu, S., Gu, X., Wu, K., Yang, J. and Qiu, S. (1992) Quantitative risk assessment for lung cancer from exposure to metal ore dust. *Biomedical and Environ. Sci.* **5**, 221–228.

Goldsmith, D. F., Beaumont, J. J., Morrin, L. A. and Schenker, M. B. (1995) Respiratory cancer and other chronic disease mortality among silicotics in California. *Am. J. ind. Med.* **28**, 459–467.

Goldsmith, D. F. and Goldsmith, J. R. (1996) Joint cancer risks among workers having silica, smoking, and radon exposures must be examined. International Conference on Radiation and Health, Beer Sheva, Israel, November 3–7.

Goldsmith, D. F. (1996) Importance of causation for interpreting occupational epidemiology research: a case study of quartz and cancer. *Occupational Medicine: State of the Art Reviews* **11**, 433–449.

Guenel, P., Hojberg, G. and Lynge, E. (1989) Cancer incidence among Danish stone workers. *Scand. J. Work, Environ. Health* **15**, 265–270.

Hill, A. B. (1965) The environment and disease: Association or causation? *Proc. Roy. Soc. Med.* **58**, 295–300.

Hnizdo, E. (1994) Risk of silicosis in relation to fraction of respirable quartz (letter). *Am. J. ind. Med.* **25**, 771–772.

Hnizdo, E. and Sluis-Cremer, G. K. (1991) Silica exposures, silicosis, and lung cancer: a mortality study of South African gold miners. *Br. J. ind. Med.* **48**, 53–60.

IARC (1987) IARC monographs on the evaluation of the carcinogenic risk of chemicals to humans. Silica and some silicates, monograph #42, Lyon. International Agency for Research on Cancer, Lyon, France.

Koskela, R.-S., Klockars, M., Jarvinen, E., Kolari, P. J. and Rossi, A. (1987) Cancer mortality of granite workers. *Scand. J. Work Environ. Health* **13**, 26–31.

Koskela, R.-S., Klockers, M., Jarvinen, E., Rossi, A. and Kolari, P. J. (1990) Cancer mortality of granite workers 1940–1985. *Occupational Exposure to Silica and Cancer Risk* (Edited by L. Simonato, A. C. Fletcher, R. Saracci and T. L. Thomas), pp. 43–53. Lyon, France, International Agency for Research on Cancer (IARC).

McLaughlin, J. K., Chen, J.-Q., Dosemeci, M., Chen, R.-A., Rexing, S. H., Zhien, W., Hearl, F. J., McCawley, M. A and Blot, W. J. (1992) A nested case-control study of lung cancer among silica exposed workers in China. *Br. J. ind. Med.* **49**, 167–171.

Merlo, F., Fontana, L., Reggiardo, G., Ceppi, M., Barisone, G., Garrone, E. and Daria, M. (1995) Mortality from lung cancer among 515 Genoa, Italy silicotics: results from the follow-up period 1961–1987. *Scand. J. Work Environ. and Health* **21**(2), 77–80.

Miller, A. B., Scarpelli, D. and Weiss, N. S. (1987) Report to the workers' compensation board on the Ontario gold mining industry of the scientific panel on mortality from cancer among Ontario gold miners 1955–1977 industrial disease standards panel Ontario Ministry of Labour, Toronto, Ontario.

Muhle, H., Takenaka, S., Mohr, U., Dasenbrock, C. and Marmelstein, R. (1989) Lung tumor induction upon long-term low-level inhalation of crystalline silica. *Am. J. ind. Med.* **15**, 343–346.

Muller, J. and Kusiak, R. A. (1986) Study of mortality of Ontario gold miners 1955–1977. Ontario Ministry of Labour, Toronto, Ontario, July.

Ng, P. N., Chan, S. L. and Lee, J. (1990) Mortality of a cohort of men in a silicosis register: further evidence of an association with lung cancer. *Am. J. ind. Med.* **17**, 163–171.

Rosenman, K.D., Stanbury, M. J. and Reilly, M. J. (1995) Mortality among persons with silicosis reported to two state based disease surveillance systems. *Scand. J. Work Environ. Health* **21** (Suppl. 2) 73–76.

Smith, A. H., Lopipera, P. A. and Barroga, V. R. (1995) Meta-analysis of studies of lung cancer among silicotics. *Epidemiology* **6**, 617–624.

Spietoff, A., Wesch, H., Wegener, K. and Klimisch, H. (1992) The effects of thoratrast and quartz on the induction of lung tumors in rats. *Health Phys.* **63**, 101–110.

Winter, P. D., Gardner, M. J., Fletcher, A. C. and Jones, R. D. (1990) A mortality follow-up study of pottery workers: preliminary findings on lung cancer. In *Occupational Exposure to Silica and Cancer Risk* (Edited by L. Simonato, A. C. Fletcher, R. Saracci and T. L. Thomas), pp. 83–94. International Agency for Research on Cancer (IARC), Lyon, France.

Wyndham, C. H., Bezuidenhout, B. N., Greenacre, M. J. and Sluis-Cremer, G. K. (1986) Mortality of middle aged white South African gold miners. *Br. J. ind. Med.* **43**, 677–684.

 Pergamon

Ann. occup. Hyg., Vol. 41, Supplement 1, pp. 480–484, 1997
© 1997 British Occupational Hygiene Society
Published by Elsevier Science Ltd. All rights reserved
Printed in Great Britain
0003–4878/97 $17.00 + 0.00
Inhaled Particles VIII

PII: S0003–4878(96)00101–9

REVISED QUANTITATIVE RISK ASSESSMENT FOR SILICOSIS AND SILICA RELATED LUNG CANCER IN AUSTRALIA

J. Leigh,† P. Macaskill† and M. Nurminen‡

†National Institute of Occupational Health and Safety, GPO Box 58, Sydney 2001, New South Wales, Australia; and ‡Finnish Institute of Occupational Health, Helsinki, Finland

INTRODUCTION

As part of a standards revision process, the purpose of this study was to revise our previous estimates of the risks of silicosis and lung cancer associated with exposure to silica dust at the current concentration levels in Australia or, at most, at a level of 0.05 mg m^{-3} in the light of new research published since our previous study. The risk estimates were then translated into predicted numbers of cases of these diseases to obtain an assessment of the magnitude and significance of the occupational health problem related to silica on a national level.

MATERIALS AND METHODS

We used the general methods previously described by Nurminen *et al.* (1992) with different risk functions and exposure unit conversion factors based on more recent data.

RESULTS

Silicosis

To estimate the parameters of the risk function we used the study of Hnizdo and Sluis-Cremer (1993) which fitted a log-logistic survival model to cohort data of 2235 white South African gold miners. The empirical risk function in this study was very similar to that obtained by Steenland and Brown (1995) in a cohort study of 3330 gold miners from South Dakota. The Hnizdo model used cumulative total respirable dust level as the "time" variable and respirable silica was assessed to constitute 30% of the total respirable dust. The model used was $R_a (LD_a) = 1 - [1 + \{LD_a\}^{1/\sigma} \exp \{- \mu/\sigma\}]^{-1}$ where: R_a = risk in age category a; L = exposure level mg m^{-3}; D_a = duration of exposure in age category a(y); LD_a = cumulative respirable dust exposure mg m^{-3}.y = cumulative respirable silica exposure/0.3; μ = 2.439, σ = 0.2199.

Table 1 shows the average lifetime risks and the expected number of silicosis cases per year computed according to the Hnizdo and Sluis-Cremer model compared with the results for three or more X-ray reader agreement according to the Canadian (Muir *et al.*, 1989, 1991) model reported by us previously. It can be seen that the Hnizdo model gives much greater dose-related risk and correspond-

Table 1.1. Average lifetime risk (%) of silicosis $\geq \frac{1}{1}$ ILO

	Current exposure	< 0.2 mg m^{-3}	< 0.1 mg m^{-3}	< 0.05 mg m^{-3}
Hnizdo model				
$I = R/D$	27	25	12	1.4
$I = 1/D\log_e (1 - R)$	60	51	16	1.5
Muir model				
$I = R/D$	0.87	0.74	0.40	0.17
$I = 1/D\log_e (1 - R)$	0.88	0.75	0.41	0.18

Table 1.2. Silicosis ($\geq \frac{1}{1}$ ILO). Cases expected in Australia in 40 years and 1 year

	Current exposure		< 0.2 mg m^{-3}		< 0.1 mg m^{-3}		< 0.05 mg m^{-3}	
	40 years	1 year	40 years	1 year	40 years	1 year	40 years	1 year
Hnizdo model								
$I = R/D$	31 953	799	24 033	600	4142	104	313	8
$I = 1/D\log_e (1 - R)$	54 595	1365	31 413	785	4227	106	313	8
Muir model								
$I = R/D$	1014	25	784	20	440	11	192	5
$I = 1/D\log_e (1 - R)$	1022	26	786	20	440	11	192	5

ingly many more predicted cases, especially at the higher exposure levels. Some reasons for the large difference in the two models have been discussed by Rice and Stayner (1995). They include physicochemical differences in silica particles; exclusion of retired miners from the Canadian study; differences in the definition of radiographic silicosis; differences in methods used to estimate exposures; errors in exposure estimates; use of aluminium dust as a prophylactic agent in Canada and possible blocking effects on quartz toxicity by other components of dust, with a greater effect in Canada.

However, the fact that the South Dakota study agrees very closely with the South African suggests that the Canadian study provides an underestimate of the risk of silicosis.

The preceding risk model assumes, first, that silica dust is a necessary cause of silicosis. In other words, there is no risk ($R = 0$) if there is no cumulative exposure ($L = 0$ or $D = 0$). Second, it is the product of level and duration (LD), that determines the risk and not the intensity of exposure in itself. However, several additional assumptions are necessary to predict realizations of risks in terms of numbers of people sustained.

Suppose that a dynamic population is being followed. The industrial population has a turnover in membership as new employees enter the labor force and others retire, but the population profile is unchanging over time (stationary) with respect to its distribution of (i) size of age groups, and (ii) duration of exposure. Suppose also that the age distribution is that of a general, employed male population, that the members of the population remain exposed to a constant intensity of silica dust while in a particular job and that the exposure ceases upon retirement, whereby the retired ex-workers are no longer at risk of silicosis progression. Finally, the hypothetical period of follow-up is extended over 40 years, corresponding to the maximum duration of employment in a worker's life (of which, on the average, 14 years is spent in exposed work).

The expected number of silicosis cases (E) was thus computed as $E = S \cdot T \cdot I$, where S stood for the size of the industrial subpopulation, T was the follow-up time (y) and I was the incidence rate of silicosis (in units of cases per year). This rate was considered a weighted average of the age-specific incidence densities, I_a ($a = 20$–24 years, . . . , 60–64 years); that is, $I = \Sigma_a W_a I_a$, where the weights were defined as $W = N_a F_a / \Sigma_a N_a F_a$ ($\Sigma_a W_a = 1$). Here N_a was the number of the male resident population of New South Wales (NSW) on 30 June 1986 in the ath age category and F_a was the fraction of the corresponding population employed in 1989–1990. I_a was solved from the relation between incidence density and cumulative incidence rate, which was used as an estimate for risk; specifically, $R_a = 1 - \exp(-I_a D_a)$ or $I_a = 1/D_a \log_e (1 - R_a)$. When silicosis risks were small, a reasonably accurate approximation was $I_a = R_a/D_a$. The total expected number of silicosis cases was then obtained by summing over the 665 industrial or occupational subpopulations associated with a nonzero exposure intensity. Results could be expressed as number of cases over 40 years or average number per year, or as lifetime risks.

Lung cancer

In the case of cancer of the bronchus, trachea and lung, the risk was assessed in terms of excess numbers over the age-specific incidences among NSW male residents in 1984 on the basis of the statistics of the NSW Central Cancer Registry. The gradient in the risk ratio (RR) for a unit increase of 10^3 respirable surface area (RSA) particle-years (Y), standardized for smoking, year of birth and age, was estimated from the South African gold miner material by a proportional hazards model as $RR = 1.023^Y$. In these data, the combined effect of dust and smoking was more than additive and it was assumed to be multiplicative so that the proportionality of hazards would apply. The 95% confidence limits for RR, again based on the gold miner data, were computed as 1.005^Y (lower limit) and 1.042^Y (upper limit).

We have revised the conversion factor used in converting the dust exposure measurement used in Hnizdo and Sluis-Cremer (1991) to give the smoking adjusted dose-response relationship between silica and lung cancer. Hnizdo and Sluis-Cremer use a cumulative dust exposure measure of RSA particle count (particles/cc) $\times$ shift length in hours $\times$ years. To convert this to respirable mass of silica in mg m^{-3} we relied largely on the data of Beadle (1971) for stopers/developers and assumed that the fraction of respirable quartz in respirable dust was 0.3. We did not divide the respirable mass figure by shift length (4–8 h) because we felt that the mean respirable quartz value for stopers/developers which would result (0.05 mg m^{-3}) would be unbelievably low and inconsistent with the mean value of 0.2 mg m^{-3} quoted by Beadle and Bradley (1970).

We also had available grouped side-by-side RSA and gravimetric silica directly measured data in 95 paired samples. Analysis of these data gave a conversion coefficient of 1 mg m^{-3} respirable silica $= 3822$ RSA (linear regression through origin), ($r = 0.96$). There was more scatter in the ungrouped data ($r = 0.34$). This reassured us somewhat that the conversion factor based on the larger data set ($n = 646$) was reasonable. It now appears that the RSA data in this set had not been multiplied by shift length as we had assumed.

Table 2.1. Average lifetime excess risk of lung cancer % (95% CI)

	Current exposure	< 0.2 mg m^{-3}	< 0.1 mg m^{-3}	< 0.05 mg m^{-3}
Hnizdo model				
1 mg m^{-3} = 8000 RSA h	1.87 (0.21–31.3)	1.34 (0.20–3.98)	0.83 (0.15–1.92)	0.47 (0.09–0.98)
1 mg m^{-3} = 3385 RSA h	0.47 (0.08– 1.09)	0.42 (0.08–0.90)	0.30 (0.06–0.59)	0.18 (0.04–0.35)

Table 2.2. Excess lung cancer cases in Australia in 40 years and 1 year (95% CI)

	Current exposure	< 0.2 mg m^{-3}	< 0.1 mg m^{-3}	< 0.05 mg m^{-3}
Hnizdo model				
1 mg m^{-3} = 8000 RSA h				
40 years	1960 (301–7763)	1524 (271–3526)	1058 (205–2195)	641 (131–1249)
1 year	49 (8–194)	38 (7–88)	26 (5–55)	16 (3–31)
1 mg m^{-3} = 3385 RSA h				
40 years	626 (122–1318)	551 (111–1092)	408 (85–779)	257 (55–480)
1 year	16 (3–33)	14 (3–27)	10 (2–20)	6 (1–12)

Further recent correspondence now suggests that Beadle's gravimetric data was too low, due to the omission of a factor of pi (3.1416 . . .) and other incorrect geometric and density assumptions in the calculation of mass from surface area. Thus our conversion, based on the Beadle data, omitting the factor of average shift length but not taking into account the errors in the opposite direction, gave an erroneously low figure for the RSA corresponding to 1 mg m^{-3} respirable silica. When corrected the conversion should become, approximately:

$$1 \text{ mg m}^{-3} \text{ respirable silica} = 8000 \text{ RSA h.}$$

If the fraction of silica in total respirable dust in the South African data is higher than 0.3 as suggested by Du Toit, this would tend to reduce the magnitude of this error and bring the conversion factor closer to that used in our previous estimates.

There are several assumptions germane to the preceding risk ratio model for lung cancer. First, the reference level of $RR = 1$ is attained with $Y = 0$ in the preceding risk function. Second, an increase in RR by a given intensity of exposure (dust level) is proportional to (average) duration of exposure and vice versa. Third, the RR increases monotonically with exposure at a constant level and remains constant after the cessation of exposure. Fourth, and most importantly, we assume that silica exposure per se is related to human lung cancer, independently of silicosis.

The excess number (E) of cancer cases was computed as $E = S.T. \Sigma_a (RR_a - 1) W_a I_a$, where I_a was the five year age-specific incidence of lung cancer and the summation was from 20 years onwards. Note that, if $RR = 1$, then $E - 0$. The total excess number was then obtained by a tally of the cases in the 665 industrial or occupational subpopulations with a nonzero exposure level. Results could also be expressed as average excess cases per year or as lifetime excess risk. Results are shown in Table 2.

REFERENCES

Beadle, D. G. (1971) The relationship between the amount of dust breathed and the development of radiological signs of silicosis: an epidemiological study of South African gold miners. In *Inhaled Particles III* (Edited by W. H. Walton), pp. 953–966. Unwin, Old Woking.

Beadle, D. G. and Bradley, A. A. (1970) The composition of airborne dust in South African gold mines. In *Pneumoconiosis*, *Proceedings of the International Conference* (Edited by H. A. Shapiro), Johannesburg, 1969, pp. 462–466. Oxford University Press, Cape Town.

Du Toit, R. S. J. (1991) The shift mean respirable mass concentration of eleven occupations of Witwatersrand gold miners. *NCOH Report* 4/91.

Hnizdo, E. and Sluis-Cremer, G. K. (1991) Silica exosure, silicosis and lung cancer: a mortality study of South African gold miners. *Br. J. ind. Med.* **48**, 53–60.

Hnizdo, E. and Sluis-Cremer, G. K. (1993) Risk of silicosis in a cohort of white South African gold miners. *Am. J. ind. Med.* **24**, 447–457.

Muir, D. C. F., Shannon, H. S., Julian, J. A., Verma, D. K., Sebestyen, A. and Bernholz, C. D. (1989) Silica exposure and silicosis among Ontario hardrock miners: III. Analysis and risk estimates. *Am. J. ind. Med.* **16**, 29–43.

Muir, D. C. F. (1991) Correction in cumulative risk in silicosis exposure assessment. *Am. J. ind. Med.* **19**, 555.

Nurminen, M., Corvalan, C., Leigh, J. and Baker, G. (1992) Prediction of silicosis and lung cancer in the Australian labor force exposed to silica. *Scand. J. Work, Environ. Health* **18**, 393–399.

Rice, F. L. and Stayner, L. T. (1994) An assessment of silicosis risk for occupational exposure to crystalline silica. *Scand. J. Work, Environ. Health* **21** (Suppl. 2), 87–90.

Steenland, K. and Brown, D. (1995) Silicosis among gold-miners: exposure-response analyses and risk assessment. *Am. J. pub. Health* **85**, 1372–1377.

 Pergamon

Ann. occup. Hyg., Vol. 41, Supplement 1, pp. 485–490, 1997
British Occupational Hygiene Society
Published by Elsevier Science Ltd
Printed in Great Britain
0003-4878/97 $17.00 + 0.00
Inhaled Particles VIII

PII: S0003–4878(96)00102–0

EFFECTS OF EXPOSURE ESTIMATION PROCEDURES ON THE EVALUATION OF EXPOSURE–RESPONSE RELATIONSHIPS FOR SILICOSIS

C. H. Rice,† H. Checkoway,‡ M. Dosemeci,§ P. Stewart§ and A. Blair§

†University of Cincinnati, Dept. of Environmental Health, PO Box 670056, Cincinnati, OH 45267–0056; ‡University of Washington, Seattle, Washington; and §National Cancer Institute, Bethesda, Maryland, U.S.A.

INTRODUCTION

The reconstruction of exposures in occupational epidemiology is an important and expanding research interest. Historically the number of environmental measurements available has been small. Methods used to estimate exposure for jobs during time periods for which there are few or no sample results incude: assign zero exposure, use the average exposure for jobs not evaluated, estimate exposure for only some jobs over specified time periods, estimate exposure for all jobs and time periods using a single-step procedure such as calculation of an average (Theriault *et al.*, 1974) or a multi-step imputation process involving considerations such as engineering controls and similarity of processes (Dement *et al.*, 1983; Seixas *et al.*, 1991; Stewart *et al.*, 1995).

The work reported here was undertaken to provide insight into possible influences of assumptions made regarding missing exposure data on measures of association, by comparing the results at each step of a multi-step imputation algorithm. Data for silica exposure and X-ray diagnosis of silicosis from the North Carolina Dusty Trades program have been used. A previous report included a strong, linear exposure–response relationship (Rice *et al.*, 1986); however, all exposure estimates were based on measurement data. If measurements were not available, exposure was assumed to be zero. For this report, odds ratios for silicosis have been calculated after each of five steps in an imputation algorithm and are contrasted with the original results.

MATERIALS AND METHODS

The North Carolina Dusty Trades program has been detailed previously (Rice *et al.*, 1984). Exposure and medical surveillance programs were initiated in 1935 for workers employed in the mining and processing of crushed stone, granite dimension stone, kaolin, lithium, mica/feldspar, pyrophyllite and talc, and hard rock mining and foundry work. Continued employment required annual documentation of a disease-free chest radiograph. The study subjects were drawn from a review of the records of all men diagnosed to have silicosis from 1935 to 1980 and up to four controls selected from the pool of disease-free workers. Controls were individually

Table 1. Description of baseline data set (A) and the criteria on which imputed values were constructed from data sets B–F

Data set identifier	Criteria on which exposure estimate based	Imputed values added to previous data set
A	industry/company/site/job	—
B	industry/company/job	median, across sites
C	industry/company	median, across all sites/jobs
D	industry/job	median, across all company/sites
E	industry	median, across all companies/sites/jobs
F	usual industry	median, across all companies/sites/jobs

matched on race and years of birth and hire within 5 years; each control was available for exposure at least as long as the corresponding case. Six hundred and seventy two noncases fulfilled the matching criteria for 216 silicotics.

For this analysis, the exposure estimates derived from impinger sampling results only formed the baseline data set. Additional exposure data sets were constructed in a step-wise fashion, building on the baseline data set, to provide estimates of exposure at jobs for which measurements were not available, as described in Table 1. Data set A (baseline) included exposure estimates derived from the industrial hygiene measurements; exposure at any job not included in this baseline was assumed equal to zero. Each successive data set included all values from the previous data set and the imputed values derived from calculating the median. In the absence of an estimate in the data set, a value of zero was assumed. The industry of longest employment was the "usual industry".

Cumulative silica exposure expressed as million particle years (mpy) and the number of years for which no exposure estimate was available were calculated for each study subject for data sets A–F. Two additional values for cumulative exposure were calculated by including a value of 0.1 (data set G) and 12.0 million particles per cubic foot (mppcf) (data set H), respectively, for employment period(s) not covered by data set A. These values were the means of the 5th and 95th percentiles of the exposure estimates in data set A and represented extreme estimates of exposure. Strata were constructed in all data sets using the following ranges of cumulative silica exposure: Group 0 (referent), < 20 mpy; Group 1, 20–59.9mpy; Group 2, 60–179.9 mpy; Group 3, 180 mpy and greater. Logistic regression was used to calculate odds ratios and 95% confidence intervals (Breslow and Day, 1980). Risk estimates were also calculated for each Data set for those subjects with exposure estimates for all dusty trades employment. An alternative approach to this algorithm based on industrial hygiene experience would be identification of the determinants of exposure, e.g. using linear models to calculate the amount of variance in the exposure accounted for by company, site, job and industry. This was done *a posteri* for the data set.

RESULTS

The amount of work time for which no exposure value was available decreased as additional imputations were made, as shown in Table 2. In the top portion of Table 3 are the odds ratios (ORs) for all 216 cases and 672 controls, showing that the ORs

Table 2. Mean (s.d.) years for which no exposure estimate was available for Dusty Trades study subjects for each data set

Data set*	All study subjects	Cases	Controls	By cumulative exposure group** Cases	Controls
A	8.8 (10.2)	7.0 (8.2)	9.4 (10.7)	9.9 (9.0)	13.8 (11.5)
				8.0 (8.6)	5.8 (7.8)
				2.4 (4.0)	2.4 (4.1)
				3.3 (5.5)	2.0 (4.2)
B	6.8 (9.2)	5.1 (7.1)	7.4 (9.7)	7.4 (8.1)	10.9 (11.1)
				5.6 (7.6)	4.9 (7.2)
				2.4 (4.1)	2.8 (4.9)
				2.9 (5.5)	1.6 (3.6)
C	3.3 (6.1)	3.9 (5.9)	3.1 (6.1)	6.3 (7.3)	4.3 (7.2)
				3.4 (5.5)	3.0 (5.8)
				2.0 (3.4)	1.4 (3.7)
				1.8 (3.8)	0.8 (1.8)
D	2.0 (5.2)	1.5 (4.4)	2.2 (5.4)	3.3 (6.5)	3.3 (6.6)
				1.2 (4.2)	2.2 (5.3)
				0.6 (2.3)	0.5 (2.4)
				0.4 (1.4)	0.4 (1.2)
E	1.9 (5.0)	1.3 (4.2)	2.1 (5.2)	3.2 (6.6)	3.1 (6.3)
				0.8 (3.1)	2.1 (5.2)
				0.5 (2.1)	0.5 (2.4)
				0.4 (1.4)	0.4 (1.2)
F	1.8 (5.0)	1.1 (4.1)	2.1 (5.2)	3.0 (6.6)	3.1 (6.3)
				0.6 (2.9)	2.1 (5.2)
				0.5 (2.1)	0.4 (2.4)
				0.1 (0.6)	0.4 (1.2)

* See Table 1.
** Values for referent, Group 1, Group 2 and Group 3, respectively.

generally increased as the amount of time for which estimates of exposure increased even though the estimates were based on increasingly less-specific data. The one exception is data set C, the only data set where the average amount of time without an exposure estimate was larger for cases than controls overall and in all four of the exposure groups. In the lower part of Table 3 are results from the analysis restricted to study subjects for whom an estimate of exposure was available for all jobs. The same general trend of increasing OR with increasing assumptions in the development of exposure estimates is seen, although several inconsistencies arc noted in data sets C, E and F.

For data set G, which includes the value of 0.1 mppcf for any unsampled job, the odds ratios were very similar to those in Table 3; however, when 12.0 mppcf was used in data set H, the increase in response with increasing exposure was not observed and no OR was statistically significant (calculations not shown). The results of linear modeling to identify the determinants of exposure followed the pattern developed using professional judgement. The highest R^2 was for industry/company/job/site (0.54), with consistently decreasing values for the criteria shown in Table 1.

Table 3. Odds ratio for three exposure groups compared with the referent group for baseline data set A and data sets with imputed (data sets B–F) for all study subjects (n = 888) and study subjects for which an exposure estimate is available for all dusty trades employment

Exposure groups compared	A	B	C	D	E	F
All study subjects*						
1 vs referent	0.82 (41/183)**	0.94 (46/196)	0.65 (44/226)†	1.14 (63/235)	1.11 (61/236)	1.23 (63/236)
2 vs referent	1.86 (49/97)†	1.96 (54/111)†	1.45 (57/132)	1.81 (60/141)†	1.84 (62/141)†	1.95 (62/141)†
3 vs referent	3.47 (29/30)†	4.20 (34/32)†	3.20 (34/32)†	4.35 (34/34)†	4.33 (34/34)†	4.61 (34/34)†
Exposure estimates available for all dusty trades jobs						
1 vs referent	0.4 (12/66)**	0.6 (15/78)	0.8 (22/126)	1.3 (53/164)	1.3 (55/168)	1.3 (60/170)
2 vs referent	1.4 (25/51)	1.4 (29/56)	1.5 (34/96)	1.7 (52/123)†	1.7 (56/126)†	1.7 (57/128)†
3 vs referent	2.0 (14/18)	3.4 (20/20)†	4.4 (23/24)†	4.5 (31/29)†	4.1 (31/30)†	4.3 (33/30)†
Number of cases/controls in referent group	(16/46)	(21/67)	(31/142)	(38/165)	(40/169)	(41/170)
Total number of cases/controls	(67/181)	(85/221)	(110/388)	(174/481)	(182/493)	(191/498)

* Total number of cases is 216; total number of controls is 672.

** Number of cases/controls in exposure group.

† $P < 0.05$, 95% confidence interval does not include unity.

DISCUSSION

The approach presented allowed identification of the contribution of each part of the imputation algorithm to changes in numbers of cases and controls in the various exposure groups and the subsequent silicosis risk estimates. Comparing data sets A and E, the imputed values resulted in 40 cases and 101 controls being recategorized from the referent to an exposure group. By aggregating the various assumptions, the proportions recategorized are more nearly equal, although step-wise comparison of the individual data sets indicated differential reclassification of cases and controls. The results showed that the exposure–response relationships were generally consistent, with the highest estimates of risk resulting from use of data set F. In the original work, it was assumed that error based on use of only measurement data would bias the estimate of risk toward the null (Copeland *et al.*, 1977). More recent work has shown that nondifferential misclassification may not always bias towards the null (Dosemeci *et al.*, 1995); however this occurred only under very extreme conditions. The generally increasing risk estimates at each step of the imputation process suggests that nondifferential misclassification did occur in the initial analysis and that the imputed values provide more accurate summary exposure estimates. However, had the imputation process ended with data set C, a statistically significant risk would have been found only for the highest exposure group.

The ORs calculated for the data sets restricted to those study subjects for which exposure estimates were available for all jobs were generally consistent with the values for all study subjects. If hygiene practice resulted in important differences in the level of effort expended for exposure assessment at jobs held by cases compared with controls, one would expect differentially higher exposures attributed to cases, resulting in higher ORs in the restricted analysis. This algorithm did not include a method to impute data for time periods for which the industry, company, site and job were all unknown. The logical extension of the process outlined might include the median across all jobs. This was not included since the proportion of a 40 year working lifetime for which no estimate was available was reduced to approximately 5%, by the imputation methods described.

Although selected *a priori*, the value of 12.0 mppcf (nearly five times the recommended exposure level in effect for many of the years during which these surveys were conducted) was probably unrealistic. A value of 2.5 mppcf, the recommended exposure level, would be a more realistic selection for the "high" value; using this value the OR when comparing the highest exposure group to the referent was statistically significant. Given the magnitude of the sampling program undertaken by the State of North Carolina to identify workplaces with a risk of silicosis it was likely that most employees at jobs or companies not sampled were at a lower risk of disease (e.g. working at lower exposure). Similar studies in other epidemiologic data sets which include quantitative exposure estimates will be useful in helping to identify the important determinants of exposure for the development of algorithms for situations where environmental data are sparse or missing.

Acknowledgements—This work was conducted while funded through an Interagency Personnel Agreement at the National Cancer Institute (C.R.). The assistance of Robert Banks, IMS, is gratefully acknowledged.

REFERENCES

Breslow, N. E. and Day, N. E. (1980) Statistical methods in cancer research. Vol. 1, *The Analysis of Case-Control Studies*. International Agency for Research on Cancer, Lyon.

Copeland, K. T., Checkoway, H., McMichael, A. J. *et al.* (1977) Bias due to misclassification in the estimation of relative risk. *Am. J. Epidemiol.* **105**, 488–495.

Dement, J. M., Harris, R. L., Symons, M. J. and Shy, C. M. (1983) Exposures and mortality among chrysotile asbestos workers. Part I: exposure estimates. *Am. J. ind. Med.* **4**, 399–419.

Dosemeci, M. and Wacholder, S. (1990) Does non-differential misclassification of exposure always bias a true effect towards the null value? *Am. J. Epidemiol.* **132**, 746–748.

Rice, C., Harris, R. L., Checkoway, H. and Symons, M. J. (1986) In *Silica, Silicosis and Lung Cancer: Controversies in Occupational Medicine*, (Edited by D. Goldsmith, D. Winn and C. Shy) pp. 77–86. Praeger, New York.

Rice, C., Harris, R. L., Lumsden, J. C. and Symons, M. J. (1984) Reconstruction of Silica Exposures in North Carolina Dusty Trades. *Am. ind. Hyg. Assoc. J.* **45**, 689–696.

Seixas, N. S., Moulton, L. H., Robins, T. G., Rice, C. H., Attfield, M. D. and Zellers, E. T. (1991) Estimation of cumulative exposures for the National study of coal workers' pneumoconiosis. *Appl. occup. Environ. Hyg.* **6**, 1032–1041.

Stewart, P. A., Triolo, H., Zey, J., White, D., Herrick, R. F., Horning, R., Dosemeci, M. and Pottern, L. M. (1995) Exposure assessment for a study of workers exposed to acrylonitrile. II. A computerized exposure assessment program. *App. occup. Environ. Hyg.* **10**, 698–706.

Theriault, G. P., Burgess, W. A., DiBerardinis, L. J. and Peters, J. M. (1974) Dust exposure in Vermont granite sheds. *Arch. Environ. Health.* **28**, 12–17.

SECTION 5

DEPOSITION, CLEARANCE AND MODELLING

 Pergamon

Ann. occup. Hyg., Vol. 41, Supplement 1, pp. 491–496, 1997
© 1997 British Occupational Hygiene Society
Published by Elsevier Science Ltd. All rights reserved
Printed in Great Britain
0003–4878/97 $17.00 + 0.00
Inhaled Particles VIII

PII: S0003–4878(96)00053–1

CHARACTERISING THE VARIABILITY IN ADULT HUMAN NASAL AIRWAY DIMENSIONS

R. A. Guilmette, Y. S. Cheng and W. C. Griffith

Inhalation Toxicology Research Institute, P.O. Box 5890, Albuquerque, NM 87185, U.S.A.

INTRODUCTION

Respiratory tract models used in calculating radiation doses from exposure to inhaled radioactive aerosols have only recently focused attention on the importance of the nasal airways (NAs). Because the NAs are the first tissues of the respiratory tract available for aerosol deposition in normally nose-breathing people, any deposition of aerosol in this anatomical structure will reduce the amounts available to be deposited in the remainder of the respiratory tract. Thus, uncertainties in estimating the deposition fractions in the NAs will propagate throughout the remainder of the respiratory tract, creating errors in the calculated dose estimates. Additionally, there is evidence that the NAs are at risk for induction of cancer from exposure to certain occupational aerosols such as wood dust, leather dusts, chromium and nickel (Roush, 1979).

Cheng *et al.* (1991, 1996) have summarised the human data on NA deposition of ultrafine and larger sized aerosols, respectively, and found substantial intersubject variability in the deposition fractions, particularly in the particle size range of 1–10 µm. Because the NAs have very complex shapes, adequate theoretical models for predicting aerosol deposition in NAs have not yet been developed. Because we hypothesize that much of the variability observed in measured NA deposition efficiencies in different humans is due to differences in the size and shape of individual NAs, we have undertaken to estimate the variability in NA size in a small population of humans.

METHODS

To assess the variabilities in NA dimensions among different adult humans, we have conducted an anatomical study of adult, nonsmoking, male and female human subjects with no notable NA disease or structural pathology. The protocol for this study was approved by institutional review boards of The Lovelace Institutes (review board for ITRI), the University of New Mexico School of Medicine and the Veterans Administration Medical Center (VAMC). Prior to magnetic resonance imaging (MRI), the procedures were described and each subject agreed to participate by signing a consent form that provided details of the procedures. Each subject received a MRI scan of their NAs using the 1.5 Tesla 55 cm bore Siemens MRI unit at the VAMC, Albuquerque, NM. Acquired MR images consisted of

3 mm contiguous coronal sections taken from the anterior end of the nostrils to the posterior pharynx. Before performing the MRI scan, several anthropometric measurements were also made on each subject: (1) height; (2) weight; (3) circumference of the head at the level of the glabella; (4) lateral head width at the glabella; and (5) anterior–posterior head width at the glabella. Normal cycling of nasal airway patency was not monitored prior to or during the MRI acquisitions. Consequently the MR images from several subjects who underwent airway cycling during the MRI were not used due to blurred images.

The perimeters of each NA were traced by hand and digitized using a Grafpen (SAC, Southport, CT) sonic digitizer (Guilmette *et al.*, 1989). The perimeter maps were stored and analyzed in a personal computer in terms of both individual airway cross-sectional areas and perimeter lengths. For the present analysis, the cross-sectional areas of all coronal sections from the anterior nares to the posterior end of the nasal airway just prior to the nasopharynx were summed, then multiplied by the thickness of each section (3 mm) to obtain a measure of the volume of both NAs. Likewise, the perimeter lengths for the same sections were summed and multiplied by the section thickness to obtain the NA surface area. In the case where left and right NA dimensions were compared, only those sections having a complete nasal septum were included. These data were then compared with the various anthropometric measurements by simple linear regression (REG procedure, SAS/STAT software, Cary, NC).

Separate comparisons of the *in vivo* deposition efficiencies for ultrafine particles in nine subjects previously described in Cheng *et al.* (1996) were made with the respective NA anatomy data from this study. Using an approach similar to that used by Cheng *et al.* (1995), the deposition efficiency for each subject was characterized by the respective coefficients, a_i, obtained by fitting each subject's deposition data to the relationship: $P_i = 1 - E_i = \exp(-a_i D^{0.5} Q^{-0.125})$, where P_i is the penetration efficiency for subject i, E_i is the deposition efficiency, D is the particle-size-specific diffusion coefficient ($cm^2 s^{-1}$) and Q is the flow rate ($cm^3 s^{-1}$). Data obtained using particles with sizes of 4, 8 and 20 nm and at flow rates of 167 and 333 $cm^3 s^{-1}$ were used.

RESULTS AND DISCUSSION

MRI scans of adequate quality for morphometric analysis have been analyzed for 21 male and 24 female subjects. The results of regressing surface area and volume on the variables height, weight, head circumference, lateral head width and dorsoventral head width are summarised in Table 1. The most statistically significant relationships for total NA volume and surface area were with subject height and dorsoventral head width (Fig. 1). No relationships were apparent for weight, height/weight ratio or lateral head width. Head circumference was only marginally correlated. This was found to be due to a significant lack of correlation for the male subjects; the correlation for females was approximately the same as that for dorsoventral head width. With the limited data available, the uncertainties on the fitted parameter values were relatively large; nevertheless, the slopes of the fits for all anthropometric parameters except lateral head width were statistically significant. The average relative standard error for height, circumference and

Table 1. Regression of anthropometric measures on NA surface area and volume

Variable	Intercept ($\pm$ SE)	Slope ($\pm$ SE)	Pr (slope = 0)	R^2
Surface area				
Height	2.4 (45)	1.13 (0.26)	0.0001	0.304
Weight	171 (13)	0.354 (0.178)	0.054	0.0839
Circumference	−56 (83)	4.42 (1.46)	0.0041	0.176
Lateral width	188 (26)	0.60 (1.69)	0.725	0.0029
Dorsoventral width	7.4 (51)	9.91 (2.66)	0.0006	0.2434
Volume				
Height	−20 (9.7)	0.25 (0.06)	0.0001	0.311
Weight	16 (2.8)	0.094 (0.038)	0.016	0.1275
Circumference	−47 (17)	1.22 (0.29)	0.0002	0.2845
Lateral width	17 (5.6)	0.37 (0.36)	0.316	0.0234
Dorsoventral width	−24 (10)	2.46 (0.55)	0.0001	0.3217

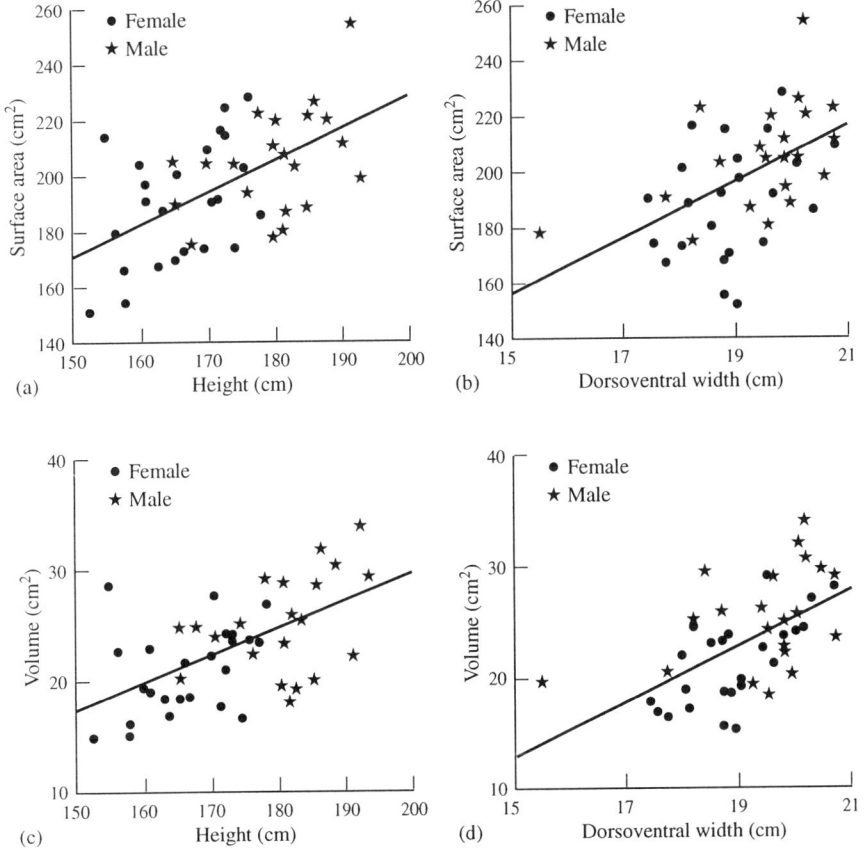

Fig. 1. Correlation of NA surface area with subject height (a) and dorsoventral head width (b), and volume with height (c) and dorsoventra head width (d).

Table 2. Regression of coefficients a_i on measures of NA size

Variable	Intercept ($\pm$ SE)	Slope ($\pm$ SE)	R^2
Height	-12 (34)	0.153 (0.190)	0.0976
Dorsoventral width	-16 (36)	1.60 (1.85)	0.1103
Surface area	-16 (8.7)	0.15 (0.041)	0.6912
Volume	-6.8 (5.4)	0.85 (0.20)	0.7451

dorsoventral head width was $25 \pm 4\%$. These data indicate that, in general, the size of human nasal airways, as represented by the coarse measures of total airway surface area and volume, are related to the height and the dorsoventral head width of the individual.

To evaluate the degree of symmetry between the left and right NAs, ratios of the larger to the smaller surface areas and volumes for the respective sides were determined. The average ratio of surface areas was 1.04 ± 0.03 (SD), and the average ratio of volumes was 1.27 ± 0.25. There were no gender-related differences that were not accounted for by subject size, and no statistically significant differences in surface areas or volumes with respect to right or left side (surface areas: left side $= 91.6 \pm 10.7$ cm^2, right side $= 91.7 \pm 11.4$ cm^2; volumes: left side $= 9.10 \pm 2.77$ cm^3, right side $= 8.69 \pm 2.11$ cm^3). The range of values of surface areas for both left and right sides was essentially a factor of two; for the volumes, it was a factor of three.

It thus appears that there is more symmetry with respect to surface area than for volume. This suggests that the size of the cell populations at risk in NAs (which are estimated to be proportional to surface area) may not vary substantially between left and right sides for different individuals. However, the more notable differences in side-specific volumes may imply that there will be unequal partitioning of airflow between the right and left NAs, which would in turn imply unequal deposition probabilities for the respective sides. This latter point requires further study, however, as it is more likely that both flow partitioning and side-to-side comparative deposition will be related to the cross-sectional areas and the shapes of the airways (and hence to the NA resistance) rather than to the more coarse volume measure of size. In any case, it cannot be reasonably assumed that both local and regional deposition of aerosols in NAs will be bilaterally symmetric.

The deposition efficiency of ultrafine aerosols (4–20 nm) in nine nonsmoking male humans (data taken from Cheng *et al.*, 1996) was compared with their respective NA surface areas and volumes, as well as with their heights and dorsoventral head widths, the parameters previously shown to be most correlated with surface area and volume (Table 2 and Fig. 2). Regression analyses showed that the deposition efficiencies were highly correlated with both surface areas ($R^2 = 0.6912$) and volumes ($R^2 = 0.7451$) of the subjects. This is consistent with the results of empirical modeling done by Cheng *et al.* (1996), in which different measures of nasal anatomy size and shape (S_f, the average NA shape factor for the turbinate region; A_s, total NA surface area; A_{min}, the minimum NA cross-sectional area) were fitted by nonlinear regression to the deposition data, and found to bring the data from their subjects into reasonable agreement. Conversely, it is interesting to note that both height and dorsoventral head width were not highly correlated

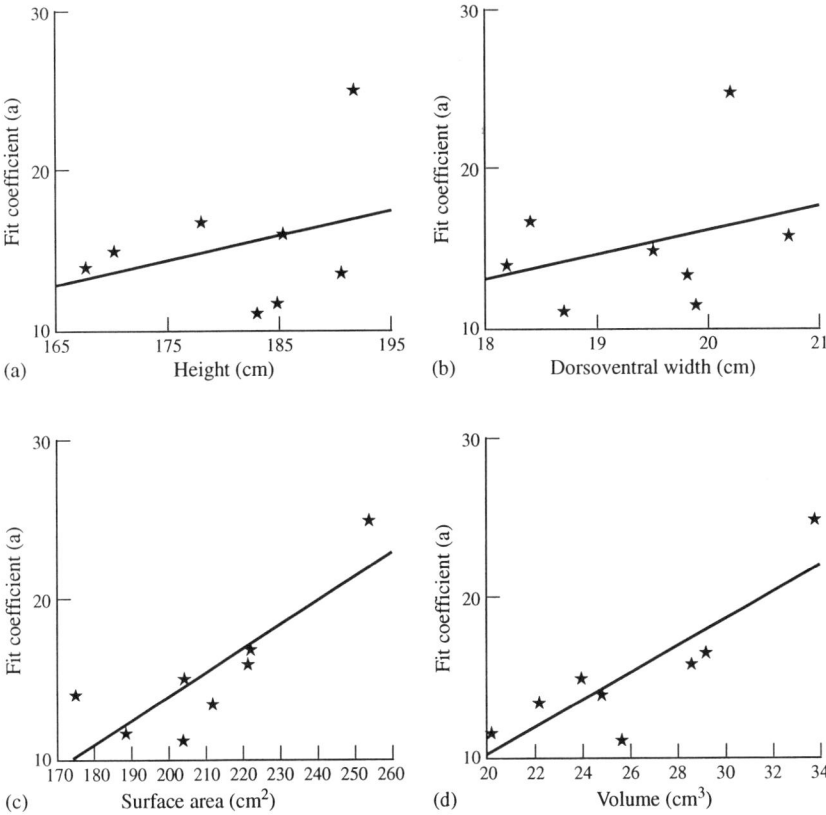

Fig. 2. Correlation of deposition parameter a_i with height (a), dorsoventral head width (b), surface area (c) and volume (d).

with the deposition data. Thus, although correlation was found between the gross anthropometric measurements of subject height and dorsoventral head width and NA surface area and volume, it appears that this correlation does not provide satisfactory predictive power for individualized deposition efficiency. Whether this would also be the case for inhalation of larger aerosol particles typical of occupational exposures (0.5–20 μm) is not known, as human deposition studies using larger aerosol particles in which NA size and shape measurements were obtained have not been done. Nevertheless, it appears that NA size and shape factors are important in explaining the large variabilities observed in studies of NA aerosol deposition.

Acknowledgements—The participation of Dr M. Williamson, radiologist, University of New Mexico School of Medicine is gratefully acknowledged. Research sponsored by the Office of Health and Environmental Research, U.S. Department of Energy, under Contract no. DE-AC04-76EV01013.

REFERENCES

Cheng, K.-H., Cheng, Y.-S., Yeh, H.-C., Guilmette, R. A., Simpson, S. Q., Yang, Y.-H. and Swift,

D. L. (1996) *In vivo* measurements of nasal airway dimensions and ultrafine aerosol deposition in the human nasal and oral airways. *J. Aerosol Sci.* **27**, 785–801.

Cheng, Y.-S., Smith, S. M., Yeh, H.-C., Kim, D.-B., Cheng, K.-H. and Swift, D. L. (1995) Deposition of ultrafine aerosols and thoron progeny in replicas of nasal airways of young children. *Aerosol Sci. Technol.* **23**, 541–552.

Cheng, Y.-S., Yeh, H.-C. and Swift, D. L. (1991) Aerosol deposition in human nasal airway for particles 1 nm to 20 μm: a model study. *Radiat. Protect. Dosim.* **38**, 41–47.

Guilmette, R. A., Wicks, J. D. and Wolff, R. K. (1989) Morphometry of human nasal airways *in vivo* using magnetic resonance imaging. *J. Aerosol med.* **2**, 365–377.

Roush, G. C. (1979) Epidemiology of cancer of the nose and paranasal sinuses: general concepts. *Head Neck Surg.* **2**, 3–11.

Pergamon

Ann. occup. Hyg., Vol. 41, Supplement 1, pp. 497–502, 1997
© 1997 British Occupational Hygiene Society
Published by Elsevier Science Ltd. All rights reserved
Printed in Great Britain
0003–4878/97 $17.00 + 0.00
Inhaled Particles VIII

PII: S0003–4878(96)00054–3

EXTRATHORACIC DEPOSITION OF INHALED, COARSE PARTICLES (4.5 μm) IN CHILDREN VS ADULTS

W. D. Bennett, K. L. Zeman, C. W. Kang and M. S. Schechter

Center for Environmental Medicine and Lung Biology, CB, 7310, Mason Farm Road, UNC,
Human Studies Division/NHEERL, USEPA and Department of Pediatrics, UNC, Chapel Hill,
NC 27599, U.S.A.

INTRODUCTION

Recent epidemiological studies suggest that children have increased morbidity from particulate air pollution (Dockery *et al.*, 1989; Schwartz *et al.*, 1994), primarily characterised by increased incidence of cough and bronchitis in the children. In general the morbidity associations with particulate pollution have been with inhalable particles (PM_{10}). It is not clear, however, which fraction of PM_{10}, the coarse (2.5–10 μm MMAD) or the fine (< 2.5 μm MMAD) mode, is most responsible for these associations. While children may be especially predisposed to acute toxic effects from airborne particulates, they may also receive an increased dose of particles to their lungs compared to adults.

Theoretical calculations by Xu and Yu (1986) predict enhanced deposition of particles (greater than 2 μm) in the head region for children when compared to adults. This model and others (Hoffman *et al.*, 1989) also predict enhanced airway (tracheobronchial) deposition for particles < 5 μm MMAD in children when compared to adults. As a result of this increased filtering in the conducting airways, a smaller fraction of particles in the size range of 2–10 μm are predicted to deposit in the pulmonary region of childrens' lungs.

There are limited data on the regional deposition of particles within the respiratory tract of children. Such measurements in humans have generally relied on the use of radiolabelled particles and gamma camera analysis. While the potential radiation risk has been deemed justified for studying normal adult volunteers, most investigators are reluctant to perform such studies in normal children. It may be that this risk is higher in children than in adults (Everard, 1994). Studies of children with cystic fibrosis lung disease, however, have been justified to assess regional deposition of inhaled therapeutic aerosols (Alderson *et al.*, 1974) or to assess mucociliary clearance following inhalation therapy (App *et al.*, 1990). In a retrospective study we analysed regional deposition of radiolabelled (Tc-99m) Fe_2O_3 particles (4.5 μm MMAD) by gamma camera analysis in a group of children and adults with mild cystic fibrosis (CF). Because the CF patients have only mild lung disease and their upper airway anatomy is likely not dissimilar from normal, intra vs extrathoracic deposition in these patients should be similar to normals.

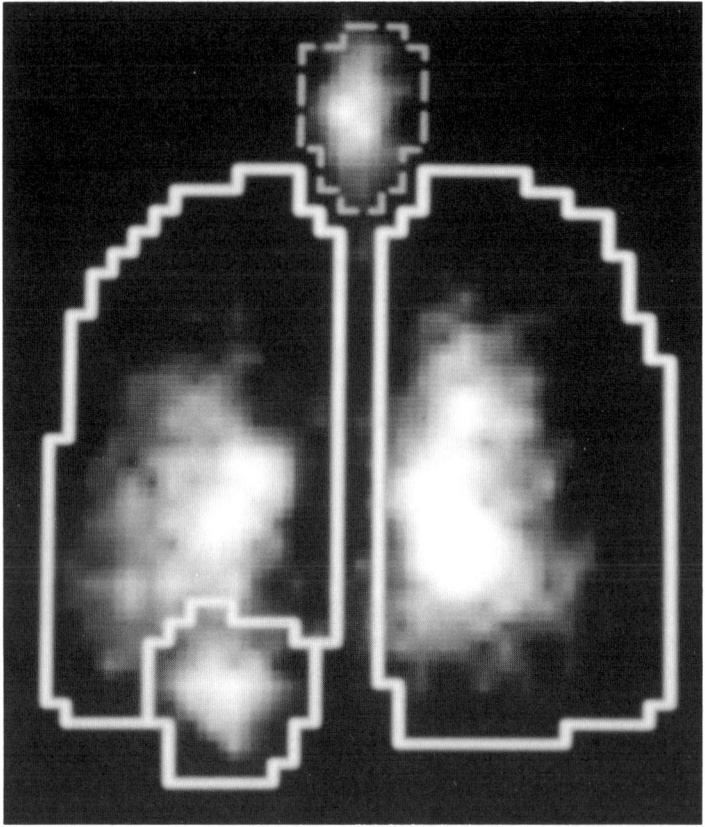

Fig. 1. Gamma camera image of coarse particle deposition in a CF child. Regions of interest (ROIs) are drawn around the lungs, head and stomach. The size of the lung ROIs are based on a Xe133 equilibrium scan in this subject.

METHODS

Regional analysis (Fig. 1) of gamma camera images was performed for a group of children and adults with mild cystic fibrosis (CF) [mean FEV1 % predicted = 80% ± 20 (SD) in children, and = 68% ± 22 in adults]. These patients had been studied as part of previous mucociliary clearance protocols. The studies were approved by the University of North Carolina Committee on the Protection of the Rights of Human Subjects and informed consent was obtained. Each subject had inhaled a 4.5 μm MMAD (geometric SD = 1.25) aerosol of insoluble iron oxide generated by spinning top from a colloidal suspension (May, 1949) and radiolabelled with Technetium 99m (Tc99m) (Wales *et al.*, 1980). For comparison to the CF adults we also analysed data from a group of normal adults. The mean of each group's age, height, body surface area (BSA), lung function and particle inhalation conditions (mean flow, tidal volume and particle size) is given in Table 1. The mean age of the children, 13.8, indicates that these patients were primarily adolescents; the youngest was age 11. The mean tidal volume and flow in the children was significantly lower than the adults, consistent with breathing parameter differences in adolescents vs adults (Xu and Yu, 1986; Schum *et al.*, 1991).

Table 1. Summary of group characteristics mean ($\pm$ SD)

	CF children (6M/7F)	CF adults (5M/10F)	Normal adults (3M/5F)	P-value
Age	13.8 (2.1)	29.1 (8.4)	23.4 (4.5)	
Height (cm)	161 (14)	167 (8)	174 (11)	0.04
BSA (m^2)	1.49 (0.24)	1.63 (0.15)	1.80 (0.21)	0.006
FEV$_1$ %pred	80 (19)	68 (22)	89 (20)	0.08
Flow$_{insp}$ (ml/s)	263 (36)	302 (52)	377 (25)	<0.001
Tidal volume (ml)	528 (83)	622 (107)	583 (285)	NS
MMAD (µm)	4.8 (0.3)	4.6 (0.4)	4.8 (0.8)	NS

P-value is for analysis of variance between three groups for each variable.
M = male, F = female.

Analysis of regional deposition

Using region-of-interest (ROI) analysis (Fig. 1), we compared the percentage of deposited particles in the lungs (intrathoracic lung activity) vs mouth and larynx (extrathoracic = head + stomach activity) in each group of subjects. For each subject we analysed a 2 minute, posterior deposition image acquired by gamma camera immediately after particle inhalation. Activity in each region (lung, head and stomach) was multiplied by an attenuation factor to correct for different gamma attenuations for the lungs, 2.5, head, 2.0 and stomach, 4.0 (Svartengren *et al.*, 1991). Per cent extrathoracic deposition (%ED) was then calculated as

$$\%ED = \frac{\text{head activity} + \text{stomach activity}}{\text{total activity}} \times 100.$$

To assess the degree of central (C) vs peripheral (P) airway deposition within the lung, we also calculated a C/P ratio of Tc99m activity (Bennett *et al.*, 1985), normalized to a Xenon 133 equilibrium scan for each subject. Only the right lung was used for C/P analysis because of interference from stomach activity on the left side.

Analysis of total and lung deposition fractions

Total and lung deposition fractions, i.e. the fraction of inhaled particles depositing in either the total respiratory tract or the lung, were also estimated by a total and lung deposition index (DI),

$$DI_{tot} = \text{Aerosol}_{\text{dep in total respiratory tract}}/\text{Aerosol}_{in}$$

and

$$DI_{lung} = \text{Aerosol}_{\text{dep in lung}}/\text{Aerosol}_{in}$$

for each subject.

Aerosol inhaled was estimated by

$$\text{Aerosol}_{in} = C \times F \times T \times Sp$$

where C = inhaled particle concentration;
F = average inhalation flow rate;
T = total inhalation time; and
Sp = particle specific activity (determined as mCi of Tc/ml of colloid soln).

Table 2. Summary of deposition data mean ($\pm$ SD)

	CF children	CF adults	Normal adults	P-value
%ED	30.7 (15.3)	20.1 (14.7)	16.0 (7.2)	<0.05
C/P	2.1 (0.5)	2.0 (0.9)	2.0 (0.4)	NS
DI_{total}	33.6 (16.9)	39.4 (24.6)	22.8 (15.2)	NS
DI_{lung}	24.1 (15.3)	30.3 (17.3)	19.1 (12.3)	NS
nDI_{lung}	58 (40)	66 (43)	38 (24)	NS

P-value is for analysis of variance between three groups for each variable.

Aerosol deposited ($Aerosol_{dep}$) was estimated as either total regional activity in the deposition image (Fig. 1) (i.e. head + stomach + lung, each corrected for different attenuations as described above) for the respiratory tract or only lung activity for lung deposition. While relative variations in gamma attenuation between head, stomach, and lung probably differ little between individuals, total attenuation has been shown to increase with increasing body surface area (BSA) (Messina and Smaldone, 1985). Based on the data of Messina and Smaldone (1985) we divided total or lung activity by a factor, 0.8 – 0.3 (BSA), for each subject to correct for the effect of variations in body size on total or lung activity detected by the gamma camera.

The calculated DIs (DI_{tot} and DI_{lung}), while incorporating a number of factors and having final units with little meaning, should be proportional to a unitless deposition fraction of the inhaled 4.5 µm particles and allowed comparisons of such between the three groups. Finally, because the children tend to have smaller lungs relative to the adults we also calculated a deposition index normalised for lung surface area, nDI_{lung}. The DI_{lung} was divided by the area (no. of pixels) for the right lung ROI (Fig. 1) which was created from a Xenon 133 equilibrium scan in that subject to obtain nDI_{lung}.

RESULTS

Table 2 summarises the regional and total deposition for the three study groups. There was a tendency for %ED to be greater in the CF children as compared to CF adults ($P = 0.12$) and normal adults ($P = 0.06$) (by about 50%). There was a large variability in %ED among all subjects. Among the children, ED tended to increase with decreasing age ($r = -0.50, p = 0.08$). When the CF children were divided into two groups, children (age < 14) ($n = 8$) vs adolescents (age > 14) ($n = 5$), the younger group had almost twice the %ED of the older group, %ED = 37 $\pm$ 14 (SD) vs 20.5 $\pm$ 11.9 respectively ($P = 0.05$). In a stepwise multiple regression analysis of %ED, which included the variables inspiratory flow, tidal volume, FEV1 %predicted, subject age and subject height, only subject height was found to significantly predict %ED. Figure 2 shows that the %ED is negatively correlated with subject height ($r- = 0.35, P = 0.04$), i.e. increasing %ED with decreasing height. The C/P ratios were not different between the three groups and among all subjects tended to correlate negative with FEV 1% pred ($P = 0.10$).

There were no significant differences in total or lung deposition (DI_{total}, DI_{lung}, or nDI_{lung}) between the three groups (Table 2), though there was considerable

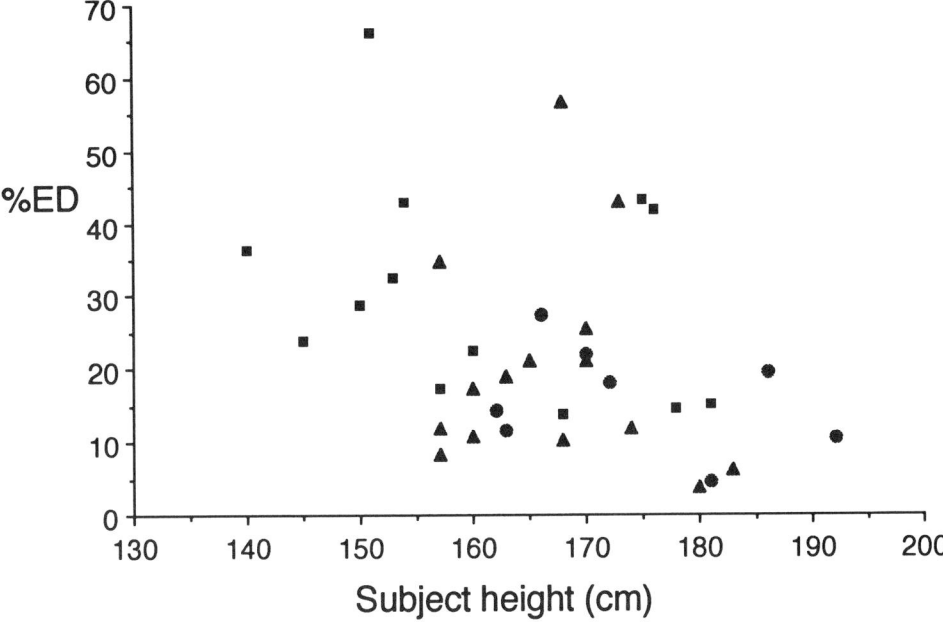

Fig. 2. Extrathoracic deposition as a % of total deposition (%ED) as a function of subject height for CF children (■), CF adults (▲), and normal adults (●).

variability. The CF children and adults tended to have greater deposition than the normals. In fact there was a tendency for nDI_{lung} (i.e. lung deposition normalised to lung surface area) to be negatively correlated with FEV 1% pred ($P = 0.13$).

DISCUSSION

Our results suggest that children have enhanced upper airway deposition of coarse particles when compared to adults. We found that the increase was correlated with decreasing height of the subject. The fact that total deposition, DI_{total}, was not different between children and adults, suggests that the upper airways (mouth and larynx) serve to filter coarse particles that might otherwise reach the lower respiratory tract. This is consistent with Alderson *et al.* (1974) who found decreased pulmonary deposition of a nebulised aerosol (3.8 µm MMAD) with decreasing age in CF children. On the other hand, children have smaller lungs than adults, so for coarse particles they still may have comparable lung deposition per surface area (nDI_{lung} from Table 2). It was interesting that the three groups had similar C/P ratios (Table 2) suggesting that intrathoracic airway deposition was not different in the children as compared to adults. It may be that the tendency for greater obstruction (diminished FEV 1%) in the CF adults tended to increase their airway deposition relative to the children. But even the normal adults with no evidence of airway obstruction had similar C/P ratios.

While our results are consistent with the predicted increase in head deposition with decreasing age of Xu and Yu (1986), the magnitude of the increase seems greater than the predictions of their model. They predict even greater enhancement of head deposition with decreasing age below age 10. The subjects in our study were not younger than age 11, yet we found %ED in some cases greater than 40%. It seems reasonable that in extrapolating these results to even younger children we might expect as much as 60–70% extrathoracic deposition for this particle size. The extrathoracic deposition in very young children may be dependent on the relative growth of the oropharyngeal dead space relative to the lung (Schum et al., 1991).

While our data are derived from subjects with CF we think they may be extrapolated to children without lung disease. The CF children we studied had mild airways obstruction, the presence of which may tend to increase deposition in the lung (Kim et al., 1988) and thus decrease %ED in these individuals compared to nonobstructed subjects. So it seems likely to expect that normal children may have even greater %ED relative to normal adults than the CF children.

What is the relevance of our results to recent epidemiological studies finding increased morbidity in children (Dockery et al., 1989; Schwartz et al., 1994)? If the enhanced effects in children are due to increased deposition of particles, then lower respiratory tract effects are less likely to be due to the action of coarse particle deposition in children. On the other hand, if the effects of airborne particulate pollution are found to be in the nature of upper respiratory problems, then enhanced upper airway deposition of coarse particles in children may play a role. Finally, our results also support the findings that clinical delivery of aerosolised drugs to the lungs of children is reduced for larger particles (Alderson et al., 1974).

REFERENCES

Alderson, P. O., Secker-Walker, R. H., Strominger, D. B., Markham, J. and Hill, R. L. (1974) Pulmonary deposition of aerosols in children with cystic fibrosis. *J. Pediatrics* **84**, 479–484.

App, E. M., King, M., Helfesrieder, R., Kohler, D. and Matthys, H. (1990) Acute and long term amiloride inhalation in cystic fibrosis lung disease. *Am. Rev. respir. Dis.* **141**, 605–12.

Bennett, W. D., Messina, M. S. and Smaldone, G. C. (1985) Effect of exercise on deposition and subsequent retention of inhaled particles. *J. appl. Physiol.* **59**, 1046–1054.

Dockery, D. W., Speizer, F. E., Stram, D. O., Ware, J. H., Spengler, J. D. and Ferris, B. G. (1989) Effects of inhalable particles on respiratory health of children. *Am. Rev. Respir. Dis.* **139**, 587–594.

Everard, M. L. (1994) Studies using radiolabelled aerosols in children. *Thorax* **49** (12), 1259–1266.

Hoffman, W., Martonen, T. B. and Graham, R. C. (1989) Predicted deposition of nonhygroscopic aerosols in the human lung as a function of subject age. *J. Aerosol Med.* **2** (1), 49–68.

Kim, C. S., Lewars, G. G. and Sackner, M. A. (1988) Measurement of total lung aerosol deposition as an index of lung abnormality. *J. appl. Physiol.* **64**, 1527–1536.

May, K. R. (1949) An improved spinning top homogeneous spray apparatus. *J. appl. Phys.* **20**, 932–938.

Messina, M. S. and Smaldone, G. C. (1985) Evaluation of quantitative aerosol techniques for use in bronchoprovocation studies. *J. Allergy clin. Immunol.* **75**, 252–257.

Schwartz, I., Dockery, D. W., Neas, L. M., Wypij, D., Ware, J. H., Spengler, J. D., Koutrakis, P., Speizer, F. E. and Ferris, B. G. (1994) Acute effects of summer air pollution on respiratory symptom reporting in children. *Am. Rev. respir. Dis.* **150**, 1234–42.

Schum, G. M., Phalen, R. F. and Oldham, M. J. (1991) The effect of dead space on inhaled particle deposition. *J. Aerosol Med.* **4**, 297–311.

Svartengren, M., Anderson, M., Bylin, G., Philipson, K. and Camner, P. (1991) Mouth and throat deposition of 3.6 μm radiolabelled particles in asthmatics. *J. Aerosol Med.* **4** (4), 313–321.

Wales, K. A., Petrow, H. and Yeats, D. B. (1980) Production of Tc99m-labeled iron oxide aerosols for human lung deposition and clearance studies. *Int. J. appl. Rad. Isotopes* **31**, 689–694.

Xu, G. B. and Yu, C. P. (1986) Effects of age on deposition of inhaled aerosols in the human lung. *Aerosol Sci. Technol.* **5**, 349–357.

 Pergamon

Ann. occup. Hyg., Vol. 41, Supplement 1, pp. 503–508, 1997
© 1997 British Occupational Hygiene Society
Published by Elsevier Science Ltd. All rights reserved
Printed in Great Britain
0003–4878/97 $17.00 + 0.00
Inhaled Particles VIII

PII: S0003–4878(96)00171–8

DEPOSITION AND CLEARANCE OF FINE PARTICLES IN THE HUMAN RESPIRATORY TRACT

C. Roth, W. G. Kreyling, G. Scheuch, B. Busch and W. Stahlhofen

National Research Centre for Environment and Health, Institute for Inhalation Biology,
Ingolstädter Landstr. 1, 85764 Neuherberg, Germany

INTRODUCTION

Total deposition of respirable particles in the human respiratory tract is well determined by experimental data and model calculations whereas regional deposition is still under discussion, particularly in the size range from 0.1–1.0 μm particle diameter. Experimental data on regional deposition of radiolabelled aerosol particles can be derived from gamma camera images on the basis of the particle clearance kinetics during the first few days after inhalation, since the limited resolution of the planar image allows no separation between the peripheral and the bronchial airways.

When particles are deposited in the human respiratory tract, two distinct phases of clearance from the thorax are usually observed: a fast cleared fraction during the first hours after inhalation followed by much slower clearance. It is generally assumed that the fast phase, completed within about 24 h, represents predominantly mucociliary clearance of particles deposited in the tracheobronchial tree, whereas the slow phase represents predominantly clearance from the alveolar region (Lippmann, 1977; Stahlhofen *et al.*, 1980). Nevertheless, during recent years the assumption that there also exists a slow cleared component from the tracheobronchial region has been confirmed by a large number of experiments (Scheuch, 1991; Stahlhofen *et al.*, 1994).

Recently, a new radio-aerosol "Technegas" (TcG; Tetley Manufacturing Ltd, Lucas Heights, Australia) was developed for lung ventilation scans in nuclear medicine. "Technegas" (TcG) is a condensation type aerosol consisting of hydrophobic inert graphite particles. For the production of TcG 20–100 μl of Na^{99m}TcO$_4$ in saline is put into the cavity of a graphite crucible which is dried at 50–70°C. Thereafter the cavity of the crucible is heated to 2500°C in an atmosphere of pure argon for 5 s such that carbon as well as the salt and ^{99m}Tc evaporates, followed by condensation during cooling. This procedure is similar to the production of the C-60 molecules, named buckminster-fullerenes (Kroto *et al.*, 1985). Therefore, it was assumed that TcG consists of ultrafine spherical particles which are formed via homogenous condensation from fullerene vapor at temperatures > 500°C. More recent investigations have shown that these ultrafine particles agglomerate rapidly to fine particles (Lemb *et al.*, 1993). The latter is in agreement with the fact that the distribution of deposited TcG applied to the lungs of normal subjects was found in the entire ventilated lungs, that is, it is transported in the lungs like a gas.

The aim of this study was the application of the new radio-aerosol TcG in order to determine its lung deposition and clearance kinetics from the lungs. Since the median diameter of this aerosol was reported to be 0.1 μm or even less it would provide regional deposition data in a size range lacking information (Roth *et al.*, 1994). A unique feature of this aerosol is its high specific ^{99m}Tc radioactivity of > 100 MBq/L aerosol reported to be tightly bound to the particle matrix (Burch *et al.*, 1986) allowing gamma camera analysis after single breath inhalation.

MATERIALS AND METHODS

The particle size distribution of the TcG-aerosol was determined with a Differential Mobility Analyser (DMA; Hauke GmbH, A-4810 Gmunden, Austria). To balance the production efficiency of the generator using 50 μl Na^{99m}TcO$_4$ saline, the whole aerosol of 6 L volume produced during one run was collected on a filter. The ^{99m}Tc activity of the initial 50 μl sample was compared to that of the filter and the crucible. To investigate the salt content of the aerosol particles, the filter was washed with distilled water and the Na$^+$-ion concentration was determined using a Na$^+$-ion sensitive electrode (PHM 95, Radiometer, Copenhagen) to calculate the sodium chloride mass.

To test the hygroscopic behaviour of the TcG-particles a Tandem Differential Mobility Analyser was used, consisting of two DMAs and controlled humidifiers for the sheath air and aerosol flow of the second DMA. With the first DMA, a monodisperse fraction of the polydisperse TcG-aerosol was separated and subsequently moistened to a relative humidity RH > 90%. The altered size distribution due to the growth of the particles related to their hygroscopic content was determined by the second DMA (Busch *et al.*, 1994).

Six healthy non-smokers (age 46 ± 9 years, 1 female, 5 males) with normal lung function volunteered for the measurements of lung deposition and retention with an inhalation apparatus described before (Scheuch *et al.*, 1989). After several breaths of clean air the subject started the inhalation of the aerosol from the "Technegas" generator for one single breath. The subject inhaled with a constant flow rate of 250 cm^{3}s^{-1} and with a tidal volume of 1000 cm^3. Three subjects performed a breath-hold of 10 s at the end of inhalation, the other three exhaled immediately after inhalation without any breath-hold.

A system of four shielded and collimated NaI(Tl) detectors (lung counter) has been used for the determination of the ^{99m}Tc activity of the particles present in the head, chest and stomach (Stahlhofen *et al.*, 1980). The activity was measured immediately after inhalation and at subsequent intervals up to 30 h. Total deposition was calculated from the lung deposit, the exhaled ^{99m}Tc activity collected on an exhalation filter and the ^{99m}Tc activity of a 250 cm^3 sample drawn from the TcG-generator onto a filter. To determine particle dissolution including TcG-particle disintegration and leakage of ^{99m}Tc from TcG, urine was measured after quantitative sampling for 24 h using the lung counter.

RESULTS

The size distribution of the TcG-aerosol is plotted in Fig. 1. Characterizing this

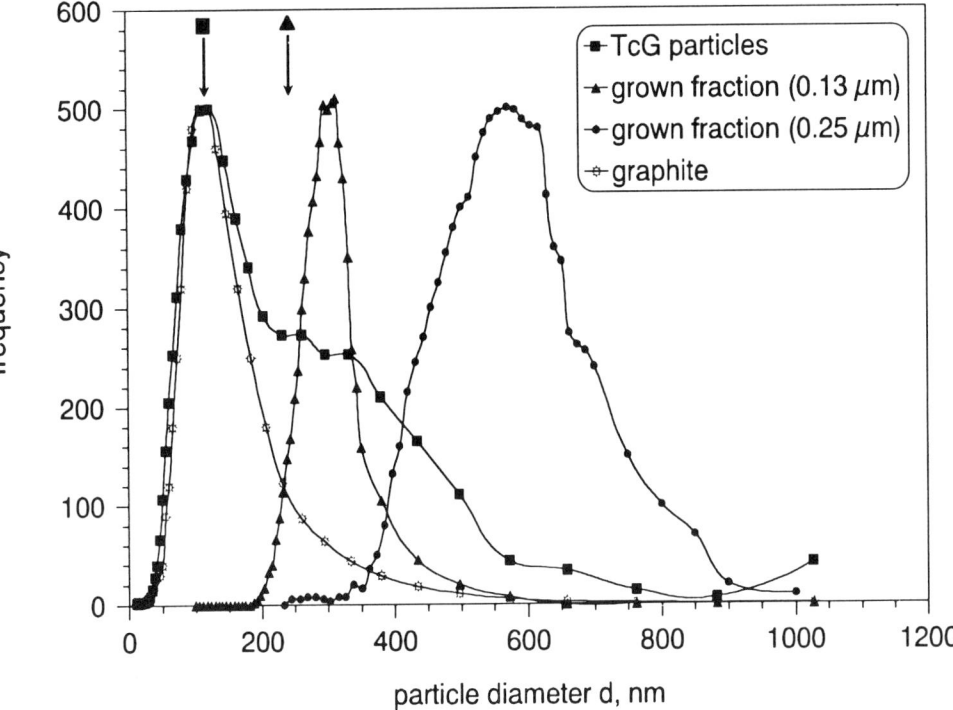

Fig. 1. Size distribution of TcG and the graphite aerosol (without $Na^{99m}TcO_4$ saline). Additionally, two distributions of 0.13 and 0.25 μm monodisperse fractions grown in the humid atmosphere (count mode diameters of the fractions selected by the first DMA are indicated by arrows).

polydisperse distribution by a log-normal distribution yielded a count median diameter (CMD) of 0.13 μm, a mass median diameter (MMD) of 0.55 μm and a geometric standard deviation (GSD) of $\sigma_g = 2.0$. For comparison the size distribution of the graphite aerosol without $Na^{99m}TcO_4$-saline in Fig. 1 presents the same distribution for particles < 200 nm, but does not show the shoulder at 300 nm. The size distribution showed the same count median diameter but a much smaller geometric standard deviation of $\sigma_g = 1.4$. Figure 1 gives also the size distributions of initially monodisperse TcG-particle fractions after hygroscopic growth in humid atmosphere (RH > 90%). Both fractions of 0.13 and 0.25 μm particles selected by the first DMA showed considerable particle growth of about a factor of 2 in diameter.

Table 1 gives the balance of radioactivity and the partition of NaCl with respect to the entire aerosol of 6 l. volume produced during one run. Sodium chloride is

Table 1. Balance of radioactivity and NaCl during TcG-aerosol production

	Relative ^{99m}Tc-activity, %	Mass of sodium chloride, μg
50 μl/sample of $Na^{99m}TcO_4$-saline	100	425
TcG-particles on filter	30	214
Crucible	12	—
Wall losses	58	211

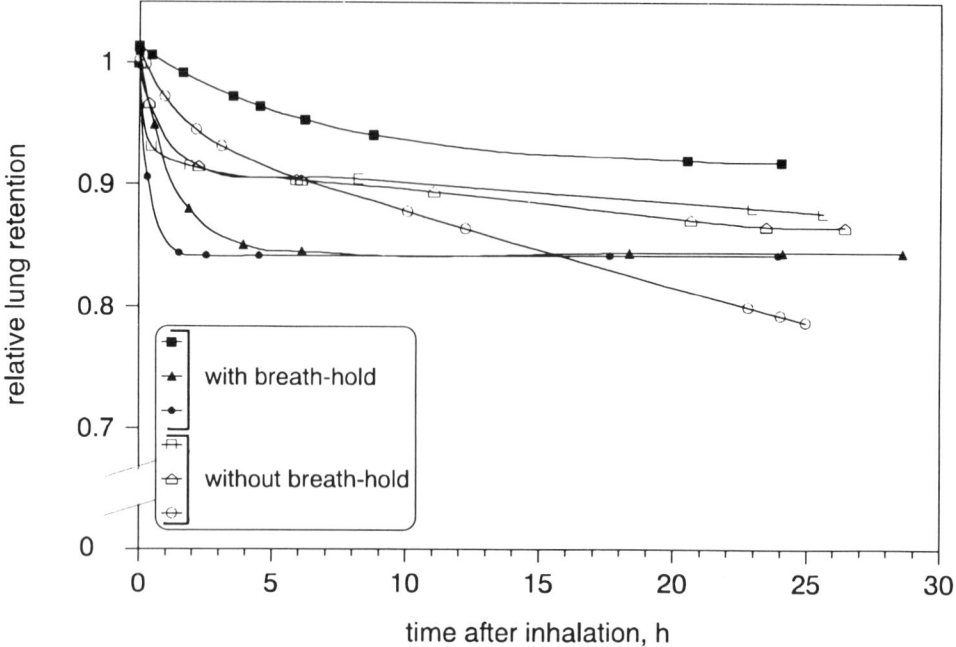

Fig. 2. Intrapulmonary retention of TcG-particles obtained from all six volunteers, data of each subject are fitted with a function of 2 exponential terms (note: the vertical axis is intersected).

completely evaporated and condensed partially on the aerosol particles. It can be washed off in distilled water. The number concentration of the TcG-aerosol of 6 l. volume was measured to be 2×10^7 cm^{-3} and, therefore, the calculated mass of carbon of the TcG particles produced during one run is about the same as the mass of NaCl given in Table 1.

Monitoring the aerosol concentration photometrically in front of the mouth of the volunteer showed an at least two-fold increase of the photometer signal of the exhaled air compared to that of the inhaled air consistent with the hygroscopic growth of the exhaled TcG aerosol due to its NaCl-content as described above.

Total deposition DE with breath-hold (DE = 47 ± 4%) is about twice the value of that without breath-hold (DE = 27 ± 4%). The new ICRP model (ICRP 66, 1994) provides DE = 29.3% for a log-normal size distribution with a MMD of 0.55 μm, a GSD of σ_g = 2.0, at the same breathing conditions.

Figure 2 shows lung retention curves obtained from all six subjects to demonstrate the rather small intersubject variability and the fast v. slow phase of clearance. All data are corrected for physical decay of [99m]Tc.

Fitting the data obtained from all six volunteers with 2 exponential terms, yielded a mean lung retention (± SEM) of:

$$\{(0.086 \pm 0.024) \exp((1.03 \pm 0.79)t)\} + \{(0.91 \pm 0.019) \exp((0.003 \pm 0.0011)t)\}$$

Rates are given in h corresponding to half-lives of 1.6 h and 228 h (9 d).

The dissolution of ^{99m}Tc activity from the particles as determined from urine samples proved to be relatively constant at 9 ± 2.1% of the inhaled ^{99m}Tc activity

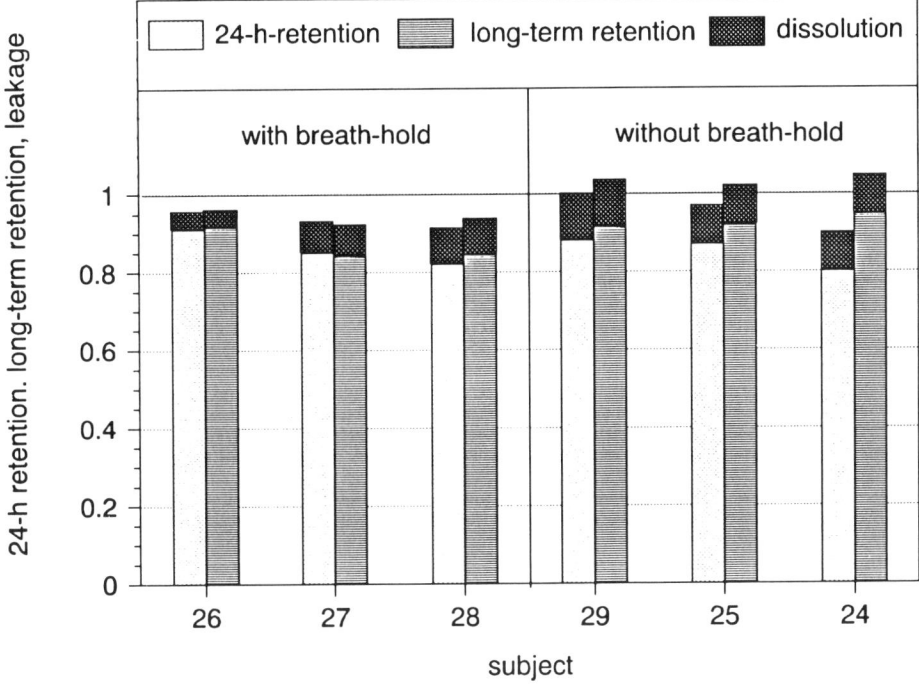

Fig. 3. 24 h retention, long-term retention and dissolution for six subjects.

for all six subjects. Dissolution occurred predominantly during the first 6 h after inhalation.

Figure 3 shows the retained fraction 24 h after inhalation and the dissolved fraction for each subject. Estimation of the fast cleared particle fraction as the difference between particle deposition v. particle dissolution and 24 h retention yielded $5.5 \pm 3.5\%$.

DISCUSSION

TcG-particles are not ultrafine but agglomerates with a polydisperse size distribution of a count median diameter of the distribution of 0.13 μm and a mass median diameter of 0.55 μm. The TcG-particles contain NaCl. As a result, they grow in the humid atmosphere of the respiratory tract at least an order of magnitude in mass (more than a factor of 2 in diameter). This was indicated by aerosol photometry of the in- and exhaled aerosol and shown by the Tandem-DMA measurements. Due to this dynamic behaviour, the distribution of the deposited TcG-particles in the lungs is difficult to predict. According to the ICRP model (ICRP 66, 1994) total deposition does not vary very much for mass median diameters between 0.5–1 μm. Therefore, the measured total deposition agrees well with the predicted values for both the dry mass median particle diameter and the estimated diameter after particle growth. Particles in the diameter range between 0.2 and 1 μm have very low intrinsic motion and are therefore considered as a nondiffusive gas. Based on these measurements and taking the growth of TcG particles into account, it is likely that particles will be deposited in all regions ventilated during the single breath maneuver. There is modest TcG particle dissolution in the lungs which either might reflect leakage of the radio-tracer or particle disintegration and dissolution.

In addition to the 24 h retention and TcG dissolution, the long-term retention is given in Fig. 3 for each subject. The latter is the long-term retention fraction of a function of 2 exponential terms fitting the data of each subject. Note that the long-term retention value is not necessarily compatible with the long-term fraction and the corresponding A-value reported earlier (Stahlhofen *et al.*, 1980) since the latter would require lung retention measurements of more than 3 days. Interestingly, the long-term retentions of those subjects performing a breath-hold were significantly smaller than those of the subjects performing no breath-hold. Accordingly, fast particle clearance was detected after breath-hold while virtually no fast particle transport was observed for the other subjects, see Fig. 3. During breath-hold, deposition of these fine TcG particles increased throughout the lungs as shown by the deposition data given above. This is true particularly in small airways contributing to the fast particle clearance. Therefore, the fast cleared fraction of 6.0 ± 2.6% is more distinct in those subjects which performed a breath-hold when compared to the others.

The ICRP model (ICRP 66, 1994) predicts that the fast cleared fraction represents only half of the bronchiolar deposition, the other half is long-term retained. Based on our data of 5.5% of fast-cleared particles, 11% deposited ^{99m}TcG particles in the bronchiolar region, a bronchiolar dose as high as 330 μSv/MBq is obtained. With the recommended loading activity of 200–400 MBq and an inhaled fraction of 10% of the produced aerosol, a dose of 7–14 mSv is delivered to the airways (Bailey, 1996). This high dose needs careful consideration.

REFERENCES

Bailey, M. R. (1996) National Radiological Protection Board, Didcot, U.K. (personal communication).
Burch, W. M., Sullivan, P. J. and McLaren, C. J. (1986) Technegas—a new ventilation agent for lung scanning. *Nucl. Med. Commun.* **7**, 865–871.
Busch, Ferron, G., Karg, E., Silberg, A. and Heyder, J. (1994) The growth of atmospheric particles in moist air. *J. Aerosol Sci.* **25** (Suppl. 1), 143–144.
International Commission of Radiological Protection (1994) *Human Respiratory Tract Model for Radiological Protection*: a report of a task group of the ICRP. Elsevier Science Ltd, Oxford (ICRP publication 66).
Kroto, H. W., Heath, J. R., O'Brien, S. C., Curl, R. F. and Smalley, R. E. (1985) C-60: buckminster-fullerene. *Nature* **318**, 162–163.
Lemb, M., Oei, T. H., Eifert, H. and Günther, B. (1993) *Eur. J. Nucl. Med.* **20**, 576–579.
Lippmann, M. (1977) *Handbook of Physiology—Reaction to Environmental Agents* (Edited by Lee, D. H. K.), p. 213. Am. Physiol. Soc., Bethesda.
Roth, C., Scheuch, G. and Stahlhofen, W. (1994) Clearance measurements with radioactively labelled ultrafine particles. *Ann. occup. Hyg.* **38** (Suppl. 1), 101–106.
Scheuch, G., Gebhart, J., Heigwer, G. and Stahlhofen, W. (1989) A new device for human inhalation studies with small aerosol boluses. *J. Aerosol Sci.* **20**, 1293–1296.
Scheuch, G. (1991) *Dispersion, deposition and clearance of aerosol particles in the human respiratory airways*. Ph.D thesis, J. W. Goethe-University, Frankfurt/M.
Stahlhofen, W., Gebhart, J. and Heyder, J. (1980) Experimental determination of the regional deposition on aerosol particles in the human respiratory tract. *Am. ind. Hyg. Assoc. J.* **41**, 385–399.
Stahlhofen, W., Scheuch, G. and Bailey, M. R. (1994) Measurement of the tracheobronchial clearance of particles after aerosol bolus inhalation. *Ann. occup. Hyg.* **38** (Suppl. 1), 189–196.

Ann. occup. Hyg., Vol. 41, Supplement 1, pp. 509–514, 1997
© 1997 British Occupational Hygiene Society
Published by Elsevier Science Ltd. All rights reserved
Printed in Great Britain
0003–4878/97 $17.00 + 0.00
Inhaled Particles VIII

PII: S0003–4878(96)00055–5

EFFECT OF MUCUS HYPERSECRETION ON INITIAL TIME COURSE OF INERT PARTICLE CLEARANCE FROM THE LUNG

N. Toms, A. Hasani, D. Pavia, S. W. Clarke and J. E. Agnew

Departments of Thoracic Medicine and Medical Physics, Royal Free Hospital and School of Medicine, Pond Street, London NW3 2QG, U.K.

INTRODUCTION

For radiation hazard estimation it is convenient to characterise retention of inhaled particles in the lung as a sum of exponentials representing clearance from different regions of the normal lung. When mucociliary transport is impaired, this characterisation may work less well. Subjects with impaired mucociliary clearance and mucus hypersecretion clear particles partly by the back-up lung defence mechanism of cough. This study aims to describe and model clearance by cough and also the underlying time-course of "non-cough" clearance of insoluble radiolabelled polystyrene particles over the first 24 h after inhalation. The model chosen was a simplified version of that proposed by the International Commission on Radiological Protection, (ICRP, 1994), with only three exponentially clearing compartments analogous to the bronchial, bronchiolar and alveolar-interstitial compartments of that model. The "alveolar interstitial" compartment was used to accommodate all the slow clearing compartments of the lung since the differences between them are negligible for this study. Most of these are effectively constant over the first 24 h. In addition we considered that a fraction of the other two compartments would be cleared exclusively by cough and could therefore be treated as a separate "cough compartment".

MATERIALS AND METHODS

This was a retrospective analysis of 30 men with mucus hypersecretion and impaired mucociliary clearance associated with COPD. The subjects were divided into two groups of 15 each matched for age, height, weight, smoking history and lung function. However, one group comprised subjects with an exacerbation of their airways disease at the time of the study. This group produced twice as much sputum in 24 h, measured by weight, as the other, baseline COPD, group. All subjects gave informed consent and the original studies had received approval from the hospital's ethical practices subcommittee.

The particle labelling, production and inhalation techniques were well established, having been in use for the last 20 years (Pavia and Thompson, 1976). A mono-disperse aerosol of polystyrene particles, mean diameter 0.5 μm, labelled with $^{99}Tc^m$ was produced by a spinning disc generator, held in a large tank and administered via a system of electronically controlled valves. For each run each patient took eight or nine inhalations of 0.45l. of aerosol, inhalation beginning at FRC and being immediately followed by an obligatory 3 s breath hold. The flow and volume of each inhalation were recorded and the "penetration index" calculated from gamma-camera images of the initial particle deposition and a Krypton ventilation image of the lungs. In this way the depth of penetration of the aerosol could be assessed (Hasani *et al.*, 1994a). These parameters were matched, within the limits of statistical significance, for all runs.

Following inhalation, lung activity was measured every half-hour for the first 6 h and then at 24 h post inhalation. Scintillation counters were used, placed anterior and posterior to the chest, aligned with the midpoint of the sternum. The detectors were collimated to exclude any activity in the oro-pharynx or the stomach. In addition the activity and time of production of any expectorated sputum were also measured. This activity was scaled to assess its previous contribution to the activity in the lung.

Cough correction and model estimation

The retention of particles in the cough compartment is considered to be essentially the inverse of the cumulative expectorated sputum activity. Each productive cough is considered to produce an instantaneous decrease in the activity retained in the cough compartment, giving a stepped function for the cough compartment retention curve. This curve is then subtracted from the raw clearance data which, after renormalisation, gives the "cough corrected clearance curve". This procedure was performed on an individual basis for each patient.

Modelling of the retention in the airways was performed using a commercial modelling program (ModelMaker version 2.0, S.B. Technology, published by Cherwell Scientific Publishing Ltd.) running on an IBM compatible PC.

RESULTS

A plot showing the activity of each productive cough as a percentage of the initial whole lung activity against its time of production shows a linearly decreasing trend, reaching zero at about 6 h after inhalation (Fig. 1). This indicates that clearance of the cough compartment is essentially complete after a period of about 6 h, although there is large variability within the group.

The retention curve for the exacerbation group's cough compartment (Fig. 2) again demonstrates large variability within the group. The plot is best fitted by a single exponential with an intercept at 11.7% initial whole lung activity, dropping to zero after approximately 6 h.

In contrast, the retention curve for the non-exacerbation group (Fig. 3) is best fitted by a straight line with an intercept at 4.4% initial whole lung activity. Again there is large variability within the group with the trend line reaching zero after 6 h. The smaller number of data points indicates the smaller number of productive coughs within this group, reflecting their lower overall sputum production.

When the cough corrected clearance curves are plotted (Fig. 4) the mucociliary clearance of the exacerbation group appears significantly retarded in comparison to the baseline COPD group over the first 6 h (retention at 6 h = $67.1 \pm 4.7\%$ and $58.1 \pm 4.1\%$, respectively). However, the retention values at 24 h for both groups

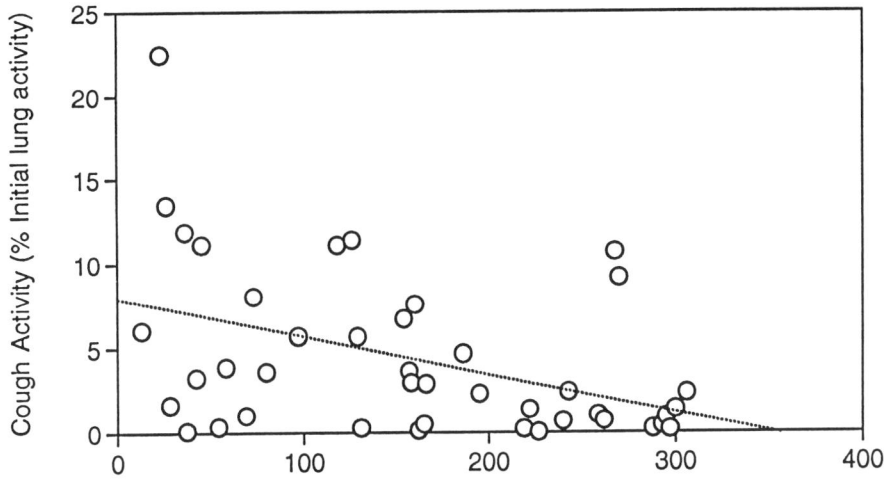

Fig. 1. Individual coughs (exacerbation group).

are not significantly different (35.3 ± 3.3% and 32.7 ± 4.2%, respectively). When a simple, three compartment model with no delay components is optimised for the first 6 h of this data, a good prediction of the 24 h retention is achieved for the non-exacerbation group (35.1% compared to 32.7 ± 4.2% measured) but not for the exacerbation group (44.6% compared to 35.3 ± 3.3% measured). The model parameters are shown in Table 1. To help explain the over-prediction for the

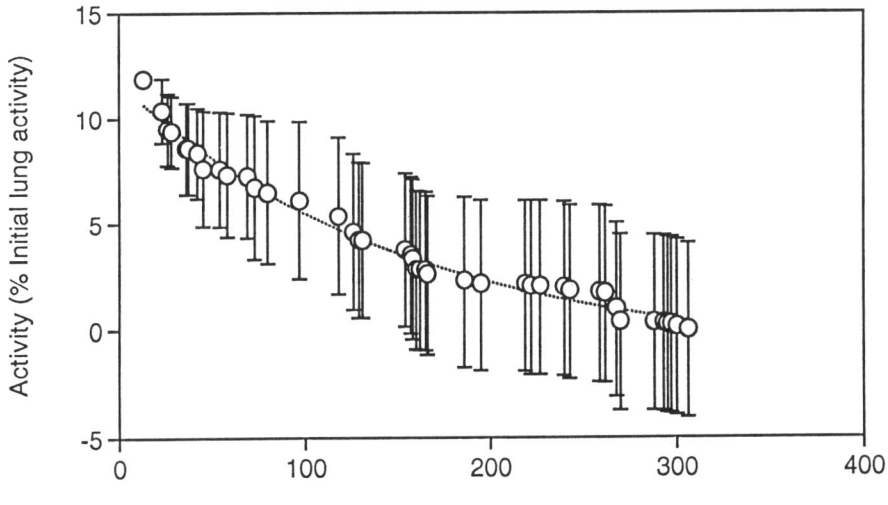

Fig. 2. Exacerbation group cough compartment.

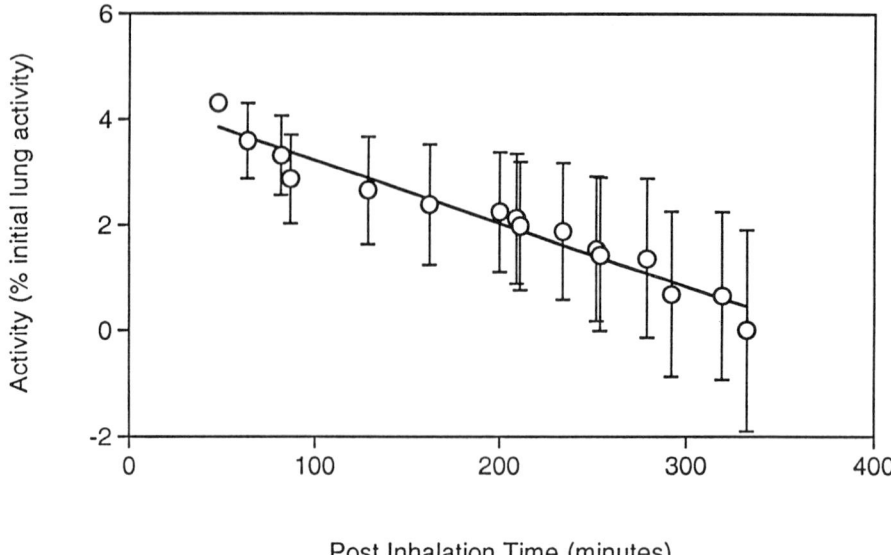

Fig. 3. Non-exacerbation group cough compartment.

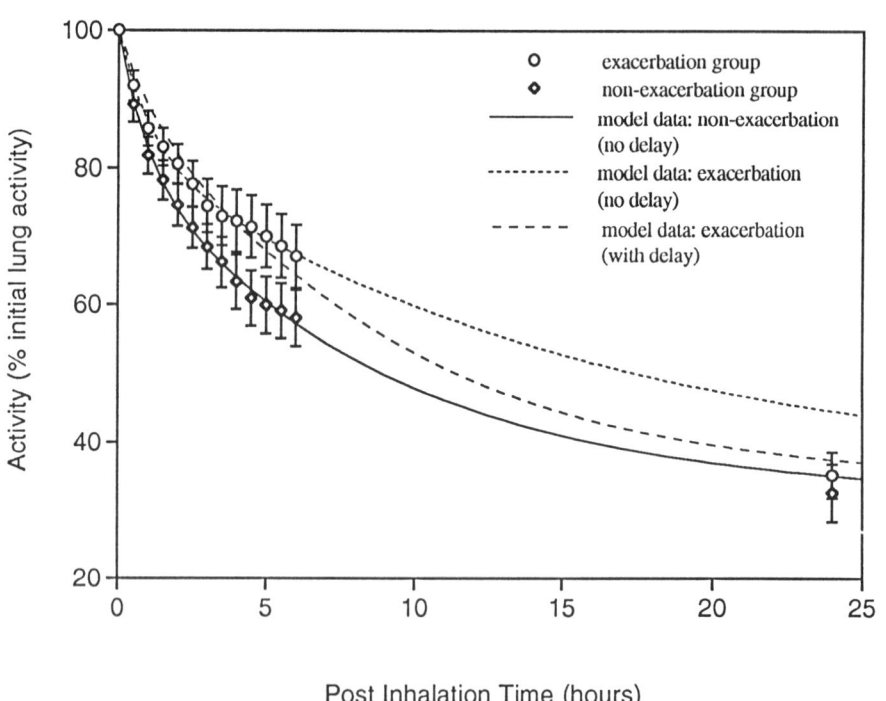

Fig. 4. Cough corrected curves and model data.

Table 1. Model parameters

| Parameter | Exacerbation group, 24 h fit | | COPD group 6 h fit without delay |
	With delay	Without delay	
B1 initial value	18.5	6.7	20.0
B2 initial value	46.1	58.1	47.0
AI initial value[†]	35.3	35.3	32.7
Rate 2	2.9	1.4	1.4
Rate 3	3.4	112.3	26.3
Rate 4	20.0	N/A	N/A

All rates are per day. All initial values are expressed as % initial lung activity.
Rate 1 was set to 0.02 d^{-1} (ICRP, 1994).
[†]AI initial value set from measured data.

exacerbation group it was hypothesised that a certain fraction of the lung clearance was delayed due to a transit time effect arising from the particles' travel time through the lung airways. A delay compartment was therefore added to the model as shown in Fig. 5. The initial values of the model's compartments were kept the same as for the 6 h fit, but the clearance rates allowed to change to optimise the model for the data over 24 h. The addition of this delay component allowed the model to fit the data with more realistic values than were obtained without it (Table 1). If the mean value of the delay and B1 compartments' clearance rates is considered as the effective clearance rate of compartment B1, then all rates generated by the model were within a factor of three of the values given in ICRP publication 66 (ICRP, 1994) for their clearance model compartments. This is in contrast to the parameters given when a delay component is not used. This agreement with the ICRP values is also true of the values obtained for the 6 h fit, without delay, of the data for the non-exacerbation group.

Examination of the initial values chosen by the modelling software reveals lower values for the proximal component (B1) in comparison to the distal component (B2) than might be expected if both compartments are cleared equally by cough. In patients with COPD, aerosol deposition tends to be centrally weighted (Dolovich *et*

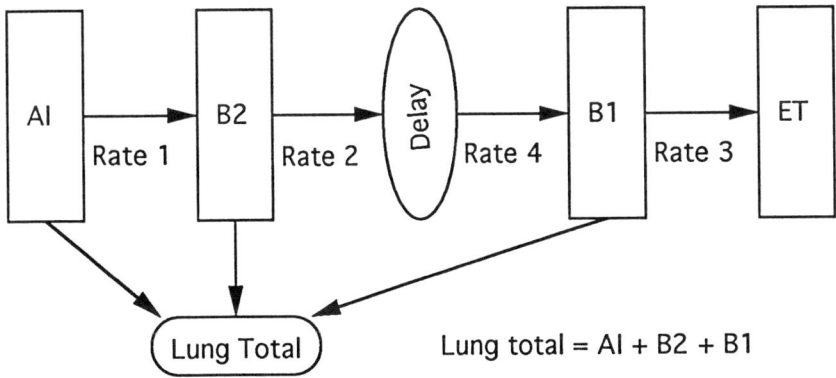

Fig. 5. Clearance model with delay compartment.

Notes: Initial value of Al compartment = measured retention at 24 h post inhalation
Rate 1 = 0.02 / d which is essentially 0 over 24 h.

al., 1976), in which case the proximal compartment (B1) should have a higher initial value than the distal compartment (B2). It would appear from these model values, derived from data in which the overt effects of cough have been removed, that cough correction affects the proximal compartment to the greatest extent. This supports the idea that cough is the predominant clearance mechanism in the central airways.

DISCUSSION

In COPD, cough is generally taken to play a vital role in maintaining airways patency (Hasani, 1994b; Ericsson, 1995). It is clear, from our own and other studies, that productive cough plays too important a role to be neglected when modelling overall lung particle clearance. This is particularly true in the case of COPD mucus hypersecretion augmented during exacerbation. Our data suggest that in such cases cough can, within the first few hours after inhalation, clear a high proportion of the particles initially deposited in the larger airways. In this circumstance a delay or "transit time" may have to be invoked when analysing mucociliary clearance rates from different airway regions. A simple sum of exponential components may not necessarily suffice.

Additionally, in subjects with moderate mucus hypersecretion, cough clearance may best be characterised by a linear rather than an exponential function. However, "non-cough" mucociliary clearance over the first 24 h could be well described by a simple sum of exponentials.

Our findings show that initial lung clearance can be treated as the sum of a small number of simple retention components. Clearance associated with productive cough can be described by one such retention component.

REFERENCES

Dolovich, M. B., Sanchis, J., Rossman, C. and Newhouse, M. T. (1976) Aerosol penetrance: a sensitive index of peripheral airways obstruction. *J. appl. Physiol.* **40**(3), 468–471.
Ericsson, C. H., Svartengren, K., Svartengren, M., Mossberg, B., Philipson, K., Blomquist, M. and Camner, P. (1995) *Eur. Respir. J.* **8**, 1886–1893.
Hasani, A., Pavia, D., Agnew, J. and Clarke, S. W. (1994a) Regional mucus transport following unproductive cough and forced expiration technique in patients with airways obstruction. *Chest* **105**, 1420–1425.
Hasani, A., Pavia, D., Agnew, J. and Clarke, S. W. (1994b) Regional lung clearance during cough and forced expiration technique (FET): effects of flow and viscoelasticity. *Thorax* **49**, 557–561.
International Commission on Radiological Protection (ICRP) (1994) Human respiratory tract model for radiological protection. *ICRP Publication 66. Ann. ICRP*, Vol. **24** (1/4).
Pavia, D. and Thomson, M. L. (1976) The fractional deposition of inhaled 2 and 5 μm particles in the alveolar and tracheobronchial regions of the healthy human lung. *Ann. occ. Hyg.* **19**, 109–114.

Pergamon

Ann. occup. Hyg., Vol. 41, Supplement 1, pp. 515–521, 1997
© 1997 British Occupational Hygiene Society
Published by Elsevier Science Ltd. All rights reserved
Printed in Great Britain
0003–4878/97 $17.00 + 0.00
Inhaled Particles VIII

PII: S0003–4878(96)00172–X

COUGH AND MUCOCILIARY TRANSPORT OF AIRWAY PARTICULATE IN CHRONIC OBSTRUCTIVE LUNG DISEASE

M. L. Groth, K. Macri and W. M. Foster

Department of Environmental Health Sciences, School of Hygiene and Public Health,
The Johns Hopkins University, Baltimore, MD 21205, U.S.A.

INTRODUCTION

Although the pathophysiology of persistent cough remains obscure, cough can be a manifestation of obstructive airway disease. We evaluated mucociliary function of the airway in elderly subjects with obstructive lung disease to determine if spontaneous cough was essential to clearance of secretions and particulate from the tracheobronchial airways.

In severe airway obstruction, failure or delayed clearance of particulate from tracheobronchial airways may be a contributing factor to the increased mortality of respiratory patients found in urban centres of the U.S. (Dockery and Pope, 1993). Voluntary cough has been demonstrated to assist mucociliary clearance of airway mucus in health and severe airway disease (Bennett *et al.*, 1990, 1993). However, the significance of involuntary or spontaneous cough or its frequency to the kinetics of mucus and particulate clearance from tracheobronchial airways in patients with chronic airflow obstruction is not well described (Camner *et al.*, 1973; Puchelle *et al.*, 1980).

METHODS

Sixteen subjects (13 men and three women), at a mean age of 67.4 ± 4.9 years ($\pm$ SD) and having obstructive airway disease were recruited for the study. Table 1 lists the pulmonary function and category of obstructive lung disease for each subject. Patients had relatively stable, moderate-to-severe airway obstruction with an FEV_1/FVC ratio of < 0.70 (mean $= 0.52$). Only two of the patients were non-smokers, four were current smokers and the remaining 10 patients were ex-smokers. Smokers were requested to refrain from smoking on the study day. Consent was obtained from the patients before admission to the study and the research was approved by the University Review Board.

Experimental protocol

Each subject had mucociliary function evaluated on a single control day. Mucociliary function was assessed by noninvasive γ-camera imaging of the lungs for clearance of radioisotopic aerosol deposited onto tracheobronchial airway surfaces. Clearance was assessed over a 4 h period and coughs during this period were recorded. The following medications were restricted: a 24 h washout for anticho-

Table 1. Characteristics of patients*

Subj.#	Age (year)	Gender	FEV$_1$/FVC	Diagnosis	Smoking status
01	68	M	0.36	CB/E	S (20/d)
02	74	M	0.47	Asth	EX
03	61	F	0.57	Asth	NS
04	75	M	0.58	CB/E	EX
05	58	M	0.59	CB	S (30/d)
06	63	M	0.70	Bron	NS
07	71	M	0.44	E	S (15/d)
08	67	F	0.45	E	EX
09	70	M	0.58	Bron	EX
10	73	M	0.40	CB	EX
11	66	M	0.46	CB	S (20/d)
12	66	M	0.47	Asth	EX
13	62	M	0.48	CB	EX
14	71	F	0.57	CB/E	EX
15	69	M	0.64	CB	EX
16	64	M	0.64	Asth	EX
Mean (± SE)	67.4 (± 1.2)		0.52 (± 0.2)		

* Diagnosis: CB, chronic bronchitis; E, emphysema; Asth, asthma; Bron, bronchiectasis.
Smoking Status: S, smoker (# cigarettes/d); EX, exsmoker; NS, non-smoker.

linergic and long-acting theophylline preparations; an 18 h washout of long-acting β-adrenergic bronchodilator; and a 12 h washout of all short-acting bronchodilator, antihistamines, short-acting theophylline preparations and oral and/or inhaled corticosteroid. Cromolyn sodium and expectorants were not permitted for 1 week prior to the evaluation of mucociliary function.

Mucociliary function measurement

To evaluate airway mucociliary function, the subjects in a seated position and with nose clips in place, inhaled by mouth an aerosol of insoluble iron oxide (Fe_2O_3) labelled with ^{99m}technetium–sulphur colloid (^{99m}Tc). To verify labelling procedures, ^{99m}Tc–sulphur colloid was assayed for unbound ^{99m}Tc with silica gel media and thin-layer chromatography. The aerosol was generated by spinning disk technique and was monodisperse in size with a diameter of 5 μm ($\sigma_g < 1.12$) and inhaled with a targeted tidal volume and breathing frequency. Immediately following aerosol inhalation, the subjects gargled and rinsed with water to clear the mouth and oropharynx of radioactivity. The initial distribution and subsequent clearance kinetics of deposited aerosol were measured with the subject in a seated and erect position by imaging the thorax with a posteriorly positioned γ-camera (Maxi Camera, General Electric Medical Systems, Pittsburgh, PA, U.S.A.). Following the initial deposition scan, lung retention of aerosol was measured (300 s scan) at 30 min intervals for the initial 90 min period post-inhalation of aerosol and at time points marking the 2nd, 3rd and 4th h post-inhalation. Subjects returned approximately 24 h post-inhalation of aerosol to have residual lung retention of ^{99m}Tc measured. Lung images were stored by computer (Sopha Med, Columbia, MD, U.S.A.) for subsequent analysis.

Analysis of radioimages

On an initial screening day, after a history was taken and a spirometric assessment of pulmonary function was made, the subjects had a [133]xenon (^{133}Xe) ventilation scan performed. The ventilation scan was acquired to evaluate regional volume and identify lung regions for subsequent analysis of radiolabeled aerosol deposition and clearance. With nose clips in place, the subjects rebreathed ^{133}Xe gas by mouth (Pulmonex, Atom Products, Shirley, NY, U.S.A.) to achieve a steady-state count rate (defined as the point at which there was no further increment in ^{133}Xe counts, a plateau in the count rate for the thorax had occurred) and a lung image was acquired and stored by computer. The steady-state image stored on a video screen enabled regions of interest to be selected by cursor manipulation and drawn to cover (1) the entire right lung field; (2) a central lung zone surrounding the hilus, and (3) a peripheral lung zone that included an outer envelope of the apical, midlung, and basal portions of the right lung. The regional area of the central and peripheral zones encompassed 33 and 67%, respectively, of the area covered by the right lung field. For a number of the patients, radiolabelled aerosol immediately after inhalation was within extrapulmonary areas (digestive tract) and made it difficult to clearly define the left lung base; therefore, deposition and clearance of ^{99m}Tc for the left lung field was not analysed.

Radiolabelled aerosol deposition and clearance were quantified with techniques previously described (Foster *et al.*, 1993; Groth and Foster, 1992).

In addition, ^{99m}Tc aerosol penetration and subsequent retention within distal regions of the lung were assessed by considering that particles initially deposited onto the tracheobronchial airways would be removed by mucociliary clearance (process believed to be completed 18–30 h postdeposition in the healthy lung) (Langenback *et al.*, 1990). Therefore, to gauge ^{99m}Tc deposition distal to the ciliated airways, post-24 h images were acquired (corrected for isotope decay and background activity) and ^{99m}Tc expressed as a percentage of the amount initially deposited.

Clearance of lung particulate was analysed from graphs of lung retention levels plotted as a function of time (Groth *et al.*, 1991). Lung retention (after decay and background correction) was normalised to lung radioactivity at time zero (100%, image acquired immediately following inhalation of radio-labelled aerosol).

Statistical analysis

Means and standard error of the mean (SE) were calculated. Analysis of variance for repeated measures was used to assess time effects on particulate retention and a Newman–Keuls *post hoc* test for significance of differences in lung retention post inhalation.

RESULTS

Patients inhaled ^{99m}Tc with an average tidal volume of 506 ± 28 ml ($\pm$ SE) and frequency of 26 ± 1.3 breaths min^{-1} for a 5 min period to provide a consistent deposition pattern of the radiolabelled aerosol. The results for the subjects are listed in Table 2 and include indices for the distribution of deposition and retention at 24 h post-inhalation of aerosol. On average the central/peripheral ratios for

Table 2. Particle deposition indices*

Subj. #	C/P ratio	24 h retention (%)
01	5.34	6.5
02	9.55	4.4
03	2.80	15.3
04	2.90	4.6
05	7.08	1.5
06	2.05	8.4
07	1.46	7.4
08	3.02	10.4
09	3.01	18.0
10	3.08	8.1
11	ND	1.5
12	2.71	4.5
13	1.65	19.0
14	2.72	40.0
15	3.57	2.0
16	6.26	2.6
Mean (± SE)	3.81 (± 0.58)	9.6 (± 2.5)

* C/P ratio, ratio of radiolabelled particle deposition in central and peripheral lung regions, adjusted for volume of regions (see Methods). ND, no data.

aerosol lung deposition had a value of 3.81 ± 0.6 (± SE), indicative of preferential deposition of aerosol within lobar and segmental bronchi. The mean ^{99m}Tc retention data observed for the patients during the 4 h measurement period are presented in Fig. 1; on average, after 240 min 63.6% of the ^{99m}Tc had cleared the lung and by the 24 h endpoint only 9.64 ± 2.4% of the ^{99m}Tc deposited (Fig. 1 and Table 2) was retained. In addition transport of the insoluble aerosol from the airways was characterised by the clearance halftime, T_{50}, the time required to attain 50% clearance and this index had a mean value for the 16 patients of 145.3 ± 23 min (± SE).

On an individual basis, using the T_{50} and the frequency of spontaneous cough during the 4 h clearance period, the subjects separated into three groups: (1) for six of the subjects the T_{50} were < 75 min and these subjects coughed on average 23 times during the 4 h measurement period; (2) three subjects had T_{50} between 75 and 150 min and averaged 13 coughs; and (3) the remaining seven subjects had T_{50} between 150 and 300 min and coughed infrequently, on average only four times during the entire measurement period (differences between groups 1 and 3 were significant, $P < 0.01$). Figure 2 graphically presents the three categories of the T_{50}. Although our laboratory has limited experience with mucociliary function studies in healthy elderly subjects (Smaldone *et al.*, 1993), based upon archived data of three subjects (59–65 years of age) with normal lung function, a mean T_{50} value of 77 min was observed for C/P aerosol distribution (mean = 3.10) and 24 h retention (mean = 17.9%) similar to those observed in the 16 patients. These normal subjects did not cough during the clearance measurements of ^{99m}Tc and thus the T_{50} represents mucociliary function, unassisted by cough.

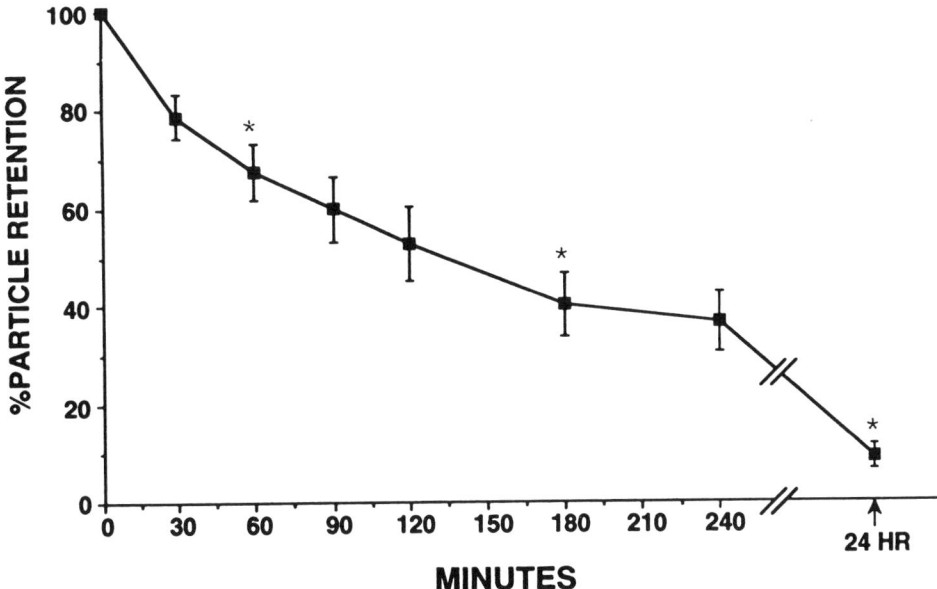

Fig. 1. Insoluble particle clearance in obstructive airway disease. Mean (± SE) lung retention post-inhalation for radiolabelled particles deposited in the right lung ($n = 16$ patients). Mean of residual retention at 24 h post-inhalation is also shown. Difference between lung retention at time zero (immediately post inhalation of labelled aerosol) and mean retention level at each subsequent time point, significant ($P < 0.01$). * Mean lung retention at 60 min, 180 min and 24 h significantly different than retention level of any preceding time point ($P < 0.01$).

DISCUSSION

These results suggest that in elderly subjects with obstructive airway disease spontaneous cough is an essential adjunct to mucociliary function for the effective clearance of epithelial secretions and deposited particulate from airway surfaces. Our results do not indicate the cause of coughing in our subjects; chronic cough is frequently associated with obstructive lung disease and for some patients is related to hypersecretory states of epithelial cells and submucosal glands. To our knowledge, only a single prior study has analysed the influence of involuntary cough on particulate clearance in elderly patients with obstructive airway disease. Camner *et al.* (1973) reported that spontaneous cough assisted airway mucus clearance; however pre-treatment of the patients with an anti-tussive agent complicated the interpretation of the results. In our present study current medications were withheld and no premedication or smoking on study days was permitted. There was a trend in our results toward an association between the number of spontaneous coughs and the speed of clearance (T_{50}) (Fig. 2); but there was no apparent relationship between baseline values of airway pulmonary function (FEV$_1$/FVC) and either the speed of particle clearance (T_{50}), or the number of spontaneous coughs observed during the 4 h measurement period.

Although the physiopathology of chronic cough is not certain, the prevalence of chronic cough is most commonly associated with four clinical disorders: nasal congestion and post-nasal drip, gastroesophageal reflux, asthma and chronic

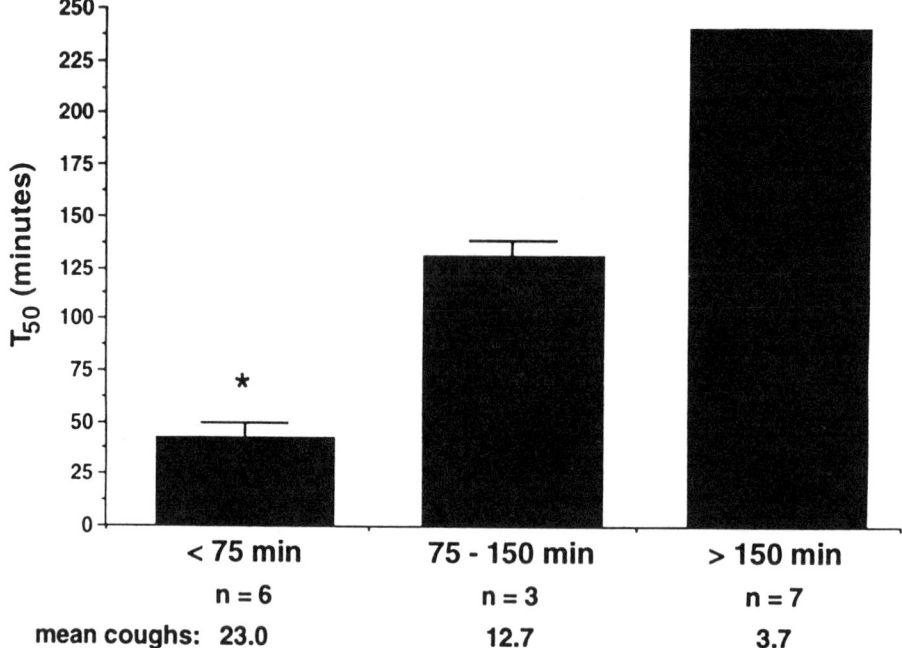

Fig. 2. Distribution of clearance halftime in obstructive airway disease. Mean clearance halftime (T_{50}) for the right lung separated into categories: $T_{50} < 75$ min, $T_{50} > 75 < 150$ min and $T_{50} > 150$ min ($n =$ number of patients). Mean coughs are number of spontaneous coughs observed during the 4 h measurement period of particle clearance. * Difference between mean T_{50} for patients with $T_{50} < 75$ min and patients with $T_{50} > 150$ min, significant ($P < 0.01$) (SE not calculated for patients with T_{50} greater than 150 min since three patients with T_{50} that exceeded the measurement endpoint, were assigned a T_{50} of 240 min).

bronchitis (Banner, 1986; Irwin *et al.*, 1981). None of the subjects studied presented with histories of post-nasal or reflux syndromes, but rather four subjects were described as asthmatic, two with bronchiectasis and the remaining 10 with chronic bronchitis and/or emphysema. Subsequently, the number of asthmatic and chronic bronchitic subjects with either normal (coughed) or delayed (seldom coughed) clearance of airway particles was approximately equal and thus neither disease category (asthma or bronchitis) was judged influential to the observed results.

In bronchial biopsies from subjects (non-asthmatic) with chronic cough a number of consistent observations have been found and include increased epithelial desquamation and the presence of inflammatory cells (particularly mononuclear), in addition to submucosal fibrosis, squamous-cell metaplasia and loss of cilia (Boulet *et al.*, 1994). We did not have biopsy information; by inference, the subjects of our study with severe airway obstruction and spontaneous cough would have been expected to exhibit cellular infiltration of the bronchial submucosa and injury of epithelial airway surfaces.

Despite similarities in age and clinical histories, and equal functional impairment from their airway disease, cough was spontaneous in only about half of our subjects. The cause of this disparity is unknown; perhaps higher levels of inflammatory mediators and cellular influx were present and led to persistent cough (Thompson *et al.*, 1989).

Acknowledgements—This research was supported by awards from the National Heart, Lung, and Blood Institute, #RO1-HL-31429 and National Institute for Environmental Health Sciences, #ES-03819, Washington, DC, U.S.A.

REFERENCES

Banner, A. S. (1986) Cough: physiology, evaluation and treatment. *Lung* **164**, 79–92.

Bennett, W. D., Foster, W. M. and Chapman, W. (1990) Cough enhanced mucus clearance in the normal lung. *J. appl. Phsyiol.* **41**, 146–152.

Bennett, W. D., Chapman, W. and Mascarella, J. M. (1993) The acute effect of ipratropium bromide bronchodilator therapy on cough clearance in COPD. *Chest* **103**, 488–495.

Boulet, L-P., Milot, J., Boutet, M., St. Georges, F. and Laviolette, M. (1994) Airway inflammation in nonasthmatic subjects with chronic cough. *Am. J. Respir. Crit. Care Med.* **149**, 482–489.

Camner, P., Mossberg, B. and Philipson, K. (1973) Tracheobronchial clearance and chronic obstructive lung disease. *Scand. J. Resp. Dis.* **54**, 272–281.

Dockery, D. W. and Pope, C. A. (1994) Acute respiratory effects of particulate pollution. *Ann. Rev. Pub. Health* **15**, 107–132.

Foster, W. M., Silver, J. and Groth, M. L. (1993) Effect of ozone on regional ventilation and particle dosimetry. *J. appl. Physiol.* **75**, 1938–1945.

Groth, M. L. and Foster, W. M. (1992) Aerosolized atropine sulfate: influence of inhalation pattern on effective blockade of vagal airway tone. *Am. Rev. Resp. Dis.* **145**, 215–219.

Groth, M. L., Langenback, E. G. and Foster, W. M. (1991) Influence of inhaled atropine on lung mucociliary function in humans. *Amer. Rev. Resp. Dis.* **144**, 1042–1047.

Irwin, R. S., Curley, F. J. and French, C. L. (1990) Chronic cough: the spectrum and frequency of causes, key components of the diagnostic evaluation and outcome of specific therapy. *Am. Rev. Resp. Dis.* **141**, 640–647.

Langenback, E. G., Bergofsky, E. H. and Foster, W. M. (1990) supramicron-sized particle clearance from alveoli: route and kinetics. *J. appl. Physiol.* **69**, 1302–1308.

Puchelle, E., Zahm, J. M., Girard, F., Bertrand, A., Polu, J. M., Aug, F. and Sadoul, P. (1980) Mucociliary transport *in vivo* and *in vitro*: relations to sputum properties in chronic bronchitis. *Eur. J. Respir. Dis.* **61**, 254–264.

Smaldone, G. C., Foster, W. M., Perry, R. J. and Langenback, E. G. (1993) Regional impairment of mucociliary clearance in chronic obstructive pulmonary disease. *Chest* **103**, 1390–1396.

Thompson, A. B., Daughton, D., Robbins, R. A., Ghafouri, M. A., Oehlerking, M. and Rennard, S. I. (1989) Intraluminal airway inflammation in chronic bronchitis: characterization and correlation with clinical parameters. *Am. Rev. Resp. Dis.* **140**, 1527–1537.

Pergamon

Ann. occup. Hyg., Vol. 41, Supplement 1, pp. 522–530, 1997
© 1997 British Occupational Hygiene Society
Published by Elsevier Science Ltd. All rights reserved
Printed in Great Britain
0003–4878/97 $17.00 + 0.00
Inhaled Particles VIII

PII: S0003–4878(96)00173–1

LUNG PATHOLOGY AND MINERALOGY ASSOCIATED WITH HIGH PULMONARY BURDEN OF METAL PARTICLES: Fe, Ti, Al AND Cr IN A PNEUMOCONIOSIS DATABASE

J. L. Abraham, A. Hunt and B. R. Burnett

Department of Pathology, SUNY Health Science Center, Syracuse, New York, U.S.A.

INTRODUCTION

The inhalation of respirable inorganic particles at low levels is a commonplace everyday occurrence. Typically, the types of particulates which are inhaled and variably retained in the lungs, are those commonly present in ambient aerosols. Much of the particulate matter is aeolian crustal material (deflationary soil) and anthropogenerated material (combustion emissions, construction products). This low or "background" level of exposure is usually orders of magnitude lower than the levels of exposure in occupational settings giving rise to lung disease. We have been studying the human lung burden of retained inorganic particulates for many years (Abraham and Burnett, 1983) and have developed an ever-expanding database (Abraham *et al.*, 1991) of information on the lung burden in currently over 700 analyses of over 600 lungs. The database contains mostly data from persons with known or suspected aerosol particulate exposures of one kind or another; it also contains data from histologically normal lungs and lungs from persons with lung disease but not known occupational or environmental exposures.

Previous studies have examined the relationship of concentrations of a few specific particles to pathologic reactions in patients with specific exposures and/or diseases, such as: sandblasters with silicosis; beryllium ceramics workers with and without chronic beryllium disease; cemented tungsten carbide workers with the unusual pathologic reaction; GIP (giant cell interstitial pneumonia); granulomatous and fibrotic lung disease in persons exposed to metals such as titanium, zirconium or aluminium; and welders. In addition, with the ongoing interest in the relationship of fine particulate aerosols in air pollution to health effects several studies have investigated the potential toxicity of various metals, especially the transition metals, Fe, Ni, V, Cu, Mn and Zn in materials such as residual oil fly ash (ROFA). The higher prevalence of such metals as V in ROFA than in particulates identified in, or recovered from lungs, is consistent with the importance of soluble metals from such particulates. Factors such as particle diameter, surface area, roughness and the presence of surface species are also likely to play critical roles. Metal particulate exposures are also of increasing interest in the etiology and pathogenesis of interstitial lung disease which might otherwise be termed "idiopathic" pulmonary fibrosis/"cryptogenic" fibrosing alveolitis (Kennedy and Chan-Yeung, 1996). Several years ago evidence of metal particulate exposure was found in a number of cases of DIP (desquamative intersitial pneumonia) (Abraham and Hertzberg, 1981). The above supports the need for further research into metal particulate exposures and lung disease.

In this study we investigate the four most commonly occurring metallic elements in the lung burden database, namely: Fe, Ti, Al and Cr. Previously, we have reported the aggregate findings in our database (Abraham *et al.*, 1991) and several groups of cases and individual cases have been investigated with comparison of the findings in selected cases to those in the database. Our aim here is to determine whether the elevated frequency of occurrence of Fe, Ti, Al and Cr is of particular significance. Furthermore, examining the cases with the highest concentrations of these metals may reveal hitherto unrecognized associations of exposures to diseases.

BACKGROUND AND METHODS

For each analysed case, an *in situ* quantitative analysis of particulates using SEM and EDXA yields total concentration as well as specific concentrations of particle types (Abraham and Burnett, 1983). The morphometric analytical technique yields concentrations in volumetric terms, numbers of particles per cubic centimeter of lung tissue. The three major groupings of particles are: silica (containing only Si), aluminium silicates (containing Si and Al at least) and metals (containing metallic elements and not containing Si). Among the metal-containing particles, the concentrations and frequencies of detection vary greatly. A summary of this data is shown in Table 1. The overall distribution of inorganic particle concentrations is log-normally distributed (Fig. 1).

Metal particulates form a diverse group. They are quite different from the silica and aluminium silicate particle groups in that the metal group includes, potentially, at least, virtually any metallic element. Nearly one-third of the elements in the periodic table have been detected in lungs in the database (Abraham *et al.*, 1991). What is difficult to present in detail is the complexity of the metal associations. Since the SEM/EDXA method permits analysis of individual particles and reveals simultaneously all those elements with atomic number greater than 9 (Na and higher) and their relative abundance (with a detection limit of approximately 1% by weight in that particle) the permutations of possible specific particle types with several different elements in the same particle (and multiple possible ratios of one element to the other(s) are practically innumerable. Therefore, initially, for the sake of manageability, the individual cases with concentrations of any of the four most prevalent metals ranked in the top 3% of all analyses were selected for study. This includes the 22 cases each ranked highest by concentrations of Fe, Ti, Al or Cr.

RESULTS AND DISCUSSION

When the top-ranked cases for Fe, Ti, Cr and Al particulate concentrations are tabulated, it becomes apparent that these metals do not always occur in isolation, and are not the result of independent exposures. Table 2 presents the 16 possible combinations of these four elements. It reveals that only a few cases contain high

Table 1. Frequency of various metal particles in 690
tissue section analyses

Metal	Number (and %) of cases where metal identified
Fe	610 (88.41%)
Ti	555 (80.43%)
Al	397 (57.53%)
Cr	315 (45.65%)
Ni	209 (30.29%)
Zn	175 (25.46%)
Sn	162 (23.48%)
Cu	146 (21.16%)
W	126 (18.26%)
Mn	120 (17.39%)
Pb	114 (16.52%)
Ba	89 (12.90%)
Zr	73 (10.58%)
Ce	73 (10.58%)
Ag	60 (8.70%)
V	51 (7.39%)
Au	41 (5.94%)
Co	34 (4.93%)
Hg	31 (4.49%)
Ta	27 (3.91%)
Sb	26 (3.77%)
Bi	25 (3.62%)
La	14 (2.03%)
Nd	14 (2.03%)
Cd	8 (1.16%)
Mo	7 (1.01%)
Br	5 (0.72%)
Nb	5 (0.72%)
Se	4 (0.58%)
Os	2 (0.29%)
As	2 (0.29%)
Ru	1 (0.14%)

concentrations of only one of these elements. Much more common are high concentrations of more than one of the four metals. Moreover, for these "frequently occurring" metals the occurrence of high concentrations of specific metal associations appears to be related to particular exposures. This is reflected in the coding of the individual cases by occupation. For example, it is easy to recognize certain combinations, such as Fe with Cr from steel working operations such as welding or grinding or foundry work. Some additional elements, not the subject of this report, may also be important in distinguishing different types of exposures and pathologic reactions. For instance, Ni is also found in many particles which contain Fe and Cr (indicating a stainless steel type of source). Although there were potentially 88 cases when the top 22 (3%) cases were tabulated by concentration for each of four elements, only 46 individual cases result, since many of the cases show increases of more than one of these elements. The median diameter for metal particles was 0.4 μm. In the analyses for which the particular metal was detected, the median concentrations (in millions of particles per cm^3 of tissue) were: Fe, 10.3; Ti, 7.5; Al, 5.6; and Cr, 3.4.

Frequency Histogram

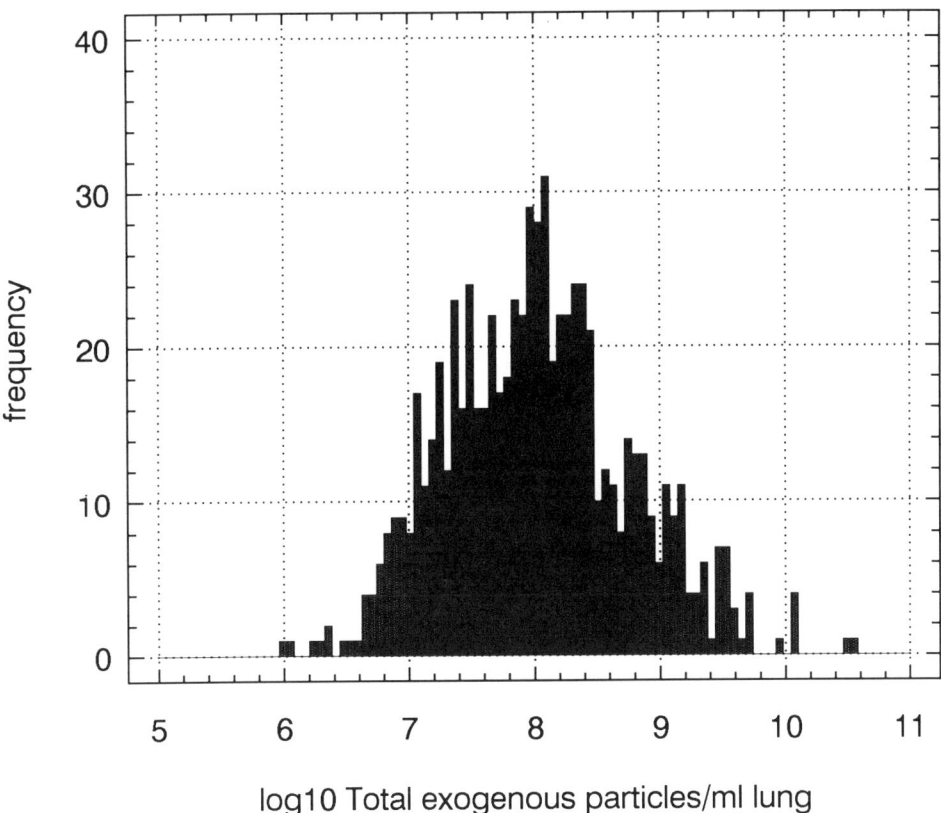

Fig. 1. Histogram showing frequency of cases by concentration of all non-fibrous inorganic particles in pneumoconiosis database. Logarithmic (base 10) scale for concentrations in particles cm^{-3} lung.

Iron

There were five cases showing high Fe but not high Ti, Al or Cr. These cases had histories of metal working and pathology revealed extensive siderosis, some with arc welder's pneumoconiosis, others with mixed dust pneumoconiosis. The other high Fe cases also had elevation of one or more of the other three major metals reported here. The most common association with Fe with Cr (12 cases) and Fe with Ti (11 cases). The histories associated with these cases included metal working and sandblasting and the pathology did not appear specific.

Titanium

There were five cases showing high Ti but not high Fe, Al or Cr. These cases had histories of spray painting and pathology revealed deposits of fine opaque, birefringent dust typical of titanium dioxide. One case had hypersensitivity pneumonitis, shown to be most likely related to toluene di-isocyanate exposure (Redline *et al.*, 1986). The other high Ti cases also had elevation of one or more of

Table 2. Cases with combinations of metals where Cr, Al, Ti, Fe concentrations singly or combined are in the top 3% of all cases

Case (occupation)	Cr Al Ti Fe	Cr Al Ti	Cr Al Fe	Cr Ti Fe	Cr Al	Cr Ti	Cr Fe	Cr	Al Fe Ti	Al Ti	Al Fe	Al	Fe Ti	Fe	Ti
1 (MW) [a]	✓	✓	✓	✓	✓	✓	✓	✓	✓	✓	✓	✓		✓	✓
2 (SB) [b]	✓	✓	✓	✓	✓	✓	✓	✓	✓	✓	✓	✓	✓	✓	✓
3 (SB)	✓	✓	✓	✓	✓	✓	✓	✓	✓	✓	✓	✓	✓	✓	✓
4 (SB)	✓	✓	✓	✓	✓	✓	✓	✓	✓	✓	✓	✓	✓	✓	✓
5 (SB)		✓		✓	✓	✓	✓	✓		✓		✓	✓		✓
6 (SB)			✓		✓			✓		✓	✓			✓	
7 (MW)			✓		✓	✓	✓	✓	✓		✓	✓			✓
8 (SB)				✓		✓	✓	✓	✓			✓	✓		✓
9 (MW)				✓		✓	✓	✓					✓		✓
10 (MW)				✓		✓	✓	✓					✓		✓
11 (SB)						✓		✓							✓
12 (F) [c]						✓		✓							✓
13 (?) [d]						✓		✓							✓
14 (SB)							✓	✓						✓	
15 (MW)							✓	✓						✓	
16 (MW)							✓	✓						✓	
17 (MW)								✓							
18 (SB)								✓							
19 (SB)								✓							
20 (SB)								✓							
21 (SB)								✓							
22 (?)								✓							
23 (DT) [e]								✓	✓			✓			✓
24 (AlW) [f]										✓	✓	✓	✓	✓	✓
25 (AlW)										✓	✓	✓			
26 (MW)												✓		✓	
27 (BeC) [g]												✓			
28 (BeC)												✓			
29 (BeC)												✓			
30 (AlW)												✓			
31 (AlW)															

continued

Table 2. *continued*

Case (occupation)	Cr Al Ti Fe	Cr Al Ti	Cr Al Fe	Cr Ti Fe	Cr Al	Cr Ti	Cr Fe	Cr	Al Fe Ti	Al Ti	Al Fe	Al	Fe Ti	Fe	Ti
32 (ECP) [h]												✓			
33 (HM) [i]												✓			
34 (HM)													✓	✓	✓
35 (HM)													✓	✓	✓
36 (MW)														✓	✓
37 (MW)													✓	✓	
38 (MW)														✓	
39 (MW)														✓	
40 (MW)														✓	
41 (NiR) [j]														✓	✓
42 (HM)														✓	✓
43 (Con) [k]															✓
44 (?)															✓
45 (P) [l]															✓
46 (P)															

(a) Metal working occupations (welding, grinding, cutting, foundry work). (b) Sand blaster. (c) Farmer. (d) Occupation unknown. (e) Dental technician. (f) Al industry/product worker. (g) Beryllium ceramic worker. (h) Electronic components polisher. (i) Hard metal (cemented tungsten carbide) worker. (j) Nickel refinery worker. (k) Construction worker. (l) Painter.

Table 3. Case (occupations), pathology and metal concentrations where Cr, Al, Ti, Fe concentrations singly or in combination are in the top 3% of all cases

Case (occupation)	Pathology	Cr conc.	Al conc.	Fe conc.	Ti conc.	Metals as % of Total
1 (MW)	MD	1079	1174	1084	200	37
2 (SB)	AcS	158	436	1160	828	27
3 (SB)	AcS	433	208	364	455	25
4 (SB)	AcS	276	281	276	241	33
5 (SB)	AcS	64	257	119	484	80
6 (SB)	AcS	380	266	251	143	17
7 (MW)	GR; MD	346	520	476	54	62
8 (SB)	AcS	335	134	328	485	21
9 (MW)	ARC; ASB	191	0	255	168	32
10 (MW)	ARC	977	122	1128	940	99
11 (SB)	AcS	431	0	21	173	21
12 (F)	MD	205	0	10	234	9
13 (?)	MD; ASB	146	142	57	329	43
14 (SB)	AcS	110	0	296	142	21
15 (MW)	MD	191	71	284	6	61
16 (MW)	MD; ASB	362	0	562	2	100
17 (MW)	GR; HSP?	157	37	167	53	60
18 (SB)	AcS	181	140	135	116	50
19 (SB)	AcS	242	0	156	154	30
20 (SB)	AcS	123	52	26	89	28
21 (SB)	AcS	152	29	34	162	24
22 (?)	MD	174	0	57	0	16
23 (DT)	AcS; MD	235	15	185	95	31
24 (AlW)	GR	8	528	327	178	61
25 (AlW)	IF	0	1239	20	378	89
26 (MW)	GR; ARC	0	1235	625	87	47
27 (BeC)	GR	0	458	9	55	45
28 (BeC)	GR	0	218	81	53	24
29 (BeC)	GR	1	213	242	1	73
30 (AlW)	IF (min)	0	1330	3	45	97
31 (AlW)	AlW	0	11 328	0	0	100
32 (ECP)	PAP; AS	0	688	0	0	31
33 (HM)	GIP	0	230	156	102	83
34 (HM)	GIP	0	167	425	821	97
35 (HM)	GIP	0	0	765	650	100
36 (MW)	MD	0	0	430	377	27
37 (MW)	ARC; ASB	0	0	286	29	30
38 (MW)	ARC; ASB	0	0	610	46	97
39 (MW)	ARC	0	14	738	89	100
40 (MW)	ARC	0	0	11 489	0	100
41 (NiR)	MD; ASB	79	9	296	36	70
42 (HM)	GIP	27	27	142	406	11
43 (Con)	MD	0	24	5	178	48
44 (?)	MD	0	80	40	180	74
45 (P)	MD (min)	14	0	20	313	64
46 (P)	MD (min)	2	0	143	226	81

Concentrations in millions of particles cm^{-3} lung; percentages = total metal/total exogenous particle concentration.

Legend for occupation: see Table 2.

Legend for pathology: AcS = accelerated silicosis; MD = mixed dust pneumoconiosis; GR = granulomatous disease; ARC = arc welder's pneumoconiosis (siderosis); GIP = hard metal disease (giant cell interstitial pneumonia); PAP = pulmonary alveolar proteinosis (AS = acute silico-proteinosis); IF = interstitial fibrosis; ASB = asbestos bodies found; AlW = aluminum arc welder's lung; HSP = hypersensitivity pneumonitis; min = minimal.

the other three major metals reported here. The most common association was Ti with Fe (see Fe, above).

Aluminum

There were eight cases showing high Al but not high Ti, Fe or Cr. These cases had histories of aluminum work or beryllium ceramics work. Their pathology revealed interstitial pneumonitis and fibrosis or granulomatous disease. The other high Al cases also had elevation of one or more of the other three major metals reported here. The most common association was Al with Ti (see Ti above). Several of these cases included a series of beryllium ceramics workers (in whose lungs Zr was also found frequently) (Abraham et al., 1995).

Chromium

There were seven cases showing high Cr but not high Ti, Al or Fe. These cases had histories of sandblasting (Wiesenfeld and Abraham, 1995), metal working and dental technician work; pathology revealed mixed dust pneumoconiosis in most. The other high Cr cases also had elevation of one or more of the other three major metals reported here. The most common association with Cr was Fe (see Fe above). Also, other dusty occupations resulted in high concentrations of silica and/or aluminium silicates in some of the cases and with Ni in cases exposed to stainless steel welding, Ni refinery work.

CONCLUSIONS

These preliminary observations further extend the study of inorganic particulates in the lung and illustrate some of the types of questions which a large database of this type can begin to ask. The complexity of the data precludes many definitive conclusions except in cases with high concentrations of single specific types of particles. There is a wide range of pathologic reactions associated with inhalation of metal particulates, very few of which are specific for a single type of exposure, with perhaps the exception of giant cell interstitial pneumonia and arc welder's pneumoconiosis. These findings of a number of granulomatous reactions and non-specific interstitial inflammatory and fibrotic processes also may indicate the need for further study of metal exposures in all such cases. It must be kept in mind that the exposures demonstrated by tissue microanalysis may reveal particles which are only markers of a more complex environment, such as the finding of tungsten in hard metal disease when cobalt is the most suspect agent (Abraham et al., 1991), or similarly, metal fume particles found and toxic gases suspected in welding (Stern et al., 1983). However, we can begin to recognize associations of unusual combinations of metallic elements with certain patterns of occupations and diseases.

REFERENCES

Abraham, J. L. and Burnett, B. R. (1983) Quantitative analysis of inorganic particulate burden *in situ* in tissue sections. *Scanning Electron Microscopy* 2, 681–696.
Abraham, J. L., Burnett, B. R. and Hunt, A. (1991) Development and use of a pneumoconiosis database of human pulmonary inorganic particulate burden in over 400 lungs. *Scanning Microscopy* 5, 95–108.

Abraham, J. L. and Hertzberg, M. A. (1981) Inorganic particulates associated with desquamative interstitial pneumonia. *Chest* **80S**, 67S–70S.

Abraham, J. L., Newman, L. S. and Burnett, B. R. (1995) Beryllium disease: pathologic and quantitative electron probe and secondary ion mass spectroscopic (SIMS) analyses of lung biopsies. *Am. J. Resp. Crit. Care Med.* **151**, A712.

Kennedy, S. and Chan-Yeung, M. (1996) Taking "cryptogenic" out of fibrosing alveolitis. *Lancet* **347**, 276–277.

Redline, S., Barna, B. P., Tomashefski, J. F. and Abraham, J. L. (1986) Granulomatous disease associated with titanium pulmonary deposition. *Br. J. ind. Med.* **43**, 652–656.

Stern, R. M., Pigott, G. H. and Abraham, J. L. (1983) Fibrogenic potential of welding fumes. *J. appl. Toxicol.* **3**, 18–30.

Wiesenfeld, S. L. and Abraham, J. L. (1995) Epidemic of accelerated silicosis (AS) in Texas sandblasters: clinical, pathological and microanalytic observations. *Am. J. Resp. Crit. Care Med.* **151**, A710.

 Pergamon

Ann. occup. Hyg., Vol. 41, Supplement 1, pp. 531–536, 1997
© 1997 British Occupational Hygiene Society
Published by Elsevier Science Ltd. All rights reserved
Printed in Great Britain
0003–4878/97 $17.00 + 0.00
Inhaled Particles VIII

PII: S0003–4878(96)00147–0

CONDUCTING AIRWAY GEOMETRY AS A FUNCTION OF AGE

M. G. Ménache* and R. C. Graham†

*Duke University Medical Center, Pulmonary, Box 3210, Durham, NC 27710, U.S.A.;
and †Man Tech Environmental, Research Triangle Park, NC 27711, U.S.A.

INTRODUCTION

Information on airway geometry (lengths, diameters and branching and gravity angles) is required to perform dosimetry calculations. Data on the dimensions of the conducting airways of the tracheobronchial region of the lung are generally obtained from measurements made on solid casts. Such measurements are time-consuming and because of the large number of conducting airways in humans (between approximately 30 000 and 60 000), only limited measurements on a few casts have been made and reported in the literature (Weibel, 1963; Horsfield and Cumming, 1968; Raabe *et al.*, 1976). More recently, Mortensen *et al.* (1983a, 1983b, 1988) have made measurements in a number of children aged from a few months to 21 years of age.

For use in dosimetry methods, the airway data must be summarized in a geometry model that may be either deterministic (Weibel, 1963; Horsfield and Cumming, 1968; Yeh and Schum, 1980; Phalen *et al.*, 1985) or stochastic (Koblinger and Hofmann, 1985). Although the Weibel (1963) geometry is widely used in dosimetry models, the cast was only described as comparing well with an average adult human lung. Horsfield and Cumming (1968) based their model on measurements made in a 25 year old male. The data of Raabe *et al.* (1976) are for two males, aged 50 and 60 years. The data of Mortensen *et al.* (1983a) provide a number of additional casts that extend the age range down to infants, include females as well as males and allow limited examination of intersubject variability. In this paper, selected descriptive statistics for 11 Mortensen casts are discussed and comparisons of diameters among the adult lung models of Weibel and Yeh and Schum with two adult (18 and 21 years) males in the Mortensen database are made.

METHODS

Eleven casts of Caucasian subjects were used for this analysis (Table 1). The details of the cast preparation and measurement techniques may be found elsewhere (Mortensen *et al.*, 1983a). Length, diameter and branching angle were measured for all airways in the first 10 generations (trachea = generation 1). No measurements were made in any airways distal to the 10th generation.

The data were evaluated qualitatively and using descriptive statistics. Both parametric (mean, standard deviation and coefficient of variation) and nonparametric (minimum, median, maximum, and selected other quantiles) statistics were

Table 1. Descriptive information for the subjects from Mortensen *et al.* (1983a)

Age (years)	Sex	Weight (kg)	Height (cm)	Chest circumference (cm)	Body surface area (m^2)	Number of airways measured
0.25	F	6	66	38	0.34	1149
1.75	M	9	71	51	0.41	1000
1.92	M	9	94	51	0.50	1124
2.33	F	12	95	51	0.58	1035
3.00	F	14	109	51	0.55	1038
8.67	M	26	118	64	0.90	1268
9.42	M	41	143	72	1.37	1270
14	F	51	175	82	1.63	980
14.08	F	56	147	82	1.48	946
18	M	52	135	85	1.48	1028
21	M	67	178	94	1.85	1024

calculated in individual generations for each cast. Where sample sizes permitted, selected statistics were calculated on a lobe basis as well as for the entire lung. The distributions were tested for normality or lognormality using the Shapiro–Wilk statistic. Quantile–quantile (Q–Q) plots were made to compare distributions. The Q–Q plot is used to determine if two samples have the same (potentially unknown) distribution. Two samples falling along the diagonal line of identity have the same distribution (i.e. same centre, spread and shape). Differences in slope are associated with similar shape but different spread and potentially, different center. Nonlinearity implies differences in shape.

RESULTS

Qualitative evaluation

Although the branching structure appeared similar in these 11 casts there were numerous differences between subjects in conducting airway geometry. Figure 1 illustrates this intersubject variability for the 18 and 21 year old subjects. The lengths of the airways are drawn to scale, however the diameters are not. In general, the airways branched dichotomously. In the children under 10 years of age, however, trifurcations were observed in the first branching in the right upper lobe (generation 4) in 5 of the 7 casts. No trifurcations were observed at this location in the 4 subjects aged 14–21.

Descriptive statistics

Box plots of the nonparametric statistics are shown for lengths and diameters in generation 10 (Fig. 2). They illustrate a slight increase in the median value with age and an increase in variability with age, but generally overlapping distributions. In the earlier generations, the increase in median dimensions with age was more pronounced although the distributions also exhibited considerable overlap. The diameter/length ratio was approximately 0.5 at all ages in all generations.

Distribution tests

A lognormal distribution described diameters in generations 6, 7, and 8 well but did not normalize diameters in generation 10. If the distributions could be

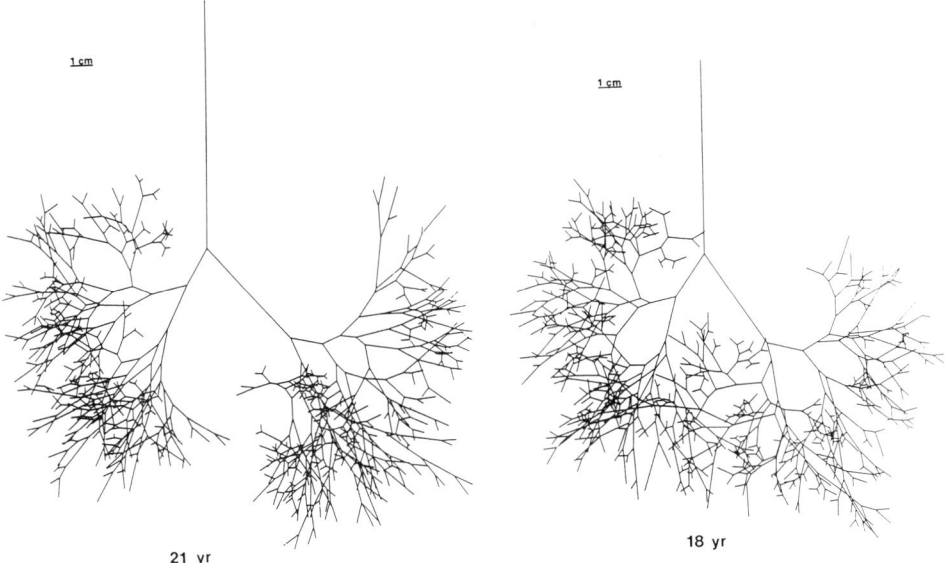

Fig. 1. The diagrams of the airways of the first 10 generations for the two young adult male subjects of the Mortensen data are drawn to scale with respect to airway length and branching angle.

evaluated within each lobe separately, however, the diameter distributions were generally well-described by a normal distribution.

Q–Q plots

In general, subjects of similar ages had similar distributions, all subjects had distributions of similar shape and as the age difference increased, the spread of the distribution associated with the older subject became greater. Figure 3 illustrates the extreme differences between the 3 month and 21 year subjects and the lack of difference between the two 14 year old subjects.

Comparison with published models

The average values for the diameters for the 18 and 21 year old subjects are plotted with the first 10 generations of the Weibel (1963) and Yeh and Schum (1980) models in Fig. 4. The data from the model of Horsfield and Cumming (1968) could not be plotted for comparison as those data are tabulated according to orders rather than generations.

DISCUSSION

Descriptive statistical analyses of airway geometry measurements made in 11 children aged 3 months to 21 years were used to examine inter- and intra-subject variability in lengths and diameters in the first 10 generations. The branching patterns were generally similar but there was substantial variability in the branching angles and airway lengths. Although the diameters were from skewed distributions that may be more extreme than lognormal, if it is possible to examine distributions

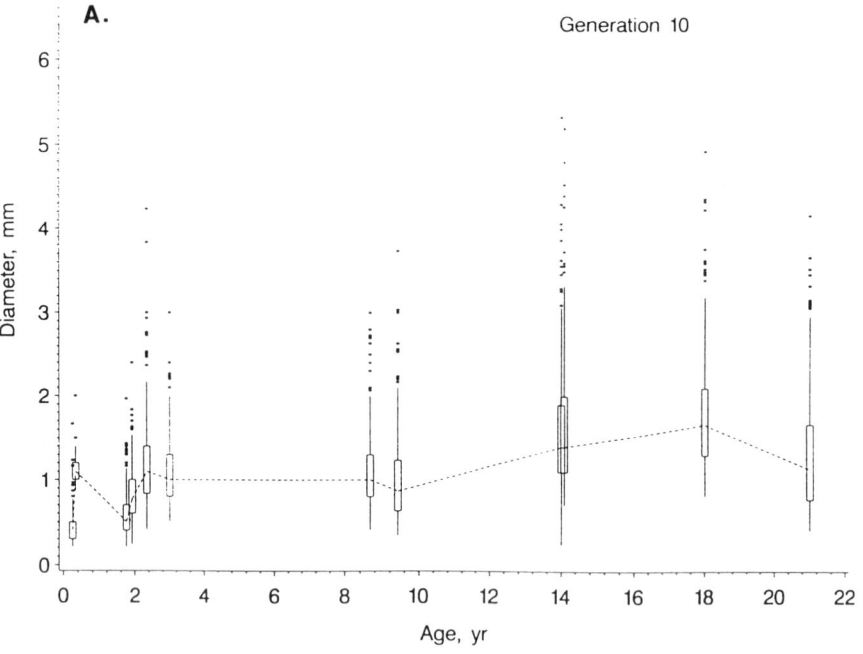

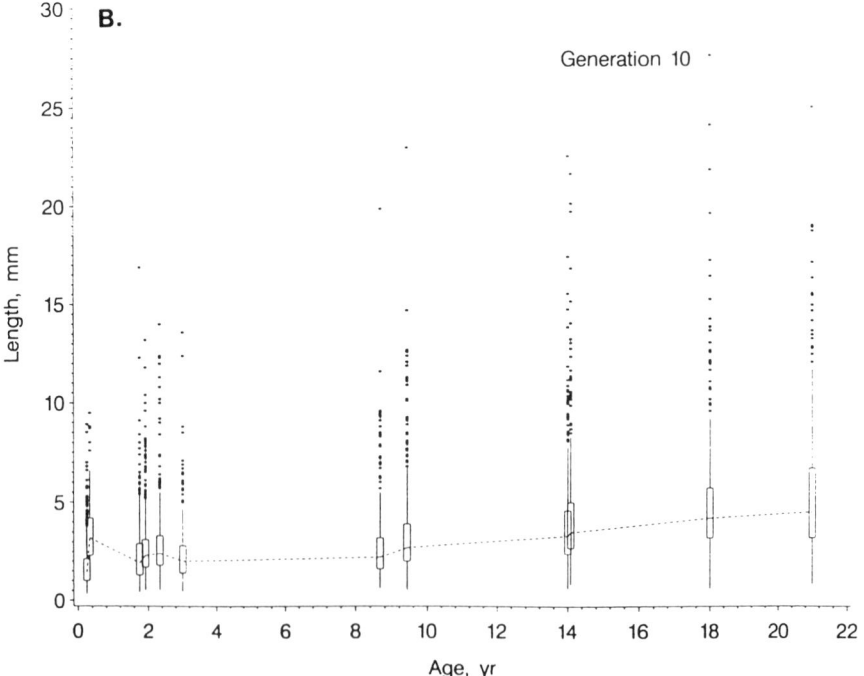

Fig. 2. Box plots of generation 10 diameters (A) and lengths (B) for the subjects as a function of age. The medians are connected by the dashed line, the box is drawn at the 25th and 75th quantiles. The hinge extends 1.5 times the length of the box. More extreme values are plotted as asterisks (*).

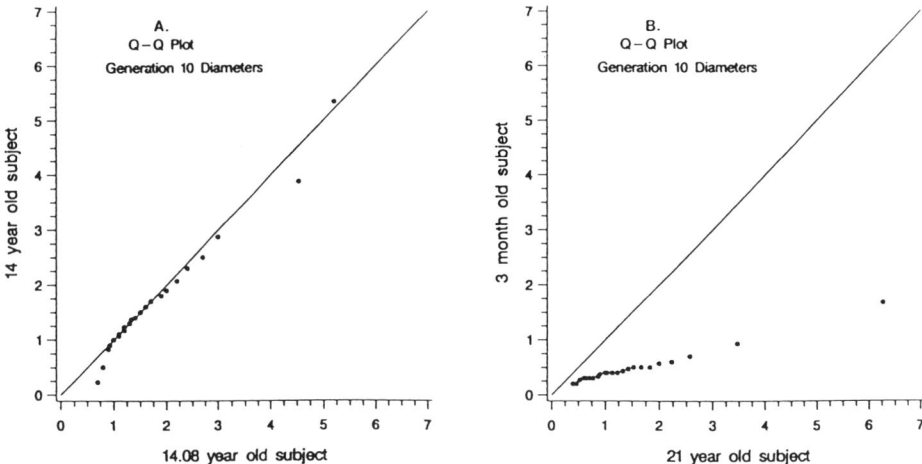

Fig. 3. The two panels are quantile–quantile (Q–Q) plots for generation 10 diameters. The diameters of defined quantiles are plotted for two subjects. The most extreme paired points represent the maximum diameter observed in each of the subjects and the minimum diameters. The intervening points correspond to quantiles 1, 5, 10, 15, . . ., 90, 95, 99. The diagonal is the line of identical distributions (i.e. same center, spread and shape). The panels illustrate that for subjects of similar age the distributions tend to be similar (A). At different ages there is a difference in spread but not shape (B).

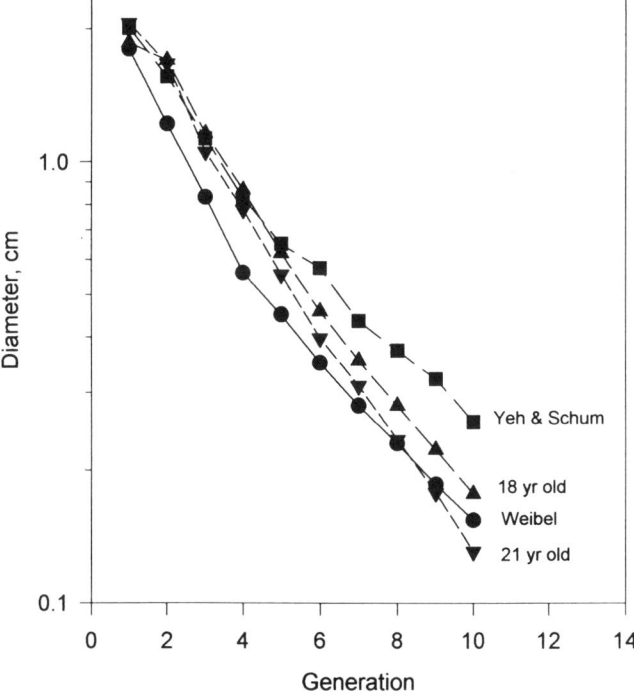

Fig. 4. The mean diameters for generations 1–10 (trachea = generation 1) are plotted for the 18 and 21 year old male subjects from the Mortensen database. Also shown are the values from the single path model of Yeh and Schum (1980) and the symmetric model of Weibel (1963).

on a lobe basis rather than entire lung, the diameters may be described by a normal distribution. Because sample sizes rapidly become small on a lobe basis, however, this observation applies only to the more distal generations. Comparisons of the 18 and 21 year old males with two published models show good agreement with the Yeh and Schum model in generations 1–4 and with the Weibel model in the more distal generations. It is possible that the expansion in the lower airway diameters observed in the Yeh and Schum model may be associated with aging.

Acknowledgements—This work was supported in part by the National Health Effects and Environmental Research Laboratory (Cooperative Agreement Number CR819093 and the National Center for Environmental Assessment of the U.S. Environmental Protection Agency. The research described in this article has been reviewed by the National Health Effects and Environmental Research Laboratory, U.S. Environmental Protection Agency, and approved for publication. Approval does not signify that the contents necessarily reflect the views and policies of the Agency nor does mention of trade names or commercial products constitute endorsement or recommendation for use.

REFERENCES

Horsfield, K. and Cumming, G. (1968) Morphology of the bronchial tree in man. *J. appl. Physiol.* **24**(3), 373–383.

Koblinger, L. and Hofmann, W. (1985) Analysis of human lung morphometric data for stochastic aerosol deposition calculations. *Phys. Med. Biol.* **30**(6), 541–556.

Mortensen, J., Schaap, R., Bagley, B., Stout, L., Young, J., Stout, A., Burkart, J. and Baker, C. (1983a) Final report: a study of age specific human respiratory morphometry. Technical Report TR 01525-010, University of Utah Research Institute, UBTL Division.

Mortensen, J., Young, J., Stout, L., Stout, A., Bagley, B. and Schaap, R. (1983b) A numerical identification system for airways in the lung. *Anat. Rec.* **206**, 103–114.

Mortensen, J., Stout, L., Bagley, B., Burkart, J. and Schaap, R. (1988) Age related morphometric analysis of human lung casts. In *Extrapolation of Dosimetric Relationships for Inhaled Particles and Gases*, (Edited by J. Crapo, E. Smolko, F. Miller, J. Graham and A. Hayes) pp. 59–68. Academic Press, San Diego.

Phalen, R., Oldham, M., Beaucage, C., Crocker, T. and Mortensen, J. (1985) Postnatal enlargement of human tracheobronchial airways and implications for particle deposition. *Anat. Rec.* **190**, 167–176.

Raabe, O., Yeh, H., Schum, G. and Phalen, R. (1976) Tracheobronchial geometry: human, dog, rat, hamster. Technical Report LF-53, Lovelace Foundation for Medical Education and Research.

Weibel, E. (1963) *Morphometry of the Human Lung*. Academic Press, New York.

Yeh, H. and Schum, G. (1980) Models of human lung airways and their application to inhaled particle deposition. *Bull. Math. Biol.* **42**, 461–480.

 Pergamon

Ann. occup. Hyg., Vol. 41, Supplement 1, pp. 537–542, 1997
© 1997 British Occupational Hygiene Society
Published by Elsevier Science Ltd. All rights reserved
Printed in Great Britain
0003–4878/97 $17.00 + 0.00
Inhaled Particles VIII

PII: S0003–4878(96)00057–9

FLOW AND DEPOSITION PATTERNS IN SUCCESSIVE AIRWAY BIFURCATIONS

T. Heistracher and W. Hofmann

Institute of Physics and Biophysics, University of Salzburg, Hellbrunner Str 34, A-5020 Salzburg, Austria

INTRODUCTION

Numerical simulation of air flow and particle deposition in lung bifurcation models has recently been improved in various aspects. For example, three-dimensional models have been used by several authors (Asgharian and Anjilvel, 1994; Balásházy and Hofmann, 1995; Yung *et al.*, 1990) and mathematical procedures have been published for the surface design of bifurcation units (Gradon and Orlicki, 1990; Heistracher and Hofmann, 1995). Despite these improvements, only scarce information on the effect of connected airway bifurcations on air flow and particle deposition within bifurcations can be found in the open literature. Indeed, only a few authors (Lee and Goo, 1993; Lee *et al.*, 1996) have theoretically dealt with double bifurcations. Since radial flow components downstream of a single branching unit can account for up to 10% of the corresponding average inlet velocity (Heistracher, 1996), such inter-bifurcation effects should be taken into account.

In the present work, the design of a symmetric double bifurcation is discussed, which is based on anatomical data. Applying the FIRE® (AVL-List, Graz, Austria) computational fluid dynamics (CFD) package for the solution of the Navier–Stokes and continuity equations, the flow field within the double bifurcation is obtained. In addition, deposition sites for 10 μm aerosol particles are calculated under consideration of the concomitant mechanisms of impaction, gravitational settling and interception.

ANATOMICAL CONSIDERATIONS

The design of the double bifurcation model is based on our recently developed mathematical description of a "Physiologically Realistic Bifurcation" (PRB) model (Heistracher and Hofmann, 1995), where the surface of a single arbitrarily shaped bifurcation is defined by a set of 11 parameters and two sigmoid functions. Anatomical data derived by Hammersley and Olson (1992) and Horsfield *et al.* (1971) have been incorporated into this model. Information on lengths and diameters is taken from the widely used Weibel (1963) and ICRP (1994) lung models.

The surface of the symmetrical double PRB model is shown in Fig. 1, with coresponding geometric parameters compiled in Table 1. The resulting numerical

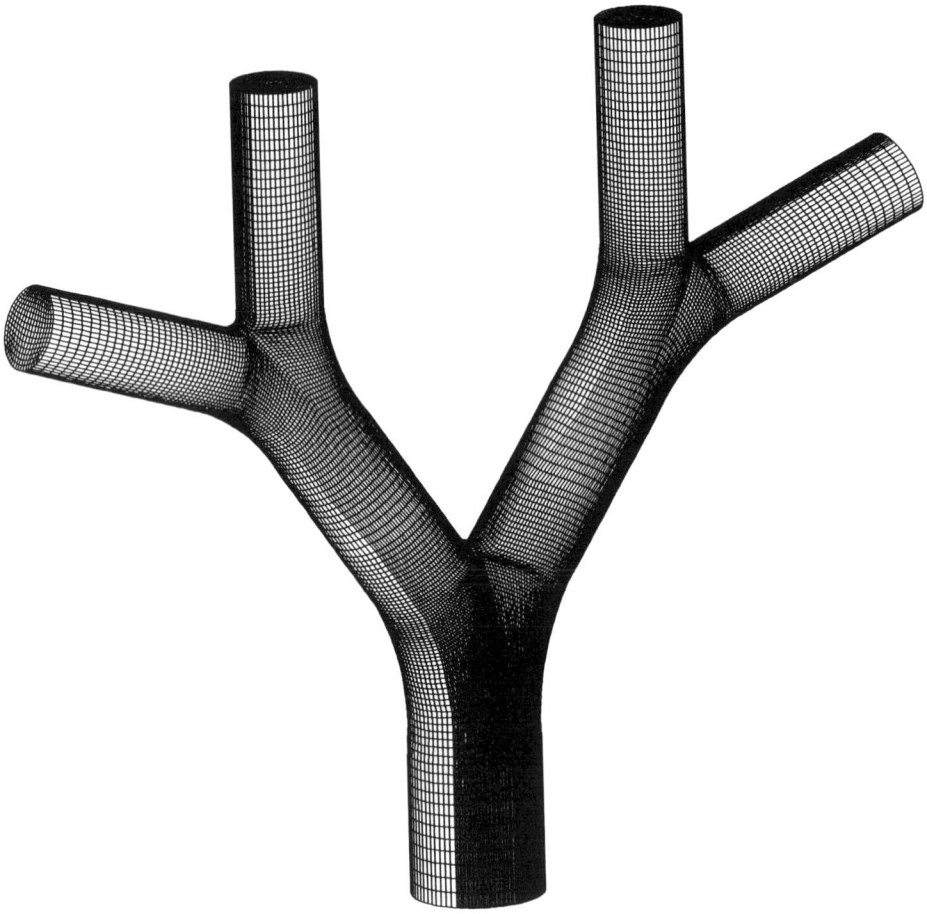

Fig. 1. Surface mesh of the double PRB model. Geometrical parameters and anatomical details
are explained in the related text and Table 1.

Table 1. Geometric parameters used for the construction of the present double bifurcation model

Geometric parameter	Dimension
Diameter of first tube	5.6 mm
Diameter of second tube	4.5 mm
Diameter of third tube	3.6 mm
Length of first tube	10.9 mm
Length of second tube	9.2 mm
Length of third tube	7.7 mm
Branching angles	35°
Outer branch curvature ratio	2.5
Carinal ridge curvature ratio	0.2

mesh consists of about 140 000 internal cells, but owing to symmetry of the present model, calculations may be performed with 70.000 internal cells only. The dimensions correspond to airway generations 3–5 (segmental bronchi), but owing to the scaling procedure mentioned below, the aspect ratios of this geometry are representative for more distal airways too. The whole geometry has a single, multiply curved surface and features three carinal ridges without any sharp edges.

The average ratio of diameters of consecutive branches in generations 0 to 15 in the Weibel (1963) and the ICRP 66 (1994) models is 1.242 and 1.246, respectively. A value of 1.245, which was used in our previous publications (Heistracher and Hofmann, 1994), was utilised for the double bifurcation modelled here.

The ratio of lengths of consecutive branches in the ICRP model varies from 0.92 to 2.53 (from 0.60 to 2.52 in the Weibel model). However, if the extreme values for the very first three generations are excluded, this value becomes rather constant with an average of 1.19 ± 0.09 (1.16 ± 0.14 in the Weibel model). Thus for the double bifurcation model, a value of 1.19 was used.

For the middle part of the double bifurcation (the second tube), the length and diameter of the 4th generation of the ICRP model was used. Starting with this airway, lengths and diameters of the proximal (first tube) and distal (third tube) generations were calculated applying the aforementioned ratios. In a last step, these ICRP based dimensions were marginally scaled by a factor of 1.023 to get the same parent tube diameter in the double bifurcation as already used in our earlier models (Heistracher and Hofmann, 1994).

RESULTS

For the CFD calculations, a parabolic inlet flow profile was chosen at the entrance of the double bifurcation, corresponding to a tracheal flow rate of $60 \, \mathrm{l \, min^{-1}}$. Constant pressure boundary conditions were applied at all outlets of the double bifurcation, which is equivalent to the assumption of symmetric filling of the lungs (for symmetric flows in a dichotomously branching network, symmetric pressure conditions would be anticipated). We will see, however, that due to secondary motions, this condition is not satisfied in the present case.

An isoline representation of the flow velocity in the main plane of the double bifurcation is illustrated in Fig. 2. The initial parabolic flow profile persists downstream up to the middle of the central zone of the proximal bifurcation; from there on, it abruptly flattens when approaching the carinal ridge. A highly skewed profile can be found in the daughter branch of the proximal bifurcation (which is identical to the parent branch of the distal bifurcation). But when comparing it with single bifurcation studies (Heistracher et al., 1995), the double bifurcation exhibits a 6.3% increase in the maximum of the exit velocity at the end of the second tube. This may be related to a minimal asymmetry in the pressure field at that site.

In the daughter branches of the distal bifurcation, considerable asymmetry in both axial and radial flow components can be observed. At the carinal ridge of the distal bifurcation, a 41.5% increase of the axial velocity can be found for the inner daughter branch when compared to the outer branch. When approaching the outlets of the daughter branches, however, this increase is reduced to 22.9%. The ratio of air mass transport in the inner and in the outer branch is 1.29. Thus the

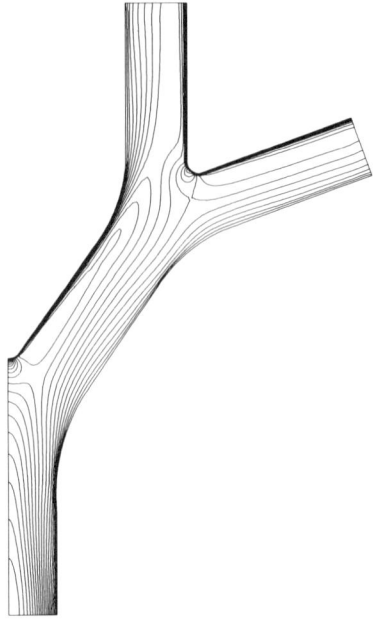

Fig. 2. Isolevel representation of the absolute value of the air velocity in the main symmetry plane of a double bifurcation (parabolic inlet flow profile with a maximum of 10.2 m s^{-1}, 20 isolines ranging from 0 m s^{-1} to 10.2 m s^{-1}, in 0.51 m s^{-1} steps.

Fig. 3. Inspiratory deposition pattern for 10 μm aerosol particles in a double PRB model corresponding to the segmental bronchi (airway generation 3–5) with a tracheal flow rate of 60 l min^{-1}.

branch with the velocity maximum in the main plane (see Fig. 2) experiences also an increased air mass transport.

For the comparison of radial (secondary) flow components, the "secondary motion intensity factor" (SMIF) is used here (Heistracher and Hofmann, 1995). It is defined as the ratio of the local radial velocity maximum (for example at the carinal ridge) and the average inlet velocity of the corresponding parent branch. Thus high SMIF values indicate strong secondary motions. In spite of the fact that the present study was conducted for a symmetric branching pattern, the SMIF values differ markedly, by 61.3%, in favour of the inner daughter branch (0.34 compared to 0.21).

Figure 3 shows the projection of deposition sites of inspired 10 μm unit relative density aerosol particles in the double bifurcation. In comparison to our earlier single bifurcation studies, marked differences in deposition patterns can be found in the double bifurcation. For symmetric flow and symmetric single bifurcation geometries, no differences in deposition in the daughter branches could be found (Heistracher, 1996; Balásházy et al., 1996). Here, however, the branch with the higher SMIF value in the double bifurcation exhibits a higher deposition density and a higher local dispersion of deposited matter compared to the low SMIF value branch. The region of deposition extends downstream of the carina in the inner daughter branch, whereas it is restricted to the vicinity of the carinal position in the outer daughter branch. The ratio of particles deposited in the inner branch to those deposited in the outer branch is 1.204 for the double bifurcation.

CONCLUSIONS

In accordance with our earlier publications (Heistracher et al., 1995), a strong correlation between secondary flow patterns and deposition density can be found in airway bifurcation studies. As shown in Figs 2 and 3, however, distinct asymmetries in air mass transport and deposition patterns can be observed in a symmetric double bifurcation. It will be noted here that the present deposition results are dominated by the mechanism of impaction, but even for very small particle sizes (10 nm), a correlation between secondary motions and local deposition patterns has been observed (Heistracher et al., 1995). In conclusion, asymmetric deposition can occur even in consecutive symmetric airway bifurcations.

Acknowledgements—This work was funded by the Austrian Fonds zur Förderung der wissenschaftlichen Forschung, Project P10426-ÖME.

REFERENCES

Asgharian, B. and Anjilvel, S. (1994) Inertial and gravitational deposition of particles in a square cross section bifurcating airway. *Aerosol Sci. Technol.* **20**, 177–193.

Balásházy, I., Heistracher, T. and Hofmann, W. (1996) Air flow and particle deposition patterns in bronchial airway bifurcations: The effect of different CFD models and bifurcation geometries. *J. Aerosol Med.* **9**, 287–301.

Balásházy, I. and Hofmann, W. (1995) Deposition of aerosols in asymmetric airway bifurcations. *J. Aerosol. Sci.* **26**, 273–292.

Gradon, L. and Orlicki, D. (1990) Deposition of inhaled aerosol particles in a generation of the tracheobronchial tree. *J. Aerosol Sci.* **21**, 3–19.

Hammersley, J. R. and Olson, D. E. (1992) Physical models of the smaller pulmonary airways. *J. appl. Physiol.* **72**, 2402–2414.

Heistracher, T. (1996) Numerical simulation of airflow and particle deposition in bronchial airway bifurcation models. Ph.D. thesis at the University of Salzburg, Austria.

Heistracher, T., Balásházy, I. and Hofmann, W. (1995) The significance of secondary flows for localized particle deposition in bronchial airway bifurcations. *J. Aerosol Sci.* **26** (Suppl. 1), S615–S616.

Heistracher, T. and Hofmann, W. (1995) Physiologically realistic models of bronchial airway bifurcations. *J. Aerosol. Sci.* **26**, 497–509.

Heistracher, T. and Hofmann, W. (1994) Simulation of airflow in a physiologically realistic airway bifurcation model. *J. Aerosol Sci.* **25** (Suppl. 1), S549–S550.

Horsfield, K., Dart, G., Olson, D. E., Filley, G. F. and Cumming, G. (1971) Models of the human bronchial tree. *J. appl. Physiol.* **31**, 207–217.

ICRP Publication 66 (1994) Human respiratory tract model for radiological protection. *Ann. ICRP*, **24**, 1–30.

Lee, J. W. and Goo, J. H. (1993) Inertial deposition of particles in a double bifurcation. *J. Aerosol Sci.* **24** (Suppl. 1), S65–S66.

Lee, J. W., Goo, J. H. and Chung, M. K. (1995) Characteristics of inertial deposition in a double bifurcation. *J. Aerosol Sci.* **27**, 119–138.

Weibel, E. R. (1963) *Morphometry of the Human Lung.* Springer, Berlin.

Yung, C. N., De Witt, K. J. and Keith, T. G. (1990) Three-dimensional steady flow through a bifurcation. *ASME J. Biomech. Engng* **112**, 189–197.

Pergamon

Ann. occup. Hyg., Vol. 41, Supplement 1, pp. 543–547, 1997
© 1997 British Occupational Hygiene Society
Published by Elsevier Science Ltd. All rights reserved
Printed in Great Britain
0003–4878/97 $17.00 + 0.00
Inhaled Particles VIII

PII: S0003–4878(96)00056–7

THE EFFECT OF VENTILATION INHOMOGENEITIES ON AEROSOL DEPOSITION AND BOLUS DISPERSION IN THE HUMAN LUNG

R. Bergmann,* W. Hofmann* and L. Koblinger†

*Institute of Physics and Biophysics, University of Salzburg, Hellbrunnerstrasse 34,
A-5020 Salzburg, Austria; and †Health Physics Department, KFKI Atomic Energy Research Institute,
H-1525 Budapest, Hungary

INTRODUCTION

The transport of aerosol particles in the human lung is influenced by different physical processes. Though some of these processes act primarily in small regions or even in bifurcations, such as convection and diffusion, they can influence the fate of particles in the whole lung. Besides these local physical transport processes, the physiological behaviour of the whole lung might also play an important role in determining aerosol deposition patterns or bolus dispersion. Studies have shown that the human lung is not ventilated homogeneously, either in time or in space (asynchrony and asymmetry effects). Differences in ventilation between topographically distributed regions of the lung have been reported (asymmetry) (Milic-Emili *et al.*, 1966) and a sequential filling of different lung parts has been observed (asynchrony) (Fukuchi *et al.*, 1980; Grant *et al.*, 1974). In the present study the effects of inhomogeneity due to asymmetry and asynchrony on deposition and bolus dispersion of inhaled aerosols in the normal lung are investigated.

METHODS

The fate of inhaled particles are simulated in a stochastic morphometric model of the human lung, using the updated and extended Monte Carlo transport and deposition code IDEAL4 (Koblinger and Hofmann, 1990; Hofmann and Koblinger, 1990). In this model, particles are traced along a randomly selected pathway based on a statistical analysis of morphometric data (Raabe *et al.*, 1976). Deposition is calculated according to the commonly applied deposition formulae. The revised code comprises a new model for alveolar ventilation, where flow splitting in the pulmonary region is calculated proportional to the volume change in the alveoli. In this paper, the airflow into each of the lobes of the lung is different and is assumed to be proportional to their volumes. Asymmetry and asynchrony are simulated by modifying the flow into the lobes by a time-dependent ventilation coefficient expressing asymmetry and a time-dependent linear function expressing asynchrony. The ventilation coefficient reflects the fact that alveoli in the upper lung regions are always partly expanded, most probably due to the weight of the lung and, therefore, alveoli in the lower regions have a higher capacity for air

R. Bergmann *et al.*

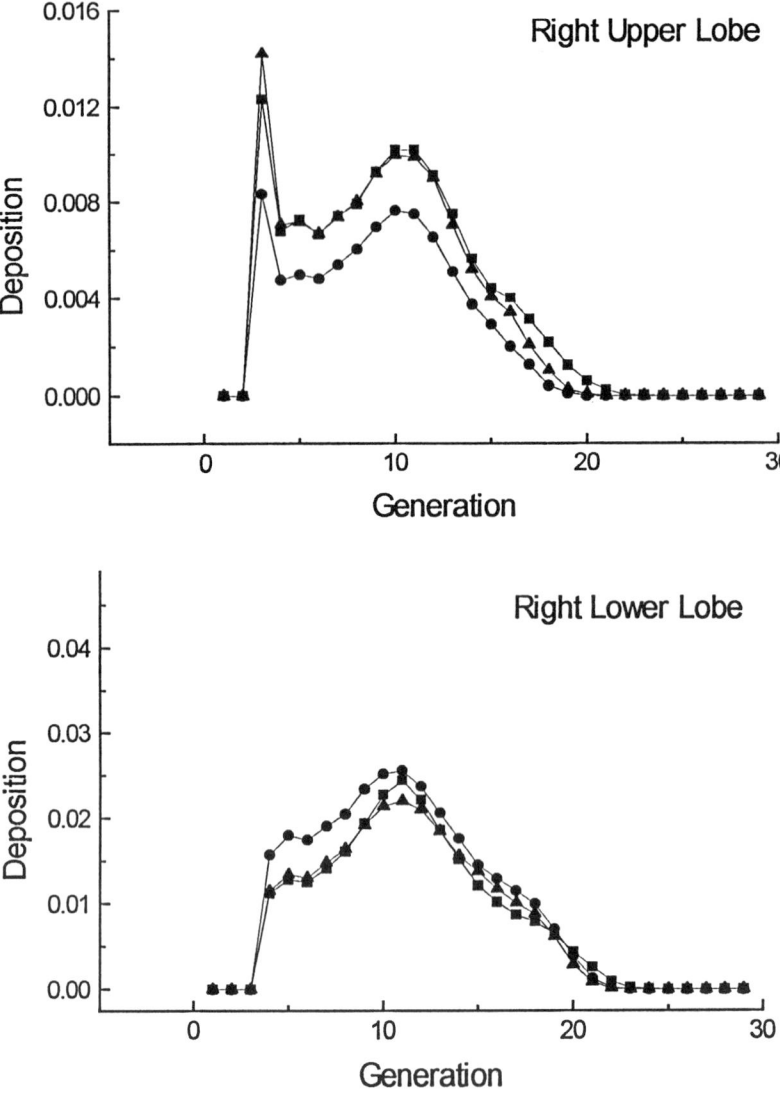

Fig. 1. Deposition of 10 μm particles vs. airway bifurcation number for various ventilation conditions (regional volumes are considered for all curves): (■) no asynchronous or asymmetric ventilation, (▲) only asynchrony, (●) asynchronous and asymmetric ventilation.

exchange. This results in a smaller flow to the upper regions and a greater flow to the lower regions. In our model the topographical regions of the lung are identified by the position of the different lobes. The quantitative value for the ventilation coefficient was 0.67, derived from Milic-Emili (1966), reflecting a 40–60 asymmetry. The properties of asynchrony were chosen according to Nixon and Egan (1987), where lowly ventilated regions are filled first and the maximum deviation from the average flow lies between 20 and 40%.

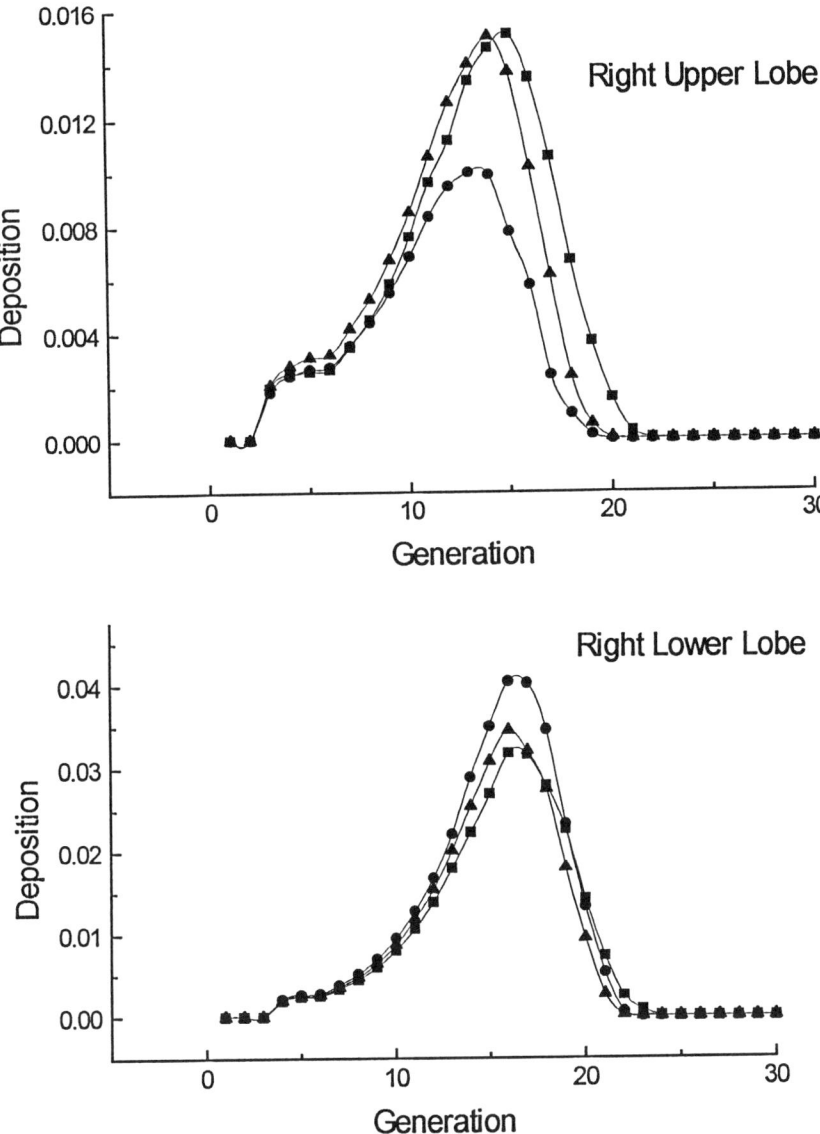

Fig. 2. Deposition of 10 nm particles vs. airway bifurcation number for various ventilation conditions (regional volumes are considered for all curves): (■) no asynchronous or asymmetric ventilation, (▲) only asynchrony, (●) asynchronous and asymmetric ventilation.

RESULTS

The results reflect three scenarios. In the first case we calculated all examined parameters without consideration of ventilation asymmetry and asynchrony. In the second case, we simulated only asynchrony and in the last case we showed the combined effect of asymmetric and asynchronic ventilation. The effects of these three ventilation scenarios on total, regional and local deposition patterns were investigated for a wide range of particle sizes (0.001–10 μm).

R. Bergmann *et al.*

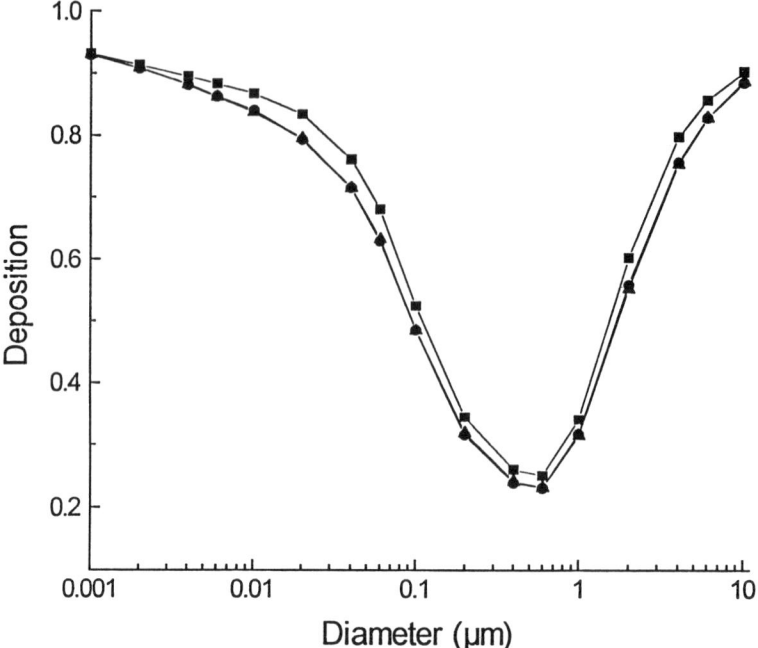

Fig. 3. Total deposition for various particle sizes under different ventilation conditions (regional volumes are considered for all curves): (■) no asynchronous or asymmetric ventilation, (▲) only asynchrony, (●) asynchronous and asymmetric ventilation.

The results displayed in the figures indicate: (i) deposition in the lower lobes is increased due to a higher flow in this region and, consequently, deposition is decreased in the upper lobes; (ii) a slight increase in bronchial deposition can be observed due to higher velocities in the highly ventilated regions. For this reason, pulmonary deposition is not increased in spite of higher velocities in the pulmonary region (Figs 1 and 2); (iii) ventilation inhomogeneities showed only minor effects on total (Fig. 3), bronchial or pulmonary deposition of the whole lung. Likewise, the exhaled bolus was not affected by asymmetry or by asynchrony (Fig. 4). A small increase in the dispersion of an inhaled bolus could be observed when the sequential filling followed the "first-in/first-out" principle with respect to inhalation and exhalation. Since only the "first-in/last-out" principle of the sequential filling is reported in the literature, those results are not presented here. Generally, it can be concluded that ventilation inhomogeneities of the lung do not appear to be an important factor for aerosol deposition or bolus dispersion for the normal lung over the parameter range considered here.

CONCLUSIONS

Though we find local differences in deposition, we can conclude that the effect of ventilation inhomogeneities plays a small role for the normal lung under resting breathing conditions. Local differences of small boli in some parts of the lung can

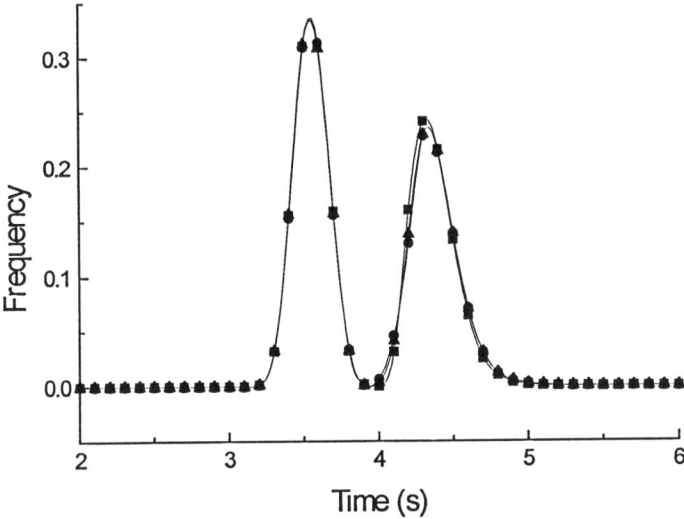

Fig. 4. Inhaled and exhaled aerosol bolus (50 cm^3 volume, 1 μm particle size), introduced 3.6 s after the start of inhalation for various ventilation conditions (regional volumes are considered for all curves): (■) no asynchronous or asymmetric ventilation, (▲) only asynchrony, (●) asynchronous and asymmetric ventilation.

be expected, consistent with the variation in local deposition patterns. Differences in bolus dispersion in a similar study of Nixon and Egan (1987) can be explained by their use of a symmetric model instead of our asymmetric Monte Carlo model where the randomness of airway geometry may compensate for the effect of ventilation inhomogeneities on bolus dispersion.

REFERENCES

Fukuchi, Y., Cosio, M., Murphy, B. and Engel, L. A. (1980) Intraregional basis for sequential filling and emptying of the lung. *Respir. Physiol.* **41**, 253.

Grant, B. J. B, Jones, H. A. and Huges, J. M. B. (1974) Sequence of regional filling during a tidal breath in man. *J. appl. Physiol.* **37**, 158–165.

Hofmann, W. and Koblinger, L. (1990) Monte Carlo modeling of aerosol deposition in human lungs. Part II: deposition fractions and their sensitivity to parameter variations. *J. Aerosol Sci.* **21**, 675–688.

Koblinger, L. and Hofmann, W. (1990) Monte Carlo modeling of aerosol deposition in human lungs. Part I: simulation of particle transport in a stochastic lung structure. *J. Aerosol Sci.* **21**, 661–674.

Milic-Emili, J., Henderson, J. A. M., Dolovich, M. B., Trop, D. and Kaneko, K. (1966) Regional distribution of inspired gas in the lung. *J. appl. Physiol.* **21**, 749–759.

Nixon, W. and Egan, M. J. (1987) Modelling study of regional reposition of inhaled aerosols with special reference to effects of ventilation asymmetry. *J. Aerosol Sci.* **18**, 563–579.

Raabe, O. G., Yeh, H. C., Schum, G. M. and Phalen, R. F. (1976) Tracheobronchial geometry: human, dog, rat, hamster, Lovelave Foundation Report LF-53, Albuquerque, NM, U.S.A.

 Pergamon

Ann. occup. Hyg., Vol. 41, Supplement 1, pp. 548–553, 1997
© 1997 British Occupational Hygiene Society
Published by Elsevier Science Ltd. All rights reserved
Printed in Great Britain
0003–4878/97 $17.00 + 0.00
Inhaled Particles VIII

PII: S0003–4878(96)00148–2

DUST EXPOSURES IN THE LUNGS OF RATS: TIME COURSE OF CHEMOATTRACTANT GENERATION AND NEUTROPHIL RECRUITMENT

I. S. Yuen, M. A. Hartsky, S. I. Snajdr and D. B. Warheit

DuPont Haskell Laboratory, PO Box 50, Elkton Road, Newark, DE 19714-0050, U.S.A.

INTRODUCTION

Inhalation of silica particles produces a chronic pulmonary inflammation response and the development of lung fibrosis and tumors in exposed rats. A variety of cytokines have been implicated in the recruitment of neutrophils into alveolar regions, including Interleukin-1 (IL-1), Tumor necrosis factor -α (TNF-α), KC and Macrophage inflammatory proteins 1 and 2 (MIP-1 and MIP-2). Alveolar macrophages (AM) secrete *ex vivo* at least two pro-inflammatory cytokines, TNF and IL-1 and these have been correlated with the development of silica-induced inflammation and fibrosis. This study was designed to elucidate the temporal sequence and mechanisms of particle-induced pulmonary inflammation following exposures to crystalline and amorphous silica, or titanium dioxide particles.

MATERIALS AND METHODS

General experimental design

Groups of male Crl:CD BR rats (7–9 weeks old, Charles River Breeding Laboratories, Kingston, NY) were intratracheally instilled with either sterile saline, a saline suspension of crystalline silica particles (Berkeley Min-U-Sil; Pennsylvania Glass and Sand Corp., Pittsburgh, PA), amorphous silica particles (Zeofree 80; JM Huber Corp., Havre de Grace, MD), or pigmentary titanium dioxide particles (mean diameter = 0.25 μm; DuPont Co., Wilmington, DE). All particles were autoclaved for 2 h for sterilization and the elimination of endotoxin contamination. Rats were lightly anesthetised with halothane (Aldrich Chemical Co., Milwaukee, WI) and instilled intratracheally according to the method described by Brain *et al.* (1976). It was determined that a silica concentration of 5 mg kg^{-1} body weight could elicit a maximum dose response. Thus, particle concentrations of 10 mg kg^{-1} body weight were used in the subsequent experiments.

Freshly lavaged fluids were centrifuged at 250 **g** to separate BAL cells and concentrated, or immediately assayed for chemotactic activity. BAL cells were used for cell differential determination and for RNA extraction and the subsequent detection of cytokine and mRNA. These experiments were designed to correlate the temporal sequence of the mRNA expression of neutrophil chemotactic cytokines with the detection of chemotactic activity and the appearance of neutrophils in lung fluids following dust exposure.

Bronchoalveolar lavage (BAL)

BAL was performed on each rat as described previously (Warheit *et al.*, 1991) with some modifications. Lavaged fluids collected from saline- or dust-exposed rats were centrifuged at 250 **g** for 10 min to separate cells from fluid. Subsequently, supernatants were recovered immediately for use in the chemotaxis assay. An aliquot of the resuspended cell pellets was used for analysis of cell number, viability and cell differential determinations according to methods described previously (Warheit *et al.*, 1991). The remaining cells were centrifuged at 250 **g**, the supernatant was removed, quick-frozen on dry ice and stored at $-70°C$ for RNA isolation.

Chemotaxis assay

Polymorphonuclear leukocytes were isolated according to the procedure described by Böyum (1976) with some modifications using a Ficoll–Paque® gradient (Pharmacia Biotech AB, Uppsala, Sweden). Peripheral blood was obtained via cardiac puncture of rats. Ten ml of blood was first mixed with 1 ml of 2.7% ethylenediaminetetraacetic acid (EDTA—to prevent coagulation) and diluted with 40 ml of saline (0.98% NaCl). The resulting erythrocyte–neutrophil pellet was mixed with 5 ml of the resulting clear plasma layer and 2 ml of 4.5% dextran (5×10^5 MW) to sediment erythrocytes. The remaining erythrocytes in the isolated polymorphonuclear cell population were removed by treatment with 0.2% hypotonic saline. The neutrophils were then suspended in HBSS supplemented with 20 mg ml^{-1} BSA to a final concentration of 2×10^6 cells ml^{-1}. Chemotaxis assays were carried out in triplicate for each sample of BAL fluids using blind well chambers. The lower wells of the chambers were filled with fresh or concentrated BAL fluids, 1×10^{-5} M fMLP (positive control), or HBSS (negative control). The upper wells of chambers were separated from the lower wells by polycarbonate filters (pore size = 3 μm) and filled with 200 μl of neutrophil suspension. The chemotaxis assay was performed for 90 min at 37°C, 5% CO_2; subsequently, the filters were removed and stained with DiffQuik® (Baxter Scientific). Migrated cells were scored as described previously (Warheit *et al.*, 1991) using light microscopy quantification of 20 predetermined 400× high-power fields (HPF). In addition, checkerboard analysis was performed. It was determined that the majority of chemotactic activity measured in the BAL fluids did not arise from random migration.

Detection of MIP-2 and KC mRNA expression

mRNA expression of two neutrophil chemotactic cytokines, MIP-2 and KC, was detected by reverse transcription–polymerase chain reaction (PT–PCR) amplification using a Stratagene RT–PCR kit followed by PCR amplification. Previously published PCR primer sequences for β-actin, MIP-2 (Driscoll *et al.*, 1993), and KC (Huang *et al.*, 1992) were utilized. The PCR reactions for MIP-2 or KC were always carried out concurrently with β-actin, which served as an internal control. In addition, an RNA sample isolated from air-exposed rats was used as a negative control. Twenty-five cycles of amplification were carried out at a denaturing temperature of 91°C for 1 min, an oligo annealing temperature of 55°C for 1 min and an extension temperature of 72°C for 2 min. Reaction products were separated

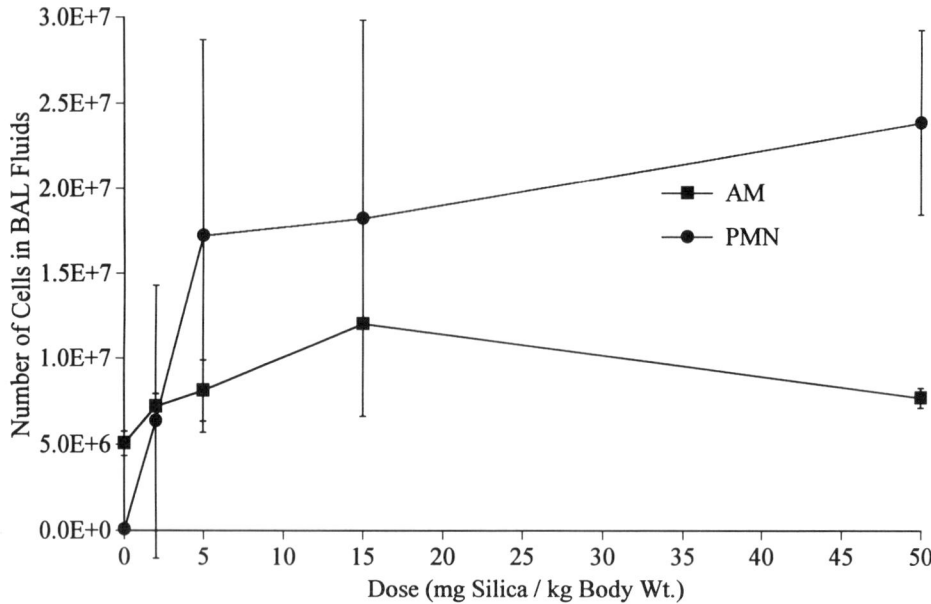

Fig. 1. Dose response of alveolar macrophages and PMN recruitment in BAL fluids of rats 5 h after instillation of silica. Based on these data, the dose of 10 mg kg^{-1} body weight was selected for study with crystalline and amorphous silica as well as TiO$_2$ particles.

by electrophoresis on a 1.5% agarose gel and visualised by ethidium bromide in Tris acetate/EDTA buffer. The sizes of MIP-2, KC and β-actin were in agreement with the published data. For semiquantification of KC or MIP-2 message, the photographs of ethidium bromide-stained gels were scanned with an Eagle Eye II Still Video System (Stratagene, La Jolla, CA) and the RNA band intensities were determined by subtracting the intensity of the β-actin bands using NIH Image 1.55 (written by Wayne Rasband at NIH and available from the Internet by anonymous ftp from zippy.nimh.nih.gov).

RESULTS

Intratracheal instillation of Min-U-Sil crystalline silica produced a marked increase of neutrophils in the lungs of rats within 4 h following dust exposure. The maximum inflammatory response was measured during the 5–8 h period following dust exposure. Thus, the 5 h postexposure time period was selected for the dose response study.

Five hours after instillation, BAL was performed and cell differentials determined. Based on these studies, an instillation dosage of 10 mg kg^{-1} body weight was chosen for subsequent experiments (Fig. 1). Groups of three rats/dust group/postexposure time period were intratracheally instilled with 10 mg kg^{-1} body weight of saline, Min-U-Sil crystalline silica, amorphous silica, or TiO$_2$ particles. Recovery periods were 0.5, 2 and 5 h, 2 and 10 days. An additional control group of three unexposed rats was also investigated. Pulmonary inflammatory responses, as evidenced by the numbers of neutrophils in BAL fluids, were measured within 5 h

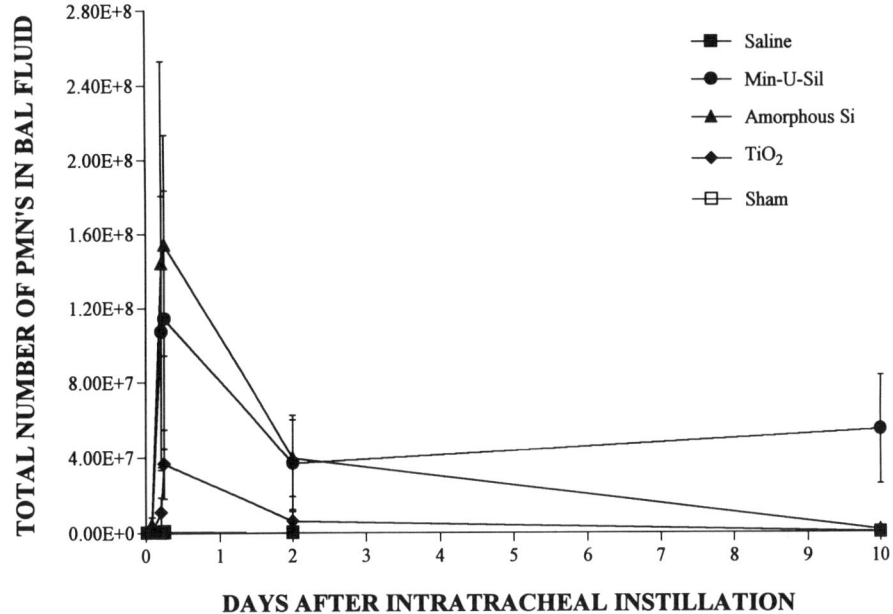

Fig. 2. Time course of neutrophil recruitment in Bal fluids after intratracheal instillation of dusts. Instillation of amorphous silica or TiO₂ particles produced a transient pulmonary inflammatory response, while instillation of crystalline silica produced a sustained inflammatory response.

following exposure to all three particle-types. Amorphous silica produced a significant neutrophilic response but returned to control levels within 10 days post-instillation (Fig. 2). Min-U-Sil silica particles produced a substantial and sustained neutrophilic inflammation, accounting for 30% of BAL cells after 10 days of recovery. In contrast, TiO₂ produced a transient but weak neutrophilic inflammation which returned to control levels after 2 days of recovery (Fig. 2).

Chemotactic activity for neutrophils in BAL fluids was detected as early as 2 h after exposure to either polymorph of silica (Fig. 3) and this was measured prior to the influx of PMNs, which were evident at 4–5 h postexposure. The generation of chemotactic factors in TiO₂-exposed rats was not measured until 5 h postexposure. Some degree of neutrophilic chemotactic activity was measured in lung fluids of crystalline silica-exposed rats after 10 days of recovery and this correlated with the continued influx of neutrophils in the lungs of exposed animals (Figs 2 and 3).

The expression of the MIP-2 gene preceded the measurable chemotactic activity and was detected in mRNA of BAL cells as early as one-half hour following exposure, although this may have been a nonspecific response to the intratracheal instillation of saline and/or particles. A semiquantitative measure of the time course of KC and MIP-2 mRNA expression using the Eagle Eye II Still Video System is presented in Table 1. The gene expression pattern of MIP-2 and KC correlated with the measurement of chemotactic activity and neutrophil recruitment in the acute pulmonary inflammatory responses following exposures to particles. However, while pulmonary inflammatory responses and associated chemotactic activity were apparent in silica-exposed rats at 2 and 10 days

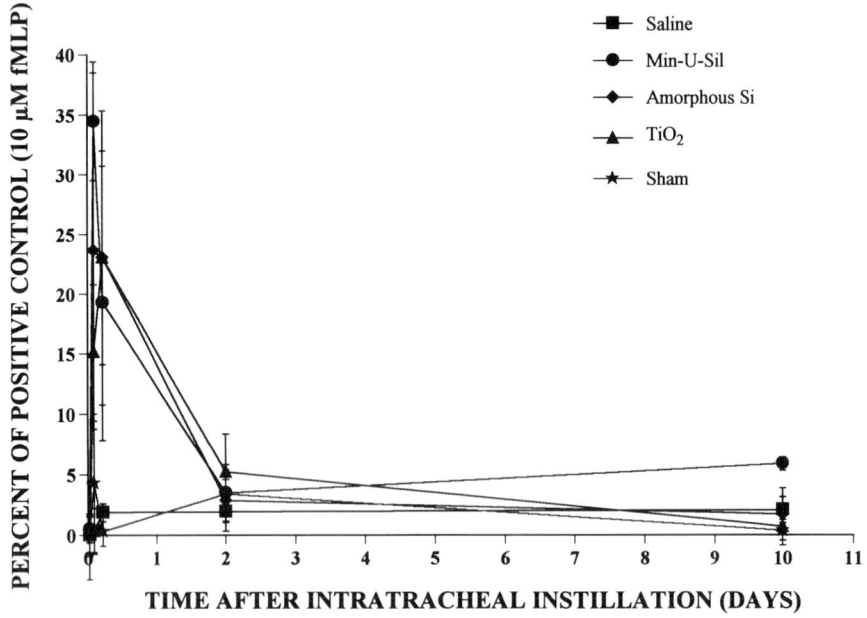

Fig. 3. Time course of neutrophil chemotactic activity in BAL fluids after instillation of dusts. Instillation of amorphous silica or TiO$_2$ particles produced a transient neutrophil chemotactic response, while instillation of crystalline silica produced a more sustained inflammatory response and mirrored the neutrophil recruitment data summarized in Fig. 2.

postexposure, this did not correlate with the expression of KC and MIP-2 mRNA in BAL cells. These results suggest that additional chemotactic factors were associated with the prolonged inflammatory response to silica.

DISCUSSION

The temporal sequence of gene expression of chemokines by BAL cells, neutrophilic chemotactic activity and neutrophil recruitment into the lung were studied following exposure to crystalline silica, amorphous silica, or titanium dioxide particles. The results demonstrated that all three dusts induced neutrophilic

Table 1. Temporal sequence of KC and MIP-2 mRNA expression by BAL cells

	0.5h	2 h	5 h	2 d	10 d
		KC			
Saline	—	—	—	—	—
Min-U-Sil silica	—	++	+++	—	—
Amorphous silica	—	+	++	—	—
TiO$_2$	—	—	—	—	—
		MIP-2			
Saline	+	—	—	—	—
Min-U-Sil silica	+	+++	+++	—	—
Amorphous silica	++	++	+++	—	—
TiO$_2$	+	+	+	—	—

+, low response; ++, moderate response; +++ high response.

inflammation as early as 5 h after exposure. Both crystalline and amorphous silica particle-types were more potent than TiO_2 particles. Maximum neutrophil influx into the lung was measured 5–6 h after exposure. Chemotactic activity for neutrophils was detected directly in bronchoalveolar lavage (BAL) fluids of dust-exposed rats within 2 h after exposure. The chemotactic activity was correlated with the influx and disappearance of neutrophils into alveolar regions of the lung in TiO_2- and amorphous silica-exposed rats. The mRNA expression of two known neutrophil chemotactic cytokines in BAL cells, MIP-2 and KC, were measured prior to the detection of chemotactic activity and the consequent pulmonary inflammatory response. However, this expression was not sustained after 2 days of recovery. It is interesting to note that although both neutrophilic chemotactic activity and inflammation remained prominent 10 days after exposure to crystalline silica, MIP-2 expression could not be detected in BAL cells. Thus, we conclude MIP-2 may be only one of several cytokines involved in directing neutrophilic recruitment after a single instillation of crystalline silica particles.

REFERENCES

Brain, J. D., Knudson, D. E., Sorokin, S. P. and Dans, M. A., (1976) Pulmonary distribution of particles given by intratracheal instillation or by aerosol inhalation. *Environ. Res.* **11**, 13–33.

Böyun, A. (1976) Isolation of mononuclear cells and granulocytes from human blood. *Scand. J. Immunol.* **5**, 77–89.

Driscoll, K. E., Hassenbein, D. G., Carter, J., Poynter, J., Asquith, T. N., Grant, R. A., Whitten, J., Purdon, M. P. and Takigiku, R. (1993) Macrophage inflammatory proteins 1 and 2: expression by rat alveolar macrophages, fibroblasts, and epithelial cells and in rat lung after mineral dust exposure. *Am. J. Respir. Cell Moll. Biol.* **8**, 311–318.

Huang, S., Paulauskis, J. D., Godleski, J. J. and Kobzik, L. (1992) Expression of macrophage inflammatory protein-2 and KC mRNA in pulmonary inflammation. *Am. J. Pathol.* **141**, 981–988.

Warheit, D. B., Carakostas, M. C., Hartsky, M. A. and Hansen, J. F. (1991) Development of a short-term inhalation bioassay to assess pulmonary toxicity of inhaled particles: comparisons of pulmonary responses to carbonyl iron and silica. *Toxicol. appl. Pharmacol.* **107**, 350–368.

Ann. occup. Hyg., Vol. 41, Supplement 1, pp. 554–560, 1997
© 1997 British Occupational Hygiene Society
Published by Elsevier Science Ltd. All rights reserved
Printed in Great Britain
0003–4878/97 $17.00 + 0.00
Inhaled Particles VIII

PII: S0003–4878(96)00058–0

ALVEOLAR MACROPHAGE CLUSTER FORMATION: A CLEARANCE MECHANISM FOR LARGE PARTICLES IN MOUSE LUNGS?

G. Oberdörster, J. Ferin,✠ R. Baggs, K. Pinkerton* and P. E. Morrow

Department of Environmental Medicine, University of Rochester, 575 Elmwood Ave, Box EHSC, Rochester, NY 14642, U.S.A.; and *Department of Pathology, University of California at Davis, Davis, California, U.S.A.

INTRODUCTION

A number of chronic inhalation studies in rats with particles of low toxicity and low solubility have shown that their retention in the lung is significantly increased when lung burdens exceed a level of $\sim$ 1–3 mg g^{-1} rat lung (Morrow, 1988). This condition has been termed "lung particle overload" and, in addition to the impairment of alveolar macrophage-mediated particle clearance, is accompanied by chronic pulmonary inflammation, lung fibrosis and even induction of lung tumours in the rat (Oberdörster, 1995). The impaired clearance function has been associated with the volume of the phagocytosed particles in alveolar macrophages (AM). Under these overload conditions clusters of particle-filled AM have been observed which are thought to represent a sequestration compartment with persistently-retained particles. Persistent retention of particles appears to be a key factor in inducing adverse chronic effects and since mice exposed chronically to high concentrations of particles show significantly lower pulmonary inflammatory and fibrotic reactions and do not induce lung tumours, we designed a study to test the volumetric particle overload hypothesis in this species using radioactively-labelled large and small particles as tracers to monitor their clearance from the lungs. In previous studies with 3 and 10.3 µm microspheres in rats, we had demonstrated that the larger particles were retained in the lungs of rats with a retention halftime ($T_{1/2}$) of $\sim$ 900 days in contrast to the small particles which were efficiently cleared with a $T_{1/2}$ of $\sim$ 70 days (Oberdörster *et al.*, 1992). The failure to clear the large particles was attributed to the volume overload of the macrophages since these particles were found to be efficiently phagocytosed by the rat alveolar macrophages. The objective of the present study was to expand on these earlier studies and to determine AM-mediated clearance rates of small (3 µm) and large (10 µm) particles in the lungs of mice to (i) obtain basic data on clearance function of mouse AM, and (ii) to evaluate a correlation between high volume loading and prolongation of particle clearance.

METHODS

We performed two studies in two different strains of mice. In the first study

B6C3F$_1$ mice (bodyweight 22.5 ± 0.8 g) were exposed by intratracheal instillation with labelled polystyrene microspheres in a 40 μl saline volume as follows: One group received a dose of 3 μg of 3.3 μm particles labelled with ^{85}Sr (3M, St Paul, MN); the second group received 4 μg of 10 μm particles labelled with ^{153}Gd (New England Nuclear, Research Products, Boston, MA). After dosing, thoracic activity of ^{85}Sr and ^{153}Gd were determined in frequent intervals up to 180 days post-exposure using a collimated counter system with two 2 × 2 in NaI detectors. The retention data were analysed after correction for physical decay and the biological halftime of particle retention in the alveolar compartment was calculated by monoexponential curve fitting. In addition, four animals each were sacrificed on days 1, 7, 14, 35 and 100 of the group 1 animals and on days 1, 7, 30, 90 and 185 of the group 2 animals for evaluation of lavage cytospins and for the retained activity in the excised lungs. Lung lavages were performed by using 1 ml of saline instilled into the lungs repeated 10 times; lavage samples were kept on ice until further analysis. Cytospins of the lavaged cells were prepared for evaluation of cell types and particle-cell association.

The second study involved C57 Black/6J mice (bodyweight 24 ± 1.2 g) which were instilled intratracheally with fluorescent polystyrene microspheres suspended in 40 μl of saline containing two particle sizes: 3 and 10 μm (Polysciences, Warrington, PA). A total of 25 μg of each particle size were instilled as a mixture into each mouse. Lung lavages were performed on days 1 and 14 post-instillation and lavage cytospins were obtained for evaluation of particle-cell association in this mouse strain. In addition, lung sections were prepared for electronmicroscopy by fixation with 10% glutaraldehyde and paraformaldehyde.

RESULTS

Figures 1 and 2 show the results of the retention measurements of the first study with B6C3F$_1$ mice dosed with either 3 or 10 μm spheres. Retention measurements of the 3 μm particle dosed mice could only be extended to 100 days since clearance of the labelled particles decreased the remaining activity to a point where it could no longer reliably be determined. The alveolar $T_{1/2}$ of the 3 μm spheres was determined to be 33 days and for the 10 μm spheres the alveolar retention halftime was 103 days as averaged from the four individual retention curves.

Examination of the cytospins of the lavaged lung cells revealed that all of the 3 μm particles were phagocytosed by AM but, in contrast, none of the large 10 μm particles were found within AM. These larger microspheres were found to be surrounded by a cluster of AM (Fig. 3) at all timepoints post-instillation consisting of up to seven AM. It appeared that the AM of the B6C3F$_1$ mice were not capable of phagocytosing these large particles, very likely because of size limitation. No signs of an inflammatory response such as increased granulocyte counts were found in the cytospins at any timepoint.

In contrast to the lavage results of the first study, the AM of C57 Black/6J mice were found to have phagocytosed the 10 μm particles. Figures 4 and 5 show examples of these findings in the lavaged cells and in a section of lung with a large particle inside AM. The retention halftime of the microspheres in this second study with C57 Black/6J mice could not be determined since the particles were not

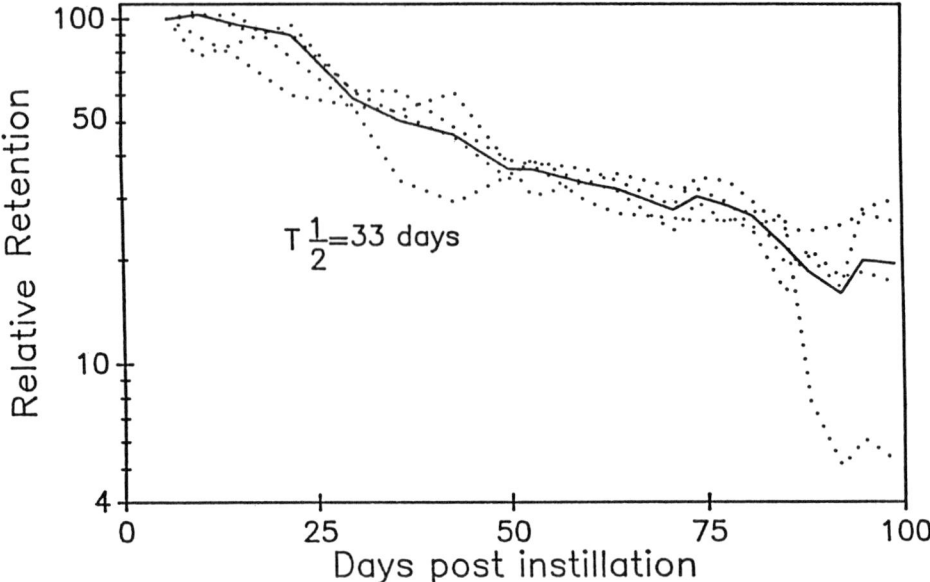

Fig. 1. Pulmonary clearance of [85]Sr-labelled 3 μm microspheres in B6C3F$_1$ mice. Retention of four individual mice (normalised to day 6 = 100%) and their mean retention is shown.

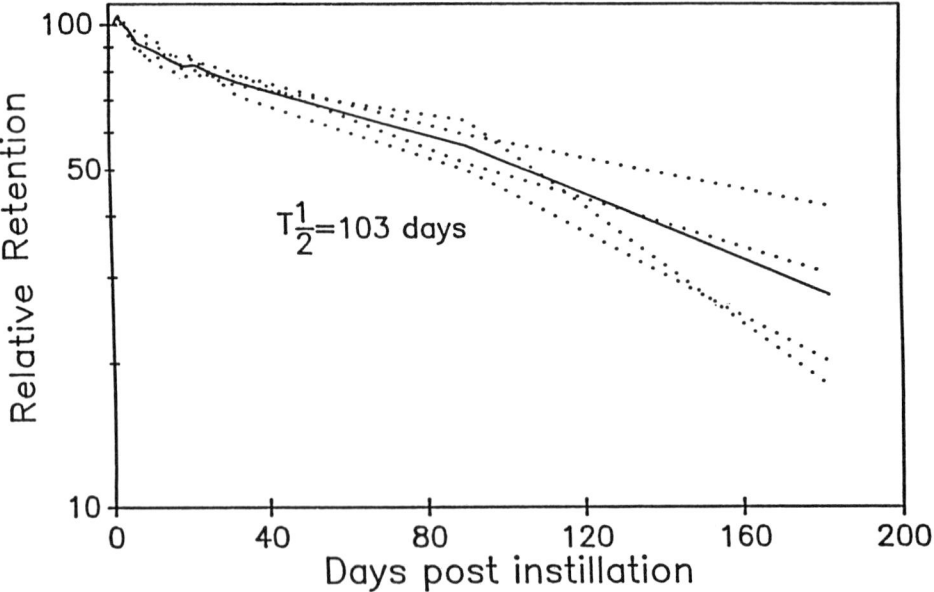

Fig. 2. Pulmonary clearance of [153]Gd-labelled 10 μm microspheres in B6C3F$_1$ mice. Retention of four individual mice and their mean retention is shown.

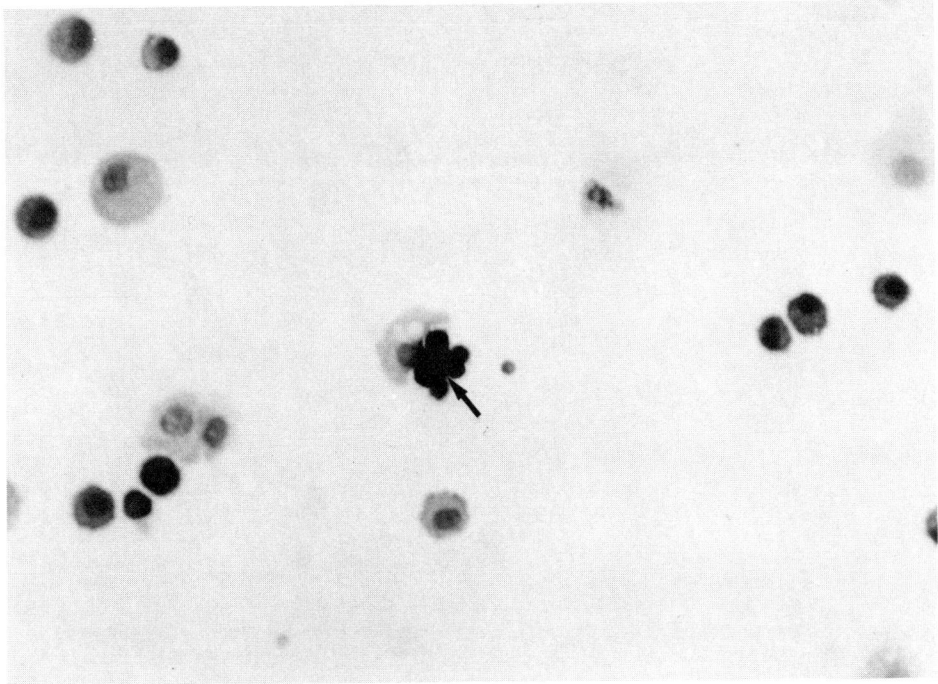

Fig. 3. AM cluster with non-phagocytosed 10 µm particle (arrow), first study with B6C3F₁ mice. No phagocytosed large particles were found.

radioactively labelled. No increase in inflammatory cells in the lavage were found at the two timepoints examined.

DISCUSSION

The alveolar $T_{1/2}$ of 33 days for the small particles found in the first study over a period of 100 days post-administration is in agreement with other data from mice reporting retention halftimes between 30 and 50 days (Kreyling, 1990). However, the finding that large 10 µm particles are eliminated from the lungs of B6C3F₁ mice with relatively good efficiency ($T_{1/2} = 103$ days) is surprising in view of the fact that in rats these large particles are persistently retained in the alveolar region (Snipes and Clem, 1981; Oberdörster et al., 1992). Perhaps even more surprising is the finding that the elimination of these large particles occurs apparently without phagocytosis by AM, but is possibly aided by the formation of clusters of AM surrounding the large particle.

A comparison of the retention halftimes of small and large particles between rats and mice is shown in Fig. 6 which summarises the present studies in the B6C3F₁ mice and our previous study in Fischer-344 rats. The large particles in the rat are retained with a > 12-fold longer $T_{1/2}$ compared to the small particles, whereas in mice the alveolar $T_{1/2}$ for large and small particles differ only by a factor of 3 indicating that AM-mediated particle clearance is not nearly as much affected as in

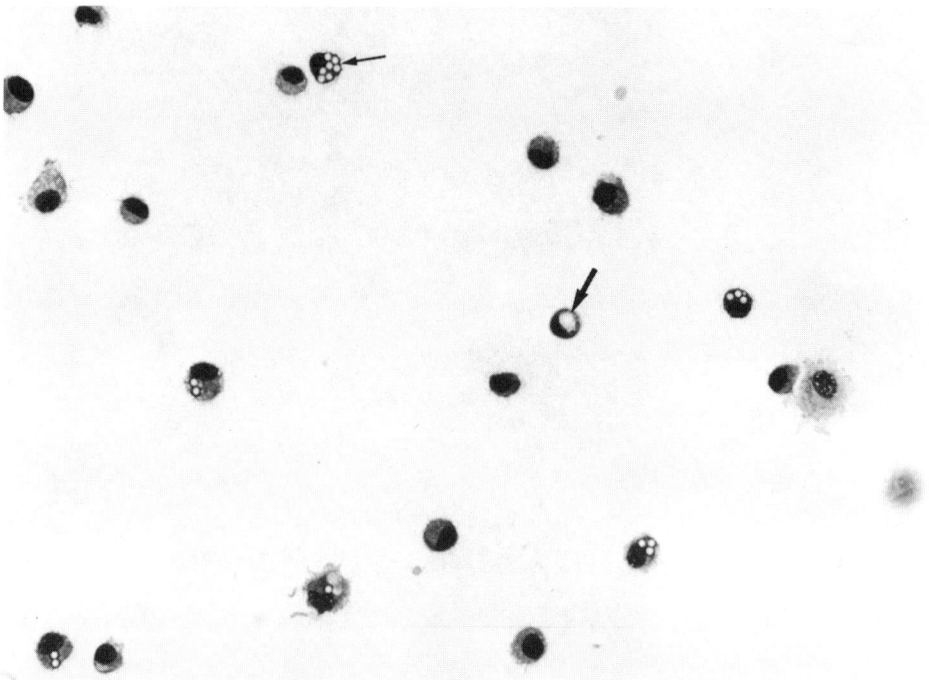

Fig. 4. AM with phagocytosed 3 μm spheres (small arrow) and phagocytosed 10 μm sphere (large arrow), second study with C57 Black/6J mice.

the rat. Thus, it appears that at least in B6C3F$_1$ mice large particles which are not phagocytosable by AM can be eliminated from the lungs rather efficiently.

Average AM diameters for mice and rats have been reported as ~ 9.5 and ~ 12 μm, respectively (Crapo *et al.*, 1983; Stone *et al.*, 1992). Thus, the administered 10 μm particles may have exceeded the capacity of the mouse macrophages to phagocytose them. However, 15 μm particles administered in an earlier study to rat lungs must also have exceeded the capacity of rat AM for phagocytosis and yet these particles were not cleared at all (Snipes and Clem, 1981). Thus, B6C3F$_1$ mice appear to have alveolar clearance mechanism which do not seem to be present in rats.

The second study reported here shows that there apparently are differences with respect to the size of microspheres being phagocytosed by AM of different strains of mice. AM of C57 Black/6J mice, in contrast to the B6C3F$_1$ mice of the first study, phagocytosed the large particles as evidenced from the cytospins at 1 and 14 days post-exposure. The clearance kinetics of these phagocytosed large particles in this mouse strain is unknown and needs to be determined in a separate study. Strains of both mice have been used in long-term inhalation studies with particles of low solubility and low cytotoxicity (NTP, 1993; Heinrich *et al.*, 1995). In those studies, both strains showed significant retardation of AM-mediated particle clearance with retention halftimes of 250–1000 days. Thus, the present finding of a retention halftime of 103 days for large particles in the B6C3F$_1$ mice is unusually

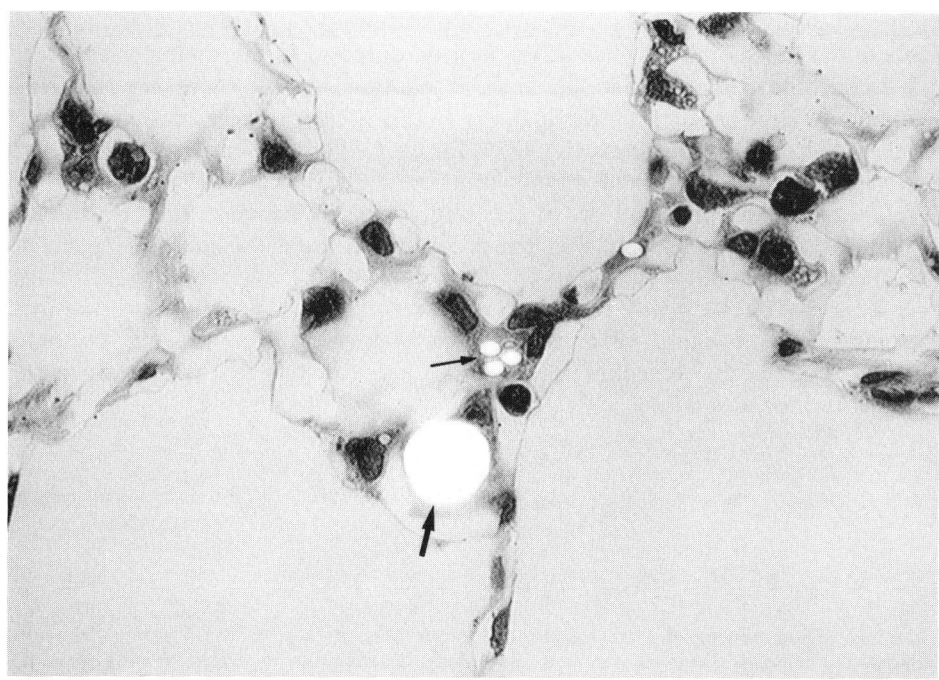

Fig. 5. Histological section of lung with AM and phagocytosed 3 μm (small arrow) and 10 μm (large arrow) microspheres.

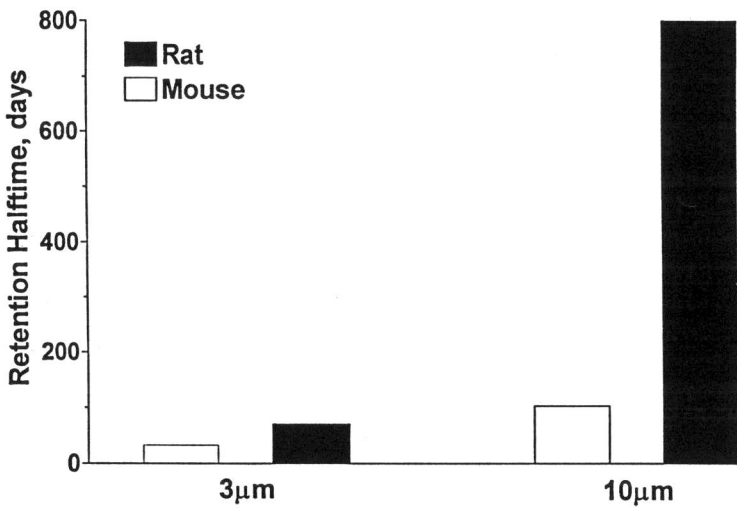

Fig. 6. Comparison of alveolar retention halftimes of 3 and 10 μm microspheres in B6C3F$_1$ mice (present study) and Fischer-344 rats (previous study, Oberdörster et al., 1992).

short, indicating that a clearance mechanism for non-phagocytosed large particles exists in the lungs of these mice which may be different from other species.

We conclude from these studies that, in contrast to rats, there is "effective" clearance of large particles in the lungs of B6C3F$_1$ mice, even though these large particles are not phagocytosed by AM. However, there appear to be significant differences between different strains of mice as is indicated in our study by differences in AM phagocytosis of the large particles between the two mouse strains. We conclude further that particle removal from the alveolar space of B6C3F$_1$ mice may be facilitated by the formation of AM clusters. These findings raise a general question with respect to the mechanism of AM movement towards the mucociliary escalator, which may be due to a chemotactic gradient or some otherwise directed movement of AM towards the conducting airways starting at the terminal bronchioles.

Acknowledgements—These studies were supported in part by NIEHS grants ESO4872 and ESO1247.

REFERENCES

Crapo, J. D., Young, S. L., Fram, E. K., *et al.* (1983) Morphometric characteristics of cells in the alveolar region of mammalian lungs. *Am. Rev. Respir. Dis.* **128**, s42.

Heinrich, U., Fuhst, R., Rittinghausen, S., Creutzenberg, O., Bellmann, B., Koch, W. and Levsen, K. (1995) Chronic inhalation exposure of Wistar rats and two different strains of mice to diesel engine exhaust, carbon black and titanium dioxide. *Inhal. Toxicol.* **7**, 533–556.

Kreyling, W. G. (1990) Interspecies comparison of lung clearance of "insoluble" particles. *J. Aerosol Med.* **3** (Suppl. 1), S93–S110.

Morrow, P. E. (1988) Possible mechanisms to explain dust overloading of the lungs. *Fundam. appl. Toxicol.* **10**, 369–384.

NTP (1993) National Toxicology Program: toxicology and carcinogenesis studies of talc in F344/N rats and B6C3F$_1$ mice. Technical Report Series no. 421; NIH Publ. no. 93–3152.

Oberdörster, G. (1995) Lung particle overload: implications for occupational exposures to particles. *Regulatory Toxicol. Pharmacol.* **27**, 123–135.

Oberdörster, G., Ferin, J. and Morrow, P. E. (1992) Volumetric loading of alveolar macrophages (AM): a possible basis for diminished AM-mediated particle clearance. *Expl. Lung Res.* **18**, 87–104.

Snipes, M. B. and Clem, M. F. (1981) Retention of microspheres in the rat lung after intratracheal instillation. *Environ. Res.* **24**, 33–41.

Stone, K. C., Mercer, R. R., Gehr, P., Stockstill, B. and Crapo, J. D. (1992) Allometric relationships of cell numbers and size in the mammalian lung. *Am. J. Respir. Cell Mol. Biol.* **6**, 235–243.

Pergamon

Ann. occup. Hyg., Vol. 41, Supplement 1, pp. 561–564, 1997
© 1997 British Occupational Hygiene Society
Published by Elsevier Science Ltd. All rights reserved
Printed in Great Britain
0003–4878/97 $17.00 + 0.00
Inhaled Particles VIII

PII: S0003–4878(96)00149–4

EFFECTS OF INHALED OVERLOAD PARTICLE CONCENTRATIONS ON ALVEOLAR MACROPHAGE (AM) CLEARANCE RESPONSES: THE ROLES OF HIGH PARTICLE BURDEN AND INFLAMMATION

D. B. Warheit, S. I. Snajdr and M. A. Hartsky

DuPont Haskell Laboratory, PO Box 50, Elkton Road, Newark, DE 19714-0050, U.S.A.

INTRODUCTION

Chronic dust overloading of the lungs is linked with macrophage-mediated particle clearance deficits. Two of the most important macrophage clearance functions are phagocytosis and cell motility or chemotaxis. In the current study we postulated that macrophage functional deficits are a response to both high particle burdens (Morrow overload theory—Morrow, 1988, 1992; Oberdoerster, 1992) as well as the influence of overload-induced pulmonary inflammatory responses. To test this hypothesis, we carried out short-term (3 day) and longer-term (4 week) inhalation studies with titanium dioxide (TiO_2) and carbonyl iron (CI) particles, particulates which are considered to be low toxicity, low solubility dusts. CD rats were exposed to aerosol concentrations ranging from 5 to 250 mg m^{-3}. In the 3 day study, rats were exposed to aerosols of TiO_2 particles (MMAD = 0.7 μm) at 0, 95 (median burden—MB) or 243 mg m^{-3} (high burden—HB). The lungs of sham and dust exposed rats were then lavaged at 0, 24 h, 48 h or 8 days postexposure. The results showed that TiO_2 produced dose-dependent transient inflammatory responses with > 95% of MB and HB recovered AM containing TiO_2 particles immediately after exposure. AM chemotaxis *in vitro* was increased over sham controls in the MB group, but cell migration in the HB group was decreased compared to sham when measured against the 5 μm but not the 8 μm pore-sized filters, indicating that the 5 μm pores prevented some of the HB AM from migrating through the filter. *In vitro* phagocytosis of carbonyl iron particles by MB and HB AM was depressed relative to controls in a dose-dependent manner. In the 4 week study, rats were exposed to 5, 50 or 250 mg m^{-3} of TiO_2 or CI particles and evaluated through 6 months postexposure. Exposures to 250 mg m^{-3} CI or TiO_2 produced an inhibition of clearance and > 98% of lavaged AM from all groups contained particles. Pulmonary inflammatory responses were measured in the high dose groups (30–50% PMNs through 3 months, 15% through 6 months) but were < 10% PMNs in the 50 mg m^{-3} group. *In vitro* phagocytosis by AM exposed to 50 and 250 mg m^{-3} were depressed relative to 0 and 5 mg m^{-3} AM through a period of 3 months postexposure, while the *in vitro* phagocytic responses of AM exposed to 50 and 250 mg m^{-3} CI particles were decreased only through 1 month postexposure. Chemotactic responses were slightly decreased in AM exposed to 50 mg m^{-3} TiO_2 or CI particles, but were significantly depressed in AM recovered from rats exposed to 250 mg m^{-3}. Whether the chemotactic impairment was related simply to particle overload or the influence of PMNs and corresponding pulmonary inflammatory effects such as protease secretion in the alveolar milieu remains to be determined. To test the hypothesis that the inflammatory response plays a role in macrophage functional deficits, we have initiated time course particle instillation studies with TiO$_2$, CI or silica dusts. Recent results demonstrate that we can separate an *in vivo* AM phagocytic response (4 h for CI and TiO$_2$) prior to the onset of particle-induced pulmonary inflammation (6–8 h). Our preliminary data indicate that both AM particle burden as well as the effects of pulmonary inflammation result in AM functional deficits and this is likely to play a role in the impairment of particle clearance measured in rats exposed to overload concentrations of dusts.

METHODS

In one study, male rats were exposed to TiO$_2$ or CI particles for 4 weeks at concentrations of 5, 50 and 250 mg m^{-3}. In another study, male rats were exposed to TiO$_2$ particles for 3 days at 95 or 243 mg m^{-3}. Only data generated from the TiO$_2$ studies are reported here. The techniques for nose-only aerosol generation of particles have been previously reported (Warheit *et al.*, 1991). Following the completion of 4 week exposures, the lungs of sham and dust-exposed animals were lavaged (Warheit *et al.*, 1984) immediately after, as well as 1 week, 30, 90 or 180 days postexposure. Following the completion of 3 day exposures, the lungs of rats were lavaged immediately after, 24 h, 48 h or 8 days postexposure. Measures of cytotoxicity and pulmonary inflammatory responses were assessed; in addition, alveolar macrophage functional parameters including morphology, phagocytosis and chemotaxis were assessed the postexposure time intervals specified above using methods described elsewhere (Warheit *et al.*, 1984).

RESULTS AND DISCUSSION

Associated with this high TiO$_2$ dust burden was the development of a persistent pulmonary inflammatory response which was evident through a period of 180 days postexposure and was manifested by significant numbers of neutrophils recovered in bronchoalveolar lavage fluids from exposed animals (Fig. 1). Three day exposures to TiO$_2$ particles at the higher exposure concentration also produced significant pulmonary inflammation (Fig. 1). Greater than 90% in the acute study and > 99% of alveolar macrophages recovered from the lung by lavage in the 4 week study contained particles. Alveolar macrophages exposed to high concentrations of TiO$_2$ particles for 4 weeks were impaired in their *in vitro* chemotactic responses. Immediately after exposure, as well as 3 months postexposure, the chemotactic responses of macrophages exposed to 50 mg m^{-3} TiO$_2$ were not significantly different from macrophages exposed to 5 mg m^{-3} and the chemotactic responses were substantially greater than macrophages exposed to 250 mg m^{-3} TiO$_2$ particles (Figs. 2 and 3). This difference between the macrophages exposed to 50 and 250 mg m^{-3} TiO$_2$ could be accounted for (a) macrophage particle burden;

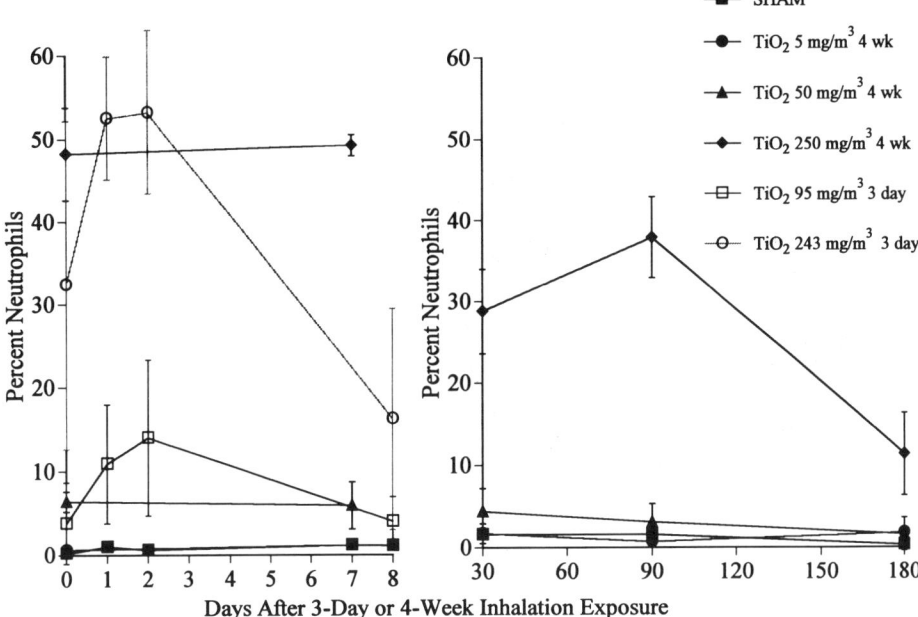

Fig. 1. Neutrophil differential percentages in BAL fluids of TiO$_2$-exposed rats. Exposures to 243 or 250 mg m^{-3} TiO$_2$ resulted in sustained pulmonary inflammatory responses which were measured through 8–90 days postexposure.

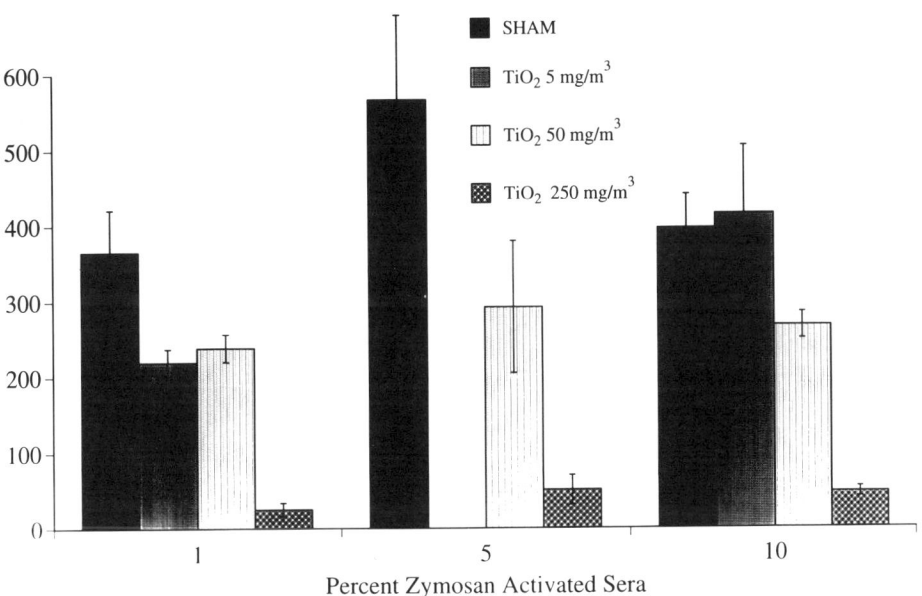

Fig. 2. Chemotaxis of alveolar macrophages recovered from TiO$_2$-exposed rats immediately after 4 week inhalation exposures. Macrophages recovered from rats exposed to 250 mg m^{-3} TiO$_2$ were significantly impaired in their *in vitro* chemotactic responses.

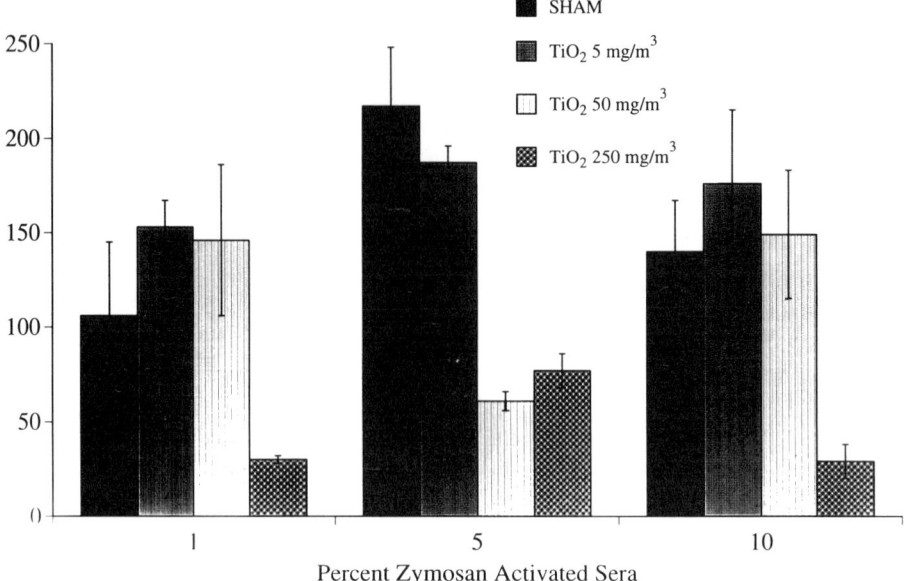

Fig. 3. Chemotaxis of alveolar macrophages recovered 3 months after a 4 week inhalation exposure to TiO$_2$ particles. Macrophages recovered from rats exposed to 250 mg m^{-3} TiO$_2$ were still significantly impaired in their *in vitro* chemotactic responses.

(b) the pulmonary inflammatory status of the lung; or (c) both factors. Efforts are currently being directed to test the hypothesis that persistent pulmonary inflammatory responses reduce the chemotactic potential of resident alveolar macrophages.

REFERENCES

Morrow, P. E. (1988) Possible mechanisms to explain dust overloading of the lungs. *Fundam. appl. Toxicol.* **10**, 369–384.

Morrow, P. E. (1992) Dust overloading of the lungs: update and appraisal. *Toxicol. appl. Pharmacol.* **113**, 1–12.

Oberdoerster, G., Ferin, J. and Morrow, P. E. (1992) Volumetric loading of alveolar macrophages (AM): a possible basis for diminished AM-mediated clearance. *Exp. Lung Res.* **18**, 87–104.

Warheit, D. B., Carakostas, M. C., Hartsky, M. A. and Hansen, J. F. (1991) Development of a short-term inhalation bioassay to assess pulmonary toxicity of inhaled particles: comparisons of pulmonary responses to carbonyl iron and silica. *Toxicol. appl. Pharmacol.* **107**, 350–368.

Warheit, D. B., Chang, L. Y., Hill, L. H., Hook, G. E. R., Crapo, J. D. and Brody, A. R. (1984) Pulmonary macrophage accumulation and asbestos-induced lesions at sites of fiber deposition. *Am. Rev. Respir. Dis.* **129**, 301–310.

 Pergamon

Ann. occup. Hyg., Vol. 41, Supplement 1, pp. 565–570, 1997
© 1997 British Occupational Hygiene Society
Published by Elsevier Science Ltd. All rights reserved
Printed in Great Britain
0003–4878/97 $17.00 + 0.00
Inhaled Particles VIII

PII: S0003–4878(96)00059–2

DEPOSITION OF FLUORESCENT PARTICLES IN REPLICA CASTS OF DISTAL HUMAN AIRWAYS: COMPARISON OF THEORY WITH EXPERIMENT

R. F. Phalen, M. J. Oldham, R. C. Mannix and G. M. Schum

Air Pollution Health Effects Laboratory, College of Medicine, University of California, Irvine,
CA 92697–1825, U.S.A.

INTRODUCTION

Bates (1989) has pointed out the vulnerability of small airways of workers who are exposed to insoluble dusts. Because small airways are nearly impossible to observe in living subjects and they are not amenable to use in bench-top studies, little is known regarding their collection efficiencies for aerosol particles. However, Brody and Roe (1983) documented enhanced deposition of inorganic particles at the bifurcations of small airways of rodents following brief inhalation exposures. Also, Cohen *et al.* (1988) observed enhanced bifurcational deposition of 0.15 μm diameter particles in generation six airways (approx. 3 mm diameter) in hollow casts of human lungs. Neither of these studies provided quantitative evaluations of theoretical models that are used for inhaled aerosol dosimetry.

The traditional predictive models of inhaled particle deposition are deterministic in that particle-size dependent deposition probabilities are calculated using anatomical and airflow information for bronchial airways on a generation-by-generation basis (Yeh and Schum, 1980). Recently, stochastic particle deposition calculations, using individual particle trajectories and finite element simulations of airflow have been described (Hofmann, 1996; Tsuda *et al.*, 1994). Both types of models can predict the deposition efficiencies of particles in small airways, but neither type has been adequately checked for its validity in these structures. Our objective was to observe the deposition patterns of particles in hollow replica casts of small human airways (generation 9–11, about 1 mm diameter) and to compare these patterns with those predicted by a deterministic mathematical model.

MATERIALS AND METHODS

Fresh lungs from a 49 year-old female (who died of brain hemorrhage) were used for replica-casting with silicone rubber using a saline-replacement technique (Kilpper and Stidd, 1973). A piece of this cast consisting of four connected complete bifurcations (generations 9–11) was removed, fitted with small tubes at each airway tip and submerged in catalysed clear plastic casting resin. After curing, the plastic block (2 × 1 × 0.5 cm) was trimmed, the silicone-rubber cast removed, and the surfaces of the block polished [Fig. 1 (a)]. A 10-power magnifier with a

Table 1. Summary of parameters used in each experiment

Experiment	Length of experiment (min)	Aerodynamic particle diameter (microns)	Total airflow rate (cc min^{-1})
1	30	2.94	233
2	60	2.94	239
3	30	0.51	236
4	30	0.51	238
5	360	0.98	39

calibrated reticle scale and protractor was used to measure the dimensions and branch angles of the hollow airway cast. For aerosol depositions, the entry of the cast was fitted with a tube that matched the diameter of the entry. This cast was placed inside a sealed cylinder with the entry tube protruding from one end and a 37 mm diameter filter covering the other (outlet) end. Air was drawn through this exit filter to introduce aerosol into the cast. Aerosols were generated by nebulizing monodisperse, fluorescently-labelled polystyrene particles (Duke Scientific Corp., Palo Alto, CA). The suspensions were diluted to produce negligible doublets during nebulisation. The aerosols were dried with purified air, deionised with a 10 millicurie 85-Kr source and held in a 20 l. steel drum. The particles used were physically 2.87, 0.96 and 0.50 μm diameter, which when corrected for density (1.05 g cm^{-3}) gave aerodynamic diameters of 2.94, 0.98 and 0.51 μm, respectively. The experimental conditions are shown in Table 1.

After the depositions of particles, the cast, back-up filter and sampling filter (used for low-flowrate sampling to determine the total number of particles cm^{-3} in the drum) were examined using a Nikon Labphot u.v. fluorescence microscope equipped with a 35 mm camera. For cast examination, a series of photographs of a bifurcation (parent and two daughter tubes, generation 10 and 11) were taken at successive depths of focus into the model [Fig. 1 (b, c)]. The filters were also examined and photographed using an asterisk-shaped pattern to obtain estimates of the number of particles that passed through the cast without depositing. Later, the photographs were projected and tracings made to mark the positions (including depths in the casts) of each particle [Fig. 1 (d)]; the total particle counts from the cast and filters were used to calculate the particle deposition efficiencies in the small airways. In addition, the particles depositing in the cast on the bifurcation were counted, as were those depositing elsewhere. These data were used to calculate the fractional deposition due to inertial impaction in relation to that from other mechanisms (presumably diffusion plus sedimentation). Predicted deposition efficiencies were calculated by inputting the aerosol aerodynamic diameters, cast dimensions and angles, and flow rates used into a deterministic computer model (Yeh and Schum, 1980; Schum *et al.*, 1991). These predictions were compared with the measured deposition efficiencies.

RESULTS

The measured and calculated deposition efficiency data are represented in Table

Table 2. Comparison of measured to predicted deposition ratios

Experiment number	Measured		Calculated	
	Impaction	Non-impaction	Impaction	Non-impaction
1	0.91	0.02	0.94	0.06
2	0.96	0.04	0.94	0.06
3	0.38	0.62	0.53	0.47
4	0.58	0.42	0.53	0.47
5	0.43	0.57	0.24	0.76

2. Within the uncertainty inherent in the study they are in agreement, providing a basic validation of the predictive model for small airways. It is not clear that the microscopic observations were able to detect every particle that deposited; the exact measured depositions could underestimate the true values. With respect to specific mechanisms producing deposition, trends in observed deposition efficiencies and calculated efficiencies were consistent. Again, the agreement with theory is satisfactory. The ratio of expected impaction efficiency as compared with other mechanisms of particle deposition in the cast is also similar to predictions by the deterministic theoretical model. Finally, the deposition patterns within the casts were examined to see if small surface features (such as ridges and cavities) on the airways, or flow inhomogeneities, led to local areas of unusually high or low particle deposition. This lack of uniformity of deposition in the small airways was seen, especially for the largest particle size.

DISCUSSION

Given the stated limitations of the microscopy in particle imaging, we conclude that the deterministic model calculations were acceptable for predicting the total and regional particle depositions in the cast. The relative mechanism-related deposition efficiencies provided by this model matched those obtained experimentally. We failed to show that the mechanistic model could not provide reasonable predictions of the efficiencies and overall locations (against bifurcations vs in airway tubes). This supports for the continued use of such models for application to aerosol dose estimates for small airways. There is a question of whether or not all of the particles deposited in the cast were in fact observable with the u.v. microscope. When the cast was turned upside down, different numbers, but similar particle patterns, were obtained, especially for the smallest particles. The reason for this effect is presumed to be optical absorption by the uneven inner surfaces of the hollow cast. Unevenness could also diffract light from the small particles in a manner that prevented their being clearly imaged and thus counted.

Future studies could involve casts that can be opened so that internal deposits can be more clearly imaged. Also, particle deposition in terminal bronchioles, respiratory bronchioles and alveolar ducts can, and should be studied. Finally, stochastic model predictions should be similarly tested using distal airway casts. These sophisticated models hold the promise of very accurately predicting sites of particle depositions within airways.

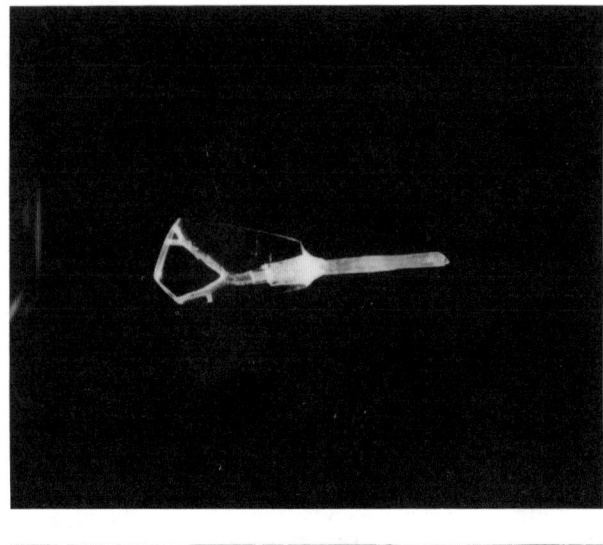

(a)

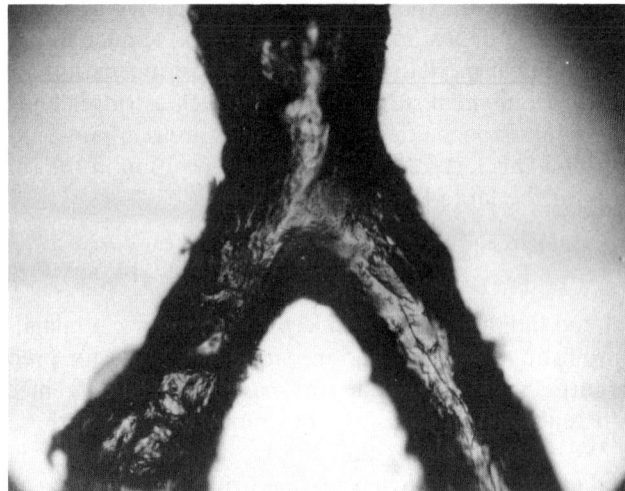

(b)

Fig. 1. a,b.

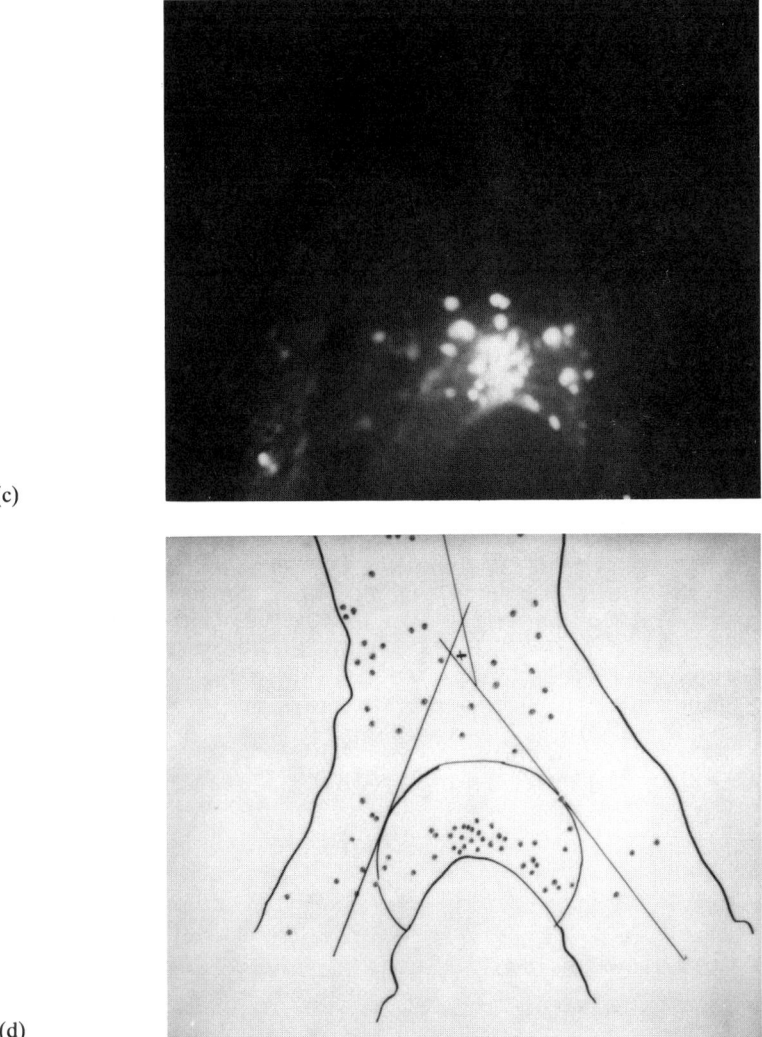

(c)

(d)

Fig. 1. (a) Distal airway cast; largest airway width is 0.15 cm. (b) Bifurcation from cast photographed through a microscope; the parent airway width is 0.11 cm, (c) Deposits 0.98 μm particles in the bifurcation seen under u.v. illumination. (d) Tracing of particle deposits in the cast made from u.v. photomicrographs.

Acknowledgements—Supported by the NIH (RO1 HL39682) and the U.C. Irvine Center for Occupational and Environmental Health. Khahn Nguyen and Thao Nguyen provided technical assistance and Marie Tonini provided secretarial support.

REFERENCES

Bates, D. A. (1989) *Respiratory Function in Disease*, 3rd edn, p. 292. Saunders, Philadelphia, PA.

Brody, A. R. and Roe, M. W. (1983) Deposition pattern of inorganic particles at the alveolar level in the lungs of rats and mice. *Am. Rev. Respir. Dis.* **128**, 724–729.

Cohen, B. S., Harley, N. H., Schlesinger, R. B. and Lippmann, M. (1988) Nonuniform particle deposition on tracheobronchial airways: Implications for lung dosimetry. *Ann. occup. Hyg.* **32** (Suppl. 1), 1045–1053.

Hofmann, W. (1996) Lung morphometry and particle transport and deposition: Overview of existing models. In *Aerosol Inhalation: Recent Research Frontiers* (Edited by J. C. M. Marijnissen and L. Gradoń), pp. 91–102. Kluwer, Dordrecht, The Netherlands.

Kilpper, R. W. and Stidd, P. J. (1973) A wet-lung technique for obtaining silastic rubber casts of the respiratory airways. *Anat. Rec.* **176**, 279–288.

Schum, G. M., Phalen, R. F. and Oldham, M. J. (1991) The effect of dead space on inhaled particle deposition. *J. Aerosol Med.* **4**, 297–311.

Tsuda, A., Butler, J. P. and Fredberg, J. J. (1994) Effects of alveolated duct structure on aerosol kinetics: I diffusional deposition in the absence of gravity. *J. appl. Physiol.* **76**, 2497–2509.

Yeh, H. C. and Schum, G. M. (1980) Models of human airways and their application to inhaled particle deposition. *Bull. Math. Biol.* **42**, 461–480.

 Pergamon

Ann. occup. Hyg., Vol. 41, Supplement 1, pp. 571–575, 1997
© 1997 British Occupational Hygiene Society
Published by Elsevier Science Ltd. All rights reserved
Printed in Great Britain
0003–4878/97 $17.00 + 0.00
Inhaled Particles VIII

PII: S0003–4878(96)00087–7

COMPARISON OF A CONVECTIVE MIXING MODEL'S PREDICTIONS TO EMPIRICAL RESULTS FOR AEROSOL DISPERSION IN THE HUMAN LUNG

G. Ganser,* M. A. McCawley,† and I. Christie*

*West Virginia University, Morgantown, WV 26506, U.S.A.; and †NIOSH, 1095 Willowdale Road, Room 111, Morgantown, WV 26505–2888 U.S.A.

INTRODUCTION

Recent modelling work has attempted to use axial diffusion to explain the observed dispersion of an aerosol bolus in transit through the lung (Edwards, 1994). This type of modelling attempts to derive from first principles the underlying mechanism of aerosol dispersion. An alternative approach has used mixing theory derived from reactor vessels to surmount problems encountered in small-scale mixing (McCawley *et al.*, 1988). Noting that Ultman (1985) had used an approach that is mathematically similar in applying network theory to this problem, it was decided to try to relate the number of vessels in series to the number of bifurcated generations in the lung.

METHODS

Aerosol dispersion data from humans were collected as previously described (Khandare, *et al.*, 1994). In brief, the apparatus consisted of an aerosol generation and detection system linked with a volume measurement system (Fig. 1). The subject was able to control both flow and volume for the breathing cycles. The subject used was an asymptomatic, white male, aged 45 who had never smoked. Data on aerosol concentration and breathing volume were collected and stored digitally at a rate of 60 Hz for the duration of the subject's test. A 0.5 μm corn oil aerosol was used. The input was normalised to a dirac delta function by a deconvolution of the input and output pulses.

THEORY

For the purposes of this theory the lung is conceived as a series of mixing tanks, in each of which complete mixing occurs. Based on the empirical data of Scherer *et al.* (1975) for a five generation glass lung, the best fit occurs when the ith generation of the lung corresponds to 2^i identical tanks. Because significant mixing is assumed to occur at the bifurcation, the generations/mixing tanks were divided so as to contain the bifurcation at the center of the mixing tank. Thus the zeroth generation begins halfway down the trachea. Other assumptions are that the transit time, t_i,

G. Ganser *et al.*

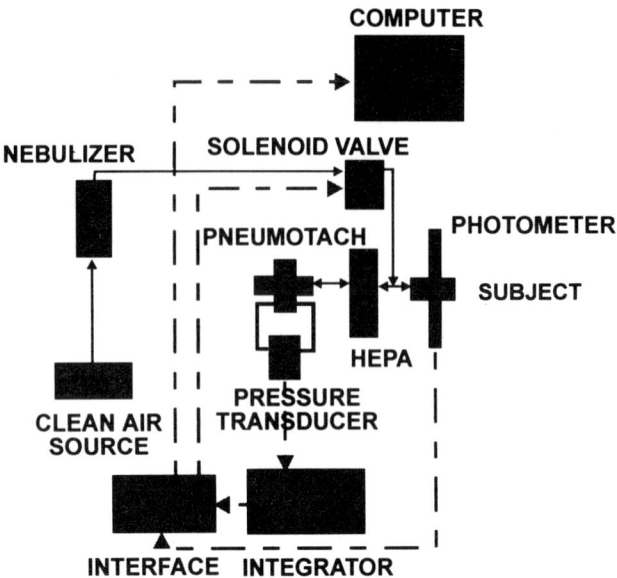

Fig. 1. Aerosol dispersion apparatus used for collecting experimental data cited in this paper.

through any generation is independent of the previous generation's; during unidirectional flow (either inspiration or expiration) any given particle makes only one entrance from the previous generation and one exit to the next generation.

Complete mixing, assumed for each mixing tank, means that all particles have the same probability of leaving the tank as stated above. The corresponding density function for t_i is

$$f(t) = (1/\bar{t}_i) \exp (-t/\bar{t}_i) \tag{1}$$

where $\bar{t}_i$ is the expected residence time.

One can then define equations for the first and second moments of the output probability function of a single completely mixed tank (having an exponential output), the first moment being $\bar{t}$, the expected mean residence time, and the second moment defined as the variance,

$$\sigma_v^2 = \bar{t}^2. \tag{2}$$

An important property of the variance is that for unidirectional flow, statistically independent tanks in series have additive variances. Statistically independent means that the history of a particle (the residence time in previous tanks) has no bearing on the residence time in the current tank. This then reduces mixing in the lung to a series of statistically independent tanks. For a given volume of penetration v_p, the lung is divided into M statistically independent mixing tanks. The total variance for a complete breathing cycle, viewed as unidirectional flow through this symmetrical lung, can then be expressed as

$$\sigma_v^2 = \sum_{i=0}^{2M-1} (\bar{t}_i)^2. \tag{3}$$

In this form it can be seen that the model has no dependence on flow rate when the output is given in terms of volume exhaled. Because it is more common to express the output in terms of time (Wen and Fan, 1974) and since the residence time in any generation is equal to the total volume of tanks in that generation, v_i^T, divided by the total flow rate, Q, equation (3) can be re-written as,

$$\sigma_v^2 = \sum_{i=0}^{2M-1} (v_i^T/Q)^2. \tag{4}$$

However, many of the published models use empirical data that express the dispersion in terms of the half-width (σ_H), which is a shortened form of the term "full width at half maximum" and is defined as the width of the expired aerosol pulse at half the maximum concentration, written as a function of expired volume. Therefore, it may be convenient to have an expression for converting the variance to the half-width. Multiplying the square root of the variance by $2Q$ gives an approximation of σ_H (designated σ_{H^*}). For a normal distribution a closer approximation can be calculated for

$$\sigma_H = (2\ln2)^{1/2} \, (2Q\sigma) \text{ or}$$

$$\sigma_H = \sqrt{2\ln2} \cdot 2 \sqrt{2} \left[\sum_{i=0}^{M-1} (v_i^T)^2 \right]^{1/2} \tag{5}$$

where the volume of penetration is

$$\sum_{i=0}^{M-1} v_i^T.$$

The conversation factor relating variance to half width is dependent on the shape of the exhaled aerosol distribution and could vary from as low a value as 0.87 to approaching $(2\ln2)^{1/2}$, that is, 1.18.

One criticism of the approach used in the statistical model above is the need for a symmetrical lung with a first-in, first-out arrangement for flow. Alternatively it is possible to again model the ith generation of the lung as 2^i identical complete mixing tanks. For each volume of penetration, v_p, all volumes are fixed, corresponding, of course, to the lung volume for that generation as specified by Weibel's Model A for regular dichotomy (Weibel, 1963). The exception to that is the last generation which expands and contracts to accommodate the changing air volume of the breathing cycle. Thus it is possible to achieve a first-in, last-out (FILO) arrangement for the aerosol flow. Because the flow is not unidirectional, equation (3) is no longer true and it is easiest to numerically solve the system of equations to find the concentration in tank T_i, over some time span, dt such that the output for the FILO Model is

$$\frac{dq_i}{dt} = \frac{r_i q_{i-1}}{v_i} - \frac{(s_i + 2r_{i+1}) \, q_i}{v_i} + \frac{2s_{i+1} \, q_{i+1}}{v_i} \, , (i = 0, \ldots, 23) \tag{6}$$

where the flow rate during inspiration into the trachea is a constant Q. The flow rate during inspiration into each of the tanks making up the ith generation is r_i, $2r_{i+1} = r_i$ and $s_i = 0$. During expiration the flow rate out of each of the tanks is s_i when $s_0 = Q$, $2s_{i+1} = s_i$ and $r_i = 0$. If this equation (6) is rearranged to give the

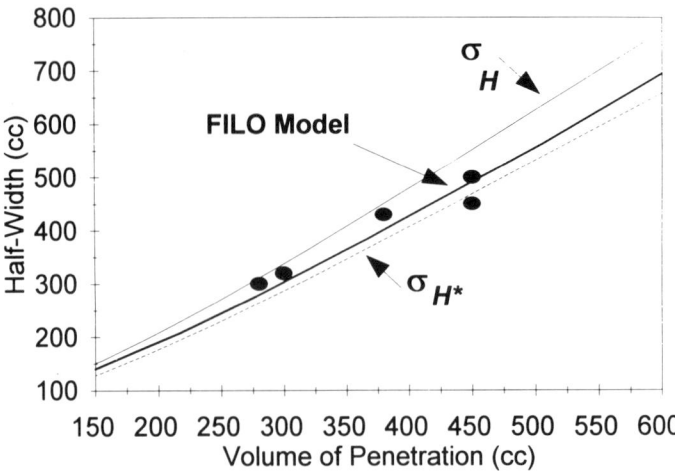

Fig. 2. Comparison of model and experimental results for a flow rate of 500 ccs^{-1}, tidal volume of 1000 cc and a bolus volume of 100 cc, for a single, typical, asymptomatic, male subject, age—44 years, height—175 cm. The values for σ_{H^*} are calculated as specified in the text, multiplying equation (4) by $2Q$; the values for σ_H are from equation (5); and the values for the FILO Model are derived from equation (6).

change in concentration with a change in volume than the FILO Model becomes independent of flow rate as was the case with the statistical model [equation (3)].

RESULTS

The output of the different approaches to half-width calculation are shown in Fig. 2. These were then compared to the data of an individual analyzed as previously noted. At penetration volumes above 400 cc, deposition loss was measureable, though less than 10%. No measurements were made above this volume since deposition losses are not accounted for by this current theory.

DISCUSSION

There is not a substantial difference in the values calculated by the different approaches taken in Fig. 2 deriving a relationship between aerosol dispersion measured by the half-width of the expired pulse and volume of penetration of the inhaled pulse. In fact, the normal distribution approximation is similar to the approach taken by Scherer *et al.* (1975) in analyzing their five generation glass model. By using only the top 10% of the output concentration they approximated a normal distribution from what was obviously a gamma distribution similar to that found in human aerosol dispersion data (McCawley *et al.*, 1988). Like Scherer we also concluded that there should be no dependence of dispersion on flow rate based on either of our model approaches. This condition is pointed to by Edwards (1994) as being a necessary model feature for agreement with data from a number of published studies.

The statistical approach taken in the first model presented in the paper used a first-in, first out sequence which is also similar to Edwards (1994). However, it has long been known (Altshuler *et al.*, 1959) that the filling and emptying sequence for aerosols cycled through the lungs is first-in, last-out. The FILO model approach also allows the model to be adjusted on a specific generation level to try to mimic changes that might occur in disease by changing the volume of any bifurcation (tank) or the flow rate through that bifurcation (tank).

One interpretation of the comparability of empirical and model results might be that the uniform residence times derived from using the regular dichotomy of Model A represent the uniformity of ventilation long recognised to occur in healthy lungs. The time constants would then be the product of the resistance and compliance of the various pathways. Therefore, the statistically significant differences between healthy subject and those with presumed mild obstruction first noted by McCawley and Lippmann (1984) and later confirmed by other authors (Blanchard, 1996) may be due to changes in the ventilation patterns due to obstruction.

CONCLUSIONS

Two separate models were derived based on Weibel's regular dichotomy for the lung. Both models were able to predict experimental results for the change in half-width with a change in penetration of an aerosol bolus. The two models also show dispersion to be independent of flow rate, in agreement with published data. One of the models uses a first-in, last-out approach and both models can be altered to mimic potential changes in the lung during obstruction. There is also no requirement to derive the flow profile in the airways since the assumed complete mixing adequately describes the empirical data.

REFERENCES

Altshuler, B., Palmes, E. D., Yarmus, L. and Nelson, N. (1959) Intrapulmonary mixing of gases studied with aerosols. *J. appl. Physiol.* **14**, 321–327.
Blanchard, J. D. (1996) Aerosol bolus dispersion and aerosol-derived airway morphometry: assessment of lung pathology and response to therapy, part I. *J. Aerosol. Med.* **9**, 183–201.
Edwards, D. A. (1994) A general theory of the macrotransport of nondepositing particles in the lung by convective dispersion. *J. Aerosol Sci.* **25**, 543–565.
Khandare, P., McCawley, M. and Zondlo, J. (1994) Validation of the mixed tank reactor model used to model aerosol dispersion in the lung. *Ann. occup. Hyg.* **38** (Suppl. 1), 151–158.
McCawley, M., Abrons, H. and Lippmann, M. (1988) Modelling an aerosol dispersion test for detecting early airway changes. *Ann. occup. Hyg.* **32** (suppl. 1), 81–89.
McCawley, M. and Lippmann, M. (1984) An aerosol dispersion test: comparison of results from smokers and non-smokers. In *Aerosols* (Edited by B. Y. H. Liu *et al.*), pp. 1007–1010. Elsevier, New York.
Scherer, P. W., Shendalman, L. H., Greene, N. M. and Bouhuys, A. (1975) Measurement of axial diffusivities in a model of the bronchial airway. *J. appl. Physiol.* **38**, 719–723.
Ultman, J. S. (1985) Gas transport in the conducting airways. In *Gas Mixing and Distribution in the Lung* (Edited by L. A. Engel and M. Paiva), pp. 63–129. Marcel Dekker, New York.
Weibel, E. R. (1963) *Morphometry of the Human Lung*. Springer-Verlag, Berlin.
Wen, C. Y. and Fan, L. T. (1975) *Models for Flow Systems and Chemical Reactors*. Marcel Dekker, New York.

Pergamon

Ann. occup. Hyg., Vol. 41, Supplement 1, pp. 576–581, 1997
© 1997 British Occupational Hygiene Society
Published by Elsevier Science Ltd. All rights reserved
Printed in Great Britain
0003–4878/97 $17.00 + 0.00
Inhaled Particles VIII

PII: S0003–4878(96)00150–0

SPATIAL AEROSOL DEPOSITION PATTERNS IN THE HUMAN LUNG: STOCHASTIC PREDICTIONS VS EXPERIMENTAL SPECT DATA

W. Hofmann,* L. Koblinger,† R. Bergmann,* J. S. Fleming,‡ A. H. Hashish§ and J. H. Conway¶

*Institute of Physics and Biophysics, University of Salzburg, Hellbrunnerstrasse 34, A-5020 Salzburg, Austria; †Health Physics Department, KFKI Atomic Energy Research Institute, H-1525 Budapest, Hungary; ‡Department of Medical Physics, Southampton General Hospital, Southampton SO16 6YD, U.K.; §Department of Electrical Engineering, Southampton University, Southampton SO17 1BJ, U.K.; and ¶University Department of Medicine, Southampton General Hospital, Southampton SO16 6YD, U.K.

INTRODUCTION

Experimental *in vivo* determination of inhaled particle deposition in human lungs is presently restricted to total and regional deposition. In contrast to these experimental efforts, lung deposition models commonly calculate particle deposition in single airway generations. So far, these theoretical predictions have not been validated by comparison with experimental data because of lack of relevant information.

Experimental data on deposition in individual airway generations are now becoming available through the application of single photon emission computed tomography (SPECT), which permits the determination of the three-dimensional distribution of radiolabelled aerosols among bronchial and pulmonary airways. To utilize these data for comparison with theoretical predictions, Fleming *et al.* (1995) have recently developed a methodology to convert the measured spatial distribution of activity to deposition fractions in hemispherical shells and single airway generations.

In this paper, we propose an opposite approach: our Monte Carlo deposition model (Koblinger and Hofmann, 1990; Hofmann and Koblinger, 1990) contains information about the spatial arrangement of the airways and thus is capable of predicting three-dimensional deposition patterns of inhaled aerosols in the human lung. The comparison between the SPECT measurements and the stochastic modelling predictions is based here on three levels: (i) total deposition; (ii) deposition per shell; and (iii) deposition per generation.

EXPERIMENTAL DATA

In a series of inhalation experiments, 12 human test subjects inhaled radiolabelled Tc-99m aerosols from six different jet nebulizers under controlled tidal breathing conditions. The present comparison between experimental data and theoretical predictions is based on SPECT measurements with subject no. 6, exposed to nebulizer no. 5. Controlled average breathing parameters of subject no. 6 for both fine and coarse aerosols are: tidal volume = 1155 ml, breathing frequency = 13 min^{-1}, inspiration time = 1.66 s and expiration time = 2.70 s. The mass median diameters (MMD) of the fine and coarse aerosols inhaled are 1.41 and 6.28 μm, respectively.

The three-dimensional distribution of radiolabelled aerosols within the lung was measured by SPECT, while aligned anatomical information was provided by magnetic resonance imaging (MRI). The spatial resolution of the measured activity distribution is determined by the size of the voxels, which is 4.67 mm in the present case. For the comparison with model predictions, the measured SPECT projections were first transformed to hemispherical shells and then converted to individual airway generations using Weibel's (1991) morphometric lung model (Fleming et al., 1995).

SPATIAL PARTICLE DEPOSITION MODELLING

While the linear dimensions of the human airway system are based on our stochastic morphometric lung model (Koblinger and Hofmann, 1985), the sizes of the two lungs are presently described by the equations proposed for the MIRD phantom (Snyder et al., 1969). The dimensions of the two lungs, their distance from each other, and the lengths and branching angles of the main bronchi leading to the two lungs, however, were adjusted to the specific lung geometry of subject no. 6 obtained from MRI data. The functional residual capacity (FRC) of our adjusted lung model volume of 2790 ml is very close to the estimated FRC of 2752 ml.

Having defined the surface of the lungs, 10 000 individual paths were constructed to compute the three-dimensional distribution of airway bifurcations. At each step of the random pathway selection, the spatial positions of a given airway bifurcation are computed. If a randomly selected daughter airway leaves the defined lung volume, the total path is rejected. Following the hemispherical transformation procedure designed by Fleming et al. (1995), shell numbers from 1 to 10 can be assigned to each computed airway bifurcation. For example, the distribution of airway bifurcations in shell numbers 3 and 8 is illustrated in Fig. 1. While the inner shell contains mostly the larger bronchial airways, the outer shell comprises a wide range of bronchial air passages (note: due to their short residual pathlengths, acinar airways always are assumed to belong to the same shell as their corresponding terminal bronchioles).

In the next step, particle deposition is computed for 10 000 inhaled particles. When deposition occurs in a given bifurcation, the spatial coordinates of the branching point of that bifurcation unit are stored, thus allowing the three-dimensional reconstruction of the particle deposition pattern. In addition, the deposition sites can be related to the corresponding shell numbers to determine the fraction of particles deposited in the various shells.

RESULTS

First, we compare total, i.e. bronchial (except trachea and first half of main bronchi) and pulmonary, deposition for mouth breathing. While excellent agree-

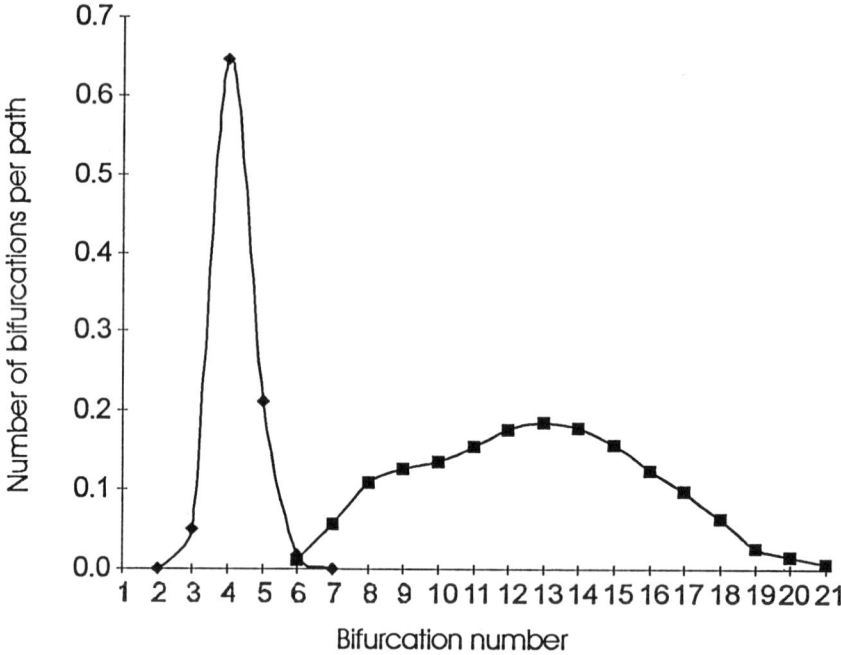

Fig. 1. Distribution of bronchial airway bifurcations in shells no. 3 (◆) and 8 (■).

ment between experiment and theory is found for the coarse aerosol particles (48.1% vs 46.3%), the predicted total deposition of 41.8% for the fine aerosols is significantly smaller than the experimentally observed value of 73.1%. Though the predicted total deposition values are supported by experimental data in human test subjects (Heyder *et al.*, 1980), it should be noted that the range of the smallest size band of the fine aerosol distribution (0.5–1.93 μm), which contains about 70% of the total mass, is too wide to allow reliable deposition calculations, considering the steep increase of total deposition in this size range (Hofmann and Koblinger, 1990).

Knowing the volume of the shells and the volume of a voxel, predicted deposition data can be converted to deposition per voxel in each shell. The computed distribution of the deposition per voxel among the 10 hemispherical shells, normalised to the inhaled activity, is plotted in Fig. 2 and compared with the corresponding data derived from the SPECT measurements. For coarse aerosols, the predicted deposition fractions are fluctuating around those based on the experimental data. For fine aerosols, however, theoretical deposition fractions are consistently smaller than the SPECT data, caused primarily by the smaller predicted total deposition. Indeed, if normalised to the same total deposition, deposition fractions are quite similar in both approaches. In addition, the stochastic deposition model correctly predicts the experimentally observed differences between the two aerosols.

Finally, the fine and coarse aerosol deposition patterns among airway generations obtained by the two approaches are displayed in Fig. 3. The comparison between the theoretical and the semi-experimental data reveals a consistent

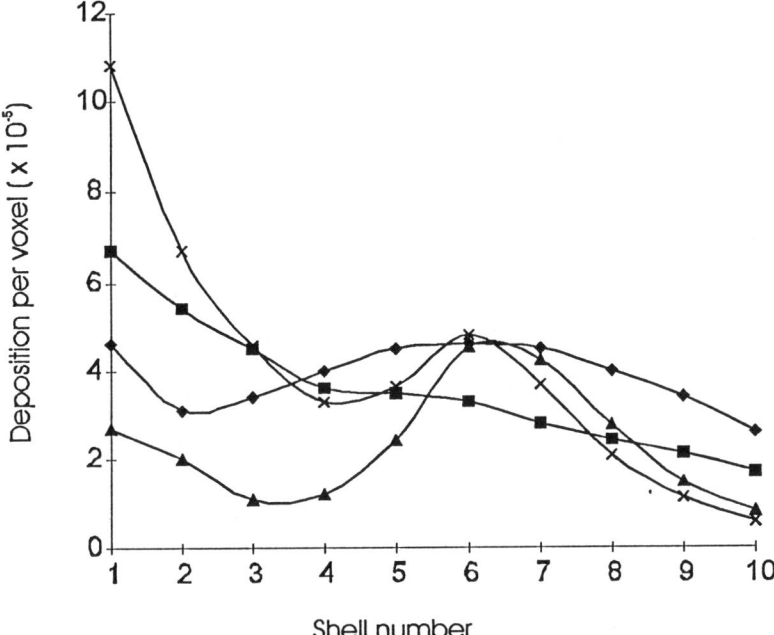

Fig. 2. Deposition per voxel of fine and coarse aerosols as functions of the shell number: comparison between data derived from SPECT measurements (fine (◆) and coarse (■) particles) and theoretical predictions (fine (▲) and coarse (×) particles). Data are expressed as a fraction of that inhaled.

behavior for both particle sizes: Below generation 17/18, the predicted deposition values are always higher than those based on the SPECT data; above this switching point, the experimentally derived data are much higher than the predicted ones. It should be noted, however, that the stochastic lung model has a variable number of airway bifurcations (here, deposition events occur in 26 bifurcations), while only 21 generations are used in the SPECT model. Thus, if all deposition fractions in airway bifurcations 22 to 26 are added to the deposition value in bifurcation 21, then the same steep increase will be obtained as observed in the experimentally derived deposition fractions.

DISCUSSION

In principle, our comparison between experiment and theory should be based on the three-dimensional activity distribution directly measured by SPECT and modelled by the stochastic deposition model. At present, however, we do not have this information. Thus the experimentally determined activity distribution was first transformed to deposition in concentric hemispherical shells and then converted to deposition in individual airway generations (Fleming *et al.*, 1995). Since these conversions are also based on modelling assumptions, the present comparison reflects not only potential differences between experiment and model, but also between two different models. For instance, two different lung models are used for the computation of the distribution of airway generations in each shell, Weibel's

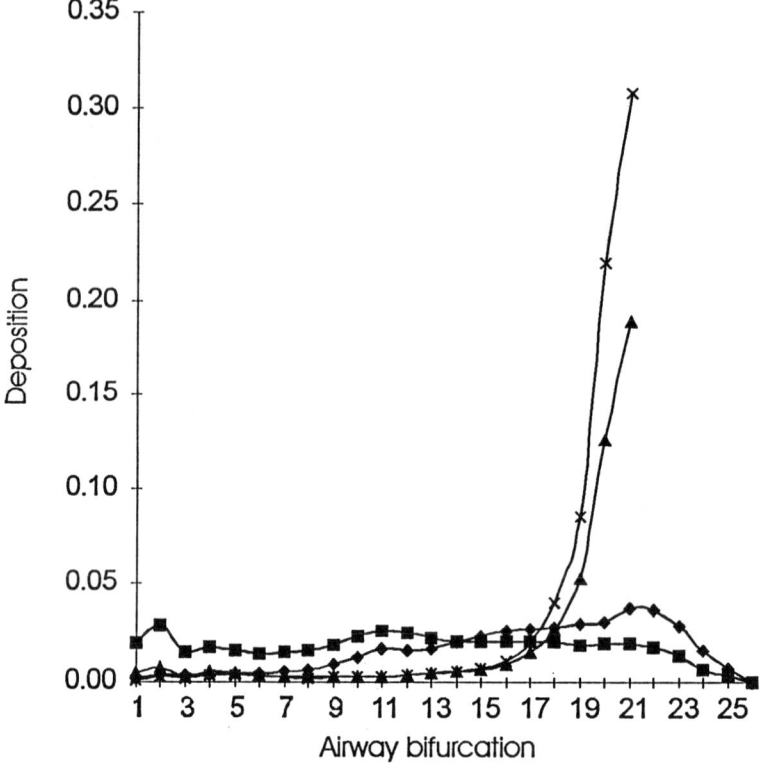

Fig. 3. Deposition of fine and coarse aerosols in human airway bifurcations: comparison between data derived from SPECT measurements (fine (×) and coarse (▲) particles) and theoretical predictions (fine (◆) and coarse (■) particles). Data are normalised to the inhaled activity.

(1991) Model A in the SPECT approach and the stochastic lung model of Koblinger and Hofmann (1985) in the modelling approach. Moreover, the experimental data refer to the right lung only, while the deposition model computes deposition patterns for the whole lung.

Inspection of the results displayed in Figs 2 and 3 reveals that the relative shapes of the deposition patterns per shell and per generation have similarities in both approaches, though there are some notable differences regarding the absolute values. Considering, however, the potential effect of inter-subject variability in lung structure and the above noted systematic differences between the two methodologies, it is not surprising at all that discrepancies still exist. Thus the first results presented here are very promising and warrant the further refinement of the computer model. In conclusion, the data derived from SPECT imaging provide valuable experimental information about the differential distribution of particles among human airways which are badly needed for model validation purposes.

Acknowledgements—This project was supported in part by the Commission of the European Communities, Contract no. FI4P-CT95-0026 and by the Austrian–Hungarian Science and Technology Cooperation Programme, Project no. A27. The experimental work was supported by Astra Draco and the Leverhulme Trust.

REFERENCES

Fleming, J. S., Nassim, M., Hashish, A. H., Bailey, A. G., Conway, J., Holgate, S., Halson, P., Moore, E. and Martonen, T. B. (1995) Description of pulmonary deposition of radiolabeled aerosol by airway generation using a conceptual three-dimensional model of lung morphology. *J. Aerosol Med.* **8**, 341–356.

Heyder, J., Gebhart, J. and Stahlhofen, W. (1980) Inhalation of aerosols—particle deposition and retention. *Generation of Aerosols* (Edited by K. Willeke), pp. 65–103. Ann Arbor Scientific Publishers, Ann Arbor.

Hofmann, W. and Koblinger, L. (1990) Monte Carlo modeling of aerosol deposition in human lungs. Part II: Deposition fractions and their sensitivity to parameter variations. *J. Aerosol Sci.* **21**, 675–688.

Koblinger, L. and Hofmann, W. (1985) Analysis of human lung morphometric data for stochastic aerosol deposition calculations. *Phys. Med. Biol.* **30**, 541–556.

Koblinger, L. and Hofmann, W. (1990) Monte Carlo modeling of aerosol deposition in human lungs. Part I: Simulation of particle transport in a stochastic lung structure. *J. Aerosol Sci.* **21**, 661–674.

Synder, W. S., Ford, M. R., Warner, G. G. and Fisher, Jr, H. L. (1969) Estimates of absorbed fractions for monoenergetic photon sources uniformly distributed in various organs of a heterogeneous phantom. MIRD Pamphlet no. 5. *J. Nucl. Med.* **10** (Suppl. 3), 7–52.

Weibel, E. R. (1991) Design of airways and blood vessels considered as branching trees. *The Lung: Scientific Foundations* (Edited by R. G. Crystal and J. B. West), pp. 711–720. Raven Press, New York.

 Pergamon

Ann. occup. Hyg., Vol. 41, Supplement 1, pp. 582–587, 1997
© 1997 British Occupational Hygiene Society
Published by Elsevier Science Ltd. All rights reserved
Printed in Great Britain
0003–4878/97 $17.00 + 0.00
Inhaled Particles VIII

PII: S0003–4878(96)00088–9

AEROSOL BOLUS DISPERSION AND RECOVERY IN A HUMAN AND DOG AIRWAY CAST

G. Scheuch, W. Stahlhofen, C. P. Fang* and M. Lippmann*

GSF-Forschungszentrum für Umwelt und Gesundheit, Inst f Inhalations Biologie, Robert Koch Allee 6, 82131 Gauting, Germany; and *Nelson Institute of Environmental Medicine, New York University Medical Center, Tuxedo, New York, U.S.A.

INTRODUCTION

In recent years methods in lung diagnostics have been developed using the aerosol bolus inhalation technique (Heyder *et al.*, 1988; Anderson *et al.*, 1989; Schulz *et al.*, 1995; Blanchard 1996). A bolus is a small volume of aerosol sandwiched in particle-free air and inhaled into different volumetric regions of the lung. During inhalation and exhalation the original small bolus is dispersed over a larger volume. This dispersion is caused by convective mixing in the airways and by intrinsic particle motion. Because particles in the size range between 0.2 and about 1 μm have low intrinsic motion, they can be used as tracers of the convective air transport in the lungs. The extent of broadening of the bolus measured in the exhaled air is a measure of the convective mixing during inhalation and exhalation, Blanchard (1996).

The bolus inhalation technique can also be used to deliver aerosol into specific volumetric regions of the lungs. If a bolus is inhaled near the end of an inhalation, the aerosol particles cannot penetrate deep into the lungs and should be delivered preferentially to the tracheobronchial airways. In order to determine where the particles from an inhaled bolus are located within the lungs at end inhalation, dispersion and recovery studies were made in a dog and a human airway cast.

MATERIAL AND METHODS

Aerosol

The monodisperse aerosol particles were produced by heterogeneous condensation of di-2-ethylhexyl sebacate (DEHS) vapour on NaCl nuclei. Two commercially available aerosol generators were used (MAGE, Lavoro E Ambiente, Bologna, Italy; TOPAS, Dresden, Germany). The terminal settling velocity (v_s) of the particles was measured with a convection-free sedimentation chamber. The aerodynamic particle diameters (d_{ae}) generated in this way ranged between 0.5 and 4 μm.

Airway cast

The lung casts were prepared from fresh lungs free of respiratory disease. A solid cast complete to the airways of 1 mm in diameter were made by filling the airways

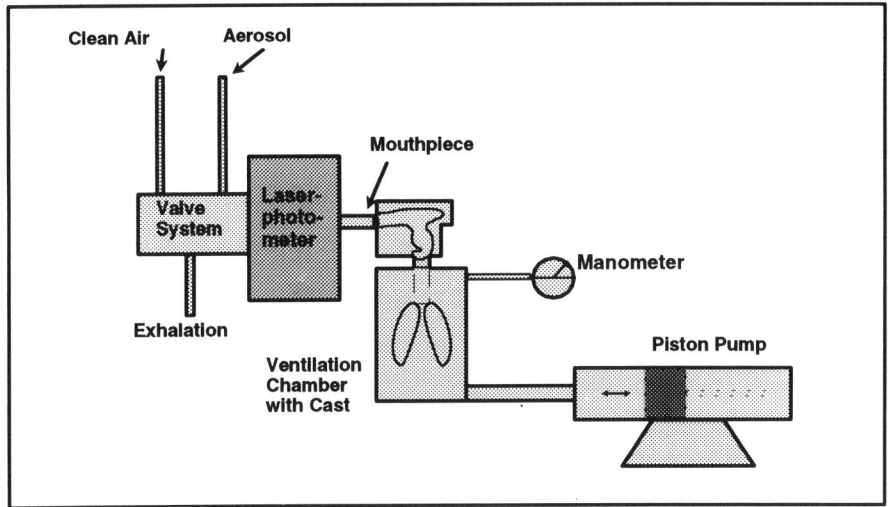

Fig. 1. Schedule of the experimental set-up.

of the inflated lung with wax. Tissue was macerated from the solid wax cast. This cast was then coated with RTV silicone material. The hollow cast was obtained after melting wax out of the cast, Briant and Lippmann (1992).

The morphometry of the casts was measured and reported. The entire volume of the dog cast was about 160 cm³, the volume of the trachea was about 70 cm³. The human cast had a total volume of 125 cm³, with a trachea volume of 30 cm³. For some measurements a model of a human mouth cavity and oropharyngeal region was connected to the human airway cast. This model had an additional volume of 54 cm³.

Aerosol administration

The cast was ventilated by a piston pump while mounted within a ventilation chamber (Fig. 1). Part of the trachea was outside of the chamber and was connected to the mouthpiece of an inhalation device. With the inhalation device the aerosol concentration could be detected during inhalation and exhalation by laser photometry. The valve system of the device allowed small volumes of aerosol to be injected into the particle free inhalation air. The penetration of each bolus into the cast could be controlled by injecting the aerosol into various predetermined volumes during the clean air inhalation. At the end of any inhalation, periods of breathholding (t_p) could be chosen.

During the experiments, the relative aerosol number concentration (c) measured with the photometer as function of the inhaled and exhaled volume (V) were recorded and stored as c (V), (Scheuch *et al.*, 1992). The influence of bolus penetration, ventilated flow rates, particle size, breathholding periods and mouth cavity were measured, (Scheuch *et al.*, 1995). In this paper a comparison between both casts is given and the influence of an additional oropharyngeal region is demonstrated.

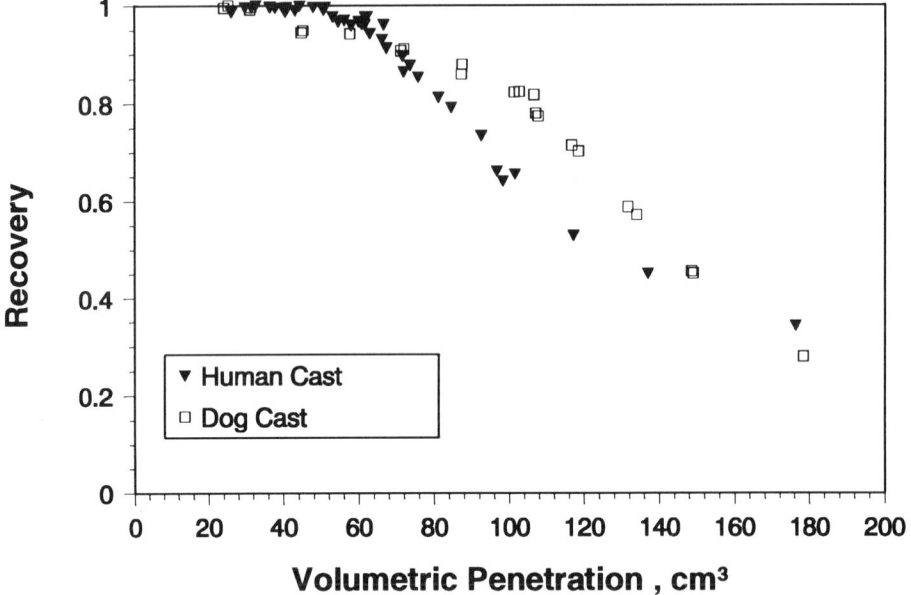

Fig. 2. Aerosol recovery as function of the volumetric penetration of the bolus. Comparison between human cast and dog cast.

RESULTS

Figure 2 shows the particle recovery after bolus ventilation in different volumetric regions of the casts with particles of $d_{ae} = 0.6–0.9$ μm. Flow rate was chosen to be 250 ml s^{-1}. Both casts were adapted to the inhalation device with the same tube (10 cm long, inner diameter 1.2 cm). While the recovery from the human cast was 100% for volumetric penetration (VP) < 50 ml, particles from the dog cast were lost earlier (VP = 35 ml). Fewer particles were lost from the dog cast between 70 and 140 ml.

The results for dispersion measurements were identical in both casts for VP < 70 ml. For VP > 70 ml the dispersion in the dog cast was significantly higher (Fig. 3). The connection between the cast and the inhalation device had a significant influence both on aerosol recovery and dispersion. In Figs 4 and 5 the recovery and dispersion after bolus injection into the human cast are given, respectively. Results of measurements are compared between connection with a mouth cavity and an oropharyngeal (MO) model and the short tubing described above. It can be seen that the MO model leads to a distinct increase in dispersion. The recovery measurement between VP = 50 and 150 yielded a higher particle recovery from the cast with the oropharyngeal model than with the human cast alone. In Fig. 5 the results of dispersion measurements of eight healthy human volunteers are also given as a comparison.

DISCUSSION

Particle losses during the ventilation cycle are caused by losses from particles into

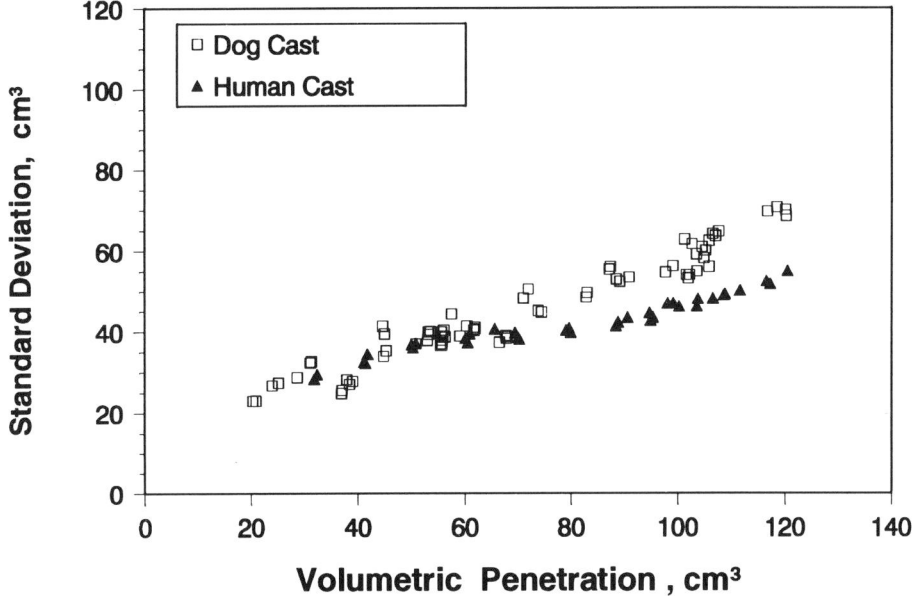

Fig. 3. Bolus dispersion (SD) as function of the volumetric penetration of the bolus. Comparison between human cast and dog cast.

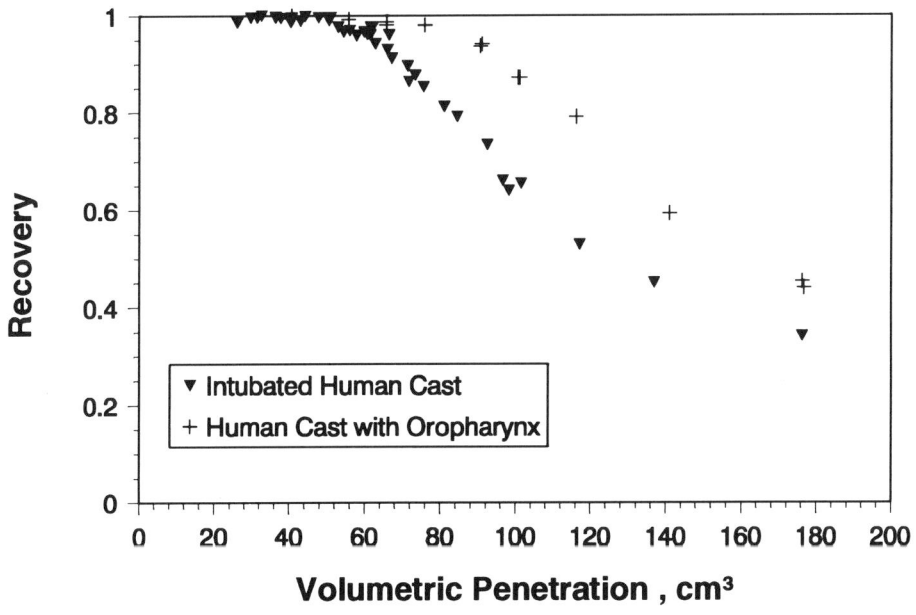

Fig. 4. Aerosol recovery as function of the volumetric penetration of the bolus. Comparison between human cast connected to the mouth piece of the inhalation device (intubated) and a measurements using an additional model for the oropharynx.

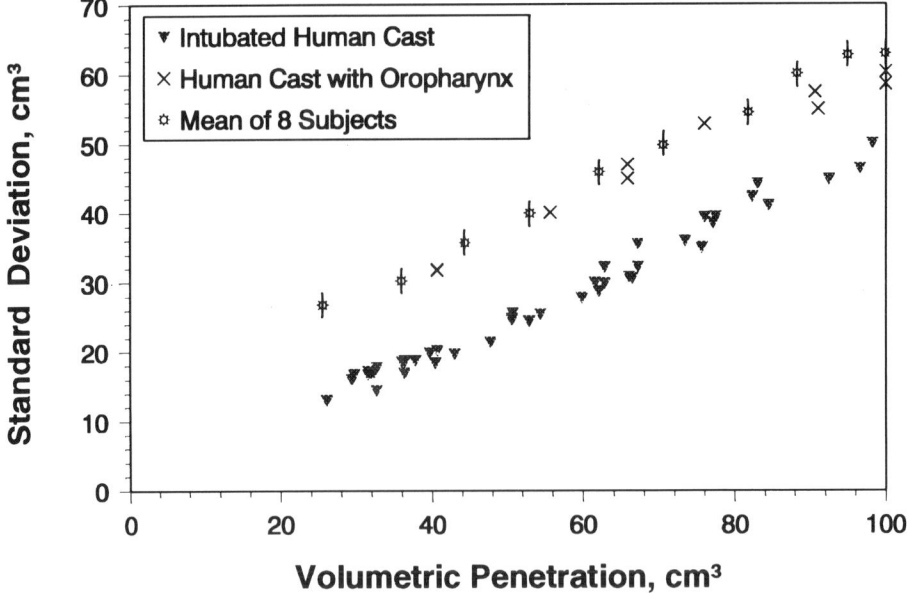

Fig. 5. Bolus dispersion (SD) as function of the volumetric penetration of the bolus. Comparison between human cast connected to the mouth piece of the inhalation device (intubated), measurements using an additional model for the oropharynx and mean values measured on eight healthy human volunteers.

the ventilation chamber and by deposition onto the walls of airways. By using particles with aerodynamic diameters between 0.6 and 0.9 μm losses by deposition are small compared to the losses into the ventilation chamber. Particles penetrating through the open ended 1 mm airways of casts reached the 5 l. ventilation chamber. Only a minor fraction (< 3%) can find the way back into the cast and will be detected with the photometer during the exhalation cycle. A recovery of 100% of the aerosol shows that no particles are deposited or reached the ventilation chamber. They are recovered from the airways of the cast. From Fig. 2 it can be seen that up to a bolus penetration volume of VP = 65 ml less than 5% of the aerosol left the cast and, therefore, could reach airways < 1 mm in diameter. For the human cast less than 2% were lost for VP < 50 ml. Because of the smaller total volume of the human cast more particles were lost at VP = 50–150 ml compared to the dog cast. An additional oropharyngeal region adapted to the human cast lead to smaller losses into the ventilation chamber. Up to a bolus penetration volume of 120 ml the losses out of the human cast airways and by deposition were smaller than 5%.

From Figs 2 and 3 it can be seen that the dispersion and recovery from human and dog cast itself showed similar results for shallow boluses. The dispersion in the dog airways are slightly higher for VP between 70 and 150 ml. This may be caused by the different branching pattern of the human and dog cast airways. The human airways have a more dichotomous branching pattern.

An additional oropharyngeal model between the trachea of the human cast and the inhalation device lead to much higher dispersion. This may be caused by mixing in low ventilated regions of the oropharynx (Scheuch *et al.*, 1995). Cast measurements and measurements in healthy human volunteers and on living dogs (Scheuch *et al.*, 1995) using the same inhalation device resulted in very similar dispersion data (Fig. 5), suggesting that the cast serve as a good model of *in vivo* behaviour of aerosol boluses.

CONCLUSION

These investigations showed that the bolus inhalation technique can be used as a tool to deposit aerosol particles in conducting airways. By inhaling boluses in volumetric regions of less than 60 ml (80 ml in human) over 95% of the aerosol will be located in airways of more than 1 mm in diameter. The dispersion measured in shallow volumetric lung depths are mainly caused by dispersion mechanisms in the oropharyngeal region. Dispersion in human airways is slightly smaller than in dog airways.

REFERENCES

Anderson, P. J., Blanchard, J. D., Brain, J. D., Feldman, H. A., McNamara, J. J. and Heyder, J. (1989) Effect of cystic fibrosis on inhaled aerosol boluses. *Am. Rev. Respir. Dis.* **140**, 1317–1324.

Blanchard, J. D. (1996) Aerosol bolus dispersion and aerosol-derived airway morphometry: assessment of lung pathology and response to therapy, Part 1. *J. Aerosol Med.* **9**, 183–205.

Briant, J. K. and Lippmann, M. (1992) Particle transport through a hollow canine airway cast by high-frequency oscillatory ventilation. *Exp. Lung Res.* **18**, 385–407.

Heyder, J., Blanchard, J. D., Feldman, H. A. and Brain, J. D. (1988) Convective mixing in the human respiratory tract: estimates with aerosol boli. *J. appl. Physiol.* **64**, 1273–1278.

Scheuch, G. and Stahlhofen, W. (1992) Deposition and dispersion of aerosols in the airways of the human respiratory tract: the effect of particle size. *Exp. Lung Res.* **18**, 343–358.

Scheuch, G., Stahlhofen, W., Fang, C. P. and Lippmann, M. (1995) Dispersion of aerosol particles in an airway cast of a dog. *Exp. Lung Res.* **21**, 519–534.

Scheuch, G., Westenberger, S., Stahlhofen, W., Fang, C. P., Cohen, B. S. and Lippmann, M. (1995) Influence of the oropharyngeal region on aerosol dispersion. *J. Aerosol Med.* **8**, 127.

Schulz, H., Schulz, A., Brand, P., Tuch, T., von Mutius, E., Erdl, R., Reinhard, D. and Heyder, J. (1995) Aerosol bolus dispersion and effective airway diameters in mild asthmatic children. *Eur. Respir. J.* **8**, 566–573.

 Pergamon

Ann. occup. Hyg., Vol. 41, Supplement 1, pp. 588–592, 1997
© 1997 British Occupational Hygiene Society
Published by Elsevier Science Ltd. All rights reserved
Printed in Great Britain
0003–4878/97 $17.00 + 0.00
Inhaled Particles VIII

PII: S0003–4878(96)00151–2

DEPOSITION AND CLEARANCE OF GLASS MICROBALLOONS: AEROSOLS OF LARGE PHYSICAL SIZE AND LOW DENSITY

P. M. Hext,* B. A. Rimmel,* A. Soames* and G. H. Pigott†

*Zeneca Central Toxicology Laboratory, Alderley Park, Macclesfield, Cheshire SK10 4TJ, U.K.; and
†27 Windermere Drive, Alderley Edge, Cheshire SK9 7UP, U.K.

INTRODUCTION

Glass microballoons (GMBs) are used in the production of emulsion explosives and to lower the bulk density of some plastics. They are hollow glass spheres with mean diameter in the range 50–60 μm (range about 4–120 μm). The walls of the spheres are very thin (wall thickness within the range 1–5 μm, but uniform for individual spheres) so the bulk density is < 0.01 g ml^{-1} and a balloon of physical diameter 50 μm has an aerodynamic diameter of approximately 5 μm. These considerations suggest that in the course of normal production and use there is a possibility that a proportion of the GMBs will be dispersed in air and their low aerodynamic diameter may result in penetration to the lower respiratory tract. The investigations reported here were primarily intended to determine whether any accumulation of such large particles could lead to the formation of agglomerates and consequent obstruction of the small airways. The investigations were conducted in rats because they are an accepted model for man and because the smaller size of the terminal airways in the rat would exacerbate any potential problem. Although the respirable limit for rats is a little below that for man (Snipes *et al.*, 1989) the size range of the GMBs is such that a significant proportion is rat-respirable.

MATERIALS AND METHODS

GMBs were obtained from Nobel's Explosive Co. Ltd as a finely divided highly mobile powder.

Male specific pathogen free Alpk:APfSD (Wistar derived) albino rats were obtained from the colony maintained by Zeneca Ltd at Alderley Park Cheshire U.K. Rats were maintained under standard laboratory conditions with food and water available *ad libitum* except during the exposures.

Atmosphere generation was by a rotating table type of dust generator with air ejection. Powder is aspirated from a groove cut into a circular rotating table. The groove is replenished by a hopper and scraper system. Concentration is adjusted by changing the combination of size of groove, speed of rotation and air flow.

Two experiments were performed. The first, of 4 h exposure duration, was designed to test the generation methods, to ensure that the integrity of the GMBs was retained and to determine whether the observed deposition pattern was as predicted from the aerodynamic diameter. For this study rats were exposed nose only in restraining tubes to an atmosphere containing 500 mg m^{-3} total particulate. Half the rats were sacrificed two days after exposure to determine the initial lung burden. The remaining rats were retained for a 14 day observation period to determine any delayed reaction.

In the second study groups of 70 rats were exposed to filtered air (control) or to atmospheres containing 10 or 60 mg m^{-3} inhalable fraction (< 15 μm AED). This was calculated initially to correspond to 25 and 125 mg m^{-3} total particulate. After 17 days the inhalable target was falling short of the desired value and total particulate concentration were raised to 50 and 250 mg m^{-3}, respectively. Exposures (whole body) were for 6 h day^{-1} for 28 consecutive days. Atmospheres were sampled for total particulate concentration at least five times daily and aerodynamic diameter distribution was determined daily by cascade impactor. Atmosphere concentration was adjusted to give the target inhalable particle concentrations assessed as cumulative total exposure.

Ten rats from each group were sacrificed two days after the last exposure (week 5) and during study weeks 9, 17, 37 and 49. All survivors were terminated at week 75. At each sacrifice half of the rats were designated for pathology with the remainder used to assess lung burden.

For the rats designated for pathology, lungs were removed and inflated with formaldehyde vapour, the trachea was ligated and the lungs immersed in formol saline. Sections of lung, trachea, tracheobroncial lymph nodes, nasal turbinates and any visually abnormal tissues were prepared for histological examination and a sample of fixed lung was taken for examination in the scanning electron microscope to give additional information on the distribution of GMBs within the lower respiratory tract.

To determine the lung burden of GMBs approximately 100 mg of fresh lung was dissolved in hydrochloric acid by digestion in a microwave "bomb". The suspension obtained was neutralised and filtered onto a vinyl filter which was dried, mounted and cleared. The GMBs were counted at 300 × magnification on 200 sequentially selected, non-overlapping fields.

RESULTS

Approximately 3% of GMBs from atmosphere samples examined in the scanning electron microscope showed evidence of damage. Intact balloons ranged from 20–120 μm diameter. For the initial (single exposure) experiment the mass median aerodynamic diameter (MMAD) of GMBs in the atmosphere was 5.8 μm with about 12% of the mass < 2.5 μm AED and over 90% of the mass < 15 μm AED. These values showed some drift over the course of the 28 day study with progressively higher MMAD recorded. This possibly represented a segregation of the balloons within the container with the lighter fraction readily rising to the surface even during short storage.

Two days after the single exposure a few GMBs were evident in the lungs of the rats examined. After 14 days only 3/5 lungs examined contained GMBs although this may reflect the limited area examined. Overall the pilot study was considered to have fulfilled the objective of demonstrating satisfactory generation and the

Table 1. Lung burden (GMBs/g lung)

Week number	Control	Exposure group 2140 mg m^{-3} h^{-1}	9900 mg m^{-3} h^{-1}
5 (post exposure)	0	760	3030
9	0	220	970
16	0	200	410
37	0	200	220
49	0	180	200
75	0	170	170

possibility of these large physical diameter spheres penetrating to the lower respiratory tract.

In the 28 day exposure study cumulative exposures of 0 (control) 2140 and 9900 mg m^{-3} h^{-1}, expressed as inhalable fraction (< 15 μm AED), were achieved. The rats remained in good condition with no evidence for an adverse effect of exposure. During the subsequent holding period the rats generally remained in good condition, there were a few deaths but the incidence was unremarkable for a study of this duration as there was no evidence from the incidence or causes that any of these were attributable to the inhalation of GMBs.

The numbers of GMBs found in the digested lung samples were small and declined rapidly following exposure (Table 1). From these values the calculated half time for clearance was approximately 21 days. A small residue remained even at final termination and the absolute numbers of this residue were apparently similar for both dose groups although the precision of measurement of such low numbers renders the value indicative rather than absolute.

In those rats examined shortly after the last exposure the majority of deposited GMBs were found in the bronchi or larger bronchioles. The median diameter of these was 15.7 ± 4 μm. These were often covered with mucus and appeared embedded in the epithelial layer. A few were also observed in the alveoli, especially in the high dose group and these were apparently free of cellular contact. Median diameter was 17.7 ± 3.7 μm, they were within the same range as those in the bronchiolar region.

At subsequent examinations no GMBs were seen in the alveolar region for either dose group. The few GMBs seen at these later time periods were firmly attached to the bronchial wall and in most cases embedded in the epithelium to at least half the diameter. The size range remained similar to that seen immediately following the exposure period. Histological reaction to the embedded GMBs was minimal, with no indications of a fibrotic response even to those remaining 75 weeks after exposure.

DISCUSSION

According to the modified ICRP lung dosimetry model (Bailey *et al.*, 1995) pulmonary deposition of about 2–3% of GMBs would be expected, although this model assumes rapid clearance of material in transit through the bronchial tree. Although the achieved deposition was not measured it was almost certainly less than this value, reflecting in part the lower size limit for respirability in the rat lung compared to man. In addition, the large physical size of GMBs would be expected to increase filtration/deposition in the upper respiratory tract due to interception, further reducing the pulmonary dose.

It is evident that alveolar clearance of deposited GMBs was rapid. This is consistent with the results of Cool and Moore (1982) who showed half times for dissolution *in vitro* of 23–280 days for three formulations of glass used to produce GMBs. GMBs are too large for phagocytosis by macrophages and dissolution would be expected to occur more rapidly at the pH of the extracellular lung fluid (Bernstein *et al.*, 1995). While this is a possible explanation, at least in part, of the rapid disappearance of GMBs from the alveoli, other mechanisms, such as transport on the surfactant to the bronchioles and ciliated airways, probably also contributed.

The behaviour of a proportion of GMBs in the conducting airways is also of interest. From the earliest stages of observation it was apparent that those GMBs visible in these regions were rapidly coated with mucus and many appeared to be immobilised on the epithelium. Eventually the remaining GMBs were found exclusively embedded in the epithelium of conducting airways. Gehr *et al.* (1993) made similar observations with polystyrene spheres, both *in vivo* and *in vitro* and suggest that the normal method of clearance from the smaller conducting airways is incorporation into the sol phase of the mucus layer, followed by macrophage phagocytosis and clearance via the mucus phase. Snipes *et al.* (1995) found that a substantial fraction of inhaled polystyrene microspheres were incorporated into the interstitium and the epithelium of the tracheobronchial region of the rat, guinea pig and dog. Similar findings have been reported for fibrous particles in man (Churg and Wright, 1988).

Schurch *et al.* (1993) suggest that this process is quantitatively important for very small particles but less so at larger diameter. The relatively few GMBs persisting in the lung even shortly following exposure is consistent with this proposal while demonstrating clearly that some incorporation into the epithelium is possible even for these very large particles.

In the case of GMBs the increased persistence, when sequestered in the epithelium did not result in any additional reaction to the particles. This is consistent with the known properties of these glasses which generally show little or no tissue reaction. A survey of 46 workers exposed to GMBs also showed no evidence for an effect on lung function or on increased prevalence of respiratory symptoms (I. P. Tams—personal communication). Thus the retention of small numbers of GMBs for extended periods in the lung tissue is without demonstrable effect in rats or humans. The consequences for similar retention of a more reactive particle are unknown, but may be less benign.

REFERENCES

Bailey, M. R., Dorrian, M-D. and Birchall, A. (1995) Implications of airway retention for radiation doses from inhaled radionuclides. *J. Aerosol Med.* **8**, 373–390.

Bernstein, D. M., Morscheidt, C., Tiesler, H., Grimm, H-G., Thevenaz, P. and Teichert, U. (1995) Evaluation of the biopersistence of commercial and experimental glass fibres. *Inhal. Toxicol.* **7**, 1031–1058.

Churg, A. and Wright, J. L. (1988) Mineral particles in airway walls in the lungs of long-term chrysotile miners. *Ann. occup. Hyg.* **32**, 173–180.

Cool, D. A. and Maillie, H. D. (1983) Dissolution of tritiated glass microballoon fragments: implications for inhalation exposure. *Health Phys.* **45**, 791–794.

Gehr, P., Geiser, M., Im Hof, V., Schurch, S., Waber, U. and Baumann, M. (1993) Surfactant and inhaled particles in the conducting airways: structural, stereological and biophysical aspects. *Microscopy Res. Tech.* **26**, 423–436.

Schurch, S., Gieser, M. and Gehr, P. (1993) Surface properties and function of alveolar and airway surfactant. In *Biosurfactants* (Edited by N. Kosaric), pp. 287–304. Marcel Dekker Inc., New York.

Snipes, M. B., McClellan, R. O., Mauderley, J. L. and Wolff, R. K. (1989) Retention patterns for inhaled particles in the lung: comparisons between laboratory animals and humans for chronic exposures. *Health Phys.* **57** (suppl. 1) 69–78.

Snipes, M. B., Guilmette, R. A. and Nikula, K. J. (1995) Microscopic distribution patterns of microspheres deposited by inhalation in the lungs of rats, guinea pigs and dogs. *Inhal. Toxicol. Res. Inst. Ann. Rep. (1994–5)*, ITRI-146, pp. 47–79.

Ann. occup. Hyg., Vol. 41, Supplement 1, pp. 593–600, 1997
© 1997 British Occupational Hygiene Society
Published by Elsevier Science Ltd. All rights reserved
Printed in Great Britain
0003-4878/97 $17.00 + 0.00
Inhaled Particles VIII

Pergamon

PII: S0003–4878(96)00089–0

DISTRIBUTION AND RETENTION OF INHALED VANADIUM ON INERT AIRBORNE PARTICLES*

O. G. Raabe and M. A. Al-Bayati

Institute for Toxicology and Environmental Health, University of California, Davis, CA 95616, U.S.A.

INTRODUCTION

Although vanadium is an essential trace element in the human body, it is highly toxic in various chemical forms at even low levels if introduced into the systemic circulation of the human body. This toxicity may be manifested by broncho-pneumonia, bronchitis, lymphatic necrosis, chronic liver damage, lung fibrosis and kidney fibrosis (Friberg *et al.*, 1986; Al-Bayati *et al.*, 1989, 1992). Most forms of vanadium are not readily absorbed via the gastrointestinal tract and less than 2% of the about 2 mg day^{-1} vanadium in the human diet is actually absorbed into the body. Workers have been exposed to potentially toxic levels of airborne vanadium (0.1–60 mgV m^{-3}) in the production of vanadium metal and vanadium compounds (Friberg *et al.*, 1986). There is also a possible risk for diseases to the general population due to the gradual increases in vanadium concentration in the atmosphere resulting from the combustion of vanadium-containing fuels, especially crude oil whose average vanadium concentration is about 600 ppm (Meish and Bielig, 1980). Inhalation and instillation experiments with animals have shown that vanadium in various forms can be quickly absorbed from the lung into the systemic circulation (Rhoads and Sanders, 1985; Knecht *et al.*, 1992).

This study was designed to provide information on lung clearance of respirable vanadium pentoxide coated onto inert particles and the distribution of the absorbed vanadium in the body of young and adult rats. Coated particles are readily produced and released to the environment by fossil fuel power plants. A total of 72 unanesthetized Sprague–Dawley male rats (36 young and 36 adult) were exposed to aerosols of respirable V_2O_5-coated on fused-clay aluminosilicate particles for up to one hour and at a vanadium concentration of about 1 mg m^{-3}. The aerosol was made by combining ^{48}V-radiolabelled V_2O_5 fume with spherical aluminosilicate particles, which were generated by nebulizing an aqueous suspension of clay and passing it through a high temperature furnace. The study showed that most vanadium was rapidly cleared from the lung, although about 5% remained for 4 weeks, and the absorbed vanadium sequestered in kidney, spleen and especially bone, sites of potential pathological effects. Higher uptake in the skeleton of the young rat may indicate the young individual has a higher risk for disease in the event of exposure to vanadium.

---

* Research supported by the United States Department of Energy.

MATERIALS AND METHODS

Use of ^{48}V labels

Radioactive ^{48}V was supplied as carrier free vanadyl chloride (VOCl$_3$) in 1 M HCl (specific activity = 74 GBq ml^{-1}; Amersham, Inc., Arlington Heights, IL). Its radioactive half-life is 16 days, decaying by positron emission (E_{max} = 0.70 MeV) and it emits four readily detected gamma photons between 0.5 and 2.2 MeV. The carrier ammonium metavanadate powder and other chemicals (purity > 99%), were purchased from J.T. Baker Chemical Co. (Phillipsburg, NJ). The carrier ammonium metavanadate was dissolved in boiling distilled water and mixed with the ^{48}V solution.

Three different sodium iodide scintillation gamma-ray counters equipped with multi-channel analyzers were used to assay the ^{48}V content of whole body, tissues, excreta and aerosol particles samples. These counters were cross calibrated and the net counts were corrected for isotope physical decay, counter efficiency and sample size geometry.

Experimental design and procedures

Two experiments were performed to study the biological distribution of inhaled vanadium in young and adult male rats at different times post exposure after inhalation of ^{48}V-radiolabelled vanadium pentoxide (V$_2$O$_5$) coated onto inert fused-clay aluminosilicate particles. The rats were respiratory disease-free Sprague–Dawley (Charles River, Wilmington, MA). Rats were maintained on a standard diet and water (pH 6) *ad libitum*. In the first experiment, 36 adult rats, 90 days of age were exposed to an aerosol of V$_2$O$_5$-coated particles. In the second experiment, 36 young rats, 35 days of age, were exposed to a similar aerosol. After exposure excreta were collected daily from each rat for radioanalysis of ^{48}V. In both experiments, the rats were serially sacrificed by anesthetic overdose at time intervals of 1 h and 1, 3, 7, 14 and 28 days after inhalation. Tissue samples collected for radioanalysis were (1) blood, (2) heart, (3) thymus, (4) right apical lobe of lung, (5) right cardiac lobe of lung, (6) right intermediate lobe of lung, (7) right diaphragmatic lobe of lung, (8) the apical portion of the left lung, (9) submaxillary salivary glands, (10) larynx and its tissue, (11) trachea, (12) oesophagus, (13), liver, (14) kidneys, (15) pancreas, (16) gastrointestinal tract and its contents, (17) testes, (18) seminal vesicles, (19) prostate, (20) urinary bladder and its contents, (21) femur, (22) head skin, (23) nasal cavity, (24) cerebrum, (25) cerebellum, (26) brain stem, (27) skull, (28) skeletal muscle samples (thigh region), (29) pelt, paws and tail and (30) carcass.

Exposure procedure

The aerosol exposure system was housed in four connected glove boxes, the first was used for aerosolisation, the second for heat treatment, the third for sampling, and the fourth contained the nose-only exposure apparatus (Raabe *et al.*, 1973). Spherical aluminosilicate fused clay particles were produced by nebulizing an aqueous suspension of pretreated montmorillonite clay obtained from Osage, Wyoming, U.S.A. (12 mg ml^{-1}), and passing it through a high temperature furnace (1150°C), which fused the clay into spherical aluminosilicate particles (Raabe *et al.*, 1971). This aerosol of spherical particles was passed through a diffusion drier and mixed with nebulized ^{48}V-labelled ammonium metavanadate solution which then passed through the second furnace at 1000°C. The slowly nebulized solution consisted of 59 GBq of ^{48}VOCl$_3$ with 2 mg NH$_4$VO$_3$ in 4 ml H$_2$O. In the second furnace the aerosol droplets were dried to small particles and then the ammonium metavanadate and the vanadyl chloride were readily transformed to vanadium pentoxide, whose melting point is 690°C and boiling or decomposition point is 1750°C. The resulting pentoxide fume rapidly diffuses to and attaches onto the surface of the aluminosilicate aerosol particles. The process forming the vanadium pentoxide fume can be described as:

$$2NH_4VO_3 \xrightarrow[\text{heat}]{1000°C} V_2O_5 + 2NH_3 \uparrow + H_2O \uparrow .$$

The resulting aerosols (Fig. 1) with 0.5 and 0.8 μm activity median aerodynamic diameters (σ_g about 1.7) were transported to the nose-only small rodent inhalation exposure apparatus which had 36 exposure ports on each side. The rats were held in tubes which fit into the inhalation ports and were exposed for forty-five minutes to about 60 mg m^{-3} of vanadium-coated airborne particles including about 1 mg Vm^{-3}. The vanadium body content for each animal was measured by whole body gamma ray spectroscopy both immediately post-inhalation and on sacrifice date.

RESULTS

The biological distribution and biokinetic results are summarized in Figs 2–4. Figure 2 shows the total retention of selected tissues with respect to the initial lung burden. Figure 3 shows the corresponding tissue concentrations of vanadium which are indicative of the toxicological dose rate. The clearance of V$_2$O$_5$ from the lungs was rapid during the first 24 h after inhalation in both experiments (Fig. 2). By 1 h, the lungs burden for all animals was about 44% of internal body burden, which was equal to 60.7 and 67.5% of initial lungs burden for young and adult rats, respectively. About 70–73% of the initial lungs burden translocated outside the pulmonary tissue by the end of the first day, 80–82% by the end of third day, but 5% remained in the lungs after 4 weeks. There was no significant difference in vanadium concentration among lung lobes in either young and adult rats, and vanadium was cleared from all lobes at about the same rate. At 1 h post-inhalation, all the tissue samples which were analysed showed measurable levels of activities except the brain. The blood showed its highest peak activity at 1 h, but lost most of that activity during the first day following the inhalation. The highest blood concentration was observed 1 h after inhalation and 83% of that activity disappeared by the end of the third day. In both young and adult rats, the skeleton showed the peak in activity at 3 days following inhalation; however, it maintained high levels of activities at four weeks throughout (Figs 2 and 3). The skeleton of the young rats showed higher vanadium uptake and retention than that of the adults.

The highest liver concentration of vanadium occurred at 1 h and 1 day in young and adult rats respectively (Fig. 3), while in all animals, the kidneys showed their peak of activity at 1 day. The spleen of rats showed the peak of activity at 3 days post-exposure and only those of the adult animals maintained most of that activity throughout the rest of the study period. Daily urinary excretion of ^{48}V is shown in

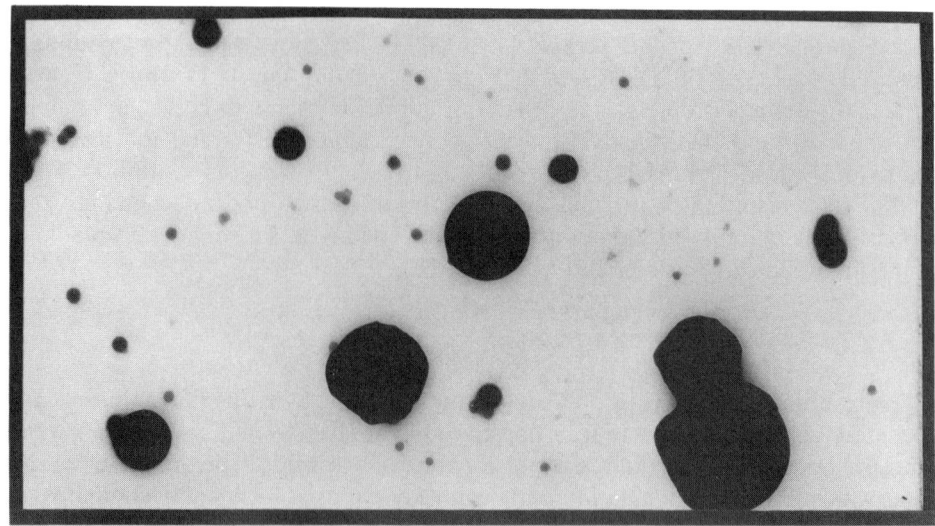

CLAY-VANADIUM AEROSOL EXPT. 1

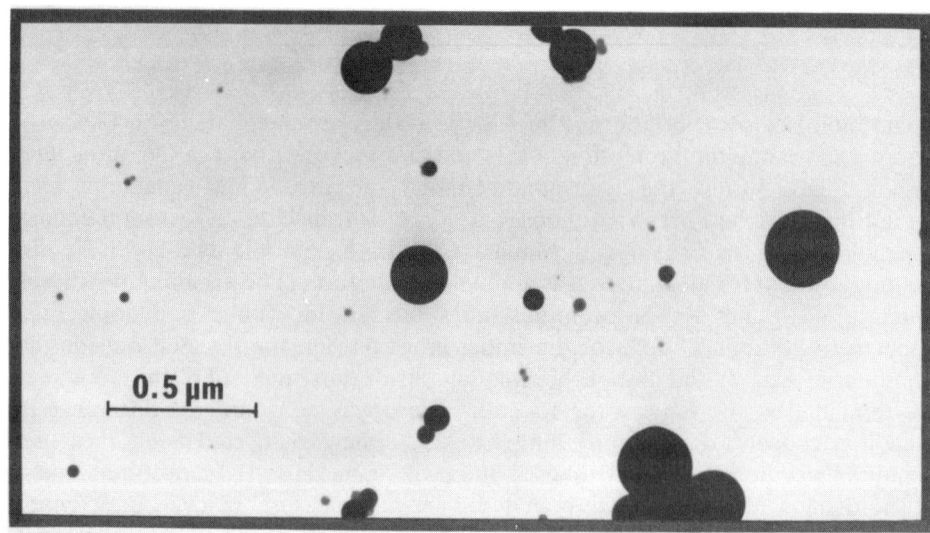

0.5 µm

CLAY-VANADIUM AEROSOL EXPT. 2

Fig. 1. Clay–vanadium aerosol, experiment 1 and 2.

Fig. 4. Most of the absorbed vanadium from the lung was excreted in the urine (about 75% in 28 days). The amounts of vanadium in the urine of the adult rats during the early stages of the study were slightly higher than those of the young rats, while in the late stages the urine of the young rats contained more vanadium than those of the adult. Fecal excretion at various times after inhalation is shown in Fig. 4.

Most of the gastrointestinal burden of [48]V was excreted through the faeces

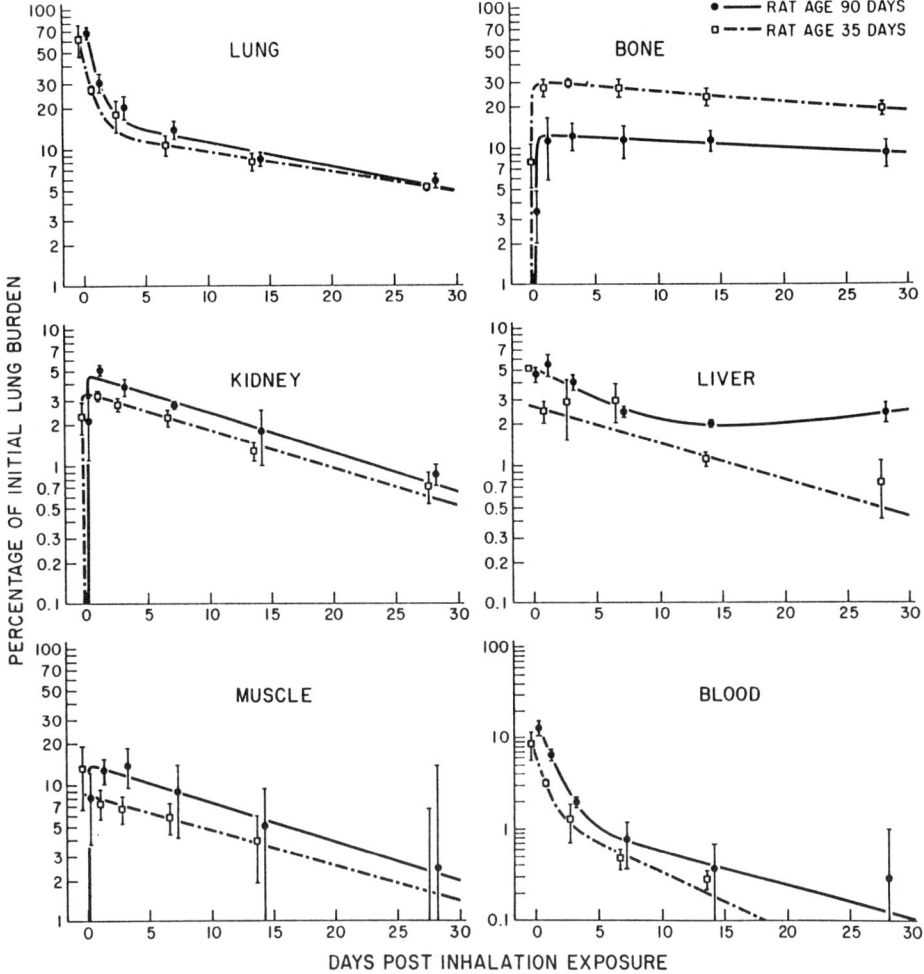

Fig. 2. Tissue retention of inhaled vanadium.

during the first 2 days. The uptake of suspended materials of [48]V-radiolabelled V_2O_5 from GI tract was also measured following gavage in the adult rats. During 7 days following gavage, 95.56% of the initial dose passed through with the faeces, 2.94% was excreted in the urine and only 1.74% retained in the body. The absorbed vanadium from the GI tract distributed in the body in same pattern as absorbed vanadium from the lungs.

DISCUSSION

The young and adult rats used in this study were exposed to [48]V-radiolabelled V_2O_5 aerosol coated on fused-clay aluminosilicate particles with activity median aerodynamic diameter of 0.5 and 0.8 μm, respectively. This study indicated that such a small difference in the size of the particles had no observed effect on the

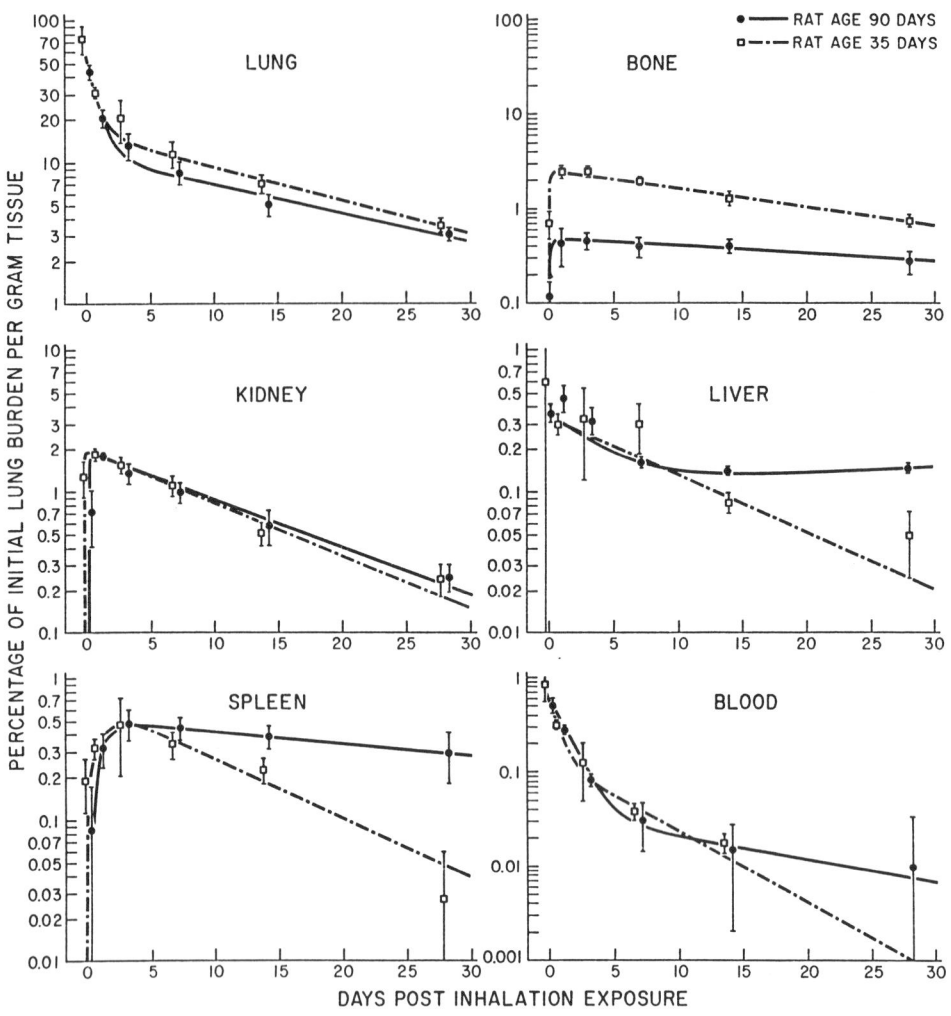

Fig. 3. Tissue concentration of inhaled vanadium.

relative pulmonary deposition, relative lobular distribution and vanadium clearance rate from the lungs throughout the study periods. In each exposure, vanadium was cleared from different lobes of lungs at about the same rate. This study showed that V_2O_5 coated on inert particles was very soluble in the alveolar fluid although 5% remained in the lung even after 28 days. A similar observation has been reported for young rats after intratracheal instillation of [48]$VOCl_3$ with about 3% of the burden remaining in the lungs after 63 days (Oberg et al., 1978).

The skeletal content reached a peak after 3 days with continued high retention at 4 weeks. The skeleton of the young rats showed higher vanadium uptake and retention than that of the adults. Because of observed higher skeletal uptake and long-term retention, children may be at higher risk than adults from inhalation of vanadium compounds.

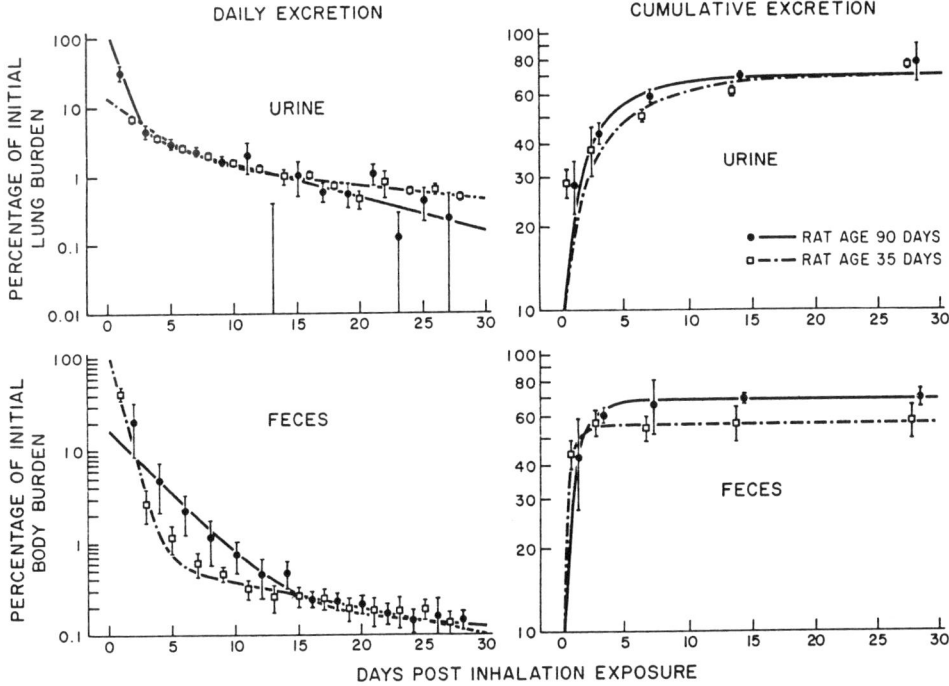

Fig. 4. Excretion of inhaled vanadium.

Acknowledgements—This work was supported by the U.S. Department of Energy and was conducted in facilities fully accredited by the American Association for the Accreditation of Laboratory Animal Care (AAALAC). The authors thank Amiram Rasolt, Stephen Teague and Victor Pietrzak for technical assistance.

REFERENCES

Al-Bayati, M. A., Culbertson, M. R., Schreider, J. P., Rosenblatt, L. S. and Raabe, O. G. (1992) The lymphotoxic action of vanadate. *J. Environ. Pathol., Toxicol., Oncol.* **11**, 19–27.

Al-Bayati, M. A., Giri, S. N., Raabe, O. G., Rosenblatt, L. S. and Shifrine, M. (1989) Time and dose–response study of the effects of vanadate in rats: morphological and biochemical changes in organs. *J. Environ. Pathol., Toxicol., Oncol.* **9**, 435–455.

Friberg, L., Nordberg, G. F. and Vouk, V. B. (1986) *Handbook on the Toxicology of Metals*, Vol. II, 2nd edn. Elsevier, Amsterdam.

Knecht, E. A., Moorman, W. J., Clark, J. C., Hull, R. D., Biagini, R. E., Lynch, D. W., Boyle, T. J. and Simon, S. D. (1992) Pulmonary reactivity to vanadium pentoxide following subchronic inhalation exposure in a non-human primate animal model. *J. appl. Toxicol.* **12**, 427–434.

Meish, H. U. and Bielig, H. J. (1980) Chemistry and biochemistry of vanadium. *Basic Res. Cardiol.* **75**, 413–417.

Oberg, S. G., Paker, R. D. R. and Sharm, R. P. (1978) Distribution and elimination of an intra-tracheally administered vanadium compound in the rat. *Toxicol.* **1**, 315–323.

Raabe, O. G., Bennick, J. E., Light, M. E., Hobbs, C. H., Thomas, R. L. and Tillery, M. I. (1973) An improved apparatus for acute inhalation exposure of rodents to radioactive aerosols. *Toixicol. appl. Pharmacol.* **26**, 264–273.

Raabe, O. G., Kanapilly, G. M. and Newton, G. J. (1971) New methods for the generation of aerosols of insoluble particles for use in inhalation studies. *Inhaled Particles III*, pp. 3–17. Unwin Bros., Surrey.

Raabe, O. G., Yeh, H-C. and Newton, G. J. (1977) Deposition of inhaled monodisperse inhaled particles. *Inhaled Particles IV*, (Edited by W. H. Walton) pp. 3–21. Pergamon Press, Oxford.

Rhoads, K. and Sanders, C. L. (1985) Lung clearance, translocation, and acute toxicity of arsenic, beryllium, cadmium, coboat, lead, selenium, vanadium and ytternium oxides following deposition in rat lung. *Environ. Res.* **36**, 359–378.

Pergamon

Ann. occup. Hyg., Vol. 41, Supplement 1, pp. 601–606, 1997
© 1997 British Occupational Hygiene Society
Published by Elsevier Science Ltd. All rights reserved
Printed in Great Britain
0003–4878/97 $17.00 + 0.00
Inhaled Particles VIII

PII: S0003–4878(96)00152–4

PARTICLE DISSOLUTION CONTRIBUTES TO LONG-TERM ALVEOLAR CLEARANCE OF COLLOIDAL GOLD IN THE RAT

G. Patrick* and C. Stirling

Medical Research Council Radiobiology Unit, Chilton, Oxfordshire OX11 0RD, U.K.

INTRODUCTION

The rate of particle clearance from the alveoli of human lung (Bailey *et al.*, 1985) Philipson *et al.*, 1996) is slower than that generally seen in rodents (Snipes *et al.*, 1983; Bailey *et al.*, 1989), although the underlying mechanisms are the same. In all species, particle dissolution can play a significant role in alveolar clearance, even for substances such as various metal oxides that are essentially insoluble in water (Kreyling *et al.*, 1988; Bailey *et al.*, 1989).

In an earlier study, alveolar microinjection was used to deposit [195]Au-labelled colloidal gold particles specifically in subpleural alveoli of rat lung, without any airway deposition (Patrick and Stirling, 1992, 1994). Since in this model the ultimate clearance rate was slow (half-time = 583 days), it was of interest to determine whether, even for such an insoluble material as metallic gold, the overall clearance rate might be partly due to dissolution and absorption to blood. For this purpose it was necessary to ascertain the pattern of excretion of gold that had dissolved in the lung and been absorbed into blood. Radiolabelled gold chloride was injected i.v. into rats and its excretion measured. Also, to estimate any urinary output of [195]Au from particles reaching the gastro-intestinal tract by clearance up the conducting airways, colloidal gold was administered to rats by gavage. From this, subject to certain assumptions, the overall alveolar clearance of colloidal gold could be analysed into particle transport and absorption to blood.

MATERIALS AND METHODS

Alveolar microinjection of colloidal gold

The procedure and experimental design of the microinjection study have been detailed elsewhere (Patrick and Stirling, 1986, 1992, 1994). The [195]Au-labelled colloidal gold particles had a count median diameter of 10.3–21.4 nm, geometric standard deviation 1.8–1.9. The amount of gold colloid injected into each animal was approx. 0.17 µg, in a volume of 0.05 µl. During four 7 day periods, six rats were housed singly in metabolism cages and the rates of urinary and faecal excretion measured. The first two periods were chosen to be during and towards the end of

---

* Present address: MRC Toxicology Unit, Hodgkin Building, University of Leicester, PO Box 138, Lancaster Road, Leicester LE1 9HN, U.K.

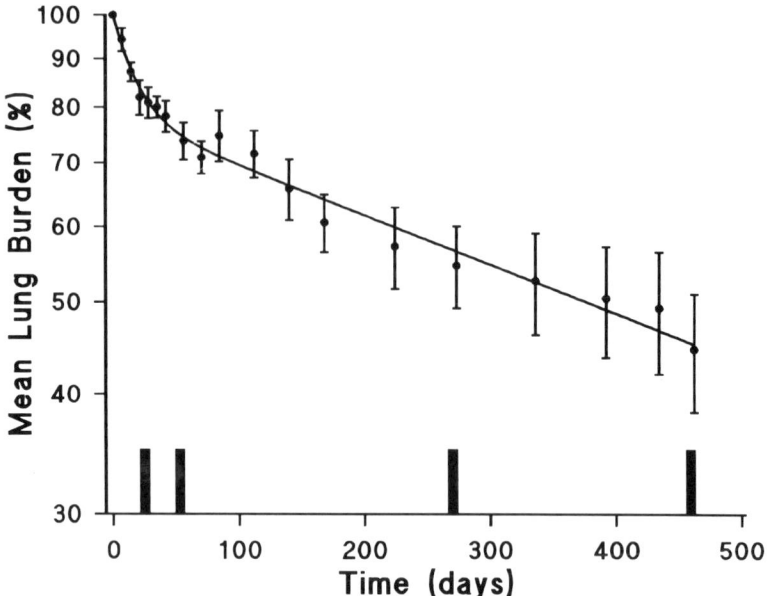

Fig. 1. Lung content of [195]Au after alveolar microinjection of gold colloid, expressed as % initial content on logarithmic scale, means ± SEM, $n = 6$. See text for fitted curve. Solid bars indicate periods of excreta collection.

the time when the first phase of alveolar clearance was still significant, and the last two periods when only the very slow phase remained (see Fig. 1).

Intravenous injection of gold chloride

Gold chloride solution ($HAuCl_4$, BDH/Merck), labelled with [195]Au ($H^{195}AuCl_4$, Dupont/NEN) was prepared for i.v. injection by making it isotonic with NaCl and adjusting the pH to 3.5 with NaOH. The resulting solution contained 1.6 MBq [195]Au and 10 µg Au per ml and was stable at this pH. 0.1 ml was injected via the sublingual vein into five 6 month old male F-344 rats. They were immediately assayed for total [195]Au content in a small animal whole-body counter and, thereafter, at intervals; meanwhile they were housed in metabolism cages for 21 days for collection and assay of excreted [195]Au. Finally they were sacrificed and dissected for radio-assay.

Intragastric administration of colloidal gold

1.7 µg [195]Au-labelled colloidal gold particles, median diameter 14.3 nm and geometric standard deviation 1.9, were administered to six male F-344 rats, instilled via a polyethylene cannula into the stomach. The rats were housed in metabolism cages and sacrificed after 7 days.

RESULTS

The kinetics of clearance of colloidal gold administered by alveolar microinjection are given in Fig. 1 (Patrick and Stirling, 1992). 22% was cleared with a

Table 1. Rates of excretion of [195]Au from colloidal gold (% contemporary lung content/day, means ± SEM, $n = 6$)

Days	Urinary	Faecal
28	0.018 ± 0.006	0.352 ± 0.057
56	0.003 ± 0.001	0.256 ± 0.051
273	0.019 ± 0.006	0.065 ± 0.023
462	0.024 ± 0.006	0.062 ± 0.035

half-time of 14 days, the remainder with a half-time of 583 days. Also shown here are the 7 day periods when urinary and faecal excretion were measured. Urinary and faecal excretion were measurable even after 462 days (Table 1). The rate of faecal excretion decreased markedly during the study. Urinary excretion, while some 20 times less than faecal excretion in the first period, remained relatively constant throughout.

In the separate experiment to determine the fate of ionic gold given i.v., some 30% of the [195]Au was cleared from the body by 21 days after injection. The kinetics of retention (R) over this period (Fig. 2) was well described by the sum of two exponentials as follows:

$$R = 12.5e^{-0.306t} + 87.5e^{-0.0085t}$$

where t = time in days. The rates of urinary and faecal excretion were similar over the 3 week period (Fig. 3). A total of 16.8 ± 0.9% (mean ± SEM) was excreted via the urine and 13.3 ± 0.7% via the faeces. 37% of the injected [195]Au was retained in the liver, while 0.14% remained in the blood (Table 2).

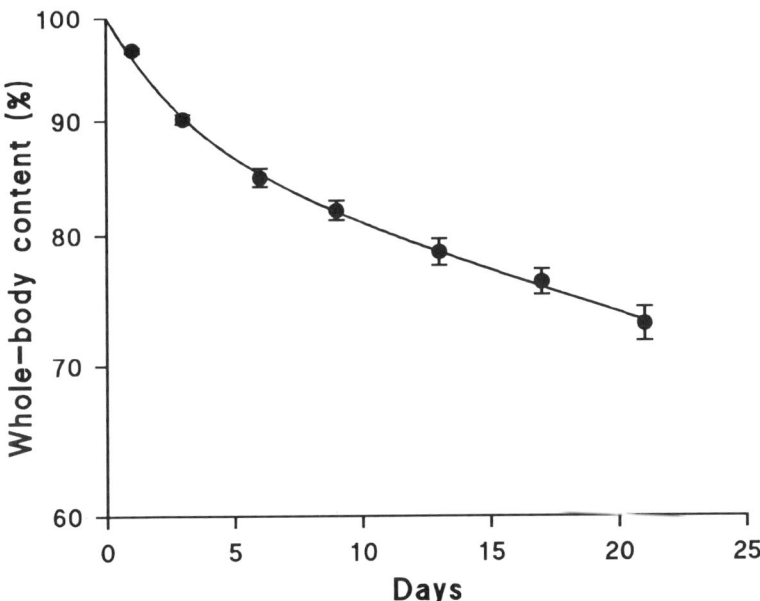

Fig. 2. Whole-body content of [195]Au after i.v. injection as $HAuCl_4$, expressed as % initial content on logarithmic scale, means ± SEM, $n = 5$. See text for fitted curve.

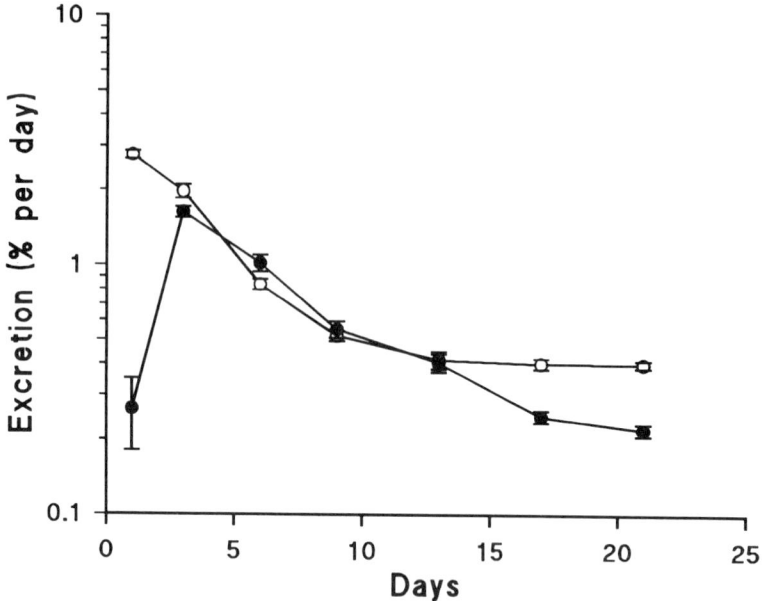

Fig. 3. Rates of excretion of ^{195}Au after i.v. injection as HAuCl$_4$, expressed as % initial content per day, means ± SEM, $n = 5$. ○ = urinary, ● = faecal excretion.

After intragastric administration of colloidal gold, 99.96 ± 0.01% and 0.020 ± 0.004% of the ^{195}Au were recovered from the faeces and urine, respectively, with only 0.019 ± 0.013% left in the carcass as a whole.

If, in the microinjection study, the ^{195}Au which appeared in the blood from dissolved gold particles were partitioned between urine and faeces in the same way as after i.v. injection of gold chloride, then the overall clearance rate for colloidal gold can be analysed into particle transport, $M(t)$ and dissolution followed by absorption to blood, $S(t)$ (see Bailey *et al.*, 1989). The calculation depends on knowing (i) the fraction of colloidal gold dissolved in the lung and absorbed to blood which is excreted via the urine, (ii) the fraction absorbed to blood which is excreted by the faeces, via the bile or from endogenous excretion, (iii) the fraction of particulate material transported via the airways to the gastro–intestinal tract

Table 2. ^{195}Au organ content at sacrifice following i.v. injection

	^{195}Au content (% initial body content, means ± SEM, $n = 5$)
Lung	0.23 ± 0.04
Thoracic lymph nodes	0.04 ± 0.00
Liver	37.43 ± 1.60
Spleen	3.02 ± 0.24
Kidney	10.49 ± 0.42
Gut	0.81 ± 0.03
Blood (approx. 8 ml)	0.14 ± 0.01
Remainder of carcass	9.49 ± 0.53

Table 3. Calculated rates of absorption to blood and particle transport of colloidal gold

Days	$S(t)^1$	$M(t)^2$	$S(t)$ (% total)^3	$M(t)$ (% total)^3
28	0.033 ± 0.010	0.338 ± 0.056	8.8	91.2
56	0.005 ± 0.002	0.254 ± 0.052	2.0	98.0
273	0.033 ± 0.011	0.051 ± 0.021	39.5	60.5
462	0.044 ± 0.011	0.043 ± 0.031	50.5	49.5

[1] Rate of absorption to blood, % contemporary lung content day^{-1}, means ± SEM, $n = 6$.
[2] Rate of particle transport, % contemporary lung content day^{-1}, means ± SEM, $n = 6$.
[3] % of combined $S(t) + M(t)$.

which is directly excreted in the faeces and (iv) the fraction transported to the gastro–intestinal tract which is excreted in the urine because of dissolution and gastro–intestinal absorption.

Values for (i) and (ii) were taken from the i.v. injection data, but modified as follows: as discussed below, the fractions excreted to urine and faeces were scaled up so that they totalled 1.0. Thus (i) and (ii) were increased from 0.168 to 0.559 and from 0.133 to 0.441, respectively. Values for (iii) and (iv), 0.9996 and 0.00020, respectively, were derived from the intragastric administration of gold colloid.

The resulting estimates of $S(t)$ and $M(t)$ are given in Table 3. $M(t)$ decreased five-fold between 56 and 273 days. In contrast, the value of $S(t)$ remained relatively constant over 28–462 days; it was equal to $M(t)$ by the end of the study.

DISCUSSION

The overall clearance of colloidal gold deposited specifically in subpleural alveoli, as shown in Fig. 1 (Patrick and Stirling, 1992), was slower than has generally been observed for long-term clearance in the rat after inhalation. The rate of particle transport was even slower in so far as particle dissolution contributed significantly to the overall clearance. To analyse the excretion data following alveolar deposition of colloidal gold, it was necessary to estimate how ^{195}Au would be excreted once it had reached the blood circulation by dissolution, or the gastro-intestinal tract by particle transport up the conducting airways. Little is known about the pattern of excretion of soluble gold from the circulation. Gold injected i.v. in humans became bound to serum albumin and was excreted via the faeces as well as the urine, some two-thirds being lost by 3 weeks after injection (Mascarenhas et al., 1972). In accord with this, similar proportions of ^{195}Au were excreted by urine and by faeces in the rat (Fig. 3).

The relative constancy of the calculated rate $S(t)$ over 462 days after alveolar deposition (Table 3) was noteworthy. The same had been observed in the F-344 rat over 180 days after the inhalation of $^{57}Co_3O_4$ particles, and also in humans and baboons though not in all the other species that were compared (Bailey et al., 1989). The decrease in $M(t)$ between the two earlier and the two later periods was as expected, consistent with the interpretation that the more rapid overall clearance at earlier times (Fig. 1) was the result of particle transport. Also as expected, particle transport was more prominent than dissolution at the earlier times. In the two later periods, when only the slow phase of clearance remained, the values of

$M(t)$ were very low, equivalent to half-times in excess of 1000 days and similar to $S(t)$ (Table 3).

Excreta were collected for periods of 7 days, yet gold chloride injected i.v. was not completely excreted even after 21 days (Fig. 2). Thus the [195]Au excreted during the 7 day collection periods included some gold dissolved earlier and lacked some gold that was dissolved then but excreted later. However, since the rate of urinary excretion was relatively constant throughout the study (Table 1), it may be assumed that these two effects approximately cancelled each other out.

The extent of [195]Au uptake by the liver after i.v. injection raises the possibility that at least some of the gold chloride had been reduced to metallic gold in the blood. In contrast, after the alveolar microinjection of colloidal gold, the mean uptake of [195]Au by the liver never exceeded 1.2% of the initial lung content (Patrick and Stirling, 1992). This may have been because gold dissolved in alveolar macrophages was released into the blood already bound, for example, to some protein. Whatever the reason, caution needs to be exercised in applying the gold chloride excretion data to the excretion rates in the colloidal gold study. Here we have assumed that the [195]Au retained in the body 21 days after i.v. injection would eventually be excreted into urine and faeces in the same proportions as were observed up to 21 days. On this basis the fractions of gold chloride excreted to urine and faeces were scaled up such that they totalled 1.0. The effect was to reduce each value of $S(t)$ by a factor of 3.3, which may represent the extreme case and so provide lower limits for $S(t)$. In any case, it may be concluded that a considerable proportion of the long-term alveolar clearance of gold colloid was the result of dissolution and absorption to blood.

Acknowledgement—This work was supported in part by the Radiation Protection Research Action of the Commission of the European Communities, under contract F13P-CT92-0064.

REFERENCES

Bailey, M. R., Fry, F. A. and James, A. C. (1985) Long-term retention of particles in the human respiratory tract. *J. Aerosol Sci.* **16**, 295–305.
Bailey, M. R., Kreyling, W. G., André, S., Batchelor, A., Collier, C. G., Drosselmeyer, E., Ferron, G. A., Foster, P., Haider, B., Hodgson, A., Masse, R., Métivier, H., Morgan, A., Müller, H.-L., Patrick, G., Pearman, I., Pickering, S., Ramsden, D., Stirling, C. and Talbot, R. J. (1989) An interspecies comparison of the lung clearance of inhaled monodisperse cobalt oxide particles—Part I: objectives and summary of results. *J. Aerosol Sci.* **20**, 169–188.
Kreyling, W. G., Schumann, G., Ortmaier, A., Ferron, G. A. and Karg, E. (1988) Particle transport from the lower respiratory tract. *J. Aerosol Med.* **1**, 351–370.
Mascarenhas, B. R., Granda, J. L. and Freyberg, R. H. (1972) Gold metabolism in patients with rheumatoid arthritis treated with gold compounds—reinvestigated. *Arthritis Rheum.* **15**, 391–402.
Patrick, G. and Stirling, C. (1986) A method for microinjection into subpleural alveoli of rat lung *in situ*. *J. appl. Physiol.* **60**, 307–310.
Patrick, G. and Stirling, C. (1992) The transport of particles of colloidal gold within and from rat lung following local deposition by alveolar microinjection. *Environ. Health Perspect.* **97**, 47–51.
Patrick, G. and Stirling, C. (1994) The redistribution of colloidal gold particles in rat lung following local deposition by alveolar microinjection. *Ann. occup. Hyg.* **38** (suppl. 1), 225–234.
Philipson, K., Falk, P., Gustafsson, J. and Camner, P. (1996) Long-term lung clearance of [195]Au-labeled teflon particles in humans. *Exp. Lung Res.* **22**, 65–83.
Snipes, M. B., Boecker, B. B. and McClellan, R. O. (1983) Retention of monodisperse or polydisperse aluminosilicate particles inhaled by dogs, rats and mice. *Toxicol. appl. Pharmacol.* **69**, 345–362.

SECTION 6

OTHER DUSTS AND TOPICS

Pergamon

Ann. occup. Hyg., Vol. 41, Supplement 1, pp. 607–614, 1997
© 1997 British Occupational Hygiene Society
Published by Elsevier Science Ltd. All rights reserved
Printed in Great Britain
0003–4878/97 $17.00 + 0.00
Inhaled Particles VIII

PII: S0003–4878(96)00086–5

CURRENT ISSUES IN EXPOSURE ASSESSMENT FOR WORKPLACE AEROSOLS

J. H. Vincent,*† L. M. Brosseau,* G. Ramachandran,* Perng-Jy Tsai,‡ T. M. Spear,§ M. A. Werner* and N. V. McCullough*

*Division of Environmental and Occupational Health, School of Public Health, University of Minnesota, Box 807, Mayo, 420 Delaware St. S.E., Minneapolis, MN 55455; ‡Department of Environmental and Occupational Health, Medical College, National Cheng-Kung University, 138 Sheng-Li Road, Tainin 70428, Taiwan; and §Montana College of Mineral Science and Technology, University of Montana, West Park Street, Butte, MT 59701, U.S.A.

INTRODUCTION

The development of new criteria for the health-related assessment of workplace aerosol exposures has stimulated interest in new or modified sampling instrumentation. In particular, samplers must now demonstrably collect aerosol fractions which are relevant to the health risk in question; namely one or more of the inhalable, thoracic and respirable fractions in the form now agreed between the International Standards Organisation (ISO, 1992), the Comité Europeén Normalisation (CEN, 1992) and the American Conference of Governmental Industrial Hygienists (ACGIH, 1995).

The primary focus of this paper is the inhalable fraction, representing airborne particles that enter through the nose and/or mouth during breathing. This will apply to those occupational exposure limits (OELs) which are currently expressed in terms of so-called "total" aerosol and is the starting point (or "envelope") for the other two fractions. It is the most difficult to simulate in a sampling instrument since it is determined by physical factors outside the body which are highly variable and largely uncontrolled. The paper first summarises the results of studies to compare measures of occupational exposure to "total" aerosol with corresponding ones obtained according to the new inhalability criterion. This leads to a discussion of the impact on OELs (and the OEL-setting process). The paper goes on to identify how the new particle size-selective criteria framework can be applied towards an improved understanding of dose arising from mixed aerosol exposures, and then discusses new methodologies for retrospective exposure assessment. Finally, there is a discussion of problems associated with the development of standards for the special case of bioaerosols.

ASSESSMENT OF EXPOSURE TO "TOTAL" AND INHALABLE AEROSOL

Wind tunnel tests of personal samplers which have been used for sampling "total" aerosol have shown that many not only fail to accurately collect true total

† To whom correspondence should be addressed.

aerosol but also poorly represent what may be inhaled by people. These results have led to the search for new samplers for inhalable aerosol and a number of instruments have been identified as suitable for given ranges of aerosol and environmental conditions (Kenny, 1995).

A number of intersampler comparison studies have been conducted in workplaces, most involving comparison between "total" aerosol exposure as measured using the 37 mm sampler (of the type widely used in the U.S. and elsewhere) and inhalable aerosol exposure as measured using the Institute of Occupational Medicine (IOM) sampler (Mark and Vincent, 1986). Industries studied include bakeries, borate processing, aluminium production, a range of nickel primary production and user industry sectors, lead primary production, machine shops and others. Altogether, many hundreds of such sample pairs have been obtained and have been widely reported (Shen *et al.*, 1993; Lillienberg and Brisman, 1994; Vincent, 1995a; Vinzents *et al.*, 1995; Tsai, 1995; Tsai *et al.*, 1995, 1996a,b; Spear, 1996; Spear *et al.*, submitted; Wilsey *et al.*, 1996; Werner, 1996). The results of these studies, when analysed by linear regression techniques, show clearly that exposure to inhalable aerosol (E_{inh}) is consistently greater than to "total" aerosol (E_{tot}). In fact, S in the regression equation $E_{inh} = S \cdot E_{tot}$ was found to take values from close to unity to as large as 4 for a range of "similarly-exposed" groups of workers. Closer inspection suggests a broad tendency for S to be greater for workplaces where the aerosol was coarsest. Although the variability in the data is such that this tendency cannot be identified as significant using stringent statistical tests, it is supported by all that is known about sampler performance based on the physics of the sampling process (Vincent, 1989, 1995b). So it is considered to be real. This body of research did reveal a number of practical problems, most notably the collection by the large sampling orifice of the IOM inhalable aerosol sampler of particulate material which should not be considered as airborne (for example "splashes" and flying "chips"). This may be a problem in certain industrial situations (for example sampling for machine shop oil mist or wood dust), and will require a technical solution. In addition, in some industries, where there is widespread usage of respirators (for example in primary lead production), exposure assessment needs also to take into account the performance of the respiratory, not just in terms of the assigned protection factor (APF) but also in terms of the more realistic effective protection factor (EPF) that applies in routine use in the workplace. Research has shown that EPF-values are consistently much lower than the corresponding APF-values, so that actual worker exposures are higher than anticipated (Spear, 1996).

The preceding may be used to guide progress towards new OELs for aerosols, in particular on the question of whether to adjust OELs to accommodate changes in measured exposure associated with the change from the "total" aerosol to the inhalability rationale (Werner *et al.*, 1996). Recognising the great variability that exists between industry processes and industries and the fact that it is not realistic to conduct intersampler comparison studies in all workplaces, the individual measured S-values are not universally applicable. However, the knowledge that has been gained can be used to arrive at a "working" matrix of S-values for industrial situations which may be broadly characterised in terms of the type of aerosol generated (Table 1).

Table 1. Suggested guidelines for use where it is deemed desirable to adjust exposure data to account for the change in exposure assessment rationale (based on generalisation of results of comparisons between "total" aerosol as measured using the 37 mm sampler and inhalable aerosol as measured using the IOM sampler)

Aerosol classification/industrial category	Approximate range of aerodynamic particle size (μm)	Suggested conversion factor
Dust: mining, ore and rock handling, handling and transportation of bulk aggregates, textiles, flour and grain handling, etc.	1–150	2.5
Mist: oil mist and other machining fluids, paint sprays, electroplating, etc.	1–50	2.0
Hot processes: metal smelting and refining, foundries, etc.	Up to 10	1.5
Welding: all	Up to 1.0	1.0
Smokes and fumes: all	Up to 1.0	1.0

The question now is: how should this information be used when the basis of the OELs is shifted from one based on "total" aerosol to one based on inhalable aerosol? Should OELs be scaled accordingly? If so, it would mean that, for a given substance, different conversion factors might need to be applied for industrial processes where that substance may be aerosolised in different ways. The alternative is to leave OELs the same, without regard for any increase in measured exposures that would ensue (and the corresponding increases in OEL exceedance fraction). Each of the preceding is an "across-the-board" approach and neither is very satisfying. Instead we propose an approach, which we believe to be the more scientific, where the knowledge gained is applied not to the OELs themselves but to the results of the individual studies that are considered during the OEL-generation process. Where, for a given substance, the studies cited contain no reference to explicit relationships between the levels of actual worker exposure observed and health effects (and, therefore, the original assignment of the OEL was made on the basis of other information, for example toxicological), there is no justification for adjusting the OEL on the basis of Table 1. On the other hand, if the original assignment of OEL was based, at least in part, on the results of studies of health effects in populations where relevant worker personal exposures to "total" aerosol had been measured (using the 37 mm sampler), then an adjustment of the OEL may be justified by choosing the appropriate S-value from Table 1. To achieve this, an adjustment would first be made to just those "total" aerosol exposure data in order to arrive at corresponding estimates of inhalable aerosol exposures. The available new exposure–response picture that results would then be considered in a new review of all the studies cited. In summary, therefore, it is important to recognise that the role of the intersampler comparisons (and the resultant conversion factors) like those discussed earlier is not in scaling of existing health-based OELs but, rather, to aid in the re-interpretation of the information which leads to the assignment of the OEL.

The actual application of the preceding ideas is scientifically consistent only in relation to OELs which are strictly health based [for example the ACGIH threshold limit values (TLVs)]. Clearly their application to OELs for which feasibility is taken into account involves additional considerations, some of which lie beyond the application of scientific knowledge.

PARTICLE SIZE-SELECTIVE SAMPLING AND DOSIMETRY

The preceding has been concerned with criteria and samplers intended primarily for use in routine exposure assessments. By contrast, "aerosol spectrometers" are sampling devices intended primarily for use in non-routine investigations and research. They provide a basis for studying the health-related dose arising from exposure to an aerosol comprised of particles of a wide range of sizes and many different chemical species (as is typical of most occupational settings). The combined distributions of particle size and species together determine how much of what is deposited where in the respiratory tract. These deposits contribute in different ways and at different rates (depending on the nature of the material deposited and the lung environment itself) towards the disease process. With this in mind, the net equivalent deposited dose (D_h) in relation to health effect h, for the amount $_iA_j$ of chemical species i depositing in region j of the respiratory tract, may be expressed in the general form (Ramachandran *et al.*, 1996).

$$D_h = \sum_{i=1}^{m} {}_i\beta_h \left\{ \sum_{j=1}^{n} {}_i\alpha_j \, {}_iA_j \right\} + S_h \tag{1}$$

where the αs are weightings which describe the relative contributions of the regional depositions, the βs represent the relative overall toxicities of the various species and S_h represents the synergisms or antagonisms that take place between the individual exposures. Basically, this equation sums the contribution to the health effect in question from particle deposition in different parts of the respiratory tract of different chemical species and allows (through S_h) for the possibility of non-additive effects.

Equation (1) is very general at this stage, and its full implementation will require physical, chemical and toxicological information most of which is not at present available. However, some progress towards its application may be indicated by some simple reasoning with respect to the α and β-values. Firstly, we may set $\alpha = 1$ for regions of the respiratory tract where particle deposition leads directly to the disease and $\alpha = 0$ elsewhere. So, for example, for pneumoconiosis, it is reasonable that $\alpha = 1$ for the alveolar region and 0 everywhere else; on the other hand, for sino-nasal cancer, $\alpha = 1$ for the extrathoracic region and 0 everywhere else. For substances which effect health regardless of where they are deposited (for example lead), the choice of the α-values is less clear-cut and will depend on bioavailability at the various sites of deposition. Secondly, for β, one approach is to consider the relative toxicity of the different substances from the point of view of their OELs. Here, for example, in order to express D_h for a mixed exposure in terms of "species 1-equivalent deposited dose", we set $_1\beta_h = OEL_1 = 1$ and subsequently $_i\beta_h = OEL_1/OEL_i$ (where it is acknowledged that the lower OEL is indicative of greater toxicity). An example of the application of this approach is given below.

Personal cascade impactors are particularly useful for the classification of occupational aerosol exposures. One instrument designed specifically to collect and

Table 2. Health-related fractions (in mg m^{-3} of air sampled and as a percentage of total ambient aerosol) for the metals contained in a typical PIDS sample taken at a nickel refinery (copper electrowinning)

	Cobalt	Copper	Nickel	Iron	Lead
Inhalable	0.0129 (62.2%)	0.1050 (63.2%)	0.0478 (61.4%)	0.0735 (59.0%)	0.0945 (56.0%)
Thoracic	0.0042 (20.4%)	0.0419 (25.2%)	0.0154 (19.7%)	0.0253 (20.3%)	0.0170 (10.1%)
Respirable	0.0019 (9.0%)	0.0188 (11.3%)	0.0065 (8.3%)	0.0123 (9.9%)	0.0049 (2.9%)

evaluate inhalable aerosol in a manner consistent with the latest particle size-selective criteria is the 2 l. min^{-1} IOM personal inhalable dust spectrometer (PIDS) (Gibson *et al.*, 1987). However, one difficulty in the use of this, and other such instruments, is the need to infer continuous size distributions from raw data which are discrete (masses collected on individual impactor stages). Thus the problem is "ill-posed", meaning that there is a large number of possible solutions that can be inferred. The optimal solution therefore requires the application of sophisticated mathematical inversion procedures. We have developed a number of such methods and applied them to PIDS data (Ramachandran *et al.*, in press; Ramachandran and Kandlikar, 1996). One has been developed for use in a Microsoft EXCEL® spreadsheet and has been shown to be quite robust for particle size distributions like those encountered in the industrial hygiene setting (Ramachandran and Vincent, in press).

The ideas outlined above have been explored for workers exposures in the nickel and lead primary production industries mentioned earlier, where interest was focussed initially on the different metals that might be present. In nickel refining, for example, it was known that exposures were occurring simultaneously not only to nickel but also to copper, cobalt, iron and lead. The samples recovered from the individual cascade impactor stages were therefore assessed for these metals. From the results of the inversion analysis to obtain the particle size distributions for the individual metals, the masses sampled in the health-related fractions were obtained directly. By way of illustration, Table 2 shows the results for a single PIDS sample taken at the nickel refinery, showing not only the mass concentrations contained in each of the health-related particle size fractions for each metal of interest, but also the relative amounts for each.

For the data shown in Table 2, both nickel and cobalt have been associated with lung cancer. This provides a useful illustration of the concepts and the rationale outlined earlier. At the copper electrowinning process at the nickel refinery referred to, nickel was present mostly in the soluble form, for which ACGIH currently lists the OEL (or TLV) as 0.1 mg m^{-3}. For cobalt, the OEL is given as 0.02 mg m^{-3}. Since it is nickel which is the primary substance of concern, we choose to relate the total mixed exposure to the nickel itself. In addition, since lung cancer has been identified as the adverse health effect of interest, we set $\alpha = 1$ for the thoracic aerosol fraction and 0 for the extrathoracic fraction, for both nickel and cobalt. Thus, from equation (1) and for the worker characterised by the sample described in Table 2, we may show that the health-related "nickel-equivalent deposited dose" for the two-component mixed aerosol of interest is 0.0364 mg m^{-3}, made up of 0.0154 mg m^{-3} directly for the nickel component itself and 0.0210 mg m^{-3} for the "secondary" cobalt component (Ramachandran *et al.*, 1996). In this illustration, it is noted that the exposure is given in terms of thoracic aerosol, since it is reasonable to assume that, for the lung cancer outcome, this is an appropriate index of exposure. However, this is not consistent with the "total" aerosol rationale for current OELs for nickel, nor indeed with the inhalable aerosol rationale expected for proposed future OELs. This is because, for nickel, there has also been a historical association with sino-nasal cancer. Nonetheless, the illustration shows how the use of a particle size-selective sampling instrument like PIDS might enable the determination of the masses of health-related species in aerodynamically-selected inhalation and lung deposition fractions in a multi-component aerosol.

RETROSPECTIVE EXPOSURE ASSESSMENT

It is recognised that, in many industries, knowledge of historical exposures is insufficient to enable quantitative estimation of the risk of disease upon which to base reliable OELs. In a recent review, Smith *et al.* (1995) underlined the importance of retrospective exposure assessment in epidemiology for long-term chronic occupational diseases, but also commented that ". . . estimation of past exposures to potential health hazards (remains) one of the most difficult problems for occupational hygiene research". In our research, we have examined historical air measurements in certain plants of the nickel primary production industry, involving data sets obtained over a 40-year period using diverse aerosol sampling instruments, both for area and personal measurements, providing different indices of exposure (particle counts, total particulate mass, nickel) and over short and long time periods (Vincent, 1996). In this work, it became clear that simple statistical correlations between these measures cannot provide reliable results. With this in mind, we are exploring a novel approach which involves analysis of the existing sparse and diverse data by a combination of formal "expert judgment" analysis and Bayesian statistical modelling.

BIOLOGICAL AEROSOLS

Whilst exposure to bioaerosols is of increasing concern in many occupational settings, there are at present no OELs for these. Although particle size-selective criteria may eventually be developed for bioaerosols, most bioaerosol disease mechanisms and sampling methodologies differ from those for inert aerosols. This problem is narrowed somewhat by focussing on a particular type of bioaerosol and the occupational setting where it might be found. The health care sector, for example, is one area where workers are potentially exposed to airborne infectious organisms. *Mycobacterium tuberculosis*, the organism which causes tuberculosis, is of major topical interest with respect to both exposure assessment and control of exposures through respiratory protection. We have conducted experiments to investigate basic problems in both sampling and control of such organisms (McCullough, 1996). In view of their pathogenic properties, an important part of the work has been the search for "safe" yet "appropriate" surrogates for use in laboratory studies. Filtration studies have shown that, similarly to inert aerosols, it is the physical properties of the organism which relate to its behavior in air (Brosseau *et al.*, 1994, in press; McCullough, 1996). Therefore, such surrogates must first match these physical properties. *Mycobacterium abscessus* (formerly *M. chelonae abscessus*) and *Bacillus subtilis*, for example, have been found to have similar rod-shaped morphometry and aerodynamic diameter as *M. tuberculosis*, which make them appropriate surrogates for certain sampler characterisation (aspiration and collection efficiency) and filtration studies. However, other assessments of bioaerosols, such as organism survival studies and characterisation of sampler bioefficiency, must also consider cell viability. Here, the ideal surrogate should react similarly to environmental stress (for example temperature, relative humidity, lack of nutrients). In this case, *M. abscessus* would be an appropriate surrogate for *M. tuberculosis* because both organisms exist in a vegetative state and are expected to be similarly affected by environmental conditions, but *B. subtilis*, a spore, would not. Currently, samplers do not exist which are designed to efficiently sample the aforementioned health-related particle size fractions (inhalable, thoracic, respirable) for viable biological aerosols (Griffiths and DeCosemo, 1994). Understanding the selection of bioaerosol surrogates with respect to research objectives is an important step in such sampler development. Once sampling methods have been developed which take into account both the physical and biological nature of the organism, establishment of exposure assessment strategies and standards for infectious organisms and other bioaerosols, can be achieved.

CONCLUDING REMARKS

A large body of work has emerged which attempts to place practical occupational aerosol exposure assessment on a sounder footing, building on the framework of particle size-selective sampling that has evolved during the past two decades. Some of the research described leads immediately to improvements in exposure assessment and standards setting. Other research will bear fruit in the longer term.

REFERENCES

American Conference of Governmental Industrial Hygienists (ACGIH) (1995) Threshold limit values for chemical substances and physical agents and biological exposure indices (1995–1996), ACGIH, Cincinnati, OH.

Brosseau, L. M., Chen, S.-K., Vesley, D. and Vincent, J. H. (1994) System design and test method for measuring respirator filter efficiency using mycobacterium aerosols. *J. Aerosol Sci.* **25**, 1567–1577.

Brosseau, L. M., McCullough, N. V. and Vesley, D. Mycobacterial aerosol collection efficiency by respirator and surgical mask filters under varying conditions of flow and humidity. *Appl. occup. Environ. Hyg.* (in press).

Comité Européen de Normalisation (CEN) (1992) Workplace atmospheres: size fraction definitions for measurement of airborne particles in the workplace, CEN Standard EN 481.

Gibson, H., Vincent, J. H. and Mark, D. (1987) A personal inspirable dust spectrometer for applications in occupational hygiene research. *Ann. occ. Hyg.* **31**, 463–479.

Griffiths, W. D. and DeCosemo, G. A. L. (1994) The assessment of bioaerosols. *J. Aerosol Sci.* **25**, 1425–1459.

International Standards Organization (ISO) (1992) Air quality—particle size fraction definitions for health-related sampling. *Technical Report ISO/TR/7708–1983 (E), ISO, Geneva*, 1983, revised 1992.

Kenny, L. C. (1995) Pilot study of CEN protocols for the performance testing of workplace aerosol sampling instruments, Report of work carried out under EC Contract MATI-CT92-0047, September 1995, Health and Safety Executive, Sheffield, U.K.

Lillienberg, L. and Brisman, J. (1994) Flour dust in bakeries—a comparison between methods. In *Inhaled Particles VII* (Edited by J. Dodgson and R. I. McCallum), pp. 571–575. Pergamon Press, Oxford.

McCullough, N. V. (1996) Testing and selection of respiratory protection for use with infectious bioaerosols. Ph.D. thesis, University of Minnesota, Minneapolis, Minnesota, U.S.A.

Mark, D. and Vincent, J. H. (1986) A new personal sampler for airborne total dust in workplaces. *Ann. occup. Hyg.* **30**, 89–102.

Ramachandran, G., Johnson, E. G. and Vincent, J. H. Inversion techniques for cascade impactor data. *J. Aerosol Sci.* (in press).

Ramachandran, G. and Kandlikar, M. (1996). Bayesian analysis for inversion of aerosol size distribution data. *J. Aerosol Sci.* **27**. 1088–1097).

Ramachandran, G. and Vincent, J. H. (1996). Evaluation of two inversion techniques for retrieving health-related aerosol fractions from personal cascade impactor measurements. *Am. ind. Hyg. Ass. J.* (in press).

Ramachandran, G., Werner, M. A. and Vincent, J. H. (1996). On the assessment of particle size distributions in workers' aerosol exposures. *The Analyst* **121**, 1225–1232.

Shen, P. T., Culver, B. D., Taylor, T. H., Granken, E. P. and Sattaur, J. (1993) A field comparison of 37 mm closed-face cassette with a personal inspirable particulate mass sampler in workplace exposure to sodium borates and boric acid dust, Paper presented at the American Hygiene Conference and Exposition, New Orleans.

Smith, T. J., Stewart, P. A. and Herrick, R. F. (1995) In *Occupational Hygiene* (Edited by J. M. Harrington and K. Gardiner), pp. 308–322. Blackwell Scientific.

Spear, T. M. (1996) Assessment of workers' exposures to lead-containing aerosol. Ph.D. thesis, University of Minnesota, Minneapolis, MN, U.S.A.

Spear, T. M., Werner, M. A., Bootland, J., Harbour, A., Murray, E. P., Rossi, R. and Vincent, J. H. Workers' exposures to inhalable and "total" lead and cadmium-containing aerosols in a primary lead smelter. *Am. ind. Hyg. Ass. J.* (submitted).

Tsai, P.-J. (1995) Health-related exposures of nickel industry workers. Ph.D. thesis, University of Minnesota, Minneapolis, MN, U.S.A.

Tsai, P.-J., Vincent, J. H., Wahl, G. and Maldonado, G. (1995) Occupational exposure to inhalable and "total" aerosol in the primary nickel production industry. *Occup. Environ. Med.* **52**, 793–799.

Tsai, P.-J., Vincent, J. H., Wahl. G. and Maldonado, G. (1996a) Worker exposures to inhalable and "total" aerosol during nickel alloy production. *Ann. occup. Hyg.* **40**, 651–669.

Tsai, P.-J., Werner, M. A., Vincent, J. H. and Maldonado, G. (1996b) Worker exposure to nickel-containing aerosol in two electroplating shops: comparison between inhalable and "total" aerosol. *Appl. occup. Environ. Hyg.* **11**, 484–492.

Vincent, J. H. (1989) *Aerosol Sampling: Science and Practice.* Wiley and Sons, Chichester, U.K.

Vincent, J. H. (1995a) *Aerosol Science for Industrial Hygienists.* Pergamon, Elsevier Science Ltd, Oxford, U.K.

Vincent, J. H. (1995b) Progress towards implementation of aerosol industrial hygiene standards. *Sci. Tot. Environ.* **163**, 3–10.

Vincent, J. H. (1996) Assessment of aerosol exposures of nickel industry workers. Final Report to the Nickel Producers Environmental Research Association (unpublished), Durham, NC, U.S.A.

Vinzents, P. S., Thomassen, Y. and Hetland, S. (1996) A method for establishing tentative occupational exposure limits for inhalable dust. *Ann. occup. Hyg.* **39**, 795–800.

Werner, M. A. (1996) Comparison of observed "total" and inhalable aerosol exposures: implications for occupational exposure standards. Ph.D. thesis, University of Minnesota, Minneapolia, MN, U.S.A.

Werner, M. A., Spear, T. M. and Vincent, J. H. (1996) Investigation into the impact of introducing workplace aerosol standards based on the inhalable fraction. *The Analyst* **121**, 1207–1214.

Wilsey, P. W., Vincent, J. H., Bishop, M. J., Brosseau, L. M. and Greaves, I. A. (1996) Workers' exposures to inhalable and "total" oil mist aerosol in a metal machining shop. *Am. ind. Hyg. Assoc. J.* **57**, 1149–1153.

Ann. occup. Hyg., Vol. 41, Supplement 1, pp. 615–620, 1997
© 1997 Published by Elsevier Science Ltd on behalf of BOHS
Printed in Great Britain. All rights reserved
0003–4878/97 $17.00 + 0.00
Inhaled Particles VIII

Pergamon

PII: S0003–4878(96)00085–3

INTRATRACHEAL EXPERIMENT ON RATS CONCERNING THE FIBROGENIC EFFECT OF RESPIRABLE DENTAL LABORATORY DUST CONTAINING PALLADIUM

A. Brammertz* and M. Augthun†

*Institute of Hygiene and Environmental Health, Department of Occupational Health; and
†Department of Prothetic Dentistry, Rheinisch-Westfälische Technische Hochschule Aachen,
Pauwelsstr. 30, D 52 057 Aachen, Germany

INTRODUCTION

Palladium, belonging to the platinum group of metals is increasingly gaining the interest of occupational and environmental medicine. The applications of palladium have multiplied over the last 10 years, particularly in catalyst techniques and dentistry (Römpp, 1992; Carson *et al.*, 1986). The use of palladium in medicine and especially in dentistry has increased, although the question of possible biological effects cannot yet be answered.

Little is known of the toxicity of palladium. Acute toxicity in humans has not yet been described. Chronic toxicity has been observed in animal experiments (Liu *et al.*, 1979). Since 1986 palladium alloys have increasingly been used in Germany instead of gold alloys for making fixed dental protheses such as crowns and bridges.

From 1986 to the start of the 1990s crown and bridgework for patients under social security had to be made of silver–palladium alloys. Nowadays their use is no longer obligatory but still widespread due to their low cost. It is estimated that 50–60% of the patients insured under the society security scheme have been provided with these alloys.

It is still not clear whether or not these newer alloys, introduced 10 years ago, cause local or systemic effects (Phielipeit and Legrum, 1986). Beyond this aspect concerning biological risks there is a possible pathological effect on the lung tissue of dental technicians during tooling or polishing (Fig. 1). Today dental technicians are exposed to growing amounts of respirable palladium-containing dusts with presently unknown biological effects. However pneumoconiosis as an occupational disease in dental laboratory technicians is widely known in the literature (Augthun, 1992; Brammertz and Augthun, 1992).

MATERIALS AND METHODS

In the first place the aim of our study was to examine what kind of dust particles are produced by polishing and tooling dental castings. Firstly dust had been collected by a personal dust sampler (PP 5 Ströhlein Instruments) while a technician was on duty (Fig. 1). Afterwards the particle size and structure was examined by SEM [electron microscopy] and physical methods (Figs 2 and 3, Table 1).

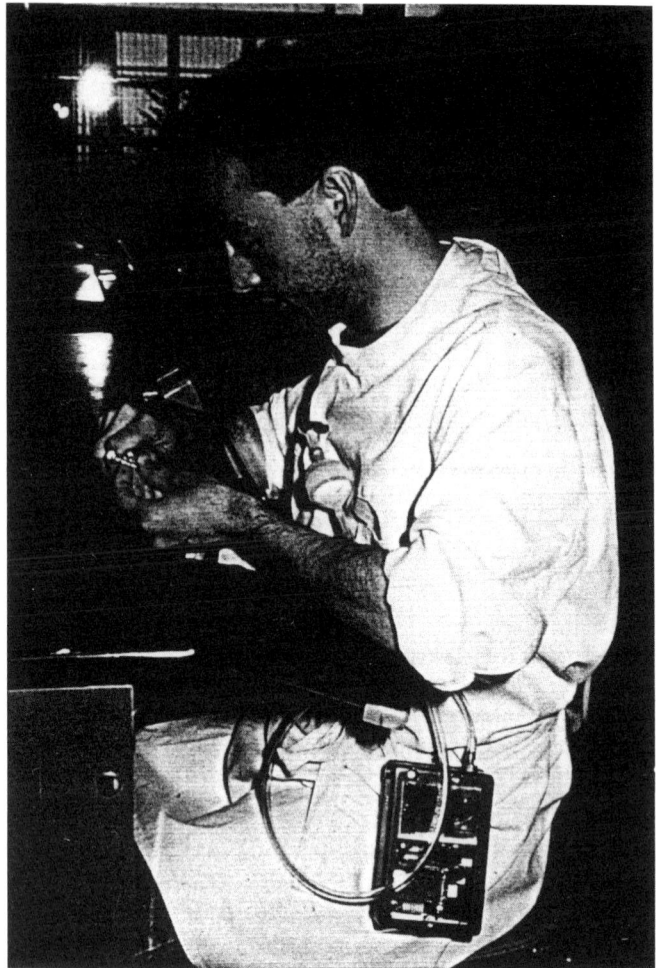

Fig. 1. Dental technician confronted with a number of dusts while producing dental prothesis.

Secondly we dealt with the question of whether there was a possible fibrogenic effect of palladium-containing dusts on the lung tissue. Dust samples of unalloyed palladium and a widely-used dental palladium alloy were studied. The particle size of both dusts amounted to about 2 μm. The dust samples were respirable even for animals like rats. Intratracheal instillation in rats, as an established standard technique was the method of choice.

Female Sprague–Dawley rats (under SPF-condition) weighing about 200 g were divided in two groups of 10 animals each and received in a non-invasive procedure one instillation of 50 mg dust suspended in 0.5 ml isotonic saline. The animals were anaesthetised with ether during the manipulation. The rats were fed on a standard diet and given water *ad libitum*. A control group of equal size underwent no instillation. The animals were sacrificed after 6 months and their lungs together

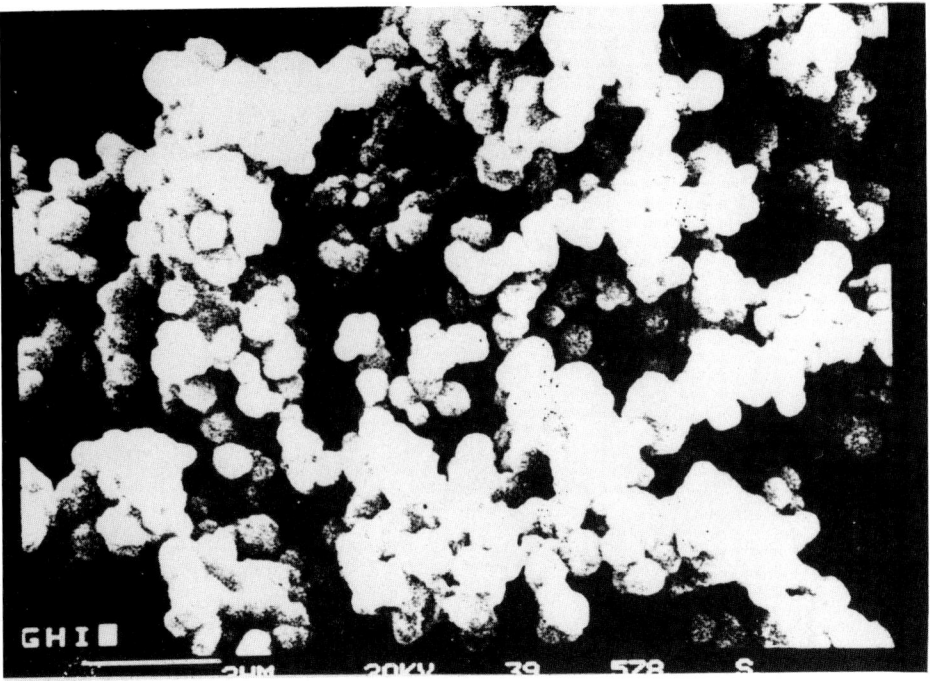

Fig. 2. Palladium dust, electronic microscopy 20KV.

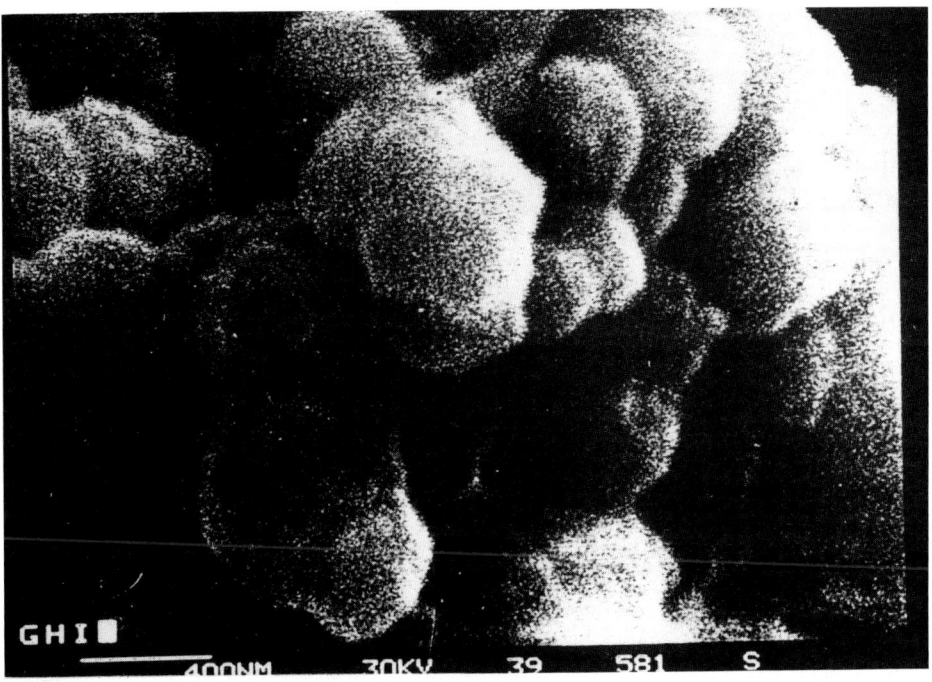

Fig. 3. Palladium dust, electronic microscopy 30KV.

Table 1. Volume %—statistics

Volume %—statistics	Palladium	Indium	Gallium
Mean	3.92 μm	8.13 μm	4.16 μm
Median	3.07 μm	9.63 μm	3.78 μm
SD	2.56 μm	1.80 μm	2.06 μm

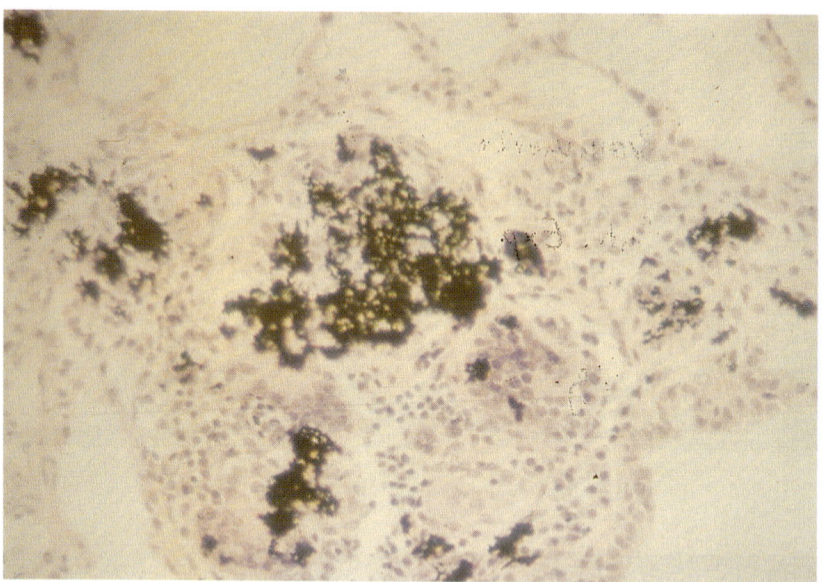

Fig. 4. Histologic results of lung tissue, 6 Month after i.t., HE-stain, magnification 10 × 20.

with pulmonary lymph nodes were dissected and examined biochemically and histologically.

RESULTS

It could be shown that the total pulmonary lipid content [determined according to Folch (1957)] as well as the total hydroxiproline content [determined according to Stegemann (1958)] were not significantly increased compared to the control group. Histological examination showed that palladium was retained in the pulmonary parenchyme and lymph nodes, but did not cause fibrosis. Only some slight inflammatory signs could be observed. The dust was mainly retained in the alveolar macrophages (Figs 4–6).

DISCUSSION

As seen from the results the risk of fibrosis for dental technicians from palladium can be classified as slight, particularly if sufficient protection in the workplace is enforced. A possible synergistic effect of some elements should be clarified. The

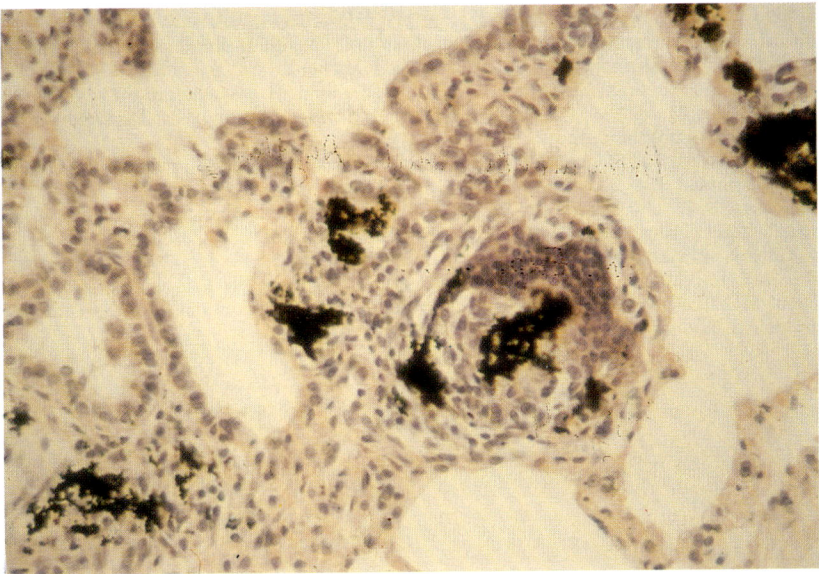

Fig. 5. Histologic results of lung tissue, 6 Month after i.t., HE-stain, magnification 10 × 40.

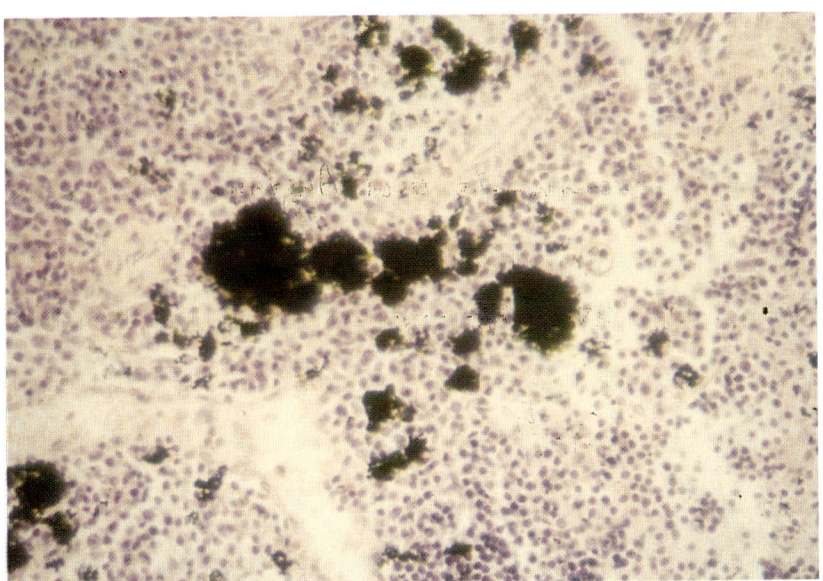

Fig. 6. Histologic results of lymph nodes, 6 Month after i.t., HE-stain, magnification 10 × 20.

multitude of partially respirable dusts in dental laboratories and the limited knowledge of the biological effects of their metallic components (for instance Pd, In, Ga) requires a simple and standardised assay of their fibrogenicity in order to decrease health hazards for dental technicians.

REFERENCES

Augthun, M. (1993) Biokompatibilität von Palladium und Palladium–Kupfer–Legie rungen. Habilitationsschrift der Medizinischen Fakultät der RWTH Aachen.

Brammertz, A. and Augthun, M. (1992) Toxicity of palladium. *Wissenschaft und Umwelt* **4**, 285–289.

Carson, B. L., Ellis, H. V. and McCann, J. L. (1986) *Toxicology and Biological Monitoring of Metals in Humans*. Lewis Publishers, Inc.

Folch, J. M., Less, M. and Sloane, S. (1957) A simple method for the isolation and purification of total lipids from animal tissues. *J. Biol. Chem.* **226**, 497–509.

Liu, T. Z., Lee, S. D. and Bhatnagar, R. S. (1979) Toxicity of palladium. *Toxicol. Lett.* **4**, 469–473.

Phielipeit, T. and Legrum, W. (1986) Zur Toxizität des palladiums Dtsch. *Zahnärztl. Z.* **41**, 1257–1260.

Römpp Chemie Lexikon (1992) (Edited by J. Falbe and M. Regitz) Vol. 9. Erweiterte und neubearbeitete Auflage, Thieme Verlag, Stuttgart.

Stegemann, H. (1958) Mikrobestimmung von hydroxiprolin mit chloramin T und p-Dimethyaminobenzaldehyd. *Hoppe Seylers Z. Physiol. Chem.* **312**, 41–42.

Pergamon

Ann. occup. Hyg., Vol. 41, Supplement 1, pp. 621–623, 1997
© 1997 British Occupational Hygiene Society
Published by Elsevier Science Ltd. All rights reserved
Printed in Great Britain
0003–4878/97 $17.00 + 0.00
Inhaled Particles VIII

PII: S0003–4878(96)00135–4

HIGH MOLECULAR WEIGHT ALLERGENS EXPOSURE IN PASTRY MAKING, ANIMAL HEALTH AND DENTAL HYGIENE IN TRAINEES

A. Dufresne,*‡ D. Gautrin,† C. Infante-Rivard* and J. L. Malo†

*McGill University, Department of Occupational Health, Faculty of Medicine, 3450 University Street, FDA Bldg Room 22, Montréal (Québec), Canada H3A 2A7; and †Centre de recherche, Hôpital du Sacré-Coeur de Montréal, 5400, boulevard Gouin-ouest, Montréal (Québec), Canada H4J 1C5

INTRODUCTION

Workplace allergens are encountered in grain elevators, bakeries, hospitals, dental offices and many others. Students in a training program for a profession or trade in the above-mentioned occupational environments may be at risk of developing diseases such as bronchial asthma, allergic rhinitis, dermatitis, etc. An epidemiological study actually conducted by our group to determine the relationship between Type I immunological sensitisation and qualitative and quantitative estimates of exposure to high molecular weight allergens taking into account atopy as well as individual and family medical history suggestive of susceptibility to allergens. We report the results of the industrial hygiene survey conducted within the scope of this study. The concentration of allergen aerosol was determined in the breathing zone of students in a training program for a profession or trade in pastry making (PM), animal health technology (AHT) and dental hygiene technology (DHT).

MATERIAL AND METHODS

The sampling train consisted of a 37 mm diameter polytetrafluoroethylene membrane filter (PTFE) with a 37 mm cellulose backing pad placed in a plastic cassette to catch the allergen aerosols. The sampling train was connected to a personal sampling pump (Gilian) functioning at $2\,l\,min^{-1}$. Each sampling train was calibrated with a Gilibrator primary flow calibrator, to determine the precise flow rate. All analyses were done at Mayo Clinic by a two-site radioimmunoassay for airborne allergens. Sampling was performed during normal laboratory work; two sites were sampled on four different days for PM; one site was sampled on two different days for AHT; and dental hygienist were sampled in one institution on three different days.

RESULTS

The industrial hygiene survey (Table 1) showed the following exposure levels in $ng\,m^{-3}$ by allergen category in the breathing zone of students. Since no α-amylase was found in the airborne particles of the PM occupational environment, what

‡ To whom all correspondence should be addressed.

Table 1. Exposure to airborne allergens (ng m^{-3})

Parameters	Wheat gluten	Animal urine	Latex
n	37	17	18
AM	64 395	4001	90.7
ASD	71 521	1833	65.9
Median	46 724	4093	91
95th percentile	182 047	7016	199
GM	37 873	3509	54.7
GSD	3.1	1.8	3.8
95% CI on GM	(25 926, 55 326)	(2607, 4722)	(28.8, 103.5)

n: number of samples; AM: arithmetic means; ASD: arithmetic standard deviation; GM: geometric mean; GSD: geometric standard deviation; CI: confidence intervals.

gluten protein was used as a marker of the exposure to wheat flour allergens. Wheat gluten and guinea pig urine (GPU) concentrations were log-normally distributed while latex aerosols were not.

DISCUSSION

Flour dust

Burdorf *et al.* (1994) showed that inhalable flour dust contains an average of 9.0% total protein, 2.3% water soluble protein and 0.03% α-amylase. Radioimmunoassay techniques are very sensitive and it was expected, even if present in very low concentration, that α-amylase would have been detected in the breathing zone samples of the students involved in this study. Since we did not detect α-amylase in any of the 37 breathing zone samples, it is suspected that α-amylase was probably absent in the flour used by the students. Although α-amylase was absent or undetectable, it does not mean that students were not exposed to allergenic dust. Indeed, Franken *et al.* (1994) suggested that a 15 kD protein is one of the three major allergens detected in a commercially available wheat flour and concluded that low molecular weight flour allergens play an important role in the pathogenesis of baker asthma. Neuwenhuijsen *et al.* (1994a) reported a range flour aeroallergen concentrations of 101.5 μg m^{-3} for transport workers and up to 1728.2 ng m^{-3} for hygiene workers. A relation between the total dust and flour aeroallergen concentrations was reported but this relation varied according to different areas and depended on the use of products other than flour.

Animal protein dust

Aerosolisation of animal proteins during their handling in laboratory sessions was confirmed in our study, although the concentration of proteins was not as high as that reported by Nieuwenhuijsen *et al.* (1995a); they reported exposure a rat urinary allergen (RUA) during different tasks such as rat handling (68.0 μg m^{-3}), cage cleaning (53.4 μg m^{-3}) and indirect contact with rats (21.3 μg m^{-3}) in 21 animal technicians. The same group of investigators reported a GM concentration of 32.4 μg m^{-1} in research establishments (Nieuwenhuijsen *et al.*, 1994b). However, our results were quite comparable to supervisors (0.6 μg m^{-3}) or scientists (0.3–6.7 μg m^{-3}) in that study. Our data are also comparable to the concentrations measured at fixed station samples and reported by Swanson *et al.* (1990) in animal care rooms. An arithmetic mean of 468 ng m^{-3} was reported at 36 in from the floor. The different concentrations reported in the previous studies are probably explained by the fact that students in training only work with animals during laboratory experiments and perform tasks that lead to less exposure than professional technicians.

Latex dust

Personal exposure levels to latex particles in a medical centre where health care workers frequently used powdered gloves (Swanson *et al.*, 1994) were found to range between 8 and 974 ng m^{-3}.

These levels are higher than our geometric mean concentration of 54.7 ng m^{-3} for dental health students. However, these concentrations seemed comparable to previously reported ones among hematopathology technicians but notably lower than that found among anesthetists (Swanson *et al.*, 1994). The difference could be explained by the less frequent changes of gloves among students. Also it has been shown that the total protein and allergen concentrations in gloves show considerable variability (Alenius *et al.*, 1994; Jones *et al.*, 1994).

CONCLUSION

If wheat gluten is a good marker of the allergenic proteins in wheat flours, one may conclude that the concentration of allergenic aerosols was highest in PM students followed by AHT and DHT students. However, the evaluation of allergen aerosols is quite complex. Many of the proteins in animal urine, flour or latex may be allergenic and the intensity of the allergic reaction may vary from one protein to another. Moreover, it was observed that AHT students were also exposed to other active contaminants such as formaldehyde and to a certain extent, to latex particles. The DHT students were also exposed to formaldehyde during their 3 years program. It is evident that further research is needed in this complex field of research.

REFERENCES

Alenius, H., Mainen-Kiljunen, S., Turnjanmaa, K., *et al.* (1994) Allergen and protein content of latex gloves. *Ann. Allergy* **73**, 315–320.
Burdorf, A., Lillienberg, L. and Brisman, J. (1994) Characterization of exposure to inhalable flour dust in Swedish bakeries. *Ann. occ. Hyg.* **38**, 67–78.
Franken, J., Stephan, U., Meyer, H. E. and Konig, W. (1994) Identification of α-amylase inhibitor as a major allergen of wheat flour. *Int. Arch. Allergy Immun.* **104**, 171–174.
Jones, R. T., Scheppman, D. L., Heilman, D. K. *et al.* (1994) Prospective study of extractable latex allergen content of disposable medical gloves. *Ann. Allergy* **73**, 321–325.
Nieuwenhuijsen, M. J., Sandiford, C. P., Lowson, C., *et al.* (1994a) Dust and flour aeroallergen exposure in flour mills and bakeries. *Occ. Environ. Med.* **51** 584–588.
Nieuwenhuijsen, M. J., Gordon, S., Tee, R. D., *et al.* (1994b) Exposure to dust and rat urinary allergens in research establishments. *Occ. Environ. Med.* **51**, 593–596.
Nieuwenhuijsen, M. J., Gordon, S., Harris, J., *et al.* (1995a) Determinants of airborne allergen-exposure in animal house. *Occ. Hyg.* **1**, 317–324.
Nieuwenhuijsen, M. J., Gordon, S., Harris, J. M., *et al.* (1995b) Variation in rat urinary aeroallergen levels explained by differences in site, task and exposure groups. *Ann. occ. Hyg.* **39**, 815–825.
Swanson, M. C., Campbell, A. R., O'Hollaren, M. T. and Reed, C. E. (1990) Role of ventilation, air filtration and allergen production rate in determining concentrations of rat allergens in the air of animal quarters. *Am. Rev. Respir. Dis.* **141**, 1578–1581.
Swanson, M. C., Bubak, M. E., Hunt, L. W., Yunginger, J. W., Warner, M. A. and Reed, C. E. (1994) Clinical aspects of allergic disease. Quantification of occupational latex aeroallergens in a medical center. *J. Allergy Clin. Immunol.* **94**, 445–451.

Pergamon

Ann. occup. Hyg., Vol. 41, Supplement 1, pp. 624–629, 1997
© 1997 British Occupational Hygiene Society
Published by Elsevier Science Ltd. All rights reserved
Printed in Great Britain
0003–4878/97 $17.00 + 0.00
Inhaled Particles VIII

PII: S0003–4878(96)00084–1

LONG-TERM EFFECTS OF COMBINED EXPOSURE TO FISSION NEUTRONS AND INHALED CADMIUM CHLORIDE IN RATS

G. Monchaux, M. Morin, J. P. Morlier and M. F. Olivier

CEA-DSV-DRR-LCE, BP 6, 92265 Fontenay aux Roses Cedex, France

INTRODUCTION

The results of several epidemiologic studies in the U.S.A. (Thun *et al.*, 1985), Sweden (Elinder *et al.*, 1985) and the U.K. (Kazantzis *et al.*, 1992; Kazantzis and Blanks, 1992) have shown a significant excess of lung cancer in workers exposed to cadmium and its inorganic compounds, but only the study of Stayner *et al.* (1992) shows clear evidence of a dose–response relationship. Moreover, in some groups heavily exposed in the past to cadmium at indoor airborne levels of about 1 mg m^{-3}, only marginal or no excess of lung cancer were observed, despite marked nephrotoxicity.

Long-term carcinogenesis studies following inhalation exposure to different cadmium inorganic compounds in different laboratory rodents showed some conflicting results according to animal species. All the experiments carried out in rats demonstrated an excess of lung cancer for cadmium oxide dust, cadmium sulphide, cadmium chloride and cadmium sulfate (Glaser *et al.*, 1990), but two experiments using cadmium oxide fumes were negative. A significant excess of lung tumours was observed in mice exposed to cadmium oxide fumes and cadmium oxide dust, but not in those exposed to cadmium sulphide (Heinrich, 1992). By contrast, all experiments following inhalation exposure in hamsters failed to demonstrate any increased occurence of lung cancer (Heinrich *et al.*, 1989), whatever the cadmium compound.

The aim of the experiment reported here was to show whether cadmium chloride ($CdCl_2$) acted, either as a complete lung carcinogen when inhaled alone during a short exposure period at dose levels similar to that encountered in industry in a recent past, or as a lung co-carcinogen in rats previously exposed to fission neutrons, irradiation acting then as an initiator, or as a lung co-carcinogen in rats exposed to cadmium chloride and treated after the end of inhalation by intramuscular injections of 5–6 benzoflavone (βNF) which has been previously demonstrated to be a specific promotor of experimentally induced squamous cell lung cancers in Sprague–Dawley rats (Morin *et al.*, 1978).

MATERIALS AND METHODS

The long-term inhalation carcinogenesis study was carried out on four experimental groups of SPF (OFA) Sprague–Dawley male rats:

- Group 1: 63 rats exposed to 0.6 Gy fission neutrons. In our series, this dose has been shown to result in a 10% incidence of lung cancers and a 3% incidence of kidney cancers (Morin and Lafuma, 1990).
- Group 2: 50 rats exposed to 0.6 Gy fission neutrons and 2 months later to CdC12 by inhalation for 1 month.
- Group 3: 50 rats exposed to $CdCl_2$ alone by inhalation for 1 month.
- Group 4: 25 rats exposed to $CdCl_2$ by inhalation for 1 month and treated 1 month after the end of exposure with six intramuscular injections of βNF.

Irradiation

The rats were exposed to fission neutrons in the "Silene" experimental reactor facilities according to the pulse operating mode with a 10 cm lead shielding. The rats were placed into individual aluminium containers suspended on a net 4 m distant from the core of the reactor. The number of fissions was: $4.82 \ 10^{16}$. The peak power was: $9.68 \ 10^{16}$ fissions s^{-1}. The dose was: 0.62 Gy neutrons and 0.118 Gy gamma rays.

Cadmium chloride inhalation

The exposure of rats to $CdCl_2$ aerosol was performed by whole body inhalation exposure in a $2 \ m^3$ dynamic type Hazleton inhalation chamber. The rats were housed continuously in individual wire cages in the chambers throughout the inhalation exposure period, provided tap water *ad libitum* and fed outside of exposure hours and maintained on a 12 h light cycle.

The aerosolization of the $CdCl_2$ solution ($3 \ g.l.^{-1}$) was achieved by a De Vilbiss 99 ultrasonic nebulizer. The rats were exposed 5 days a week, 6 h a day for 4 weeks at a nominal Cd concentration exposure of $700 \ \mu g.m^{-3}$. The Cd airborne concentration into the chamber was controlled by air sampling of aliquots of 50 air l. each on HAWP Millipore filters, 0.2 μm pore size. The Cd concentration was performed by furnace atomic absorption spectrometry, using a Perkin–Elmer 1100 B AA Spectrometer. The mean Cd concentration measured into the inhalation chambers during the 4 weeks of exposure was $685 \pm 257 \ \mu g.m^{-3}$.

Injection of the promotor (βNF)

In the group treated with βNF, 1 month after the end of the $CdCl_2$ inhalation exposure, the rats received six intramuscular injections of βNF $25 \ mg.kg^{-1}$ at fortnightly intervals. In this group all the rats were killed within 100 days after the last injection of βNF.

Animal and histologic analysis

The experiments were carried out on 10-week-old male SPF non-inbred OFA Sprague–Dawley rats. After their arrival in the laboratory, the animals were kept in quarantine for 1 week before any treatment. The rats were exposed as mentioned above. After exposure, all the rats were housed throughout their life-span in the same animal house, maintained on a 12 h light cycle. Animals were allowed to live until they died or were moribund. A full necropsy was performed in every animal. Liver, spleen, kidney, lungs and brain and all other organs with macroscopic suspicious lesions were taken systematically. Histologic samples were fixed with

Table 1. Proportion of cancers (%) in the different organs of untreated control male rats and of male
rats irradiated by fission neutrons at various doses in the Silene reactor

	Control male rats (785 rats)	Fission neutrons (0.4 Gy) (116 rats)	Fission neutrons (1.15 Gy) (80 rats)	Fission neutrons (1.73 Gy) (40 rats)
Lungs	0.64	6.0	25.0	7.5
Salivary glands	0.38	0.9	2.5	—
Intestines	0.76	1.7	—	—
Liver	0.26	0.9	—	—
Pancreas	0.63	2.6	—	2.5
Skin	2.93	9.5	13.8	10.0
Mammary tissue	0.26	1.7	7.5	5
Testes	0.51	2.6	1.3	—
Kidneys and bladder	0.64	8.6	12.5	15.0
Teratocarcinoma	0.51	—	—	—
Adrenals	2.04	3.4	7.5	7.5
Pituitary	1.67	—	6.3	2.5
Thyroid	11.34	6.9	25.0	15.0
Carcinomas	22.57	44.8	101.4	65.0
Nervous system	1.15	—	—	—
Brain	3.17	1.7	1.3	7.5
Bone	0.64	3.4	5.0	5.0
Leukemia and lymp	1.91	1.7	2.5	5.0
Spleen sarcoma	0.13	0.9	—	—
Deep tissue sarcoma	0.76	7.7	—	2.5
Angiosarcoma	1.06	13.0	7.5	5.0
Soft tissue sarcoma	2.42	12.9	26.3	42.5
Mesothelioma	—	0.9	—	—
Sarcomas	11.24	42.2	42.6	67.5
TOTAL	33.81	87.0	144.0	132.5

Bouin-Hollande fixative solution. The lungs were fixed whole by intratracheal
instillation of the fixative solution. Tissue samples were prepared in the form of 5
μm thick paraffin sections stained with haematoxylin–eosin–saffron for histopatho-
gical examination. Tumour incidence and survival time were compared with those
observed in a group of 785 untreated control rats, housed simultaneously in the
same animal house and with those observed in three historical groups of rats
exposed to fission neutrons in the Silene reactor (Morin and Lafuma, 1990).

RESULTS

The cancers observed in the different experimental groups were differentiated
into carcinomas and sarcomas. The proportion of cancers in the different organs of
untreated control rats and in the three historical groups of male rats exposed to
fission neutrons at different doses in the Silene reactor is indicated in Table 1. In
control male rats, the proportion of lung carcinomas and kidney carcinomas was
very low (0.64%).

An increase of the global incidence of cancers, including both carcinomas and
sarcomas, was observed in all the groups exposed to fission neutrons, but this
increase was higher in the group exposed to fission neutrons at 1.15 Gy than in the
other groups. In this group, the incidence of lung carcinomas was multiplied by a

Table 2. Proportion of cancers (%) observed in the organs of exposed rats

	Group 1 Fission neutrons (0.6 Gy) only (63 rats)	Group 2 Fission neutrons (0.6 Gy) + CdCl$_2$ (50 rats)	Group 3 CdCl$_2$ alone (50 rats)	Group 4 CdCl$_2$ + βNF (50 rats)
Lung	6.4	10	2	—
Salivary glands	—	4	—	—
Intestines	3.2	—	—	—
Liver	1.6	—	—	—
Pancreas	3.2	—	—	—
Skin	—	—	—	—
Mammary tissue	1.6	—	—	—
Testes	—	—	—	—
Kidneys and bladder	1.6	2	—	—
Adrenals	1.6	—	—	—
Pituitary	—	2	—	—
Thyroid	—	4	—	—
Carcinomas	20.8	22	2	0
Nervous system	6.4	2	—	4
Brain	—	4	—	—
Bone	—	—	—	—
Leukemia	3.2	2	—	—
Spleen sarcoma	1.6	—	—	—
Angiosarcoma	3.2	4	—	—
Soft tissue sarcoma	11.1	14	—	—
Deep tissue sarcoma	1.6	6	—	—
Mesothelioma	—	2	—	—
Sarcomas	27.1	34	0	4
TOTAL	47.9	56	2	4

factor of about 40 and the incidence of kidney carcinomas by a factor of about 20 as compared with controls. The incidence of cancers in the organs of rats exposed to CdCl$_2$ under different experimental conditions is indicated in Table 2.

In group 1, in which rats were exposed to fission neutrons (0.6 Gy), 30 malignant tumours (13 carcinomas and 17 sarcomas) were observed in 63 rats: four lung carcinomas (two bronchogenic adenocarcinomas and two bronchioloalveolar carcinomas), two digestive adenocarcinomas, one liver cystic cholangiocarcinoma, two pancreatic adenocarcinomas, one mammary carcinoma, two renal adenocarcinomas, one adrenal carcinoma and four malignant schwannomas, two lymphosarcomas, one spleen sarcoma, two lung angiosarcomas, seven soft tissue sarcomas and one deep tissue sarcoma (renal liposarcoma). In group 2, 28 malignant tumours (11 carcinomas and 17 sarcomas) were observed in 50 rats exposed to fission neutrons (0.6 Gy) and then to CdCl$_2$: five lung carcinomas (one squamous cell carcinoma, two bronchogenic adenocarcinoma, one bronchioloalveolar adenocarcinoma and one bronchioloalveolar carcinoma), two salivary glands carcinomas, one kidney adenocarcinoma, one pituitary carcinoma, two thyroid papillary carcinomas and one schwannoma, two meningiomas, one lymphosarcoma, two lung angiosarcomas, seven soft tissue sarcomas and three deep tissue sarcomas (two leiomyosarcomas and one liposarcoma) and one pleural mesothelioma. A slight non-significant increase of the global incidence of cancers was observed in group 2 (56%) as compared with group 1 (47.9%), which was mainly related to an increased incidence of sarcomas. For carcinomas, a slight non-significant increased incidence of lung carcinomas was observed in this group (10%) compared with group 1 (6.4%). In group 3 exposed to $CdCl_2$ alone, only one lung carcinoma, a squamous cell carcinoma was observed, but no sarcomas. In group 4, exposed to $CdCl_2$ by inhalation and treated thereafter by βNF, any lung carcinoma occured, but non-neoplastic proliferative lesions, one premetaplasia and two adenomatoses were observed in three rats.

Survival times in the different groups were compared: no significant survival time shortening was observed for the group exposed to $CdCl_2$ alone by inhalation compared with untreated controls, or in the group exposed to $CdCl_2$ after previous irradiation by fission neutrons compared with the different groups exposed to fission neutrons at different doses (Fisher's exact test).

DISCUSSION

Under the experimental conditions used, these results did not show a clear carcinogenic or co-carcinogenic effect of inhaled $CdCl_2$. They are not consistent with those reported by Takenaka *et al.* (1983) which showed a carcinogenic effect of inhaled $CdCl_2$ in rats, with a distinct dose–effect relationship. Takenaka and co-workers exposed rats by inhalation at low concentrations (from 12.5 to 50 $\mu g.m^{-3}$) 23 h a day, 7 days a week for 18 months, whereas in our experiment, rats were exposed at higher concentrations (685 $\mu g.m^{-3}$), but in a discontinuous manner and during a short period, more similar to the conditions of exposure encountered in industry in the recent past. The differences observed between these studies might be explained by a higher Cd retention in the lungs of rats exposed continuously than in the rats exposed at higher concentrations under discontinuous exposure. In the experiment of Takenaka *et al.* (1983), the first lung carcinomas were noticed 23 months after the beginning of exposure. These findings suggest that cadmium induced carcinogenicity could be restricted to aged rats. To confirm this assumption, an additional group of 25, 10 month old rats was exposed to $CdCl_2$ by inhalation in the same conditions and treated by βNF as group 4. No lung carcinomas were observed in this additional group. These results did not show clear synergistic co-carcinogenic effect of $CdCl_2$ and βNF for the induction of lung cancer. At least, they indicated that $CdCl_2$ does not act as an initiator which could be revealed by βNF. In group 2 exposed to fission neutrons 0.6 Gy and then to $CdCl_2$ by inhalation, the global incidence of cancers and especially carcinomas, was not significantly increased compared with controls. A slight non-significant in-creased incidence of lung carcinomas was observed in this group compared with rats exposed to fission neutrons at the same dose, but not of renal carcinomas. It has been shown that global irradiation by fission neutrons induced a less effective initiation of pulmonary tissues than local pulmonary irradiation by inhalation of radon and its daughters (Morin and Lafuma, 1990). In this "radon model", the strongest co-carcinogenic effect was shown by combined exposure first to radon and then to tobacco smoke which resulted in a four-fold increased incidence of lung carcinomas, mainly of the squamous cell type (Monchaux *et al.*, 1996). Using this model, it has been demonstrated that it was possible to establish the potential co-carcinogenic action, showing either multiplicative, additive or no effect of various industrial or environmental airborne pollutants. In the experiments reported here and under the experimental conditions used, these results did not demonstrate a clear synergistic effect of inhaled CdCl$_2$ after irradiation by fission neutrons.

REFERENCES

Elinder, C. J., Kjellstrom, T., Hogstedt, C., Andersson, K. and Spang, G. (1985) Cancer mortality of cadmium workers. *Br. J. ind. Med.* **42**, 651–655.

Glaser, U., Hochrainer, D., Otto, F. J. and Oldiges, H. (1990) Carcinogenicity and toxicity of four cadmium compounds inhaled by rats. *Toxicol. Environ. Chem.* **27**, 153–162.

Heinrich, U. (1992) Pulmonary carcinogenicity of cadmium by inhalation in animals. In *Cadmium in the Human Environment: Toxicity and Carcinogenicity* (Edited by G. F. Nordberg, R. F. M. Herber and L. Alessio), pp. 405–413. IARC Scientific Publications no. 118, IARC, Lyon.

Heinrich, U., Peters, L., Ernst, H., Rittinghausen, S., Dasenbrock, C. and König, H. (1989) Investigation of the carcinogenic effects of various cadmium compounds after inhalation exposure in hamsters and mice. *Exp. Pathol.* **37**, 253–258.

Kazantzis, G. and Blanks, R. G. (1992) A mortality study of cadmium exposed workers. In *Edited Proc. 7th Int. Cadmium Conf.*, New Orleans, LA, 6–8 April 1992. (Edited by M. E. Cooke, S. A. Hiscock, H. Morrow and R. A. Volpe), pp. 150–157. London-Reston, VA, Cadmium Association/ Cadmium Council.

Kazantzis, G., Blanks, R. G. and Sullivan, K. R. (1992) Is cadmium a human carcinogen? In *Cadmium in the Human Environment: Toxicity and Carcinogenicity* (Edited by G. F. Nordberg, R. F. M. Herber and L. Alessio), pp. 435–446. IARC Scientific Publications no. 118, IARC, Lyon.

Monchaux, G., Morlier, J. P., Morin, M. and Maximilien, R. (1996) An experimental model for risk assessment of combined exposure to radon and other airborne pollutants. IRPA 9, *1996 Int. Congr. on Radiation Protection*, April 14–19, 1996, Vienna, Hofburg, Austria, Proceedings/Vol. 2, pp. 369–371.

Morin, M. and Lafuma, J. (1990) Les cancers radioinduits du rat. Etude expérimentale. Rapport CEA-R-5462 Rév. 1.

Morin, M., Queval, P. and Lafuma, J. (1978) Etude expérimentale de l'action co-carcinogénique du radon-222 et de la benzo-5, 6 flavone. In *Late Biological Effects of Ionizing Radiation*, Vol. II, pp. 423–427. IAEA-SM, 224/804.

Stayner, L., Smith, R., Thun, M., Schnorr, T. M. and Lemen, R. (1992) A dose–response analysis and quantitative assessment of lung cancer risk and occupational cadmium exposure. *Ann. Epidemiol.* **2**, 177–194.

Takenaka, S., Oldiges, H., König, H., Hochrainer, D. and Oberdörster, G. (1983) Carcinogenicity of cadmium chloride aerosols in W rats. *J. Nat. Cancer Inst.* **70**, 367–373.

Thun, M. J., Schnorr, T. M., Smith, A. B., Halperin, W. E. and Lemen, R. A. (1985) Mortality among a cohort of US cadmium production workers—an update. *J. Nat. Cancer Inst.* **74**, 325–333.

Pergamon

Ann. occup. Hyg., Vol. 41, Supplement 1, pp. 630–635, 1997
© 1997 British Occupational Hygiene Society
Published by Elsevier Science Ltd. All rights reserved
Printed in Great Britain
0003–4878/97 $17.00 + 0.00
Inhaled Particles VIII

PII: S0003–4878(96)00083–X

LONG-TERM EFFECTS OF COMBINED EXPOSURE TO FISSION NEUTRONS AND INHALED LEAD OXIDE PARTICLES IN RATS

G. Monchaux, M. Morin, J. P. Morlier and M. F. Olivier

CEA-DSV-DRR-LCE, BP 6, 92265 Fontenay aux Roses Cedex, France

INTRODUCTION

The evaluation of the potential carcinogenicity of lead and lead compounds on the basis of epidemiologic and experimental data for human health risk assessment does not allow clearcut conclusions. The EPA proposed classification of lead and lead compounds as probable (B2) human carcinogens, while IARC defined lead and lead inorganic compounds as 2B carcinogens, (inadequate evidence in humans and sufficient evidence in animals).

Recent epidemiologic studies do not confirm experimental results showing that lead induces kidney cancer. However, a statistically significant increased lung cancer mortality was observed in human groups occupationally exposed to lead in battery manufacturing and smelting industries (Cooper and Gaffey, 1975; Cooper *et al.*, 1985).

A co-carcinogenic effect of lead oxide with benzo-a-pyrene was found by Kobayashi *et al.* (1974) after intratracheal instillation in hamsters, for relatively low lead lung burden (9 mg). However, animal experimental data on pulmonary carcinogenesis using the inhalation route are lacking.

The aim of this work was to ascertain whether lead oxide acts as a complete lung carcinogen when inhaled alone, or as a lung co-carcinogen in rats previously exposed to fission neutrons, irradiation acting then as an initiator; or a lung co-carcinogen in rats exposed first to lead oxide by inhalation and by the end of the inhalation period to intramuscular injections of 5–6 benzoflavone (βNF), which has been previously demonstrated to be a specific promotor of experimentally induced squamous cell lung carcinoma in Sprague–Dawley rats (Morin *et al.*, 1978).

MATERIAL AND METHODS

The most important limitation to ascertain lead pulmonary induced carcinogenesis after inhalation is the occurrence of kidney cancers in rats at cumulative doses higher than 2 g, as found by Azar *et al.* (1973). Thus, the exposure level in the inhalation chambers for the long-term inhalation carcinogenesis study had to be adjusted so that the cumulative lead kidney burden resulting from a 1 year inhalation exposure to lead oxide would be equivalent to that achieved after a 2 years exposure to 500 ppm of lead acetate in the diet. This dose is the lowest one which has been shown to be associated with lead, or lead compounds, induced kidney tumours. Accordingly, a "pilot experiment" was primarily carried out to determine the toxicokinetics of lead oxide inhaled by whole body inhalation exposure in rats, compared with those of lead oxide and lead acetate given by ingestion route (gavage). It allowed us to determine that the inhalation exposure of rats at 5 mg m^{-3}, 6 h a day, 5 days a week, during one year should be grossly equivalent to a cumulative dose of lead acetate (about 3–4 g) given by gavage. Such a dose is about two-fold lower than that inducing 10% of kidney cancers in the experiment of Azar et al. (1973).

The long-term inhalation carcinogenesis study was carried out on four experimental groups of Sprague–Dawley male rats.

• Group 1: 63 rats exposed to 0.6 Gy fission neutrons. In our series, this dose has been shown to result in a 10% incidence of lung cancers and a 3% incidence of kidney cancers (Morin and Lafuma, 1990).

• Group 2: 50 rats exposed to 0.6 Gy fission neutrons and 2 months later, exposure to lead oxide by inhalation for 1 year as previously determined.

• Group 3: 50 rats exposed to lead oxide alone by inhalation for 1 year.

• Group 4: 25 rats exposed to lead oxide by inhalation for 1 year and 1 month after the end of exposure treated with six intramuscular injections of βNF.

Irradiation

The rats were exposed to fission neutrons in the "Silene" experimental reactor facilities according to the pulse operating mode with a 10 cm thick lead shielding. The rats were placed into individual aluminium containers suspended on a net 4 m distant from the core of the reactor.

The number of fissions was: 4.88 10^{16}. The peak power was: 1.23 10^{17} fissions s^{-1}. The dose was: 0.62 Gy neutrons and 0.135 Gy gamma rays.

Lead inhalation exposure

The whole body inhalation exposure was performed into a 2 m^3 dynamic type Hazleton exposure chamber. The rats were housed continuously in individual wire cages in the chambers throughout the inhalation period, provided tap water *ad libitum* and fed outside of exposure hours. The aerosolization of dry dust lead oxide was achieved by means of a TSI Model 3433 Small-Scale Powder Disperser. The generator was calibrated so that the concentration of lead oxide aerosol into the chamber be 5 mg m^{-3}. The airborne concentration of lead oxide particles into the chamber was controlled gravimetrically using HAWP 047 0M previously weighed Millipore filters. At least three measurements were performed daily for each inhalation session. The mean lead oxide airborne concentration measured within the inhalation chamber during the 240 inhalation sessions performed was: 5.3 ± 1.7 mg m^{-3}. The aerodynamic size measurement of aerosolized particles was determined using an Andersen 1 ACFM Non-Viable Ambient Particle Sizing Sampler (Cascade Impactor). The equivalent aerodynamic diameters of lead oxide particles aerosolized into exposure chambers are listed in Table 1. These results show that the fraction respirable by rats represented more than 60% of the total mass.

The actual diameter of aerosolized particles was determined on 50 l. aliquot air samples taken from the exposure chamber and collected on 0.2 μm pore size

Table 1. Equivalent aerodynamic diameters of lead oxide particles

Particle equivalent aerodynamic diameter (μm)	Per cent in size range less than	Cumulative per cent
> 9.0	0.5	99.9
5.8–9.0	39.5	99.4
4.7–5.8	26.0	58.9
3.3–4.7	10.9	32.9
2.1–3.3	15.7	22.0
1.1–2.1	4.1	6.3
0.7–1.1	1.4	2.2
0.4–0.7	0.5	0.8
0.0–0.4	0.3	0.3

Millipore filters. The actual diameter of particles was determined by light microscopy using a Carl Zeiss Photomicroscope coupled with a video computerized image analysis system. The mean particle diameter was 0.69 μm (σg = 1.74).

Since the density of lead oxide particles is 9.5, the results obtained using the two different methods are in good agreement, the theoretical value of MMAD being 5.1 μm compared with 4.7 μm < MMAD < 5.8 μm (Table 1).

Injection of the promotor (βNF)

In the group treated with βNF, 1 month after the end of the inhalation exposure, the rats received six intramuscular injections of 25 mg kg^{-1} of βNF at fortnightly intervals.

Animal and histologic analysis

The experiments were carried out on 10-week-old male SPF outbred OFA Sprague–Dawley rats. After their arrival in the laboratory, the animals were kept in quarantine for 1 week before any treatment and acclimated to inhalation exposure chambers for 1 week before the beginning of inhalation exposure. The rats were exposed as described above.

After exposure all rats were housed throughout their life-span in the same animal house, maintained on a 12 h light cycle. All animals were kept until moribund except those of the βNF group (Group 4) which were killed 100 days after the last injection of βNF. A full necropsy was performed in every animal. Liver, spleen, kidney, lungs and brain and all organs exhibiting macroscopic lesions were taken systematically. Histologic samples were fixed with Bouin–Hollande fixative solution. The entire lungs were fixed by intratracheal instillation of the fixative solution. Tissue samples were prepared in the form of 5 μm paraffin sections stained with haematoxylin–eosin–saffron for histopathological examination.

Survival times

Tumour incidence and survival time were compared with those observed in a group of 785 untreated control rats, housed simultaneously in the same animal house and with those observed in three historical groups of rats exposed to fission neutrons in the Silene reactor (Morin and Lafuma, 1990).

Table 2. Proportion of cancers (%) in the different organs of untreated control male rats and of male rats irradiated by fission neutrons at various doses in the Silene reactor

	Control male rats (785 rats)	Fission neutrons (0.4 Gy) (116 rats)	Fission neutrons (1.15 Gy) (80 rats)	Fission neutrons (1.73 Gy) (40 rats)
Lungs	0.64	6.0	25.0	7.5
Salivary glands	0.38	0.9	2.5	—
Intestines	0.26	1.7	—	—
Liver	0.26	0.9	—	—
Pancreas	0.51	2.6	—	2.5
Skin	2.93	9.5	13.8	10.0
Mammary tissue	0.26	1.7	7.5	5.0
Testes	0.51	2.6	1.3	—
Kidneys and bladder	0.64	8.6	12.5	15.0
Teratocarcinoma	0.51	—	—	—
Adrenals	2.04	3.4	7.5	7.5
Pituitary	1.02	—	6.3	2.5
Thyroid	11.34	6.9	25.0	15.0
Carcinomas	**21.3**	**44.8**	**101.4**	**65.0**
Nervous system	1.15	—	—	—
Brain	3.17	1.7	1.3	7.5
Bone	0.64	3.4	5.0	5.0
Leukemia and lymp	1.91	1.7	2.5	5.0
Spleen sarcoma	0.13	0.9	—	—
Deep tissue sarcoma	0.76	7.7	—	2.5
Angiosarcoma	0.64	13.0	7.5	5.0
Soft tissue sarcoma	2.42	12.9	26.3	42.5
Mesothelioma	—	0.9	—	—
Sarcomas	**10.82**	**42.2**	**42.6**	**67.5**
TOTAL	**32.12**	**87.0**	**144.0**	**132.5**

RESULTS

Malignant tumours were differentiated into carcinomas and sarcomas. The proportion of cancers observed in the different organs of 785 control male rats and the proportion of cancers in the different organs of three historical groups of male rats exposed to a global irradiation by fission neutrons at different doses in the Silene reactor is indicated in Table 2. In control male rats, the spontaneous incidence of kidney and lung carcinomas was very low, 0.64%. The global incidence of cancers, including both carcinomas and sarcomas was increased in all the groups exposed to fission neutrons but this increased incidence of cancers was higher in the group exposed to fission neutrons 1.15 Gy than in the other groups. In this group the incidence of lung carcinomas was multiplied by a factor of about 40 and the incidence of kidney carcinomas by a factor of about 20 compared with controls. The proportion of cancers in the different organs of male rats exposed either to irradiation by fission neutrons 0.6 Gy alone, or to lead oxide according to the different experimental schedules is indicated in Table 3. The global incidence of cancers was lower in group 2 exposed to fission neutrons and lead oxide combined (34%) than in group 1 exposed to fission neutrons 0.6 Gy alone (47.9%).

The incidence of lung carcinomas was also lower in group 2 than in group 1. In

Table 3. Proportion of cancers (%) observed in the organs of exposed rats

	Group 1 Fission neutrons (0.6 Gy) only (63 rats)	Group 2 Fission neutrons (0.6 Gy) + PbO (50 rats)	Group 3 PbO alone (50 rats)	Group 4 PbO + βNF (25 rats)
Lung	6.4	4	—	—
Salivary glands	—	2	—	—
Intestines	3.2	—	—	—
Liver	1.6	—	—	—
Pancreas	3.2	—	—	—
Skin	—	2	2	—
Mammary tissues	1.6	2	—	—
Testes	—	—	—	—
Seminal vesicle	1.6	—	—	—
Kidneys and bladder	1.6	4	2	8
Adrenals	1.6	—	2	—
Pituitary	—	—	2	4
Thyroid	—	—	2	—
Carcinomas	**20.8**	**14**	**10**	**12**
Nervous system	6.4	2	2	—
Brain	—	—	—	4
Bone	—	—	—	—
Leukemia	3.2	4	—	4
Spleen sarcoma	1.6	2	—	—
Angiosarcoma	3.2	2	—	—
Soft tissue sarcoma	11.1	6	6	4
Deep tissue sarcoma	1.6	4	2	—
Mesothelioma	—	—	—	—
Sarcomas	**27.1**	**20**	**10**	**4**
TOTAL	**47.9**	**34**	**20**	**4**

this group, only one rat had two primitive lung carcinomas, a squamous cell carcinoma and a bronchioloalveolar adenocarcinoma, as compared with four lung carcinomas, two bronchogenic adenocarcinomas and two bronchioloalveolar carcinomas, observed in group 1.

No significant survival time shortening was observed for the group exposed to lead oxide after previous irradiation by fission neutrons compared with that exposed to fission neutrons only, nor for the group exposed to lead oxide alone compared with untreated controls (Fisher's exact test).

DISCUSSION

These results did not show a direct carcinogenic effect of inhaled lead oxide in rats, nor did they show a co-carcinogenic effect of lead oxide under the experimental conditions used. In particular, the fact that no lung carcinomas were observed either after exposure to lead oxide alone, or after treatment by βNF indicates that lead oxide is probably not a strong direct carcinogen and that it does not act as an initiator which could be revealed by βNF. They did not indicate either that lead oxide might act as a promoter after previous irradiation by fission neutrons. These results did not confirm the co-carcinogenic effect of lead oxide found by Kobayashi *et al.* (1974) with benzo-a-pyrene after intratracheal instillation in hamsters. In conclusion, the results of this study did not show any excess of cancers and especially no excess of lung or kidney cancer in the groups exposed to lead oxide by inhalation as compared with unexposed control rats or with groups irradiated by fission neutrons. They did not show a clear carcinogenic or co-carcinogenic effect of inhaled lead oxide under the different experimental conditions used.

REFERENCES

Azar, A., Trochimowicz, H. J. and Maxfield, M. E. (1973) Review of lead studies in animals carried out at Haskell Laboratory: two-year feeding study and response to hemorrage study. In *Environmental Health Aspects of Lead. Proc. Int. Symp.*, October 2–6, 1972, Amsterdam, pp. 199–210. Luxembourg: Commission of the European Communities, Directorate General for Dissemination of Knowledge, Center for Information and Documentation.

Cooper, W. C. and Gaffey, W. R. (1975) Mortality of lead workers. *J. occup. Med.* **17**, 100–107.

Cooper, W. C., Wong, O. and Kheifets, L. (1985) Mortality among employees of lead battery plants and lead-producing plants. *Scand. J. Work Environ. Health* **11**, 331–345.

Kobayashi, N. and Okamoto, T. (1974) Effects of lead oxide on the induction of lung tumors in Syrian hamsters. *J. Nat. Cancer Inst.* **52**, 1605–1610.

Morin, M. and Lafuma, J. (1990) Les cancers radioinduits du rat. Etude expérimentale. Rapport CEA-R-5462 Rév 1.

Morin, M., Queval, P. and Lafuma, J. (1978) Etude expérimentale de l'action co-carcinogénique du radon-222 et de la benzo-5,6 flavone. In *Late Biological Effects of Ionizing Radiation*, Vol. II, pp. 423–427. IAEA-SM, 224/804.

Pergamon

Ann. occup. Hyg., Vol. 41, Supplement 1, pp. 636–640, 1997
British Occupational Hygiene Society
Crown Copyright © 1997 Published by Elsevier Science Ltd
Printed in Great Britain
0003–4878/97 $17.00 + 0.00
Inhaled Particles VIII

PII: S0003–4878(96)00082–8

MEASUREMENT OF THE DEGREE OF PROTECTION AFFORDED BY RESPIRATORY PROTECTIVE EQUIPMENT AGAINST MICROBIOLOGICAL AEROSOLS

A. C. Redmayne, D. Wake, R. C. Brown and B. Crook

Health and Safety Laboratory, HSL, R111 Robens Building, Broad Lane, Sheffield S3 7HQ, U.K.

INTRODUCTION

Biological aerosols may include viable and non-viable intact microbial cells or cell components, exposure to which may cause respiratory infection, allergic lung disease or toxicosis. Respiratory protective equipment (RPE) is used widely in industry to reduce occupational exposure to a range of substances, including microbiological aerosols. Consequently, there is a need to be able to measure the performance of RPE against microbiological aerosols. This paper describes a series of tests to evaluate the performance of a range of RPE when challenged with bacterial and bacteriophage aerosols.

METHODS

Microbiological aerosols (bioaerosol)

Three bacterial species and a bacteriophage were used to provide a bioaerosol challenge.

Pseudomonas alcaligenes [National Collection of Type Cultures No. (NCTC) 10367] are Gram-negative, motile, non-sporulating rods, 0.5 μm in width and between 2.0–3.0 μm in length. They are similar in morphology to *Legionella pneumophila*, for which they may be considered as surrogates.

Bacillus subtilis sub species *globigii* (NCTC 10073) are Gram-positive, non-motile, sporulating rods. They are 0.7–0.8 μm in diameter and 2.0–3.0 μm in length and usually occur in single cells.

Micrococcus luteus (NCTC 4351) are Gram-positive, non-motile, non-sporulating cocci occurring in tetrads. Individual cocci are 0.9–1.8 μm in diameter.

The bacteriophage used was T1 coliphage wild type, American type culture collection (ATCC) No 11303B1, which has a long slender tail of 150 nm and an hexagonal head 60 nm in length (Harstad, 1965). The host bacterium is *Escherichia coli*, ATCC 12435. Aerosols were generated from aqueous suspensions of each viable organism as described below.

Test apparatus. This comprised a wind tunnel with a working section approximately 3 m in length, 1 m in height and 0.6 m in width to provide an enclosed region, maintained at negative air pressure. Within this, bacterial aerosols were generated into a perspex plenum chamber from three medical nebulisers connected in parallel via a manifold to a compressed air supply. Alternatively, a Collison atomiser was used to generate bacteriophage aerosol, as it produces a smaller primary droplet than the medical nebuliser. To challenge RPE filter material with bioaerosol, the filter under test was housed in one of two aerodynamically identical flow lines from the plenum chamber. The other flow line provided an un-filtered reference line for comparison. Perspex boxes connected with tubing to the end of these flow lines each housed an all glass impinger (AGI-30; Cox, 1987), into which the bioaerosol was collected into liquid at an air flow rate of 12.5 l.min^{-1}. The suspension was used to inoculate agar plates, from which the number of colony or plaque forming units per ml of AGI 30 collection fluid was calculated. The penetration through the mask or filter was obtained from the quotient of the count per ml passing through the filter compared to that passing through the reference line.

Non biological aerosols. The monodisperse urea aerosol tests were carried out using the automated Berglund–Liu test system (Wake, 1995) with particles of 1.5, 3, 5, 7 and 9 μm geometric diameter. The BS 4400 sodium chloride aerosol has an approximate mass median diameter of 0.6 μm and number median diameter of 0.06 μm. The procedure for carrying out both types of test has been described previously (Wake and Brown, 1988).

Penetration of a range of filters and surgical masks. A wide range of filters for full and half face respirators were tested along with a selection of available disposable nuisance dust masks and disposable surgical masks.

Leakage of masks. Although many of the devices showed acceptable filtration efficiency, in order to be useful in practice it is necessary that the extent of face-seal leakage should also be acceptable. To test the effect of face-seal leakage, the test system described above was modified to test a full-face mask respirator. The twin aerosol flow lines were replaced by a perspex chamber which accommodated the "Sheffield" manikin head wearing a respirator. Aerosol entering the chamber could pass either through the respirator and head or through the reference line. A full face mask respirator was fitted with a dust filter element (filter code C) which was shown to allow 0.01% of an aerosol of *M. luteus* to penetrate, and 0.01% of a sodium chloride aerosol and the respirator was fitted to the "Sheffield" head and then sealed on to the face to prevent unwanted leakage. To provide known leaks, five capillary tubes of different diameter were positioned on the left side of the head between the seal of the mask and the upper cheek. The assembly was sealed to the mask and head. All the capillaries were sealed and a penetration test at an air flow rate of 30 l.min^{-1} was carried out. Each capillary in turn, starting with the one of largest bore, was opened and the sodium chloride aerosol test repeated until tests had been carried out with all five opened. These tests were repeated with an aerosol of *M. luteus*.

RESULTS AND DISCUSSION

Results of penetration of filter material

The results are given in Tables 1 and 2. It can be seen that the expected high performance of the more efficient respirator filters was confirmed against both types of aerosol. The poor performance of nuisance dust masks previously

Table 1. Penetration of microbiological aerosols through filters

Filter code	Filter type	Ps. alcaligenes		B. subtilis		M. luteus	
		Pressure drop (Pa)	Penetration (%)	Pressure drop (Pa)	Penetration (%)	Pressure drop (Pa)	Penetration (%)
A	Glass fibre	41.2	< 0.01*	37.8	< 0.01*	47.1	0.88
B	Glass fibre	28.8	< 0.01*	20.0	< 0.01*	28.6	0.04
C	Glass fibre	34.8	< 0.01*	35.5	< 0.01*	45	0.01
D	Glass fibre	19.9	< 0.01*	18.9	< 0.01*	23.6	< 0.01*
E	Glass fibre	61.0	< 0.01*	49.7	< 0.01*	68.2	0.01
F	Merino			87	0.01	84	0.45
	wool			96	0.01	94	0.07
				99	< 0.01	98	0.03
G	Resin-wool			87	< 0.01	85	0.02
				81	< 0.01	78	0.22
				76	< 0.01	81	0.10
H	Filtering facepiece	48.6	0.10	—	—	—	—
I	Filtering facepiece	5.5	0.40	6.8	0.02	8.3	0.11
J	Nuisance	2.8	100	2.5	59.0	9.4	58.8
	dust mask	2.9	100	2.8	79.0	7.8	55.0
		4.0	100	2.6	56.0	7.8	63.6
K	Surgical	6.7	25.0	6.9	0.6	6.2	16.0
	mask	6.3	20.0	6.4	1.4	7.4	12.0
		7.3	15.0	6.5	0.9	7.9	5.6
L	Surgical			6	57.9	5	52.5
	mask			6	69.6	5	82.5
				7	56.1	4	82.8
M	Surgical			16	0.06	16	5.8
	mask			14	5.5	15	5.6
				15	0.24	16	4.3
N	Surgical			13	< 0.01	12	1.7
	mask			13	< 0.01	18	4.5
				12	0.01	17	0.4
O	Resuscitation			155	0.02	155	0.6
	mask			160	1.09	150	0.4
				168	7.77	141	1.4

* No organisms were detected during the counting procedure with the filtered aerosol.

observed with non-biological aerosols (Wake and Brown, 1988) was also apparent with bacterial aerosols. For example, the nuisance dust mask (code J) failed to retain a measurable number of an aerosol of *Ps. alcaligenes* and allowed up to 79% of *B. subtilis* to penetrate. Mask L was by far the worst surgical mask tested, its performance being similar to that of the nuisance dust mask J, which in appearance is closely resembled. Surgical mask K was also poor; up to 16% of *M. luteus* penetrated the mask material and up to 25.4% penetration of *Ps. alcaligenes* (Table 1) and up to 78% of NaCl aerosol (Table 2). The two surgical masks M and N, that were thought to contain electrostatically charged filter material, performed better. Although it is acknowledged that such masks do not classify as respiratory protective equipment, it is probable that they are being used widely in health care as a perceived protection of workers against possible airborne infectious agents from patients. The protection that these masks may confer should not be over-estimated.

Table 2. Penetration of monodisperse urea aerosols and BS 4400 sodium chloride aerosols through filters

Filter code	Filter type	Pressure drop (Pa)	1.5 μm	Urea aerosol Penetration (%) 3 μm	5 μm	7 μm	9 μm	NaCl aerosol Pen. (%)
A	Glass fibre	52.7	0.44	0.11	0.01	—	—	0.01
B	Glass fibre	29.0	0.01	—	—	—	—	0.01
C	Glass fibre	49.0	1.72	1.19	0.05	0.03	0.01	0.01
D	Glass fibre	24.7	0.29	0.25	0.19	0.08	0.02	0.05
E	Glass fibre	73.0	0.57	0.27	0.04	0.01	0.01	0.01
F	Merino wool	103	1.67	0.75	0.11	0.06	0.19	0.31
G	Resin wool	88	1.56	0.65	0.12	0.12	0.10	0.01
H	Filtering facepiece	63.6	1.35	1.14	0.45	0.02	0.01	0.64
I	Filtering facepiece	7.9	0.21	0.17	0.04	0.02	0.01	0.14
J	Nuisance dust mask	2.2	91.3	84.8	64.4	34.4	15.2	78.0
		3.2	90.2	75.6	57.5	23.7	9.0	—
		3.2	89.4	80.6	53.9	21.4	8.76	—
K	Surgical mask	4.7	20.2	7.10	1.89	0.54	0.17	78.0
		4.5	21.2	7.71	1.79	0.32	0.11	29.0
		4.1	18.7	7.52	2.45	1.03	0.51	32.0
L	Surgical mask	7	98.5	81.1	46.9	22.5	10.5	87.0
M	Surgical mask	16	7.32	4.93	1.95	1.29	1.15	1.00
N	Surgical mask	14	16.4	12.8	9.66	12.7	13.7	3.10
O	Resuscitation mask	173	0.43	0.29	0.11	0.77	1.59	12.0

The use of masks with a range of efficiencies enabled a comparison to be made between the penetrations of live bacterial aerosols and those of urea aerosols with a defined geometrical diameter, as in previous work (Wake *et al.*, 1992). The results obtained suggested that *Ps. alcaligenes* behaved as a particle of diameter 1.5 μm or less, *B. subtilis* behaved as 3–5 μm and *M. luteus* as 1.5–3 μm. The results of the bacteriophage aerosol penetration confirmed the large differences in performance between filter cartridge material, with low penetration and disposable dust mask-type material with much greater penetration, up to 90% and greater than in some instances. On the whole, penetration of bacteriophage through dust mask filter material was greater than that of *M. luteus*, which in turn was greater than that of *B. subtilis*. More BS 4400 sodium chloride aerosol penetrated the masks than did bacteriophage. This would place the effective particle size of the bacteriophage at smaller than bacteria but greater than sodium chloride.

Results of known leakage in a full face respirator
The results are given in Table 3. The penetration of both bacterial and sodium chloride aerosol increased with the size of leak. The observed penetration of bacteria was greater than that of the sodium chloride aerosol, which was surprising since the sodium chloride particles are smaller than the bacteria, but in general the penetrations of the two aerosols were comparable. The results provided quantifiable evidence of the influence of face seal leakage on RPE performance.

Table 3. Penetration of aerosols through known leakage in a full face respirator

Leak size (mm^2)	BS4400 Sodium chloride		*Micrococcus luteus*	
	Pressure drop (Pa)	Penetration (%)	Pressure drop (Pa)	Penetration (%)
No leak	67.5	0.28	67.3	0.92
2.26	65.3	2.33	65.5	2.96
3.91	64.4	3.29	64.3	4.21
4.52	64.2	3.81	65.0	5.31
4.87	64.2	4.83	65.0	5.19
5.11	64.5	5.92	—	5.50

The header above the aerosol subcolumns reads "Aerosol type".

CONCLUSIONS

It can be seen from the results that, generally, biological aerosols behave in a similar way to non-biological aerosols of a corresponding aerodynamic diameter. It can also be concluded that the performance of high efficiency respirator filters can be compromised by poor fit of RPE to the face.

REFERENCES

BS 4400 (1969) Method for sodium chloride particulate tests for respirator filters. British Standards Institution, London.

Cox, C. S. (1987) *The Aerobiological Pathway of Microorganisms*, pp. 172–177. Wiley, New York.

Harstad, J. D. (1965) Sampling submicron T1 bacteriophage aerosols. *Appl. Microbiol.* **13** (6), 899–908.

Wake, D. (1995) An automated system for measuring the penetration of aerosols through filters. *J. Aerosol Sci.* **26** (5), 861–865.

Wake, D. and Brown, R. C. (1988) Measurements of the filtration efficiency of nuisance dust respirators against respirable and non-respirable aerosols. *Ann. occup. Hyg.* **32**, 295–315.

Wake, D., Brown, R. C., Trottier, R. A. and Liu, Y. (1992) Measurements of the efficiency of respirator filters and filtering facepieces against radon daughter aerosols. *Ann. occup. Hyg.* **36**, 629–636.

Pergamon

Ann. occup. Hyg., Vol. 41, Supplement 1, pp. 641–646, 1997
British Occupational Hygiene Society
Crown Copyright © 1997 Published by Elsevier Science Ltd
Printed in Great Britain
0003–4878/97 $17.00 + 0.00
Inhaled Particles VIII

PII: S0003–4878(96)00081–6

THE EFFECTIVENESS OF CONTROL MEASURES TO REDUCE EXPOSURE TO HENNA AND CASTOR BEAN ALLERGENS IN A COSMETICS FACTORY

P. Griffin,* A. Redmayne,* K. Wiley* and R. G. Crane†

*Health and Safety Laboratory, Health and Safety Executive, R111 Robens Building, Broad Lane, Sheffield S3 7HQ, U.K.; and †Health and Safety Executive, Basingstoke, U.K.

INTRODUCTION

Henna is the dried and ground leaves of *Lawsonia inermis* which is used with other powdered vegetable materials in the preparation of cosmetics. One of the potential contaminants of henna is castor bean. We examined a factory where henna was processed in bulk and where workers exhibited respiratory symptoms. After our initial investigation the factory was redesigned to minimise exposure of workers to allergen. The purpose of this study was to determine the prevalence of immunological sensitisation among the work force and to quantify the airborne concentrations of henna and castor bean in the factory and thereby monitor the effectiveness of control measures instituted between our visits.

METHOD

Radio allergosorbent test (RAST)

The RAST was carried out using sera from 40 workers at the factory using the method of Topping *et al.* (1982). Total serum IgE was measured using a RIACT kit.

Measurement of airborne antigen

Static samplers were sited at positions A and B (Fig. 1). After the factory had been redesigned to minimise exposure to henna dust the factory was revisited and samplers sited at positions C–F (Fig. 2). Air was sampled, whilst the work process was being carried out, using a high volume sampler collecting onto glass microfibre sheets (22.7 × 17.7 cm). High volume 5 stage cascade impactors collecting at a flow rate of $0.46 \text{ m}^3 \text{ min}^{-1}$ were sited at positions D and E. We calculated the aerodynamic diameter (Dp50) of the particles collected to be as follows: stage 1: > 11.5 µm, stage 2: 4.8–11.5 µm, stage 3: 2.4–4.8 µm, stage 4: 1.5–2.4 µm, stage 5: 0.8–1.5 µm and base filter 0.3–0.8 µm.

Total dust was measured by gravimetric analysis. The filters were then extracted in an aqueous solution. Henna allergen concentration was measured in the extract using RAST inhibition. Castor bean antigen was measured using an inhibition ELISA.

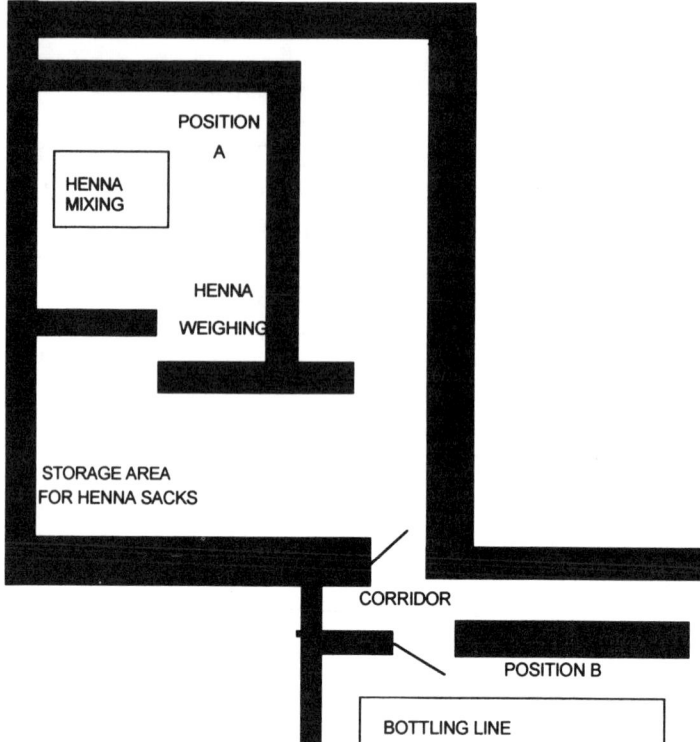

Fig. 1. Factory layout. High volume samplers were situated at the following positions: Position A; where henna and other materials were weighed, mixed and packed into drums. Position B; a separate room from the henna handling area.

Rast inhibition

RAST inhibition was performed using an adaptation of the method described by Topping *et al.* (1982). Briefly, pooled human serum with antibodies to henna was incubated with an equal volume of air sample extract or a solution containing a known concentration of henna or castor bean proteins. The RAST was then processed as described by Topping *et al.* (1982).

To demonstrate the specificity of binding to red henna discs; red henna and castor bean discs were processed as shown above with 90 µg per ml castor bean extract. To quantify henna in air sample extracts the percentage inhibition given by an air sample extract was compared to the inhibition given by a standard curve of henna extract in the range 0.8–83 µg ml^{-1}. The amount of henna allergen collected from each stage of the Anderson impactor was expressed as a percentage of the total amount of allergen collected by the sampler.

Inhibition enzyme linked immunosorbent assay (ELISA) for castor bean

Castor bean extract was coupled to a polystyrene microtitre plate. One hundred microlitres of a mixture of a 1/10 000 dilution of anti-ricin antibody (Sigma) and air sample extract or castor bean extract (range 18–180 ng ml^{-1}) was incubated in each well of a microtitre plate. The plate was then incubated with a secondary,

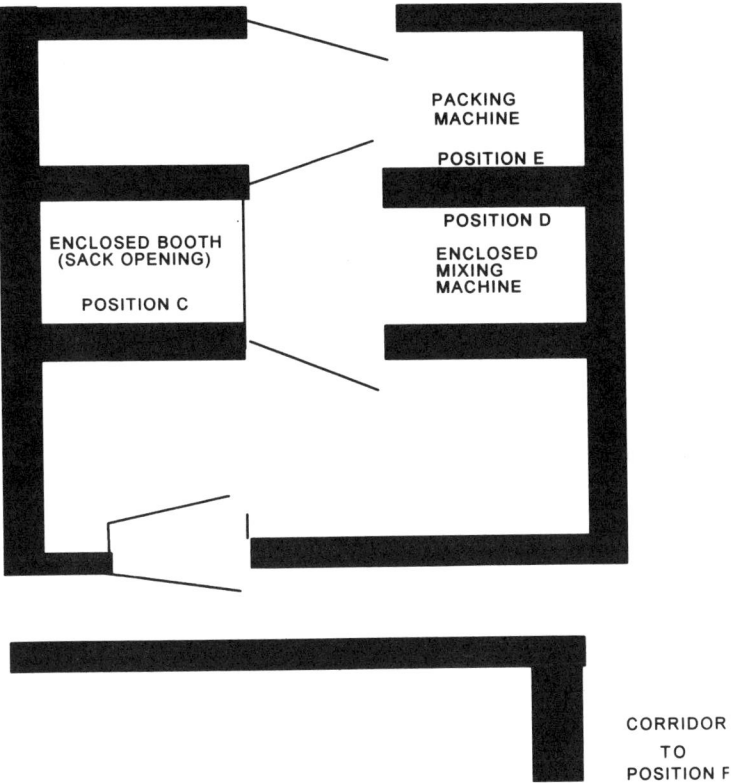

Fig. 2. Factory layout after redesign of factory. High volume background samplers were situated at the following points: Position C; a booth where sacks of henna were opened. Position D; next to an enclosed machine which mixed henna. Position E; next to a henna packing machine. Position F; away from the henna processing area.

horseradish peroxidase conjugated, antibody. Finally, an enzyme substrate solution was incubated in each well and the absorbence reading for each well was then used to calculate the % inhibition of absorbence. To quantify henna in air sample extracts the percentage inhibition given by an air sample extract was compared to the inhibition given by a standard curve of henna extract.

RESULTS

Serum IgE antibodies

The presence of specific IgE to henna and castor bean was determined by reference to 10 sera with IgE to common environmental allergens (atopic) and 10 sera without IgE to these allergens (non-atopic). Workers RAST % binding values in excess of the mean % binding + 2.5 SD of the control sera were considered positive ($p = 0.01$). There was no significant difference in the % binding to castor bean discs between the atopic and non-atopic sera.

20 workers had IgE to castor bean and 10 had IgE to henna. All workers with IgE to henna, except one, also had IgE to castor bean. Total IgE values varied from 1 to 1832 IU ml^{-1} with two results greater than 1000 IU.

Table 1. Inhibition of specific IgE binding to castor bean and red henna discs by castor bean and henna extracts

Allergen disc	Inhibitor	Per cent inhibition
Castor bean	Castor bean	87
Henna	Castor bean	4
Henna	Henna	73

Specificity of specific IgE for red henna

Table 1 shows the results obtained when castor bean or henna is used to inhibit binding to castor bean or henna coated discs. The addition of castor bean extract to the serum in this assay did not affect binding to henna discs but did inhibit binding to castor bean coated discs.

Airborne concentrations of antigen and total dust

Table 2(i) shows the airborne concentrations of total dust, henna and castor bean during our first visit and refers to the sampler positions in Fig. 1. A high concentration of henna and castor bean was detected in the room where henna was processed whilst much less was detected in an adjacent room.

Table 2(ii) show the airborne concentrations during our second visit and refers to Fig. 2. The redesign of the factory had reduced the concentration of antigen at all points measured. A transient high concentration of antigen was detected at position D and occurred when the enclosed henna processing machinery was breached for cleaning purposes. Figure 3 shows the particle size of henna allergen detected in the factory. Henna antigen was associated with particles in the size range 1.5–11.5 μm.

DISCUSSION

We found workers with specific IgE to henna and castor bean. Sera from atopics did not have elevated binding to castor bean thereby indicating that antibodies to common environmental allergens are unlikely to interfere in our RASTs. In addition most of our sera did not have elevated total IgE which could interfere with a RAST to castor bean (Topping *et al.*, 1982). We pooled sera with IgE to henna. This sample contained antibodies to henna and castor bean. Our RAST inhibition

Table 2. Airborne concentrations of total dust, henna and castor bean at various positions in the factory

Site	Sampling time (min)	Total dust (mg m^{-3})	Henna (μg m^{-3})	Castor bean (μg m^{-3})
(i) Airborne concentrations at positions A and B				
A	105	5.9	20	1.6
A	170	5.3	26	1.3
B	375	0.3	0.5	< 0.001
(ii) Airborne concentrations at positions C–F				
C	52	0.2	1.4	< 0.001
D	22	7.3	91	0.01
E	327	0.6	7	0.003
F	363	1	0.04	< 0.001

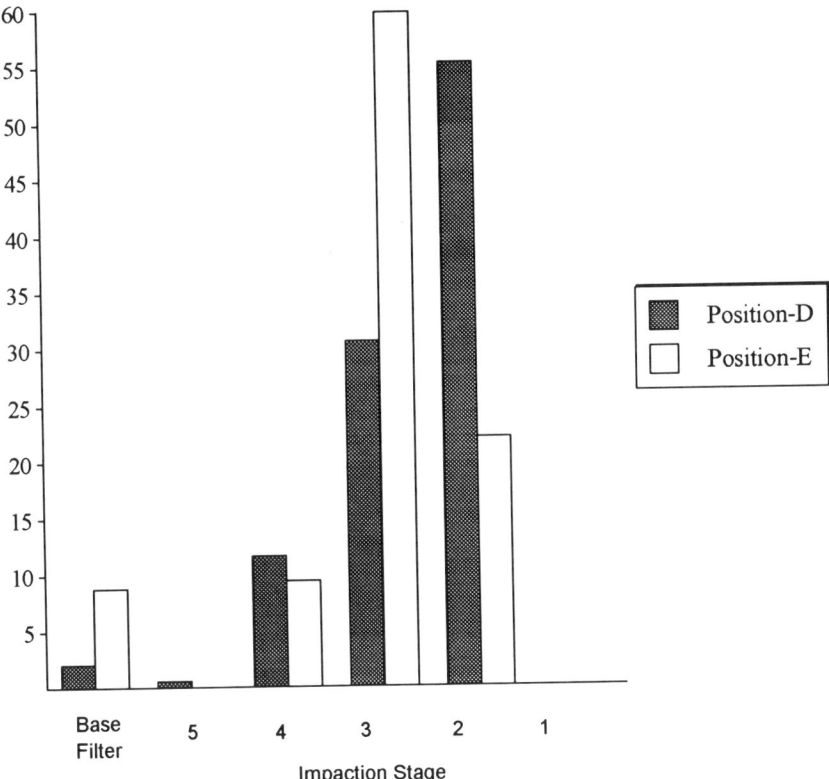

Fig. 3. Relative henna allergenic activity in different particle sized fractions. The results are expressed as a percentage of the total henna activity collected by the sampler. Henna allergen is mainly found in stages 2–4 corresponding to a particle size range of 1.5–11.5 μm.

test showed that there was no cross reactivity between these two allergens, and this material was used in the RAST inhibition assay to measure henna allergen in air sample extracts. Various studies have quantified the concentration of airborne antigen in occupational environments. Usually only one allergen is studied. In this study there were two allergens in the air sample extracts which were measured simultaneously. This enables us to monitor occupational hygiene measures which aimed to control exposure of employees to sensitising agents.

Between our visits engineering controls and, probably, the substitution of henna that was uncontaminated by castor bean resulted in a fall in the concentration of henna and castor bean antigens. However the concentration of castor bean in henna was highly variable so the significance of a fall in airborne concentration of this material would benefit from a series of measurements using different henna batches. This was outside the scope of this study.

Exposure to henna was restricted to the vicinity of the henna processing area. High concentrations of airborne henna allergen corresponded to high dust concentrations. However, at lower henna concentrations the total airborne dust level was similar to that found in control areas. This exemplifies the need for specific allergen measurement to monitor the effectiveness of aerosol control measures.

Several studies have examined the size of particles associated with allergen in occupational environments, for example Wiley and Griffin (1994) and Gordon *et al.* (1991). Similarly we have demonstrated that henna allergen is associated with respirable sized particles and confirmed an inhalatory route of allergen exposure.

In conclusion we have shown that two separate, possibly variable, constituents of an airborne allergenic material can be measured simultaneously. We have shown that it is possible to reduce the concentration of airborne allergens in the work place. Future work should entail the health surveillance of employees to assess the efficacy of allergen control measures.

REFERENCES

Gordon, S., Tee, R. D., Lawson, D., Wallace and Newman-Taylor, A. J. (1991) Reduction of airborne allergenic urinary protein from laboratory rats. *Br. J. ind. Med.* **49**, 416–422.

Topping, M. D., Henderson, R. T. S., Luczynska, C. M. and Woodmass, A. (1982) Castor bean allergy among workers in the felt industry. *Allergy* **37**, 603–608.

Wiley, K. and Griffin, P. (1994) Particle size distribution of Norwegian lobster (scampi) aerosols in seafood processing factories. *Clin. Exp. Allergy* **24**, 983.

 Pergamon

Ann. occup. Hyg., Vol. 41, Supplement 1, pp. 647–652, 1997
British Occupational Hygiene Society
Crown Copyright © 1997 Published by Elsevier Science Ltd
Printed in Great Britain
0003–4878/97 $17.00 + 0.00
Inhaled Particles VIII

PII: S0003–4878(96)00136–6

ASSESSMENT OF THE SUITABILITY OF DIFFERENT SUBSTRATE MATERIALS FOR BIOAEROSOL SAMPLING

B. Crook,* L. C. Kenny,* S. Stagg,* J. D. Stancliffe,* S. J. Futter,†
W. D. Griffiths† and I. W. Stewart†

*Health and Safety Laboratory, HSE, Broad Lane, Sheffield S3 7HQ, U.K.; and
†AEA Technology, Harwell, Oxfordshire OX11 0RA, U.K.

INTRODUCTION

Bioaerosols may comprise infectious, allergenic and microbial particles. A range of static bioaerosol sampling methods currently exist and are described in detail by Crook (1996). Most of these aim to collect the sample in a viable, culturable state so that an estimate of the bioaerosol concentration may be made by counting the number of colonies that develop on agar growth media. The majority of these methods are poorly characterised, either in terms of their physical sampling efficiency, or in terms of the recovery and survival rates of collected micro-organisms. Personal sampling for bioaerosols is generally carried out using standard aerosol samplers that collect the aerosol onto a filter. The dehydration effects caused by this method can greatly reduce the viability of the more delicate micro-organisms, especially bacteria. This is not problematic where the analysis of bioaerosol concentrations is carried out on both living and dead cells using direct counting by microscopy, biochemical or molecular detection methods. In some instances however, culturing is necessary, either to aid species identification (in the case of allergenic organisms) or to demonstrate the viability of pathogens. Existing samplers therefore have a number of disadvantages when used to assess health risks to workers arising from exposure to airborne micro-organisms. A rational, scientifically-based scheme for assessing bioaerosol exposures requires a personal sampler with the following characteristics:

- provides good rates of recovery for culturing, DNA extraction, and species identification;
- separates the bioaerosol into inhalable, thoracic and respirable size fractions according to agreed international sampling conventions, e.g. CEN/ISO;
- may be mounted in the breathing zone and be ergonomically acceptable to the wearer;
- may be operated for sampling times similar to the duration of work activities.

As a first stage in developing suitable samplers, a number of potential bioaerosol collection substrates have been selected and assessed. These substrates have since been subjected to rigorous tests for bioaerosol recovery, the results of which are reported separately in this conference. The experiments show that a number of promising solutions can be propsed for personal sampling of bioaerosols, and these will be discussed in relation to our ideal design criteria.

MATERIALS AND METHODS

Substrates considered

Size-fractionating aerosol samplers can be constructed utilising the mechanisms of inertial separation (i.e. impactors, cyclones), sedimentation (elutriators) or combinations of both mechanisms (e.g. porous foams, porous bead beds). A high efficiency selector such as a filter is needed to collect the finest aerosol fraction. Impactors have the disadvantage of high collection stresses and elutriators generally cannot be constructed on a scale suitable for personal sampling. Our initial selection of potential bioaerosol substrates included therefore porous polyurethane foams, porous beds composed of packed glass beads, a prototype miniature cyclone and conventional filters. The foams, glass beads and cyclone were tested with their surfaces uncoated and, also, coated with either gelatine or agarose.

Spike tests for bacterial recovery

Preliminary spike tests were carried out to assess the substrates' ability to maintain the viability of the collection micro-organisms. Polycarbonate, PTFE and Cellulose acetate membrane filters were tested, housed in plastic IOM personal aerosol samplers. Porous polyurethane foam plugs of nominal porosity 30 ppi (pores per inch) were housed in cowled asbestos sampling heads; both uncoated foams and foams coated with gelatine were tested. Glass beads of 1 mm diameter were sterilised, packed into a plastic 25 mm Millipore filter cassette and coated with 2% w/v agarose before testing. The biocyclone was prepared for testing by coating its internal conical surface with 2% w/v agarose gel. A 30 ppi foam plug soaked in 1/4 strength Ringers solution was placed in the cyclone "grit pot" in order to increase the humidity of the sampler chamber and covered using a PTFE filter to provide a smooth, wetted surface.

Bacteria in a liquid suspension were inoculated directly onto each substrate. Tests were repeated using three test organisms (*Bacillus subtilis*, *Pseudomonas alcaligenes* and *Escherichia coli*) chosen to represent a range of robustness, in the form of overnight, stationary phase cultures prepared in nutrient broth shaken at 35°C. The filters were inoculated with 0.5 ml of the bacterial suspensions, the foams and beads with 0.1 ml of the bacterial suspensions. The internal walls of the biocyclone were inoculated with 0.05 ml of bacterial suspension and the grit pot with 0.1 ml of bacterial suspension. Air was then drawn through the samplers at a flow rate of $2 \, l \, min^{-1}$ (except for the cyclone, operated at $1.9 \, l \, min^{-1}$) for a period of 60 min, to simulate the dehydrating effects of sampling. The bacteria were washed from each substrate and placed in PIT (peptone inositol tween), serially diluted in 1/4 strength Ringers and inoculated onto nutrient agar. The percentage recovery of the viable bacteria from each substrate was calculated by comparing it to control substrates.

Preparation of aerosol test samplers

Following the spike tests for recovery, experiments to test the recovery of aerosolised bacteria were planned (reported separately in this conference). Three types of micro-organism were selected for the aerosol tests, *Escherichia coli*, *Saccharomyces cerevisiae* and *Penicillium expansum*. These organisms were expected to provide a range of robustness and, therefore, offer a good test of the survival of organisms collected from the airborne state. Since the *E. coli* bacteria have smaller aerodynamic diameters than either *S. cerevisiae* or *P. expansum*, it was necessary to optimise the configuration of substrates for these aerosol tests, to ensure good retention rates in the test samplers. Membrane filters could obviously be used for sampling any of the three organisms and were housed in 25 mm, open-face holders (Gelman). In the case of the foam and bead substrates, different configurations were needed for the different sized organisms. For the larger organisms (*S. cerevisiae* and *P. expansum*), two foam plugs each of diameter 30 mm and depth 25 mm were placed in series in a plastic tube. These were used either coated or uncoated with 1% w/v agarose. For greater retention of the smaller *E. coli* aerosol, a foam plug 6 mm diameter and 100 mm long was placed in a plastic tube, and either coated or uncoated with 0.5% w/v agarose. For sampling *S. cerevisiae* and *P. expansum*, glass beads of diameter 0.5 mm were housed in a 25 mm Millipore cassette, with bed depth 20 mm and either coated or uncoated with 1% w/v agarose. For greater retention of the *E. coli* aerosol, glass beads of diameter 1.0 mm were packed into a plastic tube of internal diameter 6 mm and length 100 mm and either coated or uncoated with 0.5% w/v agarose. In all cases the pressure drop across the test samplers was checked to ensure that reasonable flow rates could be obtained. The retention characteristics of the biocyclone can be selected by choosing an appropriate flow rate, however it was anticipated that flow rates high enough to ensure the retention of the *E. coli* aerosol would produce excessive cell damage. For sampling *S. cerevisiae* and *P. expansum*, a sterilised foam plug (30 ppi) was situated in the "grit-pot" of the biocyclone, wetted with 1 ml of distilled water and covered using a sterilised cellulose acetate filter. Procedures for assembly, coating, disassembly and recovery of the collected aerosol were developed and documented for each test sampler.

Aerosol penetration tests

Initial aerosol penetration tests were undertaken with the larger foam and bead bed substrates and with the biocyclone, in order to establish their collection efficiency at the selected flow rates. This was to check that the sampler and substrate configurations used for the bioaerosol tests would have a sufficiently high collection efficiency for the chosen test organisms. The aerosol penetration through the substrates was measured with an aerodynamic particle sizer, TSI APS 33B, using a similar test system to that described by Maynard and Kenny (1995). Tests were carried out with the substrate inlets pointing downwards in the aerosol chamber, their outlets being connected to the APS via a transfer tube with two 90 degree bends. For the foam substrates, which are known to be subject to orientation effects, tests were repeated with the inlets pointing upwards in the chamber. The foam and bead bed samplers were tested at a flow rate of 2 l min^{-1} and the biocyclone was tested at two flow rates 3 and 4 l min^{-1}. Two specimens of each substrate type were tested.

The penetration through each substrate was measured by taking five 40 s samples

Table 1. Summary of the recovery of viable bacteria in spike tests

| Collection substrate | Micro-organisms (% recovered compared to control) | | |
	B. subtilis	*Ps. alcaligenes*	*E. coli*
Membrane filters			
Polycarbonate	13.57	66.48	0.00
PTFE	0.80	5.48	8.90
Cellulose acetate	0.47	9.81	0.00
Porous polyurethane foams			
30 ppi, gelatine coated	78.32	71.95	81.83
30 ppi, not gelatine coated	61.59	44.71	88.83
Glass bead beds			
Agarose coated	40.60	0.01	39.60
Not agarose coated	0.22	0.00	0.49

of the polydisperse aerosol, alternately with and without the substrate in the sampling line to the APS. The APS data were stored and the penetrations subsequently calculated using the APSs small particle processor (SPP) data, with the calibration corrected to allow for particle density effects. Counts from the SPP for aerodynamic diameters between 1 and 9 μm were placed into 33 size intervals, and the penetration in each interval obtained by comparing the average counts for the background aerosol in the chamber, with average counts sampled through the substrate. The resulting penetration data were analysed using curve-fitting software in order to calculate the D_{50} and other relevant parameters.

It was not possible to utilise this method to test the retention characteristics of the long thin foam and bead samplers, as the size selection curves in this case would be well below the lower limit of the range of this experimental technique (1–9 μm).

RESULTS AND DISCUSSION

The results of the initial spike tests for micro-organism viability are summarised in Table 1. Except in the case of polycarbonate filters, the collection of bioaerosols by filtration appears to greatly reduce viability and culturability. *E. coli* proved difficult to maintain in a viable state on all but the porous foam substrate and in many cases failed to be detected by culturing. The use of substrate coatings to mitigate the effects of dehydration on bioaerosol viability showed some promise, especially when used with the glass bead bed, although this needs further investigation. In general the membrane filters performed less well than any of the substrates tested whereas the polyurethane porous foams showed the most consistently high viable microbial recovery. The structure and filtration properties of porous foams have been described in detail by Wake and Brown (1991). The foam consists of a matrix of bubbles, pierced at their points of contact, thus leaving a three-dimensional network of short connected elements. This provides a large surface area per unit volume for aerosol capture whilst affording protection to the collected bioaerosol from dehydration by diffusing the air flow through the open cell structure.

Preliminary toxicity tests conducted with the nickel-plated aluminium biocyclone sampler showed that it had no harmful effect on the test bacteria. Recovery of viable bacteria from the agarose coated sampler barrel and "grit pot" filter yielded

Penetration

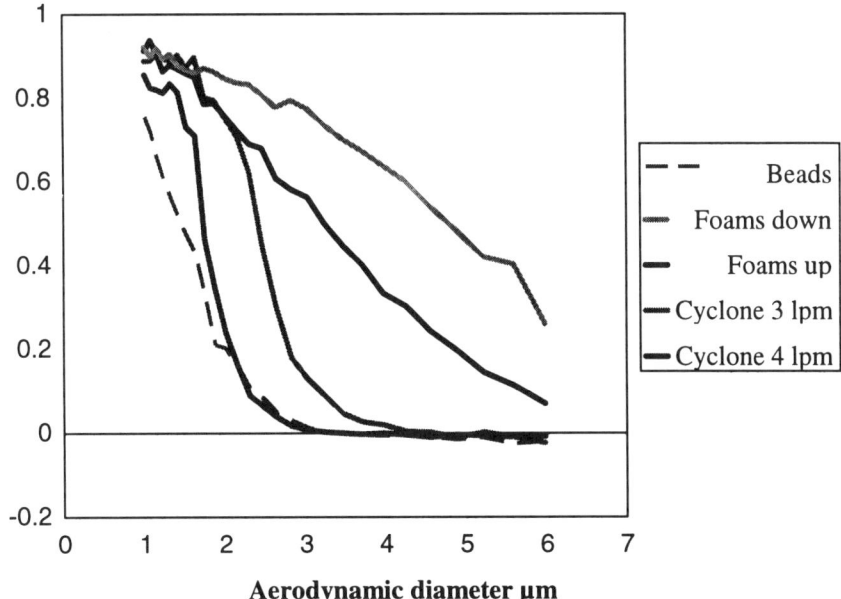

Fig. 1. Aerosol penetration through test substrates.

10.30 and 4.00%, respectively. However, inoculation of the barrel and filter proved difficult and these results may not be representative of the performance of the sampler.

The results of the aerosol penetration measurements on the substrate configurations chosen for bioaerosol testing are shown in Fig. 1. The bead bed showed the greatest collection efficiency (i.e. lowest penetration), retaining almost all particles with aerodynamic diameters greater than 2 μm. At a flow rate of 3 l min^{-1} the biocyclone retained particles with aerodynamic diameters greater than 3 μm. The foam retention depended on the orientation of the flow, with particle retention in the foam being greater when the foam inlet pointed upwards (i.e. flow in the same direction as gravity). In view of these results, a flow rate of 3 l min^{-1} was chosen for the biocyclone, and the upward pointing orientation chosen for the foams. This ensured adequate collection efficiencies in all cases for particles with aerodynamic diameters in excess of 3–4 μm.

CONCLUSIONS

A number of promising solutions to our ideal design criteria have emerged from these initial experiments to assess bioaerosol collection substrates. Of these, the use of porous polyurethane foams as collection substrates for bioaerosols appears to be the most promising. An empirical model for aerosol penetration through porous foams, developed by Vincent *et al.* (1993), facilitates the selection of suitable foam sizes, porosities and flow rates for size-selective sampling according to the CEN/ISO international sampling conventions for aerosols. Using this model, one can arrive at solutions that will allow two foam plugs with the correct size selective properties to be placed in existing personal inhalable samplers. In this way it is possible to construct a personal bioaerosol sampler that will not only size-select the bioaerosol according to inhalable, thoracic and respirable size fractions but will also provide a very economical solution to bioaerosol sampling.

The performance of glass bead beds has also been encouraging, especially since appropriate beads could be chosen to be compatible for the cell disruption method of "bead beating", a technique used in molecular microbiology to release DNA as a precursor to PCR and gene probe analysis. However, there as yet exists no model with which to predict the penetration of bioaerosols through bead beds of different bead diameter and bed depth and diameter at different flow rates. The preparation of uniform, reproducible bead beds having known size-selective characteristics has yet to be solved.

The use of long thin foams or bead beds as an alternative to filters may have some useful applications in workplace aerosol sampling. These collection elements can be included downstream of a conventional personal sampler, in the tubing between the sampler and pump. Future work will concentrate on the further examination of polydisperse aerosol penetration through the prototype porous foam plugs selected as inserts for existing personal inhalable samplers, the techniques for which have been detailed above.

REFERENCES

Crook, B. (1996) Review: methods of monitoring for process micro-organisms in biotechnology. *Ann. occup. Hyg.* **40**(3), 245–260.
Maynard, A. D. and Kenny, L. C. (1995) Performance assessment of three personal cyclone models using an aerodynamic particle sizer. *J. Aerosol Sci.* **26**(4), 671–684.
Vincent, J. H., Aitken, R. A. and Mark, D. (1993) Porous plastic foam filtration media: penetration characteristics and applications in particle size-selective sampling. *J. Aerosol Sci.* **24**(7), 929.
Wake, D. and Brown, R. C. (1991) Filtration of monodisperse aerosols and polydisperse dusts by porous foam filters. *J. Aerosol Sci.* **22**(6), 693–706.

Pergamon

Ann. occup. Hyg., Vol. 41, Supplement 1, pp. 653–658, 1997
British Occupational Hygiene Society
Crown Copyright © 1997 Published by Elsevier Science Ltd
Printed in Great Britain
0003–4878/97 $17.00 + 0.00
Inhaled Particles VIII

PII: S0003–4878(96)00137–8

AN ELECTRET-BASED PASSIVE SAMPLER USED FOR SAMPLING AIRBORNE PIGMENT DUST, RUBBER FUME AND FLOUR DUST

M. A. Hemingway,* I. Strudley,† R. C. Brown,* S. Froude* and M. M. Smith*

*Health and Safety Laboratory, Broad Lane, Sheffield S3 7HQ, U.K.; and †Health and Safety Executive, Stanley Precinct, Bootle L20 3QZ, U.K.

INTRODUCTION

Investigations have been carried out on the capture of airborne dust by an electret (a permanently charged polymer sheet), in a passive dust sampler that is simple and light-weight and, therefore, acceptable to the wearer. The device collects dust particles by electrostatic attraction. The passive sampler comprises an electret fixed to a conducting base and covered by a grid with a solid top and perforated sides. Charged particles inside the cage are attracted to the electret at a drift velocity that is proportional to the surface potential on the electret and to the electrical mobility of the particles. The grid provides physical protection for the electret and fixes the electrical field by which capture is effected. A diagram of the passive sampler is shown in Fig. 1. The electret is a polypropylene disc of 25 mm diameter, chosen to be the same shape and size as standard sampling filters and is held to the sampler base by its own electrostatic attraction.

In previous papers a passive dust sampler was subjected to preliminary laboratory investigations (Brown *et al.*, 1994) and to field trials in metal-processing industries (Brown *et al.*, 1995). The sampler was shown to perform as theory would predict and to give good correlations with measurements taken with conventional samplers. Measurements made with the passive sampler at a variety of industrial locations on a number of different occasions are presented and are compared to those made with conventional samplers.

OPERATION OF THE PASSIVE SAMPLER

The passive sampler was used to calculate the aerosol concentration, c, from:

$$c = k \times \frac{m}{Vt} \tag{1}$$

where the mass of dust collected, m, the average surface voltage, V and the exposure time of the sampler, t, were measured for each sampler. The calibration constant, k, depended upon the dimensions of the passive sampler and the average electrical mobility of the aerosol and was found by calibrating the passive sampler results with the results from conventional samplers using a linear regression constrained to pass through zero.

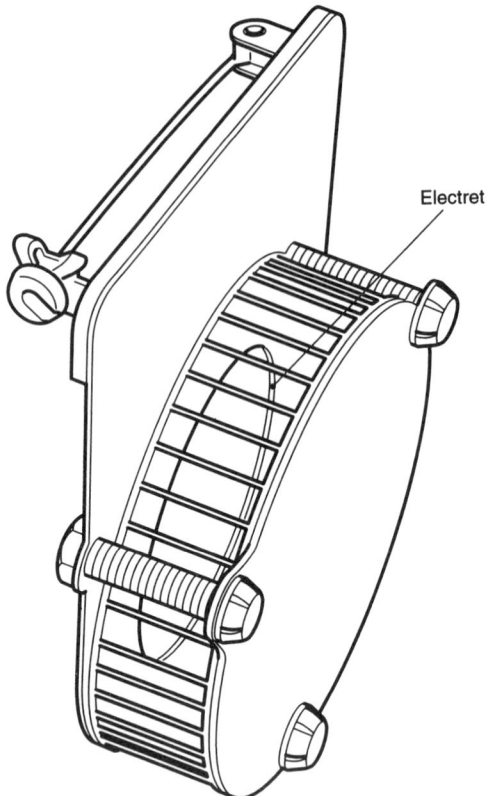

Electret

Fig. 1. Prototype passive sampler.

FIELD TRIALS OF THE PASSIVE SAMPLER

Field trials have been carried out at seven factory sites from the pigment industry of which five were visited three times; two factory sites from the rubber industry, each of which was visited twice and a bakery which was also visited twice. The factory sites from the pigment industry produced water and solvent based paints, wood varnishes and dyes for large volume plastics (for example PVC) and sampling was usually carried out in the powder weighing and mixing areas. Sampling at the factory sites from the rubber industry took place in the powder weighing and mixing areas, the rheometry laboratories and the processing belts. The bakery manufactured bread goods and sampling took place in the weighing and sieving area and along the bread line.

The electrets were weighed and then charged by a corona to a potential of approximately + 2200 V. The surface charge was allowed to settle for a period of 4 days after which their average surface potential was measured. On each visit volunteers had a conventional sampler and a passive sampler attached to the preferred side of the chest. In approximately a third of the factory sites visited from the pigment industry an additional sampler was also attached to the opposite side of the wearer's chest to obtain further data. After sampling the electrets' potential was

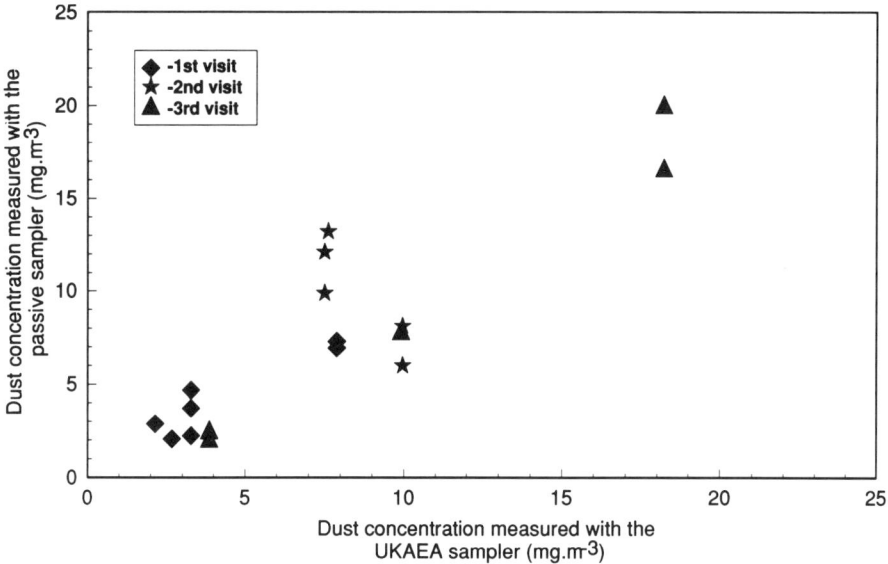

Fig. 2. Dust concentrations measured with the passive sampler and the UKAEA sampler at pigment factory site 2.

measured again, and they were reweighed. The samples taken at the bakery were also analysed for fungal alpha amylase.

The potential usefulness or otherwise of the passive sampler is shown by comparison of measurements of dust concentration made using it and those made on the same occasion with conventional dust samplers. It must, however, be borne in mind that conventional samplers are subject to error and differences between the results should not necessarily be ascribed to problems with the passive sampler. The results are plotted in Figs 2–5 and are summarised in Table 1.

The correlation coefficients, r^2, for multiple visits to single factory sites can be good. For example Fig. 2 shows $r^2 = 0.788$ for three visits to the same pigment based factory site. Indeed, $r^2 = 0.548$ for all the visits to all seven pigment based factory sites which suggests that the passive sampler can have a calibrated function which remains valid for an entire industry. However, this is not true for the rubber based factory sites visited where a good r^2 is found for visits to single factory sites but r^2 is not so good for multiple visits to a number of different factory sites. For example r^2 is 0.971 for the results from the second visit to factory site 2 of the rubber aerosols but falls to 0.151 if the results from all the visits to both factory sites are included. Figure 3 shows this is because each visit to factory site 2 has a separate calibration relationship.

The correlation of results of immunoassay for fungal alpha amylase on the bakery samplers are poor ($r^2 = 0.419$) owing to one extraneous point (Fig. 4). It is possible to analyse samples taken from the rubber industry for soluble rubber fume content, however, the amount of soluble rubber fume deposited on the passive samplers was too small for proper analysis.

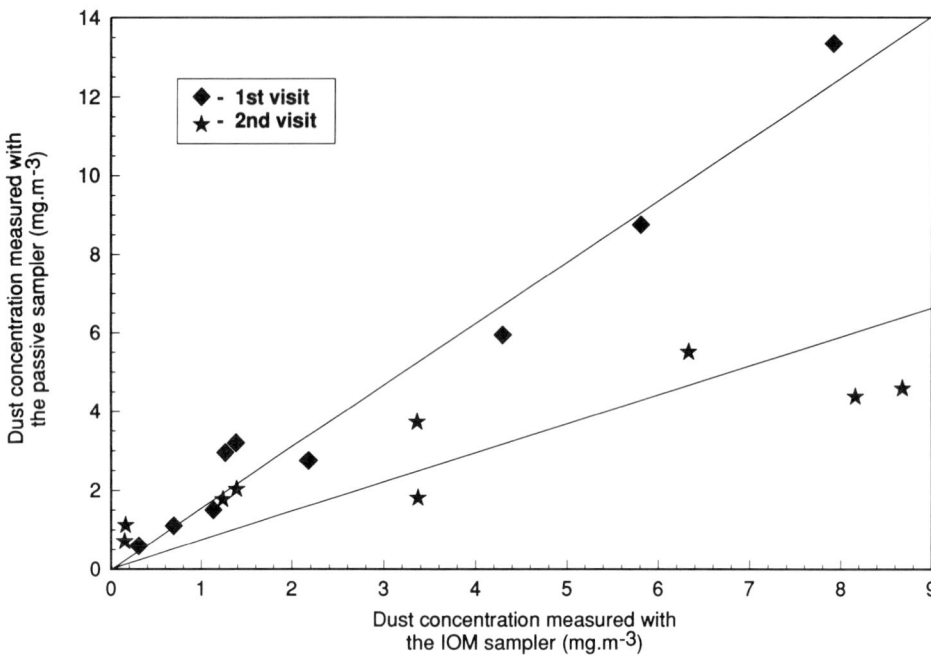

Fig. 3. Dust concentrations measured with the passive sampler and the IOM sampler at rubber factory site 2.

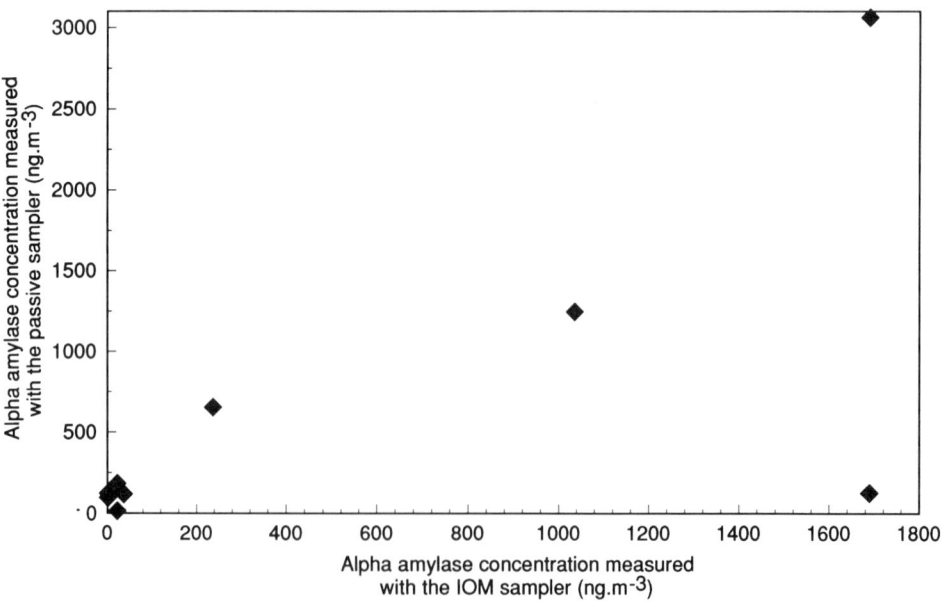

Fig. 4. Alpha amylase concentratons measured with the passive sampler and the IOM sampler at a bakery.

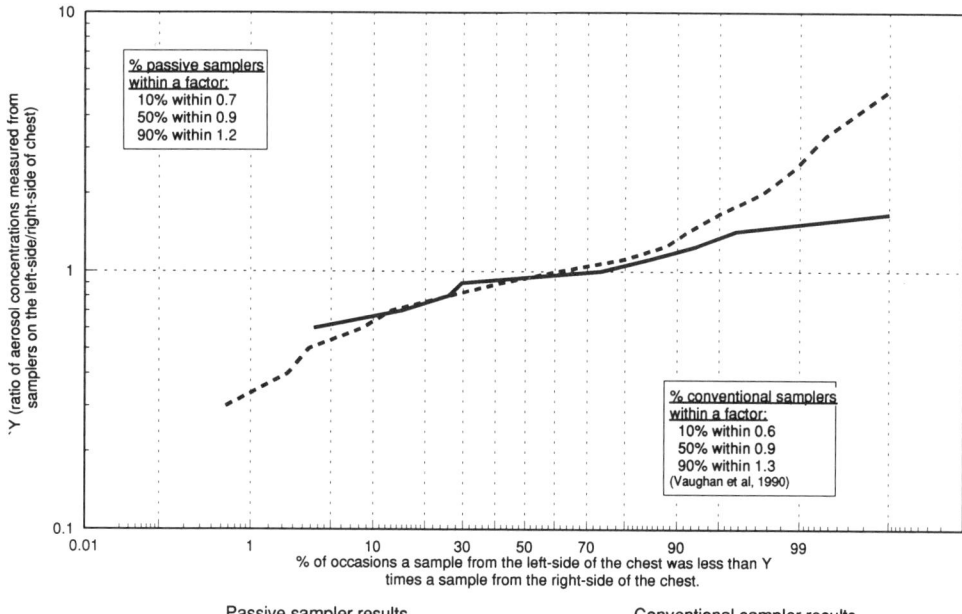

Fig. 5. Cumulative distribution of sampler aerosol concentration ratios for simultaneous passive sampling from opposite sides of the chest.

Table 1. Statistical analysis of the passive sampler results

Aerosol type	Location	Number of visits to each site	Calibration constant, k	Number of samples	Correlation coefficient, r^2	Comments
Flour dust	site 1	2	1785	13	0.904	gravimetric results
	site 1	2	8949	9	0.419	α-amylase results, see Fig. 4
Pigment dust	site 2 only	3	2286	7	0.788	see Fig. 2
	all 7 sites	3	2286	118	0.548	calibrate with site 2 results
		3		27	0.841	comparison of paired passive samplers. See Fig. 5
Rubber fume and dust	site 2 only	1	2969	9	0.971	see Fig. 3
	site 2 only	2	4723	18	0.524	see Fig. 3
	sites 1 and 2	2	2913	39	0.151	calibrated with site 2 results
	site 3*	1		32	0.773	7-hole sampler results compared to those from paired IOM samplers

* Not one of the passive sampler visits but included as a comparison of conventional sampler measurements. Data from Vaughan *et al.* (1990).

PAIRED PASSIVE SAMPLER MEASUREMENTS

A measure of sampler consistency can be obtained from the overall agreement between passive sampler readings that were taken on the same volunteer but on opposite sides of the chest and can be found from the distribution of the ratios of the concentrations measured by paired samplers. This is shown for the pigment based aerosol results in Fig. 5 in which a horizontal line at a ratio of unity would indicate perfect agreement for all the paired samplers tested and the further a line deviates from unity the worse the agreement becomes. The agreement for the passive sampler results is better than for conventional samplers (Vaughan *et al.*, 1990) which indicates that some of the difference between the two types of sampler may be attributed to the conventional samplers.

Figure 5 shows the difference between samplers mounted on the left-side and those mounted on the right-side of the chest. A ratio less than one indicates that a sampler on the left-side of the chest has sampled less than its paired sampler on the right-side. Deviations from a line passing through unity at 50% of occasions, shows an overall positional bias in sampling. Figure 5 shows that both the passive sampler and conventional sampler sample lower concentrations on the left-side of the chest.

CONCUSIONS

The electret-based passive sampler shows considerable promise in measurements of airborne dust concentrations. It has proved very acceptable to the wearers. Correlations between the passive sampler and conventional samplers have been good for single visits to single factory sites but tend to be less good for multiple visits or visits to more than one factory site. Further work will be aimed at improving the precision of the sampler and accounting for day to day variation in the electrical properties of dust.

REFERENCES

Brown, R. C., Wake, D., Thorpe, A., Hemingway, M. A. and Roff, M. W. (1994) Preliminary assessment of a device for passive sampling of airborne particulate. *Ann. occup. Hyg.* **38**, 303–318.
Brown, R. C., Hemingway, M. A., Wake, D. and Thompson, J. (1995) Field trials of an electret-based passive dust sampler in metal-processing industries. *Ann. occup. Hyg.* **39**, 603–622.
Vaughan, N. P., Chalmers, C. P. and Botham, R. A. (1990) Field comparison of personal samplers for inhalable dust. *Ann. occup. Hyg.* **34**, 553–573.

Ann. occup. Hyg., Vol. 41, Supplement 1, pp. 659–666, 1997
© 1997 British Occupational Hygiene Society
Published by Elsevier Science Ltd. All rights reserved
Printed in Great Britain
0003–4878/97 $17.00 + 0.00
Inhaled Particles VIII

PII: S0003–4878(96)00174–3

ACUTE EXPOSURE OF HUMANS TO OZONE IMPAIRS SMALL AIRWAY FUNCTION

W. M. Foster, G. G. Weinmann, E. Menkes and K. Macri

Department of Environmental Health Sciences, School of Hygiene and Public Health,
The Johns Hopkins University, Baltimore, MD 21205, U.S.A.

INTRODUCTION

Our approach has been to use noninvasive, regional analysis techniques and we have separated central from peripheral mucus membrane responses to ozone (Foster *et al.*, 1987) and characterised the effects of ozone on regional ventilation and particle dosimetry for the lung periphery (Foster *et al.*, 1993). In these prior studies we relied upon techniques that utilised radioisotopic markers to assess O_3-induced changes in regional lung function. The goal of the present study was to expand upon the earlier results using nonradioisotopic methods and multibreath washout of a resident lung gas (N_2) to measure nonhomogeneous lung function following exposure to O_3.

METHODS

Fifteen healthy male volunteers were recruited for the study. All were non-smokers and without history of lung disease nor receiving medications for any other disease. The subjects had a mean age of 25.4 ± 2 years ($\pm$ SD) and were free of respiratory infection at the time of evaluation. The forced vital capacity (FVC), forced expiratory volume at 1 s (FEV_1) and mid-maximal expiratory flow rate (FEF_{25-75}) of the group averaged $> 92 \pm 11\%$ of predicted (Crapo *et al.*, 1981). Informed consent was obtained from each individual; the study had the approval of the University's Committee on Research of Human Subjects.

Experimental protocol

The treatment plan (two treatments; crossover design) was selected to reduce within-subject variability and to facilitate comparisons between treatments to filtered air (FA) and O_3, with each subject used as his own control (FA). The order of treatments was randomized and the mean washout time between treatments was 8 weeks.

Methods of exposure of human subjects have been utilised previously (Foster *et al.*, 1993). Briefly, exposures lasted 130 min and were accomplished in a 13.6 m^3 chamber maintained at 19–23°C and 48–55% relative humidity. O_3 was generated from a 100% oxygen source by a high-frequency electric field (model G1-L, PCVI Ozone, West Caldwell, NJ, U.S.A.) and mixed with FA before being added to the chamber. Chamber O_3 was continuously monitored by an ultraviolet O_3 photometer (model 1003-AH, Dasibi, Glendale, CA, U.S.A.). Mean ($\pm$ SD) O_3 concentrations during exposures were 351 ± 6 ppb and 1 ± 1 ppb for O_3 and FA exposures, respectively. Air transport to the chamber was prefiltered and had a one-pass design with 24 changes of chamber air h^{-1}. During the initial 120 min of the exposure, the subjects alternated between 30 min periods of rest and treadmill exercise (Model Q55, Quinton Instrument, Seattle, WA, U.S.A.) and attained a cumulative ventilation per minute during exercise that was approximately ten times the volume of the FVC. The exposure ended with a final 10 min rest period. The treadmill speed and grade necessary to obtain the targeted minute ventilation were predetermined for each subject before commencement of the exposures. Exercise is frequently used during chamber exposures to increase minute ventilation to mimic an individual performing light activity under ambient conditions.

During exposures the subjects could freely choose between nasal, oral, oro-nasal breathing modes, except during assessment of minute ventilation when oral breathing and expiration into a dry gas meter was obligatory. Before and immediately following the exposures pulmonary function (at least three determinations of the FVC, FEV_1 and FEF_{25-75} by water-seal spirometer), the measurement of multibreath N_2 washout (Mauderly, 1977), and body plethysmographic measurement of thoracic gas volume (Vtg) were accomplished.

For the multibreath N_2 washout technique the subjects cleared N_2 gas from their lungs by mouth breathing 100% oxygen in a seated and erect position with a noseclip in place. Valving separated inspiratory and expiratory gases with samples of expiratory gas continuously analyzed for N_2 concentration using a mass spectrometer (Medical Gas Analyzer, Perkin–Elmer Inc., Pomona, CA, U.S.A.). Inspiratory and expiratory flow rates were measured with a heated pneumotachograph connected to the mouth piece and scaler tracings of flow were integrated to provide an instantaneous recording of tidal volume. During washout the tidal volume was presented visually to the subject on an oscilloscopic screen and simultaneous with an audio signal (for pacing the frequency of the respiratory cycle) to assist in the achievement of a targeted ventilatory pattern during the washout of N_2 from the lung. The natural logarithm of each end-tidal N_2 concentration was plotted against cumulative minute ventilation (from the start of the washout). These curves generally had two phases; an initial, steep slope followed by a more shallow slope. We defined the latter phase as the washout from 20 to 9% end-tidal N_2) concentration; flattening of the slope of this phase was used to identify nonuniform ventilation following exposure to O_3 of FA. A washout index, slope, was calculated from the best fit of the washout data to a curve generated by regression analysis.

Measures of pulmonary function were corrected to BTPS; and the trials with the highest sum of FVC and FEV_1 for both pre- and postexposure were utilised for statistical analysis.

RESULTS

The effects of O_3 on washout of lung N_2 are demonstrated in the plots presented in Fig. 1 of end-tidal % N_2 concentration v. cumulative expired volume measured during the washout. These data were acquired in a single subject pre- and

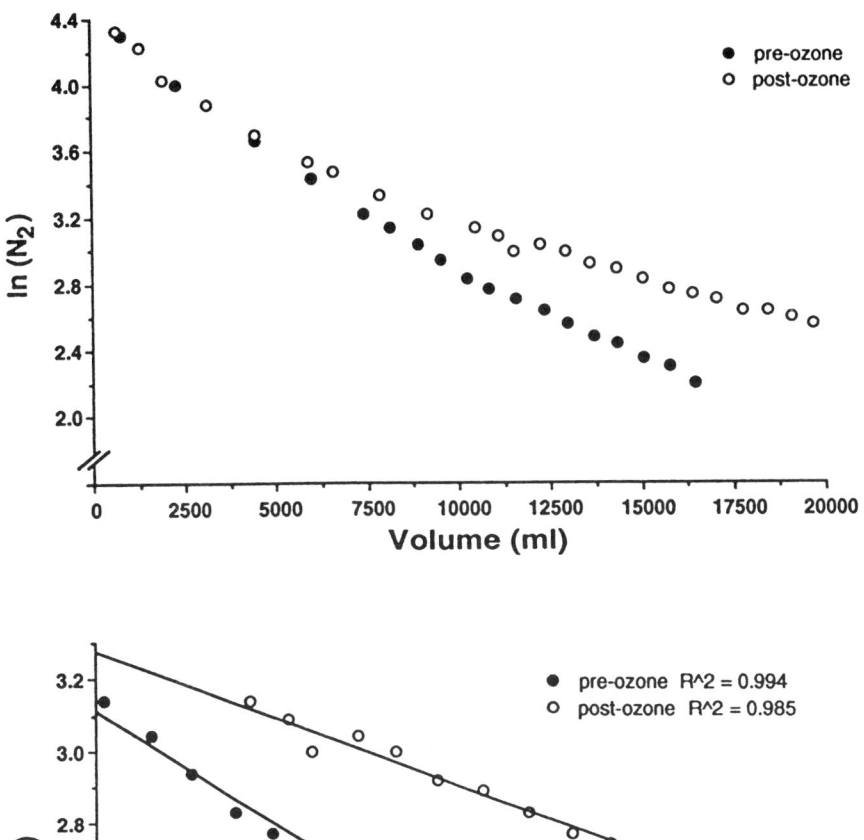

Fig. 1. Multibreath washout of lung N_2 breathing 100% O_2. Upper panel: relationship between endtidal concentration of N_2 (expressed as natural logarithm, 1n, of the % N_2 concentration) and the cumulative expired volume (ml) measured from the start of the washout; data are presented for a single subject measured pre- and post-exposure to O_3. Lower panel: data from washout of lung N_2 presented in the upper panel are fit to a linear regression (solid line, $\hat{R}2$ = coefficient) to characterize the slope of the curves during the later part of the washout between the expired lung N_2 concentrations of 25 and 9%.

post-exposure to O_3. The mean pre-post exposure, washout slopes for all 15 subjects are presented in Fig. 2 for exposures to FA and O_3. Washout slopes pre- and post-exposure to FA were similar; but the differences in washout pre-post exposure to O_3 were significant ($P < 0.05$) and represented a 24% decrease

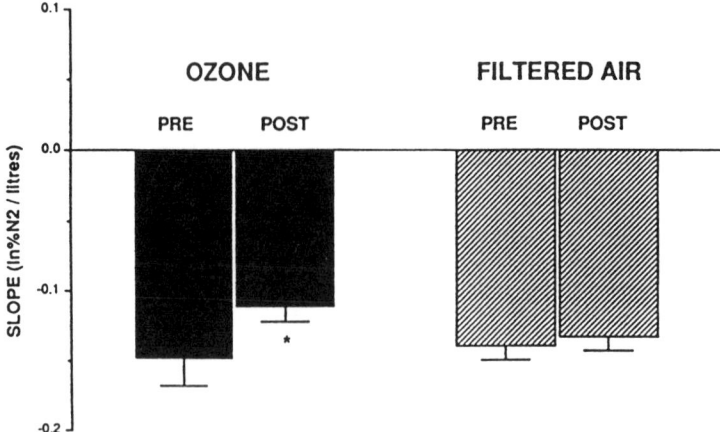

Fig. 2. Slope of lung N_2 washout pre- and post-exposure to filtered air and O_3. Mean slopes ($\pm$SE) for washouts between N_2 concentrations of 25 and 9% are presented for 15 subjects. *indicates significant as compared to mean preexposure value ($P < 0.05$).

(O_3-induced) in the mean slope of the N_2 washout, washout of lung nitrogen by breathing 100% O_2 was delayed by preexposure to O_3.

The corresponding measures of thoracic gas volume (Vtg) and minute ventilation acquired either prior to, and during, the washouts are presented in Fig. 3 and listed in Table 1, respectively. These factors were equivalent during the performance of the pre- and post-exposure washouts. Although influential to the calculation of the slope during the respective washouts, the compared values (Vtg and minute

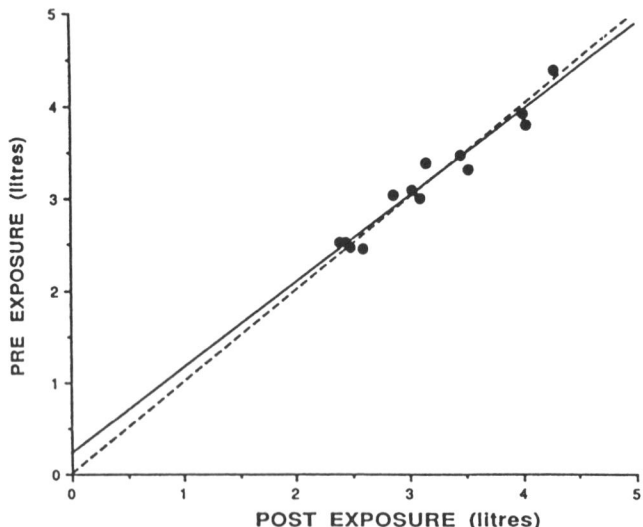

Fig. 3. Thoracic gas volume during lung washout pre- and post-exposure to O_3. Relationship between the pre- and post-exposure values of the thoracic gas volume, functional residual capacity, that were assessed in each subject at the time of the washout measurements. Data ($N = 15$) were fit to a linear regression (solid line) and for comparison is included line of identity (dashed line).

Table 1. Minute ventilation during washout of lung N_2*

Subject	Filter air exposure		Ozone exposure	
	Pre (1)	Post (1)	Pre (1)	Post (1)
1	10.20	14.86	11.37	10.26
2	11.41	15.20	11.87	13.85
3	20.93	20.61	17.00	19.50
4	11.48	10.33	8.76	9.99
5	8.95	9.31	8.52	9.95
6	12.83	12.31	9.35	9.33
7	13.30	12.85	8.61	9.38
8	11.11	11.19	8.02	8.65
9	11.54	12.60	17.68	12.86
10	8.99	9.06	12.51	11.74
11	10.72	11.06	13.11	12.18
12	11.81	11.68	8.56	11.02
13	10.61	11.10	10.38	11.06
14	11.46	11.79	9.69	8.83
15	11.53	11.98	15.45	16.56
Mean	11.79	12.40	11.39	11.68
± SE	0.72	0.73	0.82	0.78

*Average minute ventilation (1.) measured for each subject during respective washouts of lung N_2 by breathing 100% O_2.

volume) were not different and thus not factors in the effect of O_3 on the washout of lung N_2.

The effects of exposure to O_3 on spirometric indices of lung function are presented in Fig. 4. The mean values of the FVC and the FEV_1, pre- and post-exposure, are included and the influence of O_3 exposure on these indices, decrements post-exposure averaged 12 and 14%, respectively, and were significant ($P < 0.05$). The O_3-induced changes in FVC and FEV_1 (and the FEF_{25-75}, however

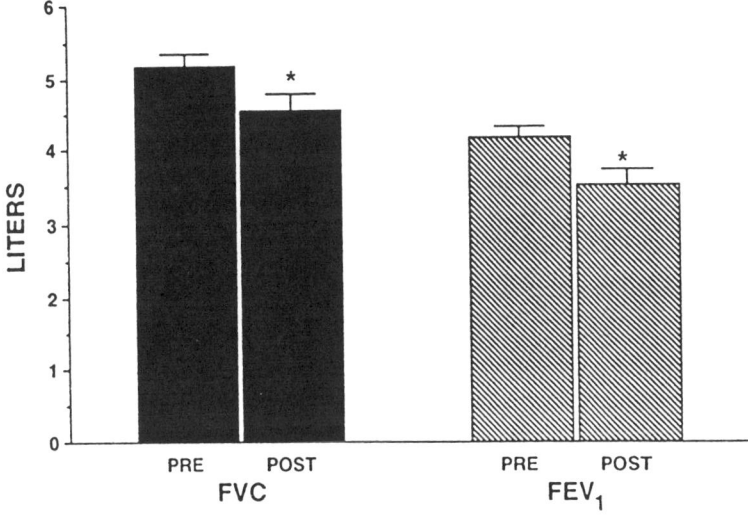

Fig. 4. Spirometric response to O_3. The mean (±SE) response data of the 15 subjects are presented for the forced vital capacity (FVC) and forced expiratory volume in 1 s (FEV_1). *indicates significantly different than mean preexposure value ($P < 0.05$).

these data are not presented) were not correlated to the observed O_3-induced changes in the slope of the N_2 washout. Although the response data have not been presented, changes in spirometric indices following exposure to FA were less than ± 2% of the pre-exposure values.

DISCUSSION

This study expands upon our prior demonstration that exposure to O_3 leads to an acute redistribution of regional ventilation (Foster *et al.*, 1993). We used a non-radioactive gas technique to evaluate the effects of O_3 on the washout kinetics of lung N_2 and small airway function. Following exposure to 350 ppb O_3, the latter phase of lung N_2 washout became prolonged. This response developed acutely after exposure to O_3. Delays in washout postexposure averaged 24% of the washout values observed preexposure and were not caused by the ventilatory pattern during the washout, nor changes in the volume of the space (Vtg) being ventilated. We suggest this response is attributed to O_3-induced alterations in bronchial tone and/or mucus secretions within smaller peripheral airways (Foster *et al.*, 1987, 1993), similar to nonuniform ventilation associated with bronchitis (Seaton and Ogilvie, 1978) and disease of the small airways (Ebert and Terracio, 1975; Wright *et al.*, 1984). The delay which occurred during the later phase of N_2 washout, did not correlate to the functional decrements observed in the FVC and FEV_1. Acute changes in lung function (FVC and FEV_1) are believed to be related to irritant and neural mechanisms (Hazucha *et al.*, 1989) within the larger airways of the lung, whereas the effect on N_2 washout we observed may represent injury, increased permeability and inflammation of more distal, smaller airways (Foster and Stetkiewicz, 1996).

Regional absorption of O_3 seems to impair small airway function and adversely affect ventilation to distal lung units, either separately or in combination with constriction of the larger bronchi. We have previously observed using radiolabelled gases that during the period immediately following exposure to O_3, lung regions with the highest ventilation per unit volume at baseline, have ventilation consistently reduced and redistributed to regions of lower ventilation (Foster *et al.*, 1993). We suggested that constriction of smooth muscle (Beckett *et al.*, 1985) and hypersecretion of mucus in peripheral airways (Foster *et al.*, 1987) within the lung regions receiving the highest local doses of O_3 could lead to uneven time constants in these airways and nonhomogeneous ventilation.

Of interest, was an additional observation that not all of the subjects exposed to O_3 exhibited delays in washout as an acute response; however when washout was remeasured 20–24 h postexposure, delays in washout of lung N_2 were found to be present. This demonstrated for a single subject in Fig. 5 and suggests that for some individuals, following oxidant exposure injury to the smaller airways may develop slowly and/or be related to the late onset of inflammatory changes known to occur with exposure to O_3 (Koren *et al.*, 1989; Foster and Stetkiewicz, 1996).

In summary the results suggest that in addition to decrements in spirometric function apparent after an acute exposure to O_3, tests of multibreath N_2 washout are also abnormal (delayed). The delays in washout are not related to changes in ventilatory pattern, nor alteration of lung volume at FRC. Changes in the washout

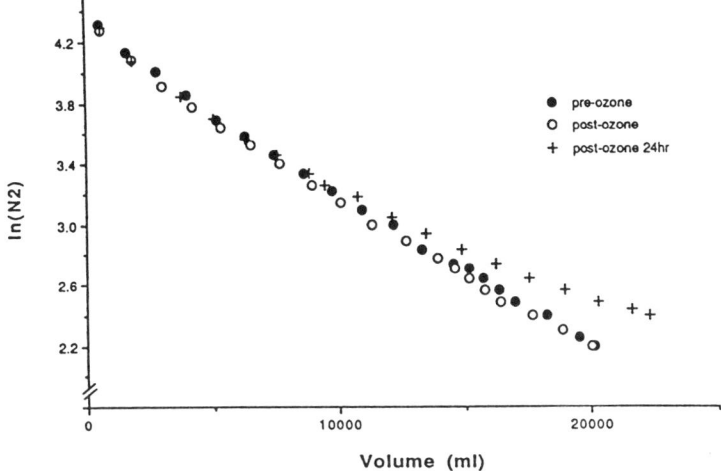

Fig. 5. Multibreath washout of lung N_2 breathing 100% O_2. Relationship between natural logarithm (1n) of endtidal % N_2 concentration and cumulative expired volume (ml) for a single subject. Washouts of lung N_2 measured pre-, immediately post- and 24 h post-exposure to O_3.

slope were most notable for the later portion of the washout and did not correlate to decreases observed in spirometric indices (FVC, FEV_1, FEF_{25-75}). Twelve of the 15 subjects were re-evaluated 24 h postexposure to O_3, and for half of these subjects the washout of lung N_2 was delayed in comparison to preexposure washout values (for two of these six subjects, the delay in washout developed at some time point after exposure and the 24 h postexposure time point).

Acknowledgments—A summary of the results was presented in part at the annual meeting of the American Thoracic Society held in San Francisco, CA, U.S.A. in 1993. The authors thank the Maryland Department of the Environment for daily reporting of ambient oxidant levels during the course of the investigation. This research was supported by awards from the National Heart, Lung and Blood Institute, #RO1-HL-31429 and National Institute for Environmental Health Sciences, #ES-03819, Washington, D.C., U.S.A..

REFERENCES

Beckett, W. S., McDonnell, W. F., Horstman, D. H. and House, D. E. (1985) Role of the parasympathetic nervous system in acute lung response to ozone. *J. appl. Physiol.* **59**, 1879–1885.

Crapo, R. O., Morris, A. H. and Gardner, R. M. (1981) Reference spirometric values using techniques and equipment that meet ATS recommendations. *Am. Rev. Resp. Dis.* **123**, 659–664.

Ebert, R. V. and Terracio, M. J. (1975) The bronchiolar epithelium in cigarette smokers. *Am. Rev. Resp. Dis.* **111**, 2–11.

Foster, W. M., Costa, D. L. and Langenback, E. G. (1987) Ozone exposure alters tracheobronchial mucociliary function in humans. *J. appl. Physiol.* **63**, 996–1002.

Foster, W. M., Silver, J. A. and Groth, M. L. (1993) Exposure to ozone alters regional function and particle dosimetry in the human lung. *J. appl. Physiol.* **75**, 1938–1945.

Foster, W. M., Jiang, L., Stetkiewicz, P. T. and Risby, T. H. (1996) Breath isoprene: temporal changes in respiratory output after exposure to ozone. *J. appl. Physiol.* **80**, 706–710.

Foster, W. M. and Stetkiewicz, P. T. (1996) Regional clearance of solute from the respiratory epithelia: 18–20 h postexposure to ozone. *J. appl. Physiol.* **81** 1143–1149).

Foster, W. M., Weinmann, G. G., Gerbase, M. W., Thomas, K. and Frank, R. (1993) Small airway function following acute exposure to 0.35 ppm ozone. *Am. Rev. Resp. Dis.* **147**, A641.

Hazucha, M. J. Bates, D. V. and Bromberg, P. A. (1989) Mechanism of action of ozone on human lung. *J. appl. Physiol.* **67**, 1535–1541.

Koren, H. S., Devlin, R. W., Graham, D. E., Mann, R., McGee, M., Horstman, D., Becker, S., McDonnell, W. F. and Bromberg, P. A. (1989) Ozone-induced inflammation in the lower airways of human subjects. *Am. Rev. Resp. Dis.* **139**, 407–415.

Mauderly, J. L. (1977) A new technique for evaluating nitrogen washout efficiency. *Am. J. Vet. Res.* **38**, 69–74.

Seaton, D. and Ogilvie, C. M. (1978) Regional lung function in asymptomatic cigarette smokers. *Am. Rev. Resp. Dis.* **118**, 265–270.

Wright, J. L., Lawson, L. W., Pare, P. D., Wiggs, B. G., Kennedy, S. and Hogg., J. C. (1984) Morphology of peripheral airways in current smokers and exsmokers. *Am. Rev. Resp. Dis.* **127**, 474–477.

Pergamon

Ann. occup. Hyg., Vol. 41, Supplement 1, pp. 667–672, 1997
© 1997 British Occupational Hygiene Society
Published by Elsevier Science Ltd. All rights reserved
Printed in Great Britain
0003–4878/97 $17.00 + 0.00
Inhaled Particles VIII

PII: S0003–4878(96)00138–X

BIOMONITORING OF A BIO-ACTIVE PARTICULATE: BLOOD PRESSURE SURVEILLANCE OF SODIUM AZIDE

S. H. Lamm,*† H. E. Rippen,* P. G. Nicoll,‡ L. Cummings,§
G. Howearth§ and D. Thayer§

*Consultants in Epidemiology and Occupational Health, Inc., 2428 Wisconsin Avenue, N.W.,
Washington, D.C. 20007, U.S.A.; ‡Arthur D. Little of Canada Ltd, Toronto, Ontario, Canada; and
§American Azide Corporation, Cedar City, Utah, U.S.A.

INTRODUCTION

Sodium azide (NaN_3) is a fine white crystalline powder that acts as the active agent for nitrogen generation in passenger car air bags. It is also used in medical laboratories as an antimicrobial agent and in the 1950s was used as a three-times-a-day therapeutic agent for hypertension, because of its vasodilatory action. Control of exposure during its manufacture is difficult. An occupational health and exposure surveillance program has been developed at the only U.S. sodium azide manufacturing plant in order to identify and assess health effects from chemical exposure (particularly sodium azide and hydrazoic acid) and to contribute to the exposure minimisation program of the plant (Rippen *et al.*, 1996). This program includes both air sampling and bio-monitoring surveillance. The bio-monitoring surveillance is based on the hypotensive effect of sodium azide exposure and examines changes in employee blood pressure measurements across the work shift. These analyses provide guidance for engineering, process and work practice changes in order to secure a safe working environment.

METHODS AND MATERIALS

Study sites and populations

This study has been conducted in the sodium azide manufacturing plant of American Azide Company in Cedar City, Utah. The plant began operations in 1993. Preliminary data were collected in 1994 and the surveillance program was established in 1995. In 1996, a comparison study was initiated at a nearby sister plant that manufactures ammonium perchlorate (a non-vasoactive substance) in order to provide a comparison occupational population. Participation in the bio-monitoring surveillance program became a routine procedure in the employment. Operating, mechanical and maintenance personnel were included. Crews worked in 12 h shifts with operating personnel assigned to processing areas for half the shift and to the control room for half the shift. The employee turnover rate was approximately 4% a year.

---

† To whom correspondence should be sent.

Industrial process and hygiene program

Sodium azide is manufactured in a two step process. Sodium is reacted with ammonia to produce sodium amide ($Na + NH_3 \rightarrow NaNH_2 + \frac{1}{2} H_2$), which is then reacted with nitrous oxide to yield sodium azide ($2 NaNH_2 + N_2O \rightarrow NaN_3 + NaOH + NH_3$). The three main processing areas are the reactor, purification, and packaging operations. Other work areas include the control room, warehousing and maintenance shop areas. Industrial hygiene air sampling has been conducted in each work area. The National Institute for Occupational Safety and Health (NIOSH) has recommended and the U.S. Occupational Safety and Health Administration (OSHA) had proposed, that azide levels not exceed 0.3 mg m^{-3}.

Bio-monitoring program

Employees in this plant take their own blood pressure measurements at the beginning and at the end of their 12 h shifts using automatic blood pressure instruments (Marshall Model 92 Digital Blood Pressure Monitor, OMRON Health Care, Inc.) The employee completes a daily report which includes their blood pressure measurements, any health incidents (such as a headache) and their sequence of job activities and location by hour throughout the shift. Changes in blood pressure measurements across the shift have served in this bio-monitoring program as an indicator of exposure. There are no adequate urinary or blood tests available to assess azide exposure (NIOSH, 1995).

Blood pressure analyses

The current analytic study data set consists of three subsets of employee daily blood pressure surveillance records: sodium azide plant (AZ) for 1995, sodium azide plant for the first two quarters of 1996 and ammonium perchlorate plant (AP) for the first two quarters of 1996. Blood pressure is usually represented as {systolic blood pressure/diastolic blood pressure} and can be summarized as a mean arterial blood pressure (MAP), derived as the {([systolic blood pressure] + [2 × diastolic blood pressure]) / 3}. Blood pressure changes across the shift can then be represented by changes in the mean arterial blood pressure (dMAP), i.e. post-shift MAP minus pre-shift MAP, producing a continuous variable for reporting the blood pressure changes in the population. Alternatively, a significant blood pressure drop may be defined as a drop equal to or exceeding 20 mm Hg in the systolic blood pressure and 10 mm Hg drop in the diastolic pressure, producing a dichotomous variable for reporting the blood pressure changes in the population. This would be equivalent to a drop in the mean arterial blood pressure (dMAP) of 13.3.

The frequency distributions of cross-shift changes in mean arterial blood pressure (dMAP) have been observed for the study and comparison populations. The average dMAP by month has been determined and compared for the study and comparison populations. For the 1995 study population, the average dMAP by month and by process area have also been determined. Odds ratio calculations for significant blood pressure drops have been determined by process areas. Results have been compared and contrasted with contemporaneous measurements of airborne particulate and vapor exposures obtained using standard industrial

Table 1. Cross-shift changes in blood pressures by plant and period

Observation	Population		
	1995 AZ (12 months)	1996 AZ (6 months)	1996 AP (6 months)
Participants	72	69	22
Matched blood pressure measurements	5654	3639	814
Cross-shift mean arterial blood pressure change (mm Hg)	−2.4	−3.3	−1.5
Number of significant blood pressure drops	272	208	33
Frequency of significant blood pressure drops (%)	4.8	5.7	4.1

hygiene methodology. Temporal trends have been observed. Analyses have been conducted to assess job–site specific associations for blood pressure changes.

RESULTS

Blood pressure changes by plant

The change in the mean arterial blood pressure (dMAP) across the work shift was calculated for each of the pairs of blood pressure measurements obtained on the study and comparison populations (Table 1). The dMAP measurements showed a normal distribution (Fig. 1) with a mean value of a few mm Hg drop across the work shift. The mean drop in the cross-shift arterial blood pressure was greater for the study population than for the comparison population. Graphic analysis of the mean dMAP by month by plant similarly shows the difference between the two groups (Fig. 2). The frequency of significant blood pressure drops among the study population and the comparison population were not significantly different.

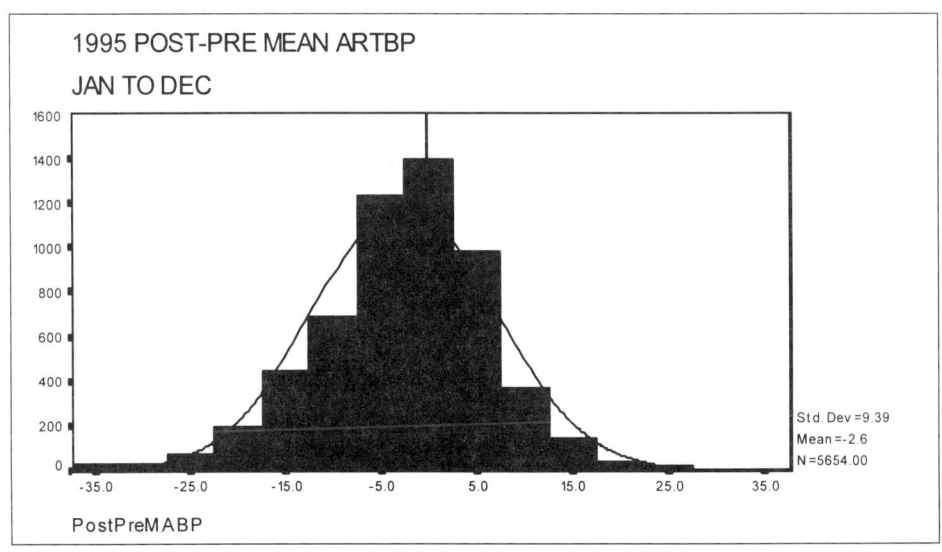

Fig. 1. Cross-shift drops of mean arterial BP among American Azide employees, January–December, 1995.

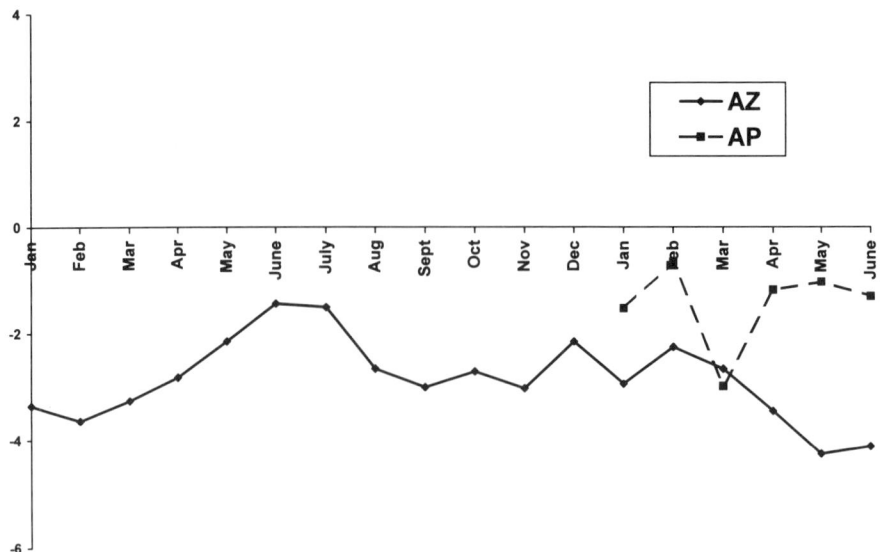

Fig. 2. Mean dMAP by month and plant.

Blood pressure changes by process area

The daily work records for 1995 for the study population were analysed to determine whether the cross-shift blood pressure change varied by work areas or work processes in which the employees had been during the work shift across which the blood pressure was measured. An odds ratio analysis was performed to observe whether the occurrence of a significant blood pressure drop was associated with having ever been in a particular work area during that shift vs not having been in that particular work area during that shift. A significant association was observed with working in the packaging area (Building 125) [odds ratio = 2.0; 95% CI, 1.6–2.6). For those who did not work in the package area, a significant association was observed with working in the purification area (Building 116), particularly with such work in the second half of the work shift [odds ratio = 1.6; 95% CI, 1.16–2.21]. Employees who worked in the control room were less likely to experience a significant blood pressure than were employees who worked elsewhere in the plant [odds ratio = 0.6; 95% CI, 0.45–0.77], probably because they were less likely to be in exposure areas. The control room is in a Building separate from the process areas. Odds ratio analysis demonstrates a hierarchy of biological response areas based on significant blood pressure changes: packaging (Building 125) > purification (Building 116) > reactor and other areas > control room (Building 106). Graphic analysis of the mean dMAP by month by process area supports the odds ratio analysis (Fig. 3). Throughout the year, the mean drop in mean arterial blood pressure was greater for those who worked in the packaging area than for those who worked in the purification area, which was greater than for those who only worked in other areas of the plant. Process area-specific analyses of the frequency of significant blood pressure drops and of monthly average drops in the mean arterial blood pressure reveal similar area-specific differences.

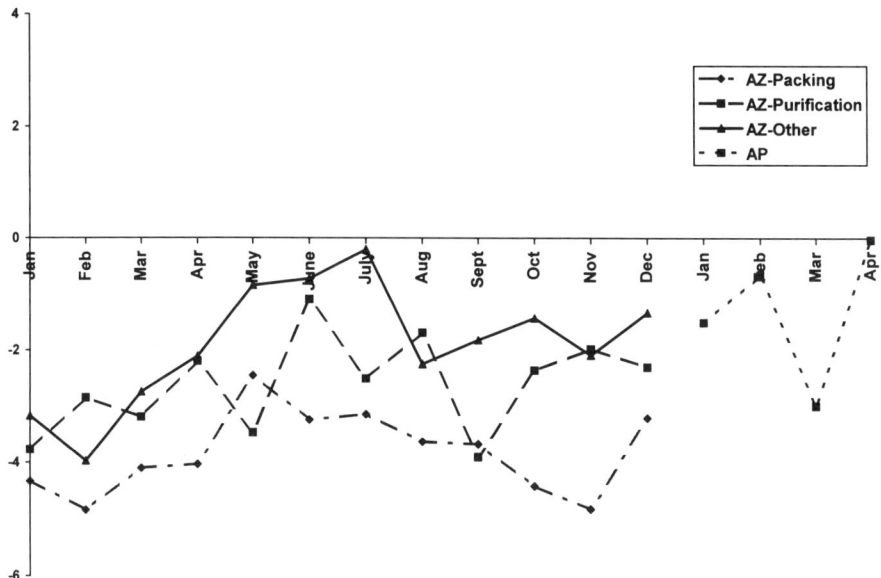

Fig. 3. Mean dMAP by month, plant and process area.

Industrial hygiene studies and reported health incidents

Five industrial hygiene surveys have been conducted at the sodium azide plant between July 1994 and June 1996 by the company, by the National Institute for Occupational Safety and Health, and by the Utah Occupational Safety and Health Division. Excess exposures appear to have been brought under control during 1995; the June 1996 air monitoring conducted by state OSHA consultation found no excessive exposure ($n = 10$ workers in the packaging and blending areas; Fig. 4). Daily surveillance of health incidents began in December 1994. The frequency of reported health incidents, particularly headaches, did not appear to have changed much in spite of apparent exposure control (Fig. 4). The air sampling procedures do not appear to have detected the exposures whose effects were detected by the bio-monitoring surveillance.

DISCUSSION

Sodium azide is a bio-active substance that causes a lowering of the blood pressure. Rigorous monitoring of changes in blood pressure across a work shift has been conducted in both an exposed and a comparison group of employees. The data from the comparison group demonstrates a drop in blood pressure across the work shift for an occupational group, a finding that has not been previously reported. The greater drop seen among the sodium azide workers shows an additional exposure-related effect whose magnitude is demonstrably different for workers with different exposure opportunities. These exposure-related effects are observed in spite of industrial hygiene air monitoring data from the same process areas and personnel showing an excessive exposure. A number of explanations may

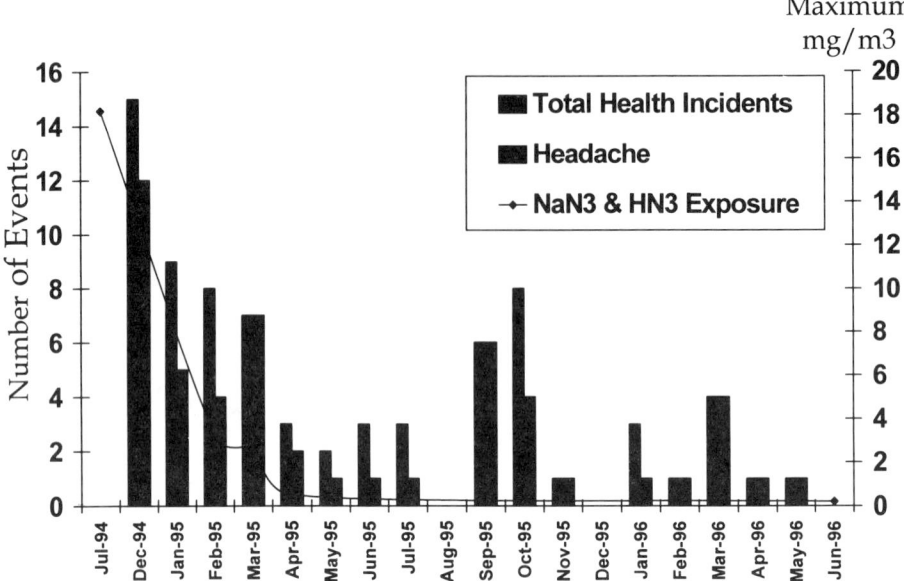

Fig. 4. Reported health incidents and industrial hygiene surveys.

be posited: (a) the exposure levels are at level at which current industrial hygiene methodology cannot effectively distinguish signal from noise and the rigorous blood pressure surveillance system has greater sensitivity; (b) the sampling strategy for air monitoring is not appropriate for actual exposures; (c) the currently proposed limits are not adequate; or (d) the routine daily blood pressure surveillance system observes a greater range of employee work practices and subsequent exposures than does a single day of sampling during which employees and management are cognizant of on-going exposure sampling. Whatever the explanation, the data from the blood pressure surveillance, as well as from the health incidents reporting, indicate that an exposure continues to exist. Recurrent analysis of the surveillance data provides guidance for engineering changes and other exposure control measures and an assessment tool for the evaluation of those control changes. The development and maintenance of a safe working environment is the goal of the program.

REFERENCES

NIOSH (1995) Hazard Evaluation and Technical Assistance Report. American Azide Corporation, Cedar City, Utah, HETA 95-0023-2531. U.S. Department of Health and Human Services, Public Health Service, Centres for Disease Control and Prevention, Cincinnati, OH.
Rippen, H. E., Lamm, S. H., Nicoll, P. G., Cummings, L., Howearth, G. and Thayer, D. (1996) Occupational health data as a basis for process engineering changes—development of a safe work environment in the sodium azide industry. *Int. Arch. occup. Hyg.* **68**(6), 459–468.

 Pergamon

Ann. occup. Hyg., Vol. 41, Supplement 1, pp. 673–676, 1997
© 1997 British Occupational Hygiene Society
Published by Elsevier Science Ltd. All rights reserved
Printed in Great Britain
0003-4878/97 $17.00 + 0.00
Inhaled Particles VIII

PII: S0003-4878(96)00080-4

EVALUATION OF THE PERFORMANCE OF ABRASIVE BLASTING RESPIRATORS

Y. Y. Hammad

University of South Florida, College of Public Health, 13201 Bruce B Downs Blvd, MDC Box 56, Tampa, FL 33612-3805, U.S.A.

INTRODUCTION

Generally, the term "protection factor" is used to describe the degree of protection that a certain respirator is expected to provide to the wearer. The protection factor is usually presented as the ratio of the concentration of the contaminant or test agent outside the respirator to the concentration measured inside the facepiece during utilization of the respirator. Thus, a respirator's maximum use concentration is determined by multiplying a contaminant's exposure limit, for example its threshold limit value (TLV), times the protection factor of the respirator. Another term, assigned protection factor (APF), has been used to define the minimum expected workplace level of protection provided by a certain class of respirators. A respirator's APF can be based on field protection factor studies or laboratory evaluations simulating work conditions. Work protection factors (WPF) are determined by measuring the protection provided by the respirator under work conditions. The values of APF and WPF are developed for properly functioning respirators used by properly fitted and trained persons. Quantitative fit testing is usually performed under controlled laboratory conditions and it represents the protection provided by a certain respirator to a given wearer. Fit testing is not usually performed for continuous flow airline respirators. It is important to note that APF is a number that is assigned to an entire class of respirators and it represents the minimum protection provided to most wearers, while, WPF is the actual protection provided by a class of respirators under work conditions.

The protection factors assigned to abrasive-blast airline respirators (ABR) have been the subject of extensive debate in the U.S.A. These difficulties stem from the fact that there are no standard procedures for the proper testing of the performance of these respirators. Generally, testing procedures performed in the laboratory tend to produce results that are not in complete agreement with those obtained under actual field conditions. One of the main problems encountered in the development of these tests is the effects of air blow-back resulting from the high pressures utilized in the blasting process, a process that cannot be easily duplicated in laboratory testing procedures. On the other hand, one of the major difficulties in field testing is the simultaneous measurement of the extremely low dust concentration inside the respirator and the very high dust concentrations prevailing outside of the respirator.

Currently, the APF assigned by the American National Standards Institute

(ANSI, 1992) is 1000. The Occupational Health and Safety Administration (OSHA) adopted an APF of 2000 for these respirators in its 29 CFR 1910.134 (OSHA, 1989). Bollinger and Schutz (1987), in a National Institute for Occupational Safety and Health (NIOSH) guide for respiratory protection, recommended a protection factor of 25 for any supplied air-respirator equipped with a hood or helmet and operated in a continuous flow mode. Recently, OSHA adopted the protection factor of 25 for ABR recommended by NIOSH in the new lead standard.

The objective of this investigation was to develop a testing procedure that can be used to determine the range of protection factors provided by ABR under actual work conditions. The ABR tested in this study were Supplied-Air Respirators, Type CE, Continuous-Flow Class, Model 77, manufactured by E.D. Bullard Company, Cynthiana, KY.

MATERIALS AND METHODS

The three abrasive blasting respirators to be tested in this investigation were selected at random by the author from the warehouse of one of the ABR distributors in New Orleans, LA. The respirator's main components include a helmet with outer window lenses, air plenum, suspension system, chin strap and a lightweight 28 in (70 cm) long cape attached to the helmet. The air supply flow rate delivered to each respirator was maintained at 8 cfm (3.8 l. s^{-1}), as measured by a mass flow meter, through a 200 feet (60 m) hose, the maximum length recommended for these respirators. The Model 77 has an air distribution plenum near the wearer's forehead. Air at a pressure of about one atmosphere (1 bar) is forced through a permeable membrane in the plenum and expelled into the interior of the hood. The respirators were modified so that respirable dust can be measured using cyclones mounted inside and outside the abrasive blasting respirators and operated in conjunction with two battery operated personal sampling pumps. Air samples were collected on Millipore, 25 mm in diameter, type AA, membrane filters.

The tests were conducted at the facilities of a paints and coating consulting firm (KTA Tator, Pittsburgh, PA), using a 12 feet long by 8 feet wide and 8 feet high ($3.6 \times 2.4 \times 2.4$ m) sand blasting booth. The exhaust air flow rate in the booth was adjusted to provide an average air velocity of 75 fpm (37 cm s^{-1}) across the booth. Preliminary tests showed that 75 fpm together with the selected abrasive blasting conditions would result in a target dust concentration of 30 mg m^{-3}. Abrasive blasting was performed using fine white sand and a ¼ in (6 mm) nozzle at a pressure of 90 psi (620 kPa).

The age, weight and height of the subjects participating in the study were as follows. Subject A is 60 years old, weighs 195 pounds (88.5 kg) and is 83 inches (2.11 m) in height; Subject B is 44 years old, weighs 185 pounds (83.9 kg) and is 71 in (1.80 m) in height and Subject C is 41 years old, weighs 230 pounds (104.3 kg) and is 69 inches (1.75 m) in height.

Each ABR was tested five times and a test lasted for an entire shift. Preparations for each test required about 1 hour. Collection of sample filters and removal of the respirators at the end of the test also required about 1 hour. As a result, actual blasting operations were performed over a period of about 6 hours. During the

Table 1. Concentration of respirable silica dust inside and outside the abrasive blasting respirators and associated work protection factors (WPF) for each subject

| Subject | Respirable silica dust concentration | | WPF |
	Inside, $\mu g\ m^{-3}$	Outside, $mg\ m^{-3}$	
Subject A:			
Mean, 5 samples	3.32	27.1	14 040
SD	1.92	8.29	14 100
Subject B:			
Mean, 5 samples	2.96	31.5	17 270
SD	1.92	8.29	12 460
Subject C:			
Mean, 5 samples	1.20	26.2	54 270
SD	1.92	8.29	43 320
All subjects:			
Mean, 15 samples	2.49	28.2	28 500
SD	1.92	6.31	31 530
Geometric mean for all subjects			16 800
Geometric SD for all subjects			2.93
5th percentile work protection factor			2800

abrasive blasting procedure in the booth, one person operated the same blasting pot, the second performed the actual blasting process and the third person tended the blasting and air supply hoses. Workers rotated their positions during the abrasive blasting activities every 10–15 min.

Total dust concentrations were determined gravimetrically and the silicon dioxide (SiO_2) fraction of each sample was determined by Proton Induced X-ray Emissions (PIXE).

RESULTS

During the testing procedures, sand was consumed at a rate of 700 lb h^{-1} (88 g s^{-1}). Average SiO_2 dust concentration in the blasting booth was 28.3 ± 8.5 mg m^{-3}. PIXE analysis showed that SiO_2 content of blank filters was 0.202 μg filter^{-1}. Respirable SiO_2 dust concentration in air supplied to the respirators (corrected for blanks) was 60 ng m^{-3}. The average respirable SiO_2 dust concentration inside the respiratorse (corrected for blanks and air supply) was 2.49 ± 1.92 $\mu g\ m^{-3}$. The average WPF obtained for the three ABR was 28 500 and the WPF ranged from 3430 to 113 000. The complete results are shown in Table 1.

DISCUSSION

At the present time, there are conflicting information, recommendations and regulations regarding the protection factors provided by abrasive blasting respirators used in the U.S.A.. The results presented in this report indicate that utilization of the PIXE for analysis of silica may revolutionize the testing protocols for determination of WPF obtained under actual field conditions. It is possible to detect dust concentrations in the ng m^{-3} range and consequently protection factors above 100 000 using the test protocol developed in this investigation.

The results also support the applicability of a protection factor of 1000 suggested by ANSI for the type of respirators tested in the study.

REFERENCES

American National Standards Institute (1992) American national standard for respiratory protection (ANSI Z88.2). American National Standards Institute, Inc., New York.

Bollinger, N. J. and Shutz, R. H. (1987) NIOSH guide to industrial respiratory protection, DHHS (NIOSH) Publication no. 87–116. Cincinnati: U.S. Department of Health and Human Services, U.S. Public Health Service, Centers for Disease Control, National Institute for Occupational Safety and Health.

Occupational Safety and Health Administration (1989) Respiratory protection, Code of Federal Regulations, Title 29, Part 1910.134, pp. 391–395.

 Pergamon

Ann. occup. Hyg., Vol. 41, Supplement 1, pp. 677–682, 1997
© 1997 British Occupational Hygiene Society
Published by Elsevier Science Ltd. All rights reserved
Printed in Great Britain
0003–4878/97 $17.00 + 0.00
Inhaled Particles VIII

PII: S0003–4878(96)00079–8

MEASUREMENT OF BIOEFFICIENCY OF A RANGE OF SUBSTRATES USED TO COLLECT BIOAEROSOLS

B. Crook,* S. J. Futter,† W. D. Griffiths,†‡ L. C. Kenny,* S. Stagg,*
J. D. Stancliffe* and I. W. Stewart†

*Health and Safety Laboratory, Broad Lane, Sheffield S3 7HQ, U.K.; and
†AEA Technology, Biotechnology Servives, 353 Harwell, Didcot, Oxfordshire OX11 0RA, U.K.

INTRODUCTION

The airborne route of worker exposure to micro-organisms in industry is regarded as potentially serious. Micro-organisms need to be viable for pathogenic effects to occur, whereas allergic responses are usually related to the total numbers of cells. Risk assessment for workers exposed to airborne micro-organisms is generally based upon culturable enumeration of a sample collected from the air within the vicinity of the worker. Griffiths and DeCosemo (1994) reported that the viable count can be affected by many factors such as the type of growth medium selected, method of aerosolisation, environmental factors like humidity, temperature, pollution, method of collection and assay. This viable count is most frequently derived from a sample collected by a static device and can at best only be regarded as providing an estimate of worker exposure. The number of micro-organisms to which the worker is actually exposed can be determined more accurately by means of a personal sampler worn by the worker. As a first step in the development of a suitable sampler, the U.K. Health and Safety Executive decided to examine the performance of a number of collecting substrates contained in a range of personal samplers, along with a small personal cyclone sampler and a passive sampler in collecting and facilitating the enumeration of selected bioaerosols (Table 1 lists the acronyms used to identify each sampler and Fig. 1 shows their cross-sections): *Penicillium expansum* spores, *Saccharomyces cerevisiae* and *Escherichia coli* cells, which are expected to have a range of robustness and aerostability. The collection substrates chosen for consideration were polycarbonate and gelatin filters; porous polyurethane foams, either uncoated or coated with an agarose gel; beds of glass beads of various diameters, again either uncoated or coated with agarose gel. The passive aerosol samplers utilise a charged electret film as the collection substrate; and the nickel-plated aluminium miniature cyclone samplers make use of their inside surfaces, coated with agarose gel to help collect the bioaerosols. Their effectiveness was evaluated in terms of culturable fraction, which is defined as the ratio of culturable to total number of micro-organisms collected by the sampler. The culturable fraction of the samples collected in the sampler and substrate

---

‡ Author to whom correspondence should be addressed.

Table 1. HSL samplers and substrates used in this study

Sampler	Description
BIOC	personal cyclone sampler
GF	gelatine filter in Gelman open-faced filter holder
LTBC	long, thin bead sampler coated with agarose gel
LTBU	long, thin bead sampler, uncoated
LTFC	long, thin foam sampler coated with agarose gel
LTFU	long, thin foam sampler, uncoated
PASS	passive sampler with charged electret film
PF	polycarbonate filter in Gelman open-faced filter holder
SFBC	short, fat bead sampler coated with agarose gel
SFBU	short, fat bead sampler, uncoated
SFFC	short, fat foam sampler coated with agarose gel
SFFU	short, fat foam sampler, uncoated

combinations were assessed against those collected in the AEAC reference sampler and from the spray suspension.

EXPERIMENTAL METHOD

The experimental procedure consisted of the following steps: cell suspensions were prepared, and then aerosolised in the bioaerosol test chamber under controlled conditions at 20°C and relative humidities of 30 and 70% (Griffiths *et al.*, 1996). Aerosolised cells were captured simultaneously in the test samplers and the AEAC, their size distributions were measured in the TSI Aerodynamic Particle Sizer. Serial dilutions of the collection fluids from the samplers were prepared and allowed to stand at room temperature for 60 min before assays were carried out. Culturable and total counts were measured on the diluted collection fluids and in the spray suspensions prior to aerosolisation. The ratio of experimental to expected airborne concentrations of the test bioaerosols were calculated for each sampler, along with the culturable fraction (ratio of culturable to total counts).

RESULTS AND DISCUSSION

Results of the experiments were recorded so that the performance of each sampler in terms of the culturable fraction of the collected sample could be compared with that of the AEAC. Examples of the results are shown in Tables 2 and 3.

Reference AEAC sampler

Griffiths *et al.* (1996, in press) reported that the AEAC reference sampler facing a 0.5 m s^{-1} wind has a 50% collection value of 0.8 μm aerodynamic diameter and a physical collection efficiency of approximately 85% for aerosol particles up to 10 μm aerodynamic diameter. The particle number and mass size distributions for the three aerosols used in this study show that more than 90% of the aerosolised material has an aerodynamic diameter greater than 0.8 μm which will be effectively captured by the AEAC. Griffiths *et al.* in press) also assessed the biological performance of the AEAC using a range of bioaerosols, and found that there is no

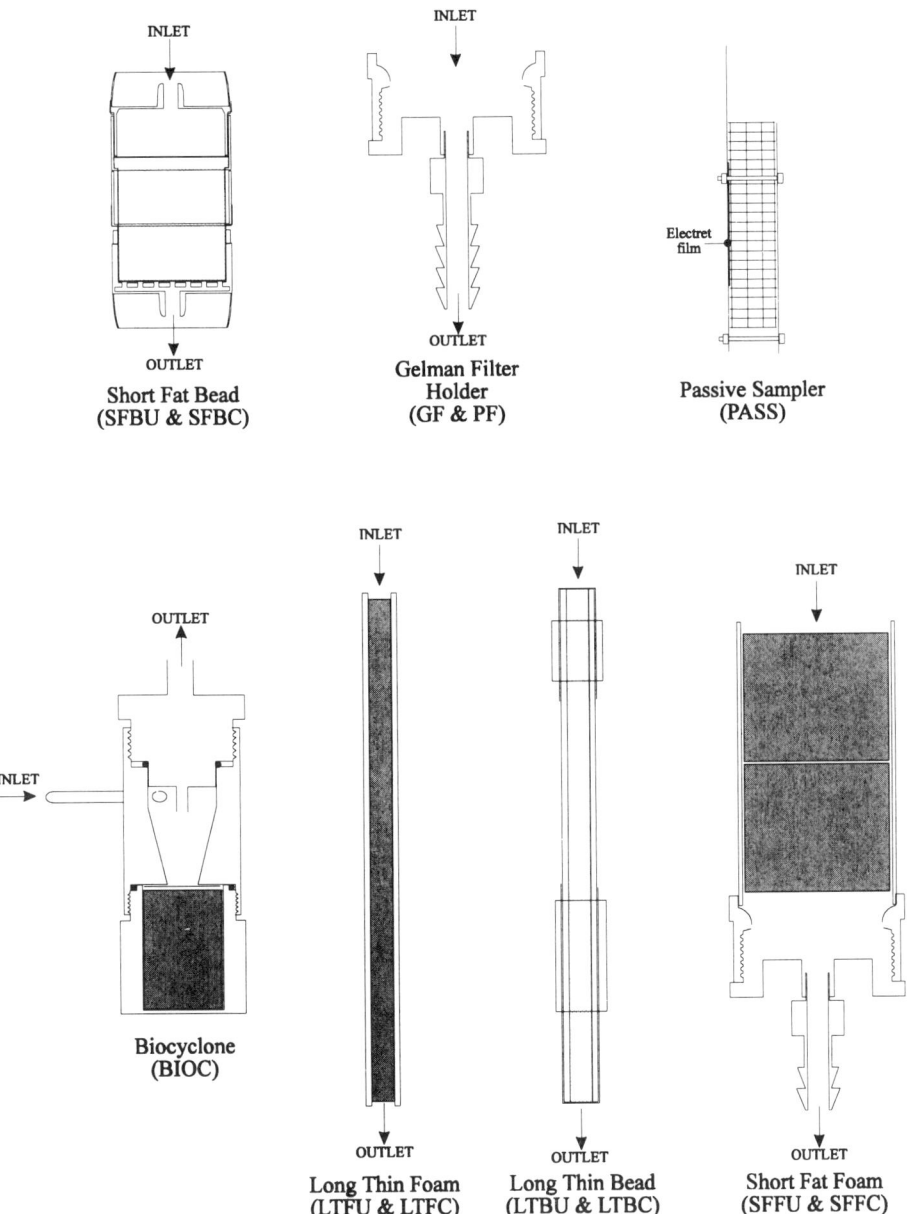

Fig. 1. Personal samplers used in this study.

significant difference between their value and the equivalent data obtained in the current study (Tables 2 and 3). Apart from *S. cerevisiae* at 30% relative humidity, there is no significant difference in the culturable fraction between the non-aerosolised and aerosolised cells captured in the AEAC (Table 3), suggesting that the AEAC provides an accurate and gentle method of capturing the cells that may have been damaged by aerosolisation. The results also indicate that the aerosolisa-

B. Crook *et al.*

Table 2. Culturable fraction at 70% relative humidity

| | Culturable fraction | | | | | | | | |
| | *E. coli* | | | *S. cerevisiae* | | | *P. expansum* | | |
Sampler	Mean	SD	*n*	Mean	SD	*n*	Mean	SD	*n*
SS	0.221	0.032	5	0.433	0.146	8	0.338	0.167	4
AEAC	0.221	0.129	5	0.268	0.108	8	0.353	0.173	4
BIOC	—	—	—	0.268	0.146	8	—	—	—
GF	—	—	—	0.527	0.353	5	—	—	—
LTBU	0.174	0.087	9	—	—	—	—	—	—
LTFU	0.157	0.041	10	—	—	—	—	—	—
PASS	0.016	0.008	4	0.196	0.142	8	0.005	0.008	8
PF	0.078	0.039	5	0.171	0.117	8	0.243	0.261	7
SFBC	—	—	—	0.363	0.250	4	—	—	—
SFBU	—	—	—	0.228	0.056	6	0.060	0.028	5
SFFC	—	—	—	0.244	0.145	5	—	—	—
SFFU	—	—	—	0.292	0.134	11	0.050	0.031	9

Table 3. Culturable fraction at 30% relative humidity

| | Culturable fraction | | | | | | | | |
| | *E. coli* | | | *S. cerevisiae* | | | *P. expansum* | | |
Sampler	Mean	SD	*n*	Mean	SD	*n*	Mean	SD	*n*
SS	0.234	0.107	5	0.348	0.017	7	0.367	0.121	3
AEAC	0.296	0.178	5	0.047	0.020	7	0.331	0.165	3
BIOC	—	—	—	0.065	0.047	7	—	—	—
GF	—	—	—	0.050	0.024	5	—	—	—
LTBU	0.037	0.024	10	—	—	—	—	—	—
LTFU	0.057	0.034	10	—	—	—	—	—	—
PASS	0.016	0.007	2	0.145	0.142	7	0.003	0.004	3
PF	0.040	0.040	5	0.104	0.116	7	0.027	0.016	6
SFBC	—	—	—	0.203	0.218	5	—	—	—
SFBU	—	—	—	0.084	0.019	4	0.003	0.002	4
SFFC	—	—	—	0.148	0.097	5	—	—	—
SFFU	—	—	—	0.094	0.072	9	0.011	0.008	7

SS—Spray suspension.

tion process and environmental conditions do not affect the AEAC performance. Collection and residence in the collecting fluid in the sampler may aid recovery.

Foam and bead substrate samplers

LTBU and LTFU were only used to capture *E. coli* aerosols. Results indicate very good agreement between the two samplers (Tables 2 and 3). These data suggest that the two samplers give the same recovery of the aerosolised cells, and have similar effects on their culturability. SFBU and SFFU showed a similar trend (SF samplers were used with *S. cerevisiae* and *P. expansum* aerosols). Culturable fractions for *S. cerevisiae* aerosols were also similar in value for the SFFU and SFFC (0.094 with a SD of 0.072 and 0.148 with a SD of 0.097 at 30% (Table 3); and 0.292 with a SD of 0.134 and 0.244 with a SD of 0.145 at 70% relative humidity (Table 2). The culturable fraction for *P. expansum* aerosols captured in the samplers are significantly lower than for *S. cerevisiae* aerosols, confirming that the *P. expansum* spores are less robust than the *S. cerevisiae* cells for SFFU (Table 3). The data in Table 2 show similar behaviour for SFBC and SFBU, at 70% relative humidity only. Coating the beads in SFBC with agarose caused recovery problems as the beads formed clumps which affect the flow of air through the sampler.

Filter samplers

The culturable fraction for micro-organisms recovered in PF at 30% relative humidity is highest for *S. cerevisiae*, followed by *E. coli* and *P. expansum* (Table 3). Culturable fractions of *S. cerevisiae* aerosols can be compared with a previous study using the same strain of micro-organism where the PF was held in an IOM Personal Inhalable Aerosol Sampler (Stewart *et al.*, in press) to give a mean and standard deviation for 36 data points of 0.100 and 0.067, respectively. For PF samples, aerosolised *P. expansum* spores have been found to be less robust (a culturable fraction of 0.027 with a SD of 0.016) than *S. cerevisiae* cells (0.104 with a SD of 0.116) at 30% relative humidity. However, at 70% relative humidity, *P. expansum* was found to have the highest culturable fraction followed by *S. cerevisiae* and *E. coli* (0.243 with a SD of 0.261; 0.171 with a SD of 0.117; and 0.078 with a SD of 0.039, respectively), quantifying the effect that relative humidity on the culturability of aerosolised micro-organisms. GF was only used to capture *S. cerevisiae* and *E. coli* aerosols. When total counts were made using optical microscopy, there are difficulties in distinguishing the captured cells in the debris of the gelatin filters. These filters contain clumps of micro-organisms which are sterilised in the production process. When the filters are dissolved, the dead micro-organisms are still visible and interfere with the total count.

Passive sampler

The ratios of measured CFU to expected airborne concentration (CFU l^{-1}) in the BTC at 70% relative humidity are identical for the three species of test micro-organism (0.135 with a SD of 0.076 for *E. coli*, 0.134 with a SD of 0.055 for *S. cerevisiae* and 0.134 with a SD of 0.048 for *P. expansum* spores). Experimentally measured CFU l^{-1} can be plotted against expected airborne concentration for the three test micro-organisms aerosolised at 30 and 70% relative humidity to give a tentative straight line relationship in each case.

CONCLUSIONS

The performance of all the foam samplers (LT and SF) and the filter samplers was compared with that of the gentle and accurate reference AEAC and shown to be dependent on the robustness of the micro-organisms and the environmental conditions. The performance of the bead and foam substrates for the capture and recovery of *E. coli* was similar in the LT samplers. Coating the bead and foam substrates proved difficult, and had no effect on the performance of the substrates.

This work indicates that an effective personal bioaerosol sampler can be constructed using polyurethane foams to provide multi-fraction, size-selective sampling. Polycarbonate filter is an effective means of recovering the finest-sized fraction. The next step is to optimise the selection of the foam plugs so that the sampling characteristics match agreed international conventions for aerosols. A prototype aerosol sampler will be constructed for further evaluation.

The need to compare the performance of new personal samplers for bioaerosols with characterised reference samplers cannot be over-emphasised. In this study, large variations have been obtained on the means of results with relatively robust and aerostable micro-organisms, even under carefully controlled experimental conditions. New methods will only be acceptable to potential users if they have significant advantages over traditional techniques. Laboratory evaluations are a necessary prerequisite to field studies to establish effective methods for the detection of bioaerosols in the workplace.

REFERENCES

Griffiths, W. D. and DeCosemo, G. A. L. (1994) An assessment of bioaerosols: a critical review. *J. Aerosol Sci.* **25**(8), 1425–1458.

Griffiths, W. D., Stewart, I. W., Futter, S. J., Upton, S. L. and Mark, D. The development of sampling methods for the assessment of indoor bioaerosols. *J. Aerosol Sci.* (in press).

Griffiths, W. D., Stewart, I. W., Reading, A. R. and Futter, S. J. (1996) Effect of aerosolisation, growth phase and residence time in spray and collection fluids on the culturability of cells and spores. *J. Aerosol Sci.* **27**(5), 803–820.

Stewart, I. W., Leaver, G. and Futter, S. J. The enumeration of aerosolised saccharomyces cerevisiae using bioluminescent assay of total adenylates. *J. Aerosol Sci.* (in press).

 Pergamon

Ann. occup. Hyg., Vol. 41, Supplement 1, pp. 683–688, 1997
© 1997 British Occupational Hygiene Society
Published by Elsevier Science Ltd. All rights reserved
Printed in Great Britain
0003–4878/97 $17.00 + 0.00
Inhaled Particles VIII

PII: S0003–4878(96)00139–1

SPORES OF THE OPPORTUNISTIC FUNGUS *ASPERGILLUS FUMIGATUS*: EVASION OF THE MACROPHAGE DEFENCE AND ADHERENCE IN INFLAMED LUNG

K. Donaldson,* J. Slight,* I. Bromley,* C. Mitchell,* P. H. Beswick,* A. Seaton†
and W. J. Nicholson*

*Biomedicine Research Group, Department of Biological Sciences, Napier University, 10 Colinton Road, Edinburgh EH10 5DY, U.K.; and †Department of Environmental and Occupational Medicine, University of Aberdeen, Aberdeen, U.K.

INTRODUCTION

Aspergillus fumigatus (Af) is a fungus that grows on dead and decaying organic matter producing spores which are present ubiquitously in the air. The 3 μm diameter of the spores renders them highly respirable and is an important factor in determining the deposition of the spores in the deep lung. Mullins *et al.* (1976) described the presence of the spores of Af in greater numbers in the lungs at autopsy compared to similar-sized spores suggesting that, over and above their respirability, they have some survival advantage in the human lung over other spores. The fungus causes a range of diseases in the human lung (Seaton *et al.*, 1989). In particular the spores appear to favour the asthmatic lung as shown by the fact that 25% of all asthmatics are sensitised to Af and allergic bronchopulmonary aspergillosis, where the fungus colonises the lung, occurs in a small proportion of asthmatics (Bateman, 1994). The spores are also more likely to germinate in the lungs of individuals with a range of chronic inflammatory lung conditions (Wagner, 1984) and this also suggests that inflamed/injured lung is a favourable environment for Af spores to gain a "foothold".

A partial explanation for the persistence of Af spores in the lungs can be found in reports of a diffusable substance from the spores that has a number of effects on macrophages including inhibition of phagocytosis, inhibition of the respiratory burst and inhibition of chemotaxis and spreading (Seaton and Robertson, 1989). We have further investigated the hypothesis that the spores have ways of avoiding the phagocytic defences of the lung and we have proposed that the spores may utilise the altered surface characteristics of the asthmatic lung lining to promote their adherence and give them the opportunity to persist.

We summarise, here, recent and novel findings towards a unified hypothesis for the pathogenicity of spores for asthmatic and inflamed lungs that embraces their ability to avoid phagocytic defences and adhere preferentially in asthmatic or inflamed lung.

MATERIALS AND METHODS

A. fumigatus *spore diffusate (AfD)*

A strain of *A. fumigatus* was isolated from the sputum of a patient with allergic bronchopulmonary aspergillosis. Spores were obtained from cultures that had been grown for 10 days at 30°C on malt agar. Spore suspensions were prepared in Hanks balanced salt solution (HBSS), at a concentration of 10^8 ml^{-1} and gently homogenised prior to incubation at 37° for 1 h, followed by filtering to remove the spores. This preparation of diffusable material from the surface of spores is referred to as *Aspergillus fumigatus diffusate* (AfD).

Alveolar macrophages

In-house-bred adult Wistar rats were used throughout and macrophages at 95% purity were obtained by conventional bronchalveolar lavage, the remainder of cells being lymphocytes.

Superoxide anion assay

Superoxide anion production by macrophages was measured by superoxide dismutase-inhibitable cytochrome C reduction.

Tumour necrosis factor

Tumour necrosis factor α (TNFα) release by macrophages over 24 h of culture was assayed by ELISA and mRMA for TNFα was assayed by reverse transcriptase polymerase chain reaction.

Adherence of spores to epithelial cells and extracellular matrix

A549 alveolar epithelial cells were seeded in 96-well tissue culture plates at 10^5 cells per well and grown overnight to confluence; spore suspension was added to each well followed by incubation for 30 min with gentle shaking. Plates were then gently washed and, after fixation, spores were counted by phase-contrast microscopy. Five fields of view were counted for each well, in triplicate wells and results were expressed as average number of spores per five high power fields. In a series of experiments A549 cells were pre-treated in interferon-gamma (IFN-γ) at 1000–2500 units per ml to activate them.

The extracellular matrix proteins fibronectin, laminin, type I collagen and type IV collagen were dissolved according to the makers instructions and then coated onto 96-well flat-bottomed plates. Plates were blocked with BSA, spore suspensions added as above and adherence determined as described above.

RESULTS

Viability

AfD was not toxic to macrophages at any concentration as measured by Trypan Blue and propidium iodide (data not shown).

Superoxide anion

AfD (1:4 dilution) inhibited the respiratory burst of rat alveolar macrophages as shown in Fig. 1.

Tumour necrosis factor

The production of TNFα was abolished by undiluted AfD treatment (Fig. 2).

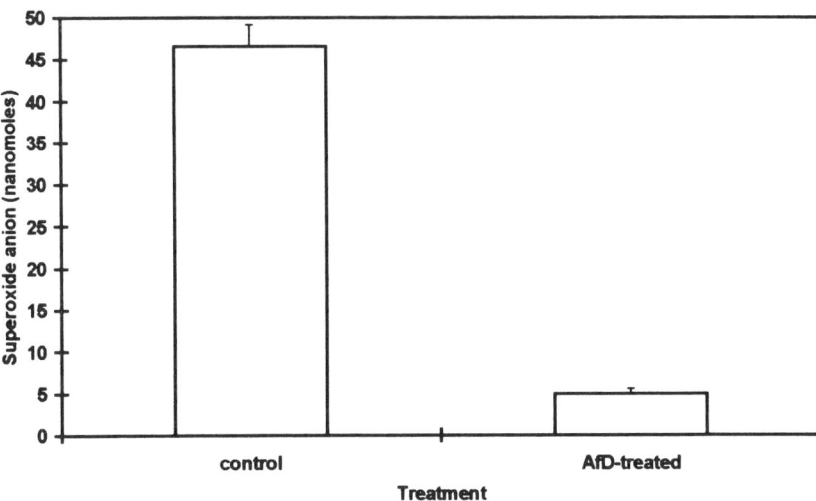

Fig. 1. Effect of AfD (1:5 dilution) on the production of superoxide by alveolar macrophages; data is mean plus standard error of three separate experiments.

AfD also substantially decreased activation of NF-κB and AP-1, transcription factors that bind to the promoter region of the TNFα gene (Fig. 4) and are necessary for its transcription.

Binding of Af spores to epithelial cells and extracellular matrix

Spores bound avidly to epithelial cells and up to twice as many bound to interferon-activated cells (Fig. 5).

Extracellular matrix proteins found in the basement membrane also bound spores to markedly greater degree than the control protein BSA (Fig. 6).

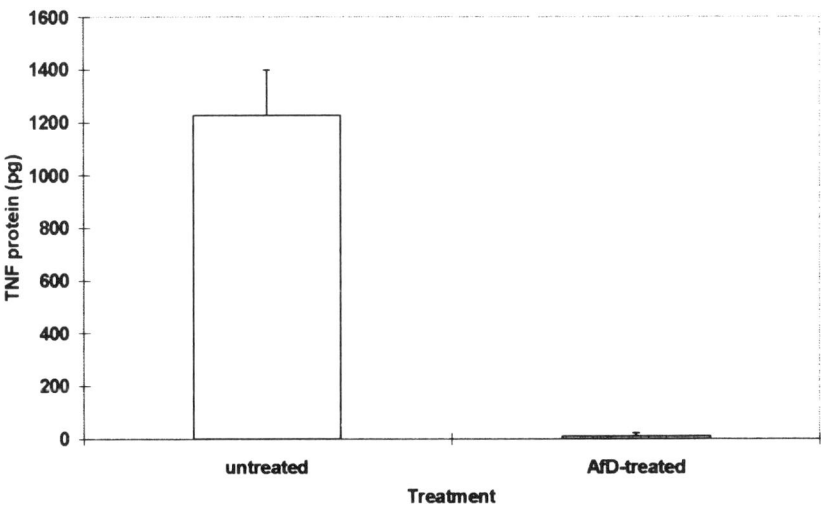

Fig. 2. Production of TNFα protein by alveolar macrophages treated with undiluted AfD; data is mean and standard error of three separate experiments.

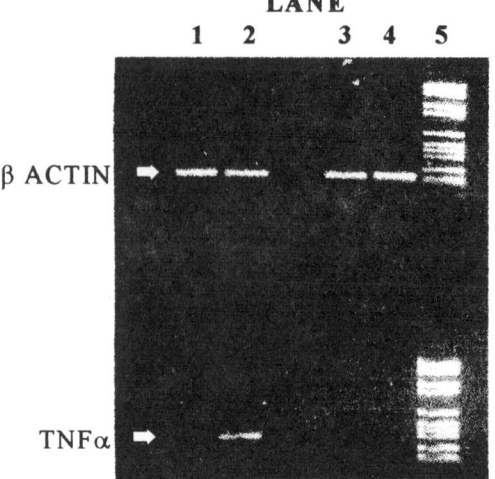

Fig. 3. Detection of cDNA made from mRNA for TNFα (lower band) and the "housekeeping" gene βactin (upper band). Lane 1—macrophages untreated; Lane 2—macrophages + LPS (50 ng ml^{-1}); Lane 3—macrophages + AfD; Lane 4—macrophages + LPS + AfD; Lane 5—DNA Ladder.

CONCLUSIONS

The data support a two stage model to explain the pathogenicity of Af. in asthmatic/inflamed lung:

(1) a diffusable product from the surface of the spores of *A. fumigatus* (AfD) renders the microbicidal defences of the alveolar macrophage largely ineffective. (2) Af spores adhere preferentially to activated epithelial cells and exposed basement membrane which are known to occur in asthmatic/inflamed lung.

Our data indicate that Af spores avoid the phagocytic defences by releasing a toxin which inhibits both the respiratory burst of macrophages and transcription of

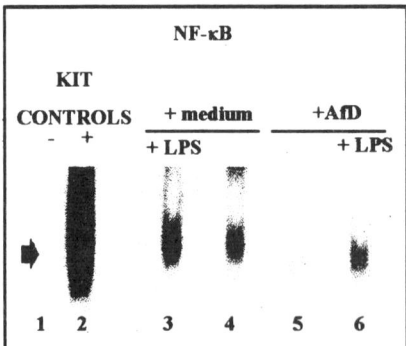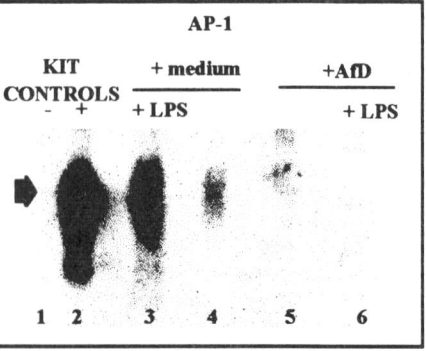

Fig. 4. Mobility shaft assays for AP-1 and NF-κB. Effect of AfD on transcription factors AP-1 (left) and NF-κB (right). There is increased nuclear transfer (arrowed bands) of the transcription factors with LPS and marked inhibition in the presence of AfD with both transcription factors. Lanes 1 and 2—kit controls; Lanes 3 and 4—normal medium, with and without LPS; Lanes 5 and 6—medium containing AfD, with and without LPS.

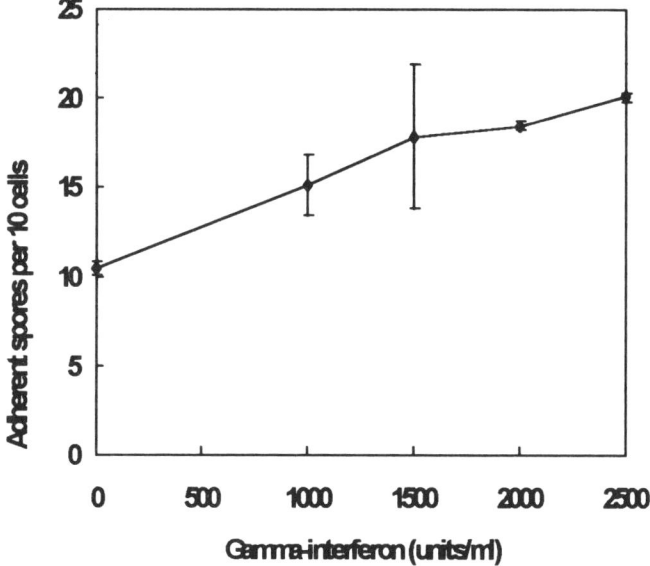

Fig. 5. Binding of spores to control and gamma interferon-treated epithelial cells; mean ± sem from three separate experiments.

the TNFα gene. Af spores are, therefore, less effectively killed and there are reports that Af spores are less effectively phagocytosed than other spores (Seaton and Robertson, 1989). Inhibition of NF-κB and AP-1 factors by AfD results in a loss of cytokine production. Subsequently the orchestration of leukocyte responses will be distorted in AfD-exposed lungs. This helps to explain the increased numbers of Af spores at autopsy (Mullins, 1976) but since these effects are reported here in normal cells, there is a need for a further virulence-enhancing factor to present in asthmatic/inflamed lung. We contend here and show evidence, that the additional factor in inflamed/asthmatic lung is:

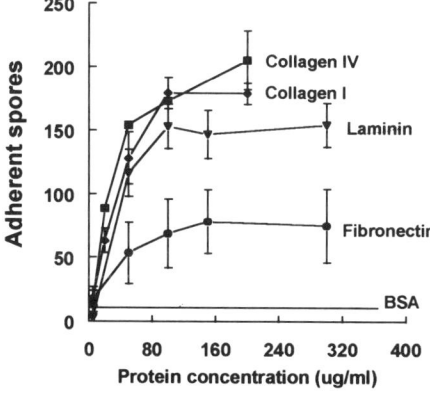

Fig. 6. Binding of Af spores to extracellular matrix proteins found in the basement membrane. Note minimal binding to the control protein BSA; data is mean ± sem from three experiments.

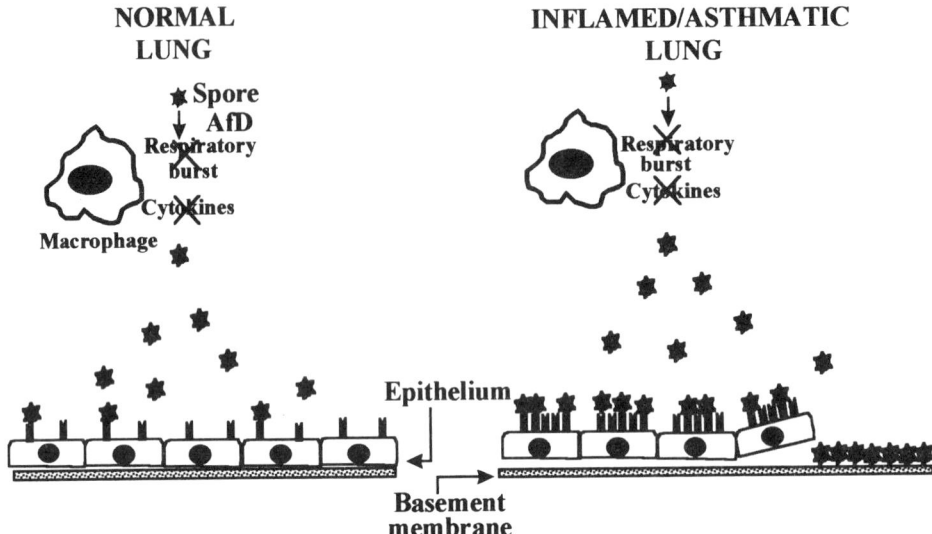

Fig. 7. Hypothetical mechanism for the increased finding of Af spores at autopsy in "normal" lungs by escape from the phagocytic defences (left) and increased persistence leading to germination of Af spores in asthmatic/inflammatory lungs by binding to an up-regulated putative receptor on activated epithelium and to areas of denuded basement membrane (right).

(1) preferential adherence to interferon-activated epithelial cells such as occur in asthmatic lung;

(2) preferential adherence to exposed basement membrane, a further characteristic of asthmatic and inflamed lung that arises through epithelial injury and detachment (Montefort *et al.*, 1993). This hypothesis is summarised in diagrammatic form in Fig. 7.

Acknowledgements—The authors acknowledge the financial assistance of the British Occupational Health Research Foundation and thank Mark Lawson and Irfan Rahman for advice and assistance with gel mobility shift assays.

REFERENCES

Bateman, E. D. (1994) A new look at the natural history of Aspergillus hypersensitivity in asthmatics. *Resp. Med.* **88**, 325–327.

Chu, H. W., Wang, J. M., Boutet, M., Boulet, L-P. and Laviolette, M. (1995) Increased expression of intercellular adhesion molecule-1 (ICAM-1) in a murine model of pulmonary eosinophilia and high IgE level. *Clin. exp. Immunol.* **100**, 319–324.

Lee, J. C. (1994) Transcription factor NF-κB: an emerging regulator of inflammation. *Ann. Rep. Med. Chem.* **29**, 235–244.

Montefort, S., Roche, W. R. and Holgate, S. T. (1993) Bronchial epithelial shedding in asthmatics and non-asthmatics. *Resp. Med.* **87**, B 9–11.

Mullins, J., Harvey, R. and Seaton, A. (1976) Sources and incidence of airborne *Aspergillus fumigatus*. *Clin. Allerg.* **6**, 209–217.

Seaton, A. and Robertson, M. D. (1989) Aspergillus, asthma and amoebae. *The Lancet*, April 22, pp. 893–894.

Wagner, G. E. (1984). Bronchopulmonary aspergillosis and aspergilloma. *Mould Allergy* (Edited by Y. Al-Doory and J. F. Domson), pp. 202–215. Lea and Febiger, Philadelphia, PA.

SECTION 7

INDOOR AIR

 Pergamon

Ann. occup. Hyg., Vol. 41, Supplement 1, pp. 689–694, 1997
© 1997 British Occupational Hygiene Society
Published by Elsevier Science Ltd. All rights reserved
Printed in Great Britain
0003–4878/97 $17.00 + 0.00
Inhaled Particles VIII

PII: S0003–4878(96)00140–8

AEROSOLS, SUBJECTIVE INDOOR AIR QUALITY AND ASTHMA IN SCHOOLS

D. Norbäck and G. Smedje

Department of Occupational and Environmental Medicine, University Hospital,
S-751 85 Uppsala, Sweden

INTRODUCTION

During the last decade, there has been an increased concern about possible human reactions to the indoor environment, including asthma, the so-called sick building syndrome (SBS), impaired mental productivity and perception of poor subjective indoor air quality (SIAQ). Hygiene measurements indicate that inadequate indoor air quality is a common problem in schools. Carbon dioxide concentration is often above current Swedish standard (Norbäck *et al.*, 1990) of 1000 μl l^{-1} (ppm) (NSBOSH, 1993) and classroom temperatures often exceed the recommended maximum value of 22°C (Norbäck *et al.*, 1990). Finally, carry-in of dust in combination with poor cleaning may cause contamination of schools by various particle pollutants such as allergens from cats, dogs and house dust mites (Munir *et al.*, 1993). Despite this knowledge, there are few studies on the health significance of ventilation rate and particle pollutants in schools. The aim of this study was to investigate possible relationships between ventilation flow, particle pollutants in air, settled dust in classrooms and human reactions among school personnel and secondary school pupils, including subjective indoor air quality, mental performance, SBS and asthmatic symptoms.

MATERIALS AND METHODS

In the county of Uppsala in mid Sweden, there were in 1992 approximately 130 public schools from which we randomly selected 40. All schools, except one, agreed to participate in the investigation. The schools varied with respect to factors such as age, construction and size. One third of the schools were situated in the city of Uppsala (117 000 inhabitants), one in the town of Enköping (19 000 inhabitants) and the other in minor communities or in the countryside. All public employees working in the school buildings were invited to participate in the study. In Sweden there are nine forms in compulsory schools and the vast majority attend public schools. Out of the 39 participating schools, 11 were secondary schools (i.e. comprising the three highest forms). For each secondary school, three classes in the 7th form were randomly chosen, except one where there were only two classes in this form. In each randomly chosen class, all pupils were also invited to participate in the investigation.

ASSESSMENT OF BUILDING CHARACTERISTICS
AND HYGIENIC MEASUREMENTS

Information on building age, building dampness, type of ventilation system, air conditioning and air humidification was gathered by inspection by an occupational hygienist. Exposure measurements were performed in March–May 1993 in 98 classrooms. They included measurement of air exchange rate, room temperature, relative air humidity, concentration of carbon dioxide, respirable dust, moulds, bacteria and the amount of settled dust in the class rooms. Temperature and air humidity were recorded by an Assman psychrometer and air exchange rate was determined by a tracer gas method. Concentration of respirable dust was measured by a direct reading instrument based on light scattering (Sibata P-5H2, Sibata Scientific Technology Ltd, Japan), calibrated by the manufacturer to 0.3 µm particles of stearic acid. Carbon dioxide (CO_2) was measured by a direct reading infrared spectrometer (Riken RI-411A, Riken Keini, Japan) calibrated by standard gases containing known concentrations of CO_2. Airborne microorganisms were sampled on a sterile polycarbonate filter (Nucleopore, Millipore Corp., U.S.A. with a pore size of 0.4 µm, the sampling rate being 1.5 l min^{-1} for 4 h. The total concentration of moulds and bacteria were determined by the CAMNEA method, utilizing staining by acridin orange and counting by epifluorescence microscopy (Palmgren *et al.*, 1986). Viable moulds and bacteria were determined by incubation on two different media.

Settled dust was collected from desks, chairs and floors by a 400 W vacuum cleaner provided with a special dust collector from ALK Laboratories, Copenhagen, containing a Millipore filter (pore size 6 µm). After passing through a sieve containing a filter with a porosity of 300 µm, the amount of fine dust was determined by weighing the filters. Cat (Fel d I) and dog (Can f I) allergens were quantified with an enzyme-linked immunosorbent assay using monoclonal antibodies. The content of the major mite allergens in the dust was also determined by enzyme immunoassay and by the semi-quantitative Acarex test. Endotoxine was analysed using the Limulus Amebocyte Lysate test.

ASSESSMENT OF SYMPTOMS AND PERSONAL FACTORS

Information on subjective indoor air quality, mental productivity, SBS symptoms and asthmatic symptoms was gathered by two versions of a self-administered questionnaire, one for school personnel and one for pupils. The questionnaire was sent by mail in January–February 1993 to the home address. Most questions were identical in both versions, including questions on personal factors such as age, smoking habits, atopy and allergy. Both questionnaires had one general question on indoor SIAQ, asking if the air quality in the school was perceived as very bad, bad, good or very good and another question on frequent respiratory infections. For the moment, however, statistical analysis for these two questions has only been performed among adults. Both questionnaires contained the same 16 questions on SBS. One question on subjective impaired mental performance was included in the questionnaire addressed to the pupils, only. Subjective mental performance due to poor indoor air quality was defined as having answered "yes" to the question: "Do you perceive that your mental performance is impaired due to poor indoor air in the school?". The questions on asthma were similar in both questionnaires and based on questions used in the European Community Respiratory Health Survey (ECRHS). In adults, self-reported asthma diagnosed by a physician was evaluated in this paper. In pupils, we studied current asthma. We defined current asthma in a pupil if he or she had reported that they ever had asthma diagnosed by a physician and gave a positive response to at least one of the questions concerning the use of asthmatic drugs, having had asthmatic attacks, attacks of shortness of breath at night or after exercise during past 12 months, or recurrent symptoms during the time they had attended the present school.

STATISTICAL METHODS

The influence of different exposure factors was analyzed by multiple logistic regression, adjusting for significant confounders. Adjusted odds ratios (OR) with 95% confidence intervals (95% CI) were calculated. In all statistical analysis, two tailed tests and a 5% level of significance were used.

RESULTS

Building characteristics and hygienic measurements

The mean age of the school buildings was 33 years (building year 1900–1992). None of the classrooms had wall-to-wall carpeting, kerosene heating or other sources of indoor combustion. In about 70% of the schools the floors were cleaned once a day with a moistened mop, while 30% were cleaned every second day. The mean room temperature was 23.5°C, but in approximately a quarter of the measurements the temperature was 25°C or more. The concentration of CO_2 was above 1000 ppm in 41% of the classrooms. The most common microorganisms in the air samples were *Cladosporium* sp., *Mycelia sterilia*, *Penicillum* sp., yeasts and *Pseudomonas* sp. Allergens from house dust mites Der p I were found in one sample only and Der f I in none. The Acarex test showed the presence of mites in three schools. Cat and dog allergens were found in all the schools but one. Arithmetic means and variations of indoor climate and pollutants are given in Tables 1 and 2. A significant relationship between low air exchange rate and increased concentrations of respirable dust ($P < 0.001$) and viable moulds ($P < 0.01$) was observed. In contrast, no significant relationship was observed between air exchange rate and concentration of viable bacteria, total bacteria, or total moulds.

Personnel and pupil characteristics

Among the school personnel, 1410 out of 1652 subjects (85%) participated. The mean age was 45 years. Current smokers were 19%, 77% were females and 54% were teachers. The prevalence of atopy was 31%, 17% had hay fever, 10% food allergy, 10% had allergy to furry animals and 6.9% had ever had asthma diagnosed by a physician. Among the secondary school pupils, 627 out of 762 participated (82%). Their age was 13–14 years. There were 3% current smokers, 52% were girls, the prevalence of atopy was 34%, 10% had hay fever, 6% had food allergy,

Table 1. Building data for 98 classrooms in 39 randomly selected schools

	Mean (min–max)
Building age (years)	33 (1–83)
General air exchange rate (h^{-1})	2.4 (< 0.1–9.8)
Outdoor air flow per classroom (l s^{-1})	120 (2–470)
Personal outdoor air flow (l $s*p^{-1}$)	5.5 (0.1–22.4)
Carbon dioxide (ppm)	990 (400–2800)
Proportion of classrooms (%) below current Swedish ventilation standard†	77
Proportion of classrooms (%) with	
Natural ventilation only	27
Mechanical exhaust air only	9
Conventional supply/exhaust ventilation	40
Displacement ventilation	24

† Carbon dioxide concentration above 1000 ppm, or outdoor air supply rate below 8 l $s*p^{-1}$.

Table 2. Indoor climate and particle pollutants in 98 classrooms

	Mean (min–max)
Room temperature (°C)	23.5 (20–28)
Relative air humidity (%)	38 (16–75)
Respirable dust ($\mu g\ m^{-3}$)	19 (6–60)
Airborne microorganisms	
Viable bacteria (10^3 cfu m^{-3})	0.92 (0.07–18)
Total bacteria ($10^3\ m^{-3}$)	52 (8–290)
Viable moulds (10^3 cfu m^{-3})	0.47 (0.07–4.5)
Total moulds ($10^3\ m^{-3}$)	40 (7–360)
Settle dust pollutants	
Cat allergen (ng g^{-1})	130 (< 20–390)
Dog allergen (ng g^{-1})	640 (< 20–3990)
Housedust mite allergens (ng g^{-1})	0.6 (< 1–31)
Endotoxin (ng g^{-1})	3.3 (2–9)
Amount of settled dust in sample (mg)	170 (30–370)

10% had allergy to furry animals and 7.7% had ever had asthma diagnosed by a physician, while 6.4% had current asthma.

Relationships between human reactions, indoor climate and particle pollutants

Among school personnel, 53% of the females and 52% of the males judged the SIAQ to be poor (bad or very bad). Poor SIAQ was more common in newer schools and in schools with exhaust air ventilation only OR = 1.8 (95% CI; 1.2–2.8) ($P < 0.01$), but less prevalent in schools with the new type of displacement ventilation system OR = 0.7 (95% CI; 0.5–0.9) ($P < 0.01$) as compared to schools with natural ventilation only. Relations between poor SIAQ and air concentration of respirable dust ($P < 0.01$), total bacteria ($P < 0.05$), total moulds ($P < 0.01$) and the amount of settled dust ($P < 0.05$) were also observed. No significant relation was observed between SIAQ and air exchange rate, CO_2, room temperature, or air humidity. Reports on frequent airway infections among school personnel were

Table 3. Effects on personnel and secondary school pupils in relation to climate, ventilation and particle pollutants in the classrooms

	Poor SIAQ (A)	Poor mental performance (C)	Airway infections (A)	SBS (A + C)	Asthma (A + C)
Climate/ventilation					
Air exchange rate	−		−		
Room temperature				+a	−c
Relative air humidity		+			+c
Airborne particles					
Respirable dust	+	+	+		
Total bacteria	+	+		+a	
Viable bacteria					+c
Total moulds	+	+			+a
Viable moulds					+c
Settled particles					
Cat allergen					+c
Dog allergen				+c	
Endotoxine				+c/−c	
Total amount of dust	+		+		

(A) indicate that relationships were investigated in adults, only.
(C) indicate that relationships were investigated in children, only.
(A + C) indicate that relationships were investigated both in adults and children.
a, Significant relation ($P < 0.05$) in adults only; c, Significant relation ($P < 0.05$) in children only.
Note: for endotoxine, a positive relationship was observed among children for airway symptoms and a negative relationship for skin symptoms.

significantly related to respirable dust, settled dust and low air exchange. Eye symptoms were significantly related to indoor concentration of total airborne bacteria ($P < 0.01$). General symptoms were more prevalent in schools with a higher room temperature ($P < 0.05$) and an increase of classroom temperature by one degree Celsius corresponded to an OR of 1.1 (95% CI 1.01–1.2) for such symptoms. No relation was observed between any type of SBS and air exchange rate, moulds, carbon dioxide concentration, or relative air humidity. Doctor diagnosed asthma among school personnel was related to total concentration of airborne moulds ($P < 0.05$), but not related to ventilation rate, CO_2, or other measured particle pollutants.

Among pupils, reports on impaired mental performance were related to low air exchange rate, respirable dust, total moulds and total bacteria. Upper airway symptoms among pupils was more common in schools with higher concentration of endotoxin in settled dust. Skin symptoms were more common in schools with more dog allergen, but less common if the dust contained more endotoxin. Current asthma among pupils was related to air concentration of viable bacteria and moulds and cat allergen concentration in settled dust. Significant statistical relationships are summarized in Table 3.

DISCUSSION

Our results indicate that the school environment may affect both pupils and school personnel. The design was cross-sectional and in such studies, selection effects may cause an underestimation of the true occurrence relation. Despite these limitations, we could demonstrate significant relations between objective measures of the school environment and symptom reporting. Selection bias can occur both because of incorrect study design and because of low response rate. Our study was performed in a random sample of schools and was designed to include all personnel in the schools and a random sample of school pupils. Thus, no selection based on symptom prevalence in the schools occurred. In addition, the response rates were high both among pupils and school personnel and it is less likely that selection bias had any major influence. The information on human reactions was gathered by means of questionnaires. Awareness of the exposure may influence symptom reporting, but the exposure measurements were done after the symptom reporting was completed.

Most of the relationships observed in our study between human reactions and the indoor environment have, to our knowledge, not been previously reported from school investigations. Our finding of a relation between general symptoms among school personnel and high room temperature, however, agrees with earlier studies, e.g. from offices (Jaakkola *et al.*, 1989). A relationship between respirable dust and eye symptoms has been reported earlier among school personnel (Norbäck *et al.*, 1990). In addition, two office studies have reported relationships between symptoms compatible with the sick building syndrome and settled dust (Skov *et al.*, 1990; Gyntelberg *et al.*, 1994). In conclusion, there is a need to improve the indoor environment in schools. We could identify certain measures that could be taken to obtain a good air quality. Room temperature should be kept below 22°C, settled dust should be removed by cleaning and the air concentration of respirable dust and airborne microorganisms should be kept as low as reasonably achievable. The new type of mechanical supply exhaust ventilation using the displacement flow ventilation seems also to be beneficial.

REFERENCES

Gyntelberg, F., Suadicani, P., Wohlfahrt Nielsen, J. *et al.* (1994) Dust and the sick building syndrome. *Indoor Air* **4**, 223–238.

Jaakkola, J. J. K., Heinomen, O. P. and Seppänen, O. (1989) Sick building syndrome, sensation of dryness and thermal comfort in relation to room temperature in an office building: need for individual control of temperature. *Environ. Int.* **15**, 163–168.

Munir, A. K. M., Einarsson, R. and Schou, C. (1993) Allergens in school dust. I. The amount of the major cat (Fel d I) and dog (Can f I) allergens in dust from Swedish schools is high enough to probably cause perennial symptoms in most children with asthma who are sensitized to cat and dog. *J. Allergy clin. Immunol.* **91**, 1067–1074.

NSBOSH. National Swedish Board of Occupational Safety and Health. Ventilation and quality of air, Stockholm, AFS 1993:5 (in Swedish).

Norbäck, D., Torgen, M. and Edling, C. (1990) Volatile organic compounds, respirable dust, and personal factors related to prevalence and incidence of sick building syndrome in primary schools. *Br. J. ind. Med.* **47**, 733–741.

Palmgren, U., Ström, G., Blomqvist, G. and Malmberg, P. (1986 Collection of Airborne Microorganisms on Nucleopore filters, estimation and analysis-CAMNEA method. *J. appl. Bacteriol.* **61**, 401–406.

Skov, P., Valbjörn, O. *et al.* (1990) Influence of indoor climate on the sick building syndrome in an office environment. *Scand. J. Work, Environ. Health* **16**, 363–371.

Pergamon

Ann. occup. Hyg., Vol. 41, Supplement 1, pp. 695–699, 1997
British Occupational Hygiene Society
Crown Copyright © 1997 Published by Elsevier Science Ltd
Printed in Great Britain.
0003–4878/97 $17.00 + 0.00
Inhaled Particles VIII

PII: S0003–4878(96)00141–X

PRELIMINARY INVESTIGATION OF AEROSOL INHALABILITY AT VERY LOW WIND SPEEDS

A. D. Maynard,* R. J. Aitken,† L. C. Kenny,* P. E. J. Baldwin*
and R. Donaldson†

*The Health and Safety Laboratory, Broad Lane, Sheffield S3 7HQ, U.K.; and
†The Institute of Occupational Medicine, 8 Roxburgh Place, Edinburgh EH8 9SU, U.K.

INTRODUCTION

International agreement on the inhalable convention was reached in the early 1990s, with the International Standards Organisation (ISO), the Comité Europeén Normalisation (CEN) and the American Conference of Government Industrial Hygienists (ACGIH) publishing standards based on the same aerosol penetration curve. This penetration curve was primarily based on the measured inhalability of breathing manikins in wind tunnels operated at between 1 and 4 m s^{-1} (Ogden and Birkett, 1977; Armbruster and Breuer, 1982; Vincent and Mark, 1982). However, evidence is beginning to accumulate indicating that typical average wind speeds in indoor work environments tend to be closer to 0.1–0.2 m s^{-1}. Recent measurements made by the Health and Safety Laboratory indicate that the wind speed in indoor workplaces is typically less than 0.2 m s^{-1} for the majority of the time, and show good agreement with earlier measurements taken by Berry and Froude (1989).

To investigate whether the current inhalable convention is a suitable model for inhalability at these low wind speeds as well as wind speeds above 1 m s^{-1}, we have devised a method of measuring inhalability in very low wind speeds, and carried out a number of preliminary measurements. Measurement methods were initially developed independently in two laboratories (the Health and Safety Laboratory (HSL) and the Institute of Occupational Medicine (IOM)) to guard against laboratory-dependent results. These were subsequently compared and a standardised approach agreed upon.

EXPERIMENTAL METHOD

Inhalability measurements were carried out in an enclosed chamber with no net through flow of air. Using this approach average air movements of significantly less than 5 cm s^{-1} were achievable, thus giving what can nominally be described as calm air conditions. An aerosol was generated and mixed above the chamber to give a spatially homogeneous distribution, then allowed to settle under gravity into the sampling area. The generation and mixing methods used by IOM and HSL differed slightly, but both were shown to give a spatially homogeneous aerosol to typically within ± 10%. All measurements were carried out using fused alumina particles

(Duralum, formerly known as Aloxite), as this has been found to be a suitable material in terms of monodisperseity and available particle sizes in similar wind tunnel experiments.

Inhalability was calculated by measuring the aerosol mass aspirated through the mouth opening of a breathing manikin and comparing it to the ambient aerosol concentration. The breathing manikin employed in both experiments was a modified "Little Annie" resuscitation dummy. Although slightly smaller than an average adult, these dummies are anatomically correct from the waist up (bar the absence of arms) and are easily adapted to connect to a breathing machine. The manikin was fitted with a mouthpiece of the same dimensions as those used in previous inhalability studies (22 mm inside diameter copper pipe flattened to a minimum internal height of 7 mm), which then constituted the mouth opening. A glass fibre filter was mounted in the rear of mouthpiece, allowing the aspirated aerosol mass to be measured by weighing the complete mouthpiece assembly. The manikin was rotated at 1 rpm in one direction throughout measurements to minimise the effect of aerosol spatial inhomogeneities. In all presented results a sinusoidal breathing pattern of 20 breaths per minute with a tidal volume of 1 litre was used.

As the nature of the calm air sampling chamber precluded isokinetic sampling, alternative methods of measuring the ambient aerosol concentration were investigated. The use of static sampling probes was rejected due to predicted reduced aspiration efficiency at large particle diameters (Grinshpun *et al.*, 1993).

In principle the mass of aerosol settling onto a surface in calm air is dependant only on the aerosol concentration, its average terminal setting velocity, the amount of time it is allowed to settle for and the area of the surface it is settling on to. Thus if the settled mass, the settling velocity, the collection time and the collection area are known, the ambient concentration may be calculated (Maynard and Baldwin, 1996). Following this approach, the ambient (reference) aerosol concentration was calculated using aerosol settling into cups of known area suspended in the sampling chamber. These were rotated at the same rate as the manikin to average over aerosol inhomogeneities. The aerodynamic diameter size distribution of the ambient aerosol was calculated using electrical sensing zone analysis of the settled particles to obtain the particle volumetric diameter, then transforming to aerodynamic diameter using data on particle density and shape factor published by Mark *et al.* (1985). The expression for particle settling at high Reynolds numbers (Hinds, 1982) was used to obtain the average terminal settling velocity from the measured aerodynamic diameter size distribution.

An alternative method of measuring the reference aerosol concentration used a rotating sharp edged probe to simulate isokinetic sampling in the calm air chamber. A 3 cm diameter sharp edged probe was rotated at 1 rpm in the chamber and the inlet velocity matched to the rotation velocity. At this rotation rate there was no noticeable turbulence in the wake of the sampler. Sampling was carried out at $1 \, l \, min^{-1}$. As the probe was circling there was an inevitable velocity mismatch across its diameter, as the outer edge travelled faster than the inner edge. However this was only of the order of 5% and was not considered a significant source of error. The mass sampled by the rotating probe was measured by weighing the complete assembly before and after sampling.

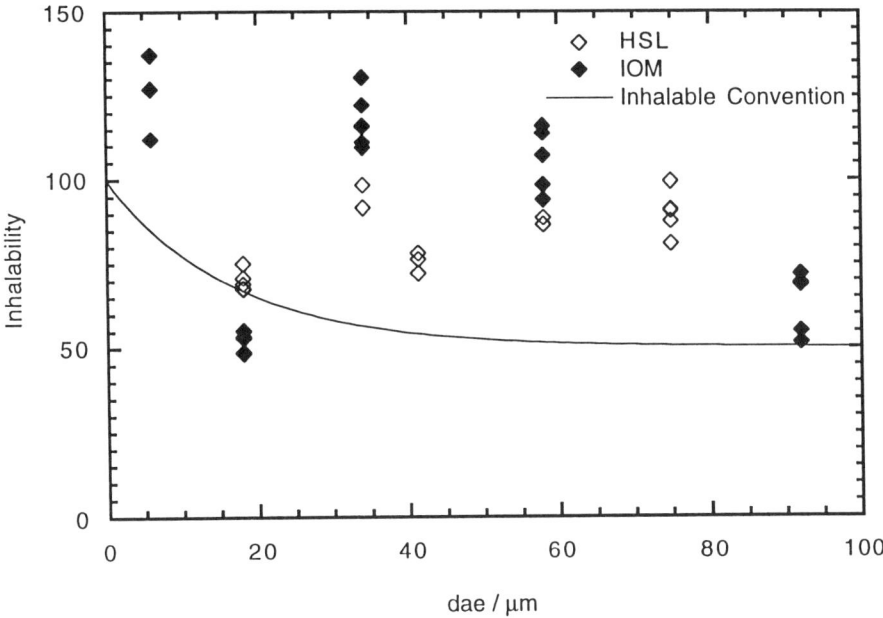

Fig. 1. Inhalability measured in calm air using gravitational settling to estimate the ambient aerosol concentration.

Both methods of measuring the ambient aerosol concentration required particles on the outside of the collection receptacles to be removed before weighing, thus introducing a potential source of error.

RESULTS

Using the experimental systems and methods above both laboratories measured inhalability in calm air at a number of particle aerodynamic diameters. The reference aerosol concentration was measured using the settling and rotating probe methods simultaneously in each case. Figure 1 shows measured inhalability at both IOM and HSL using settling to measure the reference aerosol concentration. In contrast Fig. 2 presents the measured inhalability using the rotating probe as the method of determining the reference aerosol concentration.

DISCUSSION

Both Figs 1 and 2 show a relatively good degree of agreement between the two laboratories when using both the rotating probe and the settling cups to measure the reference aerosol concentration. However, the agreement between inhalability measurements using the cups and those using the probe is poor. Before commenting on the nature of inhalability in calm air it is therefore necessary to decide which method, if any, of measuring the reference aerosol concentration gives a better estimate of the concentration.

The calculation of aerosol concentration from settled samples relies on a number

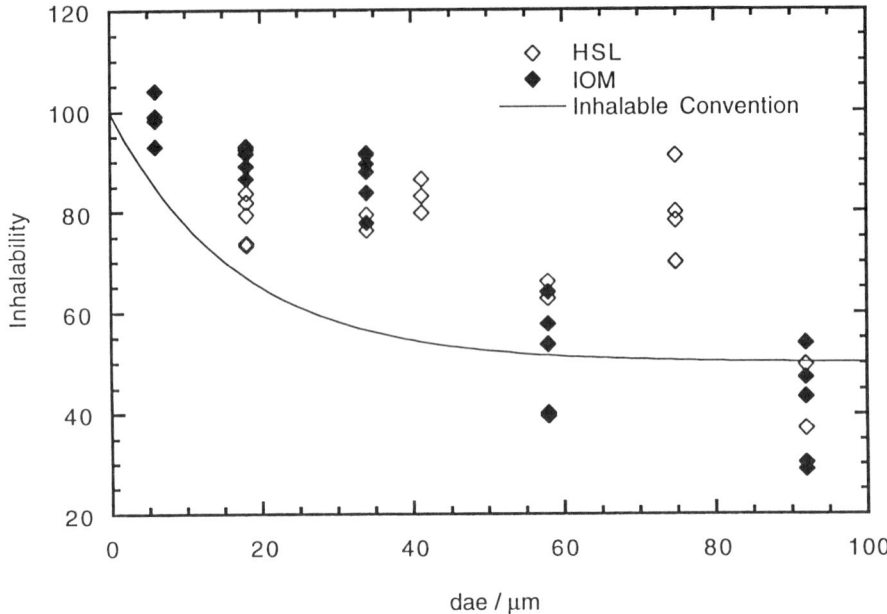

Fig. 2. Inhalability measured in calm air using a rotating 'isokinetic' probe to estimate the ambient aerosol concentration.

of assumptions: particles are travelling at their theoretical terminal settling velocity; the values of the particle shape factors and density used are correct; the rotation of the cups does not lead to significant changes in the sampled mass. In principle aerosol particles will deposit in the cups at their theoretical terminal settling velocity. However, without direct measurement this can only be an assumption. While the particle density is unlikely to differ from published values, it is possible that the shape factor of the fused alumina used will vary from batch to batch. It can be qualitatively demonstrated that rotating settling cups will over-sample small diameter particles, although quantitative analysis is not trivial. Measurements in the calm air chamber comparing static and rotating cups were inconclusive due to the spread in the results.

Errors associated with the circling probe are likely to arise from velocity mismatches across the inlet of the probe, and from turbulence induced by the passage of the probe. Using a dust lamp to visualise the aerosol there was no evidence for significant turbulence in the wake of the sampler. Velocity mismatches across the sampler were of the order of 5%, which would be unlikely to be significant at small particle diameters and of only marginal significance at larger particle diameters.

Both measurement methods are likely to have had associated handling errors, particularly as particles deposited on the outside of both the cups and the rotating probe had to be removed before weighing. However the associated errors are likely to have been similar for each sample type and thus are unlikely to account for the differences seen.

On balance there was more uncertainty over the validity of the gravitational settling method of reference aerosol concentration measurement, particularly in the light of the assumed settling velocities and the particle shape factors. Thus the rotating probe was used as the method giving the closest approximation to the reference aerosol concentration.

Comparison between the inhalability measurements with the rotating probe (Fig. 2) and the inhalable convention indicates generally higher values of inhalability at low particle diameters. Inhalability decreases with increasing particle diameter, but not as fast as the inhalable convention. Although high, these preliminary data do not lie significantly outside the envelope of previously published inhalability data. In addition the measurements carried out at IOM and HSL have indicated that the values obtained may be highly dependent on the experimental set-up. Potentially critical factors include the area of the manikin's mouth and the angle of its head. There is a clear need to investigate these factors further. Thus on the current data there is no evidence to indicate the model used to represent inhalability is inadequate at low wind speeds.

CONCLUSIONS

We have successfully developed methods for measuring inhalability in very low wind speeds at two laboratories. Comparison of inhalability measurements at a number of different aerodynamic diameters shows good agreement between them. However, these preliminary measurements have indicated that the measured inhalability is affected by experimental set-up and may vary significantly as various parameters are altered. Specifically, there are indications that the area of the manikin's mouth and the angle of the head are relevant to the measured inhalability. Although the data presented here lie above the inhalable convention they are within the envelope of published wind tunnel based measurements. In the light of this and the uncertainty over the relevance of experimental parameters there is no evidence at present indicating the current inhalable convention is not a sutiable model at very low wind speeds.

REFERENCES

Armbruster, L. and Breuer, H. (1982) Investigation into defining inhalable dust. *Inhaled Particles*. Pergamon Press, Oxford.
Berry, R. D. and Froude, S. (1989) An investigation of wind conditions in the workplace to assess their effect on the quantity of dust inhaled. HSE Internal Report, IR/L/DS/89/3. Health and Safety Executive.
Grinshpun, S. A., Killeke, K. and Kalatoor, S. (1993) A general equation for aerosol aspiration by thin walled sampling probes in calm and moving air. *Atmos. Environ.* **27A**, 1459–1470.
Hinds, W. C. (1982) *Aerosol Technology*. Wiley, New York.
Mark, D., Vincent, J. H., Gibson, H. and Witherspoon, W. A. (1985) Applications of closely graded powders of fused alumina as test dusts for aerosol studies. *J. Aerosol Sci.* **16**(2), 125–131.
Maynard, A. D. and Baldwin, P. E. J. (1996) Aerosol mass concentration measurements in calm air during sampler characterization. *10th Aerosol Society Conference*, The Aerosol Society, Swansea.
Ogden, T. L. and Birkett, J. L. (1977) The human head as a dust sampler. *Inhaled Particles IV*. Pergamon Press, Oxford.
Vincent, J. H. and Mark, D. (1982) Applications of blunt sampler theory to the definition and measurement of inhalable dust. *Ann. occup. Hyg.* **26**(1), 3–19.

Pergamon

Ann. occup. Hyg., Vol. 41, Supplement 1, pp. 700–706, 1997
© 1997 British Occupational Hygiene Society
Published by Elsevier Science Ltd. All rights reserved
Printed in Great Britain
0003–4878/97 $17.00 + 0.00
Inhaled Particles VIII

PII: S0003–4878(96)00175–5

PERSONAL EXPOSURE MEASUREMENTS OF THE GENERAL PUBLIC TO ATMOSPHERIC PARTICLES

D. Mark,* S. L. Upton,† C. P. Lyons,† R. Appleby,‡ E.J. Dyment,†
W.D. Griffith† and A.A. Fox†

*Institute of Occupational Health, University of Birmingham, Birmingham, U.K.;
†Aerosol Science Centre AEA Technology, Harwell, Didcot, U.K.; and ‡Environmental Pollution Unit,
Birmingham City Council, Birmingham, U.K.

INTRODUCTION

Airborne particles in ambient atmospheres in the U.K. are continuously monitored using tapered element oscillator microbalances (TEOM) (Patashnick and Rupprecht, 1991), equipped with inlets that select the thoracic aerosol fraction according to the PM10 definition. These are part of the Automatic Urban Network (AUN) of the U.K. Department of the Environment, which is a network of sites located in cities throughout the U.K. In order to use this information to estimate the exposure and potential dose of particles to the general public, it is necessary to understand the relationship between these measurements and actual exposure as measured with personal samplers. These personal samplers are worn on the upper part of the chest close to the nose and mouth of the wearer and are specially designed to mimic the aerodynamic behaviour of the human respiratory system. The aim of this project is to obtain a reliable assessment of the particle exposures of a representative group of the general public and to use this information to determine how effectively data from the fixed site TEOM monitors can estimate that exposure.

PREVIOUS PERSONAL EXPOSURE DATA TO PM10 PARTICLES

So far, most of the published data on the personal exposure of the general public to ambient airborne particles has been carried out in the U.S.A. Since the early 1980s, a small number of long-term studies have been initiated to estimate the exposure of the general public to airborne pollutants, including particles. One of the major studies carried out in the U.S.A. during this period was the Six City study (Spengler, 1985), carried out by investigators mainly from the Harvard School of Public Health. In this study, estimates of respirable particle exposures in six cities in the U.S.A. were obtained using both personal samplers and micro-environmental and/or outdoor fixed-site monitors. In the total human environmental exposure study (THEES) (Lioy, *et al.*, 1990) personal, indoor and outdoor concentrations of thoracic (PM10) particles were measured for the first time. The total exposure assessment methodology (TEAM) study (Wallace *et al.*, 1993) involves the measurement of PM10 particles in a large number of residents in Riverside, California.

However, despite the existence of these three large exposure studies, the number of measurements of particle exposure made with actual personal samplers is relatively small (250 respirable and 530 PM10). Instead, most of the exposure estimates were based on the use of exposure models in which particle levels in a number of micro-environments (normally five), measured using static monitors, are used in combination with information from activity diaries to build up exposure profiles for each subject. However, whilst these models may effectively estimate individual exposures to gases such as CO, NO_2 and SO_2 there is very little evidence concerning their validity for estimating particulate exposures. Consequently, many authors expressed a preference for personal sampling, but argued against large-scale personal monitoring studies on the grounds of excessive cost.

THE PERSONAL SAMPLER USED FOR THE STUDY

The choice of sampler used in this study was severely limited by the lack of suitable samplers. Consequently, a modified version of the prototype personal sampler, reported by Mark et al. (1988), was used. The sampling head has an internal cassette with an inhalable aerosol entry followed by a cylindrical plug of porous polyester foam to select the thoracic fraction of the inhaled particles, which are collected on a 37 mm diameter filter. A specially-built (small, light and quiet) sampling pump was used, capable of sucking air at $2\,l\,min^{-1}$ for at least 16 h each day with a fast battery charging system that enabled the sampler to be used the next day. Particular attention was paid to making the sampler as quiet as possible. An unobtrusive, easily removable carrier harness, based around a small camera case, was used to carry both the pump and the sampler. The sampler itself was fixed to the carrying strap and was maintained in the lapel position by a combination of removable loops and clips.

THE SAMPLING CAMPAIGN

The sampling campaign was based around the two AUN sites within the City of Birmingham. One is situated in the centre of the city in the corner of a car park close to the Centenary Square, whilst the other is in the playground of a school within a suburban area of the city.

Sampled population

The sampled population comprised 15 members of the general public at each site, although some volunteers dropped out after the first exercise and were replaced with substitutes. The main criterion for the choice of subjects was that of reliability, to ensure that the samples obtained did represent actual personal exposure measurements for the periods specified. At each site we were able to identify three main sub-groups, who have similar living and working arrangements (called lifestyle in the analysis of the results), and therefore may have similar exposures. They were: (1) retired people who live very close to the AUN stations; (2) people who also live close to the AUN stations, but who work at least 3 miles away; and (3) people who work close to the AUN stations (Centre—office workers,

East—teachers), but who live at least 3 miles from the site. The gender of the volunteers was: Centre, 15 male, 3 female; East, 12 female, 3 male.

Sampling programme

For each group of subjects, three sampling surveys were carried out: Survey 1 in August/September 1995, Survey 2 in October/November/December 1995 and Survey 3 in February/March 1996. For each survey, a target of five replicate exposure periods for each subject was set. These were of variable duration from 16 to 33 hours, with the samplers being switched off during sleep time, and at some workplaces. Subjects completed an initial questionnaire about their housing, heating, method of cooking, occupation, hobbies and kept a diary of their major activities during the exposure measurements.

Experimental analysis

Glass fibre filters were used to collect the sampled particles. The mass of particles collected on each filter was determined by weighing the filter before and after sampling twice on a six-place electronic balance. A rigid system of filter conditioning and controls was employed to enable the maximum accuracy to be achieved with the expected small particulate masses (expected to be around 50–1000 µg).

The mean outdoor PM10 mass concentration levels were calculated from the 15 min averages of continuous measurements given by the TEOM instruments at the two AUN sampling stations. The raw data have been modified following the agreed acceptance criteria for the AUN network, but must still be considered as provisional at this stage. These were calculated for each personal exposure sample for the periods during which the personal samplers were in operation.

Statistical analysis

Two main statistical analyses were carried out. Firstly, descriptive statistics were calculated for the personal exposure measurements, subdivided according to survey, smoking status and lifestyle. In order to stabilise the variances, the natural logarithms of the personal exposure measurements have been used for all models tested. Regression analyses were carried out to investigate the relationship between the personal exposure measurements and the AUN measurements. Factors such as exposure to tobacco smoke or lifestyle were treated as confounders in some analyses and their effects on the relationship between personal exposure measurements and AUN measurements were explored using analyses of covariance.

RESULTS

The results of the descriptive statistics for the personal exposure measurements for both sites are shown in the box plots of Fig. 1 (a)–(c). In Fig. 1(a), significant differences between surveys was found both between sites and between seasons (survey 1 being in the summer and 2 and 3 being in the winter). The effects of smoking on exposures can be seen in Fig. 1(b), where for both sites, the differences were highly statistically significant, with smokers exposed to almost twice as much mass of thoracic particles as the non-smokers. Levels for the passive smokers at

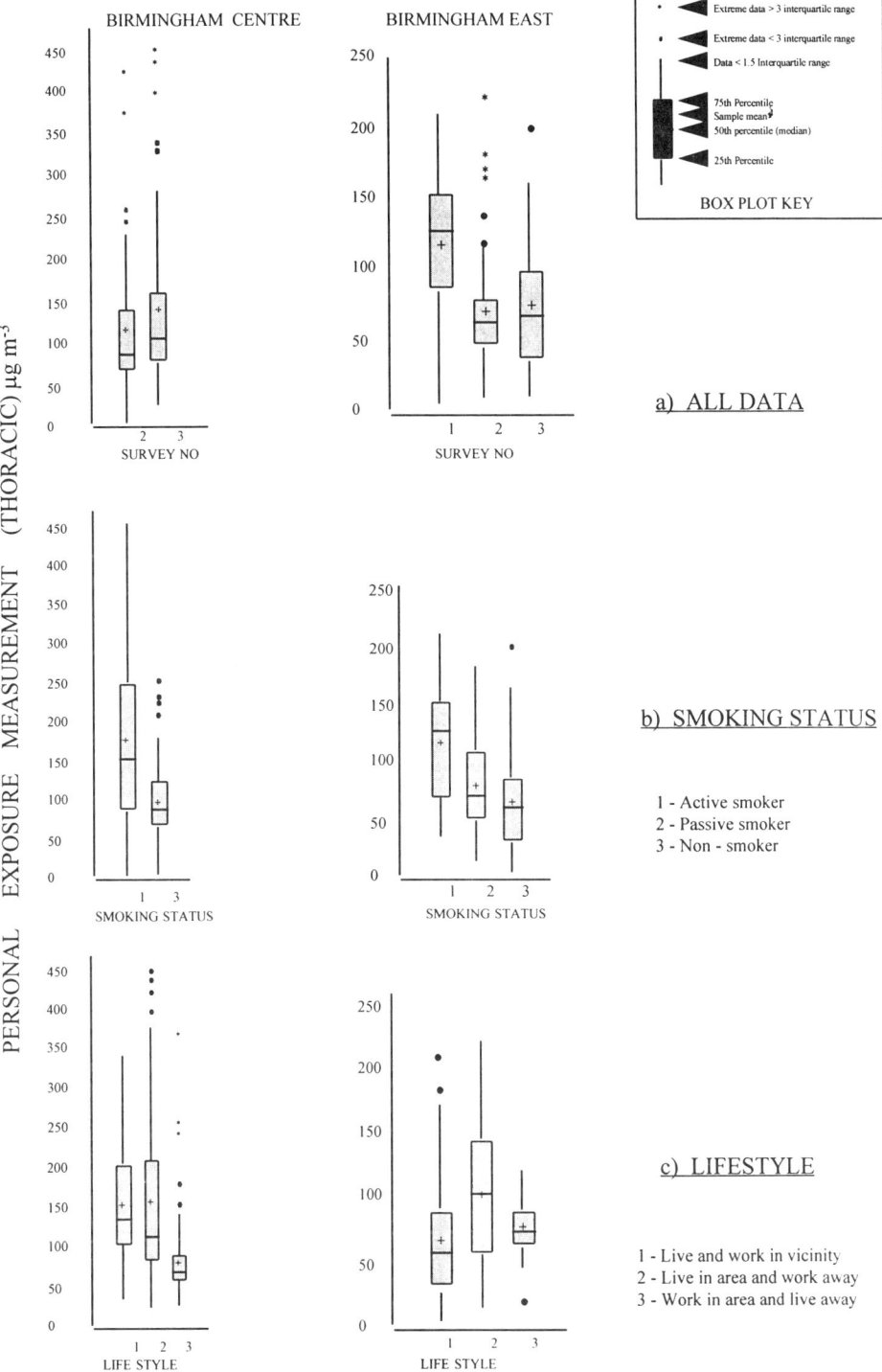

Fig. 1. Box plots of personal exposure measurements.

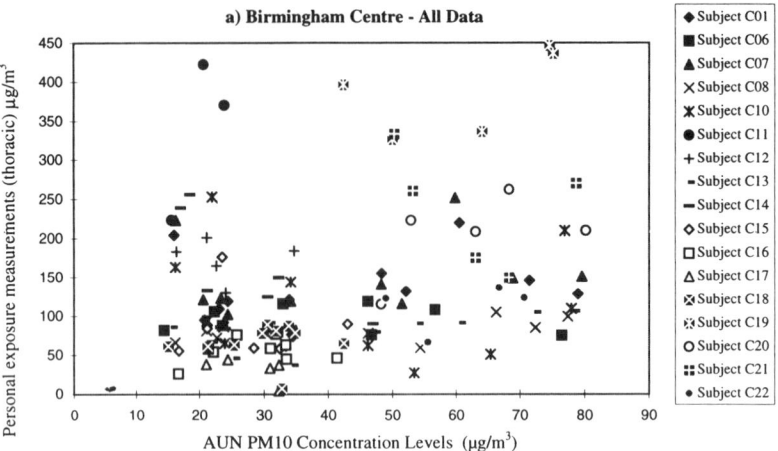

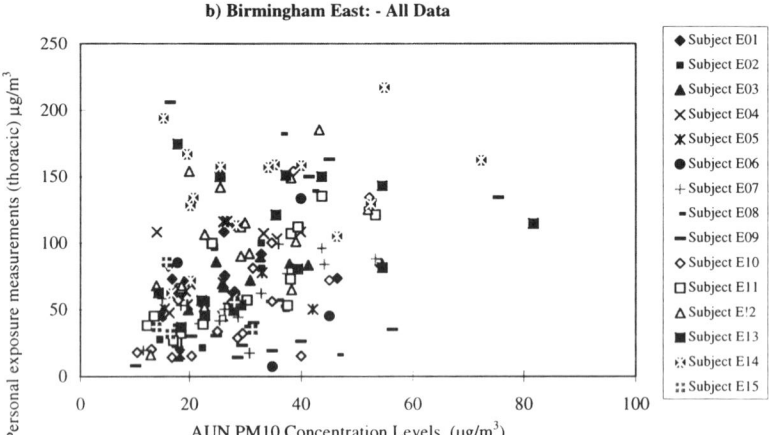

Regression equation: *ln (personal exposure) = intercept + AUN x slope*

	No of Samples	Intercept	Slope	p	R²
		ln(ug/m³)			
CENTRE					
All subjects	128	4.262	0.009	0.0035	0.066
Non smokers	76	4.077	0.009	0.0086	0.09
EAST					
All subjects	173	3.576	0.02	0.0001	0.144
Non smokers	143	3.438	0.022	0.0001	0.145

Fig. 2. Relationship between personal exposure and AUN measurements.

East were only slightly higher than those for the non-smokers. In Fig. 1(c), the effects of where people live and work in relation to the AUN sites, can be seen. For Centre, those subjects living and working in the area and those living in the area, but working away are exposed to approximately twice as much mass of thoracic particles as those working in the area and living away. For East, the picture was slightly different, with those subjects living and working in the area experiencing the lowest exposures and those living in the area but working away appeared to have the highest. These effects were statistically significant with P values < 0.001.

The results of all the measurements of personal exposure and corresponding PM10 levels from AUN monitors for Centre and East are given in Fig. 2(a) and (b), respectively. Regression analyses carried out to investigate the relationships between the data are given on the figures. For centre, the overall positive association between personal exposure measurements and AUN PM10 values, just fell short of statistical significance both for the model with all subjects and for that with smokers excluded. For East, the overall positive association between personal exposure measurements and AUN values was statistically significant ($P = 0.0001$), both for the model with all subjects and that with smokers excluded. However, there is a wide spread of results with the regression relationship explaining only 14% of the variability ($R^2 = 0.14$).

DISCUSSION

This is the first systematic study carried out in the U.K., to determine the personal exposure of members of the general public to airborne particles. It represents a 50% increase in the number of actual personal samples reported irrespective of size fraction measured and a 60% increase in measurements of the thoracic fraction. The main conclusion from this short study is that there is no simple relationship between PM10 levels obtained from the AUN particle monitors at Birmingham and the thoracic particle exposures of a small group of the general public living and/or working in the area. However, there is increasingly strong evidence of a correlation between PM10 levels in the ambient atmosphere and certain ill-health in the general public (Schwartz, 1991; Dockery et al., 1993). As people spend on average over 90% of their time indoors (confirmed in this study), the question that immediately springs to mind is: "Is there high penetration of outdoor aerosols into indoor environments?" If the answer to this question is "yes", then it is conceivable to suggest that it is the fine outdoor particles that are mainly responsible for the ill-health reported. This suggestion would lend some support to the proposition by Seaton et al. (1995) that ultra-fine particles (< 100 nm) are the main causative agents for the reported circulatory health effects. However, there are also many potentially harmful particles that are derived solely from indoor sources such as: particles from gas cookers and heaters, environmental tobacco smoke, mould spores, house dust mite faeces, dander, skin, cat, bird and insect allergens. These are variously reported to have potential allergenic, pathological and carcinogenic properties. Therefore, if the main aim of particle monitoring is to provide information for the minimisation of all particle-related ill-health, then it may not be sufficient just to monitor the ambient atmosphere via the AUN system. Further work is required.

Acknowledgements—The authors would like to acknowledge the support both financially and technically of the U.K. Department of the Environment for this work. Our special thanks go to those people in Birmingham, who willingly and responsibly participated in the exposure measurements.

REFERENCES

Dockery, D.W., Schwartz, J. and Spengler, J. D. (1992) Air pollution and daily mortality: associations with particulates and acid aerosols. *Environ. Res.* **59**, 362–373.

Lioy, P. J., Waldman, J. M., Buckley, T., Butler, J. and Pietarinen, C. (1990) The personal, indoor and outdoor concentrations of PM10 measured in an industrial community during winter. *Atmos. Environ.* **24B** (1), 57–66.

Mark, D., Borzucki, G., Lynch, G., Vincent J. H. (1988) The development of a personal sampler for inspirable, thoracic and respirable aerosol. In *Aerosols: Their Generation, Behaviour and Applications, 2nd Conference of the Aerosol Society*, Bournemouth, March 1988, pp. 183–187.

Patashnick, H. and Rupprecht, E. G. (1991) Continuous PM10 monitoring using the tapered element oscillating microbalance. *J. Air Waste Management Assoc.* **41**, 1079–1083.

Schwartz, J. (1991) Particulate air pollution and daily mortality in Detroit. *Environ. Res.* **56**, 204–213.

Seaton, A., MacNee, W., Donaldson, K. and Godden, D. (1995) Particulate air pollution and acute health effects. *The Lancet* **345**, 176–178.

Spengler, J. D., Treitman, R. D., Tosteson, T. D., Mage, D. T. and Soczek, M. L. (1985) Personal exposures to respirable particulates and implications for air pollution epidemiology. *Environ. Sci. Technol.* **19**, 700–707.

Wallace, L. A. (1993) A decade of studies of human exposure: what have we learned? *Risk Analysis* **13** (2), 135–143.

 Pergamon

Ann. occup. Hyg., Vol. 41, Supplement 1, pp. 707–713, 1997
© 1997 British Occupational Hygiene Society
Published by Elsevier Science Ltd. All rights reserved
Printed in Great Britain
0003–4878/97 $17.00 + 0.00
Inhaled Particles VIII

PII: S0003–4878(96)00142–1

SUBJECT INDOOR AND OUTDOOR PM$_{2.5}$ EXPOSURE PROFILES

A. Frøsig* and D. Sherson†

*Department of Pulmonary Medicine, Bispebjerg Hospital, Copenhagen, Denmark; and
†Department of Occupational and Environmental Medicin, Vejle Hospital, Vejle, Denmark

INTRODUCTION

Observations support the view that airborne particulate matter is an important risk factor for acute lung function changes, respiratory symptoms and chronic respiratory illness and death among high risk groups (Folinsbee, 1992; Schwartz, 1994; Pope, 1995). Taken together, these studies suggest that effects of particulate matter occur at lower concentrations than previously believed. In urban areas, information about concentrations of fine particulate matter and their deposition in human airways is important. The ability to understand the effects of inhaled particulate matter is limited at present by several factors among which is a lack of assessment of the indoor and outdoor inhaled dose.

The most widely used current method for assessing potential human exposures involves batch filter sampling over long averaging times to obtain the mass of particles below a given aero-dynamic size (particulate matter less than 2.5 µm in aerodynamic diameter, PM$_{2.5}$; or PM$_{10}$ and total suspended particulates, TSP). Subjects were followed over several days to obtain real time detailed exposure information of indoor and outdoor concentrations of PM$_{2.5}$. This paper describes preliminary exposure profiles and discusses the uptake of fine particulate matter.

MATERIALS AND METHODS

Portable nephelometers (Model M901, Radiance Research, Seattle) were used for determination of the PM$_{2.5}$ airborne particulates. The M901 nephelometer is a light weight, low power instrument designed for field use. The instrument allows measurement of the light scattering coefficient (b_{sp}) above approximately 10^{-6} m^{-1} at an averaging time of 1 min (Weiss, 1992). The light scattering coefficient is proportional to the particulate mass concentration in the size range 0.1–2.5 µm. Results have shown a very good correlation ($r^2 = 0.95$) between PM$_{2.5}$ in ambient air and b_{sp} for an averaging time of 24 h (Koenig et al., 1993).

Our results indicate that for short averaging times (½ h) the relationship between PM$_{2.5}$ and b_{sp} is dependent on wind speed (W ms^{-1}), the ambient photochemical potential P defined as the sum of ambient O$_3$ and NO$_2$ (µg m^{-3}) and the ambient relative humidity (H%):

$$PM_{2.5} = 0.356[Bsp10^4]^{0.545}[1/W]^{0.073}[P]^{0.696}[(100–H)/100]^{0.089}$$

$$(n = 1354, r^2 = 0.60). \tag{1}$$

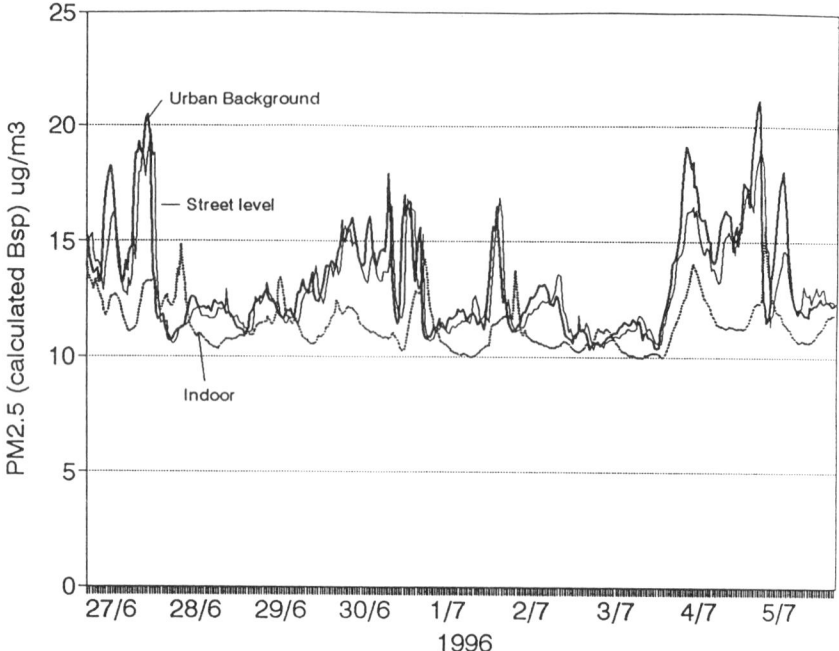

Fig. 1. PM$_{2.5}$ time series, June 27th to July 5th, Copenhagen. Urban background, street level, and indoor concentration level.

Nephelometers were set-up at indoor and outdoor sites in downtown Copenhagen, in the Copenhagen subway system, at different office environments, in welding stall at a welding school, in a pig barn and at a bar. At several of these sites significant smoking took place (about 40% of the adult Danish population are daily smokers). Exposure profiles were summarized as cumulative frequency distributions (C.F.D.).

Measured exposure concentrations can be converted to calculated depositions by the revised ICRP Human Respiratory Tract Model for Radiological Protection (ICRP, 1994). Deposition profiles were calculated for typical concentrations of PM$_{2.5}$, PM$_{10}$, TSP and Respiratory Particulates as defined by the Danish Institute of Occupational Health. Measurements of lung depositions of particulates (diameter 0.6 µm) agree very well with the ICRP model estimations (Frøsig et al., 1992).

RESULTS

Figures 1–4 show different exposure profiles. Urban background and street level PM$_{2.5}$ showed nearly identical variation over days and concentration levels. Indoor PM$_{2.5}$ levels were typical determined by outdoor concentrations and averaged 85% of the outdoor concentrations (Fig. 1). Figure 2 shows the effect of smoking among teachers during their morning and lunch breaks. It is notable that the concentration peaks in the staff room are about six times the typical urban background. Figure 3 shows the exposure environment of a subject in a pig barn. He was exposed to

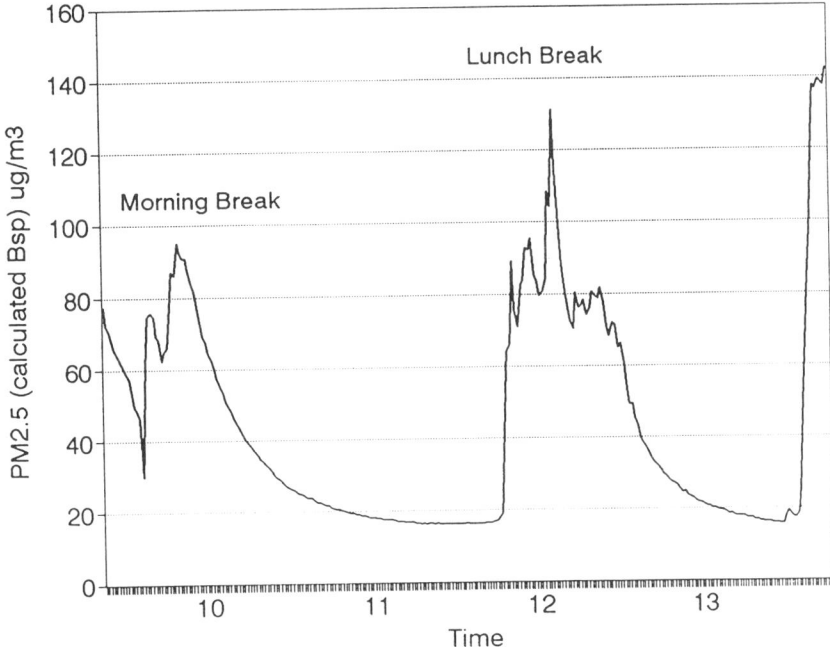

Fig. 2. PM$_{2.5}$ time series, in welding school staff room with smoking during morning and lunch breaks, May 5th, 1996.

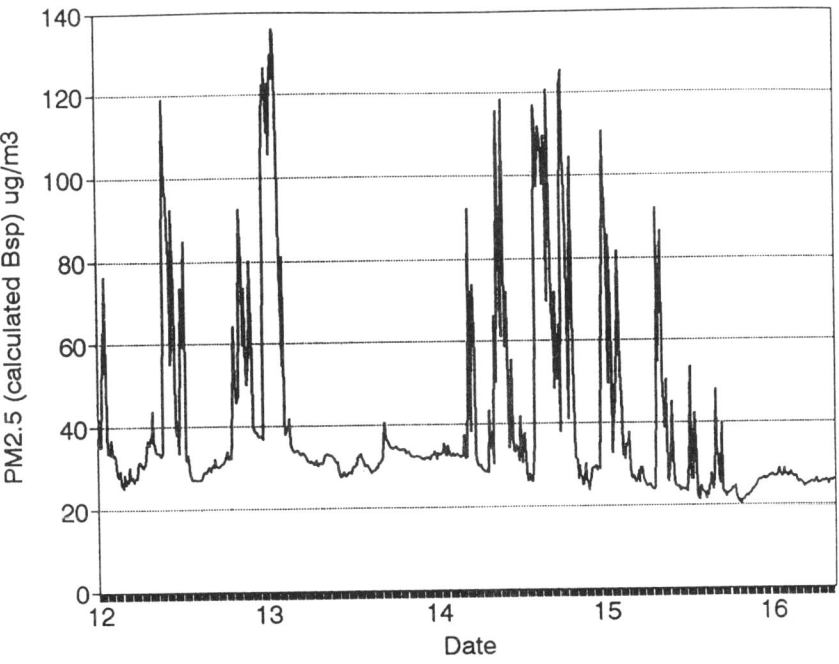

Fig. 3. PM$_{2.5}$ time series in a pig barn, January 12th to 16th, 1996.

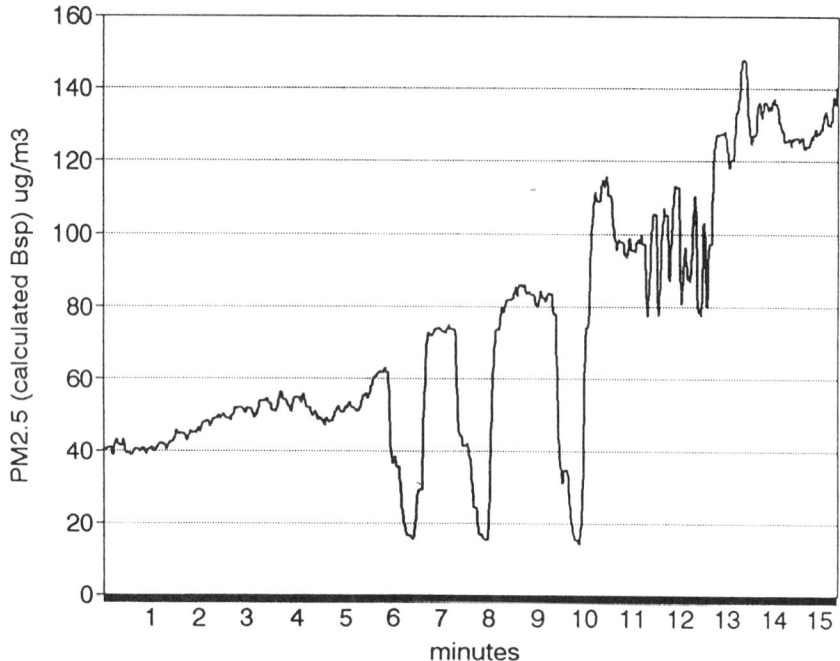

Fig. 4. PM$_{2.5}$ time series in a welding stall during breathing (please confer text for explanation).

similarly elevated concentrations as typical caused by intense smoking. Figure 4 shows PM$_{2.5}$ exposure of a subject working in a welding stall. After 6 min the subject took three deep inhalations and blew directly into a mouth piece connected to the nephelometer. The "dips" show how the PM$_{2.5}$ concentrations in his exhaled air decrease during full exhalation, i.e. most particulated matter was deposited in the airway system. The variations from 11 to 13 min show the deposition pattern during normal breathing.

Figure 5 shows the estimated fractional depositions as a function of dry particle diameter in human airways for PM$_{2.5}$ (geometric mean of dry particles, $m_g = 0.5$ µm, geometric standard deviation, $s_g = 2$, dry mass concentration $m_c = 19$ µg m^{-3}) in a healthy female performing light activity during mouth breathing (breathing frequency 21 min^{-1} with an average minute ventilation of 21 min^{-1}). The estimated cumulative alveolar mass fraction of PM$_{2.5}$ deposited was 37%. For comparison, the estimated cumulative alveolar fraction of TSP deposited is only about 5%.

Figure 6 shows PM$_{2.5}$ concentrations in a non-smoker compartment in a subway wagon during a 45 min ride. During the ride PM$_{2.5}$ due to smoking in the smoking section is significantly polluting the air in the non-smoking compartment.

Different exposure profiles are shown in Figs 7 and 8 as cumulative frequency distributions for PM$_{2.5}$. PM$_{2.5}$ exposures from residential areas with wood stoves were similar to PM$_{2.5}$ levels emitted from diesel buses in heavy traffic except for the highest percentiles (Fig. 7). The exposures in different occupational environments can be rated as follows with regard to elevated concentrations: bar with heavy smoking > welding stall > subway travel next to a smoking compartment > working in a pig barn > office environment with smoking > urban background.

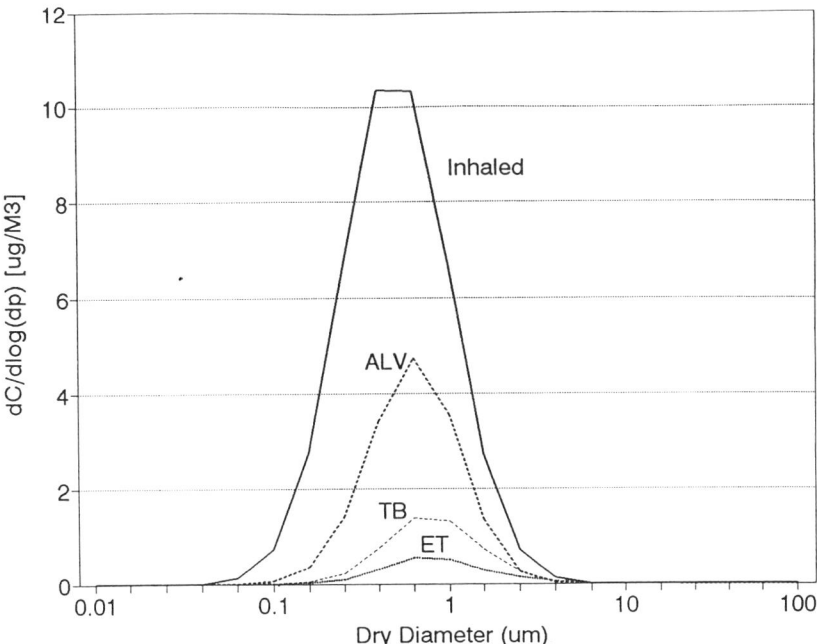

Fig. 5. Pulmonary fractional mass depositions of ambient PM$_{2.5}$ as a function of dry particulate size. Inhaled is the inhaled concentrations. ALV is the deposition in the alveolar, TB in the tracheo-bronchiolar, and ET in the extrathoracic region.

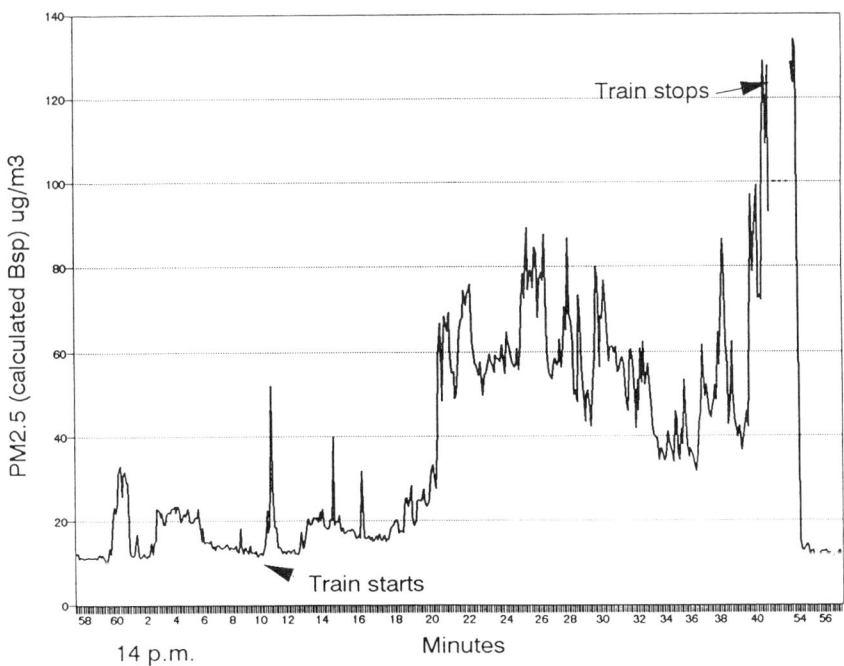

Fig. 6. PM$_{2.5}$ time series in a subway wagon during a ride from Down Town Copenhagen to Hillerød City 40 km North of Copenhagen.

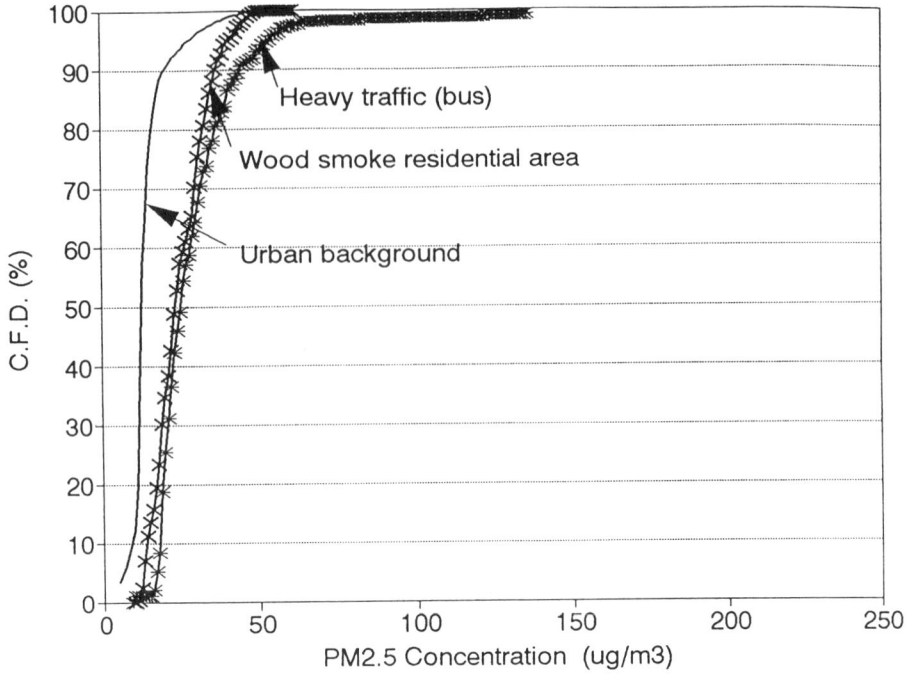

Fig. 7. PM$_{2.5}$ cumulative frequency distributions (C.F.D.) in ambient settings.

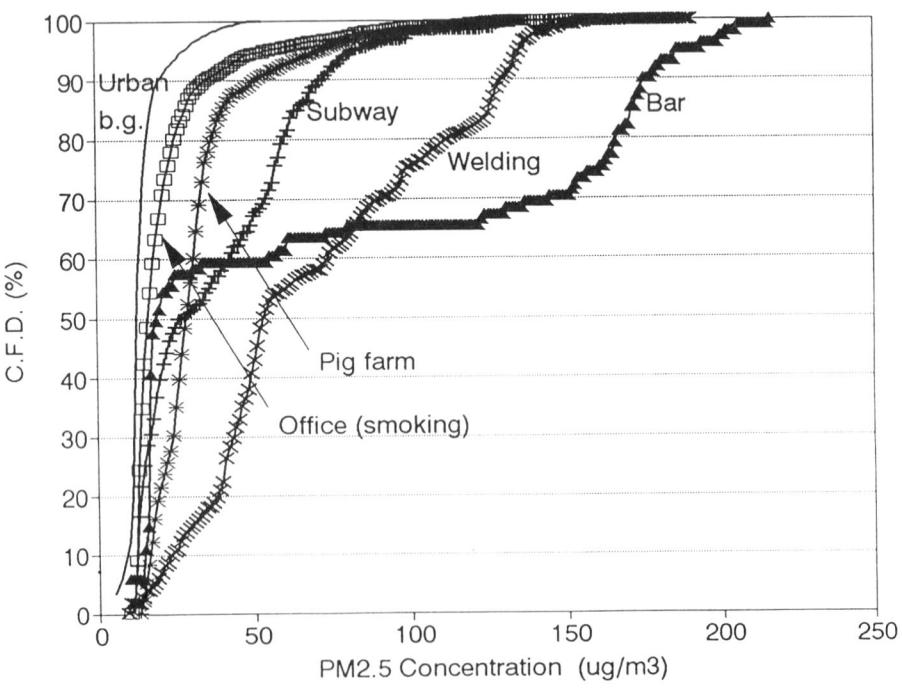

Fig. 8. PM$_{2.5}$ cumulative frequency distributions (C.F.D.) in occupational settings.

DISCUSSION

The portable nephelometer is suited for PM$_{2.5}$ assessment in different environments. Rapid and reliable real time *in situ* measurements are easily performed. The nephelometer measures the concentration profile most relevant for alveolar deposition. The fractional cumulative deposition estimations show that the nephelometer method in occupational settings can replace filter measurements with regard to determination of fine particulate alveolar deposition.

Not surprisingly, smoking in homes and other indoor places contributed to significant concentrations of fine particulate matter, such as in staff rooms, trains and offices. The concentrations reached due to heavy smoking were comparable to or higher than concentrations measured where high concentrations of fine particulates were expected, for example in welding stalls. Emissions from wood stoves can locally contribute to outdoor concentrations at a range similar to emissions from buses in heavy traffic site at downtown Copenhagen. We found significantly elevated concentrations of fine particulate matter at many places. This suggests that it is important to measure total subject exposure profiles, both indoor and outdoor.

REFERENCES

Folinsbee, L. J. (1992) Human health effects of air pollution. *Environ. Health Persp.* **100**, 45–46.
Frøsig, A., Larson, T. and Koenig, J. (1992) Development of a portable light scattering method for assessing dosimetry of fine particules. In *Proceedings from the 11th Annual Meeting*, American Association for Aerosol Research, San Francisco.
ICRP (1994) Human respiratory tract model for radiological protection. A report of a Task Group of the International Commission on Radiological Protection, ICRP Publ. 66, Pergamon Press.
Koenig, J. Q., Larson, T. V., Hanley, Q. S., Rebolledo, V., Dumler, K., Checkoway, H., Wang, D. L. and Pierson, W. (1993) Pulmonary function changes in children associated with fine particulate matter. *Environ. Res.* **63**, 26–38.
Pope, A. C., Thun, M. J., Namboodiri, M. M., Dockery, D. W., Evans, J. S., Speizer, F. E. and Heath, C. W. (1995) Particulate air pollution as a predictor of mortality in a prospective study of U.S. adults. *Am. J. respir. crit. Care Med.* **151**, 669–674.
Schwartz, J. (1994) Air pollution and daily mortality: a review and meta analysis. *Environ. Res.* **94**, 36–52.
Weiss, R. A. (1992) Operation procedures: M901 portable nephelometer. *M901 Portable Nephelometer Manual.* Published by Radian Research, Seattle, Washington, U.S.A.

Pergamon

Ann. occup. Hyg., Vol. 41, Supplement 1, pp. 714–718, 1997
© 1997 British Occupational Hygiene Society
Published by Elsevier Science Ltd. All rights reserved
Printed in Great Britain
0003-4878/97 $17.00 + 0.00
Inhaled Particles VIII

PII: S0003-4878(96)00103-2

DEPOSITION OF ULTRAFINE PARTICLES IN HUMAN TRACHEOBRONCHIAL AIRWAYS

Y.-S. Cheng, S. M. Smith and H.-C. Yeh

Inhalation Toxicology Research Institute, P.O. Box 5890 Albuquerque, NM 87111, U.S.A.

INTRODUCTION

Ultrafine particles smaller than 200 nm in diameter are produced by combustion, radioactive decay and gas-to-particle conversion in the urban air. Inhalation of ultrafine particles in both outdoor and indoor environments is a health concern. Exposure to radioactive radon progeny in the size range of 1–200 nm causes lung cancer in uranium miners and may have similar health effects in people exposed indoors. In the ambient air, ultrafine particles have a higher number concentration and may be more toxic than larger particles. Ultrafine particles deposit in the respiratory tract primarily by diffusion with increased deposition of smaller particles. Total lung deposition of ultrafine particles measured in human volunteers confirms diffusion as the mechanism (Schiller *et al.*, 1988). Higher deposition in the extrathoracic region has been determined in human nasal/oral casts and in human volunteers (Cheng *et al.*, 1993, in press). Deposition in the human tracheobronchial airways has been measured in an adult cast for particle sizes between 40 and 200 nm (Cohen *et al.*, 1990). Their data indicated higher deposition in the tracheobronchial airway than predicted by diffusional deposition in a tube, assuming a fully developed flow profile. The purpose of our study was to determine the tracheo-bronchial deposition in wider size range of particles using airway casts of adults and children. The deposition data were compared with theoretical predictions of both parabolic and plug flows.

MATERIALS AND METHODS

Hollow silicone rubber models of the tracheobronchila tree of 3, 16 and 23-year-old males were used in these studies. Each cast included the larynx, trachea and 5–8 generations of the tracheobronchial tree. Conductive material was used for the airway cast to minimize electrostatic deposition of particles. Air flow for all three casts was measured at the end of each flow path.

Molecular clusters of thoron progeny (^{212}Pb) and ^{212}Pb attached to monodisperse silver particles were used in the present study. The experimental apparatus is shown in Fig. 1. Unattached thoron progeny were generated by ^{220}Rn gas delivered from a dry ^{228}Th source to a 83.5 L stainless steel chamber. The ^{220}Rn decayed into ^{212}Pb, which is a gamma emitter with a half-life of 10.6 h. The residence time in the chamber (24 min) was sufficient to decay over 99% of thoron gas into thoron

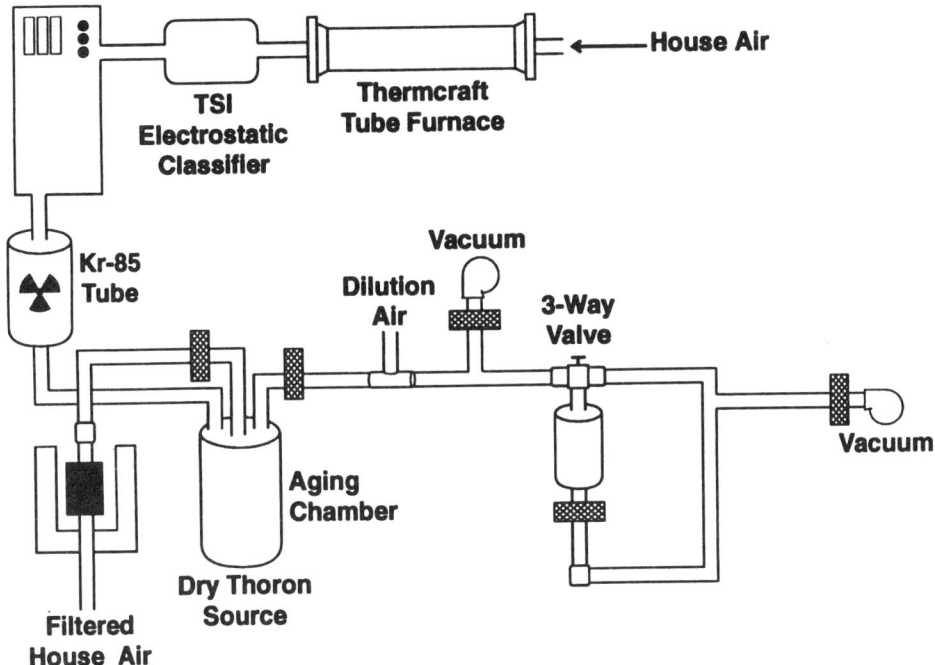

Fig. 1. Experimental apparatus for aerosol deposition study.

progeny. Polycisperse silver particles were generated using the vaporization/condensation method (Cheng *et al.*, 1990). Monodisperse Ag aerosols of 10 and 40 nm were then produced by passing the aerosol through an electrostatic classifier (Model 3071, TSI Inc., St Paul, MN). The singly charged Ag aerosol was charged–neutralized into the Boltzmann equilibrium by passing it through a ^{85}Kr tube. Silver aerosol was delivered to the stainless plenum to mix with ^{220}Rn gas. Silver particles were then radiolabelled with ^{212}Pb by coagulation (Strong *et al.*, 1994). Next, the aerosol passed through a filter holder with mesh screens (30/50/145) to remove the unattached thoron progeny. The size distribution of Ag particles was determined from the operating condition of the classifier and confirmed with a serial diffusion battery (Cheng *et al.*, 1991). The activity size distributions of the ^{212}Pb clusters and radiolabelled Ag aerosol were measured by using a graded diffusion battery (Cheng *et al.*, 1994). The aerosol charge fraction of the ^{212}Pb clusters measured using the method described by Porstendörfer and Mercer (1979) showed less than 3% were charged.

The radiolabelled aerosol was then delivered to a Lucite chamber lined with aluminum foil where an airway cast was suspended. The flow rates used for the 3-year-old cast were 10 and 20 l.min^{-1} and for the other two casts, 20 and 40 l.min^{-1}. Three particle sizes 1.7, 10 and 40 nm were deposited. Exposure times ranged from 1 h for the 1.7 nm aerosol to 4 h for 10 and 40 nm aerosols so that enough of the aerosols could be deposited in the cast. Following exposure, the outside walls of the cast were washed and the airways were cut into separate segments. In the first three generations, the bifurcation area was separated from

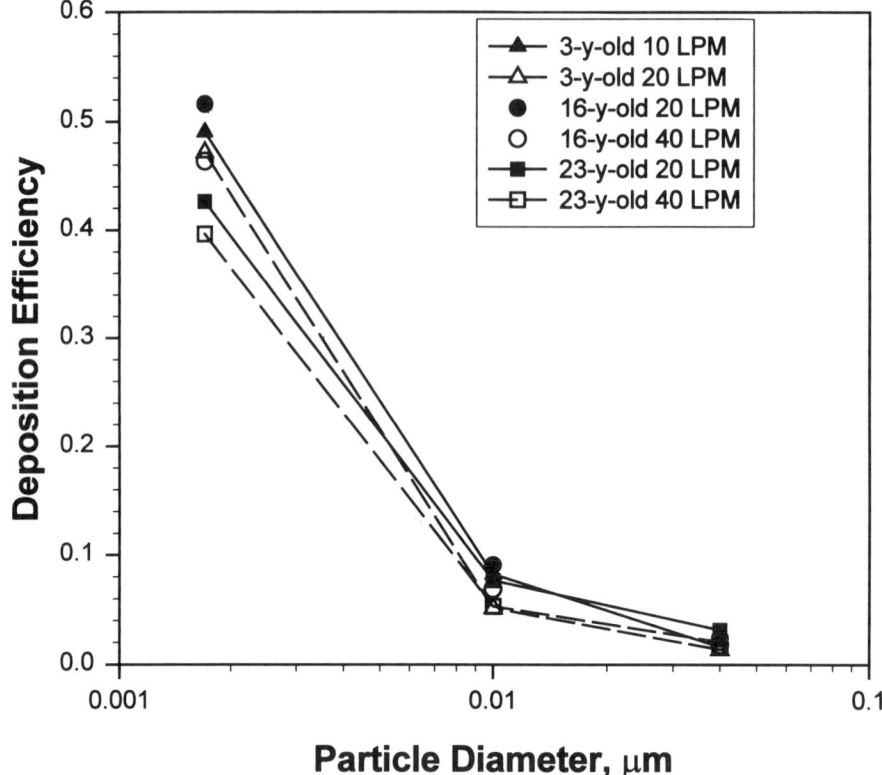

Fig. 2. Total deposition in the tracheobronchial region.

the other portion of the airway. The airway segment, foil, washes and back filter were counted for gamma activity. Based on the counting data and flow rates, the deposition was calculated for each generation of the tracheobronchial tree.

RESULTS

Figure 2 shows the tracheobronchial deposition efficiency in all three airways casts (for the flow rates used in the study). Deposition efficiency ranged from 1 to 3% for 40 nm particles to 40–53% for 1.7 nm particles. There appeared a small dependence on the flow rate, with a higher deposition for the low flow rates. These data reconfirmed that diffusional deposition is the dominate deposition mechanism. For 1.7 nm particles, deposition was higher in the 3-year-old cast than in the other casts. In the first three generations, deposition in the bifurcation was not enhanced. Experimental data were also compared with theoretical diffusion deposition in circular tubes for both parabolic flow and plug flow patterns (Ingham, 1975). Experimental data in all three casts were higher than theoretical predictions assuming a parabolic flow in tubes and are close to the prediction for plug flow. Figure 3 shows the comparison of experimental data (in the tracheobronchial region of the cast) and theoretical predictions for the 3 and 23-year-old casts in different flow rates. Similar data have been obtained for the other casts at all flow

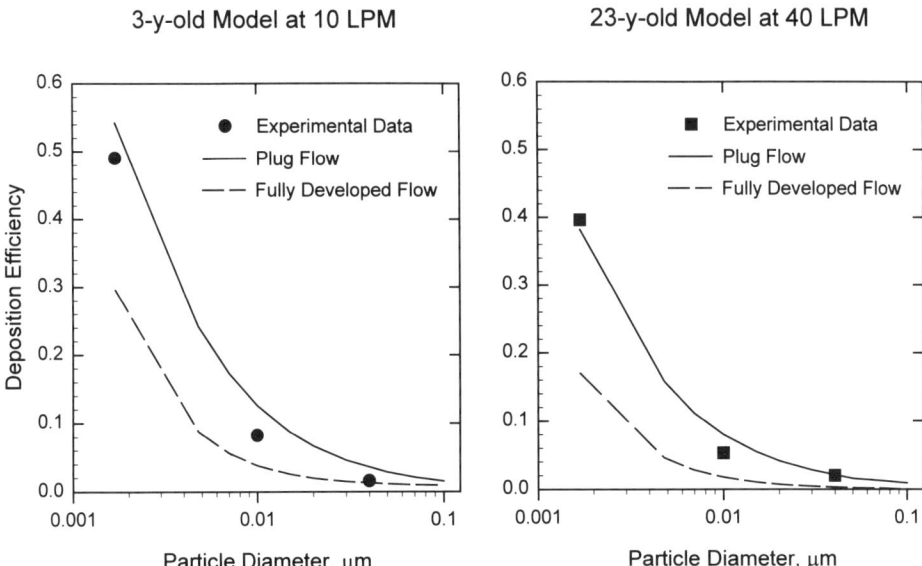

Fig. 3. Comparison of experimental data and theoretical predictions in the tracheobronchial region of the cast.

rates tested. Deposition by airway generations was also determined and compared with theoretical prediction. Figure 4 presents comparison of the deposition efficiencies of the three particle sizes in the 23-year-old cast at 20 and 40 $l.m^{-1}$, showing that the experimental results agree with the prediction based on plug flow for all three particle sizes tested. Similar results were obtained for all three casts.

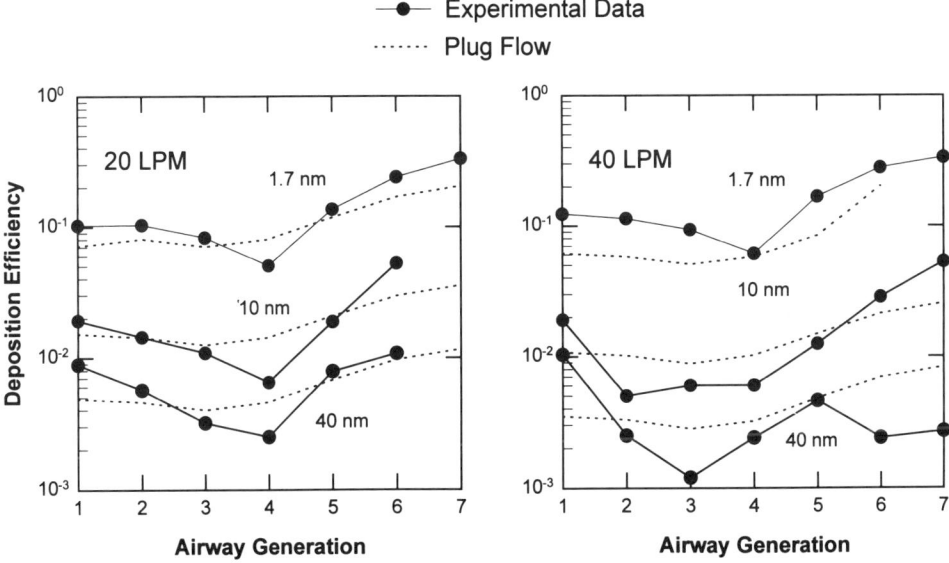

Fig. 4. Comparison of deposition by airway generation and the theoretical prediction based on plug flow in the 23-year-old cast.

DISCUSSION

Our data extend the deposition measurements of ultrafine particles to include the size of molecular clusters. The size range included attached and unattached progeny. The aerosol generation technique used can label monodisperse particles with radioactivity to quantify deposition in the airways. The use of a conducting material in the airway cast and aerosol preparation minimized the effects of electrostatic deposition, which would have increased the deposition. Consequently, our data represent deposition of aerosols in charge equilibrium. These data show that in the entire ultrafine particle size range, diffusional deposition is the dominant mechanism. Our data agree with those reported by Cohen *et al.* (1990) who showed that the measured deposition is higher than that predicted from diffusional deposition in circular tubes assuming a fully developed parabolic flow profile. Our results further show agreement with diffusional deposition predicted in circular tubes with a plug flow profile. The agreement may be due to the fact that in branched airways, the airway length is much shorter than the entrance length needed for the flow to fully develop. Therefore, deposition in a plug flow is a much closer description of the deposition process than that in a parabolic flow. The deposition in the tracheal region is higher than that predicted by the plug flow solution, indicating enhancement of deposition due to flow turbulence induced by the larynx in the region. (Research sponsored by the Office of Health and Environmental Research, U.S. Department of Energy, under Contract no. DE-AC04-76EV01013.)

REFERENCES

Cheng, Y. S., Su, Y. F., Yeh, H. C. and Swift, D. L. (1993) Deposition of thoron progeny in human head airways. *Aerosol Sci. Technol.* **18**, 359–375.
Cheng, Y. S., Yamada, Y., Yeh, H. C. and Su, Y. F. (1990) Size measurement of ultrafine particles (3 to 50 nm) generated from electrostatic classifiers. *J. Aerosol Res.* **5**, 44–51.
Cheng, Y. S., Yeh, H. C., Guilmette, R. A., Simpson, S. Q., Cheng, K. H. and Swift, D. L. Nasal deposition of ultrafine particles in human volunteers and its relationship to airway geometry. *Aerosol Sci. Technol.* (in press).
Cheng, Y. S., Yu, C. C. and Tu, K. W. (1994) Intercomparison of activity size distribution of thoron progeny by alpha- and gamma-counting methods. *Health Phys.* **66**, 72–79.
Cohen, B. S., Sussman, R. G. And Lippman, M. (199) Ultrafine particle deposition in a human tracheobronchial cast. *Aerosol Sci. Technol.* **12**, 1902–1091.
Ingham, D. B. (1975) Diffusion of aerosols for a stream flowing through a cylindrical tube. *J. Aerosol Sci.* **6**, 125–132.
Porstendörfer, J. and Mercer, T. T. (1979) Influence of electric charge and humidity upon the diffusion coefficient of radon decay products. *Health Phys.* **15**, 191–199.
Schiller, C., Gebhart, J., Heyder, J., Rudolf, G. and Stahlhofen, W. (1988) Deposition of monodisperse insoluble aerosol particles in the 0.005–0.2 um size range within the human respiratory tract. *Ann. occup. Hyg.* **32**, 41–49.
Strong, J. C., Knight, D. A. and Black, A. (1994) A novel method for the radiolabelling of the particulate phase of environmental tobacco smoke. *J. Aerosol Sci.* **25**, 199–207.

Pergamon

Ann. occup. Hyg., Vol. 41, Supplement 1, pp. 719–723, 1997
© 1997 British Occupational Hygiene Society
Published by Elsevier Science Ltd. All rights reserved
Printed in Great Britain
0003–4878/97 $17.00 + 0.00
Inhaled Particles VIII

PII: S0003–4878(96)00104–4

MEASURED TAR DEPOSITION AND COMPENSATION IN SMOKERS SWITCHING TO LOWER YIELDING PRODUCTS

J. McAughey, A. Black, J. Pritchard and D. Knight

AEA Technology, Aerosol Science Centre, 551 Harwell, Didcot, Oxfordshire OX11 0RA, U.K.

INTRODUCTION

Evidence has accumulated over the last three decades on the strong association between cigarette smoking and a range of adverse health effects (Royal College of Physicians, 1983). Thus it has been the policy of authorities responsible for public health to advise smokers to stop smoking and to discourage non-smokers from starting. However, it has been recognised that some form of product modification to reduce delivery may be of benefit to habitual smokers. In the U.K., the inception and execution of this policy (Froggatt, 1989) has achieved a great deal of practical success, with a reduction of Sales Weighted Average Tar (SWAT) in the U.K. from 20.8 mg per cigarette in 1972 to 11.0 mg per cigarette by 1993 (Waller and Froggatt, 1996). However, it was realised that a policy of progressive reduction in cigarette yield could not be continued indefinitely due to both consumer resistance and the tendency for smokers to compensate for the reduction in yields. This study has measured tar deposition and other markers of nicotine and CO intake in male and female smokers, since, over the past decades, there is an increasing prevalence of women in the population smoking while the prevalence in men is decreasing (Wald *et al.*, 1988). In view of the fact that there are known gender differences in metabolic behaviour of nicotine (Benowitz *et al.*, 1986) and deposition pattern of particles (Pritchard *et al.*, 1986), this provides a useful opportunity for comparison.

MATERIALS AND METHODS

Two groups of 13 male and 13 female middle-tar (13–16 mg yield) cigarette smokers were recruited, by advertisement. Approval was obtained from the AEA Technology Ethical Committee (ATEC) and a licence was obtained from the Administration of Radioactive Substances Advisory Committee (ARSAC) of the UK Department of Health. Suitability of the volunteers to participate was established by a medical interview including a lung function test and informed consent was obtained. Study cigarettes, a 13 mg (A) and 10 mg (B) product in which nicotine yields were maintained, were characterised by the Laboratory of the Government Chemist as described by Darrall (1988).

Volunteers smoked the 13 mg brand (A) for 9 weeks before switching to a 10 mg yield product (B) for 25 weeks. Cigarettes were issued *ad libitum* but with consumption checked by the number of butts returned. Tar retention in the

Table 1. Machine yields of study cigarette types

	Brand A	Brand B	Ratio
PMWNF* (mg)	13.0	10.0	0.77
Nicotine (mg)	1.15	1.19	1.03
CO (mg)	15.7	8.7	0.55
Nicotine filtration efficiency (%)	50	45	—

Table 2. Tar deposition markers

Tar intake	Male (mean ± SD)	Female (mean ± SD)
Total TPM—A	13.5 ± 3.8	12.5 ± 3.3
(mg.cig^{-1})—B	13.5 ± 3.7	13.2 ± 2.8
Ratio	1.02 ± 0.20**	1.08 ± 0.15**
Weekly TPM—A	3.17 ± 0.93*	1.99 ± 0.69*
(g)—B	3.56 ± 1.29*	2.09 ± 0.61*
Ratio	1.11 ± 0.21**	1.11 ± 0.21**

*($P < 0.05$) male vs female **($P < 0.05$) intake ratio vs yield ratio (0.77).

volunteers was measured using a radio-tracer, ^{123}I-1-iodohexadecane ($C_{16}H_{33}I$), as described by Pritchard et al. (1989). Deposited ^{123}I was measured (as mg tar) in one of the Harwell Laboratory whole-body monitors, consisting of six collimated 15 cm diameter NaI (T1) crystals connected via photo-multiplier tubes to a multi-channel analyser [Harwell 2000 Series (Pritchard and Black, 1988)]. Total tar particulate material (Total TPM) was estimated using a weighted combination of the signal from all six detectors to determine total body retention of the inhaled tobacco smoke. The raw data are modified using an algorithm which takes into account attenuation of the γ-ray signal by the thickness of the chest, as defined by the distance from the detector to the mid-point of chest depth. Urine and saliva samples were collected for cotinine analysis using a RIA technique based on that of Knight et al. (1985).

RESULTS

Yield data for the two types of experimental cigarettes are shown in Table 1. It is principally the poor nicotine retention of brand B that leads to the relative increase in nicotine delivery. Data for total tar deposition are shown in Table 2; in addition, weekly tar intake has been calculated by adjusting for consumption. Intake ratios indicated that compensation did occur.

Over the course of these studies, no significant increase in consumption was recorded (male: 235 ± 57 brand A vs 251 ± 68 brand B, female: 158 ± 43 brand A vs 158 ± 36 brand B): the consumption difference between genders was significant. Data for regional deposition were calculated by differential clearance; no significant differences were observed between cigarette brands or between genders. Deposition was split 35% TB: 65% P with a half-time of bronchial clearance of approximately 1.5 h. Cotinine (a metabolite of nicotine) is typically used as a steady state marker, due its slower release reflecting intake over preceding day; data are shown in Table 3.

Table 3. Cotinine markers

Cotinine	Male (mean ± SD)	Female (mean ± SD)
Saliva—A	0.40 ± 0.15	0.37 ± 0.13
(mg.l^{-1})—B	0.45 ± 0.15	0.41 ± 0.14
—Ratio	1.16 ± 0.14**	1.12 ± 0.15**
Urine—A	3.47 ± 2.00	2.82 ± 1.28
(mg.l^{-1})—B	3.92 ± 2.15	3.34 ± 1.54
—Ratio	1.17 ± 0.21**	1.20 ± 0.16**

*($P < 0.05$) male vs female **($P < 0.05$) intake vs yield ratio (1.03).

Despite an average estimated intake of nicotine which is only 55% of the male value (calculated from butt nicotine values and consumption), the female group exhibit salivary and urinary cotinine concentrations which are not significantly different from the male group. Significant compensation was observed, that is the intake ratio was significantly different from nicotine yield ratio and from unity.

DISCUSSION

It has been suggested in many reports that compensatory behaviour to maintain nicotine intake occurs in smokers switching to lower-yielding products thus negating any benefit predicted from the switch (IARC, 1986). In this study a significant yield/intake ratio was observed for tar for both men and women such that the tar intake was maintained (relative intake not significantly different from 1.0). The degree of "compensation" was consistent with an earlier study reported from this laboratory for a switch to a low-tar, low-nicotine cigarette. For nicotine and cotinine related markers, "compensation" is observed but this is not significant and the values obtained are less than in previously reported studies, where nicotine yield was reduced rather than maintained as in this study. This implies that in this group of volunteers, "compensation" is most significantly affected by maintaining tar intakes, possibly as a marker of taste. The increase in estimated nicotine intake and the markers of nicotine metabolism (10–20%) may be due to a combination of three factors, that is, as a side effect of tar intake compensation (~30–40%) or as a combination of controlling both tar and nicotine intake or alternatively, due to poorer nicotine retention efficiency of the butt in the brand B cigarette relative to brand A (~10%). In the second case, it is possible that the butt retention efficiency for nicotine differs under real smoking patterns relative to the standard yield measurements.

"Compensation" observed may have been caused by a number of factors, a major one potentially being the increase in smoke volume inhaled through the lower yield products by the volunteer, in contrast to machine smoking, where a standard smoking pattern is adopted. However, if it is assumed that yield increases linearly with puff volume, tar deposition is still greater than would have been predicted (~20%). Thus, to achieve a reduction in tar intake, yield should be reduced by the likely increase in puff volume, plus this additional 20%, i.e. for these two products 35% (puff volume increase 15%). This would suggest that yields must be substantially lowered to achieve a reduced tar intake unless product modification prevents this increase in puff volume, for example by maintaining draw resistance. Clearly, this finding has far-reaching implications for health programmes proceeding through product modification strategies where tar:nicotine ratios are being reduced and casts doubts on their effectiveness below a certain threshold.

The measurement of tar retention is of great significance in quantifying intake of the solid, particulate phase of cigarette smoke, which accounts for the majority of retention by mass. In contrast, the retention of nicotine (as vapour/particulate) and carbon monoxide gas represent special cases. A number of studies have reported high percentages of retention for cigarette particles using exhale capture techniques. Mean retention values summarised by IARC (1986) are reported in the range of 47–97% (Mean = 82 ± 17%), which as reported previously is greater than would be expected for particles in the size range 0.1–1.0 μm. As this study was designed to allow "natural" smoking, exhale capture techniques could not be deployed and hence measurements of total deposition are unavailable. However, knowledge of the radiolabel transfer under a variety of smoking procedures (Pritchard *et al.*, 1989) and the absolute mass deposited suggest that the retention percentage is consistent with the reported data.

CONCLUSIONS

In conclusion, the effectiveness of the radiolabel, ^{123}I-1-iodohexadecane as a marker for tar particulate material has again been demonstrated. Significant differences in behaviour have been observed for the particulate, vapour and gaseous phases, which may change further as smoking behaviour changes on cigarette type switching and when considered against standard deliveries from non-typical machine-smoking patterns. Thus, the commonly used markers for smoking behaviour such as cotinine and carbon monoxide are not appropriate as predictors of tar behaviour. In contrast to previous studies, there was no decrease in tar intake on switching to a lower delivery product although this switch was less pronounced. This observation may be a consequence of seeking to maintain tar intake for taste or other considerations, or due to a significant increase in the total volume of smoke drawn from the lower yielding cigarette. Since this change in smoking behaviour is observed with many low delivery products, it suggests that there must be a substantial decrease in delivery to achieve any reduction in intake. Alternatively, product modification should avoid a resultant increase in puff volume, perhaps by maintaining draw resistance. However, these results suggest that there may be a practical threshold, below which reductions in intake cannot be achieved by conventional cigarettes. It is noteworthy that the women achieved a similar deposit of tar, yet drew 30% less smoke from the cigarette. This suggests that a higher proportion of the inhaled smoke is retained. However, the regional deposition of tar showed no significant differences between men and women, despite different airway morphology. Deposition for both is higher in the tracheo-bronchial region than would be expected from equivalent sized inert particles. This reinforces the need to consider tobacco smoke aerosol as a dynamic system. It can lead to surface concentrations of deposit around the bronchial bifurcations which are three orders of magnitude greater than those in the pulmonary region of the lung. This is of importance to risk assessments from smoking behaviour, particularly as cigarette smoking is predominantly associated with bronchogenic carcinoma.

Acknowledgements—This work was supported by the Tobacco Products Research Trust. The views expressed in this work are those of the authors and do not necessarily represent those of the Trustees.

REFERENCES

Benowitz, N. L., Jacob, P., Yu, L., Talcott, R., Hall, S. and Jones, R. T. (1986) Reduced tar, nicotine and carbon monoxide exposure whilst smoking ultra-low but not low-yield cigarettes. *J. Am. Med. Assoc.* **256**, 241–246.

Darrall, K. G. (1988) Smoking machine parameters and cigarette smoke yields. *Sci. Total Environ.* **74**, 263–278.

Froggatt, P. (1989) Determinants of policy on smoking and health. *Int. J. Epidem.* **18**, 1–9.

International Agency for Research on Cancer (1986) Monographs on the evaluation of the carcinogenic risk of chemicals to humans, Vol. 38. *Tobacco Smoke*, IARC, Lyon, France.

Knight, G. J., Wylie, P., Holman, M. S. and Haddow, J. E. (1985) Improved ^{125}I-ratio-immunoassay for cotinine by selective removal of bridge antibodies. *Clin. Chem.* **31**, 118–121.

Pritchard, J. N. and Black, A. (1988) Design and calibration of a low energy (150 keV) γ-ray collimation system for the Harwell whole-body monitor. *Nucl. instr. Meth. Phys. Res.* **A264**, 453–463.

Pritchard, J. N., Jefferies, S. J. and Black, A. (1986) Sex differences in the regional deposition of inhaled particles in the 2.5–7.5 µm range. *J. Aerosol Sci.* **17**, 385–389.

Pritchard, J. N., McAughey, J. J. and Black, A. (1989) A technique for radio-labelling tar particulate material in mainstream cigarette smoke. *J. Aerosol. Sci.* **19**, 715–724.

Royal College of Physicians (1983) *Health or Smoking?* Pitman, London.

Wald, N., Kiryluk, S., Darby, S., Doll, R., Pike, M. and Peto, R. (1988) *UK Smoking Statistics.* Oxford University Press, Oxford.

Waller, R. E. and Froggatt, P. (1996) Product modification. *Br. Med. Bull.* **52**, 193–205.

Ann. occup. Hyg., Vol. 41, Supplement 1, pp. 724–727, 1997
© 1997 British Occupational Hygiene Society
Published by Elsevier Science Ltd. All rights reserved
Printed in Great Britain
0003–4878/97 $17.00 + 0.00
Inhaled Particles VIII

Pergamon

PII: S0003–4878(96)00105–6

COMPARISON OF PARTICULATE DOSE FROM EXPOSURE TO ENVIRONMENTAL TOBACCO SMOKE (ETS) AND MAINSTREAM CIGARETTE SMOKE USING RADIOTRACERS

J. McAughey, A. Black, J. Strong and C. Dickens

AEA Technology, Aerosol Science Centre, 551 Harwell, Didcot OX11 0RA, U.K.

INTRODUCTION

There has been considerable debate in recent years regarding the health risks resulting from exposure to environmental tobacco smoke (ETS). To date, there has been no direct method of measuring the amount and distribution of tar particulate from ETS deposited in the respiratory tract, or for estimating subsequent clearance. Indeed only one study by Hiller *et al.* (1982) has reported an ETS deposition fraction of 11 ± 4% in a group of five volunteers. It remains difficult to find representative chemical markers of the ETS particulate aerosol in environmental situations, as many have a significant vapour component on ageing and dilution of the smoke.

A ^{123}I-1-iodohexadecane radiolabel for mainstream smoke (Pritchard *et al.*, 1988; McAughey *et al.*, 1996) was unsuitable for sidestream tobacco smoke as 70% of the label (and particulate mass) evaporated on ageing and dilution (Pritchard *et al.*, 1988b).

As an alternative, it is known that the radioactive decay products of the naturally occurring gases ^{222}Rn (radon) and ^{220}Rn (thoron) when first formed have a high mobility and easily become attached to particles such as ETS suspended in the atmosphere. The ^{220}Rn (thoron) series decays via ^{216}Po to ^{212}Pb, with a radioactive half-life of 10.6 h (240 keV γ-ray emission) which would allow monitoring over an appropriate period.

MATERIALS AND METHODS

The methodology for the study of particulate deposition from mainstream cigarette smoke is described in the previous paper (McAughey *et al.*, 1996) with a more detailed description of the radiolabelling method for ^{123}I-1-iodohexadecane and its validation described by Pritchard *et al.* (1988). Details of the ^{212}Pb radiolabel and its validation are described elsewhere (Strong *et al.*, 1994) with the exposure system described in detail by Strong *et al.*, (1994b). In short, the exposure system comprises a ventilated 14 m^3 room, which can be maintained at constant temperature and humidity. Smoke is generated in an external glovebox and fed into the room such that steady-state atmospheres of aged and diluted sidestream smoke at various concentrations can be established and maintained. A ^{228}Th source is also maintained in a glovebox external to the room to supply ^{220}Rn to a 180 l chamber within the room. The ^{220}Rn rapidly underwent 2 α-decays to produce ^{212}Pb, which attaches principally to the smoke particles present. After a 15 min "growing-in" period, volunteers entered the room and inhaled from the chamber. The inhalation rig used consists of a shuttle valve providing separate pathways for inhalation and exhalation, allowing the collection of exhaled particulate matter; monitoring volume and flow and display of the required inhalation pattern.

Three breathing patterns were used to assess fractional and regional deposition of the labelled surrogate ETS particles; 6 × 1000 ml breaths.min^{-1} mouth breathing and nose breathing and 12 × 500 ml breaths.min^{-1} mouth breathing only.

Nine non-smoking male volunteers were recruited by advertisement, medically examined for fitness to participate in this programme and written informed consent obtained from each. Approval for the study had been granted by the local AEA ethical committee (ATEC) and the Administration of Radioactive Substances Advisory Committee (ARSAC) of the U.K. Department of Health.

Following exposure, the volunteers were monitored over the following 2 days using the collimated whole body monitoring system comprising 6 collimated 15 cm diameter NaI(Tl) detectors described previously (Pritchard and Black, 1988).

RESULTS

The particle size of the ETS aerosol (AMAD) was measured as 0.21 μm (σ_g = 1.32, Delron impactor) and 0.18 μm (σ_g = 1.5, quartz crystal microbalance). Particle concentration was 2.75 × 10^5 particle.ml^{-1} (condensation nucleus counter) and 0.86 mg.m^{-3} (gravimetric), substantially greater than reported environmental levels. However, the main criteria was to achieve high thoron progeny attachment levels to minimise radiation dose to the volunteers.

Total ^{212}Pb deposition in each volunteer was determined by comparing the inhaled with exhaled γ-activity with data shown in Table 1. Whole body measurements showed typical clearance data consisting of two exponential curves. The first curve with half times of approximately 7–10 h represents clearance from the tracheo-bronchial (TB) region of the lung. The second curve with half times of approximately 2 days represents clearance from the pulmonary (P) or alveolar region of the lung. If both curves are projected back to time zero, this gives initial deposition ratios for the respective regions. Regional deposition for the 6 × 1000 ml.min^{-1} mouth breathing pattern are shown in Table 2. Data are compared with values from the LUDEP model (Birchall et al., 1991) based on ICRP 66, 1995.

DISCUSSION

The measured deposition fraction for surrogate ETS differs from that reported by Hiller et al. (1982) of 11 ± 4% and values of 80–90% observed for mainstream smoke (U.S. Surgeon General, 1986). However, the Hiller data were for 0.41 μm mass median aerodynamic diameter (MMAD) smoke particles rather than the 0.18–0.21 μm AMAD reported here. Using the LUDEP model (Birchall et al., 1991) to predict total deposition for the experimental conditions reported by Hiller,

Table 1. Total deposition of ETS particulate

Breathing pattern	Deposition (%)	Predicted (%)
6×1000 ml.min^{-1}: Mouth	43 ± 17	33
6×1000 ml.min^{-1}: Nose	59 ± 10	33
12×500 ml.min^{-1}: Mouth	22 ± 8	21

Table 2. Regional deposition (6×1 l.min^{-1} mouth)

	Deposition (%)	Predicted
Total	43 ± 16	33
Head (ET)	2.8 ± 1.2	0.6
Thorax	37.2 ± 15	32.4
Bronchial	9 ± 4	6.6
Pulmonary	28 ± 11	25.7
$t_{1/2}$ TB (h)	8.4 ± 2.2	—

a value of 22% was obtained. It can be inferred, from these regional deposition data, that the ETS particulate is depositing in lower generations of the lung than mainstream smoke particulate. This is confirmed by the measured half-time for TB clearance of 8.4 h, which is significantly longer than the values measured by similar techniques in previous studies at Harwell. Clearance values for polystyrene particles (1.5–10 μm diameter) (Pritchard et al., 1986) and for mainstream cigarette smoke (initial particle size 0.7 μm) (McAughey et al., 1996) have all been found with a half-time of approximately 2 h.

The intake of ETS particulate mass remains low. Many authors have used the concept of "cigarette equivalents" in order to give an ETS dose relative to a single cigarette in an active smoker. Estimates of "cigarette equivalents" have varied from 0.024 to 27 cigarette equivalents per day, a range which has been seen as illogical (U.S. Surgeon General, 1986). Calculations for the highest estimate in this report divided the calculated ETS particulate intake by a mainstream smoke intake of 0.55 mg per cigarette, a minimum yield value for cigarette brands available in the U.S. However, the original authors of the cited work quote a sales weighted average mainstream smoke yield of 17 mg per cigarette (Repace and Lowrey, 1980). Use of this more realistic value would reduce the range quoted in the U.S. Surgeon General Report to 0.024–0.9 cigarette equivalents per day. While early estimates based ETS dose on ETS concentration per unit volume multiplied by volume inhaled, more recent calculations of ETS dose have taken into account deposition fraction and adjusted exposure via time-activity matrices, with ventilation volumes adjusted for activity (Holcomb, 1993). These data predict maximum intakes of 109 μg.day^{-1} for men, based on incremental exposure concentrations between smoking and non-smoking areas.

In our calculation exposure data from Holcomb (1993) and Guerin (1992) have been combined with measured data for fractional deposition and ventilation rates. Where total particulate exposure is attributed to ETS, intake in the home or work is estimated at 135 or 187 μg.day^{-1} (or 74 and 61 μg.day^{-1}, respectively for incremental exposure concentrations).

Tar deposition in cigarette smokers using radiotracers have shown an intake of

453 mg per day (range = 307–728 mg per day) in male middle-tar smokers with an average intake of 13.5 mg per cigarette (range = 7.4–22.0 mg per cigarette) (McAughey *et al.*, 1996). This mean delivery of the cigarette was equivalent to the sales weighted average tar for the U.K., consistent with the measured yield of the experimental cigarette. This suggests that daily particulate intake from ETS is of the order of 0.03–0.07% that of cigarette smokers.

However, the regional deposition patterns of mainstream tobacco smoke and ETS particulate are different, with ETS particulate depositing more deeply in the lung. Thus, direct comparisons of particulate retention on a cigarette equivalent basis may be inappropriate. This concern is supported by the U.S. Environmental Protection Agency, who chose not to consider the use of "cigarette equivalents" in their recent publication reviewing the respiratory health effects of passive smoking (USEPA, 1992) for a variety of reasons. For example, although mainstream tobacco smoke (MS) and ETS are qualitatively similar with respect to chemical composition, the absolute and proportional quantities of the smoke components, their physical state and their partitioning between phases can differ. Further differences included variations in particle size between MS and ETS and different breathing patterns in smokers and non-smokers, leading to differences in the distribution and deposition of each type of smoke in the respective populations. Subsequent metabolic differences were also discussed, suggesting dose-response associations were likely to be nonlinear.

CONCLUSIONS

These data are highly significant in terms of risk assessment of ETS particulate exposure, as they clearly demonstrate differences in the deposition pattern in the lung relative to mainstream cigarette smoke. When this is combined with the known physico–chemical differences between the two smoke types, it suggests that extrapolation of exposure and dosimetric data for the two situations is not appropriate.

Acknowledgements—Data from the studies reported were part-funded by the Tobacco Products Research Trust (in respect of mainstream smoking measurements) and the Center for Indoor Air Research (in respect of ETS deposition measurements). The views expressed are solely those of the authors.

REFERENCES

Birchall, A., Bailey, M. R. and James, A. C. (1991) LUDEP: A lung dose evaluation programme. *Rad. Prot. Dosimetry* **38**, 167–174.

Guerin, M. R., Jenkins, R. A. and Tomkins, B. A. (1992) *The Chemistry of Environmental Tobacco Smoke: Composition and Measurement.* Lewis Publishers Inc., Michigan.

Hiller, F. C., McCusker, K. T., Mazumder, M. K., Wilson, J. D., Bone, R. C. (1982) Deposition of sidestream cigarette smoke in the human respiratory tract. *Am. Rev. Respir. Dis.* **125**, 406–408.

Holcomb, L. C. (1993) Indoor air quality and environmental tobacco smoke: concentration and exposure. *Environ. Int.* **19**, 9–40.

ICRP Publication 66 (1995) International commission on radiological protection. *Human Respiratory Tract Model for Radiological Protection.* Pergamon, Oxford.

 Pergamon

Ann. occup. Hyg., Vol. 41, Supplement 1, pp. 728–735, 1997
© 1997 British Occupational Hygiene Society
Published by Elsevier Science Ltd. All rights reserved
Printed in Great Britain
0003–4878/97 $17.00 + 0.00
Inhaled Particles VIII

PII: S0003–4878(96)00176–7

CIGARETTE SMOKE-INDUCED OXIDANT/ANTIOXIDANT IMBALANCE AND ITS ROLE IN INCREASED EPITHELIAL PERMEABILITY

W. MacNee,* I. Rahman,* X. Y. Li,* D. Morrison* and K. Donaldson†

*Respiratory Medicine Unit, Department of Medicine, University of Edinburgh, Royal Infirmary; and the †Department of Biological Sciences, Napier University, Edinburgh, U.K.

INTRODUCTION

An imbalance between oxidants and antioxidants is proposed in the pathogenesis of the smoking-related lung disease—chronic obstructive pulmonary disease (COPD) (Rahman and MacNee, 1996). Cigarette smoke, contains 10^{17} oxidant molecules per puff, (Pryor *et al.*, 1993) producing an cnormous oxidant burden on the lungs. In addition to the oxidants in cigarette smoke, the numbers of inflammatory leukocytes are increased in the airspaces of cigarette smokers compared with non-smokers (Hunninghake, 1983), which release more reactive oxygen species (ROS) (Shaberg *et al.*, 1992). In addition neutrophils, sequestered in the pulmonary circulation during cigarette smoking (MacNee *et al.*, 1989), may release more ROS (Brown *et al.*, 1995). All tissues are vulnerable to oxidant damage but by virtue of its location, the airspace epithelium is particularly vulnerable. Studies in healthy smokers (Jones *et al.*, 1980) have shown an increased epithelial permeability, measured by increased clearance from the lungs of technetium 99m-labelled diethylenetriamine pentaacetic acid (^{99m}Tc-DTPA).

There is limited information on the antioxidant defences of the respiratory tract epithelial lining fluid in smokers (Cross *et al.*, 1994). The important thiol antioxidant glutathione (GSH) is increased in the epithelial lining fluid in the airways of chronic smokers (Cantin *et al.*, 1987) and appears to be related to humoral markers of inflammatory cell activity (Linden *et al.*, 1989). Glutathione may have an important role in protecting the epithelium from the effects of oxidant-induced injury produced by cigarette smoke. The aims of this study were to assess the oxidant/antioxidant imbalance produced by cigarette smoke and to determine the role of the antioxidant glutathione in the increased epithelial permeability produced by cigarette smoke.

METHODS

We studied healthy cigarette smokers, a rat model of intratracheal instillation of cigarette smoke condensate (CSC) *in vivo* and cultured human type II alveolar epithelial cells *in vitro*.

Healthy cigarette smokers/non-smokers

Fifteen regular cigarette smokers and seven life-long non-smokers, all with normal respiratory function were studied. Smokers were asked to abstain from cigarette smoking for 12 h before each study (chronic smokers) and were also studied 1 h after smoking two cigarettes (acute smokers).

Epithelial permeability was measured as the clearance of inhaled ^{99m}Tc-DTPA from the lungs, using a gamma camera. Exponential lung clearance curves were obtained from regions of interest over the lungs and the time for lung activity to fall to 50% of the initial value (t50) was calculated as a measure of epithelial permeability. Each smoker underwent bronchoscopy and bronchoalveolar lavage, eight after chronic smoking and seven after acute smoking. The results were compared with those in seven subjects who had never smoked. Blood samples were withdrawn from each subject and the trolox equivalent antioxidant capacity (TEAC), products of lipid peroxidation as thiobarbituric reactive substances and superoxide anion release from unstimulated and PMA stimulated peripheral blood neutrophils were measured using techniques which we have described previously (Rahman *et al.*, 1996).

Studies in the rat

Wistar-derived rats of the HAN strain were used in these experiments. Cigarette smoke condensate was produced by blowing smoke from three medium tar cigarettes over phosphate buffered saline (PBS) in a tonometer. Two ml of cigarette smoke condensate was instilled into the distal airspaces through a small incision in the trachea. Control rats were given 2 ml of PBS through the same route.

Rat lung epithelial permeability. Airspace epithelial permeability was measured as the passage of intratracheally instilled ^{125}I-BSA from the airspace to the blood (Li *et al.*, 1994, 1996).

Bronchoalveolar lavage (BAL) in the rat. Rats were sacrificed, the lungs removed from the thorax and bronchalveolar lavage was carried out.

In vitro studies. The permeability of monolayers of the A549 epithelial cell line was assessed as the passage of ^{125}I-BSA across the monolayer (Li *et al.*, 1994).

Glutathione assay

Reduced (GSH) and oxidise glutathione (GSSG) in BAL fluid in humans, in rats and in epithelial cells was measured using the method of Tietze (1969). In order to deplete glutathione in rat lungs, buthionine sulphoxamine (BSO, 700 mgs) was given as a single intraperitoneal injection or two injections separated by 8 h. A549 alveolar epithelial cells were also treated with 50 μM BSO to reduce intracellular GSH. The activity and gene expression of the main enzyme involved in GSH synthesis γ-glutamylcysteine synthetase (γ-GCS) was measured by reverse transcriptase-PCR (Rahman *et al.*, 1995). The transcription factor- activator protein 1 (AP1) was also assessed in epithelial cells using the gel mobility shift assay.

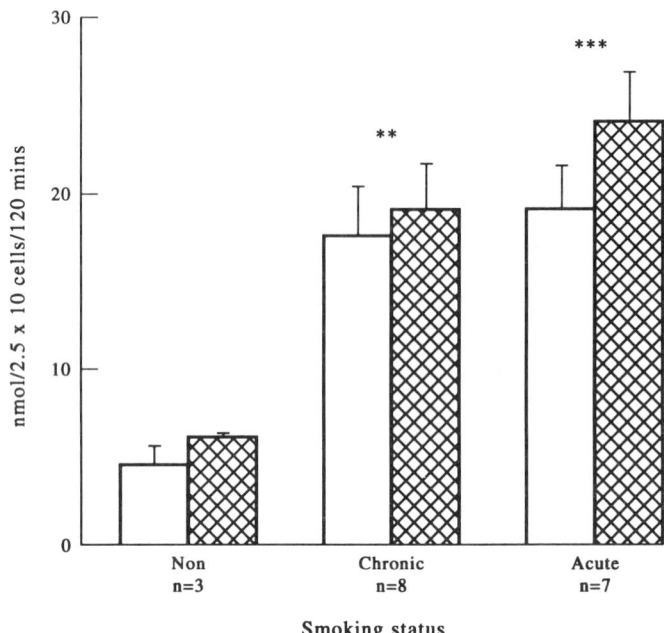

Fig. 1. Spontaneous (closed histograms) and PMA stimulated (hatched histograms) superoxide anion release from mixed leucocytes obtained from non-smokers, chronic smokers and acute smokers. $**P < 0.01$, $***P < 0.001$, compared with non-smokers.

RESULTS

In the non-smokers t50 was much longer (16.7 ± 1.3 min) than in the chronic smoking group (4.6 ± 6.2). A further reduction in t50 occured after acute smoking (14.8 ± 1.0 min, $P < 0.01$). Superoxide anion release from mixed leukocytes obtained from chronic smokers was significantly increased compared to non-smokers (Fig. 1). There was also a significant increase in products of lipid perioxidation in the plasma of both the chronic and acute smoking groups compared to the non-smokers (Fig. 2). In addition trolox equivalent antioxidant capacity (TEAC) was significantly reduced in chronic smokers compared with non-smokers (Fig. 3) and further reduced in acute smokers (Fig. 3). Reduced glutathione (GSH) was significantly increased in bronchoalveolar lavage in chronic smokers compared with non-smokers. This increase was abolished following acute smoking (Fig. 4).

In the rat, instillation of cigarette smoke condensate increased epithelial permeability (Fig. 5) associated with an initial decrease in bronchoalveolar lavage and lung GSH (Fig. 6). Depleting lung or epithelial cell GSH alone increased epithelial permeability both *in vivo* and *in vitro* (data not shown). The reason for the changes in lung GSH were investigated in more detail in A549 alveolar epithelial cells where an initial profound fall in GSH following CSC, due to the formation of CSC-GSH conjugates and a decrease in the activity of γ-GCS was followed 24 h later by an increase in γ-GCS gene expression and a rebound increase in GSH (Fig. 7). The increase in γ-GCS gene expression is associated with increased nuclear binding of the AP-1 transcription factor (data not shown).

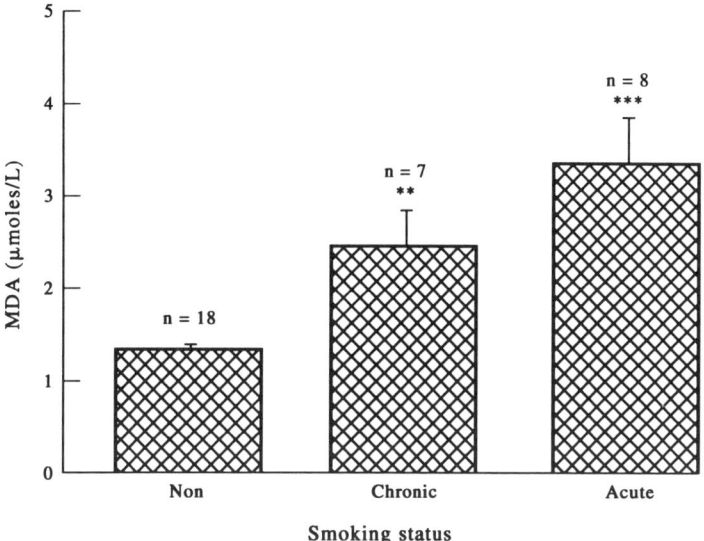

Fig. 2. Products of lipid peroxidation in plasma in non-smokers and acute and chronic smokers.
**$P < 0.01$, ***$P < 0.001$ compared with non-smokers.

DISCUSSION

These studies, both in human smokers and in a relevant animal model, show that increased epithelial permeability is an acute effect of cigarette smoking. The mechanism of the increased epithelial permeability induced by cigarette smoke is unknown. However, we hypothesise that oxidants in cigarette smoke may play a major role (Li *et al.*, 1996; Rahman and MacNee, 1996). These studies of the effects of acute and chronic cigarette smoking show profound changes in oxidant/ antioxidant balance. Both acute and chronic smoking are associated with marked

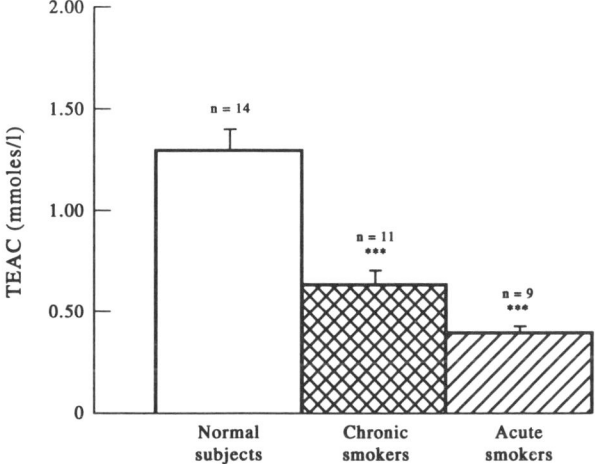

Fig. 3. Antioxidant capacity measured as the trolox equivalent antioxidant capacity (TEAC) in normal subjects, chronic and acute smokers. ***$P < 0.001$ compared with normal subjects.

W. MacNee *et al.*

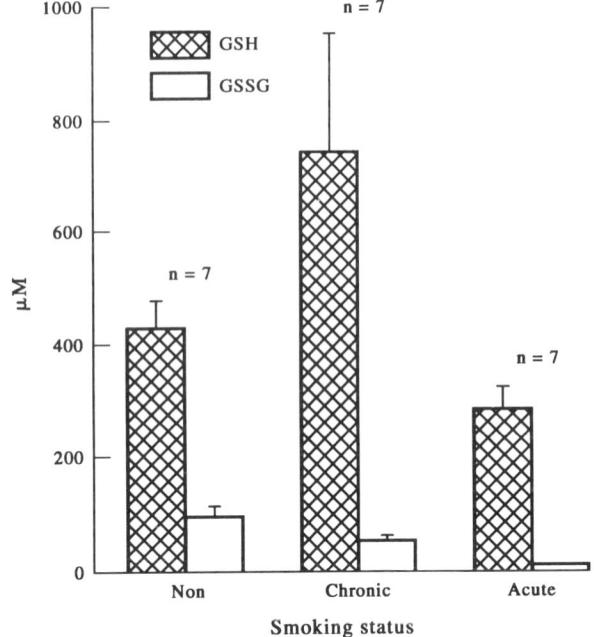

Fig. 4. Reduced (GSH) and oxidised (GSSG) glutathione levels in bronchoalveolar lavage in normal subjects, chronic and acute smokers. *$P < 0.05$.

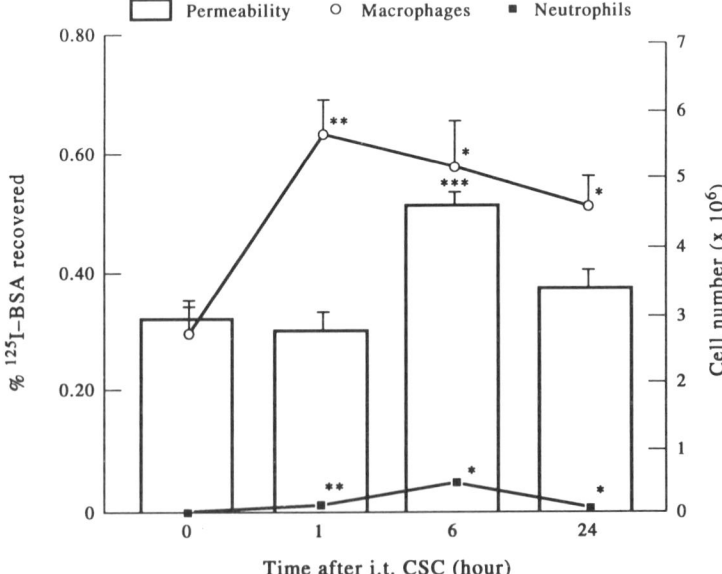

Fig. 5. The effect of instillation of cigarette smoke condensate in the rat on epithelial permeability and bronchoalveolar lavage leucocyte counts *$P < 0.05$, **$P < 0.01$ ***$P < 0.001$ compared with time zero.

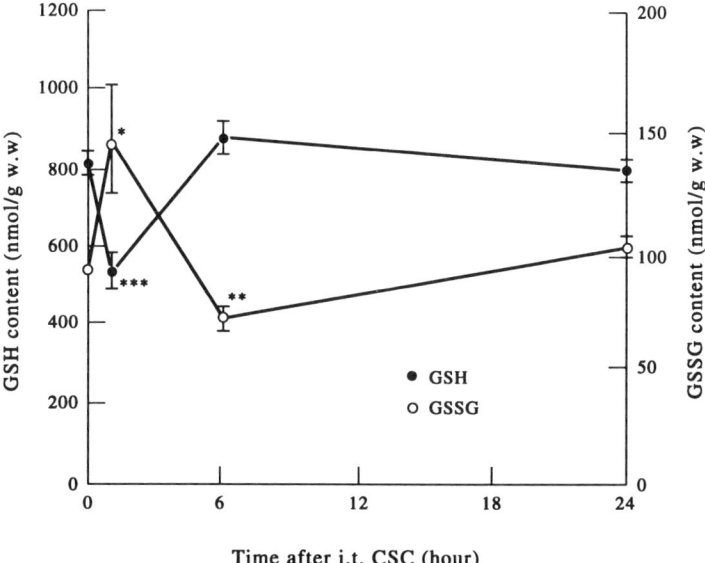

Fig. 6. Reduced glutathione (GSH; closed circle) and oxidised glutathione (GSSG; open circle) in the rat lung after intratracheal instillation of cigarette smoke condensate (CSC). *$P < 0.05$, **$P < 0.01$, ***P 0.014 0.01 compared with time zero.

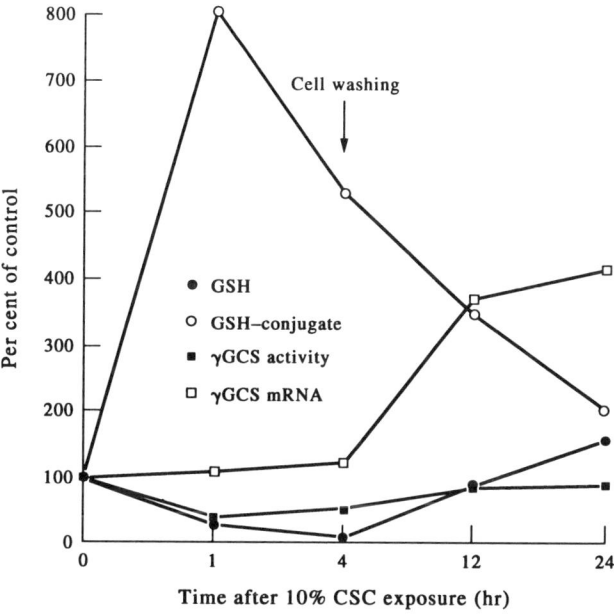

Fig. 7. The effect of cigarette smoke condensate on GSH γ-GCS activity and gene expression in A549 alveolar epithelial cells measured by RT-PCR.

changes in systemic oxidative stress, measured by a decrease in the antioxidant capacity and an increase in products of lipid peroxidation in plasma. We have previously shown increased neutrophil sequestration in the microcirculation in cigarette smokers (MacNee *et al.*, 1989). These sequestered neutrophils may release more oxygen radicals (Brown *et al.*, 1995). Thus oxygen radical release from circulating or sequestered neutrophils may result in oxidative stress in plasma. Evidence of systemic oxidative stress is also reflected by a decrease in epithelial lining fluid GSH, although this was not obvious in human studies, perhaps because of the time point chosen, since the greatest decrease in bronchoalveolar lavage GSH occurred 6 h after administration of cigarette smoke condensate in the rat, whereas the effects of smoking on epithelial permeability were measured 1 h after acute smoking in humans.

It is clear that cigarette smoke has a profound and immediate effect on lung epithelial cell glutathione. This was also reflected in studies in epithelial cells. Glutathione in levels present in epithelial lining fluid can totally protect against cigarette smoke induced injury to an epithelial cell monolayer (Lannan *et al.*, 1994). The results both animal studies and in epithelial cells, where intracellular glutathione was decreased by 90%, show that simply decreasing lung and in-tracellular glutathione can induce increased epithelial permeability. Taken together data from this and our previous studies (Li *et al.*, 1994, 1996) strongly suggest that reduced glutathione is an important protective mechanism against cigarette smoke-induced airspace perturbation. Our data also suggest that following exposure to cigarette smoke there is a profound decrease in both extra and intracellular glutathione with the formation of GSH-CSC conjugates and a decrease in the activity of the main synthesising enzyme for glutathione-γ-GCS. This is followed by a rebound increase in epithelial lining fluid GSH and in epithelial cells and this is associated with increased expression of the γ-GCS gene. Preliminary results (Rahman, 1995) also indicate that the increase in γ-GCS gene expression may be associated with activation of the AP-1 transcription factor.

In summary these data provide evidence of a marked oxidant/antioxidant imbalance systemically and in the lungs in cigarette smokers and that part of this imbalance is the depletion of the major lung antioxidant glutathione. Glutathione may also be involved in the increased epithelial permeability, which is one of the earliest injurious events produced by cigarette smoking.

REFERENCES

Brown, D., Drost, E., Donaldson, K. and MacNee, W. (1995) Deformability and CD11/CD18 expression of sequestered neutrophils in normal and inflamed lungs. *Am. J. Respir. Cell Mol. Biol.* **13**, 531–539.

Cantin, A. M., North, S. L., Hubbard, R. C. and Crystal, R. G. (1987) Normal alveolar epithelial lung fluid contains high levels of glutathione. *J. Appl. Physiol.* **63**, 152–157.

Church, T. and Pryor, W. A. (1985) Free-radical chemistry of cigarette smoke and its toxicological implications. *Environ. Health Perspect.* **64**, 111–26.

Cross, C. E., Van der Vliet, A., O'Neill, C. A., Louie, S. and Halliwell, B. (1994) Oxidants, antioxidants, and respiratory tract lining fluids. *Envion. Health Perspect.* **102**, 185–191.

Hunninghake, G. W. and Crystal, R. G. (1983) Cigarette smoking and lung destruction: accumulation of neutrophils in the lungs of cigarette smokers. *Am. Rev. Respir. Disp.* **128**, 833–38.

Jones, J. G., Lawler, P., Crawley, J. C. W., Minty, B. D., Hulands, G. and Veall, N. (1980) Increased alveolar epithelial permeability in cigarette smokers. *Lancet* **1**, 66–68.

Lannan, S., Donaldson, K., Brown, D. and MacNee, W. (1994) Effect of cigarette smoke and its condensates on alveolar epithelial cell injury *in vitro*. *Am. J. Physiol. Lung Cell Mol. Physiol.* **266**, L92–L100.

Li, X. Y., Donaldson, K., Rahman, I. and MacNee, W. (1994) An investigation of the role of glutathione in increased epithelial permeability induced by cigarette smoke *in vivo* and *in vitro*. *Am. J. Respir. Crit. Care Med.* **149**, 1518–25.

Li, X. Y., Rahman, I., Donaldson, K. and MacNee, W. (1996) Mechanisms of cigarette smoke induced increased airspace permeability. *Thorax* **51**, 465–471.

Linden, M., Hakansson, L., Ohlsson, K., Sjodin, K., Tegner, H., Tunek, A. and Venge, P. (1989) Glutathione in bronchoalveolar lavage fluid from smokers is related to humoral markers of inflammatory cell activity. *Inflammation* **13**, 651–58.

MacNee, W., Wiggs, B., Berzberg, A. S. and Hogg, J. C. (1989) The effects of cigarette smoking on neutrophil kinetics in human lungs. *N. Engl. J. Med.* **321**, 924–28.

Pryor, W. A. and Stone, K. (1993) Oxidants in cigarette smoke: radicals, hydrogen peroxides, peroxynitrate, and peroxynitrite. *Ann. N.Y. Acad. Sci* **686**, 12–28.

Rahman, I., Li, X. Y., Donaldson, K., Harrison, D. J. and MacNee, W. (1995) Glutathione homeostasis in alveolar epithelial cells *in vitro* and lung *in vivo* under oxidative stress. *Am. J. Physiol. Lung Cell Mol. Physiol.* **13**, L285–L292.

Rahman, I. and MacNee, W. (1996) Oxidant/antioxidant imbalance in smokers and chronic obstructive pulmonary disease. *Thorax* **51**, 348–350.

Rahman, I., Morrison, D., Donaldson, K. and MacNee, W. (1996) Systemic oxidative stress in asthma, COPD, and smoking. *Am. J. Respir. Crit. Care Med.* **153**.

Schaberg, T., Haller, H., Rau, M., Kaiser, D., Fassbender, M. and Lode, H. (1992) Superoxide anion release induced by platelet-activating factor is increased in human alveolar macrophages from smokers. *Eur. Rspir. J.* **5**, 387–393.

Tietze, F. (1969) Enzymic method for quantitative determination of nanogram amounts of total and oxidized glutathione: application to mammalian blood and other tissue. *Analyt. Biochem.* **27**, 502–522.

Pergamon

Ann. occup. Hyg., Vol. 41, Supplement 1, pp. 736–744, 1997
British Occupational Hygiene Society
Published by Elsevier Science Ltd
Printed in Great Britain
0003–4878/97 $17.00 + 0.00
Inhaled Particles VIII

PII: S0003–4878(96)00106–8

EXCITATORY AND INHIBITORY NEURAL REGULATION OF TRACHEAL CILIARY BEAT FREQUENCY (CBF) ACTIVATED BY AMMONIA VAPOUR AND SO_2

D. B. Yeates, S. P. Katwala, J. Daugird, A. V. Daza and L. B. Wong

Department of Medicine, University of Illinois at Chicago, 1940 W Taylor St, M/C 788, Chicago, IL 60612; and Veterans Affairs West Side Medical Center, Chicago, Illinois, U.S.A.

INTRODUCTION

Inhaled irritants have been shown to either stimulate or depress mucociliary activity dependent on the irritant, the dose and the preparation in which the irritant is studied. As the term "irritant" suggests, these compounds activate sensory nerves when delivered to the nose or the tracheobronchial airways (Masumoto *et al.*, 1993). Our proposed working model of irritant activated neural regulation of the tracheobronchial mucociliary transport system is shown in Fig. 1. We have preliminary data which shows, in dogs, that inhibition of neurotransmission via the vagosympathetic trunks and cervicothoracic ganglia results in a marked stimulation of bronchial mucociliary clearance (Yeates *et al.*, 1995). We interpret these findings in the following way. During homeostasis, the mucociliary transport system is maintained in a low basal state by inhibitory neural impulses. Abolition of the inhibitory signals allows the inherent mucosal mechanisms regulating mucociliary transport system to operate unimpeded. This results in a stimulation of the mucociliary transport system. Such an inhibitory pathway could theoretically be of central origin, or it could be part of an irritant activatible inhibitory reflex which decreases ciliary activity and thus mucociliary function. In this sense, it opposes irritant-induced excitatory reflexes that stimulate the mucociliary transport system to remove the offending irritant.

We hypothesize that irritants such as SO_2 when delivered to the bronchial airways activates the proposed inhibitory sensory nerves that results in a decrease in tracheal ciliary beat frequency and mucociliary transport, whereas ammonia vapour activates sensory nerves in the nasal passages and the bronchi that stimulate ciliary activity in the trachea via excitatory efferent nerves transmitted via the vagosympathetic trunks (VST).

METHODS

Measurement of tracheal ciliary beat frequency

Ciliary beat frequency in the distal trachea was measured using heterodyne laser light scattering (Chandra *et al.*, 1994). Briefly, a stainless steel tubular probe 30 cm long and 8 mm diameter was inserted into the trachea such that its tip was in the

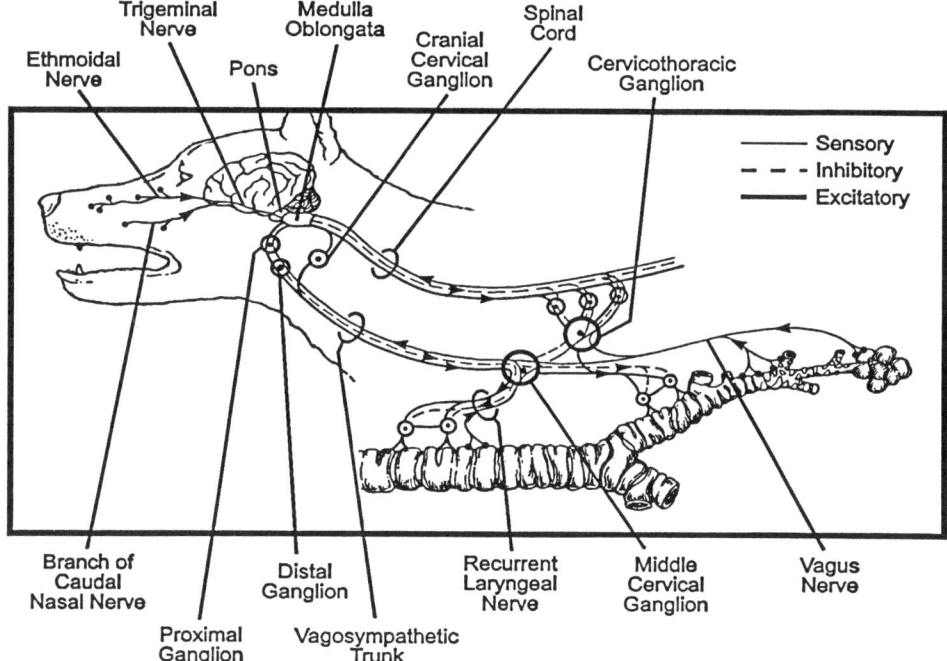

Fig. 1. The proposed neural pathways regulating tracheal and bronchial epithelial function. For simplicity the excitatory and inhibitory sensory pathways are shown as a single afferent pathway.

mid (bronchial exposures) or distal (nasopharyngeal exposures) trachea. Laser light (632.8 nm) directed through this probe exited perpendicular to the probe and was focused on the tracheal epithelium. Scattered photons from the beating cilia and the epithelium were collected through the same optics, registered by a photon counting photomultiplier tube (Hamamatsu R649) and binned at 3.3 ms intervals. Time-frequency analysis of this temporal array resulted in a measurement of ciliary beat frequency every 3–5 s. In the "ammonia" studies, three minute averaged CBF values were utilized to obtain the peak responses in the baseline, during and following each pertubation. Raw CBF in the time interval was utilized in subsequent data reduction. These raw peak CBF segments were divided into groups of three data points. The lowest CBF in each set was removed. Subsequent data analysis (averaging or mean values) was performed on this filtered data. Thus the CBF for these studies is higher than those for the "SO_2" studies.

Measurement of bronchial mucociliary clearance

Bronchial mucociliary retention was measured using radioaerosol techniques. Briefly, the anesthetized dog was placed supine such that the chest was centered on a gamma camera (Pho Gamma 4, Nuclear Chicago, or LFOV, Searle, Chicago, IL). A colloidal iron oxide sol was tagged with technetium 99m (Wales, Petrow and Yeates, 1980). This was aerosolized at 2 ml min^{-1} by a Turbotak atomizer, dried with dilution air and concentrated using a virtual impactor and maintained at a positive pressure of 20 cm water (Pillai *et al.*, 1994). About 63 µCi of this aerosol

(MMAD 10.2 μm, σ_g 1.4) was delivered to the dog for 2 min via an endotracheal tube. A series of solenoid valves ensured that the dogs inhaled for 1 s, exhaled for 2 s and rested for 3 s. This resulted a tidal volume of ~ 300 ml with a respiratory rate of 10 breaths min^{-1}. Radioactivity in the chest was sequentially monitored at 2 min intervals for 2 h and measured again for 10 min after 24 h. Following background subtraction and decay correction, the radioactivity remaining at 24 h was subtracted from each of the sequential measurements made during the first 2 h to yield bronchial retention curves. Bronchial mucociliary clearance (BMC) was ascertained from these curves.

Animal preparation

Fasted dogs were anesthetized with 2.5% thiamylal sodium; the depth of anesthesia being titrated to the animals palpebral and pedal reflexes. A percutaneous femoral artery catheter facilitated the measurement of arterial blood pressure and blood chemistry. Body temperature was maintained with heating blankets as indicated by rectal temperature. EKG, end tidal CO_2 arterial oxygen saturation and respiratory rate were monitored (Datex, Capnomac Ultima, Helsinki, Finland). Supporting fluids, (0.9% sodium chloride, 100–300 ml) were administered via an intravenous cephalic catheter.

Protocols

The responses of CBF to SO_2 were measured in three separate experiments. In part 1 each dog had its bronchial airways exposed sequentially to 0, 0.5, 2.5, 5.5, 22, 50 and 100 ppm for 20 min each. In part 2 each dog had its bronchial airways exposed to: (a) a sham SO_2 exposure i.e. dry air; (b) sequential exposures to 0, 1, 2.5, 4.5, 5.5, 10 ppm SO_2 for 20 min each; (c) sequential SO_2 exposures with prior delivery of cromolyn sodium aerosol using a peripheral aerosol deposition technique; (d) sequential SO_2 exposure with prior delivery of 2 mg kg^{-1} I.V hexamethonium. In each experiment, CBF was measured in the trachea and SO_2 (or sham) was delivered exclusively to the bronchial airways. The targeted SO_2 concentration was achieved by dilution of a 100 ppm tank of compressed SO_2-air mixture with air. The cuff of a double lumen endotracheal tube was inflated in the distal trachea. The mixture was inhaled spontaneously from a 2 l reservoir bag. The inspired and expired flows were separated using one way valves. The expired SO_2 was scrubbed (MSA 1192, Pittsburgh, PA) and vented via a fume hood. The concentration of SO_2 delivered to the dogs was validated in separate experiments using NIOSH method S308.

Six beagle dogs had their vagosympathetic trunks, VST, bilaterally elevated and enclosed in skin tubes they had their nose or bronchi independently exposed to 20 min of 10 ppm ammonia vapours, with and without their vagosympathetic trunks cooled to 0°C. Cooling the VST to 0°C was achieved by placing each of the VST between symmetrical copper blocks with half-circles machined out to form a snug fit around the skin tubes. These copper blocks were thermally coupled to thermoelectric coolers (Marlow Industries). The heat generated on the opposite side of the thermoelectric elements was dissipated through a thermally coupled copper block through which water at room temperature was circulated. A 10 ppm ammonia vapour was generated by passing air at 1 l/min through a humidifier (3M)

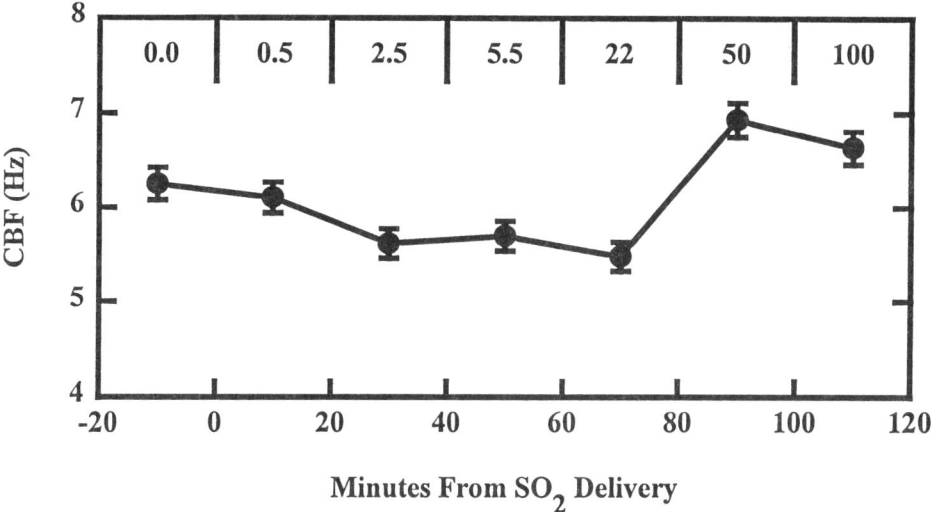

Fig. 2. The response of CBF to bronchial exposure to increasing concentrations of SO_2 in the spontaneously breathing dog. Data point represent the mean for eight dogs of 20 min averages of CBF. SO_2 concentration in ppm.

which was primed and continuously fed with a 8.8×10^{-5} M (10 ppm) ammonium hydroxide solution. The concentration at the output was validated using the NIOSH method 74–136. Duplicate measurements of the ammonia generated were 8.8 and 10.4 ppm. This ammonia vapour was delivered through short lengths of plastic tubing placed inside each of the dogs' nostrils. To facilitate delivery of ammonia vapour to the bronchi, a double lumen size 7 endotracheal tube was placed in the dogs' distal trachea and the cuff inflated. One lumen of the endotracheal tube was attached to the humidifier to which a one-way valve was attached to the input port. This enabled the dog to inhale through one lumen and exhale through the other. Ammonia vapour 10 ppm was delivered at a rate of 6–15 breaths min^{-1} depending on the spontaneous breathing rate of the dog.

RESULTS

Responses of tracheal CBF and BMC induced by SO_2

Following sequential delivery of increasing concentrations of SO_2, CBF decreased sequentially with increasing SO_2 dose levels. CBF decrease from 6.3 ± 0.2 (SE) at baseline to 5.7 ± 0.2 Hz at 5.5 ppm ($P < 0.05$). However, CBF increased to 6.9 ± 0.2 Hz and 6.6 ± 0.2 Hz at 50 ppm ($P < 0.05$) and 100 ppm ($P < 0.05$) SO_2, respectively (Fig. 2).

It can be seen in the experiments in which SO_2 was delivered with and without prior administration of either acrosolized cromolyn or hexamethonium bromide, the 5.5 ppm SO_2-induced decrease in CBF was abolished by cromolyn sodium but not by hexamethonium bromide (Fig. 3). These data suggest that the aerosolized cromolyn deposited in the alveoli attenuated the activation of sensory nerves forming the afferent arm of an inhibitory reflex. That this SO_2-induced decrease of CBF was not mediated via nicotinic ganglionic transmission is evidenced by the

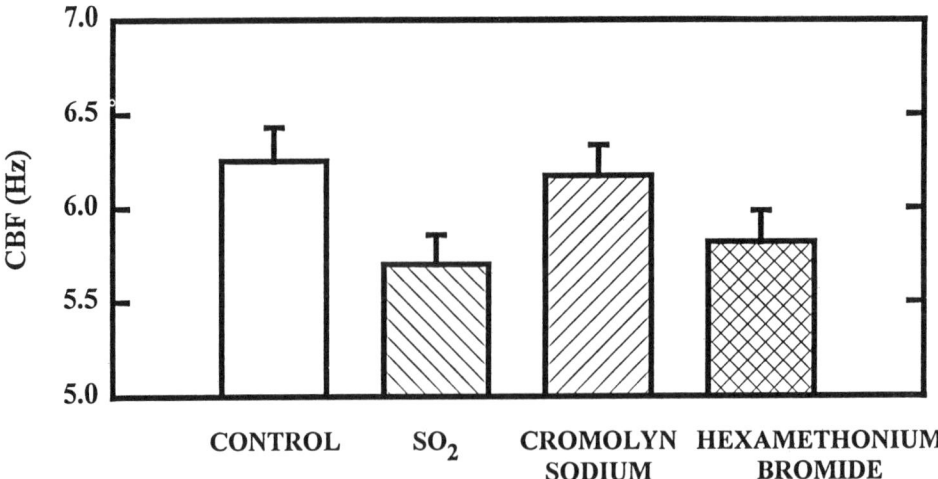

Fig. 3. Response of CBF to 5.5 ppm SO_2 alone and in conjunction with prior administration of cromolyn sodium and hexamethonium bromide. Data represent the mean values of seven dogs of 20 min averages of CBF.

observation that 2 mg kg^{-1} I.V. hexamethonium did not appear to block the SO_2-induced decrease in CBF.

It can be seen in Fig. 4 that 5.5 ppm SO_2 delivered to both the trachea and tracheobronchial airways for 20 min caused a marked decrease from 53.7 ± 5.7% to 32.8 ± 7.7% of mean bronchial mucociliary clearance after 90 min. The mean R24 for the control studies was 27.1 ± 6.7% and the SO_2 studies was 41.8 ± 2.5%.

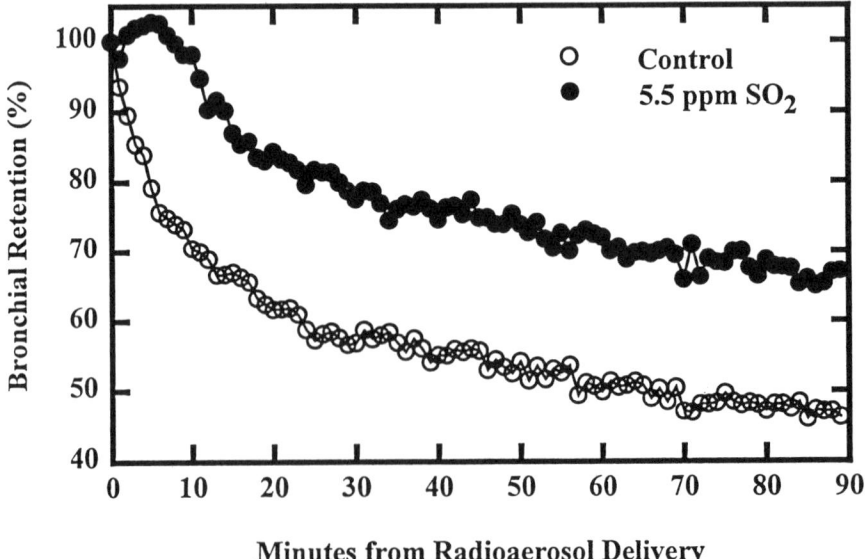

Minutes from Radioaerosol Delivery

Fig. 4. Temporal response of bronchial mucociliary retention in seven dogs with and without bronchial and tracheal exposure to 5.5 ppm SO_2.

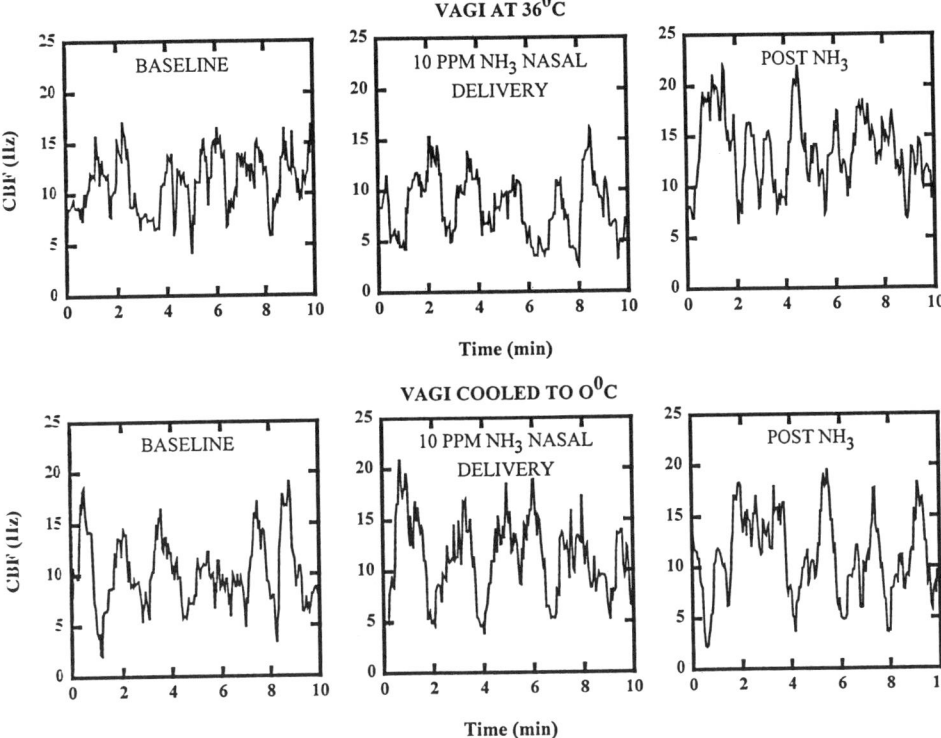

Fig. 5. Ten minute portions of CBF: (i) prior to; (ii) during; and (iii) following 10 ppm NH$_3$ exposure to the nose; (A) with the VST at 36°C; and (B) at 0°C for dog 7172. A 1 min moving average of CBF is shown.

Thus this inhibition of bronchial mucociliary clearance was not due to differences in deposition pattern.

Responses of tracheal CBF induced by ammonia vapour

The temporal sequences of CBF prior to, during and following exposure to ammonia vapour were all markedly episodic, whether or not the vagosympathetic trunks were cooled (Fig. 5). This 2 min periodicity in CBF appeared to be independent of the perturbation. It is notable that CBF did not increase during the nasal ammonia vapour delivery but rather following the ammonia exposure; an effect that was abolished when the VST were cooled to 0°C.

Nasal exposure of ammonia did not change CBF during the exposure 13.5 ± 2.2 Hz vs 13.8 ± 1.3 Hz, ($n = 7$) nor with the VST at 0°C (14.8 ± 1.2 Hz vs 15.4 ± 2.1 Hz). However, after 20 min exposure to ammonia there was a significant increase in CBF in all seven dogs ($P < 0.05$) which was not observed when the VST were cooled [Fig. 6(a)]. Exposure of the bronchi to ammonia vapour caused a significant post exposure stimulation of CBF ($P < 0.05$) that was also attenuated by cooling the VST to 0°C [Fig. 6(b)].

DISCUSSION

That SO$_2$ delivery to the bronchi inhibited CBF in the trachea, a response that

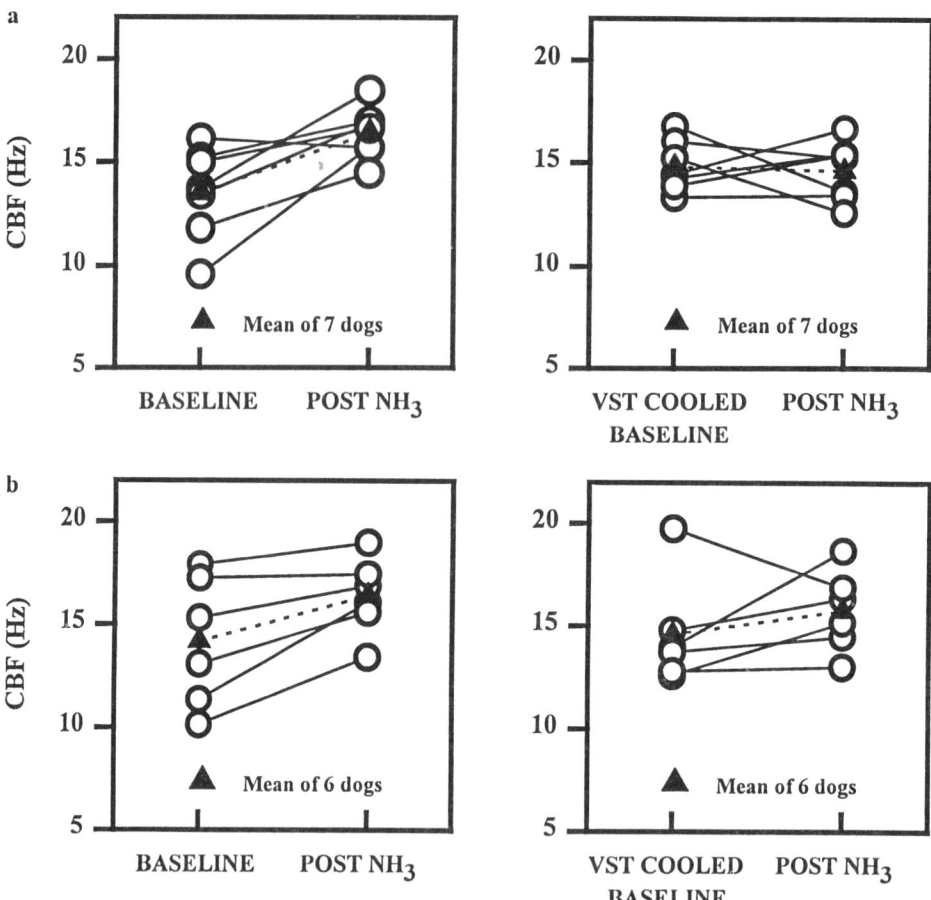

Fig. 6(a). CBF response after a 20 min exposure to 10 ppm NH$_3$ (A) VST at 36°C; (B) VST at 0°C. These data are the average of the raw peak CBF segment for baseline ans post NH$_3$ for seven dogs; (b) CBF response after a 20 min bronchial exposure to 10 ppm NH$_3$ (A) VST at 36°C; (B) VST at 0°C. These data are the average of the raw peak CBF segment for baseline ans post NH$_3$ for six dogs.

was attenuated by cromolyn (Dixon, Jackson and Richards, 1980) delivered to the afferent arm of this proposed reflex is consistent with SO$_2$ activation of an inhibitory neural reflex. That this suppression was not blocked by I.V. hexamethonium indicated that the ganglionic transmission was non-nicotinic in nature. As ammonia vapour delivered to the nasal passages as well as to the bronchi stimulated CBF in the trachea; effects that were abolished by the blockage of neural conduction in the vagosympathetic trunk suggests that the post ammonia stimulations of tracheal CBF were due to activation of excitatory naso–tracheal and broncho–tracheal reflexes, respectively.

In vitro, 3 ppm SO$_2$ (Tamaoki *et al.*, 1993), 2.5–12.5 ppm SO$_2$ (Knorst *et al.*, 1994), 30–50 ppm SO$_2$ (Blanquart *et al.*, 1995) also decreases ciliary beat frequency in the absence of neural reflexes. This toxic response may be due to the acidic nature of dissolution products resulting from the interaction between SO$_2$ and the aqueous mucus layer. This could conceivably cause a decrease in intracellular pH and Na^+, via the apical Na^+/H^+ antiporter. We have hypothesized that ciliary activity is proportional to the intracellular $[Na^+]/[Cl^-]$ ratio (Wong and Yeates, 96) a decrease in CBF may be transduced in this manner. Such a local mechanism is highly improbable in the *in vivo* experiments reported herein as the bronchi and trachea are served by separate arterial circulations. In these experiments the dilution of any diffused sulphur species in the blood would be enormous and, thus, unlikely to activate other potential reflexes or cause the release of humoral mediators.

A brief puff of ammonia vapour to the rabbit maxillary sinus, sufficient to induce a respiratory reflex, caused a stimulation of ciliary wave frequency that was in part blocked by prior administration of atropine (Lindberg *et al.*, 1987). Our results of a vagally-mediated increase of CBF following a longer low concentration of ammonia vapour are consistent with these findings. That CBF was not stimulated during the ammonia vapour exposure could be interpreted that during this period ammonia vapour stimulated both inhibitory and excitatory sensory nerves to a similar degree resulting in no net effect. Such mechanisms can be elucidated by the delivery of ammonia in the presence of specific inhibitors of the opposing pathways.

It is possible that the activation of excitatory and inhibitory neural reflexes give rise to the acceleration in bronchial clearance at very low concentrations of H_2SO_4 and the inhibition at high concentrations (Leikauf *et al.*, 1981). That neutralization of SO_x species with ammonia leads to little biological response (Schlesinger and Chen, 1994) is consistent with the balance of inhibitory and excitatory reflexes activated by each of these respective radicals.

The presence of opposing excitatory and inhibitory neural reflexes modulating airway mucosal function is consistent with the general bio-regulatory premise of neural regulation of physiological function. In the case of airway smooth muscle, this balance is largely attained through adrenergic and cholinergic mechanisms. Whereas, in the case of the mucosa, the evidence suggests that this balance is affected via excitatory and inhibitory neural reflexes, the former being transmitted via nicotinic ganglionic transmission and the latter by non-nicotinic ganglionic transmission. Consideration of the potential regulatory control of these presumably interacting reflexes leads to following possibilities. A decrease in the operation of the mucociliary transport system can be affected via either the activation of an inhibitory efferent pathway or the inhibition of an excitatory efferent pathway. Conversely, stimulation of the mucociliary tranpsort system can be affected via the activation of an excitatory efferent pathway or the inhibition of an inhibitory efferent pathway.

Acknowledgements—These data are selected from the thesis, "The effects of sulfur dioxide on the mucociliary transport system" by Joanne Daugird and the thesis, "Modulation of the mucociliary transport system via the vagosympathetic trunk and the cervicothoracic ganglia" by Sweta P. Katwala, both submitted as partial requirements for Masters Degrees in Bioengineering, University of Illinois at Chicago, 1994. This work was funded by the National Institutes of Health, Institute of Environmental Health Sciences, RO1 ESO 4137 and the Medical Service of the Department of Veterans Affairs and the Whitaker Foundation.

REFERENCES

Blanquart, C., Giuliani, O., Houcine, O., Jeulin, C., Guennou, C. and Marano, F. (1995) *In vitro* exposure of rabbit tracheal epithelium to SO_2: effects on morphology and ciliary beating. *Toxology in vitro* **9**, 123–132.

Chandra, T., Yeates, D. B., Miller, I. F. and Wong, L. B. (1994) Stationary and non-stationary correlation-frequency analysis of heterodyne mode laser light scattering: magnitude and periodicity of canine tracheal ciliary beat frequency *in vivo*. *Biophys. J.* **66**, 878–890.

Dixon, M., Jackson, D. M. and Richards, I. M. (1980) That action of sodium cromoglycate of 'C' fiber endings in the dog lung. *Br. J. Pharmacol.* **70**, 11–13.

Knorst, M. M., Kienast, K., Riechelman, H., Muller-Quernheim, J. and Ferlinz, R. (1994) Effect of sulfur dioxide on mucociliary activity and ciliary beat frequency in guinea pig trachea. *Int. Arch. occup. Environ. Health* **65**(5), 325–328.

Leikauf, G., Yeates, D. B., Wales, K. A., Albert, R. E. and Lippmann, M. (1981) Effects of sulfuric acid aerosol on respiratory mechanics and mucociliary particle clearance in healthy non-smoking adults. *Amer. ind. Hyg. Assoc.* **42**(4), 273–282.

Lindberg, S., Dolata, J. and Merke, U. (1987) Stimulation of C fibers by ammonia vapor triggers a mucociliary defence reflex. *Amer. Rev. Respir. Dis.* **135**, 1093–1098.

Matsumoto, S., Kanno, T., Yamasaki, M., Nagayama, T., Tanno, M. and Shimizu, T. (1993) H_1- and H_2-receptor influences of histamine and ammonia on rapidly adapting pulmonary stretch receptor activities. *J. Autonomic Nervous System* **43**, 17–26.

Pillai, R. S., Yeates, D. B., Eljamal, M., Miller, I. F. and Hickey, A. J. (1994) Generation of concentrated aerosols for inhalation studies. *J. Aerosol Sci.* **25**, 187–197.

Schlesinger, R. B. and Chen, L. C. (1994) Comparative biological potency of acidic aerosol: implications for the interpretation of laboratory and field studies. *Environ. Res.* **65**, 69–85.

Tamaoki, J., Chiyotani, A., Sakai, N., Takeyama, K. and Konno, K. (1993) Effect of azelastine on sulphur dioxide induced impairment of ciliary motility in airway epithelium. *Thorax* **48**, 542–546.

Wales, K. A., Petrow, H. and Yeates, D. B. (1980) Production of ^{99m}Tc-labled iron oxide aerosols for human lung deposition and clearance studies. *Int. J. appl. Radiat. Isotopes* **31**, 689–694.

Wong, L. B. and Yeates, D. B. Dynamics of the regulation of ciliary epithelium. Comments in theoretical biology (in press).

Yeates, D. B., Katwala, S. P., Daza, A. V. and Wong, L. B. (1993) Inhibition of neural transmission via the vagosympathetic trunk and cervicothoracic ganglia stimulates tracheaobronchial mucociliary clearance. *Amer. Rev. Respir. Dis.* **15**(4) (part 2), A819.

 Pergamon

Ann. occup. Hyg., Vol. 41, Supplement 1, pp. I–VIII, 1997
British Occupational Hygiene Society
Published by Elsevier Science Ltd
Printed in Great Britain
0003–4878/97 $17.00 + 0.00
Inhaled Particles VIII

LIST OF PARTICIPANTS
INHALED PARTICLES VIII—ROBINSON COLLEGE, CAMBRIDGE
26–30 AUGUST 1996

ABRAHAM, Dr J L, State University of New York, Health Science Centre, Syracuse, 750 E Adams Street, Syracuse, NY 13210, USA

AGNEW, Dr J E, Royal Free Hospital, Department Medical Physics, Pond Street, London NW3 2QG, UK

ALBIN, Dr M, Department of Occupational & Environmental Medicine, University Hospital, Lund, S-22185, Sweden

ATTFIELD, Dr M D, NIOSH, 1095 Willowdale Road, M/S 234, Morgantown, WV 26505, USA

BAILEY, Dr M R, NRPB, Chilton, Didcot, Oxfordshire OX11 0RQ, UK

BELLMANN, Dr B, Fraunhofer Institute of Toxicology & Aerosol Research, Nikolai-Fuchs-Str 1, D-30625 Hannover, Germany

BENNETT, Dr W D, University of North Carolina Chapel Hill, CB #7310 104 Mason Farm Rd, Centre for Environmental Medicine & Lung Biology, Chapel Hill, NC 27599, USA

BERNSTEIN, Dr D M, 40 chemin de la Petite Boissière, Geneva, CH 1208, Switzerland

BIGNON, Dr J, INSERM V139, Faculté de Médecine, 8 rue du General Sarrail, Creteil Cedex 94010, France

BOLTON, Dr R E, IOM, 8 Roxburgh Place, Edinburgh EH8 9SU, UK

BOYMEL, Dr P M, Unifrax Corporation, 2351 Whirlpool Street, Niagara Falls, NY 14305, USA

BRAMMERTZ, Dr A O B, Institute of Occupational Health, RWTH Aachen, Pauwelsstr 30, D-52 057 Aachen, Germany

BROWN, Prof R C, 4 Bramble Close, Uppingham, Rutland LE15 9PH, UK

BROWNE, Dr K, Leicester House, North Creake, Norfolk NR21 9JP, UK

BRUYNUIS, Mr C W, Royal Netherlands Air Force, Koninklijke Luchtmacht, Directie Personeel KLu, Staf Personeel & Organisatie, Postbus 20703, 2500 ES 's-Gravenhage, The Netherlands

BURDETT, Dr G J, HSL, Broad Lane, Sheffield, South Yorkshire S3 7HQ, UK

BURGESS, Mr G L, Centre for Occupational Health, University of Manchester, Stopford Building, Oxford Road, Manchester M13 9PT, UK

BYE, Dr E, NIOH, PO Box 8149 Dep, 0033 Oslo, Norway

CAMERON Ms K, Department of the Environment, Room A344, Romney House, 43 Marshall Street, London SW1P 3PY, UK

CASE, Dr B W, McGill University, 462 Argyle Avenue, Montreal, Quebec, Canada 43Y 3B4

CASON, Mr J E, Unifrax Corporation, 2351 Whirlpool Street, Niagara Falls, NY 14305, USA

CHENG, Dr Y S, Inhalation Toxicology Research Institute, PO Box 5890, Albuquerque, NM 87111, USA

CHERRIE, Dr J W, University of Aberdeen, c/o IOM, 8 Roxburgh Place, Edinburgh EH8 9SU, UK

CHERRY, Prof N M, Centre for Occupational Health, University of Manchester, Stopford Building, Oxford Road, Manchester M13 9PT, UK

CHURG, Mr A, University of British Columbia, 2211 Wesbrook, Vancouver, BC, Canada V6T 2B5

COGGINS, Dr C R E, RJ Reynolds Tobacco Co, Research & Development, PO Box 1487, Winston-Salem, NC 27102-1487, USA

COLLIER, Dr C G, AEA Technology plc, Biomedical Research Department, 551 Harwell, Didcot, Oxfordshire OX12 0PX, UK

COOPER, Dr A L, Manning House, 22 Carlisle Place, London SW1P 1JA, UK

COSTANTINI, Dr M G, Health Effects Institute, 955 Massachusetts Avenue, Cambridge, MA 02139, USA

CREUTZENBERG, Dr O, Fraunhofer Institute, Nikolai-Fuchs-Str 1, Hannover D-30625, Germany

CULLEN, Dr R T, IOM, 8 Roxburgh Place, Edinburgh EH8 9SU, UK

CUMMINGS, Mr L B, American Azide Company, PO Box 629, Cedar City, UT 84721, USA

DAHMANN, Dr D, Institute fur Gefahrstoff-Forschung der Bergbau-BG, Waldring 97, Bochum 44789, Germany

DALRYMPLE, Mr H L, ICL, Beaumont, Burfield Road, Old Windsor, Berkshire SL4 2JP, UK

DE KLERK, Dr N H, University of Western Australia, Department of Public Health, 6907, Western Australia, Australia

DOBSON, Dr T E, Workers Compensation Board of Nova Scotia, 5668 South Street, PO Box 1150, Halifax, Nova Scotia, Canada B3J 2YA

DONALDSON, Prof K, Biomedicine Group, Department Biological Sciences, Napier University, 10 Colinton Road, Edinburgh EH10 5DY, UK

DOUGLAS, Dr D B, Douglas Consulting Australia, PO Box 644, Kings Cross, NSW 2011, Australia

DRAHONOVSKA, Dr H, NIPH, Srobarova 48, Prague 10042, Czech Republic

DRISCOLL, Dr K E, The Procter & Gamble Company, PO Box 538707, Cincinatti, OH 45253-8707, USA

DUFRESNE, Dr A, McGill University, Department Occupational Health, Faculty of Medicine, 3450 University Street, FDA Bldg Room 22, Montreal, Quebec, Canada H3A 2A7

DUVAL-ARNOULD, Dr G, Saint Gobain, Les Miroirs, La Défense Cedex 92096, France

DWIGHT, Mrs M D, Provost & Umphrey Law Firm, PO Box 4905, Beaumont, TX 77704, USA

DYER, Dr W M, Eastman Chemical Co, Product Safety/Regulatory Programs, PO Box 1994, Kingsport, TN 37662-5394, USA

ENGHOLM, Mr B G, National Board of Health & Welfare, Socialstyrelsen, S-10630 Stockholm, Sweden

FIDALGO, Dr M M, Instituto Nacional de Silico, Department Tecnica, St/Doctor Bellmunt, Oviedo 33006, Spain

FISHER, Ms D, 14121 Heritage Lane, Silver Spring, MD 20906, USA

FLEMING, Dr J S, Southampton University Hospital Trust, Department Nuclear Medicine, Southampton General Hospital, Southampton SO16 6YD, UK

FOST, Dr U, Bundesanstalt fur Arbeitsschutz, Friedrich-Henkel-Weg 1-25, Dortmund 44149, Germany

FOSTER, Dr W M, School of Hygiene & Public Health, Johns Hopkins University, Department Environmental Health Sciences, Suite #7006, 615 North Wolfe Street, Baltimore, MD 21205, USA

FOWLER, Dr D P, Fowler Associates, 643 Bair Island Road, Suite 305, Redwood City, CA 94063, USA

FRANK, Dr A L, University of Texas, Health Center at Tyler, PO Box 2003, Tyler, TX 75710, USA

FRITSCH, Dr P, CEA, DSV/DRR/SRCA, Laboratoire de Radiotoxicologie, BP No 12, 91680 Bruyères le Châtel, France

FRØSIG, Dr A F, Selsvej 5, DK 3400 Hillerod, Denmark

GIBBS, Dr G W, Safety, Health, Environmental International Consultants Corporation 14-51221 Rge Rd 265, Spruce Grove, Alberta, Canada T7Y 1E7

GILMOUR, Mr P S, Department Biological Sciences, Napier University, 10 Colinton Road, Edinburgh EH10 5DT, UK

GLENN, Mr R E, National Industrial Sand Association, 4041 Powder Mill Road, Suite 402, Calverton, MD 20723, USA

GOLDSMITH, Dr D F, California Public Health Institute, 2001 Addison Street, Suite 210, Berkeley, CA 94704-1103, USA

GRANTHAM, Dr D L, Division of Workplace Health & Safety, GPO Box 69, Brisbane, Qld 4001, Australia

GRAY, Dr D, Sciences International Inc, 1800 Diagonal Road, Alexandra, VA 22314-2808, USA

GREEN, Dr F H Y, University of Calgary, Department of Pathology, Faculty of Medicine, 3330 Hospital Drive NW, Calgary, Alberta, Canada T2N 1N1

GREGG, Dr N, HSE, Room 149, Magdalen House, Stanley Precinct, Bootle, Merseyside L20 3QZ, UK

GRIFFITHS, Mr W D, AEA Technology, Biotechnology Services, 353 Harwell,

Didcot, Oxfordshire OX11 0RA, UK

GRIMM, Prof H-G, Technical University of Berlin, Husarenstr 32, Braunschweig D-38102, Germany

GULDBERG, Mrs M, Rockwool International A/S, Hovedgaden 584, DK-2670 Hedehusene 2670, Denmark

HADNAGY, Dr W, Institute of Hygiene, Heinrich-Heine-University, PO Box 10 1007, D-40001, Germany

HAMMAD, Dr Y Y, University of South Florida, College of Public Health, 13201 Bruce B Downs Blvd, MDC Box 56, Tampa, FL 33612-3805, USA

HANNA, Dr L M, Sciences International, 424 W Schoolhouse Lane, Philadelphia, PA 19144, USA

HARPER, Dr M, SKC Inc, 863 Valley View Road, Eighty Four, PA 15330, USA

HARRISON, Dr P T C, MRC, Institute for Environment & Health, University of Leicester, PO Box 138, Lancaster Road, Leicester LE1 9HN, UK

HEINRICH, Dr U, Fraunhofer Institute for Toxicology & Aerosol Research, Nikolai-Fuchs-Strasse 1, Hannover 30625, Germany

HEISTRACHER, Dr T, Institute of Physics & Biophysics, University of Salzburg, Austria, Hellbrunner Str 34, A-5020 Salzburg, Austria

HEMINGWAY, Dr M A, HSL, Broad Lane, Sheffield, South Yorkshire S3 7HQ, UK

HENGÉ-NAPOLI, Mrs M-H, IPSN, LEAR, BP 38, 26701 Pierrelatte, France

HENRY, Mr D C, Broken Hill Pty Ltd, PO Box 243, Warrawong, NSW 2502, Australia

HESSEL, Dr P A, University of Alberta, 13-106 Clinical Sciences Building, Edmonton, Alberta, Canada T6G 2G3

HEWSON, Mr G, Department of Minerals & Energy, 100 Plain Street, East Perth, WA 6004, Australia

HEXT, Dr P M, Zeneca CTL, Alderley Park, Macclesfield, Cheshire SK10 4TJ, UK

HEYDER, Dr J, National Centre for Environment & Health, GSF-Institute for Inhalation Biology, D-85758 Oberschleissheim, Germany

HODGSON, Mr A, NRPB, Chilton, Didcot, Oxfordshire OX11 0RQ, UK

HODGSON, Mr J T, HSE, Room 241, Magdalen House, Stanley Precinct, Bootle, Merseyside L20 3QZ, UK

HOFMANN, Dr W W, Universitat Salzburg, Institut fur Physik und Biophysik, Hellbrunnerstrasse 34, A-5020 Salzburg, Austria

HORNBERG, Mr C, Medizinisches Institut fur Umwelthygiene, Gurlittstrasse 53, D-40223 Düsseldorf, Germany

HOSKINS, Dr J A, MRC Toxicology Unit, Hodgkin Building, University of Leicester, Leicester LE1 9HN, UK

HUNT, Dr A, Department of Pathology, SUNY HSC, 750 E Adams Street, Syracuse, NY 13210, USA

HURLEY, Mr J F, IOM, 8 Roxburgh Place, Edinburgh EH8 9SU, UK

IVO, Dr M, B A D Muhlheim, Sandfuhstr 18, 44797 Bochum, Germany

JACOBSEN, Dr M, Institute fur Arbeits-und Sozialmedezin, Universitat zu Koln, Koln (Lindenthal), 50924, Germany

JANS, Dr D, IMA-Europe, Ave de la Independance Belge 75, B-1081 Brussels, Belgium

JANSSEN, Dr P Y M, Solvay Duphar BV, CJ van Houtenlaan 36, 1381 CP Weesp, The Netherlands

JONES, Dr A D, IOM, 8 Roxburgh Place, Edinburgh EH8 9SU, UK

KAINKA, Dr E, Medical Institute of Environmental Hygiene, Auf-M Hennekamp 50, D-40225 Düsseldorf, Germany

KENNY, Dr L C, HSL, Broad Lane, Sheffield, South Yorkshire S3 7HQ, UK

KNEBEL, Dr J W, Fraunhofer Institute of Toxicology & Aerosol Research, Nikolai Fuchs Str 1, D-30625 Hannover, Germany

KUEMPEL, Ms E D, National Institute for Occupational Safety & Health, 4676 Columbia Parkway MS C 32, Cincinnati, OH 45226, USA

LAMM, Dr S M, 2428 Wisconsin Avenue NW, Washington, DC 20007, USA

LEIGH, Dr J, National Institute of Occupational Health & Safety, GPO Box 58, Sydney 2001, Australia

LEVY, Dr L S, MRC, Institute for Environment & Health, University of Leicester, 94 Regent Road, Leicester LE1 7DD, UK

LINSEL, Dr G, Bundesanstatt fur Arbeitsmedizin, Noldnerstr 40142, D-10317 Berlin, Germany

LIZON, Mrs C, CEA, DSV/DRR/SRCA, Laboratoire de Radiotoxicologie, BP No 12, 91680 Bruyères le Châtel, France

LOFFLER, Mr F-W, Berufsgeenossenschaft d.keramischen u.Glas-Industrie, Reimenschneiderstrasse 2, Wurzburg 97072, Germany

MACNEE, Dr W, Respiratory Medicine Unit, Department of Medicine, Royal Infirmary, Lauriston Place, Edinburgh EH3 9YW, UK

MARK, Mr D, Institute of Occupational Health, University of Birmingham, Edgbaston, Birmingham B15 2TT, UK

MASSIOT, Mr P, CEA, DSV/DRR/SRCA, Laboratoire de Radiotoxicologie, BP No 12, 91680 Bruyères le Châtel, France

MATTON, Mrs S, CEA, DSV/DRR/SRCA, Laboratoire de Radiotoxicologie, BP No 12, 91380 Bruyères le Châtel, France

MAYNARD, Dr R L, Department of Health, Room 658 (c), Skipton House, 80 London Road, London SW1 0JU, UK

MCAUGHEY, Mr J, AEA Technology, 551 Harwell, Didcot, Oxfordshire OX11 0RA, UK

MCCAWLEY, Dr M A, NIOSH, 1095 Willowdale Road, Room 111, Morgantown, WV 26505-2888, USA

MCDONALD, Prof J C, National Heart & Lung Institute, Imperial College, Dovehouse Street, London SW3 6LY, UK

MELDRUM, Mrs M, HSE, Room 147, Magdalen House, Stanley Precinct, Bootle, Merseyside L20 3QZ, UK

MÉNACHE, Ms M G, Duke University Medical Center, Pulmonary, Box 3210, Durham, NC 27710, USA

MONCHAUX, Dr G, CEA, DSV-DRR-LCE, BP 6, 92265 Fontenay aux Roses Cedex, France

MONGAN, Dr L C, MRC Toxicology Unit, University of Leicester, Hodgkin Building, Leicester LE1 9HN, UK

MOODY, Mr J C, NRPB, Chilton, Didcot, Oxfordshire OX11 0RQ, UK

MOORE, Dr M A, Morgan Crucible Co plc, c/o Morgan Materials Technology Ltd, Newdley Road, Stourport-on-Severn, Worcestershire DY13 8QR, UK

MORFELD, Dr P, Institut fuer Arbeitswissenschaften der Ruhkohle AG, Wenge-platz 1, 44369 Dortmund, Germany

MORLIER, Dr J P, CEA, DSV-DRR-LCE, BP 6, 92265 Fontenay aux Roses Cedex, France

MORSCHEIDT, Mr C M, Isover Saint Gobain, 18 Avenue D'Alsace, Cedex 27, 92096 Paris La Défense, France

MUHLE, Dr H, Fraunhofer Institute of Toxicology & Aerosol Research, Nikolai-Fuchs-Str 1, D-30625 Hannover, Germany

MUSK, Dr A W, Department Respirable Medicine, Sir Charles Gairdner Hospital, Nedlands, Western Australia 6009, Australia

MYOJO, Dr T, National Institute of Industrial Health, Nagao 6-21-1, Tamaki, Kawasaki 214, Japan

NORBÅCK, Dr D, Department of Occupational & Environmental Medicine, University Hospital, S-75185, Sweden

OBERDORSTER, Dr G, University of Rochester, Department Environmental Medicine, 575 Elmwood Avenue Box EHSC, Rochester, NY 14642, USA

OGDEN, Dr T L, 40 Wilsham Road, Abingdon, Oxfordshire OX14 5LE, UK

OLDHAM, Mr M J, Community & Environmental Medicine, University of California, Irvine, CA 92697-1825, USA

PATRICK, Dr G, MRC Toxicology Unit, Hodgkin Building, University of Leicester, PO Box 138, Lancaster Road, Leicester LE1 9HN, UK

PETERS, Mrs A, GSF-Institute of Epidemiology, Postfach 1129, D-85758 Oberschleissheim, Germany

PHALEN, Dr R F, University of California, Community & Environmental Medicine, Irvine, CA 92717-1825, USA

PIGOTT, Dr G H, 27 Windermere Drive, Aiderley Edge, Cheshire SK9 7UP, UK

PONCY, Dr J L, CEA, DSV/DRR/SRCA, Laboratoire de Radiotoxicologie, BP No 12, 91680 Bruyères le Châtel, France

PRITCHARD, Dr J, Glaxo Group Research Ltd, Inhalation System Research, Building 5, Park Road, Ware, Hertfordshire SG12 0DP, UK

QUINN, Dr M M, University of Massachusetts Lowell, Department Work Environment, 1 University Avenue, Lowell, MA 01854, USA

RAABE, Prof O G, University of California, Davis, Institute of Toxicology & Environmental Health, CA 95616, USA

RAMOUNET, Mrs B, CEA, DSV/DRR/SRCA, Laboratoire de Radiotoxicologie, BP No 12, 91680 Bruyères le Châtel, France

REDMAYNE, Miss A C, HSL, R111 Robens Building, Broad Lane, Sheffield, South Yorkshire S3 7HQ, UK

REVELL, Mr G, HSE, Room 016, Robens Laboratory, HSL, Broad Lane, Sheffield, South Yorkshire S3 7HQ, UK

RICE, Dr C H, University of Cincinnati, Department Environmental Health, PO Box 670056, Cincinnati, OH 45267-0056, USA

ROGERS, Mr A J, Alan Rogers OH&S Pty Ltd, PO Box 2128, Clovelly, NSW 2031, Australia

ROOD, Dr A P, Davy Faraday Laboratory, Royal Institution, 21 Albemarle Street, London W1X 4BS, UK

ROSSITER, Prof C E, 10 Mynchen Road, Knotty Green, Beaconsfield, Buckinghamshire HP9 2AS, UK

ROTH, Dr C, GSF-Forschungs zentrum fur Umwelt & Gesundheit, Inst f Inhalations biologie, Postfach 1129, D-85758 Oberschleissheim, Germany

RYDMAN, Dr C, Partec Insulation, c/o Rockwool AB, 54186 Skoeyde, Sweden

SAHLE, Dr W, National Institute for Working Life, S-171 84 Solna, Sweden

SCHEUCH, Dr G, GSF-Forschungszentrum fur Umwelt und Geisundheit, Inst f Inhalations biologie, Robert-Koch-Allee 6, 82131 Gauting, Germany

SCHULTE, Dr P A, NIOSH, 4676 Columbia Parkway, Cincinnati, OH 45226, USA

SCHUM, Dr G M, California Environmental Protection Agency, Aphel, Community & Environmental Medicine, University of California, Irvine, CA 92717-1825, USA

SCHUMAN, Dr L D, Department of Labor/OSHA, 200 Constitution Avenue NW, Room N-3718, Washington, DC 20210, USA

SCHUMM, Dr M, Grunzweig & Hartmann AG, Burgermeister-Grunzweig Strasse 1, D-67059 Ludwigshafen, Germany

SEARL, Dr A, IOM, 8 Roxburgh Place, Edinburgh EH8 9SU, UK

SÉBASTIEN, Dr P, Kerlane-Saint Gobain, 31 Blvd des Bouvets, Nanterre 92000, France

SEEMAYER, Prof N H, Medical Institute of Environmental Hygiene, Gurlittstr 53, D-40223 Düsseldorf, Germany

SMITH, Dr J R H, NRPB, Chilton, Didcot, Oxforsdshire OX11 0RQ, UK

SOUTAR, Dr C A, IOM, 8 Roxburgh Place, Edinburgh EH8 9SU, UK

STANCLIFFE, Mr J D, HSL, Broad Lane, Sheffield, South Yorkshire S3 7HQ, UK

STAYNER, Dr L T, 4676 Columbia Parkway, NIOSH, Robert Taft Labs, Cincinatti, OH 45226, USA

STEWART, Mr D L, Department of Local Government & the Environment, Murray House, Mt Havelock, Douglas, Isle of Man IMI 2SF, UK

STEWART, Mr I W, AEA Technology, Biotechnology Services, 353 Harwell, Didcot, Oxfordshire OX11 0RA, UK

SWART, Dr P S, JCI Ltd, Sir Albert Medical Centre, PO Box 1794, Randfontein 1760, South Africa

TANAKA, Dr I, Department Environmental Health Engineering, University of Occupational & Environmental Health Japan, 1-1 Iseigaoka, Yahatanishi-kuk, Kitakyushu 807, Japan

TOMS, Mr N J, Royal Free Hospital, Department Thoracic Medicine, Pond Street, London NW3 2QG, UK

TRAN, Mr C L, IOM, 8 Roxburgh Place, Edinburgh EH8 9SU, UK

VALLYATHAN, Dr V, NIOSH, 1095 Willowdale Road, Morgantown, WV 26505, USA

VINCENT, Prof J H, University of Minnesota, Environmental & Occupational Health, School of Public Health, Box 807 Mayo 420, Delaware Street SE, Minneapolis, MN 55455, USA

WALKER, Mrs G S, Lucas Industries plc, 46 Park Street, London W1Y 4DJ, UK

WALLACE, Dr W E, NIOSH, 1196 Willowdale Road, Morgantown, WV 26505, USA

WALTON, Mr W H, "Ardlui", Delta Place, Inveresk, Musselburgh EH21 7TP, UK

WARHEIT, Dr D B, Du Pont Haskell Laboratory, PO Box 50, Elkton Road, Newark, DE 19714-0050, USA

WELLS, Dr A R, Owens-Corning Canada, 5140 Yonge Street, Suite 700, Toronto, Ontario, Canada M2N 6T9

WIESLANDER, Dr G, Department Occupational & Environmental Medicine, Uppsala University, University Hospital, S-751 85 Uppsala, Sweden

WINTON, Mr J S, Tioxide Europe Ltd, Tees Road, Hartlepool, Cleveland TS25 2DD, UK

WITHAM, Mr B H, CIBA, Hulley Road, Macclesfield, Cheshire SK10 2NS, UK

WOERFEL, Mrs P F, Centre for Occupational Health, University of Manchester, Stopford Building, Oxford Road, Manchester M13 9PT, UK

YEATES, Dr D B, University of Illinois at Chicago, 1940 W Taylor Street, M/C 788, Chicago, IL 60612, USA

YOUNG, Dr J, Sheffield Hallam University, School of Science & Mathematics, Pond Street, Sheffield, South Yorkshire S1 1WB, UK

AUTHOR INDEX

SUBJECT INDEX